Handbook of Perovskite Solar Cells, Volume 2

Perovskite solar cells (PSCs) have received significant attention in academia and industry due to their low cost and high-power conversion efficiency (PCE). Single- and multijunction PSCs have obtained promising certified PCEs, which suggests that PSCs are a very promising next-generation photovoltaic technology. In addition to the perovskite absorber layer, other functional layers, including electron transport layer (ETL), hole transport layer (HTL), and electrode layer (EL), have also made huge contributions to enhancing device performance. This book focuses on the development, advancement, and application of these functional layers in various PSCs.

This volume:

- Introduces ETL, HTL, and EL in efficient and stable PSCs.
- Covers material properties.
- Discusses a wide variety of PSCs including single-crystal PSCs, flexible PSCs, perovskite tandem solar cells, lead-free PSCs, inorganic PSCs, fully printable mesoscopic PSCs, electron/hole-transport-layer-free PSCs, semitransparent PSCs for building-integrated photovoltaics (BIPV), tandem solar cells, perovskite indoor photovoltaics, and inverted PSCs.
- Details potential for commercial application.

This book is aimed at researchers, advanced students, and industry professionals in materials, energy, and related areas of engineering who are interested in development and commercialization of photovoltaic technologies.

Advances in Materials Science and Engineering

Series Editor: Sam Zhang

Nanostructured and Advanced Materials for Fuel Cells
San Ping Jiang and Pei Kang Shen

Hydroxyapatite Coatings for Biomedical Applications
Sam Zhang

Carbon Nanomaterials: Modeling, Design, and Applications
Kun Zhou

Materials for Energy
Sam Zhang

Protective Thin Coatings and Functional Thin Films Technology, Two-Volume Set
Sam Zhang, Jyh-Ming Ting, Wan-Yu Wu

Fundamentals of Crystallography, Powder X-ray Diffraction, and Transmission Electron Microscopy for Materials Scientists
ZhiLi Dong

Materials for Devices
Sam Zhang

Handbook of Perovskite Solar Cells, Volume 1: Fundamentals and Absorber Layer Optimization for Efficiency and Stability
Edited by Jiangzhao Chen and Sam Zhang

Handbook of Perovskite Solar Cells, Volume 2: Functional Layer Optimization and Diverse Device Types for Efficiency and Stability
Edited by Jiangzhao Chen and Sam Zhang

Handbook of Perovskite Solar Cells, Volume 3: Critical Challenges and Opportunities in Commercialization
Edited by Jiangzhao Chen and Sam Zhang

Materials in Advanced Manufacturing
Yinquan Yu and Sam Zhang

For more information about this series, please visit: www.routledge.com/Advances-in-Materials-Science-and-Engineering/book-series/CRCADVMATSCIENG

Handbook of Perovskite Solar Cells, Volume 2

Functional Layer Optimization and Diverse Device Types for Efficiency and Stability

Edited by
Jiangzhao Chen and Sam Zhang

CRC Press
Taylor & Francis Group
Boca Raton London New York

CRC Press is an imprint of the
Taylor & Francis Group, an **informa** business

Designed cover image: Shutterstock.

First edition published 2024
by CRC Press
2385 NW Executive Center Drive, Suite 320, Boca Raton FL 33431

and by CRC Press
4 Park Square, Milton Park, Abingdon, Oxon, OX14 4RN

CRC Press is an imprint of Taylor & Francis Group, LLC

ISBN: 978-1-032-50969-3 (hbk)
ISBN: 978-1-032-50970-9 (pbk)
ISBN: 978-1-003-40049-3 (ebk)

DOI: 10.1201/9781003400493

Typeset in Times
by Apex CoVantage, LLC

Contents

About the Editors

Jiangzhao Chen (陳江照) is a professor at Faculty of Materials Science and Engineering in Kunming University of Science and Technology. He received his BS and PhD degrees in Applied Chemistry from Northeast Forestry University in July 2011 and in Optical Engineering from Huazhong University of Science and Technology in June 2016, respectively. From October 2016 to February 2019, he worked as a postdoctoral researcher at Sungkyunkwan University. From March 2019 to November 2019, he worked as a postdoctoral researcher at University of Hong Kong. From December 2019 to December 2023, he worked as a professor at School of Optoelectronic Engineering, Chongqing University. His current research interests focus on perovskite solar cells.

Sam Zhang Shanyong (張善勇), FRSC, FTFS, FIoMMM, academically known as Sam Zhang, was born and brought up in the famous "City of Mountains" Chongqing, China. He received his Bachelor of Engineering in Materials in 1982 from Northeastern University (Shenyang, China), Masters of Engineering in Materials in 1984 from Iron and Steel Research Institute (Beijing, China), and PhD degree in Ceramics in 1991 from The University of Wisconsin-Madison, USA. He was a tenured full professor (beginning 2006) at the School of Mechanical and Aerospace Engineering, Nanyang Technological University, Singapore. In January 2018, he joined the School of Materials and Energy, and served as the founding director of the Centre for Advanced Thin Films and Devices, Southwest University, China. In 2023, he joined the School of Aeronautics, Harbin Institute of Technology as Chair Professor and also served in the Harbin Institute of Technology's Zhengzhou Research Institute.

Contributors

Jingjing Chang
Faculty of Integrated Circuit
Xidian University
Xi'an, China

Chun-Hao Chen
Institute of Functional Nano and Soft Materials (FUNSOM)
Jiangsu Key Laboratory of Advanced Negative Carbon Technologies
Jiangsu Key Laboratory for Carbon-Based Functional Materials and Devices
Soochow University
Suzhou, Jiangsu, People's Republic of China

Ge Chen
Green Catalysis Center and College of Chemistry
Zhengzhou University
Zhengzhou, China

Mengran Chen
College of Materials Science and Engineering
Nanjing Tech University
Nanjing, China

Xin Chen
Institute of Functional Nano and Soft Materials (FUNSOM)
Jiangsu Key Laboratory of Advanced Negative Carbon Technologies
Jiangsu Key Laboratory for Carbon-Based Functional Materials and Devices
Soochow University
Suzhou, Jiangsu, People's Republic of China

Nian Cheng
School of Physics and Optoelectronic Engineering
Yangtze University
Jingzhou, China

Guozheng Du
College of Materials
Xiamen University
Xiamen, China

Zhuowei Du
School of New Energy and Materials
Southwest Petroleum University
Chengdu, People's Republic of China
Institute of Photovoltaic, Southwest Petroleum University
Chengdu, People's Republic of China

Junfeng Fang
School of Physics and Electronic Science
East China Normal University
Shanghai, China

Xing Guo
Academy of Advanced Interdisciplinary Research
Xidian University
Xi'an, China

Feng Hao
School of Materials and Energy
University of Electronic Science and Technology of China
Chengdu, China

Shanyue Hou
School of New Energy and Materials and Institute of Photovoltaic
Southwest Petroleum University
Chengdu, People's Republic of China

Zhangfeng Huang
School of New Energy and Materials and Institute of Photovoltaic
Southwest Petroleum University
Chengdu, People's Republic of China

Danish Khan
College of New Materials and New Energies
Shenzhen Technology University
Shenzhen, Guangdong, China

Pengwei Li
Green Catalysis Center and College of Chemistry
Zhengzhou University
Zhengzhou, China

Xiaodong Li
School of Physics and Electronic Science
East China Normal University
Shanghai, China

Yanlin Li
School of New Energy and Materials and Institute of Photovoltaic
Southwest Petroleum University
Chengdu, People's Republic of China

Yuelong Li
Institute of Photoelectronic Thin Film Devices and Technology
Key Laboratory of Photoelectronic Thin Film Devices and Technology of Tianjin
Solar Energy Research Center
Nankai University
Tianjin, China.

Zhenhua Lin
Faculty of Integrated Circuit
Xidian University
Xi'an, China

Shengzhong Liu
Dalian National Laboratory for Clean Energy
iChEM Dalian Institute of Chemical Physics Chinese Academy of Sciences
Dalian, Liaoning, China

Linchuan Ma
Institute of Photoelectronic Thin Film Devices and Technology
Key Laboratory of Photoelectronic Thin Film Devices and Technology of Tianjin
Solar Energy Research Center
Nankai University
Tianjin, China

Zhu Ma
School of New Energy and Materials and Institute of Photovoltaic
Southwest Petroleum University
Chengdu, People's Republic of China

Geping Qu
Department of Chemistry
Southern University of Science and Technology
Shenzhen, Guangdong, China

Shaowen Tang
College of Materials Science and Engineering
Nanjing Tech University
Nanjing, China

Zeguo Tang
College of New Materials and New Energies
Shenzhen Technology University
Shenzhen, Guangdong, China

Pengpeng Teng
College of Materials Science and Engineering
Nanjing Tech University
Nanjing, China

Kang Wang
Materials Engineering Technology
School of Materials Science and Engineering,
North Minzu University
Yinchuan, People's Republic of China

Shurong Wang
School of Materials and Energy
University of Electronic Science and Technology of China
Chengdu, China

Zhao-Kui Wang
Institute of Functional Nano and Soft Materials (FUNSOM)
Jiangsu Key Laboratory of Advanced Negative Carbon Technologies
Jiangsu Key Laboratory for Carbon-Based Functional Materials and Devices
Soochow University
Suzhou, Jiangsu, People's Republic of China

Wenjun Wu
Key Laboratory for Advanced Materials and Feringa Nobel Prize Scientist Joint Research Center
Shanghai Key Laboratory of Functional Materials Chemistry
Shanghai, People's Republic of China

Zong Xiang Xu
Department of Chemistry
Southern University of Science and Technology
Shenzhen, Guangdong, China

Li Yang
College of Materials
Xiamen University
Xiamen, China

Huanhuan Yao
School of Materials and Energy
University of Electronic Science and Technology of China
Chengdu, China

Chunfu Zhang
State Key Discipline Laboratory of Wide Band Gap Semiconductor Technology and Shaanxi Joint Key Laboratory of Graphene
School of Microelectronics
Xidian University
Xi'an, China

Jinbao Zhang
College of Materials
Xiamen University
Xiamen, China

Jun Zhang
College of Materials Science and Engineering
Nanjing Tech University
Nanjing, China

Yiqiang Zhang
Green Catalysis Center and College of Chemistry
Zhengzhou University
Zhengzhou, China

Zemin Zhang
Institute of Photoelectronic Thin Film Devices and Technology
Key Laboratory of Photoelectronic Thin Film Devices and Technology of Tianjin
Solar Energy Research Center

Nankai University
Tianjin, China

Long Zhou
State Key Discipline Laboratory of Wide Band Gap Semiconductor Technology and Shaanxi Joint Key Laboratory of Graphene
School of Microelectronics
Xidian University
Xi'an, China

Peng-an Zong
College of Materials Science and Engineering
Nanjing Tech University
Nanjing, China

1 The Electron Transport Layer in Perovskite Solar Cells

Jingjing Chang, Xing Guo, and Zhenhua Lin

1.1 INTRODUCTION

The Electron Transport Layer (ETL) is the key component of perovskite solar cells (PSCs), which plays an important role in realizing high-efficiency PSCs. Ideally, the ETL should satisfy the following conditions: (i) a well-matched energy band with perovskite and high electron mobility which ensure efficient electron transfer from perovskite to ETL; (ii) a wide bandgap to maximize the light absorption of perovskite film and block the hole; (iii) a uniform surface morphology and good wettability to ensure the deposition of high-quality perovskite film; (iv) good photochemical and thermal stability to guarantee the long-term stability of PSCs; (v) a simple, mild, and low-cost fabrication process.

To date, the widely reported n-type electron transport materials are typically divided into the following categories: (i) n-type metal oxide materials, such as titanium dioxide (TiO_2), tin (IV) oxide (SnO_2), zinc oxide (ZnO), indium (III) oxide (In_2O_3), niobium pentoxide (Nb_2O_5), and zinc stannate (Zn_2SnO_4); (ii) other n-type metal sulfide or nitride materials, such as cadmium sulfide (CdS) and indium sulfide (In_2S_3); (iii) n-type organic materials, such as fullerene derivatives and non-fullerene electron acceptors.

In this chapter, first, the development of inorganic and organic ETLs used in PSCs is introduced. Then some strategies for ETL optimization to improve the PSC efficiency, including doping, interface modification, and bilayer structure are summarized. Finally, an outlook on the development of ETL for large-area, high-efficiency, long-term stable commercial perovskite solar cells is presented.

1.2 INORGANIC ETLs

Due to the high stability, low cost, simple fabrication process, and excellent electronic properties, inorganic materials are considered as desirable ETLs. In this section, TiO_2, ZnO, SnO_2, and other inorganic ETLs are introduced. First, the typical photoelectric parameters of n-type inorganic ETLs are summarized in Table 1.1.

DOI: 10.1201/9781003400493-1

TABLE 1.1
Photoelectric Parameters of N-Type Inorganic ETLs

ETLs	Conduction band (eV)	Eg (eV)	Bulk mobility ($cm^2 V^{-1} s^{-1}$)	Ref.
TiO_2	–4.10	3.0–3.2	1	1
ZnO	–4.17	3.30	230	1, 2
SnO_2	–4.22	3.6–4.0	250	3, 4
In_2O_3	–4.30	3.80	20	5
Nb_2O_5	–4.02	3.41	0.2	6, 7
Cr_2O_3	–3.93	3.50	10^{-5}–1	8, 9
CeO_x	–4.00	3.50	100	10, 11
WO_x	–4.20	3.79	10–20	12
$BaSnO_3$	–3.91	3.10	150	13, 14
$SrTiO_3$	–3.65	3.25	5–8	15, 16
$SrSnO_3$	–3.92	4.16	55	17, 18
CdSe	–4.40	1.80	600	2, 19
CdS	–3.98	2.42	440	2, 20
Bi_2S_3	–3.86	1.59	49	21
In_2S_3	–3.98	2.0–2.8	17.6	22
MoS_2	–4.42	1.29	100	23
SnS_2	–4.24	2.30	7.85×10^{-4}	24

1.2.1 TiO_2

TiO_2 is the most commonly investigated ETL in PSCs due to its mature development in the field of dye-sensitized solar cells. In 2009, Miyasaka et al.[25] studied $CH_3NH_3PbBr_3$ and $CH_3NH_3PbI_3$ perovskite as visible-light sensitizers and fabricated the first PSC based on a mesoporous structure (Figure 1.1a and b) with compact TiO_2 and mesoporous TiO_2 scaffold. Among them, the compact TiO_2 can prevent direct contact between the conductive substrate and perovskite film, preventing holes transfer to conductive electrode, and the mesoporous TiO_2 can provide a framework and support for the growth of perovskite film, improving the crystallization of the perovskite film and electrons collection efficiency. In 2013, Liu et al.[26] reported the planar heterojunction PSCs by vapor deposition based on compact TiO_2 (Figure 1.1c), and they found that the PSCs without mesoporous scaffold also achieved a high efficiency of 15.6%. Due to the wide bandgap and well-matched energy level with perovskite, the TiO_2 ETL is widely used and deeply studied, and the PSCs based on TiO_2 ETL have kept the highest certified efficiency of PSCs for a long time in the past.

TiO_2 has four crystal structures: anatase, rutile, brookite, and TiO_2-B (Figure 1.1d–g).[27–30] Among them, anatase TiO_2 is the most widely used ETL in PSCs due to its wider bandgap (3.2 eV). In the mesoscopic structure PSCs, the mesoscopic layer is needed as a scaffold to assist in perovskite nucleation and increase the contact area between perovskite and ETL to improve electron extraction. The mesoscopic layer is usually

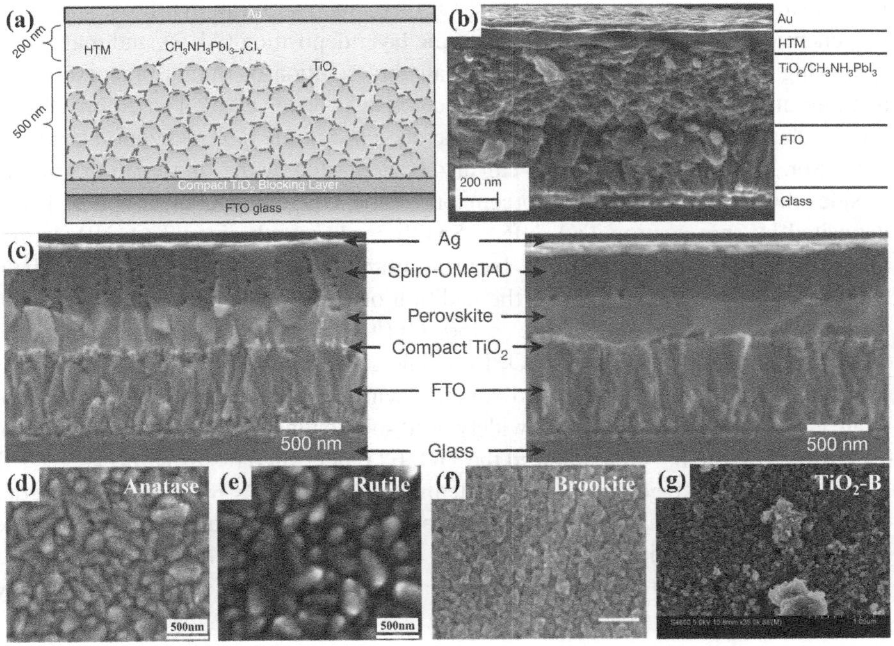

FIGURE 1.1 a) The structure of PSCs based on mesoporous TiO_2. Reproduced with permission.[33] Copyright 2013, Nature Publishing Group. b) Cross-sectional scanning electron microscopy (SEM) image of PSCs based on mesoporous TiO_2. Reproduced with permission.[34] Copyright 2013, Nature Publishing Group. c) Cross-sectional SEM image of planar heterojunction PSCs based on compact TiO_2. Reproduced with permission.[26] Copyright 2013, Nature Publishing Group. SEM images of d) Anatase TiO_2, e) Rutile TiO_2. Reproduced with permission.[28] Copyright 2019, John Wiley and Sons. f) Brookite TiO_2. Reproduced with permission.[29] Copyright 2015, Royal Society of Chemistry. g) TiO_2-B. Reproduced with permission.[30] Copyright 2019, Elsevier.

prepared by spin-coating or screen-printing a diluted TiO_2 nanoparticles (NPs) suspension. The thickness, particle size, pore size, and crystal phase of the mesoscopic layer directly affect the performance of PSCs. Yang et al.[31] studied the effect of the size of anatase TiO_2 NPs on the device performance. They found that the series resistance (R_s) and charge recombination resistance (R_{rec}) decreased with the size increase. The PSCs based on TiO_2 NPs with a diameter of 25 nm achieved the best PCE of 13.41%. Kim et al.[32] studied the effect of the thickness of mp-TiO_2 on the hysteresis of PSCs, and they found that the minimum hysteresis of PSCs could be achieved when the thickness of mp-TiO_2 increased to 220 nm.

Usually, the mesoscopic layer needs a high annealing temperature of more than 500 °C to remove the polymer templates and ensure the transfer of the amorphous layer to the crystalline phase (anatase), which is crucial for electron transport. The mesoscopic layer increases the complexity and cost of device fabrication and limits the application of PSCs in flexible configurations. The planar structure without a mesoscopic layer has been deeply researched. Many methods have been developed to deposit compact and pinholes-free TiO_2 ETL, such as spray pyrolysis, spin coating, chemical bath deposition (CBD), atomic layer deposition (ALD), and magnetron sputtering. The spray pyrolysis uses the alcohol solution of titania precursor, such as titanium diisoethpropoxide bis(acetylacetonate) (TiAcAc).[34,35] The compact TiO_2 layer is deposited on a hot substrate by the decomposition and oxidization of titania precursor. The compact TiO_2 layer can also be deposited by the spin-coating process. Besides the ethanolic solution of precursor,[36,37] the TiO_2 compact layer could also be deposited by spin-coating TiO_2 NPs.[38] Snaith et al.[39] synthesized TiO_2 NPs with a diameter of about 4.5 nm at 80 °C by using titanium tetrachloride ($TiCl_4$) and benzyl alcohol (Figure 1.2a–c). With the addition of TiAcAc, the conductivity of TiO_2 film prepared at a low temperature (< 150 °C) (lt-TiO_2) is 100 times higher than that of high-temperature processed TiO_2 film. The compact TiO_2 film prepared by this low-temperature process is flat and compact without cracks and can cover the substrate completely. CBD is another widely used solution method for the preparation of TiO_2 film. Yella et al.[40] first reported the TiO_2 ETL deposited by using the $TiCl_4$ CBD process in PSCs. They controlled the diameter of TiO_2 NPs by adjusting the concentration of $TiCl_4$. This method could prepare the rutile TiO_2 and form an intimate junction of the large interfacial area with perovskite film without further annealing process. Huang et al.[41] proposed a ligand-engineered deposited (LD) strategy

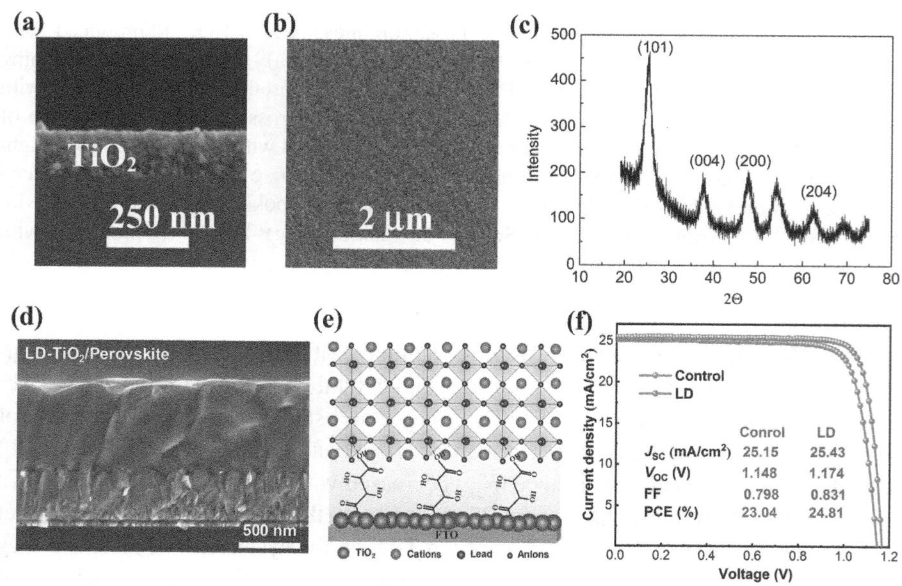

FIGURE 1.2 a) Cross-sectional SEM image of TiO_2 NPs spin-coated on glass, thickness: 100–120 nm, b) top-view SEM image of lt-TiO_2, c) X-ray diffraction (XRD) spectrum of the TiO_2 nanoparticle powder. Reproduced with permission.[39] Copyright 2014, Royal Society of Chemistry. d) Cross-sectional SEM image of perovskite films deposited on LD-TiO_2 film. e) Schematic illustration of cross-linked structure at the LD-TiO_2/perovskite interface. f) Current density-voltage (J-V) curves of the planar PSCs based on various TiO_2. Reproduced with permission.[41] Copyright 2022, Elsevier.

to precisely regulate the TiO_2 CBD deposition process using the additive ligands such as tartaric acid (TA), ethylene diamine tetraacetic acid (EDTA), phenylphosphonic acid (PA), and so on (Figure 1.2d–f). They found that the regulated LD-TiO_2 showed reduced particle aggregation and improved electron mobility. The LD-TiO_2/perovskite interface contact could be optimized to decrease contact impedance and enhance electron extraction. The PSCs based on LD-TiO_2 exhibited a PCE of 24.81% (certificated 24.5%) with an enhanced ultra-violet (UV) and humidity stability. Seok et al.[42] controlled the thickness of the TiO_2 layer by adjusting the concentration of $TiOCl_2$ aqueous solution. Except for the solution methods, the physical approaches can also be employed to prepare high-quality compact TiO_2 film. The ALD method can control the thickness of film accurately to produce thinner and conformal film. Kim et al.[43] first constructed flexible planar PSCs based on TiO_2 deposited by the ALD method. They obtained a thin TiO_2 film (about 20 nm) at a low temperature (80 °C), and the PSCs showed a champion PCE of 12.2% with a good bending property. Lv et al.[44] used the ALD method to deposit pin-hole free TiO_2 as ETL in p-i-n PSCs and achieved a PCE of 18.3%. They found that the ALD process has negligible damage to the underlying perovskite film. Moreover, the TiO_2 layer could resist water and improve the long-term stability of PSCs. Magnetron sputtering is another low-temperature physical approach for TiO_2 deposition. Chen et al.[45] prepared TiO_2 film on flexible substrates using the radio frequency magnetron sputtering method and achieved a PCE of 8.9%. Recently, the PCE of PSCs based on magnetron-sputtered TiO_2 ETL has broken through 18%.[46]

The well-matched energy level and easy and low-cost fabrication make TiO_2 the most common ETL, but TiO_2 is not the most ideal ETL in PSCs yet. First, the electron mobility of TiO_2 is relatively low, only in the range of 0.1–4 $cm^2\ V^{-1}\ s^{-1}$, which is lower than the electron mobility of perovskite film.[47,48] The low electron mobility leads to charge accumulation at TiO_2/perovskite interface, resulting in increased carrier recombination at the interface, which has been recognized as one of the origins of the *J-V* hysteresis of PSCs based on TiO_2 ETL. Second, TiO_2 contains a large density of oxygen vacancies, and the oxygen vacancies can act as electronic trap states below the conduction band, significantly influencing the efficiency of PSCs.[49] Third, TiO_2 owns high photocatalytic activity, which can decompose perovskite when the TiO_2 ETL is illuminated under ultraviolet light (Figure 1.3).[33,50] Therefore, it is important to improve the bulk or surface properties of TiO_2 ETL for achieving high-efficiency and stable PSCs.

1.2.2 ZnO

ZnO is another n-type metal oxide semiconductor, and the energy band of ZnO is similar to that of TiO_2, which is well matched with perovskite. Compared with TiO_2, ZnO has a much higher electron mobility of 230 $cm^2\ V^{-1}\ s^{-1}$, which facilitates electron transfer from perovskite to ETL. Moreover, ZnO ETL with high quality can be easily prepared at low temperatures by the solution process, which potentially enables a low cost of production and utilization of flexible devices.

Fabrication of uniform and compact ETL film is crucial for high-performance PSCs. The sol-gel method is the most widely used to prepare ZnO ETL.[51,52]

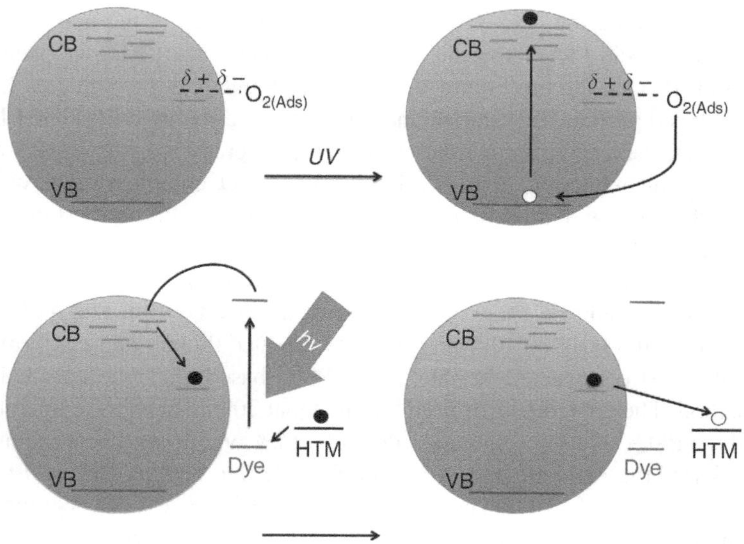

FIGURE 1.3 Proposed mechanism for UV-induced degradation. Reproduced with permission.[33] Copyright 2013, Nature Publishing Group.

Heo et al.[53] prepared ZnO ETL via a sol-gel process at 150 °C and compared the ZnO and TiO_2 ETL (Figure 1.4a and b). The PSCs based on ZnO ETL showed an average PCE of 17.6%. Due to the better electron mobility of ZnO compared to TiO_2, the electron extraction from perovskite to ETL was improved, and the PSCs based on ZnO ETLs exhibited less hysteresis. The low-temperature ZnO could be used on polyethylene naphthalate (PEN)/indium tin oxide (ITO) substrate and achieved an average PCE of 15.5%. Zheng et al.[54] applied a combustion method to prepare ZnO ETL, in which the $Zn(NO_3)_2{\cdot}6H_2O$ and acetylacetone in precursor were used as oxidizer and fuel, respectively (Figure 1.4c). The ZnO ETL prepared by the combustion process had negligible organic impurities and high crystallinity under low annealing temperature. The PSCs based on combustion-derived ZnO ETL exhibited a PCE of 19.84%.

Spin-coating ZnO NPs is another efficient solution method to prepare ZnO ETL. Liu et al.[55] synthesized ZnO NPs by dissolving Zinc acetate dihydrate in methanol (Figure 1.4d and e). The synthesized ZnO NPs could be dispersed in a mixed solvent of n-butanol, methanol, and chloroform. The ZnO ETL was deposited by spin-coating ZnO NPs directly without further annealing process. The PSCs based on low-temperature ZnO ETL yielded a PCE of 15.7% and the flexible PSCs achieved a PCE of 10.2%. Manspeaker et al.[56] prepared ZnO ETL by spin-coating a 12 wt.% water dispersion of ZnO NPs. They developed a restricted volume solvent annealing process to deposit $MAPbI_3$ film on ZnO. The PSCs achieved a PCE of 13.7%. Song et al.[57] deposited a trip cation perovskite ($Cs_x(MA_{0.17}FA_{0.83})_{(100-x)}Pb(I_{0.83}Br_{0.17})_3$) using a

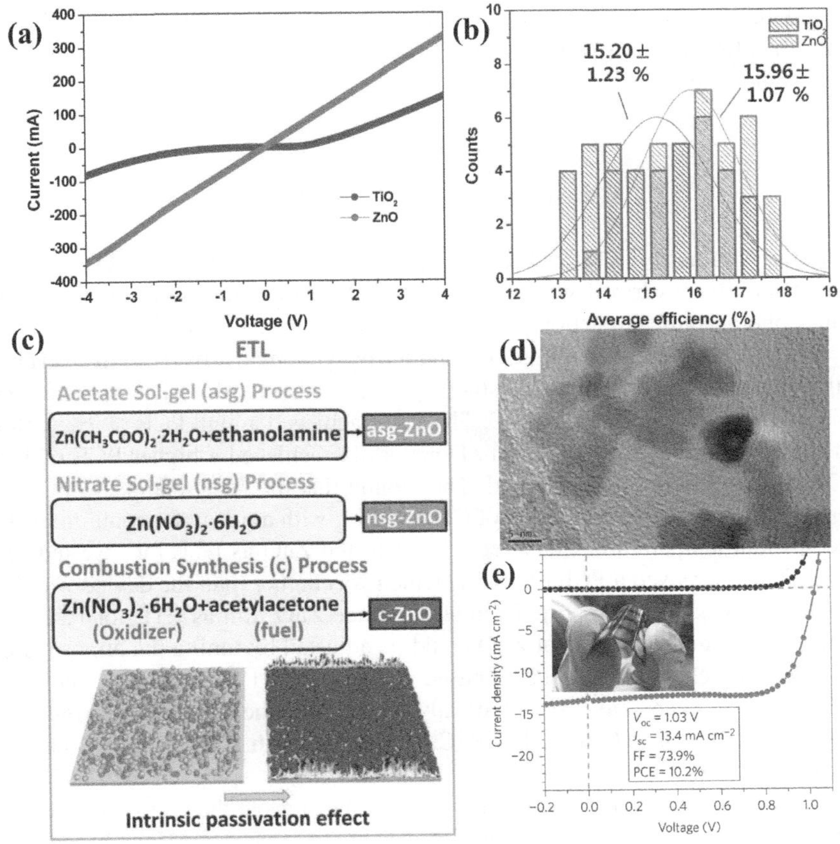

FIGURE 1.4 a) Conductivity test of TiO_2 and sol-gel ZnO films. b) PCE distribution of PSCs based on TiO_2 and sol-gel ZnO ETLs. Reproduced with permission.[53] Copyright 2016, Royal Society of Chemistry. c) Three different synthetic approaches to solution-processed ZnO ETLs. Reproduced with permission.[54] Copyright 2019, John Wiley and Sons. d) Transmission electron microscopy (TEM) micrographs of ZnO NPs. e) *J-V* characteristics for the flexible PSCs (inset) based on ZnO NPs ETLs. Reproduced with permission.[55] Copyright 2014, Nature Publishing Group.

one-step deposition method on ZnO ETL prepared by ZnO NPs. The perovskite film was annealed at 95 °C, leading to a high PCE of 18.9%.

Although ZnO has promising properties and has been widely studied, the PCE of PSCs based on ZnO is still lower than that of TiO_2-based PSCs, which is mainly due to the ZnO-perovskite interface instability. Yang et al.[58] reported that the basic surface of ZnO can induce proton-transfer reactions with $MAPbI_3$ and the decomposition of perovskite film. Cheng et al.[59] demonstrated that hydroxyl groups on ZnO NPs could accelerate $MAPbX_3$ perovskite degradation, and higher annealing temperature could reduce the surface density of hydroxyl groups and improve the stability of perovskite film (Figure 1.5a and b). However, the perovskite films still turned yellow after annealing at 100 °C for 30 min. The proton-transfer reactions can be explained by the following reaction:

$$OH^{-}+CH_3NH_3I \rightarrow CH_3NH_3OH+I^{-}$$

$$CH_3NH_3OH \rightarrow CH_3NH_2\uparrow+H_2O \text{ (heat)}$$

Moreover, similar to TiO_2, ZnO has high photocatalysis activity under UV light irradiation, and the hydroxyl groups and chemical residues on the ZnO surface will induce photochemical reactions, which affects the light stability of PSCs.[60]

The ZnO film deposited by ALD or sputtering without hydroxyl groups and chemical residues can produce higher stability. Lee et al.[61] used ALD to prepare ZnO ETL at a low temperature of 80 °C. The PSCs achieved a high PCE of 7% with the optimized ZnO thickness of 30 nm. Dong et al.[62] achieved a higher PCE of 13.1% at 70 °C by using the same method. They found that the ALD-compact ZnO could promote the single-step formation of $CH_3NH_3PbI_3$ with a precursor containing chloride at room temperature. Chang et al.[63] deposited ZnO as ETL for inverted PSCs by ALD and achieved a PCE of 16.5%, which was higher than the device with ZnO NPs film. Tseng et al.[64] reported sputtering made ZnO film as ETL for PSCs. The surface electronic properties of ZnO could be adjusted by tuning the atmosphere of the sputtering chamber. The PSCs based on the $CH_3NH_3PbI_3$ absorber achieved a PCE of 15.9%. Zhao et al.[65] systematically investigated the influence of Ar/O_2 ratios on the properties of ZnO film. The PSCs achieved a high PCE of 16.60% based on the Ar/O_2 ratio of 1:4.

Replacing the MA cation with FA or Cs cation can avoid the proton-transfer reactions between the ZnO and MA cation. Schutt et al.[66] deposited ZnO ETL by spin-coating zinc oxide solution which was prepared by dissolving zinc oxide powder in a mixed solution of ammonia hydroxide and water. Then they replaced the MA cation with lower acidity FA mixed with Cs (FACs) cations and found the FACs perovskite greatly improved the stability of PSCs based on ZnO ETL with a PCE of 21.1% (Figure 1.5c–e). Further, Chen et al.[67] supposed that the combination of aluminum-doped ZnO (AZO) ETL and MA-free perovskite could significantly improve the efficiency and stability of ZnO-based PSCs. They introduced multifunctional mussel-inspired polymer of PDA to reduce the interfacial mismatch between AZO and perovskite. The PDA could inhibit the agglomeration of AZO NPs to improve the surface morphology. PDA could warp well around AZO NPs due to its superb adhesive properties, which could significantly inhibit the deprotonation reaction between perovskite and AZO ETL. As a result, the MA-free PSCs based on PDA treated AZO achieved a PCE of 21.36%. The unencapsulated devices maintained 90% and 88% of their initial PCEs after storing at 85°C for 360 h and continuous light for 500 h, respectively. Ma et al.[68] prepared ZnO ETL for inorganic PSCs by using a low-temperature combustion solution process. They found that the proton-transfer reactions did not exist at the $ZnO/CsPbI_2Br$ interface. The all-inorganic $CsPbI_2Br$ PSCs based on ZnO exhibited a PCE of 14.78% with a V_{OC} of 1.21 V and an FF of 81.42%.

Although the vacuum method (ALD or sputtering) can be employed to prepare high-quality and stable ZnO layers, its cost is high and unsuitable for large-area

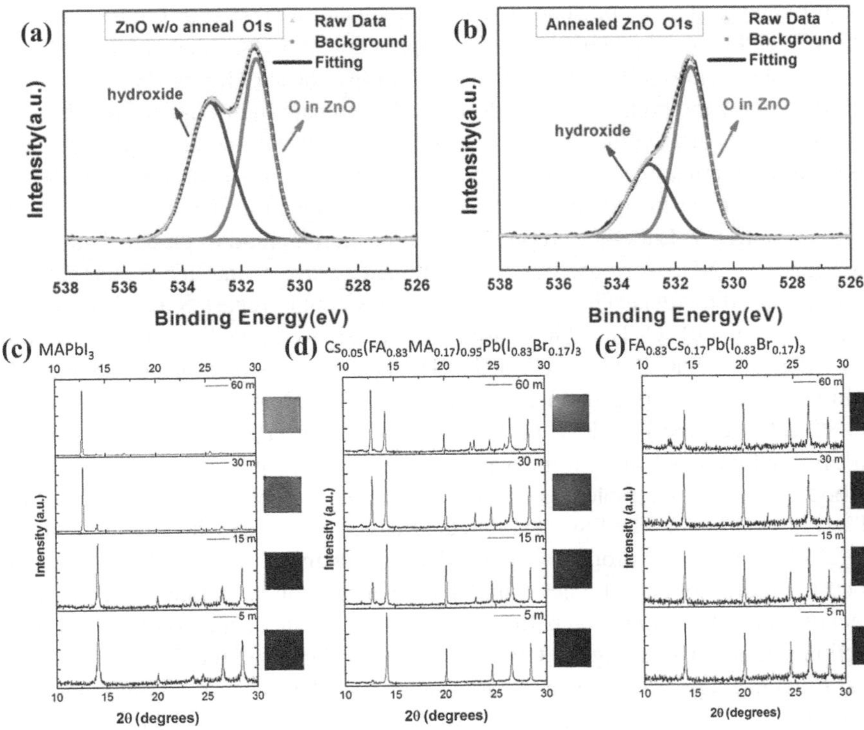

FIGURE 1.5 High-resolution X-ray photoelectron spectroscopy (XPS) of O 1s core level spectra of ZnO a) without annealing and b) with annealing at 200 °C for 1 h in air. Reproduced with permission.[59] Copyright 2015, American Chemical Society. Thermal stability (100°C) of c) $MAPbI_3$, d) CsFAMA, and e) FACs perovskites on ZnO ETLs. Reproduced with permission.[66] Copyright 2015, John Wiley and Sons.

preparation. Solution method is the simplest and lowest-cost method to prepare ZnO ETL. Direct spin coating of ZnO NPs can effectively reduce the annealing temperature and process complexity of the ZnO ETL. The solution method combined with perovskite component engineering and interface engineering is the optimal method to prepare high-performance ZnO-based PSCs. Interfacial engineering will be discussed in Section 1.4.

1.2.3 SnO_2

Recently, SnO_2 has received significant attention and is considered as the most promising ETL for PSCs due to its preferred properties. SnO_2 has a high electron mobility of up to 240 $cm^2\ V^{-1}\ s^{-1}$, which is 100 times that of TiO_2. The high electron mobility can improve electron transfer and suppress charge recombination efficiently. Moreover, SnO_2 has an appropriate energy band structure, a similar conduction band and work function to TiO_2, and a deeper valence band, which can efficiently extract electrons and block holes. The wide band gap of SnO_2 (3.6–4.0 eV) can improve light transmittance. SnO_2 can be deposited at low temperatures (< 200 °C), which is suitable for flexible and low-cost large-scale applications. Most importantly, SnO_2 has excellent chemical stability, which helps to improve the stability of PSCs.

So far, many methods have been developed to prepare high-quality SnO_2 films. In 2015, Ke et al.[69] first synthesized SnO_2 ETL for PSCs by spin-coating the $SnCl_2·2H_2O$ precursor and annealing at 180 °C. By optimizing the thickness of SnO_2 to 60 nm, the PSCs based on low-temperature SnO_2 ETL exhibited a PCE of 17.21%, with a high V_{OC} of 1.11 V. Liu et al.[70] reported a combustion method to prepare SnO_2 ETL. $SnCl_2$ and NH_4NO_3 were used as the combined oxidizer, and acetylacetonate and urea were used as fuels. The combustion process could prepare SnO_2 ETL at a low temperature of 140 °C. The PSCs based on SnO_2 showed a PCE of about 13%. Jung et al.[71] deposited SnO_2 ETL based on Tin(IV) isopropoxide/isopropanol solution. By optimizing the annealing temperature, the PSCs based on SnO_2 ETL annealed at 250 °C achieved the best PCE of 19.40%. CBD is an efficient method to prepare uniform and compact SnO_2 film. Anaraki et al.[72] combined spin-coating and CBD methods to realize a simple, fast, and low-temperature deposition method of SnO_2 ETL (Figure 1.6a–e). The PSCs exhibited a PCE close to 21% with a remarkable V_{OC} of 1.214 V. Yoo et al.[73] proposed the synthesis mechanism of SnO_2 during the CBD process (Figure 1.6f and g). They found that the formation of Sn intermediate species depends on the decomposition pathway of the $SnCl_2$, which depends on the pH of the reaction solution. By tuning the CBD process, they deposited the SnO_2 ETL with an ideal film coverage, thickness, and composition. The PSCs based on SnO_2 reached a record PCE of 25.2%. Furtherly, Min et al.[74] and Park et al.[75] achieved a record PCE of 25.5% and 25.73% based on CBD-SnO_2, respectively.

Directly spin-coating the synthesized SnO_2 nanoparticles (NPs) is an efficient method to prepare SnO_2 ETL. Song et al.[76] spin-coated the SnO_2 NPs to prepare SnO_2 ETL. The SnO_2 NPs with a size of 22–43 nm were dispersed in butanol and annealed at 200 °C. The PSCs based on SnO_2 NPs showed a PCE of 13.0%. The size of SnO_2 NPs affects the pinholes and coverage of SnO_2 films. Large-sized SnO_2 NPs are more prone to produce pinholes, resulting in charge recombination due to the direct contact between the cathode and perovskite film. Yang et al.[77] synthesized SnO_2 quantum dots (QDs) at room temperature by adding thiourea in the $SnCl_2·2H_2O$ water solution (Figure 1.7a and b). The SnO_2 QDs had a small size of 3–5 nm, which improved the contact between SnO_2 and the fluorine doped tin oxide (FTO) substrate. The PSCs based on SnO_2 QDs reached a PCE of 20.79%. You et al.[48] first used the commercial SnO_2 colloidal solution with a SnO_2 NPs size of 3–4 nm to prepare SnO_2 ETL at a low temperature of 150 °C (Figure 1.7c–f). The PSCs exhibited a PCE of 20.54% (19.9% certificated). Recently, they achieved a record certificated PCE of 25.6% based on SnO_2 NPs.[78,79] The use of commercial SnO_2 NPs has greatly promoted the development of SnO_2-based PSCs. SnO_2 NPs were also used to prepare ETL in inverted PSCs. Zhu et al.[80] prepared highly crystalline SnO_2 NPs with a size of 5–10 nm by hydrothermal method. The SnO_2 NPs were dispersed in isopropyl and spin-coated on C_{60} to form a bilayer ETL. They got a PCE of 18.8% in inverted PSCs. In addition, the compact SnO_2 could resist oxygen and water and improve the stability of PSCs, which retained over 90% of its initial PCE after being stored for 30 d in ambient with ~70% relative humidity.

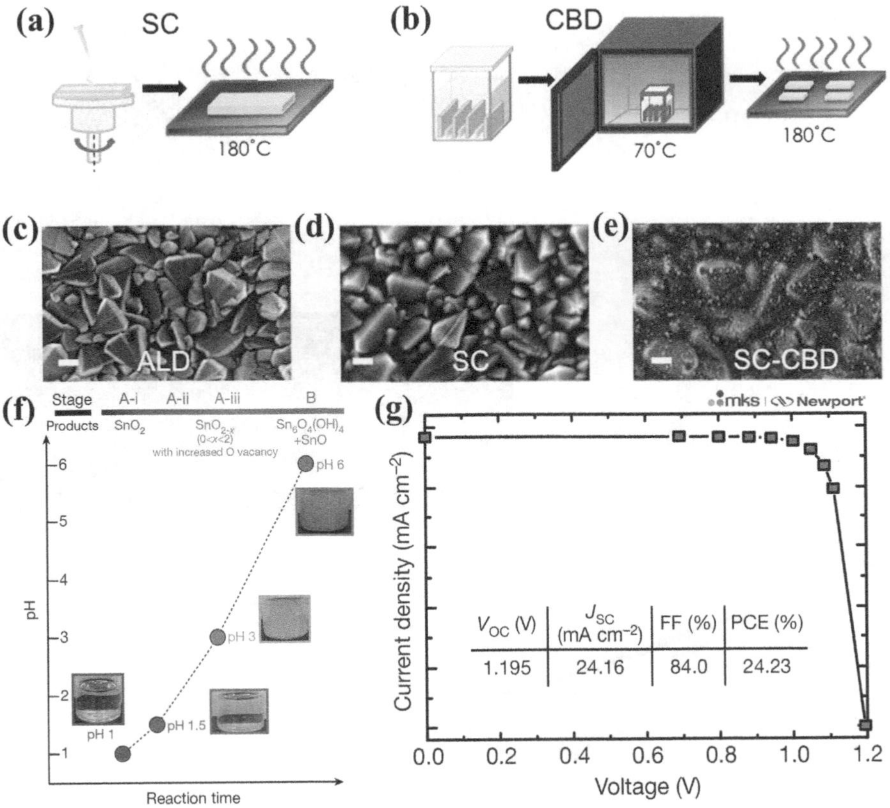

FIGURE 1.6 Schematic illustration of a) spin coating (SC) and b) chemical bath deposition (CBD) methods of SnO_2 thin-film deposition. Top view SEM images of SnO_2 films deposited by c) ALD, d) spin coating, and e) spin coating and CBD method, respectively. Reproduced with permission.[72] Copyright 2016, Royal Society of Chemistry. f) Schematic illustration of the reaction progress of CBD method. g) *J-V* curves of the champion device based on CBD SnO_2 ETL. Reproduced with permission.[73] Copyright 2021, Nature Publishing Group.

SnO_2 can also be deposited by other vapor-based processes. In 2015, Baena et al.[81] deposited the SnO_2 ETL by the ALD method at 118 °C. The PSCs based on SnO_2 ETL achieved a barrier-free energetic configuration and obtained a PCE of over 18% with a record high V_{OC} of up to 1.19 V and negligible hysteresis. Furtherly, they have promoted the PCE of PSCs based on ALD SnO_2 to over 20% and V_{OC} to 1.23 V.[82] Chen et al.[83] prepared SnO_2 ETL by the pulsed laser deposition (PLD) method at room temperature and achieved a PCE of 17.29%. Ma et al.[84] used e-beam deposition to prepare SnO_2 ETL, which is a low-cost and large-scale manufacturing technology to form uniform and thickness-controlled films. The PSCs based on e-beam deposition exhibited a PCE of 18.2%.

In addition to the improvement and innovation in the preparation process, it is also necessary to combine various interface engineering methods to further improve the device efficiency, which will be further discussed in Section 1.4.

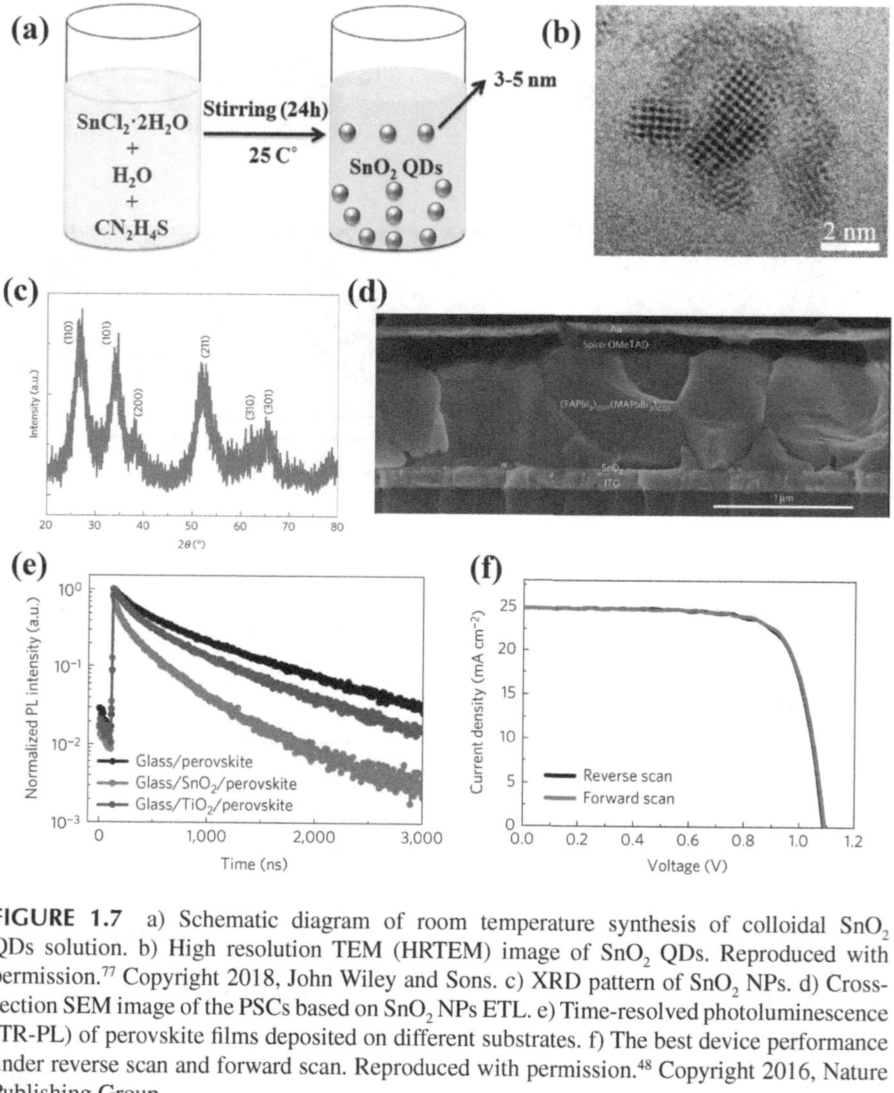

FIGURE 1.7 a) Schematic diagram of room temperature synthesis of colloidal SnO_2 QDs solution. b) High resolution TEM (HRTEM) image of SnO_2 QDs. Reproduced with permission.[77] Copyright 2018, John Wiley and Sons. c) XRD pattern of SnO_2 NPs. d) Cross-section SEM image of the PSCs based on SnO_2 NPs ETL. e) Time-resolved photoluminescence (TR-PL) of perovskite films deposited on different substrates. f) The best device performance under reverse scan and forward scan. Reproduced with permission.[48] Copyright 2016, Nature Publishing Group.

1.2.4 Other Inorganic ETLs

In_2O_3 is an n-type semiconductor with high electron mobility (10 $cm^2\ V^{-1}\ s^{-1}$), wide band gap (3.8 eV), and good thermal stability. Yoon et al.[85] directly spin-coated $In(NO_3)_3{\cdot}xH_2O$ solution to prepare In_2O_3 film, and they found the precursors rapidly decomposed at 270 °C. The PSC with optimized In_2O_3 at 300 °C showed the highest PCE of 14.63%. Qin et al.[86] deposited In_2O_3 ETL using the sol-gel method, combined with PCBM interface modification, and the PSCs based on In_2O_3 ETL achieved a PCE of 14.83%. Chen et al.[87] introduced acetylacetone into In_2O_3 sol-gel precursors to improve the surface morphology of In_2O_3 ETL, and the PSCs based on In_2O_3 ETL without PCBM achieved a PCE of 15.3%. Due to the good photostability of In_2O_3, the PSCs based on In_2O_3 ETL showed improved stability compared to those based on TiO_2 and ZnO ETL. To reduce the annealing temperature of In_2O_3 ETL, Guo et al.[5] reported a low-temperature solution combustion approach to deposit In_2O_3 ETL (Figure 1.8a–d). The indium nitrate hydrate and acetylacetone in precursors acted as oxidizers and fuel, which could cause an exothermic reaction during annealing to lower the crystal temperature. The PSCs based on In_2O_3 ETL at 200 °C exhibited the highest PCE of 18.12% with impressive stability.

Another n-type metal oxide, niobium oxide (Nb_2O_5), exhibits similar optical band gap and electronic properties, importantly, improved chemical stability relative to that of TiO_2. Kogo et al.[88] used a solution process to prepare Nb_2O_5 compact layer, which was found to work as an effective blocking layer in mesoporous Al_2O_3-based PSCs. As a result, the PSCs achieved a high V_{OC} of 1.13 V. Feng et al.[89] directly deposited Nb_2O_5 ETL by using E-beam evaporation without any further treatment.

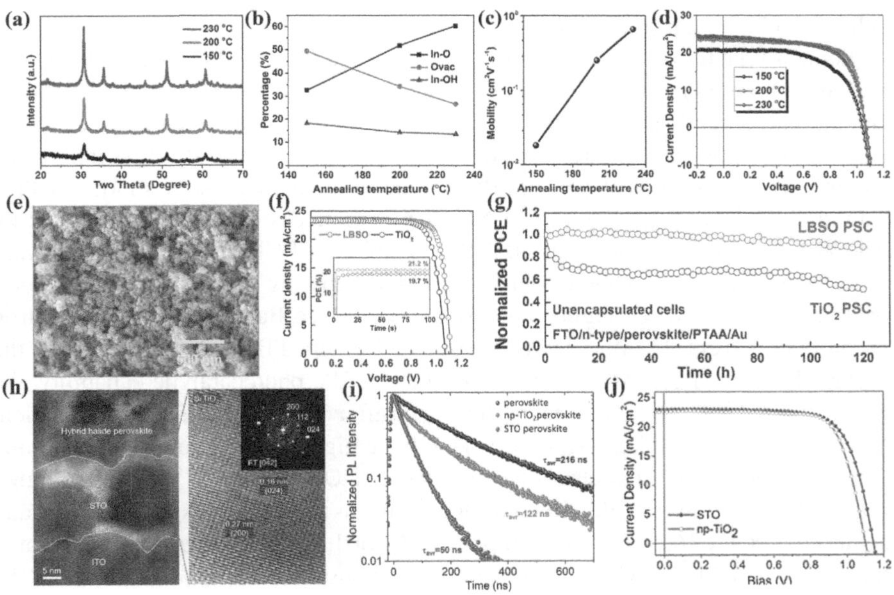

FIGURE 1.8 a) XRD patterns of In_2O_3 films annealed at 150–230 °C on glass substrates. b) Oxygen binding state percentage of In_2O_3 films annealed at different temperatures. c) The charge carrier mobilities of In_2O_3 TFTs as a function of annealing temperatures. d) *J-V* curves of In_2O_3-based devices. Reproduced with permission.[5] Copyright 2019, Elsevier. e) SEM image of the as-prepared $BaSnO_3$ powder synthesized at 50 °C. f) *J-V* curves, and (inset) stabilized PCEs at a maximum power point for the best-performing La-$BaSnO_{3-}$ and TiO_2-based PSCs. g) Photostability tests under constant AM 1.5G illumination with a xenon lamp, including UV radiation for two unencapsulated devices. Reproduced with permission.[91] Copyright 2017, AAAS. h) TEM images of the cross-sectional interface of ITO/$SrTiO_3$/perovskite samples. i) TR-PL of pristine perovskite, np-TiO_2/perovskite, and $SrTiO_3$/perovskite samples. j) *J-V* characteristics of the $SrTiO_3$ and np-TiO_2 devices. Reproduced with permission.[16] Copyright 2019, American Chemical Society.

It was found that 60 nm thick Nb_2O_5 film delivered the best performance of PSCs (18.59%). Dong et al.[8] synthesized chromium oxide (Cr_2O_3) nanosheets by a facile solvothermal reaction. Then they spin-coated the Cr_2O_3 ink on FTO substrates without further annealing process to prepare Cr_2O_3 ETL. The as-prepared Cr_2O_3 layer had high optical transmittance and superb electron extraction and carrier transport property. The PSCs base on Cr_2O_3 ETL achieved a PCE of 16.23%. Cerium oxide (CeO_x) is one of the most important rare-earth oxides, with a wide bandgap, large dielectric constant, high ionic conductivity, and high thermal and chemical stability. Wang et al.[10] reported a simple low-temperature solution process to prepare CeO_x ETL. By optimizing the concentration of the precursor solution, the PSCs base on CeO_x achieved a champion PCE of 14.32% and 17.04% with $PC_{61}BM$ modification. Tungsten oxide (WO_x) is also a promising ETL material due to its wide bandgap (2–3 eV), high electron mobility (10–20 $cm^2\ V^{-1}\ s^{-1}$), good chemical stability, and low processing temperature. Chen et al.[12] developed a record low temperature process to prepare nanocrystalline WO_x film at 50 °C. They also found that hexanol could help reduce charge recombination and increase the Fermi level. The corresponding PSCs achieved a high PCE of 20.77%.

Some n-type ternary metal oxides are also used as ETL in PSCs. $BaSnO_3$, $SrTiO_3$, and $SrSnO_3$ are perovskite-structured semiconductors with good chemical stability. $BaSnO_3$ has a bandgap of 3.1 eV and high electron mobility, which makes $BaSnO_3$ a successful application in dye-sensitized solar cells.[13,14] Zhu et al.[90] first introduced $BnSnO_3$ as a mesoporous layer into PSCs. Due to the higher electron mobility than TiO_2 and the same structure as $CH_3NH_3PbX_3$, the electron transfer was enhanced. The PSCs based on mesoporous $BaSnO_3$ exhibited a PCE of 12.3%. Seok et al.[91] reported a superoxide colloidal solution route for preparing a La-doped $BaSnO_3$ compact layer under 300 °C (Figure 1.8e–g). The La-doped $BaSnO_3$ has a high electron mobility of 320 $cm^2\ V^{-1}\ s^{-1}$. The PSCs based on La-doped $BaSnO_3$ ETL exhibited the highest PCE of 21.2%. Moreover, due to the low UV photocatalytic capability, the PSCs retained 93% of their initial performance after 1,000 h of full-sun illumination. $SrTiO_3$ has a wide band gap similar to TiO_2 but a higher conduction band, matching better with perovskite. The electron mobility of $SrTiO_3$ (5–8 $cm^2\ V^{-1}\ s^{-1}$) is also higher than that of TiO_2. Bera et al.[15] first reported $SrTiO_3$ as a mesoporous layer in PSCs and achieved a PCE of 6.97% with an improved V_{OC} of 1 V. Neophytou et al.[16] fabricated the $SrTiO_3$ as ETL using a low-temperature combustion process at 150 °C, and the PSCs based on $SrTiO_3$ ETL exhibited a PCE of 19% with good stability, retaining 80% of its initial efficiency after 1,000 h of constant illumination (Figure 1.8h–j). Luo et al.[92] spin-coated modified $SrSnO_3$ nanocrystals to prepare $SrSnO_3$ ETL. The $SrSnO_3$ ETL was highly transparent and conductive, and most importantly its lattice parameters were well matched with the perovskite film. The high lattice matching enables a more ordered beginning of the growth of halide perovskite on the electron transport layer, avoiding the formation of a deteriorated buried interface. The PSCs based on $SrSnO_3$ ETLs achieved a high PCE of 25.17% with improved stability. The PSCs maintained 90% of its initial PCE (24.4%) after operating at the maximum power point (MPP) for 1,000 h.

In addition to metal oxides, other n-type inorganic materials are also used as ETL in PSCs, especially metallic sulfides. Cadmium sulfide (CdS) has a high electron mobility of 440 $cm^2\ V^{-1}\ s^{-1}$. Liu et al.[93] prepared metal sulfide (CdS and ZnS) ETLs using CBD process. The ultraviolet photoelectron spectra (UPS) results showed that the conduction band offsets at CdS/perovskite and ZnS/perovskite interfaces were −0.1 eV and +1.3 eV, respectively. This result indicated that CdS/perovskite interface may have better charge transfer than ZnS/perovskite interface. The PSCs based on ZnS showed a low PCE of 0.98%, which was much lower than that based on CdS (11.17%). Peng et al.[20] prepared CdS film by CBD process at 100 °C and achieved a PCE of 16.1%. It is worth noting that Shohl et al.[94] found Cd could diffuse into perovskite film and react with $MAPbI_3$ to form MA_2CdI_4, which acted as an electrical barrier at the interface and affected the performance of PSCs, while Tong et al.[95] found that CdS could lead to an in-situ Cd-doping in perovskite film and improve the device performance. The mechanism of the influence of CdS on perovskite film still needs to be deeply studied. Apart from the CBD process, Dong et al.[96] directly spin-coated CdS NPs to deposit CdS ETL without any further high-temperature annealing process, and finally the PSCs achieved a PCE of 16.5%. Li et al.[21] used thermal evaporation to fabricate a compact and smooth amorphous Bi_2S_3 film with high conductivity to replace the PCBM in inverted PSCs. The PSCs with the structure of $NiO/CH_3NH_3PbI_3/Bi_2S_3$ exhibited a PCE of 13%, which was comparable with the PSCs based on PCBM ETL. Moreover, due to the hydrophobic and hermetic encapsulation of Bi_2S_3, the PSCs showed improved ambient stability. In_2S_3 has an electron mobility of 17.6 $cm^2\ V^{-1}\ s^{-1}$ and a band gap of 2.0–2.8 eV. Hou et al.[22] developed a low-temperature CBD method to prepare In_2S_3 films (Figure 1.9a–c). They prepared a well-organized In_2S_3 nanoflake array by optimizing the reaction time. The PSCs based on In_2S_3 ETL achieved a PCE of 18.22% with negligible hysteresis, which was higher than that based on TiO_2 ETL (15.70%). Yang et al.[97] constructed $CsPbIBr_2$ PSCs based on In_2S_3 ETL deposited by low-temperature CBD methods, and the PSCs exhibited a PCE of 5.59%. Chen et al.[98] proposed a facial and controllable reflux condensation technique to prepare In_2S_3 film as ETL for $CsPbBr_3$ PSCs. The PSCs with the structure of $FTO/In_2S_3/CsPbBr_3/Carbon$ exhibited a PCE of 6.54%. Singh et al.[23] first reported a MoS_2 thin film as the ETL for PSCs. They directly fabricated thin uniform MoS_2 film on an FTO substrate by using microwave-assisted low-temperature synthesis in the solution phase. The MoS_2 thin film could form a thermal stable interface with perovskite without annealing. The PSCs based on MoS_2 ETL exhibited a PCE of 13% with a high current density of 21.7 mA/cm^2, comparable to those based on TiO_2 and SnO_2 ETL. Zhao et al.[24] developed a two-dimensional (2D) SnS_2 ETL deposited by self-assembly stacking method (Figure 1.9d–f). The large-scaled 2D multilayer SnS_2 sheet structure triggered a heterogeneous nucleation of perovskite film. The intermolecular interactions between Pb and S could passivate the interfacial trap states, which suppressed charge recombination and facilitated electron extraction at interface. The PSCs based on SnS_2 achieved a PCE of 20.12% with a high V_{OC} of 1.161 V and a high J_{SC} of 23.55 mA/cm^2.

Gallium nitride (GaN) has a direct bandgap of 3.4 eV, exceeding 82% transmittance over the entire visible wavelengths and high electron mobility (~1,000 $cm^2\ V^{-1}\ s^{-1}$). These excellent optical and electrical properties have attracted researchers to

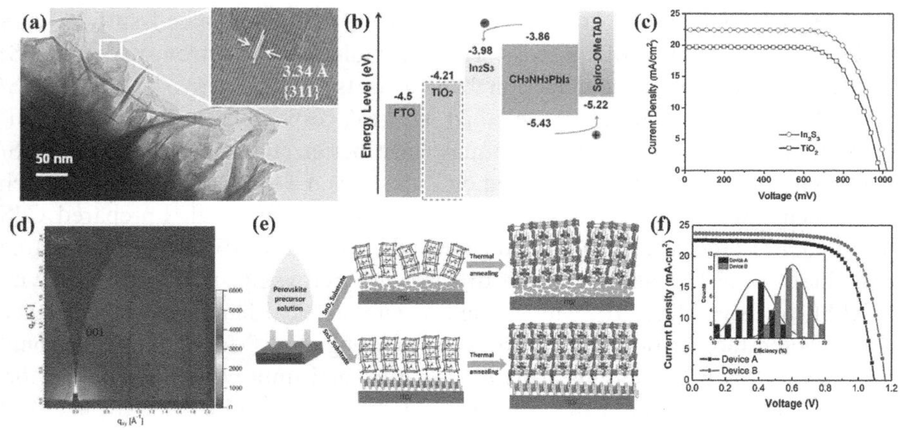

FIGURE 1.9 a) High-resolution TEM image of In_2S_3 nanoflake array, inset showing the lattice fringes with interplanar spacing. b) Energy level of In_2S_3 based PSCs. c) *J-V* curves of PSCs based on In_2S_3 and TiO_2 ETLs. Reproduced with permission.[22] Copyright 2017, Elsevier. d) Grazing incidence wide-angle X-ray scattering (GIWAXS) patterns of 2D multilayer SnS_2 film prepared by self-assembly stacking deposition method. e) The one-step spin-coating forming process of perovskite films from the precursor solution onto SnO_2 and SnS_2 substrates. f) *J-V* curves and histogram of efficiency (inset) of devices based on SnO_2 (Device A) and SnS_2 (Device B) ETL. Reproduced with permission.[24] Copyright 2019, John Wiley and Sons.

study the combination of GaN with perovskites. Lee et al.[99] introduced nanoporous (NP) GaN epitaxially grown on n-type GaN (n-GaN) (NP GaN/n-GaN) to form a dual-function cathode structure replacing transparent conductive layers and ETL. The $MAPbI_3$ PSCs based on NP GaN/n-GaN cathode structure (without any other ETLs) achieved a high PCE of 18.79%. MXenes are a large family of 2D materials with unique optoelectronic properties and tunable surface termination. Yang et al.[100] first used $Ti_3C_2T_x$ MXenes in PSCs as ETL. $Ti_3C_2T_x$ ETL was prepared by spin-coating colloidal solutions onto ITO substrates, followed by UV-ozone treatment. UV-ozone treatment could form oxide-like Ti-O bonds, which could improve the properties of $Ti_3C_2T_x$/perovskite interface. The PSCs based on $Ti_3C_2T_x$ achieved a PCE of 17.17%.

1.3 ORGANIC ETLs

N-type inorganic materials are usually prepared from precursors based on polar solvents such as water or alcohol and require high annealing temperatures to obtain good crystallinity and electrical properties, which is suitable for n-i-p structures. In p-i-n PSCs, the ETL is deposited on perovskite film, the n-type organic materials with low fabrication temperature, suitable energy level, and high electron mobility are more suitable.

TABLE 1.2
Device Structure and Performance of PSCs Based on Other ETLs

ETLs	Device structure	V_{OC} (V)	J_{SC} (mA/cm²)	FF	PCE (%)	Ref.
In_2O_3	$ITO/In_2O_3/PCBM/MA_{1-y}FA_yPbI_{3-x}Cl_x/$Spiro-OMeTAD/Ag	1.06	23.40	0.73	18.12	5
Nb_2O_5	$FTO/Nb_2O_5/(FAPbI_3)_{1-x}(MAPbBr_3)_x/$Spiro-OMeTAD/Au	1.08	24.00	0.74	19.2	101
Cr_2O_3	$FTO/Cr_2O_3/Cs_{0.05}(MA_{0.17}FA_{0.83})_{0.95}Pb(I_{0.83}Br_{0.17})_3/$Spiro-OMeTAD/Au	1.10	21.28	0.70	16.23	8
CeO_x	$FTO/CeO_x/PC_{61}BM/MAPbI_3/$Spiro-OMeTAD/Ag	1.06	23.25	0.69	17.04	10
WO_x	$FTO/WO_x/MAPbI_3/$Spiro-OMeTAD/Ag	1.08	22.71	0.78	19.16	12
$BaSnO_3$	FTO/La doped $BaSnO_3/MAPbI_3/PTAA/Au$	1.12	23.40	0.81	21.30	91
$SrTiO_3$	$ITO/SrTiO_3/Cs_{0.07}FA_{0.73}MA_{0.20}PbI_{2.53}Br_{0.47}/$Spiro-OMeTAD/Au	1.14	23.02	0.72	19.00	16
$SrSnO_3$	$ITO/SrSnO_3/Cs_{0.05}(FAMA)_{0.95}PbI_3/$Spiro-OMeTAD/Au	1.17	25.72	0.84	25.17	92
CdS	$FTO/CdS/Cs_{0.05}(MA_{0.17}FA_{0.83})_{0.95}Pb(I_{0.83}Br_{0.17})_3/$Spiro-OMeTAD/Au	1.09	21.58	0.70	16.50	96
In_2S_3	$FTO/In_2S_3/(FAPbI_3)_{0.85}(MAPbBr_3)_{0.15}/$Spiro-OMeTAD/Ag	1.03	22.54	0.78	18.22	22
MoS_2	$FTO/MoS_2/CH_3NH_3PbI_3/$Spiro-OMeTAD/Au	0.89	21.70	0.64	13.14	23
SnS_2	$ITO/SnS_2/FA_{0.75}MA_{0.15}Cs_{0.1}PbI_{2.65}Br_{0.35}/$Spiro-OMeTAD/Au	1.16	23.55	0.73	20.12	24
GaN	Sapphir/n-GaN/NP $GaN/MAPbI_3/$Spiro-OMeATD/Au	1.07	22.87	0.77	18.79	99
$Ti_3C_2T_x$	$ITO/Ti_3C_2TX/MAPbI_3/$Spiro-OMeTAD/Ag	1.08	22.63	0.70	17.17	100

1.3.1 Fullerenes and Their Derivatives

Fullerenes and their derivatives are commonly used as ETL in PSCs (Figure 1.10a–d). The low-cost C_{60} can be packed densely to enhance the charge transfer, which shows an electron mobility of 1.6 $cm^2\ V^{-1}\ s^{-1}$ and conductivity of 2.3×10^{-3} S cm^{-1}.[102] Due to the low solubility in generally used solvents, most of the compact C_{60} film is prepared by a thermal evaporation method. Ke et al.[103] deposited C_{60} layer as ETL in n-i-p structure by employing a vacuum deposition process. The PSCs based on C_{60} layer achieved a maximum PCE of 15.14%. Their results suggested that low-temperature and vacuum deposited fullerene could be used as an efficient ETL for flexible PSCs.

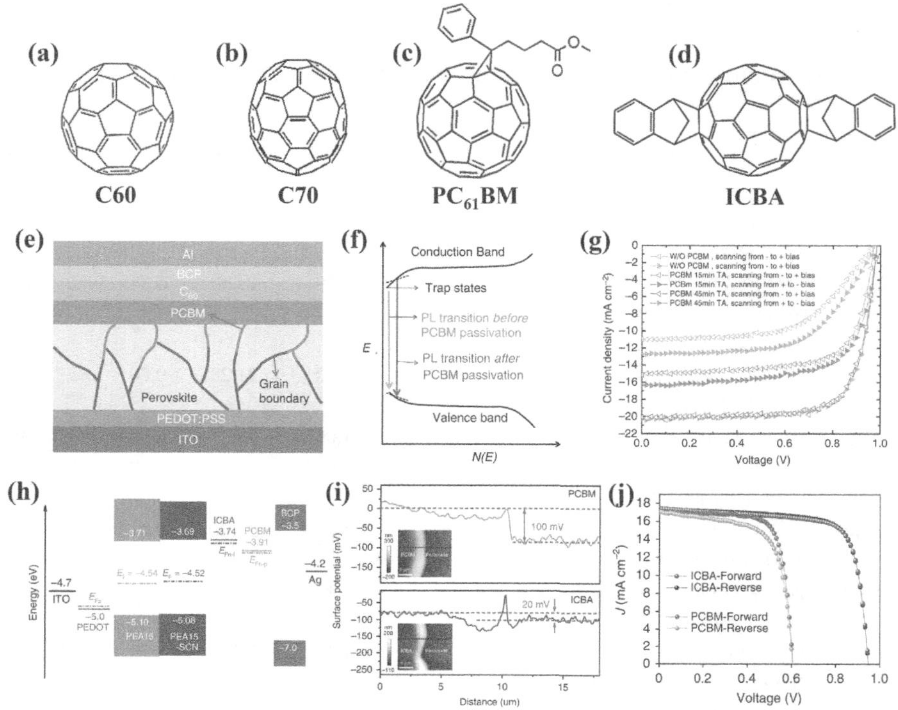

FIGURE 1.10 Chemical structures of fullerenes and their derivatives discussed in this section. a) C_{60}, b) C_{70}, c) $PC_{61}BM$, and d) ICBA. a–b) Reproduced with permission.[107] Copyright 2018, American Chemical Society. c–d) Reproduced with permission.[114] Copyright 2018, John Wiley and Sons. e) Device structure of PSC with PCBM layer. f) Schematic of steady-state photoluminescent (PL) peaks blue-shift from passivation effect. g) *J-V* curves of devices with PCBM layer. Reproduced with permission.[108] Copyright 2014, Nature Publishing Group. h) Schematic illustration of energy levels. i) Surface potential distribution of perovskite film with NH_4SCN (PEA15-SCN)/PCBM and PEA15-SCN/ICBA; the insert images are AFM topography images for the corresponding samples. j) *J-V* curves of the device with ICBA and PCBM. Reproduced with permission.[113] Copyright 2020, Nature Publishing Group.

Yoon et al.[104] found that a 35 nm-thick C_{60} layer showed a homogenously compact layer, which could improve the electron extraction and eliminate the hysteresis of PSCs. The PSCs exhibited a PCE of 19.1% and the flexible PSCs exhibited a PCE of 16.0%. Recently, Jiang et al.[105] achieved a PCE of 25.37% based on C_{60} ETL, which was the highest PCE of the p-i-n PSCs up to now. Similar to C_{60}, Nodoushan et al.[106] prepared C_{70} films as ETL for PSCs by organic molecular beam deposition. The PSCs showed the highest PCE of 12% with an optimized C_{70} thickness of 10 nm. Lin et al.[107] studied the reasons for the low performance of fullerene-based PSCs prepared by spin-coating compared with thermal deposition. They found the crystallinity of the fullerene influenced the performance of ETL. They mixed C_{70} into C_{60} to reduce the crystallinity of the spin-coated film, and the mixed-fullerene-based PSCs showed a PCE of 16.7%. Combined with vacuum-dried spin-coated mixed-fullerene, the PCE of PSCs was further improved to 18.0%.

Pristine fullerene materials exhibit low solubility and insufficient interaction with perovskite film. It is necessary to introduce chemical modification to improve the solubility and regulate the electronic configurations of fullerene. $PC_{61}BM$ has been widely used as ETL in PSCs due to its low-temperature solution process and highly efficient electron transport. Most importantly, Shao et al.[108] found that $PC_{61}BM$ could efficiently passivate charge trap states at $CH_3NH_3PbI_3$ perovskite film surface and grain boundaries and suppress the hysteresis of PSCs (Figure 1.10e–g). The PCE of p-i-n PSCs based on $PC_{61}BM$ ETL was close to that of n-i-p structure PSCs. Notably, $PC_{61}BM$ could be partially dissolved by the solvent of perovskite precursor, and in most reports, $PC_{61}BM$ are used as top ETL in p-i-n PSCs. Wu et al.[109] reported a solution-processed and annealing-free $PC_{61}BM$ ETL in n-i-p PSCs. They doped perovskite precursor solution with $PC_{61}BM$ to prepare a presaturated solution. As a result, the underlying $PC_{61}BM$ layer could be reserved and the defects in perovskite grain boundaries could be passivated. The $CH_3NH_3PbI_{3-x}Cl_x$-based champion device showed the highest PCE of 20.85%. Organic ETLs have a great advantage in continuous printing process. Lee et al.[110] developed a self-organized organic ETL based on $PC_{61}BM$. They combined polyethyleneimine (PEI) as a surface work function modifier and $PC_{61}BM$ as an electron acceptor, using a spontaneous vertical phase separation driven by the surface energy difference of the PEI and $PC_{61}BM$ to form the ETLs with electron acceptor and surface work function modifier. The printed PEI:$PC_{61}BM$ film was comparably uniform and smooth to the spin-coated film. The PSCs based on printed PEI:$PC_{61}BM$ ETLs achieved a PCE of 17.3%.

$PC_{61}BM$ is widely used as ETL in p-i-n PSCs and achieved a relatively higher efficiency. But the high degree of energy disorder and severe self-aggregation of $PC_{61}BM$ limit the efficiency and lifespan of devices. Jiang et al.[111] designed a series of fullerene dyads FP-Cn (n = 4, 8, 12) to replace $PC_{61}BM$. Among them, FP-C8 showed the most enhanced molecule ordering and adhesion with the perovskite surface due to the balanced decoupling between the chelation effect from terpyridine and the self-assembly of fullerene. The PF-C8/C_{60}-based PSCs reached a PCE of 21.69%, which was higher than that of $PC_{61}BM/C_{60}$ (20.09%). Moreover, the FP-C8-based PSCs showed better moisture and thermal stability than PCBM/C_{60}-based PSCs and maintained 96% of their original efficiency after 1,200 h of operation. Xie et al.[112] reported a multi-adduct fullerene ETL named hexakis[di(ethoxycarbonyl)methano]-C60 (HEMC). HEMC has twelve ester groups onto the fullerene cage to form a star-like molecule, leading to high solubility in polar solvents. HEMC could bond with perovskite films by Lewis base-acid interactions between carbonyl groups and unsaturated Pb^{2+} to enhance the electron extraction at interface. The PSCs based on HEMC ETL achieved a champion PCE of 20.02%. Indene-C_{60} bisadduct (ICBA) is another fullerene derivative used as ETL for PSCs. ICBA can be easily synthesized from C_{60} and has high solubility in common organic solvents. ICBA is similar in size to PCBM and has high electron affinity. The lowest unoccupied molecular orbital (LUMO) of ICBA is about 0.17 eV higher than that of PCBM, which is closer to the conduction band of perovskite film, leading to more efficient charge extraction from perovskite film. Jiang et al.[113] replaced PCBM with ICBM as ETL for tin PSCs (Figure 1.10h–j). ICBA with a shallower energy level significantly improved the V_{OC} of PSCs to 0.94 V, which was much higher than that of PSCs based on PCBM ETL (0.6 V). Table 1.3 summarized the photovoltaic parameters of PSCs based on different fullerene derivatives.

TABLE 1.3
Photovoltaic Parameters of Devices Based on Different Fullerene Derivatives

Device structure	Fullerene derivatives LUMO (eV)	Fullerene derivatives μ $cm^2 V^{-1} s^{-1}$	V_{OC} (V)	J_{SC} (mA/cm^2)	FF	PCE (%)	Ref.
ITO/CPTA/$CH_3NH_3PbI_3$/Spiro-OMeTAD/Au	-3.90	5.4×10^{-3}	1.10	22.06	0.76	18.39	115
FTO/PyCEE/$CH_3NH_3PbI_3$/Spiro-OMeTAD/Ag	-3.94	\	1.05	22.95	0.76	18.27	116
FTO/Sil-C_{60}/ $Cs_{0.05}FA_{0.75}MA_{0.2}PbBr_{0.3}I_{2.7}$/Spiro-OMeTAD/Au	-3.51	\	1.04	19.40	0.74	15.20	117
FTO/Sol-gel C_{60}/ $MAPbI_{3-x}Cl_x$/Spiro-OMeTAD/Au	\	3.8×10^{-4}	1.07	23.00	0.73	17.90	118
ITO/PTAA/PFN-Br/$Cs_{0.05}FA_{0.90}MA_{0.05}PbI_{2.85}Br_{0.15}$/FP-C8/BCP/Ag	-4.15	4.9×10^{-3}	1.10	23.83	0.83	21.69	111
FTO/NiO_x/$MAPbI_xCl_{3-x}$(HEMC)/ HEMC/BCP/Ag	-3.94	2.0×10^{-4}	1.09	22.31	0.79	19.14	112
ITO/PEDOT/$PEA_xFA_{1-x}SnI_3$-SCN/ICBA/BCP/Ag	-3.74	\	0.94	17.40	0.75	12.40	113
ITO/2PACz/$CsPbI_2Br$/ICBA/BCP/Ag	\	\	1.29	15.25	0.82	16.00	119
FTO/PEDOT:PSS/$CH_3NH_3PbI_3$/ $PC_{61}BEH$/Au	-3.89	4.76×10^{-4}	0.95	22.05	0.78	16.26	120
FTO/NiO_x/perovskite/PCBB/BCP/Ag	-3.90	3.09×10^{-4}	1.08	22.85	0.80	19.84	121
PET/IZO/PTAA/$Cs_{0.04}(MA_{0.17}FA_{0.83})_{0.96}Pb(I_{0.83}Br_{0.17})_3$/PCBC6/BCP/Ag	-3.51	\	1.05	23.22	0.76	18.41	122
FTO/NiO_x/ $MAPbI_3$/ C5-NCMA/Ag	-3.87	1.59×10^{-3}	1.08	20.68	0.79	17.60	123
FTO)/NiO_x/$MAPbI_3$/C60–3-Py/ BCP/Ag	-3.80	2.95×10^{-3}	1.02	22.46	0.76	17.57	124
ITO/PTAA/$Cs_5(MA_{0.17}FA_{0.83})_{95}Pb(I_{0.83}Br_{0.17})_3$/AHF-2/BCP/Au	-4.13	1.23×10^{-6}	1.11	23.03	0.76	20.21	[125]
ITO/P3HT-COOH/$Cs_{0.05}FA_{0.78}MA_{0.17}PbI_{2.53}Br_{0.47}$/5F-PCBP/PEI/Ag	-4.24	2.27×10^{-4}	1.11	23.53	0.82	21.27	[126]

1.3.2 Other n-Type Organic Materials

Various non-fullerene electron acceptors not only retain the properties of fullerenes but also have some advantages compared with fullerenes, such as easily tunable electronic energy levels and good thermal and chemical stability. Recently, several well-known organic molecules, including perylenediimide (PDI), naphthalene imide (NDI), and hexaazatrinaphthylene (HATNA) derivatives have been developed as ETLs in PSCs. Their electronic, optical, and morphological characteristics can be designed and adjusted accurately.

PDI has high electrical conductivity (3.43×10^{-5}–4.4×10^{-4} S cm^{-1}) and electron mobility (0.2–1.3 cm^2 V^{-1} s^{-1}), and a relatively low LUMO (3.5–4.3 eV), which has been widely studied as ETL for PSCs. The properties of PDI could be controlled by molecular design and morphological regulation. Huang et al.[127] introduced PDI with two fluorine groups (FPDI) as ETL for PSCs, and they studied the influences of FPDI morphology controlled by different solvent treatments on perovskite morphology. By optimizing the FPDI morphology, the FF of PSCs based on FPDI was significantly increased, leading a PCE of 7.93%. Zhang et al.[128] reported amino-substituted PDI (N-PDI) as ETL. The PDI backbone provided suitable energy levels and good electron-transport mobility (Figure 1.11a). The terminal amino group enhanced the surface wetting capability of N-PDI, reduced the surface work function of the FTO, and passivated the surface trap states of the perovskite films. The PSCs based on N-PDI ETL achieved PCEs of 17.66% and 14.32% for rigid and flexible substrates, respectively. Wu et al.[129] developed three solution processable perylene diimides (H-PDI, F-PDI, and Br-PDI) as ETLs. Among them, the PSCs based on F-PDI did not show any photovoltaic effect due to the poor thin-film quality caused by the low solubility. The solubility of H-PDI was better than F-PDI, but the performance of PSCs based on H-PDI was relatively worse due to its inferior electron mobility and conductivity. The Br-PDI had the highest solubility, electron mobility, and conductivity, and the PSCs based on Br-PDI ETLs achieved the highest PCE, which was almost as efficient as $PC_{61}BM$-based PSCs. Zou et al.[130] used tetrachloroperylene diimide (TCl-PDI) to replace TiO_2 as ETL for PSCs. The PSCs based on TCl-PDI exhibited higher V_{OC} and dramatically improved FF compared to that of TiO_2-based PSCs, which maybe come from the better energy band alignment and reduced nonradiative recombination at TCl-PDI/perovskite interface. The PSCs based on TCl-PDI ETL achieved a PCE of 14.73%. Moreover, TCl-PDI based PSCs displayed higher stability than TiO_2 based PSCs due to the improved perovskite film quality.

It is worth noting that the strong π-π stacking of PDI molecules could cause aggregation, which decreased the solubility and film quality. Introducing molecular twisting can prevent PDIs from forming excessively large aggregates. Jiang et al.[131] demonstrated a 3D type PDI-based molecule ($TPE-PDI_4$), comprising a tetraphenylethylene core surrounded by four PDIs, which was an efficient ETL in inverted PSCs (Figure 1.11b). The electron mobility of $TPE-PDI_4$ measured by space-charge-limited current (SCLC) method was about 1.73×10^{-2} cm^2 V^{-1} s^{-1}. Moreover, the energy levels of $TPE-PDI_4$ and $PC_{61}BM$ were similar. The highly twisted 3D molecular structure could provide a good solubility in common organic solvents. The PSCs based on $TPE-PDI_4$ ETLs achieved a PCE of 16.29%, which was slightly higher than that

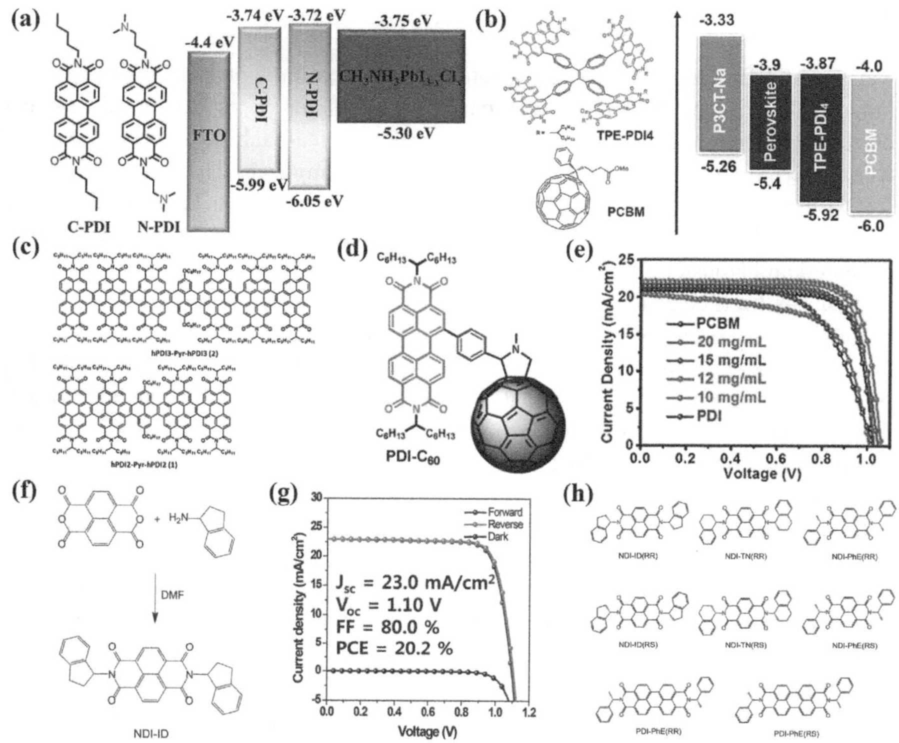

FIGURE 1.11 a) Molecular structures of the C-PDI and N-PDI used as ETL, as well as the corresponding energy band diagram. Reproduced with permission.[128] Copyright 2016, Royal Society of Chemistry. b) Chemical structure of TPE-PDI4 and PCBM and corresponding energy levels. Reproduced with permission.[131] Copyright 2018, Royal Society of Chemistry. c) Molecular structures of hPDI2-Pyr-hPDI2 and hPDI3-Pyr-hPDI3. Reproduced with permission.[132] Copyright 2018, John Wiley and Sons. d) Molecular structure of PDI-C_{60}. e) *J-V* curves of the PSCs based on different ETLs. Reproduced with permission.[133] Copyright 2018, John Wiley and Sons. f) Synthetic route of NDI-ID. g) *J-V* curves of the PSCs based NDI-ID ETLs. Reproduced with permission.[136] Copyright 2018, John Wiley and Sons. h) Chemical structures of NDI-based and PDI-based materials. Reproduced with permission.[138] Copyright 2020, John Wiley and Sons.

of $PC_{61}BM$-based PSCs (15.27%). Castro et al.[132] reported two cove-edge graphene nanoribbons hPDI2-Pyr-hPDI2 and hPDI3-Pyr-hPDI3 as ETLs for inverted PSCs (Figure 1.11c). HPDI3-Pyr-hPDI3 had a higher electron mobility and more matched energy levels than hPDI2-Pyr-hPDI2 due to its larger conjugate plane. The PSCs based on hPDI3-Pyr-hPDI3 exhibited a PCE of 16.5%, which was higher than that of hPDI2-Pyr-hPDI2 (15.6%) and $PC_{61}BM$- (14.9%) based PSCs. Importantly, the PSCs based on hPDI3-Pyr-hPDI3 and hPDI2-Pyr-hPDI2 ETLs showed improved stability compared to $PC_{61}BM$-based PSCs due to their more hydrophobic surfaces. The PSCs based on PDI derivatives ETLs usually exhibit lower PCE but higher stability, while the PSCs based on fullerene derivative ETL exhibit higher PCE but lower stability. To combine the advantages of PDI and fullerene, Luo et al.[133] reported a PDI/fullerene hybrid (PDI-C_{60}), which had an improved electron mobility and molecular ordering compared to $PC_{61}BM$ due to the introduction of PDI unit with large π-conjugated structure (Figure 1.11d). The N and C=O in PDI could provide lone pair electrons to form coordination bonds with under-coordinated Pb atoms to enhance the stability of perovskite film. The PSCs based on PDI-C_{60} achieved a champion PCE of 18.6%, which was significantly higher than that of $PC_{61}BM$ (16.6%) and PDI (13.8%; Figure 1.11e). Meanwhile, the higher hydrophobicity of PDI-C_{60} resulted in better ambient stability of the PDI-C_{60}-based PSCs compared to $PC_{61}BM$-based PSCs. Zheng et al.[134] developed a PDI-fullerene dimer by incorporating PDI with two fullerenes. To further increase solubility and conductivity, halogen anions were introduced to induce an intramolecular self-doping. The final product was named C60-PDI-I. Compared with $PC_{61}BM$, C60-PDI-I exhibited dramatically enhanced electron mobility and reduced molecular self-aggregation. The PSCs based on C60-PDI-I ETLs achieved a champion PCE of 18.29%, which was higher than that of $PC_{61}BM$ (15.60%) based PSCs. Moreover, the ambient stability was significantly improved due to the increased surface hydrophobic and morphology of C60-PDI-I films.

NDI is another n-type non-fullerene electron acceptor, which can be synthesized in short steps with high yields and has been demonstrated to show high electron mobility and thermal stability. Kwon et al.[135] introduced N-substituted phenylmethyl into the NDI core to form NDI-PM as ETL for p-i-n PSCs. NDI-PM showed comparable electron mobility, charge extraction, and solubility to $PC_{61}BM$. The NDI-PM-based PSCs reached a high PCE of 18.4%, which was close to the $PC_{61}BM$-based PSCs (18.9%). Although the PCE of NDI-PM-based PSCs was slightly lower than that of $PC_{61}BM$-based PSCs, the NDI-PM-based PSCs provided improved stability than $PC_{61}BM$-based PSCs. They also synthesized a new NDI-ID consisting of an N-substituted indane group and having simultaneous alicyclic and aromatic characteristics by a low-cost, one-step reaction and facile purification method (Figure 1.11f).[136] The NDI-ID had a high electron mobility of 0.2 $cm^2\ V^{-1}\ s^{-1}$. The PSCs based on NDI-ID ETL reached a champion PCE of 20.2% (Figure 1.11g). Meanwhile, the NDI-ID-based PSCs exhibited high long-term stability, which retained 90% of their initial PCE after 500 h at 100 °C with 1 sun illumination without encapsulation. Further, they added two methyl groups into NDI-PM to form two asymmetric-shaped chiral (R)-1-phenylethyl (PhE) groups, named NDI-PhE (Figure 1.11h).[137] The PhE groups could reduce the molecular symmetry and increase the free volume. The NDI-PhE exhibited high solubility (about two times higher than $PC_{61}BM$ in o-dichlorobenzene) and excellent film-forming ability. The inverted PSCs based on NDI-PhE ETLs achieved a high PCE of 20.5%, which was higher than $PC_{61}BM$-based PSCs (20.2%). To clarify the influence of stereoisomerism on the properties of NDI. They designed a series of chiral stereoisomers of NDI derivatives with two chiral substituents.[138] It was found that enantiomeric NDI derivatives with homochiral side groups had more asymmetrical crystal characteristics and higher solubility, compared to mesomeric NDI derivatives. The PSCs based on NDI enantiomer-based ETLs exhibited a high PCE of 19.07 ± 0.65% and high thermal stability with less than 10% degradation of initial PCE at 100 °C for 1,000 h in air without encapsulation. It is also worth noting the aggregation of NDIs would decrease the solubility and film quality. Similar to PDIs, building distorted spatial structures could enhance the isotropy of charge transport and inhibit the aggregation of NDIs. Ge et al.[139] designed two core-expanded NDI derivatives NDI3HU-DTYM2 and (DTYM-NDI-DTYA)2 with planar and twisted molecular backbones, respectively. The twisted molecular skeleton weakened the aggregation effect of molecules and enhanced the isotropic charge transport. The PSCs based on (DTYM-NDI-DTYA)2 exhibited a PCE of 12.9%, which was higher than that of NDI3HU-DTYM2-based PSCs (8.7%).

Other n-type small molecules are also developed as efficient ETLs for PSCs. Jen et al.[140] designed a serious of HATNA-based molecules by replacing three F groups on HATNA-F_6 with three alkylsulfanyl chains bearing different alkyl-chain lengths (Figure 1.12a). This modification could increase solubility and improve energy level alignment. The PSCs based on HATNASOC7-Cs ETLs exhibited a higher PCE of 17.62% compared to $PC_{61}BM$-based PSCs (15.9%; Figure 1.12b). The stability of PSCs was also improved due to the significantly increased surface hydrophobicity of HATNASOC7-Cs. Wang et al.[141] designed a new n-type sulfur-containing small molecule hexaazatrinaphtho[2,3-c][1,2,5]thiadiazole (HATNT) as ETL for PSCs, which has a high electron mobility of 1.73×10^{-2} cm^2 V^{-1} s^{-1}. The PSCs based on HATHT ETL exhibited a PCE of 18.1%, which was higher than that of PSCs based on $PC_{61}BM$. The improvement was due to the more efficient suppression of charge recombination at the perovskite/HATNT interface than at the perovskite/$PC_{61}BM$ interface. Zhan et al.[142] reported a fused-ring electron acceptor, indacenodithiophene (IDIC), as ETL for n-i-p PSCs (Figure 1.12c). IDIC had good solution processability, a high electron mobility of 1.1×10^{-3} cm^2 V^{-1} s^{-1} and a suitable LUMO energy level matched with the CBM of the perovskite film. The PSCs based on IDIC ETLs showed a champion PCE of 19.1%, which was higher than that of TiO_2-based PSCs (17.4%). Furthermore, the stability and hysteresis of the IDIC-based PSCs had been significantly improved against the TiO_2-based devices.

Except for small molecules, abundant acceptor polymer materials are also promising ETLs for PSCs due to their easily tunable properties, good morphological stability, strong solvent resistance (DMF, DMSO and GBL etc.), and film-forming properties. Tan et al.[143] designed a series of PDI-based polymers (PX-PDIs) containing different copolymerized units (X), including vinylene (V), thiophene (T), selenophene (Se), dibenzosilole (DBS), and cyclopentadithiophene (CPDT; Figure 1.12d). They studied the influence of energy alignment, electron mobility, and film morphology of PX-PDIs on PSC performance. The PSCs based on PV-PDI ETLs reached the highest PCE of 10.14% due to the deeper LUMO energy level, the highly delocalized LUMO electron density, and better planar structure of PV-PDI. Sun et al.[144] used an amino-functionalized copolymer (PFN-2TNDI) with a conjugated backbone composed of fluorene, naphthalene diimide, and thiophene spacers as ETL for PSCs (Figure 1.12e and f). It was found that the amines on the polymer side chains could passivate the surface traps of perovskite film and reduce the work function of metal cathode simultaneously. The PSCs based on PFN-2TNDI ETL exhibited a higher PCE of 16.7% than that based on PCBM ETL (12.9%). In addition, they also found that a wide range of thicknesses of PFN-2-TNDI can be applied to produce high efficiency due to its good electron transport property. He et al.[145] reported two

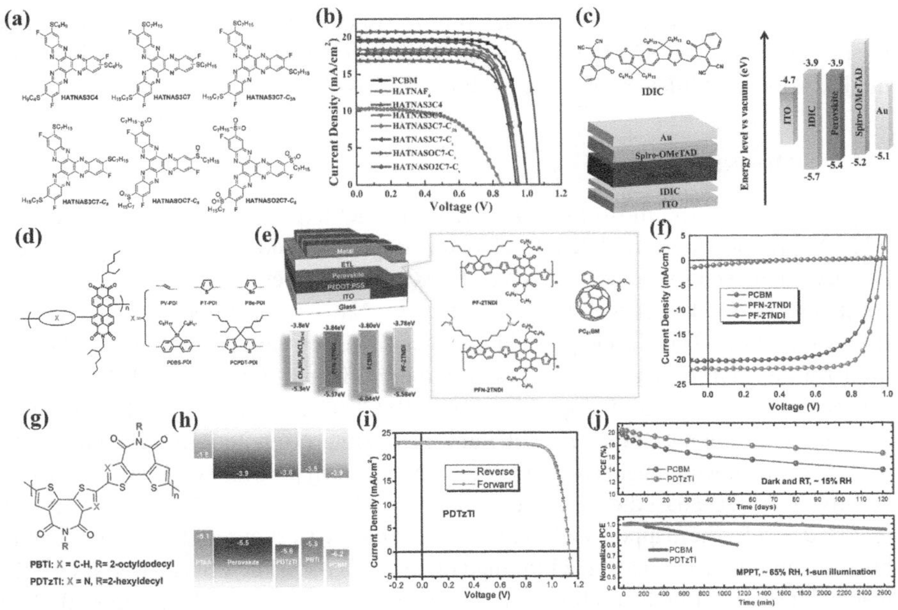

FIGURE 1.12 a) Chemical structures of HATNA-derived ETLs. b) *J-V* curves of the PSCs based on HATNA derivatives ETLs. Reproduced with permission.[140] Copyright 2016, John Wiley and Sons. c) Chemical structure of IDIC and corresponding device structure and energy level diagram. Reproduced with permission.[142] Copyright 2017, Royal Society of Chemistry. d) Molecular structures of PX-PDIs. Reproduced with permission.[143] Copyright 2017, American Chemical Society. e) Molecular structures of PFN-2TNDI and PF-2TNDI ETLs and the corresponding energy band diagram. f) *J-V* curves of the PSCs based on PFN-2TNDI and PF-2TNDI ETLs. Reproduced with permission.[144] Copyright 2018, John Wiley and Sons. g) Molecular structures of the n-type polymers PDTzTI and PBTI. h) Schematic illustration of the band alignment, i) *J-V* curves, and j) stability of the PSCs based on PBTI, PDTzTI and PCBM. Reproduced with permission.[145] Copyright 2020, Elsevier.

n-type polymers PBTI and PDTzTI as ETLs for inverted PSCs, which were based on bithiophene imide and thienylthiazole imide, respectively (Figure 1.12g–j). The PSCs based on PDTzTI ETLs exhibited a high PCE of 20.86%, which was higher than that of PCBM (20.14%) and PBTI- (10.60%) based PSCs. Due to the highly hydrophobic properties as well as the mobile ion blocking capability of polymer, the PSCs based on PDTzTI ETLs also exhibited excellent long-term and operational stability as compared with the PCBM-based PSCs.

1.4 OPTIMIZATION OF ETLs

The properties of ETLs highly affect the performance of PSCs. The pristine ETLs usually demonstrate unavoidable disadvantages, such as low electron mobility, defects, poor surface morphology, and so on, which can cause charge recombination and energy loss, leading to the low efficiency of PSCs. The properties of ETL also affect the stability of PSCs. Thus, the optimization of ETLs is required to maximize the performance of PSCs. In this section, typical strategies, such as doping, interface modification, and bilayer structures will be discussed.

1.4.1 Doping

Doping is one of the most direct and efficient methods to adjust the properties of ETL. Many additives have been developed to dope ETL and significant improvement has been achieved in the performance of PSCs. Alkali metals are the most common additives. Ma et al.[146] successfully introduced a NaCl surface doping to TiO_2 ETL. It was found that NaCl doping could enhance the electronic properties of TiO_2 and improve the charge transfer. As a result, a more efficient PSC with a PCE of 18.3% was achieved. Tang et al.[147] doped TiO_2 with RbCl. The similar polyhedral structure and lattice parameters between RbCl and perovskite promoted seed-mediated perovskite growth, which helps to form dense and uniform perovskite film with low trap density. The PSCs based on RbCl-doped TiO_2 yielded a PCE of 20.75%. Tan et al.[38] reported a chlorine-capped TiO_2 nanocrystal strategy to suppress interfacial recombination and improve interface binding (Figure 1.13a–c). The PSCs base on Cl-TiO_2 reached a certified PCE of 20.1% and retained 90% of its initial PCE after 500 h of continuous AM 1.5g illumination. Azmi et al.[148] dipped ZnO into various solutions of alkali metal hydroxide (LiOH, NaOH, and KOH) to prepare alkali metal doped ZnO ETLs (Figure 1.13d). They found that alkali metal doping significantly improved the electron mobility of ZnO and induced a more favorable energy band. The PSCs based on K-doped ZnO showed the highest PCE of 19.9%, which was higher than that of pristine PSCs (16.10%). Additionally, the decomposition reaction at ZnO/perovskite interface was suppressed. The PSCs based on K-doped ZnO retained 91% after being stored in air for 800 h, much higher than 36% of the pristine PSCs. Park et al.[3] introduced Li element by adding bis(trifluoromethane)sulfonimide lithium salt (Li-TFSI) into SnO_2 ETL. They found that Li doping could enhance the conductivity and induce the downshift of the conduction band of SnO_2. The modified properties of Li-doped SnO_2 improved the electron extraction and transfer from perovskite to SnO_2 ETL, leading to an impressive PCE of 18.2% with suppressed *J-V* hysteresis. Guo et al.[149] doped SnO_2 with lithium fluoride (LiF; Figure 1.13e and f). They found that LiF could enhance the conductivity of SnO_2. Moreover, LiF doping enlarged the lattice constant of SnO_2 and reduced the lattice mismatch between SnO_2 and perovskite, which reduced the trap densities at the SnO_2/perovskite interface and improved the crystal quality of perovskite film. The PSCs based on SnO_2:LiF showed a PCE of 21.33%, which was higher than the pristine PSCs. The alkali cations can diffuse into perovskite film to induce double-layer doping of ETL and perovskite film. Xie et al.[150] doped SnO_2 ETL with various alkali chlorides (LiCl, NaCl, KCl, RbCl, and CsCl). It was found that the oxygen vacancies at the SnO_2 surface and under-coordinated Pb^{2+} in $CsPbBr_3$ film could be passivated due to the diffusion of alkali and chloridion. The PSCs based on SnO_2-RbCl ETL exhibited the champion PCE of 10.04%. Min et al.[74] reported a Cl-bonded SnO_2 (Cl-bSO) ETL. It was found that a coherent interlayer could be formed at Cl-bSO/perovskite interface by coating a Cl-containing $FAPbI_3$ perovskite precursor on Cl-bSO. They inferred that the atomically coherent interlayer was a crystalline $FASnCl_x$ phase, which could reduce the interfacial charge recombination and contact resistance, leading to a high PCE of 25.8%.

Other metal cations are also reported as efficient additives in ETLs. Zhou et al.[151] reported PSCs with a PCE of 19.3% based on yttrium-doped TiO_2 ETL. The conductivity of Y-doped TiO_2 increased 3.3 times than that of the pristine TiO_2. The Y-doped devices exhibited a faster photocurrent decay and shorter PL decay time compared to pristine devices. Sandhu et al.[152] doped TiO_2 compact layer with Zr by adding ZrO_2 precursors. They demonstrated that Zr doping significantly improved the conductivity and reduced the electron trap-state density, leading to enhanced electron extraction and suppressed non-radiative recombination. The PSCs based on Zr-doped TiO_2 showed an improved PCE of 18.16% with decreased hysteresis.

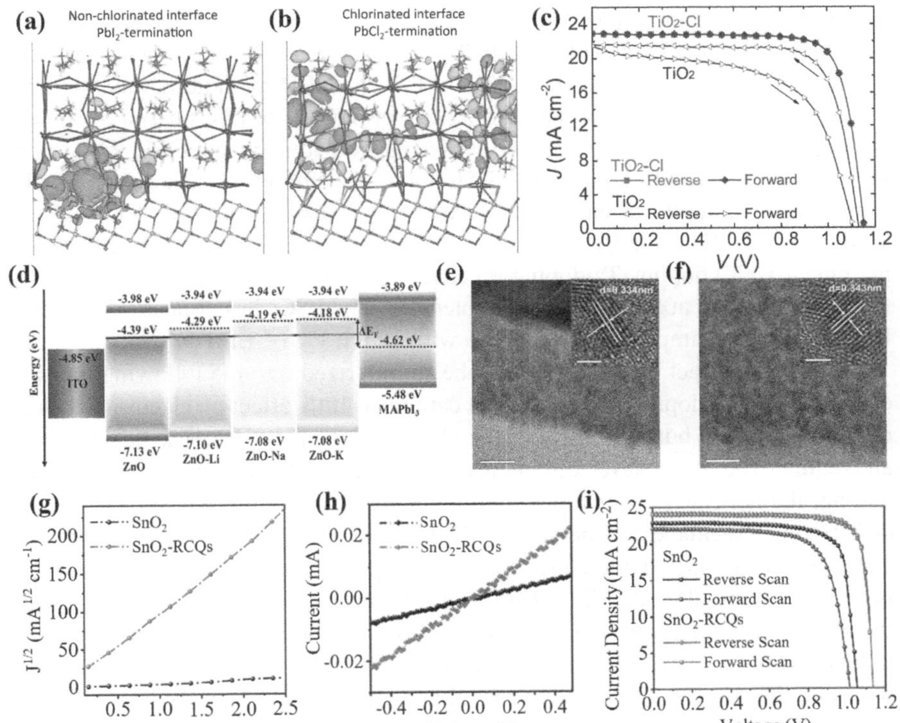

FIGURE 1.13 a) Trap-like localized antisite defects form near the valence band edge for the PbI_2-terminated TiO_2/perovskite interface. b) Shallow and delocalized Pb-Cl antisite defects are seen for the $PbCl_2$-terminated interface. c) *J-V* curves of PSCs with TiO_2 and TiO_2-Cl as ETLs. Reproduced with permission.[38] Copyright 2017, AAAS. d) Schematic illustration of energy levels of ZnO-ETLs and perovskite film. Reproduced with permission.[148] Copyright 2018, American Chemical Society. The HR-TEM images of e) SnO_2 and f) SnO_2:LiF. Reproduced with permission.[149] Copyright 2021, Elsevier. g) *J-V* characteristics of the electron-only devices based on the SnO_2-RCQs and SnO_2 ETLs. h) Conduction properties of the SnO_2 films with and without RCQs. i) *J-V* curves of PSCs based on SnO_2-RCQs and SnO_2 ETLs. Reproduced with permission.[167] Copyright 2020, John Wiley and Sons.

Metal cations like Mg^{2+},[153] Sb^{3+},[154] Sn^{4+},[155] Nb^{5+},[156] and so on are also reported to dope TiO_2 and achieved an improved performance of PSCs. Tseng et al.[157] prepared high-quality, full-covered Al-doped ZnO (AZO) ETL by magnetron sputtering. Compared to ZnO, the PSCs based on AZO had better stability and comparable J_{SC} and higher V_{OC} and FF. The best AZO-based PSCs achieved a PCE of 17.6% with a V_{OC} of 1.07 V. Song et al.[158] developed Mg-doped ZnO (ZMO) ETL for PSCs. Due to the improved optical properties, suitable energy levels, excellent charge extraction efficiency, and suppressed charge recombination, the PSCs based on ZMO ETL achieved a PCE of 16.5%. Most importantly, the $CH_3NH_3PbI_3$ film deposited on ZMO ETL exhibited better thermal stability and photo-stability. Huang et al.[159] demonstrated a low-temperature solution-processed SnO_2 ETL with Sb doping. The electrical conductivity was remarkably enhanced by Sb-doping, due to the increased carrier concentration in Sb-doped SnO_2. Moreover, Sb-doping could move up the Fermi level and improve the energy level alignment, leading to reduced charge recombination. The PSCs based on Sb-doped SnO_2 ETL showed a steady state efficiency of 17.7%. Anaraki et al.[160] prepared Nb-doped SnO_2 ETL by CBD method. It was found that Nb doping would not change the morphology of SnO_2. The PSCs based on 5 mol.% Nb-doped SnO_2 yielded an improved performance with a champion PCE of 20.5%. Ren et al.[161] found Nb-doped SnO_2 ETL showed a smooth surface and high electron mobility, leading to enhanced electron extraction. Zhan et al.[162] developed a Ta-doped SnO_2 ETL for PSCs. The conductivity of SnO_2 increased from 2.4 to 4.7 mS cm^{-1} after Ta doping. Ta-doping accelerated the electron transfer and suppressed the recombination at SnO_2/perovskite interface. The PSCs based on Ta-doped SnO_2 ETL reached a champion PCE of 20.8% with a high V_{OC} of 1.161 V.

The doping effect of metal ions can be summarized into the following types: (1) equivalent cation doping; the equivalent cation has little effect on electron density of metal oxide ETLs but may improve the electron dynamics, decrease defect density, and regulate the energy level alignment of metal oxide ETL. (2) Nonequivalent cation doping; the nonequivalent cation could regulate the concentration of defects (especially oxygen vacancies) in metal oxide ETLs, which could significantly improve the conductivity of metal oxide ETLs, leading to highly efficient electron transfer from perovskite to metal oxide ETLs. (3) Double doping of ETL and perovskite; the diffusible alkali cations could diffuse from underlying ETL to perovskite film to induce a double doping of ETL and perovskite film. The diffused alkali cations not only passivate the surface oxygen vacancies of ETL but also improve the crystallinity of perovskite film.

In addition to metal cations, other inorganic compounds are also used as additives in ETL. Liu et al.[163] doped commercial SnO_2 aqueous colloidal dispersion with ammonium chloride (NH_4Cl). It was found that NH_4Cl doping increased the electron mobility of SnO_2 and improved the energy band alignment at SnO_2/perovskite interface. In addition, the NH_4^+ and Cl^- at SnO_2/perovskite interface could suppress the formation of deep-level defects. The PCE of PSCs significantly increased from 18.71% to 21.38% after NH_4Cl doping due to improved charge transport and reduced charge recombination. Other ammonium halides are also reported as additives in different ETLs.[164] Mahmood et al.[165] reported a nitrogen-doped ZnO nanorod (N:ZnO) as ETL for PSCs. It was found that nitrogen could increase the electron concentration and mobility of ZnO nanorods and improve the aspect ratio of ZnO nanorod to enhance the infiltration of the perovskite into ETL. A conformal polyelectrolyte polyethylenimine (PEI) was coated on the electron-rich high aspect ratio N:ZnO nanorod arrays to induce a favorable work function shift. These factors resulted in a significantly increased PCE from < 10% to > 16%.

Quantum dots (QDs) have many excellent properties and have also been used to dope ETL. Li et al.[166] used the carbon QDs (CQDs) to dope TiO_2 ETL. It was found that CQDs could enhance electron extraction and injection between TiO_2 and perovskite. The increased electron transport at TiO_2/perovskite interface yielded an improved PCE of 18.89% with reduced *J-V* hysteresis. Hui et al.[167] reported a low-temperature solution-processed SnO_2 ETL doped by carboxylic-acid and hydroxyl-rich red-carbon QDs (RCQDs; Figure 1.13g–i). RCQDs doping increased the electron mobility of SnO_2 by ~20 times from 9.32×10^{-4} to 1.73×10^{-2} $cm^2\ V^{-1}\ s^{-1}$, and the corresponding PCE of PSCs increased from 19.15% to 22.77%. The PSCs based on RCQDs-doped SnO_2 also had enhanced long-term stability, retaining over 95% of its initial PCE after 1,000 h under 40–60% humidity at 25 °C. Xie et al.[168] prepared graphene quantum dots (GQDs) doped SnO_2 as ETL. They demonstrated that the photogenerated electrons in GQDs could transport to the conduction band of SnO_2 to fill the electron trap states, which could improve the Fermi level and conductivity of SnO_2. The PSCs based on GQDs-doped SnO_2 reached a steady-state PCE of 20.23% with little hysteresis. Tavakoli et al.[169] doped ZnO ETL by reduced graphene oxide quantum dots (rGOQDs). They found that rGOQDs not only improved the charge transfer from perovskite to ZnO but also passivated the ZnO surface to prevent the interface reaction. The PSCs based on rGOQDs doped ZnO showed a PCE of 15.2%. Some other QDs, such as PbS[170], CdSe[171], black phosphorus[172], and so on are also reported as additives for ETLs and significantly improved PCE of PSCs have been achieved.

Except for the inorganic additive, organic materials are also used as a common additive for ETLs due to its diverse properties. Peng et al.[173] doped mesoporous-TiO_2 with deoxyribonucleic acid (DNA). They found that DNA doping decreased the trap states in mesoporous-TiO_2. The optimized DNA-doped TiO_2 showed improved energy band alignment with perovskite. The PSCs based on DNA-doped TiO_2 exhibited a PCE of 17.59%. Ethylene diamine tetraacetic acid (EDTA) is an excellent additive material due to its strong chelation function. Yang et al.[174] prepared an EDTA-complexed SnO_2 ETL for PSCs (Figure 1.14a and b). It was found that EDTA-complexed SnO_2 showed three times higher electron mobility than pristine SnO_2. Moreover, EDTA-complexed SnO_2 provided a well-matched Fermi level with the conduction band of perovskite, leading to a higher V_{OC}. The PSCs based on EDTA-complexed SnO_2 ETL achieved a high PCE of 21.60% (certified 21.52%) with negligible hysteresis. Wang et al.[175] developed an EDTA-complexed ZnO ETL. Except for the more suitable energy levels and improved electron extraction, the EDTA-complexed ZnO could avoid the decomposition of perovskite efficiently due to the strong chelation between EDTA and organic ligands on the ZnO surface. Huang et al.[176] doped SnO_2 ETL with ethoxylated polyethyleneimine (PEIE) polyelectrolyte (Figure 1.14c and d). It was found that PEIE doping could improve the conductivity and surface morphology of SnO_2. PEIE doping also reduced the work function of

SnO_2 ETL. Thus, the charge transfer at SnO_2/perovskite interface was significantly improved, leading to an improved PCE of 20.61%. Tu et al.[177] used the electron donor triphenylphosphine oxide (TPPO) to realize the n-type doping of SnO_2. The DFT simulation revealed that strong interaction between TPPO and SnO_2 and the electrons were transferred from the R_3P^+-O–σ-bond mainly to the peripheral thin atoms, resulting in the presence of delocalized electrons at the surface, which increased the conductivity and decreased the work function of SnO_2. The built-in field increased from 0.01 to 0.07 eV, leading to an increased PCE from 19.01% to 20.69%. Dong et al.[178] presented an in-situ passivation method by doping SnO_2 ETL with hydroxyacid potassium synergist (Figure 1.14e and f). They found that the orbital overlap of the carboxyl oxygen with the Sn atom along with the nano particle deposition effectively suppressed the interfacial defects and released the internal residual strains in the perovskite. The in-situ PSCs achieved a high PCE of 24.91% with a high FF of 85.2%. Yin et al.[179] demonstrated two functionalized $Ti_3C_2T_x$ MXene molecules with dodecyltrimethoxysilane (MXene-H) and fluoroalkylsilane (MXene-F) as efficient additives of SnO_2 ETL. Functionalized MXene doping could enlarge the crystal size of perovskite by regulating the suitable surface energy of SnO_2. The high conductivity and mobility of functionalized MXene could form zero Schottky barrier heterojunction with SnO_2 to enhance charge transfer. The PSCs based on MXene-H-doped SnO_2 achieved a PCE of 24.12% with a high FF of 0.84.

Doping is also an efficient strategy to adjust the conductivity, electron mobility, energy band, and other properties of organic ETLs. Wang et al.[180] demonstrated a "air-stable" doping of n-type C_{60} ETL by 4-(1,3-dimethyl-2,3-dihydro-1H-benzimidazol-2-yl)-N,N-diphenylaniline (N-DPBI; Figure 1.14g and h). N-DPBI doping improved the conductivity and reduced the work function of the C_{60} layer. N-DPBI also tuned the surface wettability of C_{60} film, which helped to fabricate a "void-free" perovskite/C_{60} interface. The $MAPbI_xCl_{3-x}$ PSCs based on N-DPBI doped C_{60} reached the highest PCE of 18.3%. The $FA_{0.83}Cs_{0.17}Pb(I_{0.6}Br_{0.4})_3$ device achieved a T_{80} lifetime of over 600 h without encapsulation and over 3,420 h when sealed. Liu et al.[181] proposed a co-doping strategy by using tris(pentafluorophenyl)borane (TPFPB) and nonhygroscopic lithium salt ($LiClO_4$) to prepare C_{60} ETL. It was found that TPFPB could decrease the LUMO of C_{60} by forming a Lewis acidic adduct. The enlarge free energy difference could enhance the electron injection and reduce the charge recombination. $LiClO_4$ doping could improve the electron mobility and conductivity of C_{60} film, reducing the hysteresis of devices. The $CsPbI_2Br$ PSCs based on a co-doped C_{60} layer achieved a PCE of 15.19% and a PCE of 14.44% for large-area PSCs (1.0 cm^2). Zhou et al.[182] doped the PCBM ETL with a surfactant Triton X-100 to improve the coverage of PCBM on the hydrophobic perovskite surface treated by the $-CF_3$ group (Figure 1.14i and j). In addition, Triton X-100 could enhance the charge transfer from perovskite film to PCBM ETL, yielding a PCE of 21.9% with a significantly increased J_{SC}. Kakavelakis et al.[183] added reduced graphene oxide (rGO) into PCBM ETL for inverted PSCs. The rGO-doped PCBM showed increased conductivity and reduced series resistance and surface roughness, which produced a faster charge injection. The PSCs with rGO-doped PCBM exhibited a PCE of 14.5%. Zhu et al.[184] reported CQDs-doped PCBM as ETL for inverted PSCs. It was found that CQDs doping could not only improve the conductivity of PCBM but also prevent the

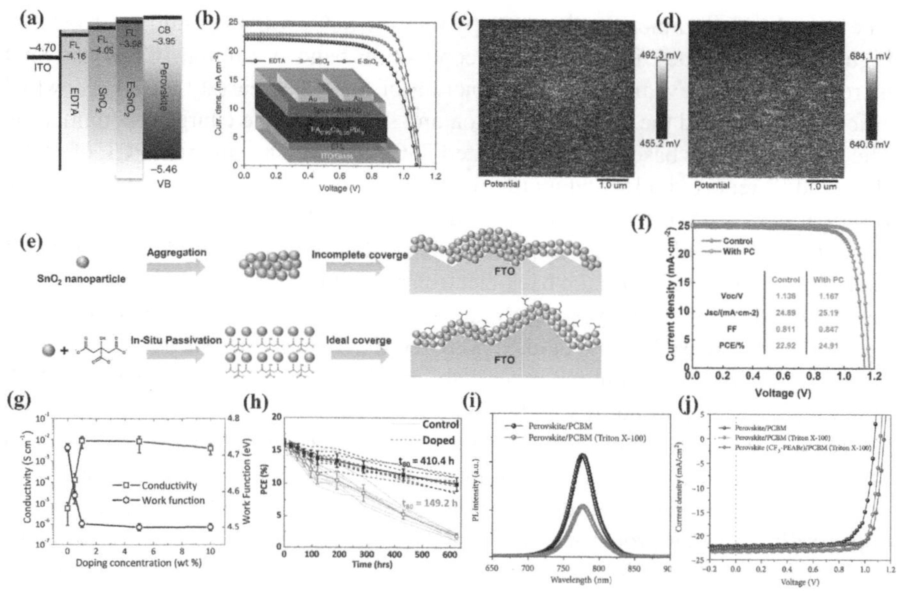

FIGURE 1.14 a) Schematic illustration of Fermi levels of EDTA, SnO_2, and E-SnO_2 relative to the conduction band of the perovskite layer. b) *J-V* curves with the inset showing device configuration. Reproduced with permission.[174] Copyright 2018, Nature Publishing Group. KPFM images of c) SnO_2 and d) SnO_2:PEIE ETLs. Reproduced with permission.[176] Copyright 2020, John Wiley and Sons. e) Schematic illustration of in situ passivation strategy of SnO_2. f) *J-V* curves of PSCs based on SnO_2 ETL and PC-SnO_2 ETL. Reproduced with permission.[178] Copyright 2023, John Wiley and Sons. g) Conductivity and work function of N-DPBI-doped C60 films at different doping concentrations. h) Aging for 624 h of encapsulated PSCs based on pristine and N-DPBI doped C_{60} ETLs under full spectrum simulated. Reproduced with permission.[180] Copyright 2017, John Wiley and Sons. i) PL spectra of perovskite films with PCBM or PCBM (Triton X-100) layers. j) *J-V* curves of PSCs with different conditions. Reproduced with permission.[182] Copyright 2021, AAAS.

diffusion of I^- to Ag electrode to form AgI, which would suppress the dimerization in fullerene and improve the stability of PSCs. The PCE of PSCs was improved from 16.1% to 18.1% after CQDs doping.

1.4.2 Interface Modification

UV-ozone treatment is a widely used surface treatment strategy for metal oxide, which can remove surface residuals, reduce oxygen vacancies, and improve surface wettability. Dong et al.[185] demonstrated the UV-ozone treatment is an efficient method to remove the surface contaminants or residues and improve wettability when the low annealing temperature is not able to eliminate surface contaminants or residuals. Bu et al.[186] found that UV-ozone treatment could improve the quality of the as-sintered SnO_2 films and reduce the electron transporting resistance of the ETLs and the charge recombination at the interface of ETL/perovskite. Zhou et al.[187] reported an energy band alignment engineering by treating SnO_2 with deep-ultraviolet (DUV; Figure 1.15a–c). They found that the conduction band minimum (CBM) of SnO_2 increased by 0.26 eV and the work function of SnO_2 decreased by 0.31 eV, which efficiently improved the charge extraction and suppressed the charge recombination. The $CsPbI_2Br$ PSCs based on DUV-treated SnO_2 achieved a champion PCE of 15.1%. Han et al.[188] reported a low-temperature modification of ZnO NPs by UV exposure. UV treatment could effectively remove the surface hydroxyl groups on the ZnO surface, combining with HNO_3 treatment could further reduce the content of the residual hydroxyl ions and retard the back-electron recombination at the interface between the ZnO and perovskite layers, and the stability of PSCs based on ZnO ETL was significantly increased. Wang et al.[189] prepared the $Ti_3C_2T_x$ MXene as ETL for PSCs via a room-temperature solution process followed by oxygen plasma treatment. They found that oxygen plasma treatment could break parts of Ti-C bonds and generated abundant Ti-O bonds randomly distributed on MXene, resulting in tunable work functions and reduced trap states to improve that charge transfer at interface.

$TiCl_4$ treatment has been usually used to modify the surface morphology and improve the electron properties of TiO_2 ETL. Enhanced photovoltaic performance has been observed for PSCs based on compact and mesoporous TiO_2 ETL with $TiCl_4$ treatment.[37,190] $TiCl_4$ treatment is also used to modify other metal oxides. Ma et al.[191] introduced a TiO_x interlayer at ZnO/perovskite interface by $TiCl_4$ treatment, which could efficiently suppress the decomposition reaction at ZnO/perovskite interface. The PSCs based on $TiCl_4$-treated ZnO achieved a PCE of 20.71%. $TiCl_4$ treatment is also used to tune the interface energy band alignment. The conduction band of SnO_2 is lower than that of TiO_2, which induces V_{OC} loss of PSCs. Rao et al.[192] treated SnO_2 ETL with $TiCl_4$. They found that $TiCl_4$ treatment could not only enhance the wettability of SnO_2 to form a uniform perovskite film but also improve the interface energy band alignment. The PSCs based on $TiCl_4$-treated SnO_2 reached a PCE of 14.69%.

Metal oxide ETLs contain a large amount of charged defects, such as hydroxyl groups, oxygen vacancies, exposed metal atoms, and dangling bonds. The alkali metal halide, containing metal cation and halide anion, can efficiently passivate charge defects of ETL and perovskite film. Liu et al.[193] modified SnO_2 with KCl, which could passivate both types of charge defects by K^+ cation and Cl^- anion. After introducing a KCl interlayer at the SnO_2/perovskite interface, the interfacial trap defects were reduced and carrier lifetime was increased. The PSCs with KCl passivation exhibited a PCE of 20.5% with negligible hysteresis. The penetrated moisture and oxygen could react with defect sites at interface/grain boundary, leading to the degradation of perovskite film. After KCl passivation, the PSCs maintained 90% of their initial PCEs after being stored in a dark N_2 box for over 30 d, while the pristine PSCs demonstrated significant degradation after just 10 d. Li et al.[194] reported a CsBr modification for TiO_2 ETL. It was found that CsBr modification passivated the interface defects and suppressed the UV-induced degradation of perovskite film. The mechanism for the improved UV stability was that CsBr passivation reduced the chemical reactivity of TiO_2 and reduced the defect density at TiO_2/perovskite interface. Ye et al.[195] developed an inorganic shunt-blocking layer LiF between SnO_2 and $CsPbI_{3-x}Br_x$ perovskite (Figure 1.15d and e). They found that the LiF interlayer

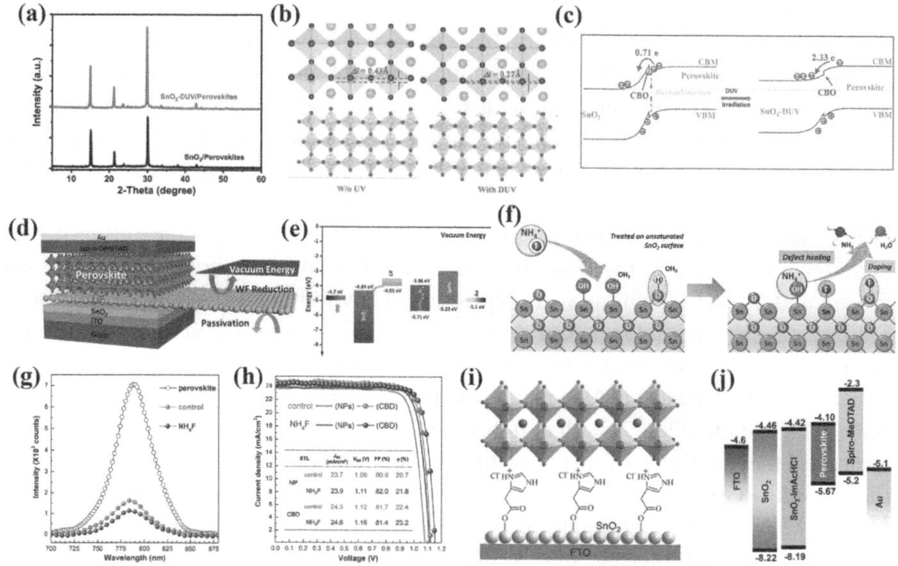

FIGURE 1.15 a) XRD spectra of $CsPbI_2Br$ films deposited on SnO_2 without and with DUV treatment. b) Optimized structures of $CsPbI_2Br$ on the pristine and hydroxylated SnO_2 surfaces, and the marked Δl reflects the degree of octahedral distortion. c) Schematic illustration of charge transfer and recombination at interface of $SnO_2/CsPbI_2Br$. Reproduced with permission.[187] Copyright 2020, John Wiley and Sons. d) The scheme of device architecture of the inorganic perovskite solar cells. e) Energy band alignment for each layer in $CsPbI_{3-x}Br_x$ inorganic solar cells. Reproduced with permission.[195] Copyright 2019, John Wiley and Sons. f) Schematic illustration of NH_4F treatment on SnO_2 surface. g) PL spectra of perovskite films. h) *J-V* curves of PSCs based different ETLs. Reproduced with permission.[197] Copyright 2020, American Chemical Society. i) Schematic illustration of the formation of ImAcHCl on SnO_2 layer. j) Energy level diagram of PSCs. Reproduced with permission.[205] Copyright 2019, John Wiley and Sons.

could passivate interface defects and adjust the energy bang alignment. The CBM of SnO_2 shifted from 4.30 to 4.01 eV after LiF modification, leading to a better energy band alignment with $CsPbI_{3-x}Br_x$. Moreover, the insulated LiF suppressed the charge recombination at the interface. The PSCs with LiF modification exhibited a significantly increased V_{OC} from 1.08 to 1.18 V and FF from 67.89% to 78.26%. Zhang et al.[196] modified SnO_2 with InX_3 (X=F, Cl, Br, and I) to optimize the energy band alignment, improve light absorption and utilization, enhance charge extraction at the interface, and inhibit charge recombination. The InF_3-treated SnO_2 showed a champion PCE of 21.39%.

Other materials containing covalently bonded cations and anions on the same molecules, called zwitterionic molecules, are also widely used to modify the ETL/perovskite interface. Zwitterionic molecules can efficiently enhance charge extraction and injection by forming interfacial dipoles, and their cationic and anionic groups enable them to form chemical interaction with perovskite and ETL to passivate interface defects. Jung et al.[197] reported a bifunctional surface treatment that used ammonium fluoride (NH_4F) to reduce defect sites and adjust the Fermi level of SnO_2 ETL (Figure 1.15f–h). They found that the ammonium cation with weak acidity could react with terminal-hydroxyls on the SnO_2 surface to form ammonia gas and water vapor, and the fluoride anions could substitute into the defect sites to tune the energy level of SnO_2. The PSCs based on NH_4F-treated SnO_2 reached a high PCE of 23.2% with a significantly increased V_{OC} from 1.12 to 1.16 V. Kim et al.[198] reported an approach to enhance passivation effects by controlling the Cl^- and NH_4^+ of NH_4Cl on the front and back side of SnO_2, respectively. The ITO was pre-treated by NH_4Cl before SnO_2 deposition. They found NH_4^+ could diffuse to the SnO_2 surface during SnO_2 crystal growth. This approach improved the crystallinities of SnO_2 and perovskite and reduced the Schottky barrier of the ITO/SnO_2 interface to enhance electron transfer. As a result, the PCE of PSCs based on NH_4Cl-pre-treated SnO_2 increased to 22.25% with a high V_{OC} of 1.18 V. NH_4^+ pretreatment significantly improved the stability of PSCs, and the unencapsulated PSCs maintained 86.8% of their initial PCE after being stored in 50% relative humidity at room temperature for 1200 h. Bi et al.[199] modified SnO_2/perovskite interface with KPF_6. It was found that PF_6^- located at interface and K^+ diffused into perovskite films. PF_6^- could bond with perovskite and SnO_2 by hydrogen and coordination bond, respectively, which could improve the interfacial contact. KPF_6 modification could improve perovskite film quality, passivate interfacial defects, and release interfacial stress, leading to a PCE improvement from 19.66% to 21.39%. The unencapsulated modified device maintained 80.1% of the initial PCE after aging at 60 °C for 960 h and 57.2% after aging under one sun illumination for 960 h, respectively. Liu et al.[200] introduced MACl to eliminate the deprotonation ability of ZnO. It was found that ZnO preferred to react with MACl than MAI due to the different electronegativities of chlorine and iodine. The MA cation could pre-react with ZnO during annealing process to avoid the further deprotonation of organic cations in perovskite. Cl could also perform an effective defect passivation for perovskite film. The perovskite film deposited on ZnO/MACl could maintain strong light absorbance and stable crystal structure. The efficiency of PSCs based on ZnO decreased rapidly after storing for 60 d, while the stability of PSCs based on MACl-modified ZnO was significantly enhanced, indicating MACl could effectively protect the perovskite film from degeneration.

Yang et al.[201] modified TiO_2 ETL with an ionic-liquid (IL) [BMIM]BF_4. [BMIM]BF_4 has a high charge mobility and conductivity and superior optical transparency. Moreover, [BMIM]BF_4 could form a surface dipole to tune the energy level of TiO_2. The PSCs based on [BMIM]BF_4 modification achieved a PCE of 19.62% with eliminated hysteresis. Further, Noel et al.[202] and Xia et al.[203] introduced [BMIM]BF_4 to modify SnO_2 and PCBM ETL, respectively. [BMIM]BF_4 could tune the work function of both SnO_2 and PCBM and passivate interfacial defects to enhance the charge transfer at ETL/perovskite interface. Xu et al.[204] employed a novel IL 3-sulphonatopropyl acrylate potassium salt (SPA) as an effective multifunctional modifier to modify TiO_2/$CsPbI_3$ interface. It was found that the SPA could not only donate lone-pair electrons to undercoordinated Ti^{4+} to passivate defects on TiO_2 and increase the conductivity by strong electron-donating sulfonate and carbonyl groups of SPA but also suppress the vacancies and undercoordinated Pb^{2+} defects at the perovskite buried interface by forming coordination interaction and electrostatic binding. Most importantly, SPA could regulate the perovskite crystallization and optimize interfacial energy level alignment. The $CsPbI_3$ PSCs based on SPA modification reached an excellent PCE of 20.98% and a record-low V_{OC} deficit of 0.451 V. The PSCs with SPA modification maintained 91% of the initial efficiency after 1,500 h, while less than 60% was retained for the control devices under the same condition. The improved stability was mainly caused by the reduced interfacial defects. Chen et al.[205] introduced 4-imidazoleacetic acid hydrochloride (ImAcHCl) IL to passivate SnO_2/perovskite interface (Figure 1.15i and j). They found that ImAcHCl could bond with SnO_2 through an ester via esterification reaction and form electrostatic interaction with perovskite via imidazolium cation in ImAcHCl and iodide anion in perovskite, which could provide a chemical bridge between SnO_2 and perovskite. The Cl^- in ImAcHCl could improve the crystallinity of perovskite film. Moreover, ImAcHCl modification could improve the interface energy band alignment. The PCE of PSCs with ImAcHCl modification increased from 18.60% ± 0.50% to 20.22% ± 0.34%. ImAcHCl treatment significantly improved the thermal and moisture stability of PSCs, which was mainly due to the reduced interface defects. Wang et al.[206] reported a functionalized TiO_2 surface treated by polar molecules of amino naphthalene sulfonates. The $CsPbI_3$ perovskite crystal grains could be anchored by the strong ion-dipole interaction between the molecule-level polar interlayer and the ionic perovskite film, resulting in high-quality $CsPbI_3$ films with the merit of defect-immunity and large shunt resistance. The corresponding PSCs achieved an indoor PCE of up to 41.2% under illumination from a commonly used indoor light-emitting diode light source (2956 K, 1062 lux). The treated PSCs reserved ~91% (83% for the pristine) of its initial efficiency after 540 h under maximum power point (MPP) tracking using LED@1062 lux in an N_2 glovebox. The improved perovskite film quality and reduced charge accumulation at interface were the main reasons for the improved stability of PSCs.

Some wide band gap materials are introduced at ETL/perovskite interface to suppress the charge recombination. Al_2O_3 was first reported as a scaffold for perovskite film by Lee et al,[207] and the insulating Al_2O_3 did not take part in electron transfer but induced a high efficiency. Si et al.[208] separated ZnO and perovskite film with an ultra-thin Al_2O_3 insulating layer. The Al_2O_3 insulating layer could suppress the degradation of perovskite. The compact ZnO layer and Al_2O_3 layer were prepared by ALD method. They optimized the thickness of the Al_2O_3 layer by a quantum tunneling model based on effective-mass approximation theory. Due to the suppressed carrier recombination, the PSCs based on ZnO ETL reached a PCE of 15.55% with an Al_2O_3 thickness of 4 Å. The PSCs with Al_2O_3 protective layer kept almost invariable PCE after 12 d exposure at room temperature with the relative humidity about 20%. While the PCE of PSCs without Al_2O_3 protective layer decreased seriously, which may result from the degradation of perovskite and ZnO compact layer. Tsvetkov et al.[209] found that PbO-covered metal oxide ETL could form a zipper-like interdigitated interface with Pb-based halide perovskite (Figure 1.16a–c). The PbO films were prepared by dipping the TiO_2 film layers into the aqueous $PbCl_2$ solution at 90 °C for 60 s and washed with water and ethanol. They found that the controlled substitution of interfacial Ti atoms of TiO_2 ETL by Pb atoms at the $TiO_2/MAPbI_3$ interface

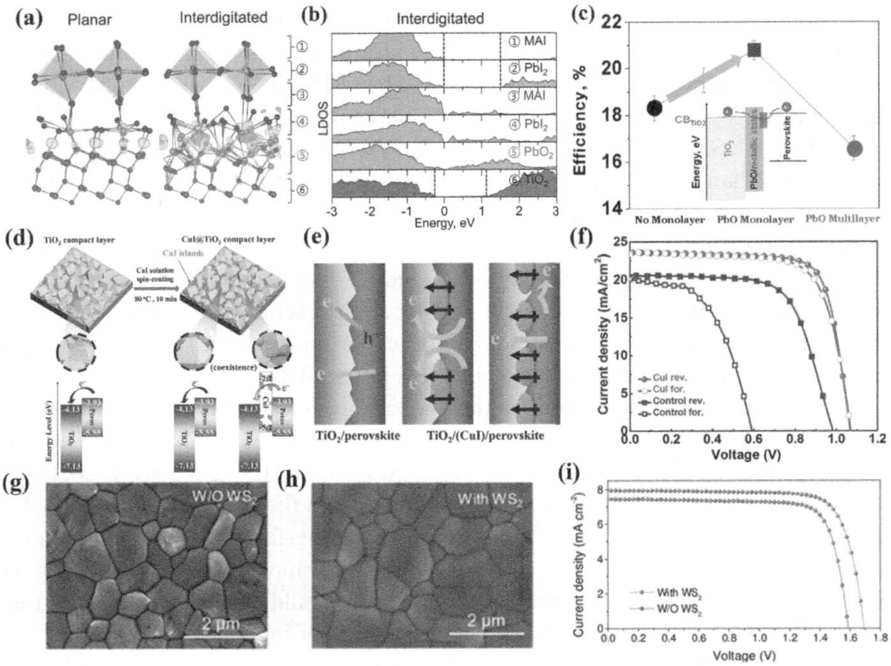

FIGURE 1.16 a) DFT-derived atomic structures of the planar and interdigitated (110)-TiO_2/(110)-$MAPbI_3$ interfaces. b) Layer-by-layer projected DOS for the Pb-interdigitated TiO_2/$MAPbI_3$ model. c) Mechanism of the enhancement of charge transfer across the interface between perovskite and a PbO-capped TiO_2 layer having metallic states. Reproduced with permission.[209] Copyright 2021, American Chemical Society. d) Schematic illustration of the formation of CuI islands on TiO_2 compact layer and deposition of perovskite layer subsequently. e) The figure of interface with various amount of CuI islands. f) *J-V* curves of device with bare TiO_2 and CuI@TiO_2 ETL. Reproduced with permission.[214] Copyright 2018, John Wiley and Sons. Top-view SEM images of $CsPbBr_3$ perovskite films g) without and h) with the modification of WS_2 nanoflakes. i) *J-V* curves of the PSCs with and without WS_2 modification. Reproduced with permission.[217] Copyright 2020, John Wiley and Sons.

or the establishment of Pb-I-Pb networks bridging $MAPbI_3$ and TiO_2 produced an atomically thin two-dimensional metallic layer to enhance the charge extraction at the interface. They also demonstrated that PbO modification was also applicable to ZnO and $SrTiO_3$ to form a similar zipper-like interdigitated metallic interface, which improved the charge extraction rate at the interface about two times. Dagar et al.[210] deposited a thin MgO overlayer on SnO_2 ETL by spin-coating the magnesium acetate tetrahydrate precursor in ethanol. It was found that the thin MgO layer led more uniform perovskite film and reduced interfacial carrier recombination. The PSCs based on SnO_2/MgO ETL showed a high PCE under indoor illumination conditions. Wang et al.[211] introduced an ultrathin MgO interlayer at SnO_2/$CsPbIBr_2$ interface by sol-gel method. The MgO layer provided a better substrate for $CsPbIBr_2$ growth to reduce the δ-phase perovskite at the interface. The tunneling effect and better band alignment effectively blocked holes and accelerated electrons to the electrode. The MgO interlayer significantly promoted the V_{OC} of $CsPbIBr_2$ PSCs from 1.25 to 1.36 V. The PSCs with MgO layer presented a significantly enhanced air stability, which was due to faster charge extraction and decreased grain boundaries in perovskite films. Cao et al.[212] used a thin MgO layer to inhibit the interfacial charge recombination and enhance the stability of PSCs based on ZnO ETL. The ZnO and MgO layer were prepared by solution process. Combining with protonated ethanolamine, the PSCs with the best PCE of 21.1% and no hysteresis were successfully fabricated.

Other inorganic materials have also been reported for ETL surface modification. Chen et al.[213] reported an effective strategy to convert the ZnO surface into ZnS at ZnO/perovskite interface by sulfidation. The ZnS layer was prepared by dipping ZnO layer into an ethanol solution of 1 M thiourea. It was found that the sulfide on the ZnO-ZnS surface bound strongly with Pb^{2+} and created a novel electron transport pathway to enhance charge transfer and reduce charge recombination at the interface, yielding a PCE of 20.7%. Most importantly, sulfide modification significantly improved the stability of PSC based on ZnO, which retained 88% and 87% of its initial PCE after 1,000 h storage and 500 h UV radiation, respectively. Byranvand et al.[214] reported an innovative surface modification method by preparing p-type CuI island-modified TiO_2 ETL (Figure 1.16d–f). The CuI island was prepared by spin-coating the CuI solution of different concentrations. It was found that the polarity of the CuI-modified TiO_2 surface could pull electrons to TiO_2/perovskite interface, resulting in a shift of the energy levels and superior electron extraction with a low trap density. TiO_2 surface was only partially coated by CuI, and the p-type CuI consisting of small-sized Cu^+ and large-sized I^- could act as PbI_2 at interface, which could prevent the back-recombination and facilitate electron extraction. The higher conduction and valence band energy levels of CuI could induce a shift of TiO_2 band edge and reduce the density of trap states. The PSCs based on CuI-modified TiO_2 ETL reached a PCE of 19.0%. 2D materials are also used to modify the ETL/perovskite interface due to their remarkable optoelectronic characteristics. Huang et al.[215] introduced a thin 2D TiS_2 at SnO_2/perovskite interface and achieved a PCE of 21.73% with suppressed hysteresis. The water-soluble 2D TiS_2 with different concentrations was spin-coated on SnO_2 to form the TiS_2 film. The improved PCE could be attributed to better energy level alignment and passivation of trap states by forming bonds between TiS_2 and perovskite via SO_4^{2-} groups. Tavakoli et al.[216] demonstrated a modification method for ZnO/perovskite interface by introducing monolayer graphene (MLG). It was found that MLG not only enhanced the charge extraction but also improved the stability of perovskite at high temperature. Zhou et al.[217] introduced a 2D WS_2 nanoflake into SnO_2/perovskite interface (Figure 1.16g–i) by spin-coating the WS_2 solution. Similar to the layer-translational feature of 2D materials, the weak van der Waals interaction between WS_2 and perovskite allowed the uncommitted movement of the interfacial perovskite lattice during annealing and cooling processes, playing a role of lubricant at SnO_2/perovskite interface to reduce the interfacial tensile strain significantly. The $CsPbBr_3$ PSCs with WS_2 interlayer reached a high PCE of 10.65% with an ultrahigh V_{OC} of 1.70 V. Arising from the lubricant feature of WS_2/$CsPbBr_3$ van der Waals heterostructure, the tensile strain was released, and the ion migration was significantly reduced. The WS_2 treated device maintained over 80% of the initial PCE over 120 d storage under ambient condition (80% humidity at 25 °C). Zhu et al.[218] synthesized carbon quantum dots with bromine functional groups (Br-CQDs) as buried interface modifiers and an additive to enhance $CsPbBr_3$ film quality. The Br-CQDs could improve the energy band alignment and enhance the charge transfer at $TiO_2/CsPbBr_3$ interface. The all-inorganic $CsPbBr_3$ PSCs with Br-CQDs achieved a PCE of 10.79% with an ultrahigh V_{OC} of 1.651 V.

Organic materials have abundant functional groups, which are also used to improve the surface electronic properties of ETL and passivate defects at ETL/perovskite interface. Fullerene C_{60} and its derivatives are widely used to passivate ETL/perovskite interface. Zhou et al.[219] prepared pristine C_{60} on TiO_2 ETL by a low-temperature all-solution process (Figure 1.17a and b). They found that C_{60} interlayer greatly benefitted the ETL/perovskite interface by improving perovskite film quality, promoting interfacial charge transfer and suppressing interfacial charge recombination. The flexible PSCs based on C_{60}-modified TiO_2 exhibited a PCE of over 16% with high bending resistance (≈80% after 1500 cycles) and high light-soaking retention (≈100% after 100 min). Cao et al.[220] introduced a hydroxylated C_{60}, fullerenol, at TiO_2/perovskite interface by spin-coating the fullerenol solution. The fullerenol interlayer could improve the interfacial electron transfer and improve the crystallinity of perovskite film. PCBM has been reported as an efficient surface modifier for many metal oxides ETLs and achieved significant improvement of PCE.[5,37,191] Wang et al.[221] introduced an n-type doped, 1,3-dimethyl-2-phenyl-2,3-dihydro-1H-benzoimidazole (DMBI) doped PCBM as an effective modification layer at $TiO_2/MAPbI_3$ perovskite interface. DMBI-doped PCBM showed multiple beneficial effects, including enhanced charge conductivity, smaller energy offset, surface defect passivation, enhanced charge transfer efficiency, and large perovskite crystal grain size. The DMBI-doped PCBM interlayer realized an average PCE of 20.14% with an improved stability. Wang et al.[222] compared the passivation effect of three representative fullerenes, C_{60}, PCBM, and [6,6]-phenyl-C61-butyric acid (PCBA) at SnO_2/perovskite interface. The thermally evaporated C_{60} could be mostly retained after solvent treatment, while the low efficiency indicated that no efficient passivation was developed at SnO_2 interface. The PCBM treatment showed inefficient charge transfer at SnO_2/perovskite interface unless thermal annealing was applied after PCBM deposition, because the thermal annealing induced reaction between PCBM and hydroxyl groups on SnO_2 surface to form covalent bonds. In contrast, PCBA could easily bond with SnO_2 due to the high reactivity of carboxylic acid side chain to passivate defects at SnO_2/perovskite interface, which yielded the highest PCE of 18.8%.

The terminal functional groups of organic compounds can directly interact with metal or oxygen dangling bonds on the metal oxide ETL surface, which could form a permanent dipole moment to adjust the Fermi level and conduction band of ETL. Self-assembled monolayers usually contain multiple groups to form chemical bonding with the substrates and ordered arrangement on substrates, which is a benefit for charge transfer at interface. Wojciechowski et al.[223] reported C_{60}-SAM-modified TiO_2 ETL for PSCs (Figure 1.17c). C_{60}-SAM could bind with the hydroxyl groups on TiO_2 surface by the carboxylic acid anchoring group. C_{60}-SAM modification significantly improved the electron transfer at TiO_2/perovskite interface. Guo et al.[36] compared the modification effect between PCBM and C_{60}-SAM. They found that C_{60}-SAM

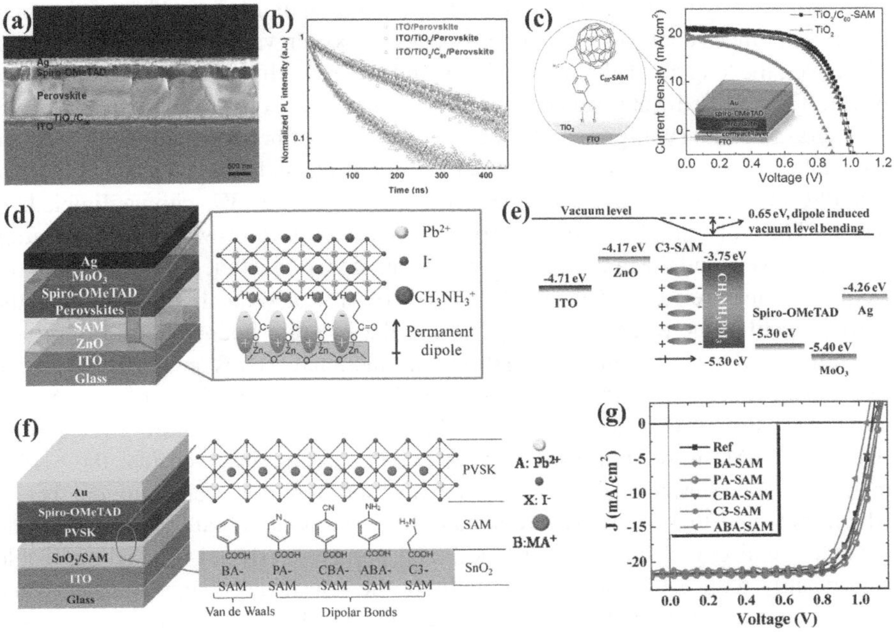

FIGURE 1.17 a) Cross-sectional SEM images of TiO_2/C_{60}-based PSC. b) TR-PL spectra of perovskite films on ITO, ITO/TiO_2, and ITO/TiO_2/C_{60}. Reproduced with permission.[219] Copyright 2018, John Wiley and Sons. c) *J-V* characteristic curves of PSCs without or with C_{60}-SAMs modification. Inset is the schematic of device structure and the binding of SAMs to TiO_2 surface. Reproduced with permission.[223] Copyright 2014, American Chemical Society. d) Schematic diagram of perovskite solar cell device structure, SAM induced permanent dipole formation, and involvement of the SAM in the crystalline structure of perovskite crystals. e) Schematic energy level of each layer in perovskite solar cell. Reproduced with permission.[224] Copyright 2015, American Chemical Society. f) Schematic diagram of the PSCs structure and the SAM between the SnO_2 and perovskite film. g) *J-V* characteristic curves of PSCs without or with different SAMs modification. Reproduced with permission.[226] Copyright 2017, American Chemical Society.

modification provided an improved perovskite film quality and enhanced interface charge transfer compared to PCBM modification due to its compact and ordered arrangement at TiO_2/perovskite interface. As a result, the PSCs based on C_{60}-SAM modification showed improved efficiency and suppressed hysteresis compared to that based on PCBM modification. Zuo et al.[224] introduced a C_3-SAM interfacial modification at ZnO/perovskite interface (Figure 1.17d and e). They found that C_3-SAM modification resulted in improved morphology and increased light absorption of perovskite film, the lowered cathode work function, and the enhanced electronic coupling between ZnO and perovskite. Liu et al.[225] used a long chain fullerene derivative C_9 with two hydroxyl anchoring groups on the side chain to modify the SnO_2 surface. C_9 has a suitable LUMO energy level and high electron affinity, which was beneficial to the electron extraction and transfer from perovskite to SnO_2 ETL. It was found that the Lewis base hydroxyl groups at the end of the long chain in C_9 could form chemical bonds with under-coordinated Sn to passivate the oxygen vacancies on SnO_2 surface. Moreover, the long alkyl chain in C_9 facilitated ordered self-assembly, and the hydrophobic C_9-modified SnO_2 surface suppressed heterogeneous nucleation and facilitated grain boundary migration during the deposition of perovskite films to improve the quality of perovskite films with enlarged grain size and reduced grain boundary. The PSCs based on C_9-modified SnO_2 achieved a PCE of 21.3% with negligible hysteresis. The ambient, photo, and electric field stability of PSCs with C_9 modification were improved, and the main reason could be attributed to efficient passivation of the oxygen vacancy-related defects on the surface of SnO_2 and much lower tDOS in the perovskite film due to enlarged grain size, reduced grain boundary, and improved crystallinity. Zuo et al.[226] introduced different functional groups onto SnO_2 surface to form various chemical interactions with perovskite layer by depositing different SAMs (BA-SAM, PA-SAM, CBA-SAM, ABA-SAM, and C_3-SAM; Figure 1.17f and g). All these SAMs were spin-coated on SnO_2 ETL. It was observed that the performance of PSCs showed an opposite trend to that of the energy level alignment theory, which indicated that chemical interactions were the predominant factor governing the interfacial optoelectronic properties. The proper interfacial interactions could significantly reduce trap states and facilitate the interfacial charge transfer. Yang et al.[227] modified SnO_2/perovskite interface with APTES SAM. The APTES SAM was prepared by immersing the as-prepared SnO_2 in isopropanol solution of APTES for several hours. It was found that APTES modification had multiple functions, including increasing the surface energy induced the formation of high-quality perovskite films with improved morphology, forming dipoles on the SnO_2 surface to decrease the work function of SnO_2 and enlarge built-in potential of SnO_2/perovskite heterojunctions, and passivating the trap states at the perovskite surface via hydrogen bonding to reduce the charge recombination at the interface. The PSCs with APTES modification exhibited a PCE of 17.54% with a high V_{OC} of 1.16 V. Du et al.[228] introduced glycine to modify SnO_2/perovskite interface. They found that the carboxylic acid group of glycine could react with the hydroxyl groups on SnO_2 surface and its outward amino group could assist the formation of perovskite film, which acted as a linker between SnO_2 and perovskite to enhance the charge transfer. As a result, the PSCs based on glycine modification achieved a PCE of 20.68%.

Other small molecules have been extensively reported to modify ETL surface. Lou et al.[229] reported a π-conjugated small molecules ($BTAC_4$) to modify SnO_2/perovskite interface. They found that a compact and ordered molecular packing of the organic passivator was advantageous to improve the SnO_2/perovskite interfacial quality, including lower trap density, increased electron conduction, and suppressed charge recombination. The PSCs based on $BTAC_4$ modification showed a PCE of over 23% with a high V_{OC} of 1.252 V. Moreover, the PSCs based on $BTAC_4$ modification exhibited improved stability, maintaining 90% of the initial PCE after being stored under relative humidity of 35% and room temperature (25 °C) for 30 d, which was higher than that of pristine PSCs (70%). Wu et al.[230] developed P-biguanylbenzoic acid hydrochloride (PBGH) to modify SnO_2/perovskite interface. It was found that the PBGH could effectively passivate the trap states of Sn dangling bonds and O vacancies on the SnO_2 surface via Lewis acid/base coordination, which could improve the conductivity of SnO_2 film and accelerate the electron extraction. Moreover, PBGH modification assisted the formation of high-quality perovskite film with low defect density due to its strong interaction with PbI_2. The PBGH-modified $FACsPbI_3$-PSCs reached a high PCE of 24.79%.

1.4.3 Bilayer Structures

Bilayer ETL structure is another efficient strategy. Different from interface modification, the upper layer of the bilayer structure is usually thicker and can also be used as an ETL alone. This structure can efficiently remove the pinholes and cracks in ETL and utilize the advantages of respective constituent ETL materials to realize improved energy band alignment and enhanced charge transfer to improve the performance of PSCs. Combined TiO_2 and SnO_2 to form bilayer ETL are most popular for PSCs. The compact TiO_2 has outstanding advantage in dense structure and coverage capability but a relatively low electron extraction capability and interface stability. SnO_2 prepared by nanoparticles has excellent electron extraction capability but a relatively loose structure and low coverage capability. Kim et al.[231] deposited a conformal ultrathin SnO_2 layer on compact TiO_2 using a polyacrylic acid-stabilized SnO_2 QDs (paa-QD-SnO_2@TiO_2; Figure 1.18a–d). The uniform bilayer of paa-QD-SnO_2@TiO_2 largely improved the absorption of perovskite film and formed an outstanding electron-selective contact with perovskite film. The PSCs achieved a high PCE of 25.7% (certified 25.4%). It should be noted that the relatively high CBM of TiO_2 compared with SnO_2 may introduce an energy barrier, leading to charge recombination at TiO_2/SnO_2 interface. Shu et al.[232] proposed to build an effective built-in electric field at TiO_2/SnO_2 interface to drive electrons to transfer from SnO_2 to TiO_2. They spin-coated a Li-TFSI solution on TiO_2 surface to realize an n-type doping of TiO_2 to build an interface electric field. The PSCs achieved a PCE of 23.2% with a high V_{OC} of 1.201 V. Hu et al.[233] prepared SnO_2/TiO_2 bilayer by using commercially SnO_2 NPs and as-prepared TiO_2 NPs. The CBMs of SnO_2, TiO_2, and perovskite are -4.33, -4.10, and -3.60 eV, respectively, suggesting that the TiO_2 intercalation could form a cascade structure in terms of the energy-level alignment, where the drastic energy change of electrons moving from perovskite to SnO_2 could be mitigated with a "stair". The energy loss was significantly reduced owing to the improved energy band alignment. The PCE of PSCs based on SnO_2/TiO_2 increased from 18.09% to 20.50%. You et al.[234] inserted a compact TiO_2 layer between FTO and WO_3 to form a TiO_2/WO_3 bilayer ETL by ALD method. It was found that TiO_2/WO_3 bilayer ETL showed better coverage on FTO than single WO_3 layer to avoid the direct contact between perovskite and FTO. Moreover, the lower valence band of TiO_2 could block holes to suppress charge recombination. The PSCs based on TiO_2/WO_3 bilayer ETL exhibited a PCE of 20.14% with a high FF of 78.55% and J_{SC} of 23.54 mA/cm^2.

Ma et al.[235] introduced a low-work function ZnO interlayer between ITO and SnO_2 (Figure 1.18e–g). It was found that the Fermi level of bilayer ZnO/SnO_2 upshifted compared to that of SnO_2, leading to an improved energy band alignment at ETL/perovskite interface. Moreover, the ZnO interlayer could improve the surface morphology of SnO_2 to further improve the morphology and crystal quality of the $CsPbI_2Br$ film. Combining with GAI treatment on $CsPbI_2Br$ surface, the PSCs showed a PCE

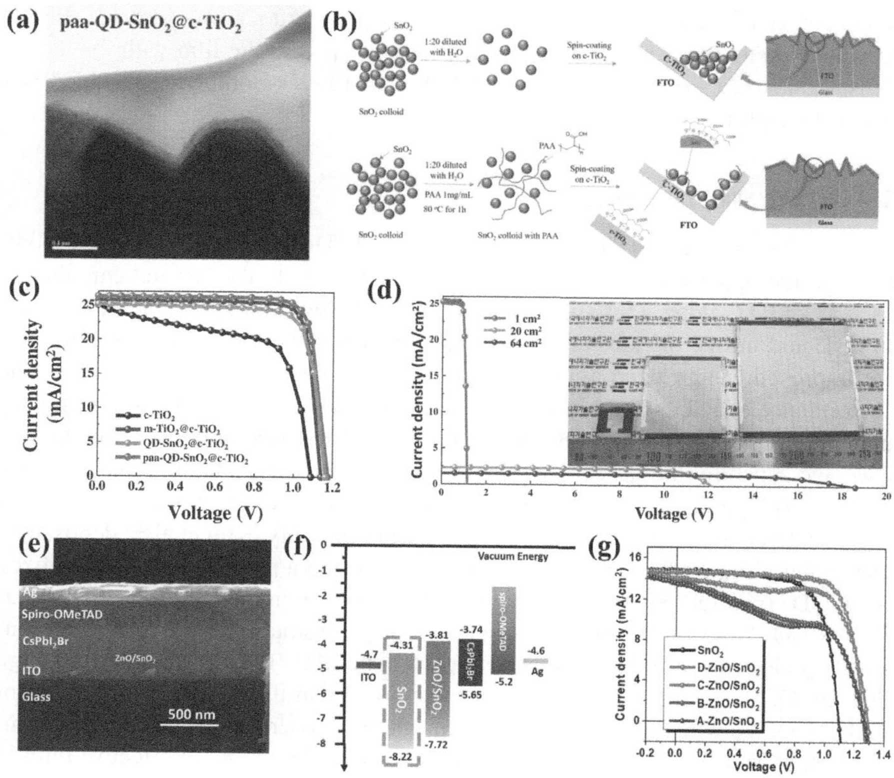

FIGURE 1.18 a) The cross-sectional TEM image of paa-QD-SnO_2@c-TiO_2 over the FTO substrates. b) A schematic diagram of the formation of QD-SnO_2 and paa-QD-SnO_2 layers on c-TiO_2 from solutions. c) The *J-V* curves of the PSCs with different ETLs. d) *J-V* curves of the large-size PSCs based on paa-QD-SnO_2@c-TiO_2 ETL. (Inset) Photo of the large-size PSCs. Reproduced with permission.[231] Copyright 2022, AAAS. e) Cross-section SEM image of the completed device. f) Energy band alignment of $CsPbI_2Br$ inorganic PSCs. g) *J-V* characteristics of inorganic $CsPbI_2Br$ PSCs with and without ZnO interlayer. Reproduced with permission.[235] Copyright 2022, John Wiley and Sons.

of 16.25% with an excellent FF of 82.15% and a V_{OC} of 1.27 V. The similar bilayer structure was also reported by Noh et al.[236] and He et al.[237] in $MAPbI_3$-PSCs and PSCs module and significant performance improvement was achieved. Su et al.[238] prepared ZnO/SnO_2 bilayer ETL by ALD method. Combining the high conductivity of ZnO with fast electron extraction of SnO_2, the ZnO/SnO_2 bilayer showed an efficient electron transport than the single layer. The wide-bandgap PSCs achieved a PCE of 18.1% with a high V_{OC} of 1.233 V. Interestingly, Yan et al.[239] demonstrated that SnO_2/ZnO bilayer was also an efficient bilayer ETL for PSCs. The high-quality $CsPbI_2Br$ film with regular crystal grains and full coverage could be obtained on SnO_2/ZnO surface. Due to the higher CBM of ZnO, the energy band alignment of SnO_2/ZnO/$CsPbI_2Br$ interface was improved to enhance the electron extraction and suppress the charge recombination. The PSCs based on SnO_2/ZnO bilayer ETL reached a PCE of 14.6%

with a V_{OC} of 1.23 V. Wang et al.[240] reported a In_2O_3/SnO_2 bilayer ETL for PSCs. It was found that the presence of In_2O_3 resulted in a uniform and compact perovskite film with low trap density. Moreover, the CBM of In_2O_3 was higher than that of ITO, leading to an enhanced charge transfer from perovskite to ETL. The PSCs reached a PCE of 23.24% with a high V_{OC} of 1.17 V, corresponding to a low V_{OC} loss of 0.36 V. The same metal oxide prepared by different processes could exhibit different characteristics, which could also construct bilayer ETL. Lin et al.[241] prepared a bilayer SnO_2 as ETL. They spin-coated colloidal nanoparticulate SnO_2 on sol-gel $SnCl_2{\cdot}2H_2O$ derived compact SnO_2 layer. It was found that the bilayer SnO_2 provided a better energy band alignment than the single layer, leading to faster charge extraction and lower perovskite trap-density. The PSCs based on bilayer SnO_2 exhibited a PCE of 20.3% with a V_{OC} of 1.13 V. Liu et al.[242] prepared an intermeshing SnO_2 (Im-SnO_2) by combining two types of SnO_2 (NP-SnO_2 and Col-SnO_2) with complementary electronic/physical properties. They found that the Im-SnO_2 had compact and smooth surface with favorable wettability and high electron mobility, which was benefit for $CsPbI_2Br$ film growth. The gradually decreased conduction band of Im-SnO_2 provided desirable cascade energy levels between perovskite and ITO for fluent electron transport, leading to the improved charge extraction and suppressed charge recombination. The $CsPbI_2Br$ PSCs showed a PCE of 16.10% with a V_{OC} of 1.31 V. Gong et al.[243] prepared a SnO_2 bilayer by adjusting the F doping concentration of SnO_2. They demonstrated that gradual substitution of F^- into the SnO_2 ETL could effectively reduce the band offset and result in a substantial increase in device V_{OC}. The PSCs based on F-doped bilayer SnO_2 ETL achieved a PCE of 20.2% with a V_{OC} of 1.13 V.

Except the metal oxide bilayer ETLs, metal oxides and organic materials could also form an efficient bilayer ETL for PSCs. Liu et al.[244] reported a ZnO/C_{60} bilayer ETL for an inverted $CsPbI_2Br$ PSCs (Figure 1.19a–e). The ZnO/C_{60} bilayer ETL exhibited superior electron-extracting performance compared to either one of them alone. The $CsPbI_2Br$ PSCs reached a PCE of 13.3%. Moreover, the PSCs without encapsulation exhibited highly long-term thermal stability and retained 80% of its initial PCE after being heated at 85 °C for 360 h. Zhu et al.[80] successfully deposited SnO_2 NCs on C_{60} to form a bilayer ETL for inverted PSCs (Figure 1.19f–i). The SnO_2 NCs prepared by hydrothermal could afford efficient electron transport at a relatively thick thickness of about 120 nm due to its high crystallinity, leading to a high PCE of 18.8%. Importantly, the compact bilayer ETL could form a protecting layer to enhance the stability of PSCs, the PSCs retained over 90% of its initial PCE after 30 d storage in ambient with > 70% relative humidity. Jiang et al.[245] prepared $PC_{61}BM$/ZnO NPs bilayer ETL for inverted PSCs, which combined the advantages of efficient charge extraction ability of $PC_{61}BM$ and excellent blocking ability of ZnO NPs against oxygen, water, and electrodes. Zheng et al.[246] reported a bilayer $PC_{61}BM/CrO_x$ as ETL for inverted PSCs. It was found that CrO_x interfacial layer could resist the diffusion water and oxygen into perovskite and prevent the reaction of metal electrode with perovskite layer, which could enhance the stability of PSCs in air without encapsulation by more than ten times. The CrO_x interlayer could also improve the energy level alignment between the active layer and metal electronic to enhance the electron collection. The PSCs based on $PC_{61}BM/CrO_x$ bilayer exhibited a PCE of 17.5%.

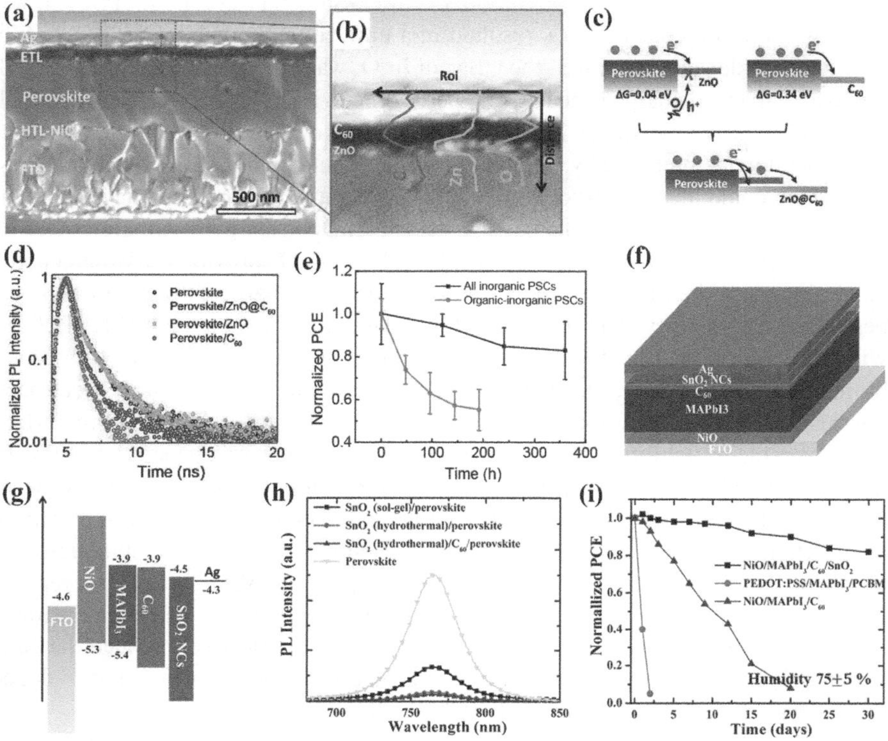

FIGURE 1.19 a) Cross-sectional SEM image of the ZnO/C_{60} bilayer-based PSC. b) The corresponding EDS line scanning of the selected area marked within a red rectangle. c) Schematic view of different ETL with band gap alignment. d) TRPL spectra of $CsPbI_2Br$ thin films based on different ETLs. e) Normalized PCEs of the $CsPbI_2Br$ all inorganic PSCs and $Cs_{0.04}FA_{0.8}MA_{0.16}PbI_{0.85}Br_{0.15}$ organic-inorganic PSCs continuously heated at 85 °C in N_2 atmosphere without encapsulation and tested in air (relative humidity 30–40%). Reproduced with permission.[244] Copyright 2018, American Chemical Society. f) Illustration of the PSCs with a structure of FTO/NiO/$MAPbI_3$/C_{60}/SnO_2 NCs/Ag. g) The corresponding energy level diagram of the PSCs. h) Steady-state PL spectra of the perovskite film capped with different ETLs. i) The stability test of devices in ambient with a relative humidity 75 ± 5%. Reproduced with permission.[80] Copyright 2018, John Wiley and Sons.

TABLE 1.4

Device Structure and Performance of PSCs Based on Bilayer ETLs

Device structure	V_{OC} (V)	J_{SC} (mA/cm^2)	FF	PCE	Ref.
FTO/c-TiO_2/paa-QD-SnO_2/$FAPbI_3$/ OAI/Spiro-MeOTAD/Au	1.18	26.09	0.84	25.72	231
FTO/C-TiO_2(Li)/SnO_2/$FAPbI_3$/ Spiro-MeOTAD/Ag	1.17	26.66	0.78	24.30	232

TABLE 1.4 *(Continued)*
Device Structure and Performance of PSCs Based on Bilayer ETLs

Device structure	V_{OC} (V)	J_{SC} (mA/cm²)	FF	PCE	Ref.
ITO/SnO_2/TiO_2/$(FAPbI_3)_{1-x}$ $(MAPbBr_3)_x$/Spiro-OMeTAD/Au	1.10	24.20	0.77	20.50	233
FTO/TiO_2/WO_3/$MAPbI_3$/ Spiro-OMeTAD)/Ag	1.09	23.54	0.79	20.14	234
ITO/ZnO/SnO_2/$CsPbI_2Br$(GAI)/ Spiro-OMeTAD/Ag	1.27	15.58	0.82	16.25	235
ITO/ZnO/SnO_2/$MAPbI_3$/ Spiro-OMeTAD/Au	1.20	22.10	0.77	20.43	236
FTO/ZnO/SnO_2/$Cs_{0.05}FA_{0.95}PbI_3$/ Spiro-OMeTAD/Au	1.08	22.92	0.80	19.78	237
FTO/ZnO/SnO_2/$Cs_{0.05}FA_{0.8}MA_{0.15}Pb$ $(I_{0.75}Br_{0.25})_3$/Spiro-OMeTAD/Au	1.23	20.50	0.72	18.10	238
ITO/SnO_2/ZnO/$CsPbI_2Br$/ Spiro-OMeTAD/MoO_3/Ag	1.23	15.00	0.79	14.60	239
ITO/In_2O_3/SnO_2/$FA_{1-x}MA_xPbI_3$/ Spiro-OMeTAD/Au	1.17	24.45	0.81	23.24	240
FTO/sol-gel SnO_2/SnO_2 colloid/ $(FAPbI_3)_{1-x}(MAPbBr_3)_x$/ Spiro-OMeTAD/Au	1.13	23.10	0.78	20.30	241
ITO/NP-SnO_2/Col-SnO_2/$CsPbI_2Br$/ Spiro-OMeTAD/Au	1.31	15.53	0.79	16.10	242
FTO/$(F{:}SnO_2)_{380}$/$(F{:}SnO_2)_{180}$/ $(FAPbI_3)_{0.85}(MAPbBr_3)_{0.15}$/Spiro-OMeTAD/Au	1.13	22.92	0.78	20.20	243
FTO/NiO_x/$CsPbI_2Br$/ZnO/C_{60}/Ag	1.14	15.20	0.77	13.30	244
FTO/NiO/$MAPbI_3$/C_{60}/hydrothermal SnO_2/Ag	1.12	21.80	0.77	18.80	80
ITO/PTAA/$MAPbI_3$/PCBM/ ZnO NPs/Al	1.11	19.60	0.79	17.20	245
FTO/PEDOT:PSS/$MAPbI_{3-x}Cl_x$/ $PC_{61}BM$/Cr_2O_3/Ag	1.01	16.43	0.70	17.50	246

1.5 CONCLUSION AND OUTLOOK

ETL is one of vital components in PSCs that facilitates selective charge extraction and has a significant influence on charge recombination. In this chapter, we summarized the recent developments of ETLs for highly efficient PSCs, including the preparation of inorganic and organic ETL, doping and interface engineering to promote the properties ETL. The main features of ETL which need to be taken into account are high electron conductivity, high transparency, well-matched conduction band and deep valence band, high stability, and low production cost.

For the inorganic ETLs, n-type metal oxides are the most widely used due to their excellent electrical properties and low fabrication cost. TiO_2 is considered as the most significant contender as inorganic ETLs due to its well-matched energy level and easy and low-cost fabrication process, but the high annealing temperature, low electron mobility, and photocatalytic activity hinder its journey in PSCs. Another common ETL is ZnO, which has low sintering temperature and high electron mobility, but the interface instability of ZnO can induce the decomposition of $MAPbI_3$ perovskite, and therefore it shows great application in inorganic PSCs. SnO_2 is considered as the most promising ETL due to its high electron mobility, appropriate energy band structure, low annealing temperature, and excellent chemical stability. The PSCs based on SnO_2 ETL has achieved the highest record PCE of PSCs. Furthermore, various studies have also identified different n-type inorganic materials, such as In_2O_3, $SrTiO_3$, CdS, and so on, which could be used as ETL in PSCs and achieve good results. Organic ETLs have low fabrication temperature, suitable energy level, and high electron mobility, which is suitable for flexible or p-i-n PSCs. However, the synthetic cost of organic ETLs is higher than that of inorganic ETLs. Moreover, the organic ETLs also exhibit lower thermal stability than inorganic ETLs.

High-quality ETLs and ETL/perovskite interface are crucial to promote the large-scale commercial application of high-efficiency and low-cost PSCs. The properties of ETLs materials could be adjusted by optimizing the preparation process and doping. In addition, large-scale fabrication desires more attention with the booming development of PSCs. Various deposition technologies (inkjet printing, screen printing and etc.) should get further development to promote the commercial application of PSCs. Flexibility is another major direction in the development of PSCs. Flexible PSCs have higher requirements for ETLs than rigid PSCs. The high annealing temperature of inorganic ETLs would destroy the flexible substrates. Moreover, compared with rigid substrates, moisture and oxygen are more likely to penetrate the polymer substrates. It is necessary to develop low-temperature processes to deposit high-quality and compact ETLs.

Except the properties of ETLs, the properties of ETL/perovskite interface, such as energy band alignment, interface defects, and interface stability, should also be considered for highly efficient PSCs. Interface engineering is an efficient strategy to improve the interface contact, which could suppress the non-radiative recombination and enhance that charge transfer at ETL/perovskite interface. Moreover, the stability of PSCs should be also deserved more attention. It is necessary to develop functional interfacial modification materials with high stability to improve the long-term stability of PSCs. For flexible PSCs, due to the different Young's modulus between ETLs and perovskite films, the displacement of the ETL and perovskite film is detrimental to the photovoltaic performance and mechanical flexibility of flexible PSCs under bending operation. It is necessary to introduce an interlayer to enhance the adhesive property between ETLs and perovskite films. Organic materials have rich functional groups and can interact with ETL and perovskite simultaneously to form an adhesive interface. However, considering that most inorganic materials have better stability and lower cost than organic materials, it is necessary to develop inorganic interface materials for strengthening interfacial adhesion.

REFERENCES

1. P. Tiwana, P. Docampo, M. B. Johnston, H. J. Snaith, L. M. Herz, Electron mobility and injection dynamics in mesoporous ZnO, SnO_2, and TiO_2 films used in dye-sensitized solar cells. *ACS Nano* **2011**, *5*, 5158.
2. X. Yang, C. Xu, N. C. Giles, Intrinsic electron mobilities in CdSe, CdS, ZnO, and ZnS and their use in analysis of temperature-dependent Hall measurements. *J. Appl. Phys.* **2008**, *104*, 073727.
3. M. Park, J.-Y. Kim, H. J. Son, C.-H. Lee, S. S. Jang, M. J. Ko, Low-temperature solution-processed Li-doped SnO_2 as an effective electron transporting layer for high-performance flexible and wearable perovskite solar cells. *Nano Energy* **2016**, *26*, 208.
4. L. Xiong, Y. Guo, J. Wen, H. Liu, G. Yang, P. Qin, G. Fang, Review on the application of SnO_2 in perovskite solar cells. *Adv. Funct. Mater.* **2018**, *28*, 1802757.
5. X. Guo, Z. Lin, J. Ma, Z. Hu, J. Su, C. Zhang, J. Zhang, J. Chang, Y. Hao, Low temperature combustion synthesized indium oxide electron transport layer for high performance and stable perovskite solar cells. *J. Power Sources* **2019**, *438*, 226981.
6. B. Gu, Y. Zhu, H. Lu, W. Tian, L. Li, Efficient planar perovskite solar cells based on low-cost spin-coated ultrathin Nb_2O_5 films. *Sol. Energy* **2018**, *166*, 187.
7. R. F. Janninck, D. H. Whitmore, Electrical conduction in nonstoichiometric α-Nb_2O_5. *J. Chem. Phys.* **1962**, *37*, 2750.
8. J. Dong, J. Wu, J. Jia, X. He, Z. Lan, L. Fan, J. Lin, M. Huang, Annealing-free Cr_2O_3 electron-selective layer for efficient hybrid perovskite solar cells. *ChemSusChem* **2018**, *11*, 619.
9. J. A. Crawford, R. W. Vest, Electrical conductivity of single-crystal Cr_2O_3. *J. Appl. Phys.* **1964**, *35*, 2413.
10. X. Wang, L. Deng, L. Wang, S. Dai, Z. Xing, X. Zhan, X. Lu, S. Xie, R. Huang, L. Zheng, Cerium oxide standing out as an electron transport layer for efficient and stable perovskite solar cells processed at low temperature. *J. Mater. Chem. A* **2017**, *5*, 1706.
11. S. Ahmmed, A. Aktar, M. H. Rahman, J. Hossain, A. B. M. Ismail, Design and simulation of a high-performance $CH_3NH_3Pb(I_{1-x}Cl_x)_3$-based perovskite solar cell using a CeO_x electron transport layer and NiO hole transport layer. *Semicond. Sci. Technol.* **2021**, *36*, 035002.
12. C. Chen, Y. Jiang, Y. Wu, J. Guo, X. Kong, X. Wu, Y. Li, D. Zheng, S. Wu, X. Gao, Z. Hou, G. Zhou, Y. Chen, J.-M. Liu, K. Kempa, J. Gao, Low-temperature-processed WO_x as electron transfer layer for planar perovskite solar cells exceeding 20% efficiency. *Sol. RRL* **2020**, *4*, 1900499.
13. D. W. Kim, S. S. Shin, S. Lee, I. S. Cho, D. H. Kim, C. W. Lee, H. S. Jung, K. S. Hong, $BaSnO_3$ perovskite nanoparticles for high efficiency dye-sensitized solar cells. *ChemSusChem* **2013**, *6*, 449.
14. S. S. Shin, J. S. Kim, J. H. Suk, K. D. Lee, D. W. Kim, J. H. Park, I. S. Cho, K. S. Hong, J. Y. Kim, Improved quantum efficiency of highly efficient perovskite $BaSnO_3$-based dye-sensitized solar cells. *ACS Nano* **2013**, *7*, 1027.
15. A. Bera, K. Wu, A. Sheikh, E. Alarousu, O. F. Mohammed, T. Wu, Perovskite oxide $SrTiO_3$ as an efficient electron transporter for hybrid perovskite solar cells. *J. Phys. Chem. C* **2014**, *118*, 28494.
16. M. Neophytou, M. De Bastiani, N. Gasparini, E. Aydin, E. Ugur, A. Seitkhan, F. Moruzzi, Y. Choaie, A. J. Ramadan, J. R. Troughton, R. Hallani, A. Savva, L. Tsetseris, S. Inal, D. Baran, F. Laquai, T. D. Anthopoulos, H. J. Snaith, S. De Wolf, I. McCulloch, Enhancing the charge extraction and stability of perovskite solar cells using strontium titanate ($SrTiO_3$) electron transport layer. *ACS Appl. Energy Mater.* **2019**, *2*, 8090.
17. H. Guo, H. Chen, H. Zhang, X. Huang, J. Yang, B. Wang, Y. Li, L. Wang, X. Niu, Z. Wang, Low-temperature processed yttrium-doped $SrSnO_3$ perovskite electron transport layer for planar heterojunction perovskite solar cells with high efficiency. *Nano Energy* **2019**, *59*, 1.

18. T. Wang, A. Prakash, Y. Dong, T. Truttmann, A. Bucsek, R. James, D. D. Fong, J.-W. Kim, P. J. Ryan, H. Zhou, T. Birol, B. Jalan, Engineering $SrSnO_3$ phases and electron mobility at room temperature using epitaxial strain. *ACS Appl. Mater. Interfaces* **2018**, *10*, 43802.
19. L. Wang, W. Fu, Z. Gu, C. Fan, X. Yang, H. Li, H. Chen, Low temperature solution processed planar heterojunction perovskite solar cells with a CdSe nanocrystal as an electron transport/extraction layer. *J. Mater. Chem. C* **2014**, *2*, 9087.
20. H. Peng, W. Sun, Y. Li, W. Yan, P. Yu, H. Zhou, Z. Bian, C. Huang, High-performance cadmium sulphide-based planar perovskite solar cell and the cadmium sulphide/perovskite interfaces. *J. Photonics Energy* **2016**, *6*, 022002.
21. D.-B. Li, L. Hu, Y. Xie, G. Niu, T. Liu, Y. Zhou, L. Gao, B. Yang, J. Tang, Low-temperature-processed amorphous Bi_2S_3 film as an inorganic electron transport layer for perovskite solar cells. *ACS Photonics* **2016**, *3*, 2122.
22. Y. Hou, X. Chen, S. Yang, Y. L. Zhong, C. Li, H. Zhao, H. G. Yang, Low-temperature processed In_2S_3 electron transport layer for efficient hybrid perovskite solar cells. *Nano Energy* **2017**, *36*, 102.
23. R. Singh, A. Giri, M. Pal, K. Thiyagarajan, J. Kwak, J.-J. Lee, U. Jeong, K. Cho, Perovskite solar cells with an MoS_2 electron transport layer. *J. Mater. Chem. A* **2019**, *7*, 7151.
24. X. Zhao, S. Liu, H. Zhang, S.-Y. Chang, W. Huang, B. Zhu, Y. Shen, C. Shen, D. Wang, Y. Yang, M. Wang, 20% Efficient perovskite solar cells with 2D electron transporting layer. *Adv. Funct. Mater.* **2019**, *29*, 1805168.
25. A. Kojima, K. Teshima, Y. Shirai, T. Miyasaka, Organometal halide perovskites as visible-light sensitizers for photovoltaic cells. *J. Am. Chem. Soc.* **2009**, *131*, 6050.
26. M. Liu, M. B. Johnston, H. J. Snaith, Efficient planar heterojunction perovskite solar cells by vapour deposition. *Nature* **2013**, *501*, 395.
27. S.-D. Mo, W. Y. Ching, Electronic and optical properties of three phases of titanium dioxide: Rutile, anatase, and brookite. *Phys. Rev. B* **1995**, *51*, 13023.
28. Y. Wang, J. Wan, J. Ding, J. Hu, D. Wang, A rutile TiO_2 electron transport layer for the enhancement of charge collection for efficient perovskite solar cells. *Angew. Chemie Int. Ed.* **2019**, *58*, 9414.
29. A. Kogo, Y. Sanehira, M. Ikegami, T. Miyasaka, Brookite TiO_2 as a low-temperature solution-processed mesoporous layer for hybrid perovskite solar cells. *J. Mater. Chem. A* **2015**, *3*, 20952.
30. F. Xie, J. Zhu, Y. Li, D. Shen, A. Abate, M. Wei, TiO_2-B as an electron transporting material for highly efficient perovskite solar cells. *J. Power Sources* **2019**, *415*, 8.
31. Y. Yang, K. Ri, A. Mei, L. Liu, M. Hu, T. Liu, X. Li, H. Han, The size effect of TiO_2 nanoparticles on a printable mesoscopic perovskite solar cell. *J. Mater. Chem. A* **2015**, *3*, 9103.
32. H.-S. Kim, N.-G. Park, Parameters affecting I–V hysteresis of $CH_3NH_3PbI_3$ perovskite solar cells: effects of perovskite crystal size and mesoporous TiO_2 layer. *J. Phys. Chem. Lett.* **2014**, *5*, 2927.
33. T. Leijtens, G. E. Eperon, S. Pathak, A. Abate, M. M. Lee, H. J. Snaith, Overcoming ultraviolet light instability of sensitized TiO_2 with meso-superstructured organometal tri-halide perovskite solar cells. *Nat. Commun.* **2013**, *4*, 2885.
34. J. Burschka, N. Pellet, S.-J. Moon, R. Humphry-Baker, P. Gao, M. K. Nazeeruddin, M. Grätzel, Sequential deposition as a route to high-performance perovskite-sensitized solar cells. *Nature* **2013**, *499*, 316.
35. H. Choi, S. Paek, N. Lim, Y. H. Lee, M. K. Nazeeruddin, J. Ko, Efficient perovskite solar cells with 13.63% efficiency based on planar triphenylamine hole conductors. *Chem.—A Eur. J.* **2014**, *20*, 10894.
36. X. Guo, B. Zhang, Z. Lin, J. Ma, J. Su, W. Zhu, C. Zhang, J. Zhang, J. Chang, Y. Hao, Interface engineering of TiO_2/perovskite interface via fullerene derivatives for high performance planar perovskite solar cells. *Org. Electron.* **2018**, *62*, 459.

37. J. Ma, J. Chang, Z. Lin, X. Guo, L. Zhou, Z. Liu, H. Xi, D. Chen, C. Zhang, Y. Hao, Elucidating the roles of $TiCl_4$ and PCBM fullerene treatment on TiO_2 electron transporting layer for highly efficient planar perovskite solar cells. *J. Phys. Chem. C* **2018**, *122*, 1044.
38. H. Tan, A. Jain, O. Voznyy, X. Lan, F. P. García de Arquer, J. Z. Fan, R. Quintero-Bermudez, M. Yuan, B. Zhang, Y. Zhao, F. Fan, P. Li, L. N. Quan, Y. Zhao, Z.-H. Lu, Z. Yang, S. Hoogland, E. H. Sargent, Efficient and stable solution-processed planar perovskite solar cells via contact passivation. *Science* **2017**, *355*, 722.
39. K. Wojciechowski, M. Saliba, T. Leijtens, A. Abate, H. J. Snaith, Sub-150 °C processed meso-superstructured perovskite solar cells with enhanced efficiency. *Energy Environ. Sci.* **2014**, *7*, 1142.
40. A. Yella, L.-P. Heiniger, P. Gao, M. K. Nazeeruddin, M. Grätzel, Nanocrystalline rutile electron extraction layer enables low-temperature solution processed perovskite photovoltaics with 13.7% efficiency. *Nano Lett.* **2014**, *14*, 2591.
41. H. Huang, P. Cui, Y. Chen, L. Yan, X. Yue, S. Qu, X. Wang, S. Du, B. Liu, Q. Zhang, Z. Lan, Y. Yang, J. Ji, X. Zhao, Y. Li, X. Wang, X. Ding, M. Li, 24.8%-efficient planar perovskite solar cells via ligand-engineered TiO_2 deposition. *Joule* **2022**, *6*, 2186.
42. M. J. Paik, Y. Lee, H. Yun, S. Lee, S. Hong, S. Il Seok, TiO_2 colloid-spray coated electron-transporting layers for efficient perovskite solar cells. *Adv. Energy Mater.* **2020**, *10*, 2001799.
43. B. J. Kim, D. H. Kim, Y.-Y. Lee, H.-W. Shin, G. S. Han, J. S. Hong, K. Mahmood, T. K. Ahn, Y.-C. Joo, K. S. Hong, N.-G. Park, S. Lee, H. S. Jung, Highly efficient and bending durable perovskite solar cells: toward a wearable power source. *Energy Environ. Sci.* **2015**, *8*, 916.
44. Y. Lv, P. Xu, G. Ren, F. Chen, H. Nan, R. Liu, D. Wang, X. Tan, X. Liu, H. Zhang, Z.-K. Chen, Low-temperature atomic layer deposition of metal oxide layers for perovskite solar cells with high efficiency and stability under harsh environmental conditions. *ACS Appl. Mater. Interfaces* **2018**, *10*, 23928.
45. C. Chen, Y. Cheng, Q. Dai, H. Song, Radio frequency magnetron sputtering deposition of TiO_2 thin films and their perovskite solar cell applications. *Sci. Rep.* **2015**, *5*, 17684.
46. F. Shahvaranfard, N. Li, S. Hosseinpour, S. Hejazi, K. Zhang, M. Altomare, P. Schmuki, C. J. Brabec, Comparison of the sputtered TiO_2 anatase and rutile thin films as electron transporting layers in perovskite solar cells. *Nano Sel.* **2022**, *3*, 990.
47. H. S. Jung, N. Park, Perovskite solar cells: from materials to devices. *Small* **2015**, *11*, 10.
48. Q. Jiang, L. Zhang, H. Wang, X. Yang, J. Meng, H. Liu, Z. Yin, J. Wu, X. Zhang, J. You, Enhanced electron extraction using SnO_2 for high-efficiency planar-structure $HC(NH_2)_2PbI_3$-based perovskite solar cells. *Nat. Energy* **2016**, *2*, 16177.
49. J. Lian, B. Lu, F. Niu, P. Zeng, X. Zhan, Electron-transport materials in perovskite solar cells. *Small Methods* **2018**, *2*, 1800082.
50. B. Roose, J.-P. C. Baena, K. C. Gödel, M. Graetzel, A. Hagfeldt, U. Steiner, A. Abate, Mesoporous SnO_2 electron selective contact enables UV-stable perovskite solar cells. *Nano Energy* **2016**, *30*, 517.
51. J. Zhou, X. Meng, X. Zhang, X. Tao, Z. Zhang, J. Hu, C. Wang, Y. Li, S. Yang, Low-temperature aqueous solution processed ZnO as an electron transporting layer for efficient perovskite solar cells. *Mater. Chem. Front.* **2017**, *1*, 802.
52. J. Zhang, T. Pauporté, Effects of oxide contact layer on the preparation and properties of $CH_3NH_3PbI_3$ for perovskite solar cell application. *J. Phys. Chem. C* **2015**, *119*, 14919.
53. J. H. Heo, M. H. Lee, H. J. Han, B. R. Patil, J. S. Yu, S. H. Im, Highly efficient low temperature solution processable planar type $CH_3NH_3PbI_3$ perovskite flexible solar cells. *J. Mater. Chem. A* **2016**, *4*, 1572.
54. D. Zheng, G. Wang, W. Huang, B. Wang, W. Ke, J. L. Logsdon, H. Wang, Z. Wang, W. Zhu, J. Yu, M. R. Wasielewski, M. G. Kanatzidis, T. J. Marks, A. Facchetti, Combustion synthesized zinc oxide electron-transport layers for efficient and stable perovskite solar cells. *Adv. Funct. Mater.* **2019**, *29*, 1900265.

55. D. Liu, T. L. Kelly, Perovskite solar cells with a planar heterojunction structure prepared using room-temperature solution processing techniques. *Nat. Photonics* **2014**, *8*, 133.
56. C. Manspeaker, P. Scruggs, J. Preiss, D. A. Lyashenko, A. A. Zakhidov, Reliable annealing of $CH_3NH_3PbI_3$ films deposited on ZnO. *J. Phys. Chem. C* **2016**, *120*, 6377.
57. J. Song, L. Liu, X.-F. Wang, G. Chen, W. Tian, T. Miyasaka, Highly efficient and stable low-temperature processed ZnO solar cells with triple cation perovskite absorber. *J. Mater. Chem. A* **2017**, *5*, 13439.
58. J. Yang, B. D. Siempelkamp, E. Mosconi, F. De Angelis, T. L. Kelly, Origin of the thermal instability in $CH_3NH_3PbI_3$ thin films deposited on ZnO. *Chem. Mater.* **2015**, *27*, 4229.
59. Y. Cheng, Q.-D. Yang, J. Xiao, Q. Xue, H.-W. Li, Z. Guan, H.-L. Yip, S.-W. Tsang, Decomposition of organometal halide perovskite films on zinc oxide nanoparticles. *ACS Appl. Mater. Interfaces* **2015**, *7*, 19986.
60. P. Zhang, J. Wu, Y. Wang, H. Sarvari, D. Liu, Z. D. Chen, S. Li, Enhanced efficiency and environmental stability of planar perovskite solar cells by suppressing photocatalytic decomposition. *J. Mater. Chem. A* **2017**, *5*, 17368.
61. K.-M. Lee, S. H. Chang, K.-H. Wang, C.-M. Chang, H.-M. Cheng, C.-C. Kei, Z.-L. Tseng, C.-G. Wu, Thickness effects of ZnO thin film on the performance of tri-iodide perovskite absorber based photovoltaics. *Sol. Energy* **2015**, *120*, 117.
62. X. Dong, H. Hu, B. Lin, J. Ding, N. Yuan, The effect of ALD-ZnO layers on the formation of $CH_3NH_3PbI_3$ with different perovskite precursors and sintering temperatures. *Chem. Commun.* **2014**, *50*, 14405.
63. C.-Y. Chang, K.-T. Lee, W.-K. Huang, H.-Y. Siao, Y.-C. Chang, High-performance, air-stable, low-temperature processed semitransparent perovskite solar cells enabled by atomic layer deposition. *Chem. Mater.* **2015**, *27*, 5122.
64. Z.-L. Tseng, C.-H. Chiang, C.-G. Wu, Surface engineering of ZnO thin film for high efficiency planar perovskite solar cells. *Sci. Rep.* **2015**, *5*, 13211.
65. W. Zhao, K. Wang, H. Li, Z. Yang, Z. Liu, J. Sun, D. Wang, S. (Frank) Liu, Stoichiometry control of sputtered zinc oxide films by adjusting Ar/O_2 gas ratios as electron transport layers for efficient planar perovskite solar cells. *Sol. Energy Mater. Sol. Cells* **2018**, *178*, 200.
66. K. Schutt, P. K. Nayak, A. J. Ramadan, B. Wenger, Y.-H. Lin, H. J. Snaith, Overcoming zinc oxide interface instability with a methylammonium-free perovskite for high-performance solar cells. *Adv. Funct. Mater.* **2019**, *29*, 1900466.
67. G. Liu, Y. Zhong, H. Mao, J. Yang, R. Dai, X. Hu, Z. Xing, W. Sheng, L. Tan, Y. Chen, Highly efficient and stable ZnO-based MA-free perovskite solar cells via overcoming interfacial mismatch and deprotonation reaction. *Chem. Eng. J.* **2022**, *431*, 134235.
68. J. Ma, J. Su, Z. Lin, L. Zhou, J. He, J. Zhang, S. Liu, J. Chang, Y. Hao, Improve the oxide/perovskite heterojunction contact for low temperature high efficiency and stable all-inorganic $CsPbI_2Br$ perovskite solar cells. *Nano Energy* **2020**, *67*, 104241.
69. W. Ke, G. Fang, Q. Liu, L. Xiong, P. Qin, H. Tao, J. Wang, H. Lei, B. Li, J. Wan, G. Yang, Y. Yan, Low-temperature solution-processed tin oxide as an alternative electron transporting layer for efficient perovskite solar cells. *J. Am. Chem. Soc.* **2015**, *137*, 6730.
70. X. Liu, K. Tsai, Z. Zhu, Y. Sun, C. Chueh, A. K.-Y. Jen, A low-temperature, solution processable tin oxide electron-transporting layer prepared by the dual-fuel combustion method for efficient perovskite solar cells. *Adv. Mater. Interfaces* **2016**, *3*, 1600122.
71. K.-H. Jung, J.-Y. Seo, S. Lee, H. Shin, N.-G. Park, Solution-processed SnO_2 thin film for a hysteresis-free planar perovskite solar cell with a power conversion efficiency of 19.2%. *J. Mater. Chem. A* **2017**, *5*, 24790.
72. E. H. Anaraki, A. Kermanpur, L. Steier, K. Domanski, T. Matsui, W. Tress, M. Saliba, A. Abate, M. Grätzel, A. Hagfeldt, J.-P. Correa-Baena, Highly efficient and stable planar perovskite solar cells by solution-processed tin oxide. *Energy Environ. Sci.* **2016**, *9*, 3128.

73. J. J. Yoo, G. Seo, M. R. Chua, T. G. Park, Y. Lu, F. Rotermund, Y.-K. Kim, C. S. Moon, N. J. Jeon, J.-P. Correa-Baena, V. Bulović, S. S. Shin, M. G. Bawendi, J. Seo, Efficient perovskite solar cells via improved carrier management. *Nature* **2021**, *590*, 587.
74. H. Min, D. Y. Lee, J. Kim, G. Kim, K. S. Lee, J. Kim, M. J. Paik, Y. K. Kim, K. S. Kim, M. G. Kim, T. J. Shin, S. Il Seok, Perovskite solar cells with atomically coherent interlayers on SnO_2 electrodes. *Nature* **2021**, *598*, 444.
75. J. Park, J. Kim, H.-S. Yun, M. J. Paik, E. Noh, H. J. Mun, M. G. Kim, T. J. Shin, S. Il Seok, Controlled growth of perovskite layers with volatile alkylammonium chlorides. *Nature* **2023**, *616*, 724.
76. J. Song, E. Zheng, J. Bian, X.-F. Wang, W. Tian, Y. Sanehira, T. Miyasaka, Low-temperature SnO_2-based electron selective contact for efficient and stable perovskite solar cells. *J. Mater. Chem. A* **2015**, *3*, 10837.
77. G. Yang, C. Chen, F. Yao, Z. Chen, Q. Zhang, X. Zheng, J. Ma, H. Lei, P. Qin, L. Xiong, W. Ke, G. Li, Y. Yan, G. Fang, Effective carrier-concentration tuning of SnO_2 quantum dot electron-selective layers for high-performance planar perovskite solar cells. *Adv. Mater.* **2018**, *30*, 1706023.
78. Q. Jiang, Y. Zhao, X. Zhang, X. Yang, Y. Chen, Z. Chu, Q. Ye, X. Li, Z. Yin, J. You, Surface passivation of perovskite film for efficient solar cells. *Nat. Photonics* **2019**, *13*, 460.
79. Y. Zhao, F. Ma, Z. Qu, S. Yu, T. Shen, H. Deng, X. Chu, X. Peng, Y. Yuan, X. Zhang, J. You, Inactive $(PbI_2)_2RbCl$ stabilizes perovskite films for efficient solar cells. *Science* **2022**, *377*, 531.
80. Z. Zhu, Y. Bai, X. Liu, C.-C. Chueh, S. Yang, A. K.-Y. Jen, Enhanced efficiency and stability of inverted perovskite solar cells using highly crystalline SnO_2 nanocrystals as the robust electron-transporting layer. *Adv. Mater.* **2016**, *28*, 6478.
81. J. P. Correa Baena, L. Steier, W. Tress, M. Saliba, S. Neutzner, T. Matsui, F. Giordano, T. J. Jacobsson, A. R. Srimath Kandada, S. M. Zakeeruddin, A. Petrozza, A. Abate, M. K. Nazeeruddin, M. Grätzel, A. Hagfeldt, Highly efficient planar perovskite solar cells through band alignment engineering. *Energy Environ. Sci.* **2015**, *8*, 2928.
82. J.-P. Correa-Baena, W. Tress, K. Domanski, E. H. Anaraki, S.-H. Turren-Cruz, B. Roose, P. P. Boix, M. Grätzel, M. Saliba, A. Abate, A. Hagfeldt, Identifying and suppressing interfacial recombination to achieve high open-circuit voltage in perovskite solar cells. *Energy Environ. Sci.* **2017**, *10*, 1207.
83. Z. Chen, G. Yang, X. Zheng, H. Lei, C. Chen, J. Ma, H. Wang, G. Fang, Bulk heterojunction perovskite solar cells based on room temperature deposited hole-blocking layer: Suppressed hysteresis and flexible photovoltaic application. *J. Power Sources* **2017**, *351*, 123.
84. J. Ma, X. Zheng, H. Lei, W. Ke, C. Chen, Z. Chen, G. Yang, G. Fang, Highly efficient and stable planar perovskite solar cells with large-scale manufacture of E-beam evaporated SnO_2 toward commercialization. *Sol. RRL* **2017**, *1*, 1700118.
85. S. Yoon, S. J. Kim, H. S. Kim, J.-S. Park, I. K. Han, J. W. Jung, M. Park, Solution-processed indium oxide electron transporting layers for high-performance and photo-stable perovskite and organic solar cells. *Nanoscale* **2017**, *9*, 16305.
86. M. Qin, J. Ma, W. Ke, P. Qin, H. Lei, H. Tao, X. Zheng, L. Xiong, Q. Liu, Z. Chen, J. Lu, G. Yang, G. Fang, Perovskite solar cells based on low-temperature processed indium oxide electron selective layers. *ACS Appl. Mater. Interfaces* **2016**, *8*, 8460.
87. P. Chen, X. Yin, M. Que, X. Liu, W. Que, Low temperature solution processed indium oxide thin films with reliable photoelectrochemical stability for efficient and stable planar perovskite solar cells. *J. Mater. Chem. A* **2017**, *5*, 9641.
88. A. Kogo, Y. Numata, M. Ikegami, T. Miyasaka, Nb_2O_5 blocking layer for high open-circuit voltage perovskite solar cells. *Chem. Lett.* **2015**, *44*, 829.
89. J. Feng, Z. Yang, D. Yang, X. Ren, X. Zhu, Z. Jin, W. Zi, Q. Wei, S. (Frank) Liu, E-beam evaporated Nb_2O_5 as an effective electron transport layer for large flexible perovskite solar cells. *Nano Energy* **2017**, *36*, 1.

90. L. Zhu, Z. Shao, J. Ye, X. Zhang, X. Pan, S. Dai, Mesoporous $BaSnO_3$ layer based perovskite solar cells. *Chem. Commun.* **2016**, *52*, 970.
91. S. S. Shin, E. J. Yeom, W. S. Yang, S. Hur, M. G. Kim, J. Im, J. Seo, J. H. Noh, S. Il Seok, Colloidally prepared La-doped $BaSnO_3$ electrodes for efficient, photostable perovskite solar cells. *Science* **2017**, *356*, 167.
92. C. Luo, G. Zheng, F. Gao, X. Wang, C. Zhan, X. Gao, Q. Zhao, Engineering the buried interface in perovskite solar cells via lattice-matched electron transport layer. *Nat. Photonics* **2023**, *17*, 856.
93. J. Liu, C. Gao, L. Luo, Q. Ye, X. He, L. Ouyang, X. Guo, D. Zhuang, C. Liao, J. Mei, W. Lau, Low-temperature, solution processed metal sulfide as an electron transport layer for efficient planar perovskite solar cells. *J. Mater. Chem. A* **2015**, *3*, 11750.
94. W. A. Dunlap-Shohl, R. Younts, B. Gautam, K. Gundogdu, D. B. Mitzi, Effects of Cd diffusion and doping in high-performance perovskite solar cells using CdS as electron transport layer. *J. Phys. Chem. C* **2016**, *120*, 16437.
95. G. Tong, Z. Song, C. Li, Y. Zhao, L. Yu, J. Xu, Y. Jiang, Y. Sheng, Y. Shi, K. Chen, Cadmium-doped flexible perovskite solar cells with a low-cost and low-temperature-processed CdS electron transport layer. *RSC Adv.* **2017**, *7*, 19457.
96. J. Dong, J. Wu, J. Jia, L. Fan, Y. Lin, J. Lin, M. Huang, Efficient perovskite solar cells employing a simply-processed CdS electron transport layer. *J. Mater. Chem. C* **2017**, *5*, 10023.
97. B. Yang, M. Wang, X. Hu, T. Zhou, Z. Zang, Highly efficient semitransparent $CsPbIBr_2$ perovskite solar cells via low-temperature processed In_2S_3 as electron-transport-layer. *Nano Energy* **2019**, *57*, 718.
98. J. Chen, W. Qiu, C. Huang, L. Wu, C. Liu, Q. Tian, Z. Peng, J. Chen, A novel solvent for multistep solution-processed planar $CsPbBr_3$ perovskite solar cells using In_2S_3 as electron transport layer. *Energy Technol.* **2022**, *10*, 2200054.
99. K. J. Lee, J.-W. Min, B. Turedi, A. Y. Alsalloum, J.-H. Min, Y. J. Kim, Y. J. Yoo, S. Oh, N. Cho, R. C. Subedi, S. Mitra, S. E. Yoon, J. H. Kim, K. Park, T.-H. Chung, S. H. Jung, J. H. Baek, Y. M. Song, I. S. Roqan, T. K. Ng, B. S. Ooi, O. M. Bakr, Nanoporous GaN/n-type GaN: a cathode structure for ITO-free perovskite solar cells. *ACS Energy Lett.* **2020**, *5*, 3295.
100. L. Yang, C. Dall'Agnese, Y. Dall'Agnese, G. Chen, Y. Gao, Y. Sanehira, A. K. Jena, X. Wang, Y. Gogotsi, T. Miyasaka, Surface-modified metallic $Ti_3C_2T_x$ MXene as electron transport layer for planar heterojunction perovskite solar cells. *Adv. Funct. Mater.* **2019**, *29*, 1905694.
101. D. Shen, W. Zhang, Y. Li, A. Abate, M. Wei, Facile deposition of Nb_2O_5 thin film as an electron-transporting layer for highly efficient perovskite solar cells. *ACS Appl. Nano Mater.* **2018**, *1*, 4101.
102. X. Liu, Z. Liu, H. Ye, Y. Tu, B. Sun, X. Tan, T. Shi, Z. Tang, G. Liao, Novel efficient C_{60}-based inverted perovskite solar cells with negligible hysteresis. *Electrochim. Acta* **2018**, *288*, 115.
103. W. Ke, D. Zhao, C. R. Grice, A. J. Cimaroli, J. Ge, H. Tao, H. Lei, G. Fang, Y. Yan, Efficient planar perovskite solar cells using room-temperature vacuum-processed C_{60} electron selective layers. *J. Mater. Chem. A* **2015**, *3*, 17971.
104. H. Yoon, S. M. Kang, J.-K. Lee, M. Choi, Hysteresis-free low-temperature-processed planar perovskite solar cells with 19.1% efficiency. *Energy Environ. Sci.* **2016**, *9*, 2262.
105. Q. Jiang, J. Tong, Y. Xian, R. A. Kerner, S. P. Dunfield, C. Xiao, R. A. Scheidt, D. Kuciauskas, X. Wang, M. P. Hautzinger, R. Tirawat, M. C. Beard, D. P. Fenning, J. J. Berry, B. W. Larson, Y. Yan, K. Zhu, Surface reaction for efficient and stable inverted perovskite solar cells. *Nature* **2022**, *611*, 278.
106. F. Jafari Nodoushan, A. Behjat, N. Torabi, M. Ahmadpour, B. Ramesh Patil, V. Turkovic, M. Madsen, Planar perovskite solar cells using fullerene C_{70} as electron selective transport layer. *Int. J. Opt. Photonics* **2020**, *14*, 15.

107. H.-S. Lin, I. Jeon, R. Xiang, S. Seo, J.-W. Lee, C. Li, A. Pal, S. Manzhos, M. S. Goorsky, Y. Yang, S. Maruyama, Y. Matsuo, Achieving high efficiency in solution-processed perovskite solar cells using C_{60}/C_{70} mixed fullerenes. *ACS Appl. Mater. Interfaces* **2018**, *10*, 39590.
108. Y. Shao, Z. Xiao, C. Bi, Y. Yuan, J. Huang, Origin and elimination of photocurrent hysteresis by fullerene passivation in $CH_3NH_3PbI_3$ planar heterojunction solar cells. *Nat. Commun.* **2014**, *5*, 5784.
109. F. Wu, S. Lu, C. Hu, H. Lu, C. Chen, J. Tang, S. Yang, L. Zhu, A smart way to prepare solution-processed and annealing-free PCBM electron transporting layer for perovskite solar cells. *Adv. Sustain. Syst.* **2022**, *6*, 2200212.
110. J. Lee, J. Kim, C. Lee, G. Kim, T. K. Kim, H. Back, S. Jung, K. Yu, S. Hong, S. Lee, S. Kim, S. Jeong, H. Kang, K. Lee, A printable organic electron transport layer for low-temperature-processed, hysteresis-free, and stable planar perovskite solar cells. *Adv. Energy Mater.* **2017**, *7*, 1700226.
111. Y. Jiang, J. Wang, H. Zai, D. Ni, J. Wang, P. Xue, N. Li, B. Jia, H. Lu, Y. Zhang, F. Wang, Z. Guo, Z. Bi, H. Xie, Q. Wang, W. Ma, Y. Tu, H. Zhou, X. Zhan, Reducing energy disorder in perovskite solar cells by chelation. *J. Am. Chem. Soc.* **2022**, *144*, 5400.
112. Z. Xing, S.-H. Li, Y. Hui, B.-S. Wu, Z.-C. Chen, D.-Q. Yun, L.-L. Deng, M.-L. Zhang, B.-W. Mao, S.-Y. Xie, R.-B. Huang, L.-S. Zheng, Star-like hexakis[di(ethoxycarbonyl) methano]-C_{60} with higher electron mobility: An unexpected electron extractor interfaced in photovoltaic perovskites. *Nano Energy* **2020**, *74*, 104859.
113. X. Jiang, F. Wang, Q. Wei, H. Li, Y. Shang, W. Zhou, C. Wang, P. Cheng, Q. Chen, L. Chen, Z. Ning, Ultra-high open-circuit voltage of tin perovskite solar cells via an electron transporting layer design. *Nat. Commun.* **2020**, *11*, 1245.
114. P.-W. Liang, C.-C. Chueh, S. T. Williams, A. K.-Y. Jen, Roles of fullerene-based interlayers in enhancing the performance of organometal perovskite thin-film solar cells. *Adv. Energy Mater.* **2015**, *5*, 1402321.
115. Y.-C. Wang, X. Li, L. Zhu, X. Liu, W. Zhang, J. Fang, Efficient and hysteresis-free perovskite solar cells based on a solution processable polar fullerene electron transport layer. *Adv. Energy Mater.* **2017**, *7*, 1701144.
116. H.-R. Liu, S.-H. Li, L.-L. Deng, Z.-Y. Wang, Z. Xing, X. Rong, H.-R. Tian, X. Li, S.-Y. Xie, R.-B. Huang, L.-S. Zheng, Pyridine-functionalized fullerene electron transport layer for efficient planar perovskite solar cells. *ACS Appl. Mater. Interfaces* **2019**, *11*, 23982.
117. P. Topolovsek, F. Lamberti, T. Gatti, A. Cito, J. M. Ball, E. Menna, C. Gadermaier, A. Petrozza, Functionalization of transparent conductive oxide electrode for TiO_2-free perovskite solar cells. *J. Mater. Chem. A* **2017**, *5*, 11882.
118. K. Wojciechowski, I. Ramirez, T. Gorisse, O. Dautel, R. Dasari, N. Sakai, J. M. Hardigree, S. Song, S. Marder, M. Riede, G. Wantz, H. J. Snaith, Cross-linkable fullerene derivatives for solution-processed n-i-p perovskite solar cells. *ACS Energy Lett.* **2016**, *1*, 648.
119. Z. Peng, J. Tian, K. Zhang, A. These, Z. Xie, Y. Zhao, A. Osvet, F. Guo, L. Lüer, N. Li, C. J. Brabec, Reducing open-circuit voltage losses in all-inorganic perovskite cells by dedoping. *ACS Energy Lett.* **2023**, *8*, 2077.
120. C. Tian, E. Castro, G. Betancourt-Solis, Z. Nan, O. Fernandez-Delgado, S. Jankuru, L. Echegoyen, Fullerene derivative with a branched alkyl chain exhibits enhanced charge extraction and stability in inverted planar perovskite solar cells. *New J. Chem.* **2018**, *42*, 2896.
121. C. Tian, G. Betancourt-Solis, Z. Nan, K. Liu, K. Lin, J. Lu, L. Xie, L. Echegoyen, Z. Wei, Efficient and stable inverted perovskite solar cells enabled by inhibition of self-aggregation of fullerene electron-transporting compounds. *Sci. Bull.* **2021**, *66*, 339.
122. T. Ahmad, B. Wilk, E. Radicchi, R. Fuentes Pineda, P. Spinelli, J. Herterich, L. A. Castriotta, S. Dasgupta, E. Mosconi, F. De Angelis, M. Kohlstädt, U. Würfel, A. Di Carlo, K. Wojciechowski, New fullerene derivative as an n-type material for highly efficient, flexible perovskite solar cells of a p-i-n configuration. *Adv. Funct. Mater.* **2020**, *30*, 2004357.

123. X. Meng, Y. Bai, S. Xiao, T. Zhang, C. Hu, Y. Yang, X. Zheng, S. Yang, Designing new fullerene derivatives as electron transporting materials for efficient perovskite solar cells with improved moisture resistance. *Nano Energy* **2016**, *30*, 341.
124. B. Li, J. Zhen, Y. Wan, X. Lei, L. Jia, X. Wu, H. Zeng, M. Chen, G.-W. Wang, S. Yang, Steering the electron transport properties of pyridine-functionalized fullerene derivatives in inverted perovskite solar cells: the nitrogen site matters. *J. Mater. Chem. A* **2020**, *8*, 3872.
125. R. D. Chavan, D. Prochowicz, B. Bończak, M. Fiałkowski, M. M. Tavakoli, P. Yadav, M. J. Patel, S. K. Gupta, P. N. Gajjar, C. K. Hong, Azahomofullerenes as new n-type acceptor materials for efficient and stable inverted planar perovskite solar cells. *ACS Appl. Mater. Interfaces* **2021**, *13*, 20296.
126. H.-H. Huang, H. Tsai, R. Raja, S.-L. Lin, D. Ghosh, C.-H. Hou, J.-J. Shyue, S. Tretiak, W. Chen, K.-F. Lin, W. Nie, L. Wang, Robust unencapsulated perovskite solar cells protected by a fluorinated fullerene electron transporting layer. *ACS Energy Lett.* **2021**, *6*, 3376.
127. J. Huang, Z. Gu, L. Zuo, T. Ye, H. Chen, Morphology control of planar heterojunction perovskite solar cells with fluorinated PDI films as organic electron transport layer. *Sol. Energy* **2016**, *133*, 331.
128. H. Zhang, L. Xue, J. Han, Y. Q. Fu, Y. Shen, Z. Zhang, Y. Li, M. Wang, New generation perovskite solar cells with solution-processed amino-substituted perylene diimide derivative as electron-transport layer. *J. Mater. Chem. A* **2016**, *4*, 8724.
129. J. Wu, W.-K. Huang, Y.-C. Chang, B.-C. Tsai, Y.-C. Hsiao, C.-Y. Chang, C.-T. Chen, C.-T. Chen, Simple mono-halogenated perylene diimides as non-fullerene electron transporting materials in inverted perovskite solar cells with ZnO nanoparticle cathode buffer layers. *J. Mater. Chem. A* **2017**, *5*, 12811.
130. D. Zou, F. Yang, Q. Zhuang, M. Zhu, Y. Chen, G. You, Z. Lin, H. Zhen, Q. Ling, Perylene diimide-based electron-transporting material for perovskite solar cells with undoped poly(3-hexylthiophene) as hole-transporting material. *ChemSusChem* **2019**, *12*, 1155.
131. K. Jiang, F. Wu, H. Yu, Y. Yao, G. Zhang, L. Zhu, H. Yan, A perylene diimide-based electron transport layer enabling efficient inverted perovskite solar cells. *J. Mater. Chem. A* **2018**, *6*, 16868.
132. E. Castro, T. J. Sisto, E. L. Romero, F. Liu, S. R. Peurifoy, J. Wang, X. Zhu, C. Nuckolls, L. Echegoyen, Cove-edge nanoribbon materials for efficient inverted halide perovskite solar cells. *Angew. Chemie Int. Ed.* **2017**, *56*, 14648.
133. Z. Luo, F. Wu, T. Zhang, X. Zeng, Y. Xiao, T. Liu, C. Zhong, X. Lu, L. Zhu, S. Yang, C. Yang, Designing a perylene diimide/fullerene hybrid as effective electron transporting material in inverted perovskite solar cells with enhanced efficiency and stability. *Angew. Chemie Int. Ed.* **2019**, *58*, 8520.
134. T. Zheng, B. Fan, Y. Zhao, B. Jin, L. Fan, R. Peng, Rational design of anion-doped perylene diimide-fullerene dimer as effective electron transporting material in inverted perovskite solar cells. *Chem. Eng. J.* **2021**, *420*, 129730.
135. J. H. Heo, S.-C. Lee, S.-K. Jung, O.-P. Kwon, S. H. Im, Efficient and thermally stable inverted perovskite solar cells by introduction of non-fullerene electron transporting materials. *J. Mater. Chem. A* **2017**, *5*, 20615.
136. S. Jung, J. H. Heo, D. W. Lee, S. Lee, S. Lee, W. Yoon, H. Yun, S. H. Im, J. H. Kim, O. Kwon, Nonfullerene electron transporting material based on naphthalene diimide small molecule for highly stable perovskite solar cells with efficiency exceeding 20%. *Adv. Funct. Mater.* **2018**, *28*, 1800346.
137. S. Jung, J. H. Heo, D. W. Lee, S. Lee, S. Lee, W. Yoon, H. Yun, D. Kim, J. H. Kim, S. H. Im, O. Kwon, Homochiral asymmetric-shaped electron-transporting materials for efficient non-fullerene perovskite solar cells. *ChemSusChem* **2019**, *12*, 224.

138. S. Jung, J. H. Heo, B. M. Oh, J. B. Lee, S. Park, W. Yoon, Y. Song, H. Yun, J. H. Kim, S. H. Im, O. Kwon, Chiral stereoisomer engineering of electron transporting materials for efficient and stable perovskite solar cells. *Adv. Funct. Mater.* **2020**, *30*, 1905951.
139. C. Ge, W. Wu, L. Hu, Y. Hu, Y. Zhou, W.-S. Li, X. Gao, Core-expanded naphthalene-diimide derivatives as non-fullerene electron transport materials for inverted perovskite solar cells. *Org. Electron.* **2018**, *61*, 113.
140. D. Zhao, Z. Zhu, M. Kuo, C. Chueh, A. K. -Y. Jen, Hexaazatrinaphthylene derivatives: efficient electron-transporting materials with tunable energy levels for inverted perovskite solar cells. *Angew. Chemie Int. Ed.* **2016**, *55*, 8999.
141. N. Wang, K. Zhao, T. Ding, W. Liu, A. S. Ahmed, Z. Wang, M. Tian, X. W. Sun, Q. Zhang, Improving interfacial charge recombination in planar heterojunction perovskite photovoltaics with small molecule as electron transport layer. *Adv. Energy Mater.* **2017**, *7*, 1700522.
142. M. Zhang, J. Zhu, K. Liu, G. Zheng, G. Zhao, L. Li, Y. Meng, T. Guo, H. Zhou, X. Zhan, A low temperature processed fused-ring electron transport material for efficient planar perovskite solar cells. *J. Mater. Chem. A* **2017**, *5*, 24820.
143. Q. Guo, Y. Xu, B. Xiao, B. Zhang, E. Zhou, F. Wang, Y. Bai, T. Hayat, A. Alsaedi, Z. Tan, Effect of energy alignment, electron mobility, and film morphology of perylene diimide based polymers as electron transport layer on the performance of perovskite solar cells. *ACS Appl. Mater. Interfaces* **2017**, *9*, 10983.
144. C. Sun, Z. Wu, H.-L. Yip, H. Zhang, X.-F. Jiang, Q. Xue, Z. Hu, Z. Hu, Y. Shen, M. Wang, F. Huang, Y. Cao, Amino-functionalized conjugated polymer as an efficient electron transport layer for high-performance planar-heterojunction perovskite solar cells. *Adv. Energy Mater.* **2016**, *6*, 1501534.
145. W. Chen, Y. Shi, Y. Wang, X. Feng, A. B. Djurišić, H. Y. Woo, X. Guo, Z. He, N-type conjugated polymer as efficient electron transport layer for planar inverted perovskite solar cells with power conversion efficiency of 20.86%. *Nano Energy* **2020**, *68*, 104363.
146. J. Ma, X. Guo, L. Zhou, Z. Lin, C. Zhang, Z. Yang, G. Lu, J. Chang, Y. Hao, Enhanced planar perovskite solar cell performance via contact passivation of TiO_2/perovskite interface with NaCl doping approach. *ACS Appl. Energy Mater.* **2018**, *1*, 3826.
147. Y. Tang, R. Roy, Z. Zhang, Y. Hu, F. Yang, C. Qin, L. Jiang, H. Liu, Rubidium chloride doping TiO_2 for efficient and hysteresis-free perovskite solar cells with decreasing traps. *Sol. Energy* **2022**, *231*, 440.
148. R. Azmi, S. Hwang, W. Yin, T.-W. Kim, T. K. Ahn, S.-Y. Jang, High efficiency low-temperature processed perovskite solar cells integrated with alkali metal doped ZnO electron transport layers. *ACS Energy Lett.* **2018**, *3*, 1241.
149. X. Guo, X. Huang, J. Su, Z. Lin, J. Ma, J. Chang, Y. Hao, Reducing the interfacial energy loss via oxide/perovskite heterojunction engineering for high efficient and stable perovskite solar cells. *Chem. Eng. J.* **2021**, *417*, 129184.
150. G. Xie, X. Lu, J. Duan, Y. Dong, X. Jiang, F. Tu, Y. Duan, Q. Tang, Alkali chloride doped SnO_2 electron-transporting layers for boosting charge transfer and passivating defects in all-inorganic $CsPbBr_3$ perovskite solar cells. *J. Mater. Chem. A* **2021**, *9*, 15003.
151. H. Zhou, Q. Chen, G. Li, S. Luo, T. Song, H.-S. Duan, Z. Hong, J. You, Y. Liu, Y. Yang, Interface engineering of highly efficient perovskite solar cells. *Science* **2014**, *345*, 542.
152. S. Sandhu, C. Saharan, S. K. Buruga, S. A. Kumar, P. S. Rana, P. C. Nagajyothi, S. D. Mane, Micro structurally engineered hysteresis-free high efficiency perovskite solar cell using Zr-doped TiO_2 electron transport layer. *Ceram. Int.* **2021**, *47*, 14665.
153. H. Zhang, J. Shi, X. Xu, L. Zhu, Y. Luo, D. Li, Q. Meng, Mg-doped TiO_2 boosts the efficiency of planar perovskite solar cells to exceed 19%. *J. Mater. Chem. A* **2016**, *4*, 15383.
154. Y. Xu, J. Duan, X. Yang, J. Du, Y. Wang, Y. Duan, Q. Tang, Lattice-tailored low-temperature processed electron transporting materials boost the open-circuit voltage of planar $CsPbBr_3$ perovskite solar cells up to 1.654 V. *J. Mater. Chem. A* **2020**, *8*, 11859.

155. T. Wu, C. Zhen, H. Zhu, J. Wu, C. Jia, L. Wang, G. Liu, N.-G. Park, H.-M. Cheng, Gradient Sn-doped heteroepitaxial film of faceted rutile TiO_2 as an electron selective layer for efficient perovskite solar cells. *ACS Appl. Mater. Interfaces* **2019**, *11*, 19638.
156. D. H. Kim, G. S. Han, W. M. Seong, J.-W. Lee, B. J. Kim, N.-G. Park, K. S. Hong, S. Lee, H. S. Jung, Niobium doping effects on TiO_2 mesoscopic electron transport layer-based perovskite solar cells. *ChemSusChem* **2015**, *8*, 2392.
157. Z.-L. Tseng, C.-H. Chiang, S.-H. Chang, C.-G. Wu, Surface engineering of ZnO electron transporting layer via Al doping for high efficiency planar perovskite solar cells. *Nano Energy* **2016**, *28*, 311.
158. J. Song, E. Zheng, L. Liu, X. Wang, G. Chen, W. Tian, T. Miyasaka, Magnesium-doped zinc oxide as electron selective contact layers for efficient perovskite solar cells. *ChemSusChem* **2016**, *9*, 2640.
159. Y. Bai, Y. Fang, Y. Deng, Q. Wang, J. Zhao, X. Zheng, Y. Zhang, J. Huang, Low temperature solution-processed Sb:SnO_2 nanocrystals for efficient planar perovskite solar cells. *ChemSusChem* **2016**, *9*, 2686.
160. E. Halvani Anaraki, A. Kermanpur, M. T. Mayer, L. Steier, T. Ahmed, S.-H. Turren-Cruz, J. Seo, J. Luo, S. M. Zakeeruddin, W. R. Tress, T. Edvinsson, M. Grätzel, A. Hagfeldt, J.-P. Correa-Baena, Low-temperature Nb-doped SnO_2 electron-selective contact yields over 20% efficiency in planar perovskite solar cells. *ACS Energy Lett.* **2018**, *3*, 773.
161. X. Ren, D. Yang, Z. Yang, J. Feng, X. Zhu, J. Niu, Y. Liu, W. Zhao, S. F. Liu, Solution-processed Nb:SnO_2 electron transport layer for efficient planar perovskite solar cells. *ACS Appl. Mater. Interfaces* **2017**, *9*, 2421.
162. Q. Liu, X. Zhang, C. Li, H. Lu, Z. Weng, Y. Pan, W. Chen, X.-C. Hang, Z. Sun, Y. Zhan, Effect of tantalum doping on SnO_2 electron transport layer via low temperature process for perovskite solar cells. *Appl. Phys. Lett.* **2019**, *115*, 143903.
163. Z. Liu, K. Deng, J. Hu, L. Li, Coagulated SnO_2 colloids for high-performance planar perovskite solar cells with negligible hysteresis and improved stability. *Angew. Chemie* **2019**, *131*, 11621.
164. H. Wang, S. Cao, B. Yang, H. Li, M. Wang, X. Hu, K. Sun, Z. Zang, NH_4Cl-modified ZnO for high-performance $CsPbIBr_2$ perovskite solar cells via low-temperature process. *Sol. RRL* **2020**, *4*, 1900363.
165. K. Mahmood, B. S. Swain, A. Amassian, 16.1% efficient hysteresis-free mesostructured perovskite solar cells based on synergistically improved ZnO nanorod arrays. *Adv. Energy Mater.* **2015**, *5*, 1500568.
166. H. Li, W. Shi, W. Huang, E.-P. Yao, J. Han, Z. Chen, S. Liu, Y. Shen, M. Wang, Y. Yang, Carbon quantum dots/TiO_x electron transport layer boosts efficiency of planar heterojunction perovskite solar cells to 19%. *Nano Lett.* **2017**, *17*, 2328.
167. W. Hui, Y. Yang, Q. Xu, H. Gu, S. Feng, Z. Su, M. Zhang, J. Wang, X. Li, J. Fang, F. Xia, Y. Xia, Y. Chen, X. Gao, W. Huang, Red-carbon-quantum-dot-doped SnO_2 composite with enhanced electron mobility for efficient and stable perovskite solar cells. *Adv. Mater.* **2020**, *32*, 1906374.
168. J. Xie, K. Huang, X. Yu, Z. Yang, K. Xiao, Y. Qiang, X. Zhu, L. Xu, P. Wang, C. Cui, D. Yang, Enhanced electronic properties of SnO_2 via electron transfer from graphene quantum dots for efficient perovskite solar cells. *ACS Nano* **2017**, *11*, 9176.
169. M. M. Tavakoli, R. Tavakoli, Z. Nourbakhsh, A. Waleed, U. S. Virk, Z. Fan, High efficiency and stable perovskite solar cell using ZnO/rGO QDs as an electron transfer layer. *Adv. Mater. Interfaces* **2016**, *3*, 1500790.
170. Z. Pang, S. Yang, Y. Sun, L. He, F. Wang, L. Fan, S. Chi, X. Sun, L. Yang, J. Yang, Hydrophobic PbS QDs layer decorated ZnO electron transport layer to boost photovoltaic performance of perovskite solar cells. *Chem. Eng. J.* **2022**, *439*, 135701.
171. X. Zeng, T. Zhou, C. Leng, Z. Zang, M. Wang, W. Hu, X. Tang, S. Lu, L. Fang, M. Zhou, Performance improvement of perovskite solar cells by employing a CdSe quantum dot/PCBM composite as an electron transport layer. *J. Mater. Chem. A* **2017**, *5*, 17499.

172. Y. Zhang, L. Xu, Y. Wu, Q. Zhou, Z. Shi, X. Zhuang, B. Liu, B. Dong, X. Bai, W. Xu, D. Zhou, H. Song, Double-layer synergistic optimization by functional black phosphorus quantum dots for high-efficiency and stable planar perovskite solar cells. *Nano Energy* **2021**, *90*, 106610.
173. X. Peng, H. Lu, J. Zhuang, X. Liu, Z. Ma, H. Wang, Z. Guo, Q. Wang, H. Zhang, S. Zhao, Enhanced performance of perovskite solar cells using DNA-doped mesoporous-TiO_2 as electron transporting layer. *Sol. Energy* **2020**, *206*, 855.
174. D. Yang, R. Yang, K. Wang, C. Wu, X. Zhu, J. Feng, X. Ren, G. Fang, S. Priya, S. Liu, High efficiency planar-type perovskite solar cells with negligible hysteresis using EDTA-complexed SnO_2. *Nat. Commun.* **2018**, *9*, 3239.
175. Z. Wang, X. Zhu, J. Feng, C. Wang, C. Zhang, X. Ren, S. Priya, S. (Frank) Liu, D. Yang, Antisolvent- and annealing-free deposition for highly stable efficient perovskite solar cells via modified ZnO. *Adv. Sci.* **2021**, *8*, 2002860.
176. X. Huang, J. Du, X. Guo, Z. Lin, J. Ma, J. Su, L. Feng, C. Zhang, J. Zhang, J. Chang, Y. Hao, Polyelectrolyte-doped SnO_2 as a tunable electron transport layer for high-efficiency and stable perovskite solar cells. *Sol. RRL* **2020**, *4*, 1900336.
177. B. Tu, Y. Shao, W. Chen, Y. Wu, X. Li, Y. He, J. Li, F. Liu, Z. Zhang, Y. Lin, X. Lan, L. Xu, X. Shi, A. M. C. Ng, H. Li, L. W. Chung, A. B. Djurišić, Z. He, Novel molecular doping mechanism for n-doping of SnO_2 via triphenylphosphine oxide and its effect on perovskite solar cells. *Adv. Mater.* **2019**, *31*, 1805944.
178. W. Dong, C. Zhu, C. Bai, Y. Ma, L. Lv, J. Zhao, F. Huang, Y.-B. Cheng, J. Zhong, Low-cost hydroxyacid potassium synergists as an efficient in situ defect passivator for high performance tin-oxide-based perovskite solar cells. *Angew. Chemie Int. Ed.* **2023**, e202302507.
179. L. Yin, C. Liu, C. Ding, C. Zhao, I. Z. Mitrovic, E. G. Lim, H. Wang, Y. Sun, Y. Han, Z. Li, L. Yang, C.-Q. Ma, C. Zhao, Functionalized-MXene-nanosheet-doped tin oxide enhances the electrical properties in perovskite solar cells. *Cell Reports Phys. Sci.* **2022**, *3*, 100905.
180. Z. Wang, D. P. McMeekin, N. Sakai, S. van Reenen, K. Wojciechowski, J. B. Patel, M. B. Johnston, H. J. Snaith, Efficient and air-stable mixed-cation lead mixed-halide perovskite solar cells with n-doped organic electron extraction layers. *Adv. Mater.* **2017**, *29*, 1604186.
181. C. Liu, Y. Yang, C. Zhang, S. Wu, L. Wei, F. Guo, G. M. Arumugam, J. Hu, X. Liu, J. Lin, R. E. I. Schropp, Y. Mai, Tailoring C_{60} for efficient inorganic $CsPbI_2Br$ perovskite solar cells and modules. *Adv. Mater.* **2020**, *32*, 1907361.
182. L. Zhou, J. Su, Z. Lin, X. Guo, J. Ma, T. Li, J. Zhang, J. Chang, Y. Hao, Synergistic interface layer optimization and surface passivation with fluorocarbon molecules toward efficient and stable inverted planar perovskite solar cells. *Research* **2021**, *2021*, 1.
183. G. Kakavelakis, T. Maksudov, D. Konios, I. Paradisanos, G. Kioseoglou, E. Stratakis, E. Kymakis, G. Kakavelakis, T. Maksudov, D. Konios, E. Kymakis, G. Kioseoglou, E. Stratakis, I. Paradisanos, Efficient and highly air stable planar inverted perovskite solar cells with reduced graphene oxide doped PCBM electron transporting layer. *Adv. Energy Mater.* **2017**, *7*, 1602120.
184. X. Zhu, J. Sun, S. Yuan, N. Li, Z. Qiu, J. Jia, Y. Liu, J. Dong, P. Lv, B. Cao, Efficient and stable planar perovskite solar cells with carbon quantum dots-doped PCBM electron transport layer. *New J. Chem.* **2019**, *43*, 7130.
185. Q. Dong, Y. Shi, C. Zhang, Y. Wu, L. Wang, Energetically favored formation of SnO_2 nanocrystals as electron transfer layer in perovskite solar cells with high efficiency exceeding 19%. *Nano Energy* **2017**, *40*, 336.
186. T. Bu, S. Shi, J. Li, Y. Liu, J. Shi, L. Chen, X. Liu, J. Qiu, Z. Ku, Y. Peng, J. Zhong, Y.-B. Cheng, F. Huang, Low-temperature presynthesized crystalline tin oxide for efficient flexible perovskite solar cells and modules. *ACS Appl. Mater. Interfaces* **2018**, *10*, 14922.

187. L. Zhou, J. Su, Z. Lin, X. Guo, J. Ma, L. Feng, J. Zhang, S. Wang, S. F. Liu, J. Chang, Y. Hao, Deep-ultraviolet photoactivation-assisted contact engineering toward high-efficiency and stable all-inorganic $CsPbI_2Br$ perovskite solar cells. *Sol. RRL* **2020**, *4*, 2000001.
188. G. S. Han, H. Shim, S. Lee, M. L. Duff, J. Lee, Low-temperature modification of ZnO nanoparticles film for electron-transport layers in perovskite solar cells. *ChemSusChem* **2017**, *10*, 2425.
189. J. Wang, Z. Cai, D. Lin, K. Chen, L. Zhao, F. Xie, R. Su, W. Xie, P. Liu, R. Zhu, Plasma oxidized $Ti_3C_2T_x$ MXene as electron transport layer for efficient perovskite solar cells. *ACS Appl. Mater. Interfaces* **2021**, *13*, 32495.
190. H. Tao, W. Ke, J. Wang, Q. Liu, J. Wan, G. Yang, G. Fang, Perovskite solar cell based on network nanoporous layer consisted of TiO_2 nanowires and its interface optimization. *J. Power Sources* **2015**, *290*, 144.
191. J. Ma, Z. Lin, X. Guo, L. Zhou, J. Su, C. Zhang, Z. Yang, J. Chang, S. Liu, Y. Hao, Low-temperature solution-processed ZnO electron transport layer for highly efficient and stable planar perovskite solar cells with efficiency over 20%. *Sol. RRL* **2019**, *3*, 1.
192. H.-S. Rao, B.-X. Chen, W.-G. Li, Y.-F. Xu, H.-Y. Chen, D.-B. Kuang, C.-Y. Su, Improving the extraction of photogenerated electrons with SnO_2 nanocolloids for efficient planar perovskite solar cells. *Adv. Funct. Mater.* **2015**, *25*, 7200.
193. X. Liu, Y. Zhang, L. Shi, Z. Liu, J. Huang, J. S. Yun, Y. Zeng, A. Pu, K. Sun, Z. Hameiri, J. A. Stride, J. Seidel, M. A. Green, X. Hao, Exploring inorganic binary alkaline halide to passivate defects in low-temperature-processed planar-structure hybrid perovskite solar cells. *Adv. Energy Mater.* **2018**, *8*, 1800138.
194. W. Li, W. Zhang, S. Van Reenen, R. J. Sutton, J. Fan, A. A. Haghighirad, M. B. Johnston, L. Wang, H. J. Snaith, Enhanced UV-light stability of planar heterojunction perovskite solar cells with caesium bromide interface modification. *Energy Environ. Sci.* **2016**, *9*, 490.
195. Q. Ye, Y. Zhao, S. Mu, F. Ma, F. Gao, Z. Chu, Z. Yin, P. Gao, X. Zhang, J. You, Cesium lead inorganic solar cell with efficiency beyond 18% via reduced charge recombination. *Adv. Mater.* **2019**, *31*, 1905143.
196. W. Zhang, Z. Lin, Q. Cai, X. Xu, H. Dong, C. Mu, J. Zhang, Electron transport assisted by transparent conductive oxide elements in perovskite solar cells. *ChemSusChem* **2022**, *15*, e202102002.
197. E. H. Jung, B. Chen, K. Bertens, M. Vafaie, S. Teale, A. Proppe, Y. Hou, T. Zhu, C. Zheng, E. H. Sargent, Bifunctional surface engineering on SnO_2 reduces energy loss in perovskite solar cells. *ACS Energy Lett.* **2020**, *5*, 2796.
198. J. Kim, J. Park, Y. Kim, W. Jo, Improvement of open-circuit voltage deficit via pre-treated NH_4+ ion modification of interface between SnO_2 and perovskite solar cells. *Small* **2022**, *18*, 2204173.
199. H. Bi, B. Liu, D. He, L. Bai, W. Wang, Z. Zang, J. Chen, Interfacial defect passivation and stress release by multifunctional KPF_6 modification for planar perovskite solar cells with enhanced efficiency and stability. *Chem. Eng. J.* **2021**, *418*, 129375.
200. C. Liu, W. Wu, D. Zhang, Z. Li, G. Ren, W. Han, W. Guo, Effective stability enhancement in ZnO-based perovskite solar cells by MACl modification. *J. Mater. Chem. A* **2021**, *9*, 12161.
201. D. Yang, X. Zhou, R. Yang, Z. Yang, W. Yu, X. Wang, C. Li, S. (Frank) Liu, R. P. H. Chang, Surface optimization to eliminate hysteresis for record efficiency planar perovskite solar cells. *Energy Environ. Sci.* **2016**, *9*, 3071.
202. N. K. Noel, S. N. Habisreutinger, B. Wenger, Y. Lin, F. Zhang, J. B. Patel, A. Kahn, M. B. Johnston, H. J. Snaith, Elucidating the role of a tetrafluoroborate-based ionic liquid at the n-type oxide/perovskite interface. *Adv. Energy Mater.* **2020**, *10*, 1903231.

203. J. Xia, J. Luo, H. Yang, C. Sun, Z. Wan, H. A. Malik, H. Zhang, Y. Shi, C. Jia, Ionic selective contact controls the charge accumulation for efficient and intrinsic stable planar homo-junction perovskite solar cells. *Nano Energy* **2019**, *66*, 104098.
204. C. Xu, S. Zhang, W. Fan, F. Cheng, H. Sun, Z. Kang, Y. Zhang, Pushing the limit of open-circuit voltage deficit via modifying buried interface in $CsPbI_3$ perovskite solar cells. *Adv. Mater.* **2023**, *35*, 2207172.
205. J. Chen, X. Zhao, S. Kim, N. Park, Multifunctional chemical linker imidazoleacetic acid hydrochloride for 21% efficient and stable planar perovskite solar cells. *Adv. Mater.* **2019**, *31*, 1902902.
206. K. Wang, H. Lu, M. Li, C. Chen, D. Bo Zhang, J. Chen, J. Wu, Y. Zhou, X. Wang, Z. Su, Y. Shi, Q. Tian, Y. Ni, X. Gao, S. M. Zakeeruddin, M. Grätzel, Z. Wang, L. Liao, Ion-dipole interaction enabling highly efficient $CsPbI_3$ perovskite indoor photovoltaics. *Adv. Mater*. **2023**, 2210106.
207. M. M. Lee, J. Teuscher, T. Miyasaka, T. N. Murakami, H. J. Snaith, Efficient hybrid solar cells based on meso-superstructured organometal halide perovskites. *Science* **2012**, *338*, 643.
208. H. Si, Q. Liao, Z. Zhang, Y. Li, X. Yang, G. Zhang, Z. Kang, Y. Zhang, An innovative design of perovskite solar cells with Al_2O_3 inserting at ZnO/perovskite interface for improving the performance and stability. *Nano Energy* **2016**, *22*, 223.
209. N. Tsvetkov, M. E. Khan, B. C. Moon, Y.-H. Kim, J. K. Kang, Strain-induced metallization and defect suppression at zipper-like interdigitated atomically thin interfaces enabling high-efficiency halide perovskite solar cells. *ACS Nano* **2021**, *15*, 1805.
210. J. Dagar, S. Castro-Hermosa, G. Lucarelli, F. Cacialli, T. M. Brown, Highly efficient perovskite solar cells for light harvesting under indoor illumination via solution processed SnO_2/MgO composite electron transport layers. *Nano Energy* **2018**, *49*, 290.
211. H. Wang, H. Li, S. Cao, M. Wang, J. Chen, Z. Zang, Interface modulator of ultrathin magnesium oxide for low-temperature-processed inorganic $CsPbIBr_2$ perovskite solar cells with efficiency over 11%. *Sol. RRL* **2020**, *4*, 2000226.
212. J. Cao, B. Wu, R. Chen, Y. Wu, Y. Hui, B.-W. Mao, N. Zheng, Efficient, hysteresis-free, and stable perovskite solar cells with ZnO as electron-transport layer: effect of surface passivation. *Adv. Mater.* **2018**, *30*, 1705596.
213. R. Chen, J. Cao, Y. Duan, Y. Hui, T. T. Chuong, D. Ou, F. Han, F. Cheng, X. Huang, B. Wu, N. Zheng, High-efficiency, hysteresis-less, UV-stable perovskite solar cells with cascade ZnO-ZnS electron transport layer. *J. Am. Chem. Soc.* **2019**, *141*, 541.
214. M. M. Byranvand, T. Kim, S. Song, G. Kang, S. U. Ryu, T. Park, P-type CuI islands on TiO_2 electron transport layer for a highly efficient planar-perovskite solar cell with negligible hysteresis. *Adv. Energy Mater.* **2018**, *8*, 1702235.
215. P. Huang, Q. Chen, K. Zhang, L. Yuan, Y. Zhou, B. Song, Y. Li, 21.7% efficiency achieved in planar n-i-p perovskite solar cells via interface engineering with water-soluble 2D TiS_2. *J. Mater. Chem. A* **2019**, *7*, 6213.
216. M. M. Tavakoli, R. Tavakoli, P. Yadav, J. Kong, A graphene/ZnO electron transfer layer together with perovskite passivation enables highly efficient and stable perovskite solar cells. *J. Mater. Chem. A* **2019**, *7*, 679.
217. Q. Zhou, J. Duan, X. Yang, Y. Duan, Q. Tang, Interfacial strain release from the WS_2/$CsPbBr_3$ van der waals heterostructure for 1.7 V voltage all-inorganic perovskite solar cells. *Angew. Chemie Int. Ed.* **2020**, *59*, 21997.
218. J. Zhu, B. He, M. Wang, X. Yao, H. Huang, C. Chen, H. Chen, Y. Duan, Q. Tang, Elimination of defect and strain by functionalized CQDs dual-engineering for all-inorganic HTMs-free perovskite solar cells with an ultrahigh voltage of 1.651 V. *Nano Energy* **2022**, *104*, 107920.

219. Y.-Q. Zhou, B.-S. Wu, G.-H. Lin, Z. Xing, S.-H. Li, L.-L. Deng, D.-C. Chen, D.-Q. Yun, S.-Y. Xie, Interfacing pristine C_{60} onto TiO_2 for viable flexibility in perovskite solar cells by a low-temperature all-solution process. *Adv. Energy Mater.* **2018**, *8*, 1800399.
220. T. Cao, Z. Wang, Y. Xia, B. Song, Y. Zhou, N. Chen, Y. Li, Facilitating electron transportation in perovskite solar cells via water-soluble fullerenol interlayers. *ACS Appl. Mater. Interfaces* **2016**, *8*, 18284.
221. B. Wang, J. Yang, L. Lu, W. Xiao, H. Wu, S. Xiong, J. Tang, C. Duan, Q. Bao, Interface engineering of air-stable n-doping fullerene-modified TiO_2 electron transport layer for highly efficient and stable perovskite solar cells. *Adv. Mater. Interfaces* **2020**, *7*, 1901964.
222. J. Wang, K. Datta, C. H. L. Weijtens, M. M. Wienk, R. A. J. Janssen, Insights into fullerene passivation of SnO_2 electron transport layers in perovskite solar cells. *Adv. Funct. Mater.* **2019**, *29*, 1905883.
223. K. Wojciechowski, S. D. Stranks, A. Abate, G. Sadoughi, A. Sadhanala, N. Kopidakis, G. Rumbles, C.-Z. Li, R. H. Friend, A. K.-Y. Jen, H. J. Snaith, Heterojunction modification for highly efficient organic-inorganic perovskite solar cells. *ACS Nano* **2014**, *8*, 12701.
224. L. Zuo, Z. Gu, T. Ye, W. Fu, G. Wu, H. Li, H. Chen, Enhanced photovoltaic performance of $CH_3NH_3PbI_3$ perovskite solar cells through interfacial engineering using self-assembling monolayer. *J. Am. Chem. Soc.* **2015**, *137*, 2674.
225. K. Liu, S. Chen, J. Wu, H. Zhang, M. Qin, X. Lu, Y. Tu, Q. Meng, X. Zhan, Fullerene derivative anchored SnO_2 for high-performance perovskite solar cells. *Energy Environ. Sci.* **2018**, *11*, 3463.
226. L. Zuo, Q. Chen, N. De Marco, Y.-T. Hsieh, H. Chen, P. Sun, S. Chang, H. Zhao, S. Dong, Y. Yang, Tailoring the interfacial chemical interaction for high-efficiency perovskite solar cells. *Nano Lett.* **2017**, *17*, 269.
227. G. Yang, C. Wang, H. Lei, X. Zheng, P. Qin, L. Xiong, X. Zhao, Y. Yan, G. Fang, Interface engineering in planar perovskite solar cells: energy level alignment, perovskite morphology control and high performance achievement. *J. Mater. Chem. A* **2017**, *5*, 1658.
228. J. Du, L. Feng, X. Guo, X. Huang, Z. Lin, J. Su, Z. Hu, J. Zhang, J. Chang, Y. Hao, Enhanced efficiency and stability of planar perovskite solar cells by introducing amino acid to SnO_2/perovskite interface. *J. Power Sources* **2020**, *455*, 227974.
229. Q. Lou, Y. Han, C. Liu, K. Zheng, J. Zhang, X. Chen, Q. Du, C. Chen, Z. Ge, π-Conjugated small molecules modified SnO_2 layer for perovskite solar cells with over 23% efficiency. *Adv. Energy Mater.* **2021**, *11*, 2101416.
230. M. Wu, Y. Duan, L. Yang, P. You, Z. Li, J. Wang, H. Zhou, S. Yang, D. Xu, H. Zou, Z. Liu, Multifunctional small molecule as buried interface passivator for efficient planar perovskite solar cells. *Adv. Funct. Mater.* **2023**, 2300128.
231. M. Kim, J. Jeong, H. Lu, T. K. Lee, F. T. Eickemeyer, Y. Liu, I. W. Choi, S. J. Choi, Y. Jo, H.-B. Kim, S.-I. Mo, Y.-K. Kim, H. Lee, N. G. An, S. Cho, W. R. Tress, S. M. Zakeeruddin, A. Hagfeldt, J. Y. Kim, M. Grätzel, D. S. Kim, Conformal quantum dot-SnO_2 layers as electron transporters for efficient perovskite solar cells. *Science* **2022**, *375*, 302.
232. H. Shu, C. Peng, Q. Chen, Z. Huang, C. Deng, W. Luo, H. Li, W. Zhang, W. Zhang, Y. Huang, Strategy of enhancing built-in field to promote the application of C-TiO_2/SnO_2 bilayer electron transport layer in high-efficiency perovskite solar cells (24.3%). *Small* **2022**, *18*, 2204446.
233. M. Hu, L. Zhang, S. She, J. Wu, X. Zhou, X. Li, D. Wang, J. Miao, G. Mi, H. Chen, Y. Tian, B. Xu, C. Cheng, Electron transporting bilayer of SnO_2 and TiO_2 nanocolloid enables highly efficient planar perovskite solar cells. *Sol. RRL* **2020**, *4*, 2070014.
234. Y. You, W. Tian, L. Min, F. Cao, K. Deng, L. Li, TiO_2/WO_3 bilayer as electron transport layer for efficient planar perovskite solar cell with efficiency exceeding 20%. *Adv. Mater. Interfaces* **2020**, *7*, 1901406.

235. J. Ma, J. Su, Z. Lin, J. He, L. Zhou, T. Li, J. Zhang, S. Liu, J. Chang, Y. Hao, Double side interfacial optimization for low-temperature stable $CsPbI_2Br$ perovskite solar cells with high efficiency beyond 16%. *Energy Environ. Mater.* **2022**, *5*, 637.
236. Y. W. Noh, I. S. Jin, K. S. Kim, S. H. Park, J. W. Jung, Reduced energy loss in SnO_2/ZnO bilayer electron transport layer-based perovskite solar cells for achieving high efficiencies in outdoor/indoor environments. *J. Mater. Chem. A* **2020**, *8*, 17163.
237. R. He, S. Nie, X. Huang, Y. Wu, R. Chen, J. Yin, B. Wu, J. Li, N. Zheng, Scalable preparation of high-performance ZnO-SnO_2 cascaded electron transport layer for efficient perovskite solar modules. *Sol. RRL* **2022**, *6*, 2100639.
238. Z. Su, D. Xu, Q. Ma, K. Gao, C. Zhang, C. Xing, S. Wang, W. Shi, X. Wang, K. Li, J. Hui, X. Yang, Atomic layer deposited ZnO-SnO_2 electron transport bilayer for wide-bandgap perovskite solar cells. *Sol. RRL* **2023**, *7*, 2201026.
239. L. Yan, Q. Xue, M. Liu, Z. Zhu, J. Tian, Z. Li, Z. Chen, Z. Chen, H. Yan, H. L. Yip, Y. Cao, Interface engineering for all-inorganic $CsPbI_2Br$ perovskite solar cells with efficiency over 14%. *Adv. Mater.* **2018**, *30*, 1.
240. P. Wang, R. Li, B. Chen, F. Hou, J. Zhang, Y. Zhao, X. Zhang, Gradient energy alignment engineering for planar perovskite solar cells with efficiency over 23%. *Adv. Mater.* **2020**, 1905766.
241. L. Lin, T. W. Jones, J. T. Wang, A. Cook, N. D. Pham, N. W. Duffy, B. Mihaylov, M. Grigore, K. F. Anderson, B. C. Duck, H. Wang, J. Pu, J. Li, B. Chi, G. J. Wilson, Strategically constructed bilayer tin (IV) oxide as electron transport layer boosts performance and reduces hysteresis in perovskite solar cells. *Small* **2020**, *16*, 1901466.
242. S. Liu, W. Chen, Y. Shen, S. Wang, M. Zhang, Y. Li, Y. Li, An intermeshing electron transporting layer for efficient and stable $CsPbI_2Br$ perovskite solar cells with open circuit voltage over 1.3 V. *J. Mater. Chem. A* **2020**, *8*, 14555.
243. X. Gong, Q. Sun, S. Liu, P. Liao, Y. Shen, C. Grätzel, S. M. Zakeeruddin, M. Grätzel, M. Wang, Highly efficient perovskite solar cells with gradient bilayer electron transport materials. *Nano Lett.* **2018**, *18*, 3969.
244. C. Liu, W. Li, C. Zhang, Y. Ma, J. Fan, Y. Mai, All-inorganic $CsPbI_2Br$ perovskite solar cells with high efficiency exceeding 13%. *J. Am. Chem. Soc.* **2018**, *140*, 3825.
245. T. Jiang, W. Fu, Improved performance and stability of perovskite solar cells with bilayer electron-transporting layers. *RSC Adv.* **2018**, *8*, 5897.
246. S. Zheng, W. Li, T. Su, F. Xie, J. Chen, Z. Yang, Y. Zhang, S. Liu, M. P. Aldred, K. Y. Wong, J. Xu, Z. Chi, Metal oxide CrO_x as a promising bilayer electron transport material for enhancing the performance stability of planar perovskite solar cells. *Sol. RRL* **2018**, *2*, 1700245.

2 Hole Transport Materials for Perovskite Solar Cells

Danish Khan, Zeguo Tang, Geping Qu, and Zong-Xiang Xu

2.1 INTRODUCTION

The average rise in the earth's temperature has doubled in the last 50 years, implying the increase in greenhouse gases emitted by the energy conversion machines made by human beings.[1] Everything from small gasoline cars to huge industries or thermal power stations emit these gases in massive amounts. To save the planet earth, it is necessary to shift energy production to sustainable and renewable energy resources. However, the high initial cost for renewable energy production, especially the expensive photovoltaics (PV) power plants, is one of the major hurdles to get 100% renewable power generation. The conventional silicon solar modules, so-called first-generation PV, acquire a large area of land and are fabricated in expensive ways, adding up the initial costs. In this era of nanotechnology, where cheaper thin-film PV is introduced, it is quite possible to cut this cost. Furthermore, the future applications of energy harvesting and building-integrated PVs are compelling scientists and researchers to think about cheaper, mechanically flexible, lighter, and charming thin-film PVs. At present, the second generation PVs, copper indium gallium selenide (CIGS), cadmium telluride (CdTe), amorphous silicon (a-Si), and nanocrystalline silicon (NC-Si) occupy the market of thin-film PVs.[2] In the emerging PV thin-film technology, so-called third generation PV, the cost and efficiency both are especially enforced. Dye-sensitized solar cells (DSSCs), quantum dot-sensitized solar cells (QDSSCs), organic solar cells (OSCs), and perovskite solar cells (PSCs) are part of third-generation PV. Among these solar cells, most of the PSCs-based research groups and major PV market players are in pursuit of commercializing PSCs as power conversion efficiency (PCE) in PSCs broke all the previous PCE records achieved by single-junction solar cells, including the silicon one. Hitting 26% of record PCE in such a short period is extraordinary.[3] This surge in the PCE enhancement is the specialty of PSCs, which comes from some remarkable properties such as low exciton binding energy, intrinsic ferroelectric polarization, long-range ambipolar transport characteristics, and a high dielectric constant.[4] Nonetheless, stability and scalability are the remaining obstacles to PSCs' commercialization.

Halide perovskite is generally written as ABX_3, where A is a cation ($CH_3NH_3^+$ (MA^+), $NH_2CH{=}NH_2^+$ (FA^+), Cs^+, etc.), B is a metal cation (Pb^{2+}, Sn^{2+}, Ge^{2+}, etc.), and X is a halogen anion (I^-, Br^-, Cl^-). Everything that we have in our surroundings, such as oxygen, moisture, light, and heat, boosts the degradation of PSCs.[5] Perovskites lattice

 DOI: 10.1201/9781003400493-2

is made from weak ionic and hydrogen bonds and van der Waals forces. On top of that, the components such as I^-, MA^+, and Pb^{2+} are large and can be polarized easily. These issues speed up under ambient conditions. In short, the instability of PSCs is caused by extrinsic environmental factors as well as intrinsic factors inhibited in the devices by material and interfaces.[6] On the contrary, intrinsic stability issues are also mainly initiated by extrinsic environmental factors. Consequently, a keen interest in the stability enhancement of PSCs has been seen recently. The extrinsic stability issues can be controlled by encapsulation, while intrinsic ones are critical. The major intrinsic issues can be categorized as hygroscopicity, ion migration, and thermal instability.[7] The most common type of halide perovskite is $MAPbI_3$. Still, this perovskite structure is the most unstable due to the smaller size of MA, which indirectly allows the perovskite to adopt a low symmetrical tetragonal phase. As a result, MA was mixed with FA, which enhances the PCE due to the low band gap, but it degrades the perovskite structure under the light.[8–10] This degradation is due to the instable yellow phase of FA-based perovskite.[11] Cs-based perovskite is relatively stable under light and has high thermal stability as it is inorganic. Nevertheless, the phase conversion issues in $CsPbI_3$ result in low PCEs.[12] To enhance its PCE, the band gap was optimized by mixing it with FA and MA, where different research groups reported the PCEs above 20%.[8,13,14] Mix cations perovskites became too famous due to their high PCE and decent stability. Now, most of the research groups use the mix cation organic, inorganic halide PSCs as their reference device. X site of perovskite has also been thoroughly studied. Mostly, iodine (I) holds its position at X sites of perovskite, but I^- ions suffer from oxidation under the light. Halides are mixed to control this issue while keeping a low band gap. For example, the replacement of I^- with Cl^- enhances the stability and PCE through efficient charge transfer and carrier diffusion length. Furthermore, Br^- is also introduced to symmetrize the tetragonal phase as the combination of Br and I increase MA–Pb's bonding strength. In this way, both the PCE and thermal stability can be enhanced as it enlarges the crystal grain size. In short, the optimized compositional engineering of halides plays a vital role in the thermal and light stability of perovskites. Conclusively, the simultaneous optimization of A and X sites of perovskite can play a massive role in the improvement of thermal and light stability. Cs at the A site form a better cubic phase, and Br at the X site leads to a reduction of octahedral tilting and lattice distortion. $CsPbBr_3$ seems perfect after considering all this, which is relatively stable, but its high band gap reduces the PCE of the device.[15–17] For the B site in ABX_3, Pb is incorporated chiefly, which also retards stability due to the soft nature of Pb–I bonding and Pb^{2+} ions. On exposure to high temperature and light, Pb^{2+} ions are converted into Pb atoms, generating deep defect traps. On top of that, Pb is one of the most environmentally unfriendly materials. The replacement of Pb with Sn, Ge, Cd, etc. can be an option. Nevertheless, Sn faces the stability while other Pb replacements (Ge, Cd, Ge, etc.) exhibit low PCEs. [18]

Generally, the PSC device structure consists of a transparent conductive oxide (TCO), a compact electron transport layer (ETL), perovskite layer, hole transportation layer (HTL), and metal cathode (Ag or Au). TCO generally can be indium tin oxide (ITO) or fluorine-doped tin oxide (FTO), which is chosen because of its suitable work function and transparency. ETL facilitates the electron flow, and HTL stimulates the holes, while the charge separation happens on the light striking at the perovskite layer. In the end, separated charges are collected at the metal cathode and TCO. According to device architecture, PSCs have three main categories, planar n-i-p, mesoporous n-i-p, and inverted planar p-i-n, as shown in Figure 2.1 (a), (b), and (c), respectively. ETL and HTL are in contact with the absorber layer, and these layers should not only play the role of carrier mobility enhancement but also protect it from the environment. In p-i-n devices, the role of HTM is more crucial as it acts as the substrate to the perovskite film, which could govern the perovskite film growth and crystallization kinetics. In contrast, HTMs could be designed to play an essential part in perovskite defect passivation and/or as a guardian layer against harsh environmental conditions in n-i-p devices. Nonetheless, there are also some limitations on HTMs owing to which n-i-p devices HTMs cannot be used in p-i-n devices or vice versa. These limitations include their energy levels or appropriate solvents used during wet-chemistry techniques. Both types of materials, organic and inorganic, have been employed as hole transporting materials (HTMs) in PSCs, having their pros and cons (Figure 2.1d and e).

Whenever the inorganic HTMs such as CuI, CuO_X, CuSCN, MoO_X, and NiO_X are deposited through solution processing techniques, the required solvents dissolve the perovskite.[19] On top of that, doping is required to get higher conductivity and p-type properties in inorganic HTMs.[20] Though inorganic HTM-based PCE recorded values are somehow less than the organic HTM-based PSCs, the chemical stabilities and versatility in the fabrication methods are the two main reasons that make inorganic HTMs more attractive than organic ones. On the other side, conventional organic HTMs possess low chemical stabilities and hygroscopic nature but high PCE values. Recently, researchers have been aiming for organic HTMs, divided into two main

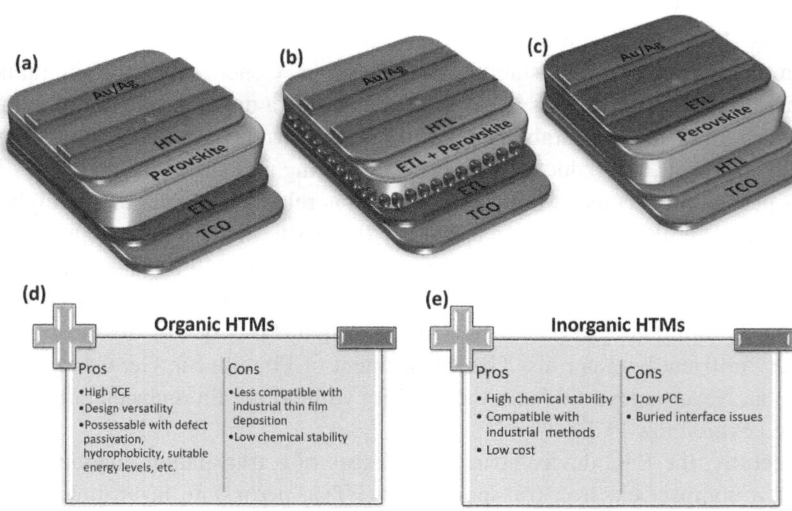

FIGURE 2.1 (a) Planar n-i-p structure; (b) mesoporous n-i-p structure; (c) planar p-i-n structure; (d) Organic HTMs and (e) Inorganic HTMs' pros and cons.

categories: polymers and small molecules (SMs). But, if commercialization is concerned, the batch-to-batch variation in polymer molecular weight is an issue.[21] Small molecules-based HTMs (SM-HTMs) seem a better option than polymeric HTMs due to reproducibility and accessible designs. Therefore, recently some dopant-free SM-HTMs with new molecular designs have been reported with moderately higher PCEs.[22]

2.1.1 Selection Criteria for HTM in PSCs

For efficient PSC devices, reducing charge recombination is a basic necessity, and HTM is responsible for performing this job systematically for hole carriers. Furthermore, some other properties are needed for HTMs to fit in the device structure (Figure 2.2). Depositing smooth and thin HTL through wet chemistry techniques requires high solubility of the HTMs in common solvents. Similarly, chemical or thermal stability is also required for depositing high-quality morphological thin films. The dense, compact, and stable (chemically and thermally) films can guard the perovskite layer against the harsh environment in the n-i-p device. Besides that, it can also assist in high-quality perovskite deposition in p-i-n PSCs. If the material is soluble and stable, it should have high hole mobility and conductivity afterward.[23] HTMs are p-type materials, and their primary function is to extract holes and transfer them efficiently. Similarly, highest occupied molecular orbital (HOMO) and lowest unoccupied molecular orbital (LUMO) energy levels should fit in the range of perovskite and electrodes to avoid interface recombination losses. Defect passivation can be an extra property that can be inherited in HTMs to enhance the stability further and tackle the ion vacancies.[24]

As PCE reached a record level, and stability is greatly concerned, the new issue is adopting industrial fabrication. Commonly, the PCE reduces whenever we scale up the PSC device area or the fabrication process.[25,26] Spin coating is the conventional

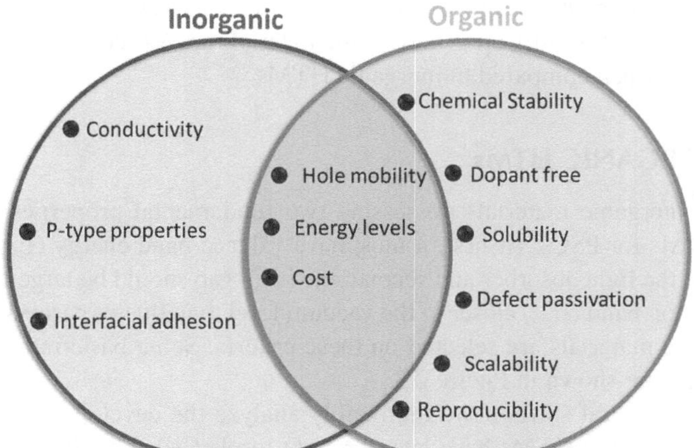

FIGURE 2.2 Properties which are practicable for Inorganic and Organic HTMs.

technique to fabricate thin-film devices, which provides an effective and easy way to fabricate solar cells. The automatic quenching, controlled spinning speed, solvent evaporation, and antisolvent dripping allow the formation of dense and smooth thin films. However, it is well known that spin coating is not practical and economical on a large scale. Learning from other thin-film technologies, researchers have adopted industrial methods such as blade coating, slot die coating, laser printing, evaporation, etc.[27] Nevertheless, the perovskite or organic layers are generally challenging to make thin and dense; whereas the inorganic HTM-based PSCs could be more efficient, they can adopt all the industrial scale fabrication techniques. Interestingly, organic HTMs are efficient and can effectively solve instability issues due to their design universality.

Moreover, the commonly used organic HTMs are poly (3,4-ethylenedioxythio phene)-poly (styrenesulfonate) (PEDOT:PSS), poly (triaryl amine) (PTAA), poly (3-hexylthiophene-2,5-diyl) (P3HT), and Spiro-OMeTAD. Nevertheless, the high cost and requirement of hygroscopic dopants are the main reasons researchers are looking for dopant-free organic materials with similar properties and low cost. The commonly used dopants in organic HTMs are lithium bis(trifluoromethanesulfonyl) imide (LiTFSI), cobalt complexes, and tert-butylpyridine (t-BP). These p-type dopants enhance the PCE due to high hole mobility and conductivity. But on the other side, they are responsible for hindering long-term stability due to their hygroscopic nature. The hygroscopic nature of LiTFSI, FK209, and t-BP speeds up perovskite degradation.[28] All these three dopants retard the PSCs' performance. For example, LiTFSI doping may result in immobile acceptor anions and deep traps, affecting free carriers' transportation. Furthermore, it decreases the adhesion on substrates when it is doped with HTMs for p-i-n PSCs.[29] On the other side, the evaporation rate of t-BP is slow, which results in the corrosion of perovskite and HTL.[30,31] t-BP also drops the glass transition temperature of the HTM, which reduces its thermal stability.[32] Furthermore, it decomposes the perovskite chemically while reacting with PbI_2.[33,34] As a consequence, the fabrication process will become more complicated and expensive on an industrial scale if conventional HTMs are not replaced with dopant-free HTMs. Several dopant-free polymers and small molecules are synthesized to replace the traditional HTMs. Still, up to now, these materials are less compatible with industrial methods when compared to inorganic HTMs.[26]

2.2 INORGANIC HTMs

Generally, inorganic materials possessing two fundamental properties are considered as HTMs for PSCs. At first, it must have valence band energy (E_{VB}) matching with that of the light absorber and second, the band gap should be large enough with the conduction band (E_{CB}) closer to the vacuum level than that of perovskite. Most of the inorganic materials are selected on these criteria. Some basic inorganic HTMs energy levels are shown in Figure 2.3.

There are several shifts if we thoroughly analyze the development of HTMs in PSCs. First, organic HTMs have been brought to play, such as Spiro-OMeTAD or PEDOT:PSS. However, by considering the cost and instability of PSCs, the research trends were shifted towards inorganic HTMs, where different inorganic HTMs were

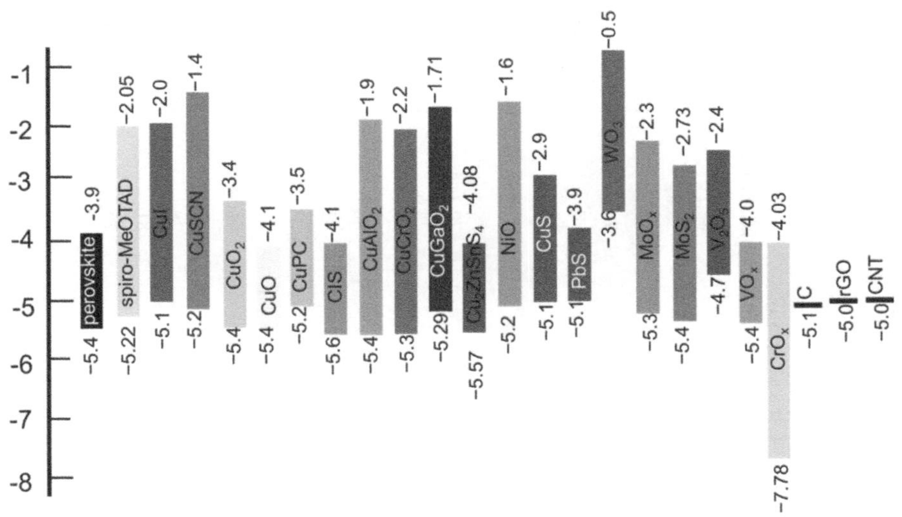

FIGURE 2.3 Energy levels comparison of commonly used inorganic HTMs. Reproduced with permission.[35] Copyright, 2022 MDPI.

introduced, such as NiO_X or CuO_X. After tremendous efforts of doping through Li, Zn, Sr, Cd, etc., the performance was enhanced up to certain levels, but still, the inorganic HTMs couldn't defeat the organic HTMs in terms of high PCE.[36,37] After many studies on organic HTMs, the inorganic HTMs are in the game again due to their scalable fabrication ease. On top of that, the inorganic HTMs show better performances if we deposit these through industrial methods.[38–40] The efficient PSCs (PCE above 17%) with inorganic HTMs are collected in Table 2.1. In Table 2.1, it can be noticed that more than 90% of PSCs are p-i-n PSCs, where the HTM is deposited on FTO or ITO. Depositing HTM on ITO or FTO provides versatility in fabrication techniques. While in n-i-p devices, HTM is deposited on top of perovskite, which is the most sensitive layer of PSCs. Though the higher PCEs in organic HTL-based PSCs are fascinating, the deposition of organic thin films on an industrial scale is quite an issue.

2.2.1 Copper Derivatives

The low-temperature γ-phase (below 350 °C) in Copper(I) iodide (CuI) behaves like a p-type semiconductor.[92] Kamat and coworkers used it for the first time in mesoscopic PSCs through an automated drop-casting method.[93] The PCE was only 6% due to low V_{OC} and highly complex deposition technique made it inattentive. Later on, Spiccia et al. used doctor blade coating to deposit CuI in p-i-n PSC, where the PCE of 7.5% was achieved, and the charge separation was ineffective.[94] Deng and coworkers made a $MAPbI_3$-based p-i-n device and achieved a PCE of 13.6%.[95] The deeper VB position in CuI enhanced the V_{OC}, and the high conductivity of CuI helped in high J_{SC}, but the CuI-based devices lag in FF, which was attributed to the imperfect coverage

TABLE 2.1
Photovoltaic Parameters of Inorganic HTM-Based PSCs

HTM	HTL deposition method	Device structure	Open circuit voltage (V_{OC}) V	Current density (J_{SC}) mA/cm^2	Fill factor (FF)	PCE	Ref
(Li,Mg):NiO	Spray pyrolysis	Glass/FTO/NiMgLiO/$MAPbI_3$/Ti(Nb)O_X/Ag	1.08	20.4	82.7	18.3	41
CuO_X	Spin coating	Glass/ITO/HTL/$MAPbI_{3-x}Cl_x$/PCBM/C60/BCP/Ag	1.11	22.5	75.8	19	42
Cu:NiO_X	Spin coating	Glass/ITO/HTL/$MAPbI_3$/PCBM/BCP/Ag	1.11	20.76	81	18.66	43
CuSCN	Spin coating	Glass/ITO/HTL/$MAPbI_3$/PCBM/Ag	1.10	22.7	71	17.5	44
$CuGaO_2$	Spin coating	Glass/FTO/c-TiO_2/$MAPbI_{3-x}Cl_x$/$CuGaO_2$/Au	1.11	21.66	77	18.51	45
CuO_X	Spin coating	Glass/ITO/HTL/$MAPbI_3$/ZnO(NP)/Al	1.03	22.42	76	17.43	46
Cu:NiO_X	Spin coating	Glass/FTO/HTL/$MAPbI_{3-x}Cl_x$/PCBM/ZrAczc/Al	1.13	23.4	77.5	20.5	47
Cs:NiO_X	Spin coating	Glass/ITO/HTL/$MAPbI_3$/PCBM/ZrAcac/Ag	1.12	21.77	79.3	19.35	48
NiO	Spin coating	Glass/FTO/HTL/$CsFAMAPbBr_zI_{3-z}$/PCBM/bis-C60/Ag	1.09	22.8	74.6	18.6	49
$NiMgO_X$	magnetron sputtering	Glass/FTO/HTL/$MAPbI_3$/PCBM/ZnMgO/Al	1.08	21.3	80	18.5	50
Mesoporous Cu:NiO_X	Spin coating	Glass/FTO/HTL/$MAPbI_3$/PCBM/C_{60}/Ag	1.11	21.66	82.3	19.79	51
CuSCN	Spin coating	Glass/FTO/m-TiO_2/$CsFAMAPbI_{3-x}Br_x$/CuSCN/GO/Au	1.10	23.1	77	20.2	52
Mesoporous NiO_X	Electrochemical deposition	Glass/FTO/HTL/$MAPbI_3$/PCBM/BCP/Ag	1.01	21.55	81.34	17.77	53
$NiCo_2O_4$	Spin coating	Glass/ITO/HTL/$MAPbI_{3-x}Cl_x$/ZrAcac/PCBM:C_{60}/Ag	1.07	21.83	78	18.23	54
$CuCrO_2$	Spin coating	Glass/ITO/HTL/$MAPbI_3$/PCBM/BCP/Ag	1.07	21.94	81	19	55
NiO	Solution–combustion-based method	Glass/FTO/HTL/$MA_{1-y}FA_yPbI_{3-x}Cl_x$/PCBM/BCP/Ag	1.12	23.7	76	20.2	56
NiO	ALD	Glass/FTO/HTL/$Cs_{0.05}MA_{0.95}PbI_3$/PCBM/BCP/TiO_2/Ag	1.05	22.56	78	18.45	57

NiO_X	Spin coating	Glass/ITO/HTL/$Cs_{0.05}(MA_{0.17}FA_{0.83})_{0.95}Pb(I_{2.7}Br_{0.3})$/PCBM/ZnO/Al	1.02	22.2	82	18.6	58
(Li,Ag):NiO_X	Spin coating	Glass/ITO/HTL/$MAPbI_3$/PCBM/BCP/Ag	1.13	21.29	80	19.24	59
Zn:NiO_X	Spin coating	Glass/ITO/HTL/$MAPbI_3$/PCBM/BCP/Ag	1.1	22.8	78	19.6	60
Sr:NiO_X	Spin coating	Glass/FTO/HTL/$FA_{0.15}MA_{0.85}Pb(I_{0.95}Br_{0.05})_3$/PCBM/AgAl	1.11	22.68	79.1	20.07	61
Cu:NiO	Spin coating	Glass/ITO/HTL/$MAPbI_3$/C_{60}/BCP/Ag	1.12	22.28	81.2	20.26	62
Cu_3PS_4	Spin coating	Glass/FTO/HTL/$MA_{0.7}FA_{0.3}PbI_3$/PCBM/BCP/Ag	1.06	20.83	81.6	18.17	63
Mn-S	Vacuum vapor deposition	Glass/FTO/TiO_2/$FA_{0.85}MA_{0.15}Pb(_{I0.85}Br_{0.15})_3$/MnS/Au	1.11	23.40	76.4	19.86	64
surface-modified Cu_2O QDs	Spin coating	Glass/FTO/c-TiO_2/m-TiO_2/$Cs_{0.05}FA_{0.81}MA_{0.14}PbI_{2.55}Br_{0.45}$/HTL/Au	1.15	22.2	74.2	18.9	65
Fe:NiO_X	Spin coating	Glass/ITO/HTL/$MAPbI_3$/PCBM/BCP/Ag	1.07	18.9	85.6	17.57	66
Co:NiO_X	Spin coating	Glass/ITO/HTL/$MAPbI_3$/PCBM/BCP/Ag	1.04	22.46	75	17.52	67
K:NiO_X	Spin coating	Glass/FTO/HTL/$Cs_{0.05}FA_{0.81}MA_{0.14}Pb(Br_{0.15}I_{0.85})_3$/PCBM+$TBABF_4$/TIPD/Ag	1.13	20.53	74	17.05	68
Rb:NiO_X	Spin coating	Glass/ITO/HTL/$Cs_{0.1}FA_{0.7}MA_{0.2}I_{3-x}Br_x$/PCBM/BCP/Ag	1.05	21.77	75.3	17.21	69
(Li,Co):NiO_X	Spin coating	Glass/ITO/HTL/$MA_{1-y}FA_yPbI_xCl_{3-x}$/PCBM/BCP/Ag	1.1	24.3	80	20.1	70
Au:NiO_X	Spin coating	FTO/HTL/$MAPbI_3$/PCBM/PPDIN6/Ag	1.11	23.1	79	20.2	71
NiO	Plasma-assisted ALD	Glass/ITO/HTL/$Cs_{0.05}(MA_{0.17}FA_{0.83})Pb_{1.1}(I_{0.83}Br_{0.17})_3$/C60/BCP/Cu	1.07	21.75	73.3	17.07	72
NiO_X	Spin coating	Glass/ITO/HTL/$Cs_{0.1}MA_{0.2}FA_{0.7}Pb_{0.5}Sn_{0.5}Br_3$/C60/BCP/Au	0.77	31	73.3	17.6	73
NiO_X	Electron beam evaporative (EBE)	Glass/ITO/HTL/$Cs_x(FA_{0.83}MA_{0.17})_{1-x}PbBr_{0.15}I_{0.85})_3$/C60/BCP/Au	1.11	23.1	82	20.7	74
KCL treated NiO_X	Spin coating	Glass/ITO/HTL/CsMAFAPbIBr/PCBM/ZrAcac/Ag	1.15	22.89	79.5	20.96	75

(Continued)

TABLE 2.1 *(Continued)*
Photovoltaic Parameters of Inorganic HTM-Based PSCs

HTM	HTL deposition method	Device structure	Open circuit voltage (V_{OC}) V	Current density (J_{SC}) mA/cm²	Fill factor (FF)	PCE	Ref
n-Butyamine treated NiO_X	Spin coating	Glass/FTO/HTL/$MAPbI_3$/C_{60}/BCP/Ag	1.08	22.8	77.9	18.9	76
UVO-NiO_X	Electrochemical deposition (ECD)	Glass/FTO/NiOx/$MAPbI_3$/PCBM/BCP/Ag	1.11	23.39	75.5	19.67	77
NiO_X	EBE	Glass/FTO/HTL/$MAPbI_{3-x}Cl_x$/C_{60}/BCP/Ag	1.08	21.37	78.3	18.13	78
(Li,Pb):NiO_x	Spin coating	Glass/FTO/HTL/$MAPbI_3$/PCBM/BCP/Ag	1.01	21.31	78.8	17.44	79
Ni^{3+}:NiO	Spin coating + hydrothermal treatment	Glass/FTO/HTL/CsMAFAPbIBr/PCBM/ZnO/Al	1.11	22.87	70	17.75	80
$Cu_2Cd_xZn_{1-x}SnS_4$	Thermal evaporation	Glass/ITO/SnO_2/$MAPbI_3$/HTL/Au	1.15	23.15	78.1	20.62	81
Cu:NiO_X	pulsed laser deposition (PLD)	Glass/FTO/HTL/$FA_{0.2}MA_{0.8}PbI_xCl_{3-x}$/PCBM:RHB/LiF/Ag	1.11	23	79.6	20.41	82
Ni_2O_3	Spin coating	Glass/FTO/HTL/$MA_{0.85}FA_{0.15}PbI_{0.9}Cl_{0.1}$/PCBM/BCP/Ag	1.10	19.67	82.6	17.89	83
NiO_X	magnetron sputtering	Glass/FTO/HTL/$MA_{0.65}FA_{0.35}PbI_3$/PCBM/BCP/Ag	1.04	23.21	78.8	18.7	84
NiO_X	Spin coating	Glass/ITO/HTL/$Cs_{0.25}FA_{0.75}Pb(Br_{0.2}I_{0.8})_3$ LiF/C_{60}/BCP/Ag	1.14	24	85	19.66	85
NiO_X	NIR	FTO/NiO_X/$MAPbI_3$/$PC_{61}BM$/PEI/Ag	1.09	20.46	79.8	17.77	86
NiO_X	EBE	Glass/ITO/HTL/$Cs_{0.1}MA_{0.15}FA_{0.75}Pb(I_{0.85}Br_{0.15})_3$/$C_{60}$/BCP/Au	1.11	24.6	80	21.6	87
Sm:NiO_X	Spin coating	Glass/ITO/HTL/$Cs_{0.05}FA_{0.81}MA_{0.14}Pb(I_{0.95}Br_{0.05})_3$/PCBM/BCP/Ag	1.10	23.62	79.3	20.71	88
(Li,Cu):NiO_X	Spin coating	Glass/ITO/HTL/$MAPbI_{3-x}Cl_x$/PCBM:C60/ZrAczc/Ag	1.11	23.15	81	20.83	89
Cd:NiO_X	Spin coating	Glass/ITO/HTL/$FA_{0.2}MA_{0.8}PbI_xCl_{3-x}$/PCBM:RHB101/LiF/Ag	1.11	23	79.6	20.47	90
NiO_X	Surface redox + EBE	Glass/FTO/HTL/CsMAFAPbIBr/PCBM/BCP/Ag	1.16	24.8	81.4	23.4	91

of CuI. Sun et al. achieved a PCE of 16.8% where the CuI was spin-coated as HTM, which provided the non-uniform thin film.[96] In fact, the film's smoothness can be improved through industrial-scale fabrication techniques, but depositing a thin film through scalable methods is quite a challenge. Yu et al. deposited the CuI layer in p-i-n PSCs through a solid-gas reaction, where the thermally evaporated solid copper film was exposed to iodine vapor.[97] The thickness of CuI was well controlled, and PCE of 14.7% was executed.

Copper(I) thiocyanate (CuSCN) was first introduced in several thin-film devices due to its high transparency and high conductivity,[98–102] and later on, it was included as HTM in PSCs. Ito et al. coated the CuSCN through doctor blade coating and achieved a PCE of just 4.85% in mesoscopic PSCs.[103] Later on, the sequential deposition was used to deposit the thick layer of CuSCN in n-i-p PSC devices, where a PCE of 12.4% was executed.[104] Nazeeruddin and co-workers fabricated the PSC with an absorber layer of $(FAPbI_3)_{0.85}(MAPbBr_3)_{0.15}$ and achieved 16.6% of PCE through doctor blade coating.[105] Seok and coworkers fabricated the mesoscopic PSCs through spin coating with a PCE of 18%.[106] The uniform layer of CuSCN has also been tried through spray coating, where a PCE of 17.1% was achieved.[107] Later on, Sarkar et al. executed the PCE of 16.6% in p-i-n PSCs, where the CuSCN was deposited through the electrodeposition method.[108] In 2017, Wijeyasinghe et al. executed the PCE of 17.5%, where aqueous CuSCN–ammine solution was spin-casted as HTL in p-i-n PSCs.[44] To hit a higher PCE with solution-based bottom-up approaches are more accessible, but the solvents in which the CuSCN is mixed degrade the perovskite layer.[52] That is one of the main reasons for not incorporating CuSCN in n-i-p devices. Arora and co-workers published an article in *Science* that introduced a fast solvent removal technique, which executed the PCE of 20.2% with high stabilities.[52] Bian et al. incorporated CuS as HTM in $MAPbI_3$-based p-i-n PSCs through the spin coating of CuS nanoparticles and achieved a PCE of 16.2%.[109] PCE of 18.51% was achieved in n-i-p devices by incorporating $CuGaO_2$ nanosheets as inorganic HTM.[45] The copper-based HTMs provide higher conductivity than other inorganic HTMs. Zhang et al. enhanced the photostability of p-i-n PSCs by using solution-processed $CuCrO_2$ nanocrystals as HTM, which delivered a PCE of 19%.[55] Cu_3PS_4, with a small effective mass for holes and a wide-enough bandgap, is used as HTM in p-i-n devices, where the PCE of 18.17% is achieved.[63] Cu_2ZnSnS_4 (CZTS), used as an absorber in CZTS solar cells, is applied as HTM in PSCs by Wu et al. They increased the bandgap of CZTS with Cd doping ($Cu_2Cd_xZn_{1-x}SnS_4$) to get the dual functionality of HTM (light harvesting + hole transportation) and achieved the PCE of 20.62%, which was further enhanced up to 21.11% by incorporating the Au nanorods below Au electrode.[81]

Some other copper derivatives, such as Cu_2O and CuO, have also been included in PSCs as HTM. Cu_2O was used as HTM in p-i-n PSCs by Chatterjee and Pal through successive ionic layer adsorption and reaction (SILAR) method with the PCE value of 8.3%.[110] Thermal evaporation of Cu on ITO in the air has been performed to get Cu_2O as HTM thin layer in p-i-n PSCs and the PCE up to 11% is reported.[111] Later, Cu_2O quantum dots were deposited on perovskite as HTM in n-i-p PSCs, where the non-polar inorganic nanoparticle dispersions were achieved by ethenyltriethyloxysilane.

The modified hydrophobic Cu_2O quantum dots as HTM enhanced the PCE from 11.9 to% 18.9%.[65]

CuO_X thin film was first introduced by Bian et al. through the solution-processing method, and later on, it was employed as HTM in PSCs.[112] CuO_X was deposited on ITO through the conventional spin coating and annealing method, and UV treatment helped to adjust the band gap. It is well known that the UV treatment can increase the oxygen and Cu^{2+} content, which enhances the hole mobility of the CuO_X film. This breakthrough UVO optimization delivered a PCE of 17.1%.[113] Bian and coworkers improved the perovskite layer absorption composition and achieved 19% in similar devices.[42] A similar approach in CuO_X based PSCs has been adopted by Yu et al., who reported a PCE of 17.43%.[46]

2.2.2 Nickel Derivatives

The work on Nickel oxide (NiO) has been greatly studied as compared to other inorganic HTMs, due to its deep valance band and wide bandgap. Furthermore, it possesses higher chemical stabilities and a very cost-effective inorganic HTM. Nonstoichiometric phases of NiO_X enhance the p-type conductivity, generally generated through the oxygen exchange or UV treatment.[114] In NiO, the p-type conductivity is attributed to nickel vacancies V_{Ni}.[91] As long as the NiO is undoped, its hole density is limited due to the high ionization energy of Ni vacancies, which is the reason behind the high PCE of doped NiO_X or NiO. The doping of NiO is generally needed with extrinsic dopants with shallower acceptor levels. Therefore, doping or codoping NiO with Cu, Mg, Co, or Cs is advantageous in enhancing the conductivity. However, it is common that doping also enhances complexities and cost whenever it is involved in any part of the PSC device. The second approach is the post-treatment of the deposited layer or modification of the fabrication process, which can enhance the Ni vacancies. For example, in vacuum-processed NiO_X films, oxygen and metal deficiency are formed. As a result, the hole mobility and p-type characteristics are enhanced. The hole mobility and energy level alignments are done by doping NiO_X with several materials. For the sake of p-type properties enhancement, NiO_X is also doped with Co, where the PCE of 17.52% was achieved.[67] Samarium-doped Nickel Oxide (Sm: NiO_X) as HTM in p-i-n PSC delivered a PCE of 20.71% because the Sm^{3+} doping significantly boosted the electrical conductivity and deepened the work function.[88] Similarly, in a recently published report, NiO_X was doped with Cd to get more nickel vacancies and low formation energy, where a PCE of 20.47% was obtained.[90] Chen et al. doped NiO_X with Cs and executed the PCE of 19.35% in p-i-n PSCs.[48] NiO_X has also been doped by Rb, which increased the conductivity of HTL and optimized the energy level and a PCE of 17.21% was achieved.[69] Moreover, other metals are also doped with NiO or NiO_X to achieve higher conductivities. When Fe-doped NiO_X was included as HTM in p-i-n PSCs, 17.57% of PCE was recorded.[66] Synergistic effects in codoping of NiO_X with Li and Ag achieved high conductivities, and the PCE up to 19.24% was recorded.[59] NiO_X is also co-doped with Li and Co, which enhanced the conductivity and especially the hole mobility of the HTM layer.[70] The PCE of 20.1% was achieved when NiO_X HTM was doped with 10% Li and 5% Co in p-i-n

PSCs. The optical transmittance and conductivity are also improved by codoping of NiO_X with Li and Cu, where the PCE of 20.8% was executed.[89] Wan et al. reported Zn-doped NiO_X for the first time and got 19.6% of PCE due to improved transparency and conductivity in HTL.[60] In addition, Zn-doped NiOx were more ambient stable than the pristine NiOx-based PSCs. Sr doping in NiO_X also boosted conductivity, and in Sr:NiO_X-based PSCs PCE was recorded above 20%.[61] However, among all the dopants, Cu stayed prominent and was doped in NiO_X through different strategies. A high annealing temperature above 400 °C was required in NiO_X deposition or doped NiO_X deposition. Jen et al. introduced a low-temperature solution-based combustion process in solution-processed Cu-NiO_X and achieved a PCE of 17.7%.[115] NiO_X nanocrystal ink was prepared in water and fabricated on ITO substrates at room temperature, and a PCE of 17.6% was achieved.[116] In the same report, flexible PSCs with a PCE of 14.5% were also designed with a similar procedure. Cu:NiO_X HTM-based p-i-n PSCs are also reported by He et al., and a PCE of 18.66% was achieved.[43] For the purpose of high stability and low hysteresis, mesoporous Cu:NiO_X is applied to achieve an efficiency of 19.79%.[51] Furthermore, Feng et al. achieved a PCE of 20.41% through Cu doping of NiO_X.[82] The doping of Cu with NiO or NiO_X has been reported more than a dozen times because the optimized doping tunes the band gap and enhances the conductivity, especially the p-type conductivity. Chen et al. noticed that the Cu:NiO HTM possesses the Cu^+ and Cu^{2+} states, and the substitution of Ni^{2+} with Cu^{2+} boost both carrier concentration and carrier mobility.[62] As a result, Chen et al. achieved a PCE of 20.26% in Cu: NiO HTM-based PSCs.

2.2.2.1 Buried Interface Issues

Ni vacancies have a considerable role in enhancing the p-type behavior of NiO. Thakur et al. grew the NiO layer with Ni^{3+} defects through a hydrothermal process on FTO and achieved a PCE of 17.75%.[80] However, the interface chemistry of NiO_X/perovskite is quite complicated. Most of the reports published on the nickel derivatives suggest that the reason behind low PCE is the V_{OC} loss caused by the imperfect band alignment and low work function. But in reality, low V_{OC} loss is noticed in organic-based HTMs with similar ionization potential (= 5.2 eV) and band gaps.[117,118] Furthermore, some researchers also convicted the organic counterparts of NiO_X. Nevertheless, some reports indicate that surface recombination velocities of perovskite films on NiO_X and PTAA are equivalent.[119] Boyd et al. closely analyzed the chemical and electronic structure of the NiO_X affecting the chemistry between perovskite precursors and the oxide surface.[85] The under-coordinated metal cation sites ($Ni^{\geq 3+}$) act as Lewis acid or oxidant towards dissociated lead halide precursors (i.e., I^-) and as a Brønsted base (proton-acceptor) with protonated amine precursors. As a result, a defective perovskite with deficient A-site and enriched $PbI_{2-x}Br_x$ is formed at the perovskite-NiO_X interface. These interface defects are tackled by including an excess amount of A site in precursor, and PCE of 19.66% was achieved with V_{OC} of 1.15 V. The experiments performed by Boyd et al. suggest that the reaction between the perovskite and NiO_X proceeded as shown in the reactions below ((1) to (5)), where HA^+ is a protonated A-site cation ($CH_3NH_3^+$ or MA^+, $CH_5N_2^+$ or FA^+) and A is the deprotonated A site cation (CH_3NH_2, CH_4N_2).[85]

$$HAPbI_3 \leftrightarrow PbI_2 + HA^+ + I^- \tag{1}$$

$$HA^+ \leftrightarrow A_{(g)} + H^+ \tag{2}$$

$$I^- \leftrightarrow \frac{1}{2} I_{2(g)} + e^-. \tag{3}$$

$$Ni^{\geq 3+}O_x + e^- + H^+ \leftrightarrow Ni^{\geq 2+}O_xH \tag{4}$$

$$HAPbI_3 + Ni^{\geq 3+}O_x \leftrightarrow Ni^{\geq 2+}O_xH + PbI_2 + A_{(g)} + \frac{1}{2} I_{2(g)} \tag{5}$$

Du et al. applied surface redox engineering on EBE-deposited NiO_X to enhance the Ni^{3+} and Ni^{4+} vacancies and drop the percentage of NiO and $Ni(OH)_2$ species (Figure 2.4a).[91] In addition, Brønsted acid treatment (surface oxidation engineering (SOE)) eliminates the detrimental Ni^{4+} and -OH but reserves the beneficial Ni^{3+} and NiO backbone that results in one of the highest PCEs (23.4%). Figure 2.4b represents the change in the content of different Ni species. After SOE, both the concentrations of Ni^{3+} and Ni^{4+} enhanced, while there was a drop off in the percentage of NiO and $Ni(OH)_2$. It is well known that the Ni vacancies are recurring defects in NiO, as evidenced by its low formation energy compared with O- vacancies under either Ni- or O-rich conditions (Figure 2.4c). Figure 2.4d shows the order of chemical reaction

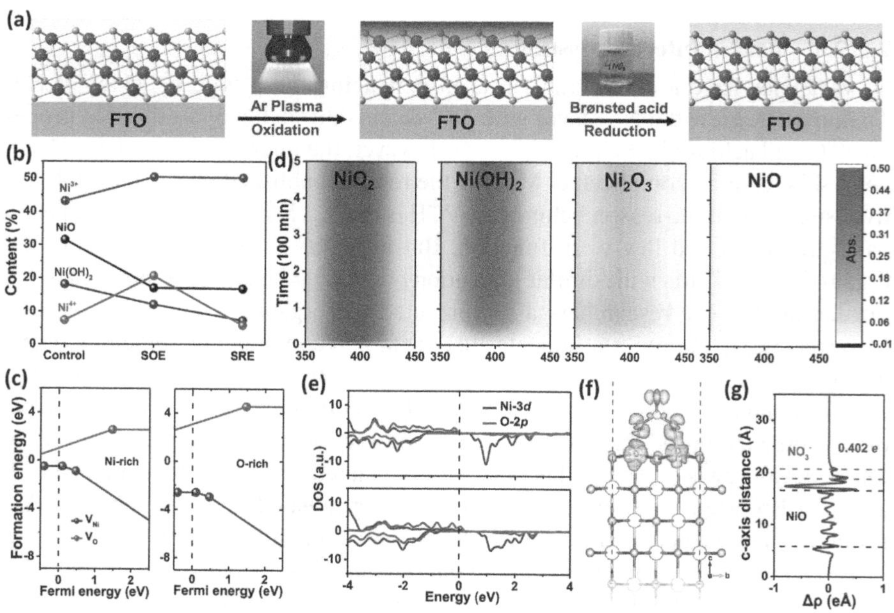

FIGURE 2.4 (a) Schematic of SRE process; (b) Ni species alteration due to SRE; (c) the formation energies of metal (Ni) and oxygen vacancies as a function of the Fermi energy in Ni-rich and O-rich conditions; (d) UV-vis absorption values for NiO_2, $Ni(OH)_2$, Ni_2O_3, NiO powder in HNO_3 solution; (e) density of states; (f) optimized NiO surface structure; (g) planar-averaged charge density difference of NiO surface with NO_{3-} adsorption. Reproduced with permission. Copyright 2022, Elsevier.[91]

activity, i.e., Ni^{4+} > Ni(OH) > Ni^{3+} > NiO. Figure 2.4e shows the density states. The pristine NiO surface has a semi-conductive electronic structure, similar to the bulk NiO and on adsorbing NO_3^- (Figure 2.4f), the d electrons of surface Ni^{2+} cations transfer to the NO_3^- group, which decreased the fermi energy level. Furthermore, the NO_3^- adsorption triggers charge reorganization, whereby almost 0.402 electron transfers from the NiO to the NO_3^- (Figure 2.4g), which could increase the conductivity of the NiO layer and induce an interfacial dipole and ease the hole injection process. The following equations explain this whole process.

$$NiO \overset{\text{Ar plasma}}{\longrightarrow} Ni^{\geq 3+}\left(Ni_2O_3 + NiO_2\right) + Ni_{\text{vacancy}} + Ni^{\leq 2+} \tag{6}$$

$$Ni^{\leq 2+} + O_2 + H_2O \overset{\text{In air}}{\longrightarrow} NiO + Ni\left(OH\right)_2 \tag{7}$$

$$NiO_2 + HNO_3 \rightarrow Ni^{4+} + H_2O + NO_3^{\,-} \rightarrow Ni^{\leq 3+} + O_2 + NO_3^{\,-}\left(\text{fast}\right) \tag{8}$$

$$NiO\ \text{or}\ Ni_2O_3 + HNO_3 \rightarrow Ni^{\leq 3+} + H_2O + NO_3^{\,-}\left(\text{slow}\right) \tag{9}$$

$$Ni\left(OH\right)_2\ \text{or}\ NiOOH + HNO_3 \rightarrow Ni^{\leq 3+} + H_2O + NO_3^{\,-}\left(\text{fast}\right) \tag{10}$$

Seldom have other treatments also executed high V_{OC} through curing the NiO_X/perovskite interface defects. For example, Chen et al. treated the spin-casted NiO_X films with alkali chlorides and achieved a PCE of 20.96% with a V_{OC} of 1.15 V.[75] Schematic diagram of the device is shown in Figure 2.5a and a cross-sectional SEM image of the similar device is shown in Figure 2.5b. The ions (Na, K) were properly diffused

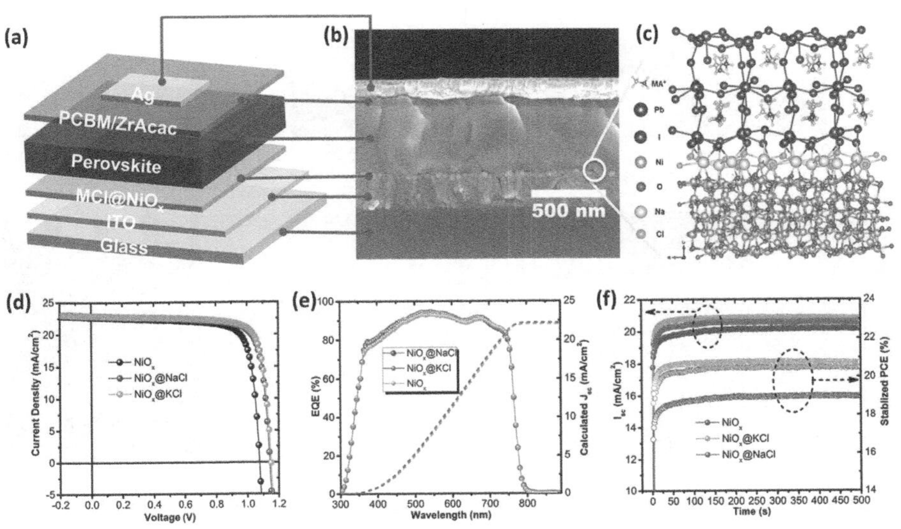

FIGURE 2.5 (a) Device structure (PSCs) used in this work; (b) cross-section scanning electron microscopy (SEM) image of the device; (c) DFT optimized structural details of perovskite/NaCl/NiO_X; (d) I-V curves of the modified and pristine NiO_X-based PSCs; (e) EQE curves of the corresponding PSCs in (d); (f) stabilized PCE of the corresponding PSCs in (d). Reproduced with permission. Copyright 2019, Wiley.[75]

in perovskite and cured the interface defects, which were experimentally and theoretically proved (Figure 2.5c). The device in which KCl diffused in the perovskite got the highest PCE of 20.96% (Figure 2.5d) with the outstanding reduction in V_{OC} loss as the EQE curves are similar, i.e., showing the same amount of FF and J_{SC} (Figure 2.5e and f). Therefore, PCE enhancement is attributed to high V_{OC}. These common interfacial defects have also been fixed by the treatment of NiO_X with a dipolar molecule n-Butylamine. This particular treatment created the dipole-ion attachment at the NiO_X/perovskite interface, which suppressed the interface defects, and a PCE of 18.9% was recorded in $MAPbI_3$-based p-i-n PSCs.[76]

Xu and coworkers added a π-conjugated ionic compound ($QAPyBF_4$) in perovskite precursor (Figure 2.6a) where $[BF_4]^-$ was found at NiO_X/perovskite interface

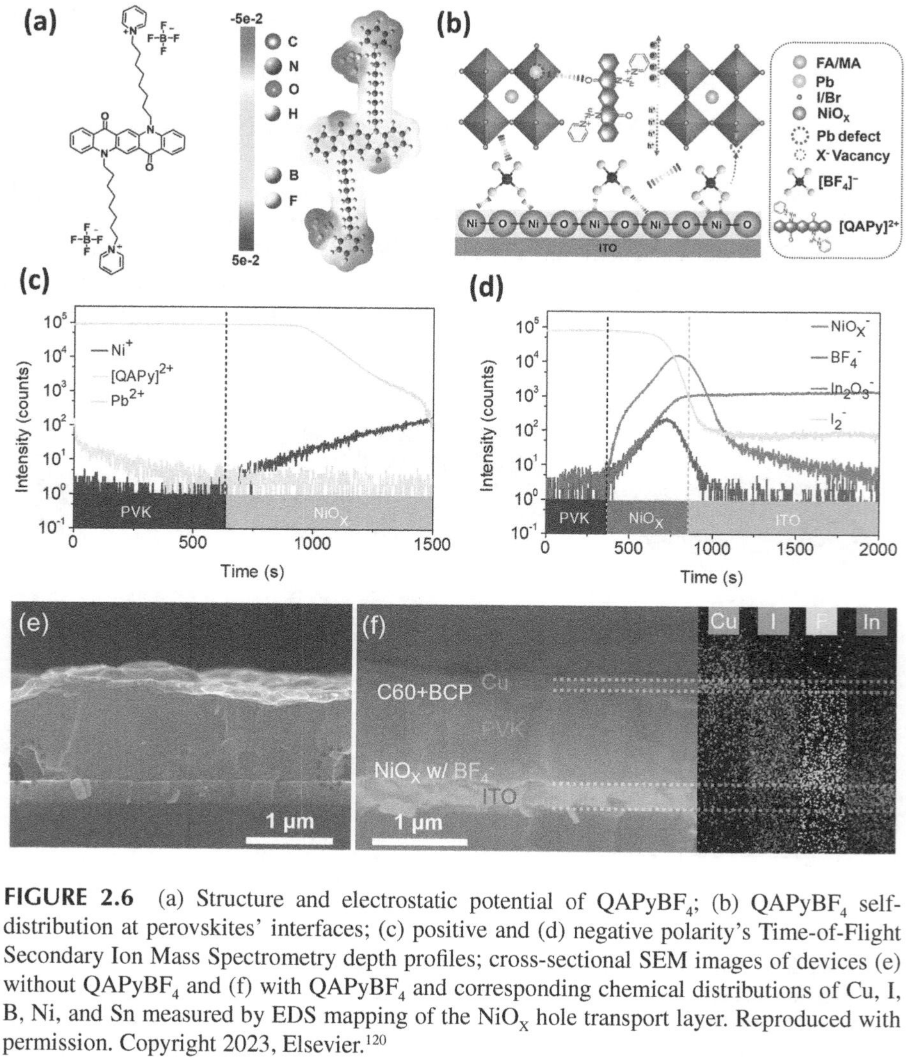

FIGURE 2.6 (a) Structure and electrostatic potential of $QAPyBF_4$; (b) $QAPyBF_4$ self-distribution at perovskites' interfaces; (c) positive and (d) negative polarity's Time-of-Flight Secondary Ion Mass Spectrometry depth profiles; cross-sectional SEM images of devices (e) without $QAPyBF_4$ and (f) with $QAPyBF_4$ and corresponding chemical distributions of Cu, I, B, Ni, and Sn measured by EDS mapping of the NiO_X hole transport layer. Reproduced with permission. Copyright 2023, Elsevier.[120]

(Figure 2.6b, (c)).[120] $[BF_4]^-$ suppressed the $Ni^{\geq 3+}$ species, and PCE of 23.38% was reported with an FF of 85%. In the cross-sectional SEM image of the device, the suppression of defects is noticed after the incorporation of $QAPyBF_4$ in the perovskite layer (Figure 2.6e and f).

2.2.2.2 Interlayers and NiO_X-based Composites for Buried Interface

Though every heterojunction interface faces defects and charge recombination issues, among all the inorganic materials, the NiO_X layer is mostly used as HTM for p-i-n PSCs. In this section, we discussed the interlayers designed to handle these issues. As explained earlier, some Ni species are responsible for the buried interface issues, and different types of organic materials are used as interlayers or dopants in the NiO_X layer to suppress the unwanted Ni species at the interface. These species destroy the perovskite crystal present near the interface, which results in hysteresis losses. Different buffer layers are used to avoid the direct contact of NiO_X with perovskite. For example, ferrocene dicarboxylic acid (FDA), as an interfacial modified layer, enhanced the PCE to 18.20% from 15.13%, and the hysteresis losses were significantly suppressed.[121] Girolamo et al. applied a hybrid (organic/inorganic) Mg^{2+}-based interlayer named MgO(R), reducing the hysteresis losses and improving the PCE from 16.8 to 18%.[122] In fact, the organic interface with the absorber layer improves the recombination losses and possesses fewer defects than inorganic ones. Lee et al. deposited the layer based on organic donor–acceptor–donor (D–A–D)-typed small molecule on NiO_X, which not only bypasses the buried interface but also provides the defect passivation through Lewis bases interaction method via methoxy (OMe) moieties available at the molecule's end. The PCE of over 20% is executed with an outstandingly long life span.[123] Different polymers, such as PTAA, poly(methyl methacrylate) PMMA, and polystyrene (PS), are also used as the buffer layer between NiO_X and perovskite by Wu and coworkers.[124] They achieved high V_{OC} (1.19 V) and PCE (21.56%) with NiO_X/PTAA/Perovskite-based devices and claimed that high V_{OC} is due to the PTAA's perfect energy level match with NiO_X. However, PTAA is commonly used organic HTM in PSCs with seldom high PCE records. Anyhow, the organic contacts with perovskite always provide high PCE and fewer interface losses than the inorganic ones. Ru et al. used 3,6-diffuoro-2,5,7,7,8,8-hexacyanoquinodimethane molecules to achieve high V_{OC} and FF.[125] As a result, 22.13% of PCE was achieved in small-area devices and 20.01% in large-area devices. But, for organic molecules, the spin coating was used. The original aim in focusing on NiO_X was to provide scalable deposition through industrial techniques, which is somehow less convenient with organic HTMs, especially the small molecules.

2.2.2.3 NiO_X for Scalable PSCs

The original aim of developing inorganic HTM is to meet the requirements of scalable PSCs, and solution processing with the spin coating is not the answer to this issue. Therefore, some scalable deposition techniques have been applied while depositing NiO_X. Han and coworkers used the spray-pyrolysis method on Mg^{2+} and Li-doped NiO_X and achieved a PCE of 18.3% on small devices and 15% on large-area devices.[41] Later on, using similar deposition methods, a PCE of 19.6% was achieved through perovskite modification with an aperture area of 1.025 cm^2.[126] Pulsed-laser-deposition

(PLD) method was also applied to deposit NiO on ITO substrate by improving morphology under controlled oxygen partial pressure, and PCE of 17.3% was achieved.[127] PLD deposition of Cu-doped NiO_X has also been done by Feng et al., where a PCE of 20.41% was obtained.[82] For scalable PSCs, the deposition of the thin and dense large-area film is crucial. The atomic layer deposition (ALD) method has been applied with O_3 as an oxidizer, and PCE of 16.4% was recorded in $MAPbI_3$-based p-i-n devices.[128] Koushik et al. used low-temperature plasma-assisted ALD to coat NiO_X on ITO and achieved a PCE of 17.07%.[72] Seo et al. also applied ALD on ETL and HTL (NiO), where a PCE of 18.45% was achieved.[57] The electrodeposition method has also been used to get mesoporous NiO_X HTL in p-i-n PSC, where the PCE of 17.77% was achieved.[53] Liu et al. applied a solution combustion-based approach to coat NiOx on FTO and annealed the HTL at low temperatures (150–250 °C).[56] The PCE above 20% was recorded due to smooth and dense HTL films of NiO_X. Magnetron sputtering of NiO_X is applied to achieve 18.7%.[84] A similar approach is used to get the doping effect of Mg where $NiMgO_X$ was coated on FTO by the co-sputtering of the ceramic MgO target, and the metallic nickel target and PCE of 18.5% were delivered.[50] Electron beam evaporation (EBE) has also been applied by different researchers in n-i-p PSCs. Pang et al. evaporated the Ni metal on the FTO substrate and annealed it in the air to get NiO_X, where the PCE of 18.13% was achieved.[78] Abzieher et al. advanced this EBE process by adding a controlled level of oxygen during the evaporation process and achieved a PCE of 20.7%.[74] One of the significant supremacies of the EBM method is that it fits the scalable solution-processed PSCs. Du et al. applied surface redox engineering on the EBE evaporated NiO_X films and got the PCE (23.4%), among the highest PCEs reported in inorganic HTM-based PSCs.[91]

2.2.3 Other Inorganic P-Type Semiconductors

Nickel and copper derivatives stayed quite effective as HTM materials for PSCs; some other inorganic p-type semiconductors have also been used. Vanadium (V)-oxytriisopropoxide (V_2O_5) is included in p-i-n PSCs deposited in a conventional spin coating and high temperature-based annealing procedure.[129] The hole mobility was impressive, but poor interfacial contacts and poor morphology resulted in the low performance of PSCs. Spiccia and co-workers applied glass assisted deposition technique on V_2O_X to achieve a PCE of 14.8%, where FF and J_{SC} were relatively low due to the low conductivity of V_2O_X.[130] The vacuum vapor deposited MnS has been reported as HTL in n-i-p PSCs, and the PCE of 19.86% was executed.[64] MnS is much cheaper ($4/g) than other inorganic HTMs. Furthermore, the well-dispersed $NiCo_2O_4$ 5 nm-sized nanoparticles were spin-coated on ITO substrate to achieve the PCE of 18.23%.[54] Several other inorganic HTMs were tried for inclusion in the PSCs, such as CIS, PbS, WO_3, MoO_X, MoS_2, CrO_X, carbon nanotubes, graphene, etc., but NiO_X came as the most prominent among all.[35]

2.3 POLYMERIC HTMs

Whenever polymeric or small molecule-based HTMs are discussed, the most demanding property is the dopant-freeness in HTMs. As discussed in the first section of this chapter, the dopants significantly damage the performance of the devices. Therefore, this section mainly focused on the dopant-free polymeric HTMs and state-of-the-art polymers such as P3HT, PEDOT:PSS, PTAA, etc. Polymers consist of a molecular structure consisting of many similar units bonded together. Therefore, in respect of mechanical properties, they are lesser than the inorganic and better than small molecules. As a consequence, the solution printing of polymer-based layers is possible and has been reported in more than a few dozen research papers.[131] Furthermore, the long chains of polymers provide high hole mobility and impressive film morphology owing to the spreadable π-conjugation.

2.3.1 State-of-the-Art Polymeric HTMs

2.3.1.1 PEDOT:PSS

Poly(3,4-ethylenedioxythiophene):poly(styrene sulfonate) (PEDOT:PSS) consists of two parts, i.e., PEDOT and PSS (Figure 2.7). PEDOT is positively charged with a conjugated backbone, while PSS is negatively charged. PEDOT helps in high conductivity, and PSS facilitates solubility enhancement as it is stably dispersible in common solvents. The work function (5.0–5.2 eV) of PEDOT:PSS enables charge mobility and extraction whenever it is used with photoactive layers. On top of that, it is a well-engineered polymer and commercially available with high mechanical properties, which makes it quite feasible for industrial film deposition techniques. Above all else, the high transparency makes it more than feasible to include it in solar cells. At the start of its development, it was greatly used in the charge transport and main layers of organic solar cells and light-emitting diodes (LEDs). In recent years, it has become one of the state-of-the-art HTMs in PSCs.

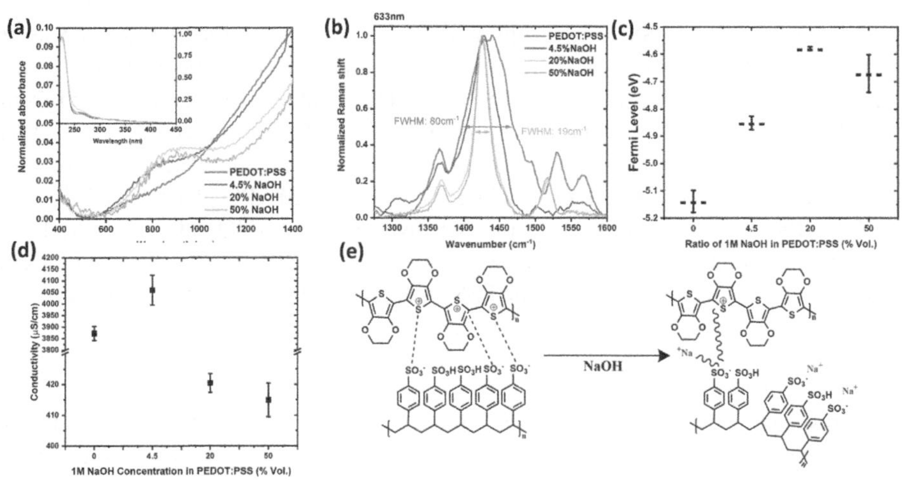

FIGURE 2.7 (a) Optical/electrical properties of pristine and dedoped PEDOT:PSS with different ratios of NaOH; (b) normalized UV-vis spectra (220 nm) with the inset of full spectra; (c) Fermi levels; (d) conductivity, and (e) PEDOT:PSS chemical structure and dedoping mechanism. Reproduced with permission. Copyright 2022, ACS. [135]

In 2016, Huang et al. used DMSO-doped PEDOT:PSS in PSCs and achieved a PCE of 16.7%.[132] As a result, light harvesting, charge extraction, and photocurrent generation are extraordinarily enhanced with improved morphological and crystalline perovskite layer properties. However, PSS is a part of PEDOT:PSS, due to its insulating property, often leads to inadequate conductivity and hole mobility. Furthermore, some perovskite layers have deeper HOMO levels. As a result, the band offset between the HOMO of perovskite and HTM results in interface recombination losses. Therefore, several dopants are incorporated to enhance carrier mobility and optimize energy levels. Liu et al. employed a commonly used F4-TCNQ dopant that enhanced the PCE from 13.3% to 17.2% due to high carrier mobility and optimized alignment of energy levels.[133] PEDOT:PSS has also been doped with sodium citrate to improve the work function of HTM, which enhanced the V_{OC} from 1.057 to 1.134 V, and the modified devices exhibited a maximum PCE of 18.39%.[134] Recently, Chin et al. de-doped the PEDOT:PSS with 4.5, 20, and 50% (vol%) of 1 M NaOH.[135] The normalized absorption spectra show typical bipolaron peaks at wavelengths > 1100 nm, showing a reduction in increasing the NaOH ratio, while the polaronic peak enhancement was noticed at around 800 nm (Figure 2.7a). Raman spectroscopy further confirms the de-doping of PEDOT:PSS, where the central peak (at 1,425 cm^{-1}) could be assigned to PEDOT interring double bond oscillation, alters more narrowly with an increased NaOH ratio (Figure 2.7b). Incorporating NaOH in PEDOT:PSS decreased the fermi level from -5.15 to -4.55 eV (Figure 2.7c). For de-doped PEDOT:PSS, lower conductivity was noticed as expected. The field effect transistor measurements reveal a significant drop in conductivity between the pure and 20% samples from around 3,875 to 420 μS/cm (Figure 2.7d). As a result, a reduction of recombination losses at the de-doped PEDOT:PSS/perovskite interface is noticed through a longer photoluminescence lifetime and higher surface photovoltage, which leads to an increased device V_{OC}, FF, and PCE.

Other than doping techniques, surface treatment techniques were applied to enhance the PEDOT:PSS electrical properties. Li et al. treated the PEDOT:PSS layer with dimethylformamide (DMF), methanol and their mixture solvents to enhance the electrical properties of PEDOT:PSS film. The conductivity was enhanced from 10^{-3} S/cm to around 10^{-2} S/cm where the PCE of 17.5% was executed.[136] Reza et al. removed the prevailing PSS by using ethylene glycol and methanol-based solvent engineering methods and achieved a PCE of 18.18% in $MAPbI_3$-based PSCs.[137] The PEDOT:PSS surface modification through hydroquinone (HQ) also improved the PCE by 5.7% compared to PEDOT:PSS-based devices.[138] Very recently, ammonium carbamate (AC) was used to strip the PSS shell around PEDOT, which enhanced the conductivity and improved the PCE in lead-free PSCs.[139]

2.3.1.2 PTAA

Poly[bis(4-phenyl)(2,4,6-trimethylphenyl)amine](PTAA) is enriched with electrons, which naturally has a property to block electrons and accept holes. Compared to other conventional HTMs, the PTAA has higher hole mobility (4×10^{-3} $cm^2V^{-1}s^{-1}$).[140] Initially, it was used as HTM for organic LEDs, but with the emergence of PSCs, it became one of the most commonly used HTMs in PSCs. In most of the flexible PSCs or large area n-i-p PSCs, PTAA is used as HTM. However, further enhancement in hole mobility is required to get higher PCEs. Therefore, PTAA also requires the necessary dopants. PTAA is modified to cure the low hole mobility and shallower HOMO levels. For example, Ryu et al. introduced PF8-TAA (HOMO = 5.44 eV), PIF-TAA (HOMO = 5.51 eV) to compete with the PTAA (HOMO = 5.14 eV) to enhance the V_{OC}.[141] PTAA is also modified to PTA, which has fewer methyl groups than PTA, which deepened the HOMO level and enhanced the V_{OC}.[142] Other than the HOMO deepening, the PTAA is modified to enhance the carrier mobility too. Kim et al. modified PTAA twice to improve both the parameters (HOMO deepening and mobility enhancement) by introducing the fluorinated PTTA, i.e., 1F-PTAA. 1F-PTAA improved the PCE from 21.3 to 22% [143] and later the CH_3O-PTAA, where PCE of 21% was achieved.[144] The methoxy part in CH_3O-PTAA improved the molecular interactions between HTM and its dopant. Matsui et al. improved the mobility by modifying the PTAA through methylphenylethenyl functional groups and achieved a PCE of 15%, but it still was incompatible with the doped PTAA.

As the dopants are doing significant damage in terms of device stability, some researchers introduced new dopants to replace the conventional dopants (Li-TFSI and t-BP). Luo et al. were the pioneers in introducing the replacement of traditional dopants.[145] They introduced a single dopant named fluorine-containing hydrophobic Lewis acid dopant (LAD) tris(pentafluorophenyl)boron (35FB) to replace both of the conventional dopants (Figure 2.8). This LAD dopant improved the HOMO mismatch and hole mobility, thus executing the PCE beyond 20%. Later on, same research group introduced another dopant with high hydrophobicity named as fluorinated iron III porphine and boosted the PCE from 21% to nearly 22%.[146]

2.3.2 Dopant-Free Polymeric HTMs

Two advantages of polymeric HTMs on small molecules are the adoption of solution printing for scalable PSCs and high hole mobility. These advantages mainly come from their substantial interchange charge transfer and amorphous nature. From a design perspective, polymeric HTMs can be divided into donor-acceptor (D-A) type, non-D-A type, and other designs.

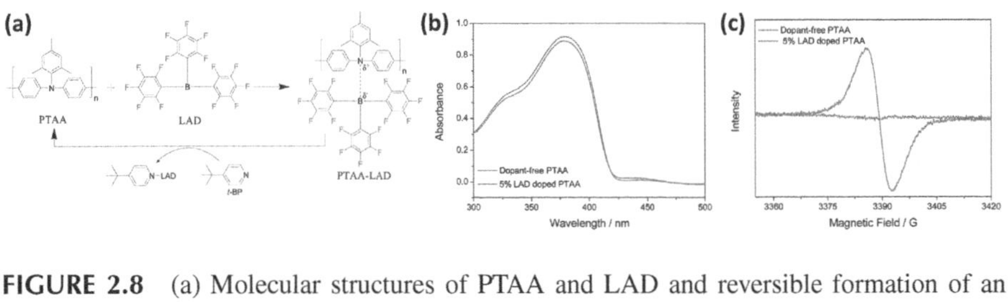

FIGURE 2.8 (a) Molecular structures of PTAA and LAD and reversible formation of an adduct between Lewis basic PTAA and Lewis acidic LAD in solution; (b) UV-vis absorption spectra and; (c) ESR spectra of dopant-free PTAA and 5% LAD doped PTAA. Reproduced with permission. Copyright 2018, RSC.[145]

2.3.2.1 D-A Type Polymeric HTMs

The concept of the D-A-type polymers is adopted from the organic photovoltaics (OPVs), where these materials have been widely employed as p-type donor materials.[147,148] By combining different donor and acceptor designs, the properties such as energy levels, hole mobilities, and optical bands can be optimized. The representative D-A-type conjugated polymeric HTMs are shown in Figure 2.9, and their properties are listed in Table 2.2. Diketopyrrolopyrrole (DPP) and isoindigo are famous acceptor units for designing D-A-type conjugated polymers. In 2018, Chang et al. copolymerized the DPP unit with thieno[3,2-b]thiophene (TT) unit and used it as dopant free HTM in PSCs where the PCE of 17.8% was achieved.[149] Later in the same year, a isoindigo-based polymer named PII2T8T was designed, composed of a biaxially extended structure. This extended structure helped in UV blockage and deepened the HOMO level closer to the perovskite to suppress the nonradiative recombination. As a result, a PCE of 18.3% with higher photostability was reported.[150]

Carbazole has also been considered an efficient donor for polymeric HTMs. Wang et al. attached different acceptors with carbazole. For example, benzothiadiazole (BT), an acceptor with varying lengths of alkoxy side chains, obtained a PCE of 17.1%.[152] The attached chains don't take part in charge flow. Therefore, performance

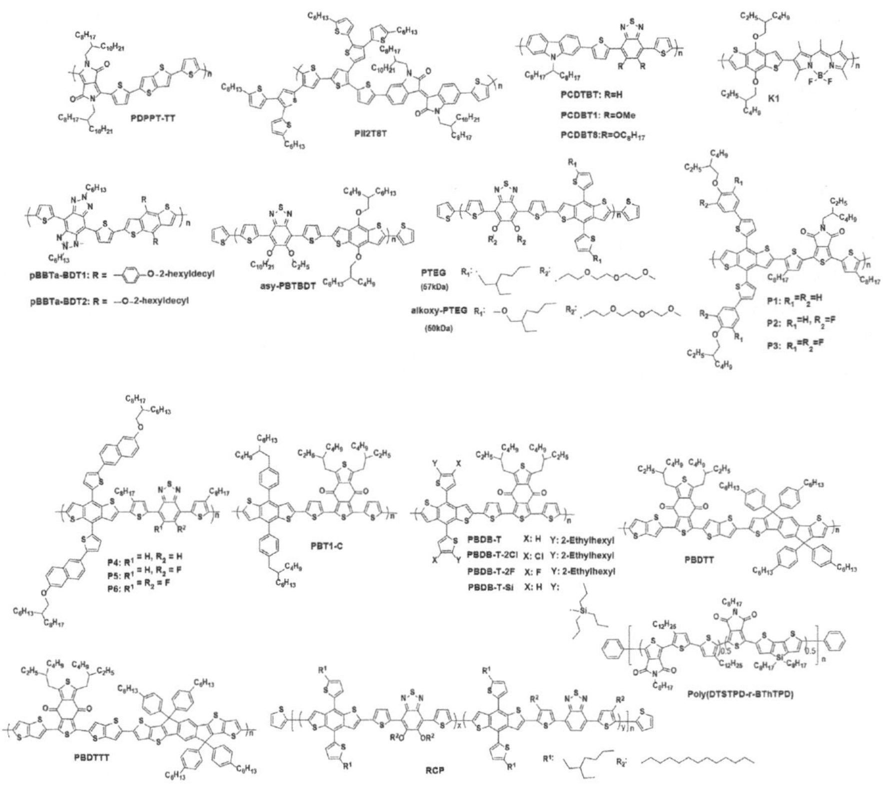

FIGURE 2.9 D-A-type polymeric HTMs structure. Reproduced with permission. Copyright 2020, ACS.[151]

TABLE 2.2
Photovoltaic Parameters of the Devices Based on D-A-Type Polymeric-Based PSCs

HTM	Device structure	V_{OC} (V)	J_{SC} (mA/cm²)	FF (%)	PCE (%)	Ref
PDPPT-TT	FTO/TiO_2/PCBM/$MA_{1-y}FA_yPbI_{3-x}Cl_x$/HTM/Ag	1.07	22.8	72	17.8	149
PII2T8T	ITO/HTM/$(FAPbI_3)_{0.85}(MAPbBr_3)_{0.15}$/$PC_{61}BM$/BCP/Ag	1.09	21.5	78.1	18.3	150
PCDTBT	ITO/TiO_2/$PC_{60}BM$/$MAPbI_3$/HTM/MoO_3/Au	1.05	21.6	72.3	16.4	152
PCDTBT1	ITO/TiO_2/$PC_{60}BM$/$MAPbI_3$/HTM/MoO_3/Au	1.10	21.8	77.6	18.7	152
PCDTBT8	ITO/TiO_2/$PC_{60}BM$/$MAPbI_3$/HTM/MoO_3/Au	1.02	21.2	67.3	14.8	152
pBBTa-BDT1	FTO/TiO_2/$MAPbCl_{3-x}I_x$/HTM/Au	0.77	18.1	50.6	7	156
pBBTa-BDT2	FTO/TiO_2/$MAPbCl_{3-x}I_x$/HTM/Au	0.95	20.3	75.2	14.5	156
asy-PBTBDT	FTO/TiO_2/m-TiO_2/$Cs_{0.05}FA_{0.81}MA_{0.14}PbI_{2.55}Br_{0.45}$/HTM/Au	1.11	22.4	73.2	18.3	157
2DP-TDB	ITO/SnO_2/$FA_{0.85}MA_{0.15}PbI_3$/HTMs/MoO_3/Ag	1.16	24.13	76.9	22.1	158
PTEG	FTO/SnO_2/$Cs_{0.05}FA_{0.81}MA_{0.14}PbI_{2.55}Br_{0.45}$/HTM/Au	1.14	22.5	77	19.8	159
alkoxy-PTEG (2-MA)	FTO/SnO_2/$Cs_{0.05}FA_{0.81}MA_{0.14}PbI_{2.55}Br_{0.45}$/HTM/Au	1.14	23.2	79.8	21.2	160
alkoxy-PTEG (3-MC)	FTO/SnO_2/$Cs_{0.05}FA_{0.81}MA_{0.14}PbI_{2.55}Br_{0.45}$/HTM/Au	1.13	23.3	75.7	19.9	160
P1	FTO/TiO_2/m-TiO_2/$MAPbI_3$/HTM/Ag	1	17.2	54.7	9.8	161
P2	FTO/TiO_2/m-TiO_2/$MAPbI_3$/HTM/Ag	1.04	20.2	70.7	14.9	161
P3	FTO/TiO_2/m-TiO_2/$MAPbI_3$/HTM/Ag	1.03	18.6	53.8	10.4	161
P4	FTO/TiO_2/m-TiO_2/$MAPbI_3$/HTM/Ag	1.03	16.1	67.7	11.2	162
P5	FTO/TiO_2/m-TiO_2/$MAPbI_3$/HTM/Ag	1.01	16.8	72	12.3	162
P6	FTO/TiO_2/m-TiO_2/$MAPbI_3$/HTM/Ag	1.09	20.7	75.8	17.3	162
PBT1-C	FTO/TiO2/m-TiO2/Cs-perovskite/HTM/MoO_3/Ag	1.05	22.4	81.2	19.1	163

(Continued)

TABLE 2.2 *(Continued)*
Photovoltaic Parameters of the Devices Based on D-A-Type Polymeric-Based PSCs

HTM	Device structure	V_{OC} (V)	J_{SC} (mA/cm^2)	FF (%)	PCE (%)	Ref
PBDB-T	ITO/ZnO/$CsPbI_2Br$/HTM/MoO_3/Ag	1.11	15.5	82.5	14.2	164
PBDB-T-2Cl	ITO/ZnO/$CsPbI_2Br$/HTM/MoO_3/Ag	1.20	14.61	79.7	14.0	164
PBDB-T-2F	ITO/ZnO/$CsPbI_2Br$/HTM/MoO_3/Ag	1.20	14.8	83.7	14.9	164
PBDB-T-2Si	ITO/ZnO/$CsPbI_2Br$/HTM/MoO_3/Ag	1.20	15.5	84	15.6	164
PBDB-O	ITO/PEDOT:PSS/perovskite/HTM/MoO_3/Ag	1.07	16.98	56.2	10.2	165
PBDB-T	ITO/PEDOT:PSS/perovskite/HTM/MoO_3/Ag	1.12	22.75	74.7	19.0	165
PBDB-Cz	ITO/PEDOT:PSS/perovskite/HTM/MoO_3/Ag	1.14	24.41	79.0	22.0	165
PBDTT	ITO/SnO_2/Cs-perovskite/HTM/MoO_3/Ag	1.12	23.6	76.7	20.3	166
PBTTT	ITO/SnO_2/Cs-perovskite/HTM/MoO_3/Ag	1.10	22.3	79.6	19.5	166
RCP	FTO/SnO_2/$MAPbI_3$/HTM/Au	1.08	21.9	75	17.3	167
poly(DTSTPD-r-BThTPD)	FTO/SnO_2/$CsPbI_2Br$/HTM/Au	1.15	15.7	81	14.6	168
PBZ-1Si	ITO/SnO_2/CSFAMAperovskite/polymer/Au	1.03	21.37	78.3	17.2	169
PBZ-2Si	ITO/SnO_2/CSFAMAperovskite/polymer/Au	1.06	22.22	81.3	19.1	169
PBZ-3Si	ITO/SnO_2/CSFAMAperovskite/polymer/Au	1.09	21.8	84.9	20.1	169
PC-DPP	ITO/SnO_2/$MA_{0.16}FA_{0.84}PbI_3$/HTMs/MoO_3/Ag	1.15	24.74	79.2	22.67	170
Nap-SiBTA	FTO/compact-TiO_2/mesoporous-TiO_2/$FAPbI_3$/HTM/Au	1.12	25.45	80.8	23.07	171

is enhanced without disturbance in the HOMO level. In the same work, they further decorated the polymer (PCDTBT) with OMe (PCDTBT-1), which passivated the perovskite, and a PCE of 18.7% was achieved. In comparison, another similar polymer was designed with extra-long chains (PCDTBT8), which reduced the performance. Hong and coworkers developed a series of polymers based on Benzo[1,2-b:4,5-b′]dithiophene (BDT) as a donor.[153] BDT is coplanar with prominent electron donating capability.[154,155] They used 4,4- difluoro-4-bora-3a,4a-diaza-s-indacene (BODIPY) as an acceptor and tuned the energy levels by changing the positions of alkyl chains at BDT and meso-substituents on BODIPY and obtained a PCE of 16%. Marks and coworkers also used BDT as an acceptor and designed the polymer named pBBTa-BDT1 and pBBTa-BDT2.[156] They used 2,6-dialkylated-benzo[1,2-d;4,5-d′]-bistriazole (BBTa) as an acceptor unit. They compared the pBBTa-BDT1 with pBBTa-BDT2 and observed that the pBBTa-BDT2 was with deeper HOMO level (5.21 eV) and high hole mobility owing to its high coplanarity as compared to its counterpart; as a consequence it exhibited more than twice amount of PCE (14.5%) than the pBBTa-BDT1.

In 2017, Park and coworkers decorated the BDT based polymeric HTM (asy-PBTBDT) with asymmetric alkyl chains on BT units to make it soluble in the green solvent (2-methylanisole (2-MA)).[157] It was the first ever HTM material soluble in a green solvent. Though asy-PBTBDT exhibited low crystallinity, it possessed high hole mobility. As a result, a PCE of 18.3% was executed. In 2018, Park and coworkers designed another efficient polymer (PTEG) by decorating tetraethylene glycol (TEG) groups as the side chains of the BT unit that enhanced the solubility through dipole–dipole interactions between the polymer's chains and solvent's molecules and also the hole mobility was obtained.[159] As a consequence, they achieved a PCE of 18.9%. In 2020, they modified the PTEG design by introducing alkoxyl side chains into the BDT units and named this polymeric HTM alkoxy-PTEG.[160] This polymer was soluble in the green solvent (2-MA), which they used three years back, and a new solvent called 3-methylcyclohexanone (3-MC). The efficient devices based on 2MA and 3-MC delivered PCEs of 21.2% and 19.9%. BDT-based polymers were also designed with two-dimensional (2D) structures for the sake of high hole mobility and high-quality film morphology. In 2021, Fu et al. compared 2DP-TDB with a small molecule (TB-DPP).[158] 2DP-TDB exhibited extended π-electron delocalization and achieved a PCE of 22.17%. The face-on orientation enhanced the dense molecular packing, increasing hole mobility. They further performed the temperature-dependent UV-vis analyses of the 2DP-TDB solution in chloroform (Figure 2.10a and b). They noticed that the polymers are stacked in a face-on orientation at room temperature and dispersed under high temperatures. The schematic of packing patterns is shown in Figure 2.10c and d at high and room temperatures, respectively. TB-DPP's crystallinity was found to be poorer than the 2DP-TDB.[158]

Three BDT-based 2D polymeric HTMs (P1, P2, and P3) were designed by Song et al. by attaching the conjugated chains with different numbers of fluorine substituents.[161] In comparison, the P2 (with only one fluorine substituent) exhibited high hole mobility and a matchable HOMO level of perovskite, where a PCE of 14.9% was achieved. Jin et al. designed a series of 2D D-A type fluorinated polymeric HTMs (P4, P5, and P6) and noticed that the polymer (P6) with high fluorination exhibited

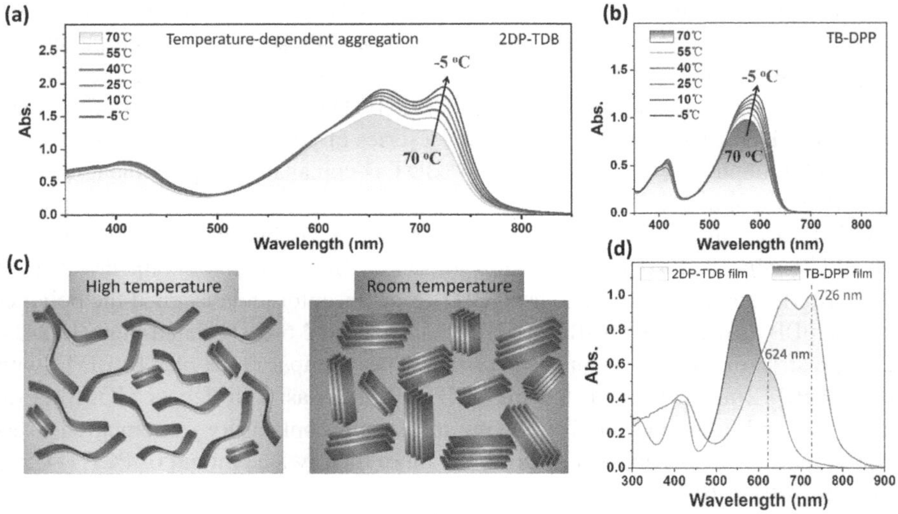

FIGURE 2.10 (a) UV-vis absorption graph of 2DP-TDB and 2D SM (b) chloroform-based solution at numerous temperature values. (c) Schematic illustration of 2DP-TDB's packing at different temperatures (c) and room temperature (d) UV-vis absorption spectra comparison of 2DP-TDB with 2D small molecule. Reproduced with permission. Copyright 2021, ACS.[158]

high mobility with enhanced hydrophobicity.[162] Therefore P6-based devices revealed comparably higher PCE (17%) with 30 days of life span. Jen et al. designed a polymer named PBT1-C by copolymerizing a phenyl-substituted 2D BDT unit with a 2D benzodithiophene-4,8-dione (BDD) acceptor.[163] The perovskite surface defect passivation through carbonyl groups and a deeper HOMO level resulted in a PCE of 19.1% with outstandingly improved FF (81.2%). 2D-BDT-BDD polymeric HTMs were also prepared by Wang et al.[164] They compared the chlorinated 2D-BDT-BDD (PBDB-T-2Cl) polymer with the fluorinated one (PBDB-T-2F), and both polymeric HTMs performed with almost equal values of PCE in $CsPbI_2Br$ PSCs. Hence, either the fluorination or chlorination of 2D-BDT, both designs can help in the performance enhancement of PSCs by optimizing HOMO levels, surface passivation, and hydrophobicity. Furthermore, two BDD-based polymeric HTMs (PBDTT and PBTTT) are designed by Ling et al. by using indacenodithiophene (IDT) and indacenodithieno[3,2-b] thiophene (IDTT) comonomers.[166] They noticed that PBDTT-based PSCs stability and PCE (20.3%) were higher than PBTT (19.5%) owing to its high solubility and improved interfacial contact. Liang et al. synthesized three polymers based on BDTT as the D unit and difluorobenzotriazole (FTAZ) as the A unit, linked through thiophene based π-bridge and terminated with siloxane (PBZ-1Si, PBZ-2Si, and PBZ-3Si; Figure 2.11a).[169] PBZ-1Si, PBZ-2Si, and PBZ-3Si, achieved PCEs of 17.24, 19.15, and 20.18 %, respectively. These polymeric-based HTM layers were smooth due to face-on orientation and dense molecular packing. On top of that, the siloxane functionality provides defect passivation, hydrophobicity, and high solubility. The dihedral angles between repeating units of each polymer indicate the steric hindrance of side chains and the co-planarity (Figure 2.11b). The clear diffraction point

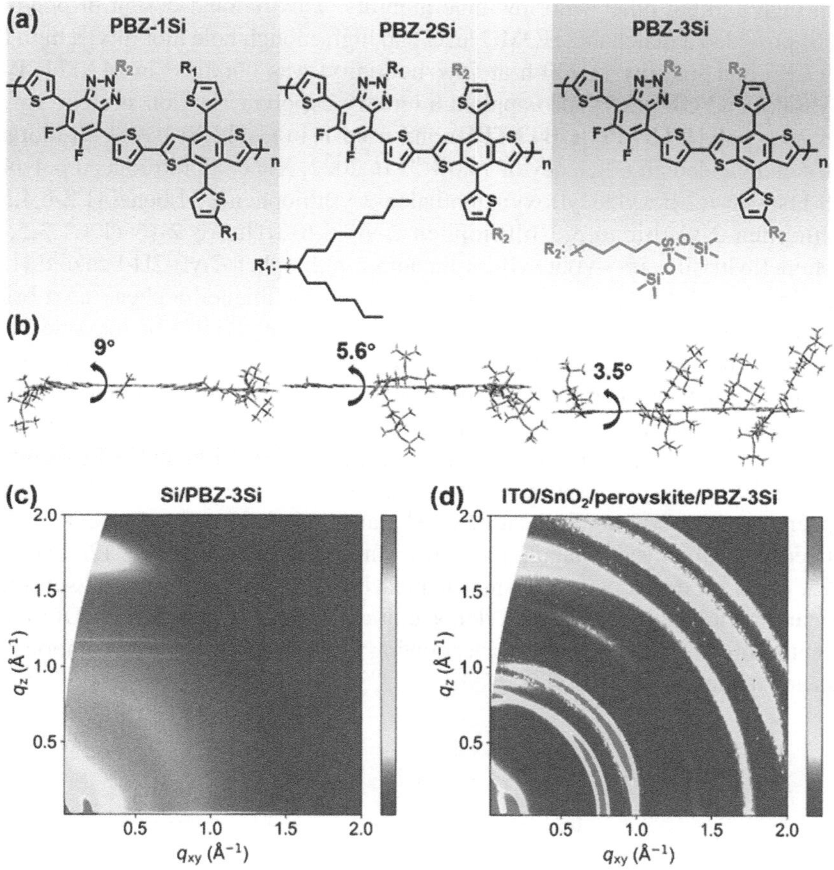

FIGURE 2.11 (a) Molecular structures of PBZ-1Si, PBZ-2Si, and PBZ-3Si; (b) Side view of the optimized geometries from DFT calculations; GIWAXS patterns conducted at (c) Si/PBZ-3Si and (d) ITO/SnO_2/perovskite/PBZ-3Si; Reproduced with permission. Copyright 2022, RSC.[169]

at $qz \sim 1.75$ Å^{-1} can be assigned to the π–π stacking of PBZ-3Si and meets with that noticed on the silicon substrates (Figure 2.11c and d), proving the same preferred face-on orientation along the out-of-plane direction.

You et al. also reported BDT–BDD-based polymeric HTMs by employing carbazole as the conjugated side chains of BDT named PBDB-Cz.[165] They compared PBDB-Cz with PBDB-O and PBDB-T to know the effect of side chains on hole transportation. PBDB-Cz performs best as a dopant-free HTM of PVSCs owing to the high hole mobility and defect passivation, delivering an outstanding PCE of 22.06% with high stability. At the same time, PBDB-O and PBDB-T-based devices achieved PCEs of 10.23% and 19.07%, respectively. Park and coworkers developed a BDT-based terpolymer named RCP by combining two similar structured copolymers.[167] Two combined polymers have different HOMO levels and hole mobilities, i.e., one with a deeper HOMO level while the other with the shallower one and, similarly, one with high and the other with low hole mobility. This rational design of polymeric HTM provided a matchable HOMO level and high enough hole mobility; a high PCE of 17.3% and stability (1,400 h at 75% humidity) was obtained in $MAPbI_3$ PSCs. Miyasaka and coworkers also applied a ternary copolymerization strategy by synthesizing **poly(DTSTPD-r-BThTPD)** and used it in $CsPbI_2Br$-based all-inorganic PSCs and yielded an efficiency of 14.6%.[168] In 2022, Xie et al. introduced poly[4-(5-(4,8-bis(5-(6-((2-hexyldecyl)oxy)naphthalen-2-yl)thiophen-2-yl)benzo[1,2-b:4,5-b'] dithiophen-2-yl)thieno[3,2-b]thiophen-2-yl)-5,6-difluoro-2-(6-(1,1,1,3,5,5,5-heptamethyltrisiloxan-3-yl)hexyl)-7-(thieno[3,2-b]thiophen-2-yl)-2H-benzo[d][1,2,3] triazole] (Nap-SiBTA), consisting of fused aromatic thienothiophene as π-bridge, which provided electron delocalization, enhanced the rigidity of the side blocks to reduce the twisting disorder, increased planarity, and improved π–π stacking (Figure 2.12a).[171] Furthermore, the thermogravimetric analysis (TGA) and differential scanning calorimetry (DSC) graphs indicate that Nap-SiBTA possesses a high decomposition temperature (394 °C) as shown in Figure 2.12b and c. From ultraviolet-visible (UV-vis) absorption spectra (Figure 2.12d), it was noticed that the peak appeared at 500–600 nm, indicating π–π* transition. A favorable HOMO level was noticed from the cyclic voltammetry (CV) measurements (Figure 2.12e and f). In 2023, Fu et al. developed a polymer named PC-DPP, with a thiophene-substituted phenanthrocarbazole (PC-T) as a donor unit and diketopyrrolopyrrole (DPP) as an acceptor unit and coated smooth layers owing to the dense face-on orientation packings and achieved the PCE of 22.67%.[170]

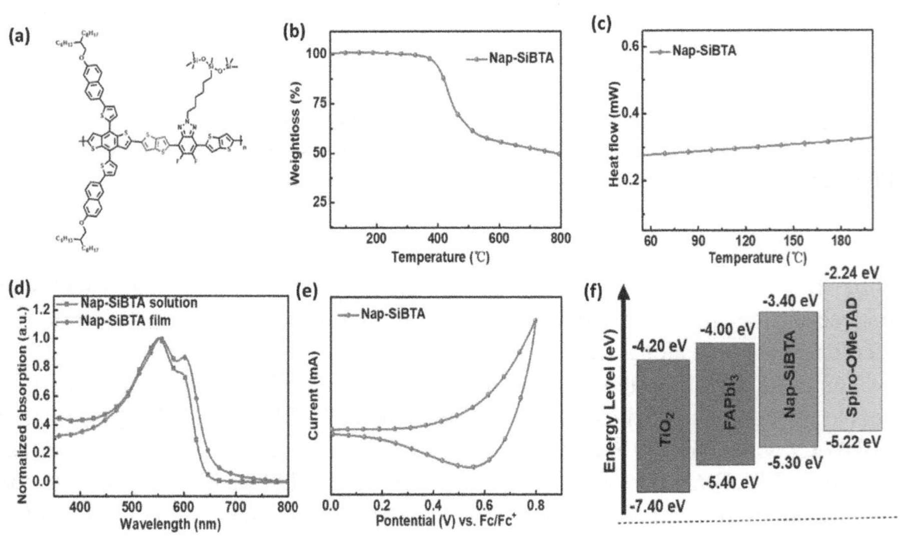

FIGURE 2.12 (a) Molecular structure of Nap-SiBTA; (b) TGA graph; (c) DSC graph; (d) UV-vis absorption curves; (e) cyclic voltammetry of Nap-SiBTA; (f) energy levels. Reproduced with permission. Copyright 2022, Wiley.[171]

2.3.2.2 Non D-A-Type Polymeric HTMs

Although dopant-free D-A type polymers are primarily focused on the design of enhancing push-pull property where the hole mobility is enhanced through electrons delocalization, the non-D-A type polymeric HTMs are based on donor unit and π -bridge forming quinoid resonance structure along the polymeric chain in which the electron delocalization is increased in an excited state.[172] The big difference between D-A and non-D-A polymers is the low conjugation, making them more soluble in non-polar solvents. The non-polar solvents are robust to perovskite precursor, especially in the p-i-n structure where the perovskite is deposited on the HTM, and, furthermore, these solvents don't bring any harm to the n-i-p structure where HTM materials are deposited on the perovskite. On top of that, non-D-A types are also more feasible for commercialization owing to their facile synthesis. P3HT is one of the most well-known non-D-A type polymers and the most feasible for large-scale production. Dopant-free P3HT has been used several times as HTM along with the buffer layers for PSCs with PCEs as high as 23%.[173], [174] Fang and coworkers reported a new type of polyelectrolytes with polythiophene backbone, i.e., P3CT-Na and P3CT-CH_3NH_2.[175] P3CT-CH_3NH_2 performance was found to be higher than P3CT-Na on both flexible and rigid substrates, 19.6% and 18.2%, respectively. However, later on, they tailored P3CT-Na with the butylammonium to design another P3CT-Na-based polyelectrolyte HTM which increased the hydrophobicity of the material and, as a result, high PCE with lengthened long-term stability was achieved.[176] Feng and coworkers copolymerized the fluorine with carboxyl acids and reported two HTMs, i.e., PFDT-COOH and PFDT-2F-COOH.[177] In comparison, PFDT-2F-COOH exhibited higher PCE (21.7%), attributed to its deeper HOMO level and improved work function. The optimized design based on fluorination always enhances the performance of dopant-free polymeric HTMs (Figure 2.13).

For perovskite/HTM interface optimization and defect passivation, Xu and coworkers prepared the DBT through copolymerization through a 2,5- di-2-ethylhexyloxy-1,4-phenylene unit and afterward arranged the polymers with the edge on molecular orientation.[178] The thiophene helped passivate the perovskite's defects, and molecular orientation helped gain high hole mobility. As a result, 19.7% of PCE with a high J_{SC} (25.5 mA cm^{-2}) was obtained. Later on, in 2019, DTB was further modified by introducing diethylene glycol (DEG) with different ratios, i.e., DTB (*x*DEG), where *x* was 3% and 10%.[179] These two randomly designed terpolymers exhibited PCEs of 20.2% ($x = 3\%$) and 17.8% ($x = 10\%$) when used as HTMs in PSC devices. DTB(0.03DEG) hole mobility was two times higher than the DTB(0.10DEG) due to molecular packing difference. Furthermore, both the polymeric HTMs showed high crystallinity and hydrophobicity and achieved impressive long-term stability. In 2020, Xu's research group further designed two anionic conjugated polyelectrolytes based on DTB; TB(K), and TB(Na), decorated with alkyl sulfonate side chains.[180] TB(K) performance (PCE = 20%) was higher than TB(Na) (PCE = 18.3%), as it helped in charge extraction and defect passivation (Table 2.3). In 2019, Bo and coworkers used a BDT block to prepare three polymers, P1, P2, and P3.[181] Bis(alkyloxy)benzene was incorporated as the monomeric unit, and through S–O interactions, the backbone of polymers was locked at a planar position. P1 and P2 structure was ladder-like

TABLE 2.3
Photovoltaic Parameters of the Devices Based on non-D-A-Type Polymeric-Based PSCs

HTM	Device structure	V_{OC} (V)	J_{SC} (mA/cm²)	FF (%)	PCE (%)	Ref
P3HT	ITO/SnO_2/$CsPbI_2Br$/TFB/HTM/Au	1.26	15.6	78.7	15.5	172
P3HT	FTO/c-TiO_2/m-TiO_2/$(FAPbI3)_{0.95}(MAPbBr_3)_{0.05}$/ $HTAB_{0.3}(FAPbI3)_{0.95}(MAPbBr3)_{0.05}$/HTM/Au	1.15	24.9	81.4	23.3	173
P3CT-CH_3NH_2	ITO/HTM/$MAPbI_3$/$PC_{60}BM$/C_{60}/BCP/Ag	1.09	22.2	81	19.6	175
PFDT-COOH	ITO/HTM/$FA_{1-x}MA_xPbI3$/PCBM/C_{60}/BCP/Cu	1.08	24.8	77.2	20.6	177
PFDT-2F-COOH	ITO/HTM/$FA_{1-x}MA_xPbI3$/PCBM/C_{60}/BCP/Cu	1.11	24.9	78.5	21.7	177
DTB	FTO/c-TiO_2/m-TiO_2/$FA_{0.85}MA_{0.15}PbI_{2.55}Br_{0.45}$/HTM/Au	1.10	25.5	69.9	19.7	178
DTB(0.03DEG)	FTO/c-TiO_2/m-TiO_2/$Cs_{0.05}FA_{0.81}MA_{0.14}PbI_{2.55}Br_{0.45}$/HTM/Au	1.14	23.6	75.1	20.2	179
DTB(0.1DEG)	FTO/c-TiO_2/m-TiO_2/$Cs_{0.05}FA_{0.81}MA_{0.14}PbI_{2.55}Br_{0.45}$/HTM/Au	1.05	23.5	72.4	17.8	179
TB(Na)	ITO/HTM/$FA_{0.85}MA_{0.15}Pb(Br_{0.15}I_{0.85})_3$/PCBM/$C_{60}$/BCP/Ag	1.06	22.0	78.1	18.3	180
TB(K)	ITO/HTM/$FA_{0.85}MA_{0.15}Pb(Br_{0.15}I_{0.85})_3$/PCBM/$C_{60}$/BCP/Ag	1.10	22.7	80	20	180
P1	FTO/c-TiO_2/PCBA/$(FAPbI_3)_{0.85}(MAPbBr_3)_{0.15}$/HTM/Au	1.06	22.4	75	17.8	181
P2	FTO/c-TiO_2/PCBA/$(FAPbI_3)_{0.85}(MAPbBr_3)_{0.15}$/HTM/Au	0.81	21.8	42.9	7.6	181
P3	FTO/c-TiO_2/PCBA/$(FAPbI_3)_{0.85}(MAPbBr_3)_{0.15}$/HTM/Au	0.93	22	63.3	12.9	181
PBDT[2F]T	FTO/SnO_2/$MAPbI_3$/HTM/Ag	1.06	22.6	72.6	17.5	182
PBDT(T)[2F]T	FTO/SnO_2/$MAPbI_3$/HTM/Ag	1.03	16.6	68	11.6	182
P4	FTO/m-SnO_2/$MAPbI_3$/HTM/Au	1.01	21.8	68	14.9	183
P5	FTO/m-SnO_2/$MAPbI_3$/HTM/Au	1.08	22.8	79	19.6	183
P6	FTO/m-SnO_2/$MAPbI_3$/HTM/Au	1.11	22.8	80	20.3	183

3,6-PCzTPA	ITO/HTM/$(FAPbI_3)_{0.85}(MAPbBr_3)_{0.15}$/$PC_{61}BM$/$C_{60}$/BCP/Ag	0.96	21.9	77.9	16	184
2,7-PCzTPA	ITO/HTM/$(FAPbI_3)_{0.85}(MAPbBr_3)_{0.15}$/$PC_{61}BM$/$C_{60}$/BCP/Ag	1.07	21.8	73.2	17.1	184
3,6 – 2,7-PCzTPA	ITO/HTM/$(FAPbI_3)_{0.85}(MAPbBr_3)_{0.15}$/$PC_{61}BM$/$C_{60}$/BCP/Ag	1.04	22.7	77.9	18.4	184
PC	ITO/SnO_2/perovskite ($FA_{0.85}MA_{0.15}PbI_3$)/HTM/MoO_3/Ag	1.16	24.2	79.6	22.2	185
PE7	FTO/SnO_2/$MAPbI_3$/HTM/Au	1.07	22.8	76	18.5	186
PE8	FTO/SnO_2/$MAPbI_3$/HTM/Au	1.09	22.7	76.6	19	186
PE9	FTO/SnO_2/$MAPbI_3$/HTM/Au	1.11	23	77.6	19.8	186
PE10	FTO/SnO_2/$MAPbI_3$/HTM/Au	1.13	23.2	80.1	21	186
PE10	ITO/SnO2/$FA_{0.85}MA_{0.15}PbI_3$/HTM/MoO_3/Ag	1.16	24.1	79.8	22.3	186

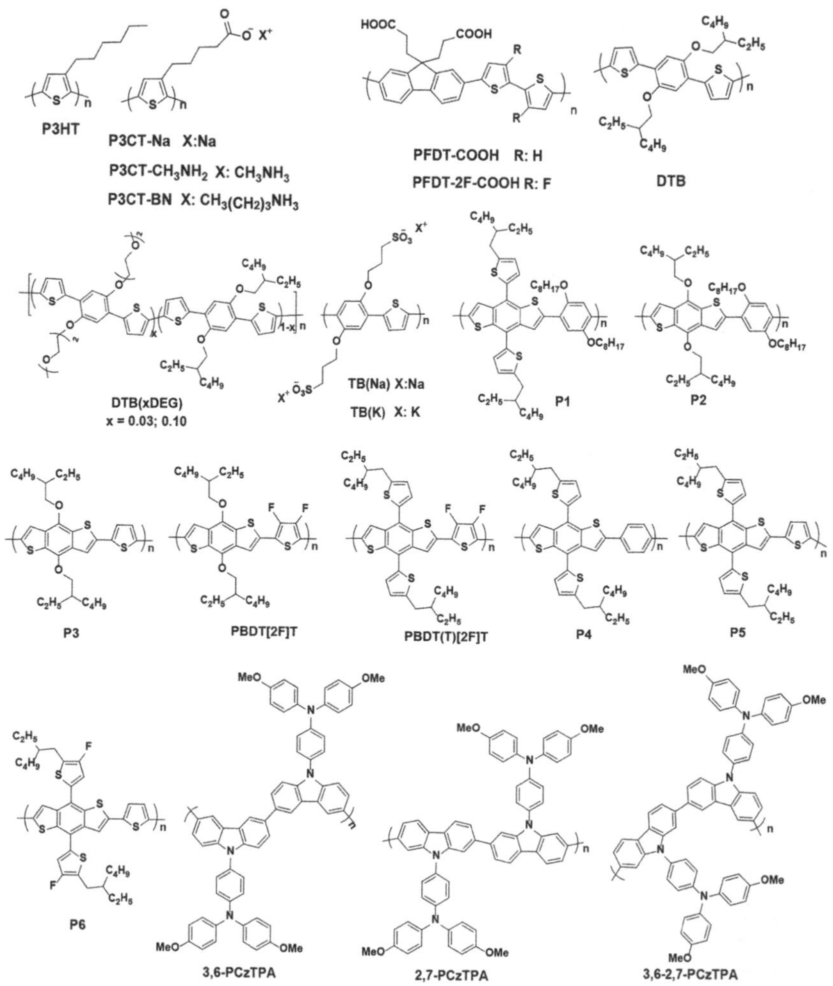

FIGURE 2.13 Non-D-A type polymeric HTMs structure. Reproduced with permission. Copyright 2020, ACS.[151]

planar, and the highest PCE of 18.3% was yielded. Alkyl thienyl-substituted 2D-BDT was employed in a series of polymers, P4, P5, and P6, and it was noticed that the P5 and P6 were more planar due to the thiophene bridge as compared to P4, which was with phenyl bridge.[183] On top of that, the thiophene, due to Pb-S interaction, helps in charge extraction and defect passivation. Furthermore, P6 possessed non-covalent interaction through F···H, and F···S. Therefore, P6 was superior among all in terms of planarity and hence executed the PCE of 20.3%, higher as compared to P4 (14.9%) and P5 (19.6%). However, the whole series of these polymers exhibited outstanding morphological and life stabilities in PSCs. Guo et al. synthesized a phenanthrocarbazole (PC) based polymeric HTM, PC6, where S···O secondary interaction was applied to fix the molecular geometry and reduce the reorganization energy.[185]

Furthermore, the planar backbone of PC enhances the π-π stacking between adjacent polymers. As a result, 22.2% of efficiency with significantly improved longevity was reported. In 2022, Yao et al. synthesized a series of polymers (PE7–PE10) by incorporating pyrene into thiophene and selenophene polymeric chains.[186] In addition, pyrene was fluorinated, and selenium was used to coordinate the Pb^{2+}. PE-10 reached a PCE of as high as 22.3%, while PE7, PE8, and PE9 delivered PCEs of 18.5, 19, and 19.8%, respectively.

2.3.2.3 Other Types of Polymeric HTMs

Other than π-conjugated polymers or D-A-based polymers, there are some specific designs, such as cross-conjugated or p–π-conjugated polymers (Figure 2.14). Although these polymers exhibit lower mobilities than their counterparts, they have some extra advantages, such as facile synthesis, more transparency, wettability, fewer batch-to-batch complications, etc. A large amount of research reports has been published to enhance hole mobility. For example, PVCz-OMeDAD was synthesized through a facile method, i.e., free-radical polymerization of vinyl monomers.[187] Though the designed polymer was non-conjugated due to its flexible design, it exhibited high morphological properties with improved hole mobility (3.44×10^{-4} $cm^2V^{-1}s^{-1}$) and

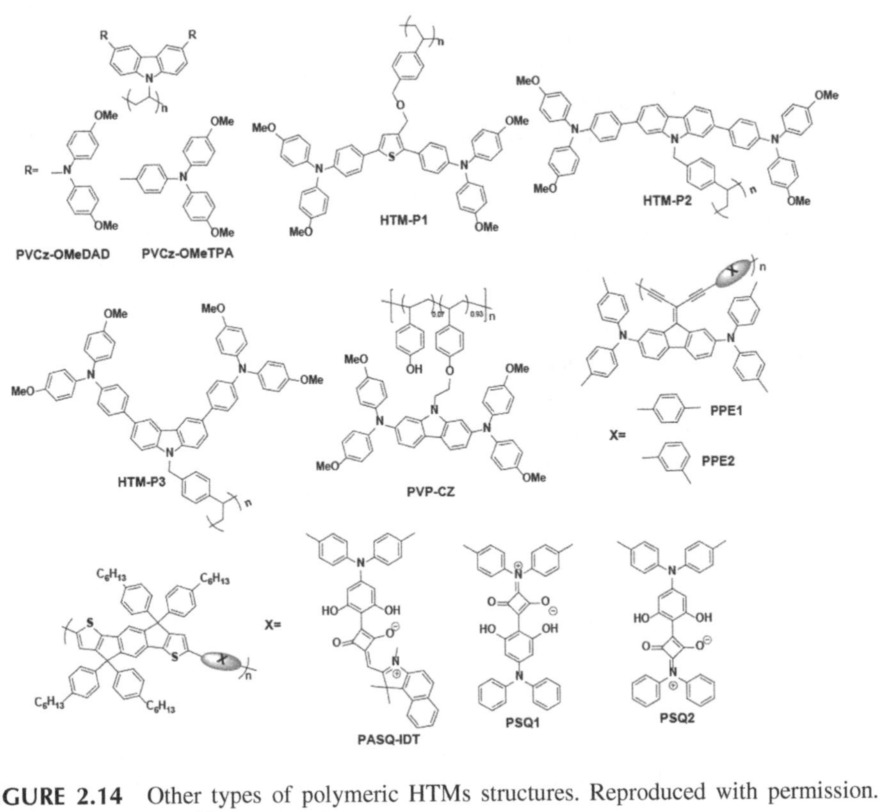

FIGURE 2.14 Other types of polymeric HTMs structures. Reproduced with permission. Copyright 2020, ACS.[151]

demonstrated a PCE of 16%. Furthermore, the PVCz-OMeDAD has been modified to PVCz-OMeTPA, which enhances the hole mobility, and a PCE of 17.2% was reported in PVCz-OMeTPA-based PSCs.[188] Xu and coworkers followed a similar synthesis methodology to synthesize the polymeric HTM P1.[189] He designed P1 by attaching the conjugated methoxytriphenylamine to the backbone of conjugated polystyrene (PS) with the hole mobility of 1.6×10^{-4} $cm^2V^{-1}s^{-1}$, which was higher than the monomer-based polymers, and executed the average PCE of 16.8% when incorporated in PSCs. Furthermore, they reported two more PS-based HTMs, HTM-P2 and HTM-P3.[190] They differentiated both the HTMs by altering the methoxy triphenylamine substitution positions on carbazole. Both of the polymers exhibited high solubility and film quality HTM-P3-based devices exhibited a PCE of 18.5%, which was higher than the HTM-P2 (16.8%). As both polymeric HTMs were hydrophobic, the PCEs retained over 80% after 30 d in ambient air with a 30% humidity level. Instead of free radical polymerization, post-functionalization is considered better for product quality and batch-to-batch variations. On top of that, the well-optimized post-functionalization also enhances the solvent tolerance, which further helps in improving film-forming capability. This strategy was adopted by Jen and Luo et al. to synthesize PS-based polymeric HTM, named (PVP-CZ); when utilized as HTM in PSCs, a PCE of 17.8% with a high FF of 81.1% was obtained. [191]

The cross-conjugated polymers have several advantages over π-conjugated, such as good film processibility and transparency, facile methodology for synthesis, and high solubility in common solvents. Li and coworkers synthesized two cross-conjugated polymers, PPE1 and PPE2, and devices based on these polymers exhibited PCEs of 11.1 and 19.3%, respectively. Furthermore, squaraines (SQs) blocks are found to be efficient as polymeric HTMs due to their NIR absorption capability and high photochemical and moisture stabilities. Li and coworkers synthesized pseudo-2D p–π-conjugated polymers (PASQ-IDT) through the copolymerization of SQ and IDT.[192] This polymeric HTM acted as a zwitterionic conjugated polyelectrolyte, which increased the surface wettability, and perovskite with enhanced film quality was grown. As a result, a PCE of 18.3% was yielded. However, low hole mobility (4.11×10^{-5} $cm^2V^{-1}s^{-1}$) was noticed in the PASQ-IDT. Later, in 2019, they introduced two new polysquaraines-based polymeric HTMs (PSQ1 and PSQ2).[193] The hole mobility was improved by introducing N,N-diarylanilinosquaraines, which provided π–π overlapping. PSQ1 and PSQ2 delivered the hole mobilities of 1.09×10^{-3} and 9.87×10^{-3} $cm^2V^{-1}s^{-1}$, respectively. And on incorporating PSQ2 as HTM, a PCE of 15.5% was achieved in $CsPbI_2Br$ solar cells with a life span of 300 h (PCE retained up to 83% of its initial value under illumination). In 2022, Chen et al. reported pyridine anchoring group functionalized poly(triarylamine) (p-PY), with 22.8% (certified 22.3%) and maintained 97.5% of the initial efficiency after operation under one sun illumination at 45°C for 1,000 h. Pyridine anchoring group enhanced the interaction with TCO as well as with perovskite.[194] In the same year, Fu et al. mixed two conjugated polymers PM6 (previously reported as donors in organic solar cells) and newly synthesized PMSe (alloy) and applied as dopant-free HTM to achieve a PCE of 24.53%.[195] PMSe provided π–π stacking and C=O···Se noncovalent interactions, while PM6 passivated the perovskite through its abundant Lewis base groups. GIWAXS was used to investigate the molecular packing and crystallinity in polymeric films (Figure 2.15a–c). The pure PM6 film exhibited a dominant π–π stacking diffraction peak at q = 1.66 $Å^{-1}$, corresponding to a d-spacing of 3.78 Å. The poly-alloy-based layer also delivered a face-on orientation (~63% ratio) and a π–π stacking diffraction peak at q = 1.67 $Å^{-1}$ with a less d-spacing of 3.75 Å, referring to a stronger π–π stacking owing to a flatter backbone of PMSe and formation of polymer alloys (Figure 2.15g and h). The resulting orientation distribution in films based on PM6, mixed (poly-alloy), and PMSe are schematically shown in Figure 2.15d, e, and f, respectively. Figure 2.15c shows that mixed polymers and PM6 had similar diffraction profiles, but the π–π stacking reflection of PMSe is less oriented as compared to PM6 and poly-alloy, suggesting a more random orientation (Figure 2.15i).

Moreover, cross-linkable polymeric HTMs can improve hole extraction and act as a protective barrier owing to their excellent solvent and thermal resistance. Nonetheless, crosslinking generally requires high temperatures, which could damage the absorber layer. In 2016, Li et al. improved the thermal stability of $MAPbI_3$-based PSCs by functionalizing 4,4′,4″-tris(N-carbazolyl)triphenylamine (TCTA) with two vinyl benzyl ether (VB) groups, named as TCTA-BVP (Figure 2.16a).[196] They lowered the crosslinking temperature by introducing an aliphatic crosslinker

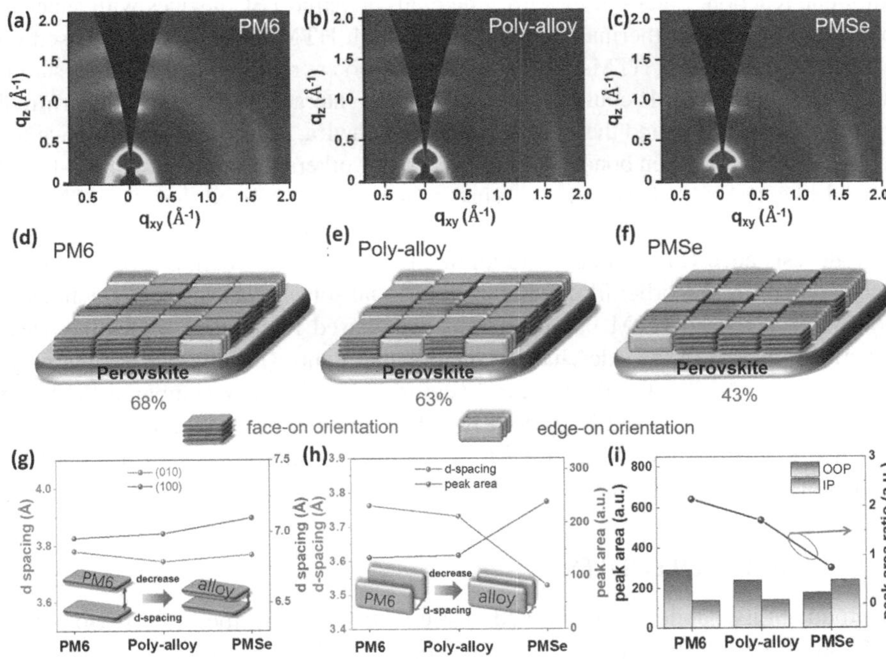

FIGURE 2.15 (a)–(c) GIWAXS scattering patterns of films based on PM6, mixed, and PMSe; (d)–(f) Schematic illustration of the orientation distribution in films based on PM6, mixed, and PMSe; (g) d-spacing of π–π stacking diffraction of out-of-plane; (h) d-spacing of π–π stacking diffraction in the in-plane direction for PM6, poly-alloy, and PMSe; (i) peak area and peak area ratio of π–π stacking diffraction in out-of-plane and in-plane directions for PM6, poly-alloy, and PMSe. Reproduced with permission. Copyright 2022, ACS.[195]

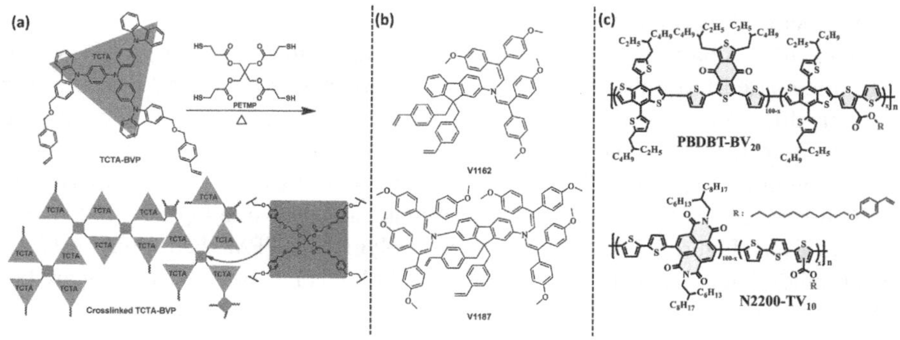

FIGURE 2.16 Cross-linkable polymeric HTMs. (a) Synthesis of crosslinked TCTA-BVP via the thiol-ene thermal polymerization. Reproduced with permission. Copyright 2016, Wiley.[196] (b) V1162 and V1187. Reproduced with permission. Copyright 2021, Wiley.[199] (c) PBDBT-BV_{20} and N2200-TV_{10}. Reproduced with permission. Copyright 2021, ACS.[200]

consisting of thiol groups, pentaerythritol tetrakis(3-mercaptopropionate) (PETMP), which can easily react with the two vinyl groups in TCTA-BVP via the well-known thiol-ene reaction. Later on, in 2018, Zhu et al. prepared PSC devices with superior photo, moisture, and thermal stability, where both HTM and ETM were based on crosslinked polymeric HTMs.[197] Brooks and coworkers revealed that the high stability in crosslinker HTMs could be due to enhanced interactions between the terminal groups of crosslinker and the I of the $[PbI_6]^{4-}$ octahedra, as terminal groups have the ability to form hydrogen bonds.[198] Getautis and coworkers introduced enamine-based cross-linkable HTMs based on two vinyl groups. They realized that the HTM layers are solvent-resistant films, improving p-i-n device stability (Figure 2.16b).[199] The solvent-resistance HTM films could allow the use of a wet-coating process for the deposition of the absorber film, without orthogonal solvents. Furthermore, Choi and coworkers designed HTM based on thiophene-based monomers possessing vinyl functional groups in the side-chain of the polymer donor (PBDBT-BV20) and polymer acceptor (N2200-TV10) structures (Figure 2.16c).[200] They compared the blend films of PBDBT-BV20 and N2200-TV10 before and after thermal crosslinking and revealed that the crosslinking could lock the internal morphology of films, which results in an excellent shelf-life of PSCs over 1,200 h.

2.4 SMALL MOLECULES

Reproducibility, facile synthesis, and cost-effectiveness are the main reasons that SMs are more attentive than polymers.[23] Spiro-OMeTAD is also a small molecule based on a spiro molecular core and exhibits an outstanding performance when used along with dopants. However, the main idea is to develop those kinds of SMs that don't need dopants to gain high hole mobility.[201] Therefore, SMs are designed with numerous strategies, such as enhancing planarity, introducing donor-acceptor units, and launching rigidness and molecular packings. These molecular strategies affect film formation, perovskite defect passivation, charge extraction, energy levels, and stability.[23] In the past six years, over a thousand SMs have been presented to compete with the state-of-the-art HTMs, and the dopant freeness stayed the principal target. Therefore, in this section, we explained the dopant-free SMs which are applied in PSCs to get high PCEs and stabilities. The necessity of developing SMs is generated from the drawbacks of the Spiro-OMeTAD structure, where the spiro 3D core and triphenylamine (TPA) together make a propeller-like structure, enhancing intermolecular distances and reducing hole mobility. TPA has been seen in more than 80% of SMs presented up to date due to its charge recombination reduction, electron coupling reduction, and TPA-perovskite interaction. Furthermore, other essential molecular blocks have been seen in most SM-HTMs, such as thiophene, alkyl chains, cynovinylene (CN), Lewis bases, ketonic groups, etc.[202] All of these molecular blocks have positive effects on the hole mobility, energy levels, defect passivation, film morphology, stability, etc. Recently, we drew a connection between the molecular blocks and their roles in the device when applied as HTMs (Figure 2.17; Table 2.4).

2.4.1 Planarity and Molecular Packing

Planarity has a close relation with hole mobility and film smoothness, generally generated from enhanced intermolecular interactions.[159] Linking different molecular blocks decides whether the molecule is planar or not. For example, Car [1,3] and Car [2,3] are two molecules in which the TPA is attached to the molecular core at

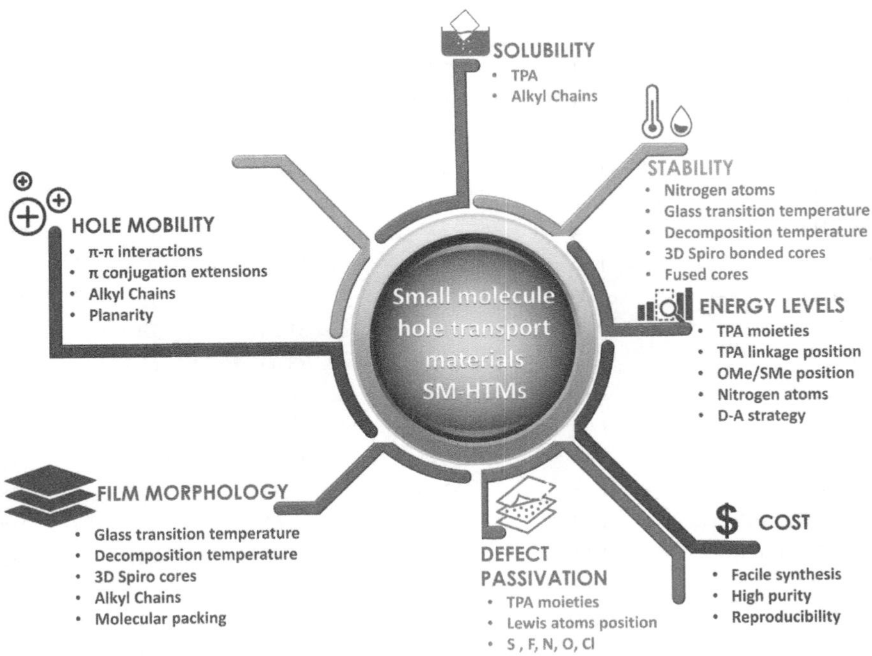

FIGURE 2.17 Chemical designs links to the essential properties of SM-HTMs; Reproduced with permission. Copyright 2023, Wiley.[23]

TABLE 2.4
Photovoltaic Parameters of Dopant-Free SM-HTMs

SM-HTMs	V_{OC} (V)	J_{SC} (mA/cm^2)	FF (%)	PCE (%)	Ref
DTP-C6Th	1.15	22.76	79.9	21.04	203
TTP-OMeTAD	1	18.56	79	14.6	204
TTP-SMeTAD	1.07	20.15	77	16.6	204
TPAC0M	0.91	20.73	74	13.92	205
TPAC2M	0.99	22.58	71	15.77	205
TPAC3M	1	22.79	78	17.54	205
BTF5	0.8	21.31	67	11.42	206
BTF6	1.13	22.23	81	20.34	206
TTE-1	1.03	18.46	71.68	13.68	207
TTE-2	1.11	23.26	77.53	20.04	207
ZT-H1	1	21.21	47.6	10.03	208
ZT-H2[	1.11	24.15	73.42	19.63	208
TQ4	1.12	23.71	79	21.03	209
TQ3	0.95	21.1	53.5	10.8	209
T-CPDT	0.88	21.37	60.9	11.45	210
C-CPDT	1.14	23.27	74.2	19.68	210
PYR-16	1.11	21.56	71	17	211
PYR-27	1.07	20.53	67	14.67	211
TPA-2,7-FLTPA-TPA	1.05	20.82	78	17.1	212
TPA-3,6-FLTPA-TPA	1.05	19.85	67	13.9	212
MeOTTVT	1.11	23.89	80.3	21.3	213
AI110	1.01	19.77	74.42	14.9	213
PTZ-TPA	0.98	21.5	67.9	14.3	214
DMZ	1.02	22.5	81.05	18.61	215
ZnPH13	1.03	20	75	15.5	216
ZnPH14	1.04	21	76	16.6	216
ZnPH15	1.05	21.6	75	17	216
ZnPH22	1.06	22.5	77	18.3	216
CoPc	0.96	11.03	51.28	5.41	217
NiPc	1.13	23.92	78.66	21.23	217
CuPc	1.1	23.96	74.05	19.61	217
ZnPc	1.06	22.91	67.7	16.43	217
H2Pc	1.11	23.85	74.36	19.68	217
OMe-DPA CuPc	1.05	22.44	70.74	16.73	218
OMe-TPA CuPc	1.09	23.41	76.67	19.67	218
SMe-TPA CuPc	1.14	23.71	79.30	21.51	219
TPADPP	1.12	23.27	83	21.62	220
TPADPP-Boc	1.02	22.22	72	16.36	220
TPA-BPFN-TPA	1.04	22.7	78	18.4	221
BT-MTP	0.97	20.87	73.6	14.95	222
DFBT-MTP	1.10	23.01	79.7	20.15	222

TABLE 2.4 *(Continued)*
Photovoltaic Parameters of Dopant-Free SM-HTMs

SM-HTMs	V_{OC} (V)	J_{SC} (mA/cm²)	FF (%)	PCE (%)	Ref
DFBT-PMTP	1.16	22.15	82.28	21.23	223
BDT-TPA	1.11	24.30	75	20.31	224
BDT-TPA-F	1.15	24.45	77.18	21.63	224
BDT-DPA-F	1.18	24.66	79.57	23.12	224
YJ01	1.08	22.53	81.5	19.85	225
YJ02	1.04	22.22	75.5	17.5	225
DTB-FL	1.14	23.8	79.1	21.5	226
ACE-QA-ACE	1.06	22.41	77	18.2	227
TPA-QA-TPA	0.99	22.4	75.1	16.6	227
DPA-QA-DPA	0.95	22.38	73.2	15.5	227
YZ18	1.03	24.1	73	18.1	228
YZ22	1.10	25.1	81	22.4	228
ACE-ANT-ACE	1.03	18.7	67.1	13.1	229
TPA-ANT-TPA	1.03	21.07	79.6	17.5	229
MPA-BTI	1.09	21.58	73.1	17.2	22
MPA-BTTI	1.12	23.23	81.4	21.17	22
CB	1.15	23.6	77.7	21.09	230
tpa-t-FBDT	1.08	22.4	78.1	18.9	231
sDTG-tpa	1.06	19.58	70.34	14.67	232
DFH	1.09	22.7	83	20.6	233
SFDT-TDM	1.13	24.1	79.5	21.7	234
TPE-S	1.26	15.6	78.5	15.4	235
XY1	1.11	22.21	76.18	18.78	236
H0	1.02	22.87	57.3	13.37	237
H2	1.06	22.93	76.9	18.69	237
Trux-OMeTAD	1.02	23.4	79	18.6	238

different positions.[239] Car [2,3] is more planar, which possessed high hole mobility and achieved a PCE of 19.23%, which was higher than Car [1,3] (16.2%). Similarly, in TPA-2,7-FLTPA-TPA and TPA-3,6-FLTPA-TPA, the TPA linkage made TPA-2,7-FLTPA-TPA, hole mobility, and PCE enhanced.[240] Two molecules, PYR16 and PYR27, are designed with different TPA linkage positions, but the high performance was noticed in less planar SM-HTM, i.e., PYR16.[211] On close comparison of PYR16 with PYR17, it can be noticed that the π-conjugated areas of PYR16 are more exposable to the adjacent similar molecules. These areas enhance the π-π interactions between the molecules. The difference between TQ3 and TQ4 is one extra C-C bond, which makes the orthogonal part of TQ4 fixed.[209] This minute rigidness fixes the molecule's planarity and enhances the intermolecular interactions. Similarly, in the comparison of T-CPDT and C-CPDT, C-CPDT arms are fixed while T-CPDT arms are free.[210] Therefore, hole mobility is enhanced, and T-CPDT delivered higher PCE

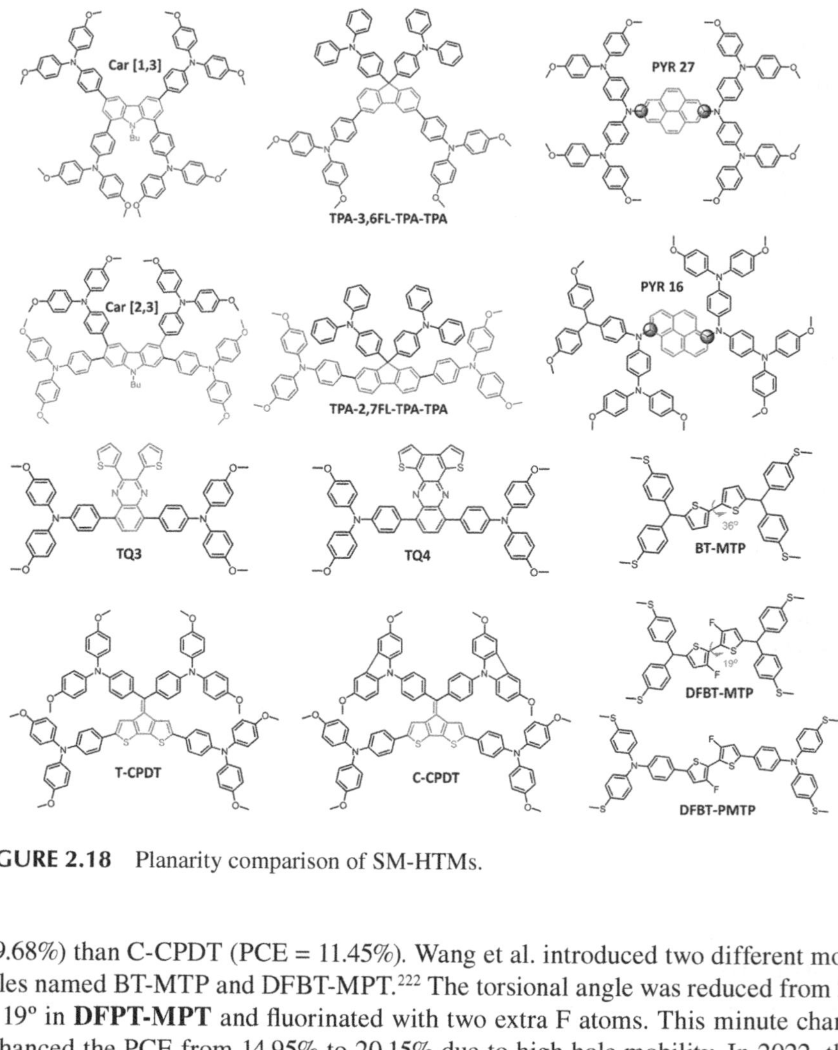

FIGURE 2.18 Planarity comparison of SM-HTMs.

(19.68%) than C-CPDT (PCE = 11.45%). Wang et al. introduced two different molecules named BT-MTP and DFBT-MPT.[222] The torsional angle was reduced from 36° to 19° in **DFPT-MPT** and fluorinated with two extra F atoms. This minute change enhanced the PCE from 14.95% to 20.15% due to high hole mobility. In 2022, they extended the conjugation of **DFPT-MPT** through π-bridge, which further enhanced the planarity of the molecule and yielded a PCE of 21.23%.[223] The high PCE in dopant-free molecules is a general trend in achieving high performances. But, the planarity can become more advantageous if molecular packing is adopted along with the planarity (Figure 2.18).

How can the molecules be designed to make the interactions more effective in establishing inline-oriented molecular packing? The answer to this question is to make the molecular blocks, whether these blocks are core, bridge, or side-arms, openly conjugated to their adjacent molecules. For this purpose, the molecular block linkage throughout the molecule can be optimized. Let us take the example of the molecules with spreadable conjugation systems, i.e., phthalocyanines (Pcs), significantly more flatly conjugated ones. Pcs with Cu metal at the center are mainly applied as HTMs because they come with high thermal stability.[241] Back in 2019, we designed two Cu-Pcs (OMe-DPA-CuPc and OMe-TPA-CuPc) with TPA

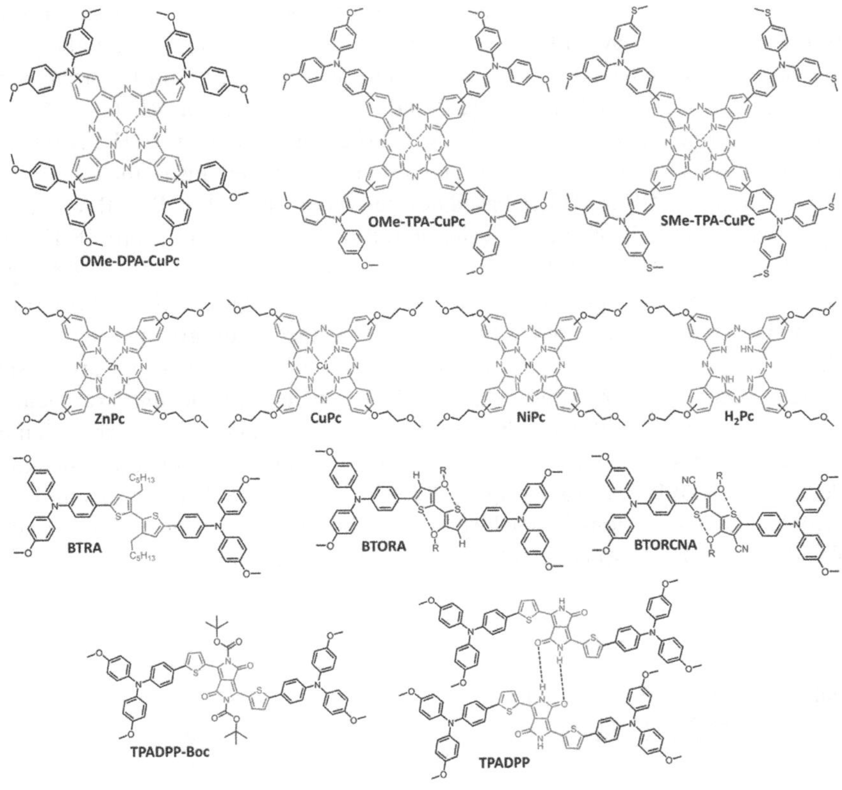

FIGURE 2.19 Design affecting the intermolecular and intramolecular interactions of SM-HTMs.

having OMe moieties.[218] OMe moiety at the TPA arm can help increase the hydrophobicity and act as defect passivation due to the Lewis-base interaction of O with Pb^{2+}. We noticed that OMe-TPA-CuPc outperformed OMe-DPA-CuPc (PCE = 19.67%). It was due to enhancement in the planarity because the molecular blocks make it more planar and exposable, which enhanced the hole mobility too. In 2022, we replaced the OMe moieties with SMe moieties, and through the diffusion of Pc molecules in bulk perovskite through grain boundaries, we hit the PCE of 23%, which is one of the highest among all the CuPc-based SM-HTMs (Figure 2.19).[219] Pcs are also decorated with some other molecular blocks to achieve the dopant-free HTMs. Chen et al. extended the conjugation of Zn-Pcs and measured the performances by substituting different oxygen and sulfur molecules at the Pc arms.[216] They noticed that the one with dioxole moiety exhibited high PCE. In 2021, Yu et al. introduced the alkylated oxygen atoms at the arms of different centered (H_2, Cu, Zn, Ni) Pcs. [217] He noticed the interaction between the oxygen atoms and metal centers, which helped in charge transfer within the molecule. This intramolecular interaction, along with the intermolecular interaction, exhibited high hole mobility, and a PCE of 21.3% was reported while using NiPc-based molecule as a dopant-free HTM.

It is worth noticing that the intramolecular interactions of two different atoms within the molecule or with the atom of the adjacent molecule are quite helpful in achieving molecular packing and, hence, high hole mobility. Similar strategy has been applied in in 4,4′-(3,3′-dihexyl-[2,2′-bithiophene]-5,5′-diyl)bis(N,N-bis(4-methoxyphenyl)aniline (BTRA) based HTMs.[242] When the methylene (CH_2) is replaced with O in BTRA, the intramolecular interaction is enhanced due to the (S···O) interaction. Therefore, PCE in BTRA-based devices was 18.4%, while BTORA (one with oxygen) yielded a PCE of 20.27%. And on introducing NC at the core, PCE is further enhanced to 20.27% (in BTORCNA molecule). As a result, the BTORCNA possessed higher hole mobility than BTRA and BTORA (Figure 2.20a). Furthermore, these molecular changings angled the BTORCNA, further improving the molecular packings, as seen in Figure 2.20b and c. The 2D-GIWAXS patterns further show that the molecular packing is inline, and these symmetrical molecular arrangements make an angle of 35° with the substrate (Figure 2.20d and e). A similar design can also enhance non-covalent interaction between the atoms of two separate but similar molecules. TPADPP-Boc was designed by Li et al., and on thermal annealing, the intermolecular interactions, from NH of one molecule to CO of a closely packed molecule, are created, which resulted in a high PCE of 21.62%.

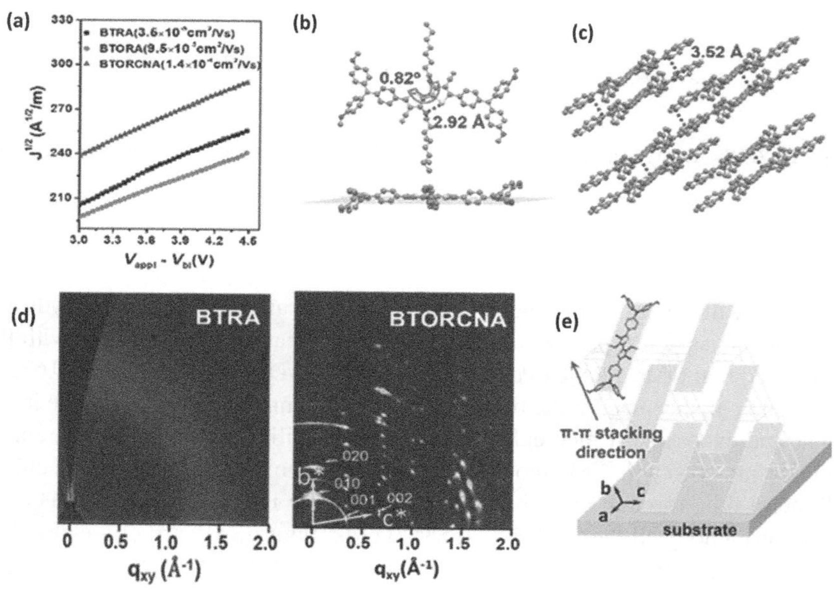

FIGURE 2.20 (a) SCLC model calculated hole mobility for BTRA, BTORA, and BTORCNA; (b) Top view (upper)/side view (down) of molecular configuration; (c) molecular packing of BTORCNA as revealed through single-crystallographic analysis; (d) 2D GIWAXS patterns of neat films of BTRA and BTORCNA. (e) Packing orientation of BTORCNA molecules. Reproduced with permission. Copyright 2022, Wiley.[242]

It is evident that the π–π intermolecular interactions govern the molecular packing. Generally, these interactions result in different molecular packing styles that generate amorphous films, dependent on intramolecular and intermolecular interactions. The optimized molecular design in terms of intramolecular or intermolecular interactions generally decides the sort of packing. However, no matter in which style the molecules are packed, packing must be in-plane. Both orientation styles, face-on or edge-on, enhance the interactions if the molecules are packed in-plane. For example, previously, we optimally designed a Pcs-based molecule (CuPrPc) to enhance the intermolecular interactions and noticed that the in-plane face-on orientation is produced.[243] SFDTM molecule is another example in which the packing was neither face-on nor edge-on, just like BTORCNA, but the molecular packing was in-plane.[234] In-plane molecular packing also enhances the film's smoothness and denseness. As a result, especially in p-i-n PSCs, the perovskite film can improve its growth on these dense and smooth SM-HTMs-based films. Polymeric-based films are smoother and denser than SMs due to their extended structures, while in SM-HTMs, optimized molecular packing can play this role.

2.4.2 D-A Strategy

Like polymers, the SMs can also be designed while focusing on donor and acceptor blocks to enhance hole mobility (Figure 2.21). If we closely watch all the efficient SM-HTMs, we can see that every efficient molecule is based on the donor-acceptor property. Primarily, the TPA block at the end plays the role of donor block, while the central core plays the acceptance part. If this push-pull concept is specially chosen for designing SMs, these SMs are named D-A-based SMs. Furthermore, the planarity through π–conjugation introduction and defect passivation through properly decorated Lewis bases (N, O, S, or F atoms) can further enhance the performance of D-A-based SM-HTMs. For example, the introduction of cynovinylene (CN) at the acceptor part not only enhances the push-pull property but also passivates the perovskite through Lewis base interactions, i.e., the interaction between the N and Pb^{2+}. CN-based D–A-type chromophore (BTPA-TCNE) was synthesized by Li et al. and obtained a PCE of 16.94%.[244] In BTPA-TCNE, TPA was used as a donor. A similar but planar molecule (TPA-BPFN-TPA) was synthesized, where TPA at both sides performed the function of donors, CN helped in acceptance, and a PCE of 18.4% was achieved. YJ01 and YJ02 possess π-bridges between the D-A units to enhance the π-conjugation, i.e., D–π–A–π–A–π–D SMs.[225] Both molecules were decorated with two CN units at different positions of acceptor blocks, and PCEs of 17.5% and 19.85% were achieved in YJ02 and YJ01, respectively. High PCE and stability in YJ01 were attributed to the extended distance between two CN units than YJ02. Similarly, DT-BT was modified to DTB-FL by inserting π-bridges between the D-A blocks and alkyl chains at the backbone of the acceptor unit.[226] As a result, PCE was enhanced from up to 21.5% due to increased intermolecular interactions and organized molecular packings.

The ketonic group (C=O) also performs the A-block role and possesses the Pb-O interaction property. Pham et al. compared three D-A-D molecules, where ketonic decorated quinacridone (QA) was tried with three different acceptors (DPA, TPA,

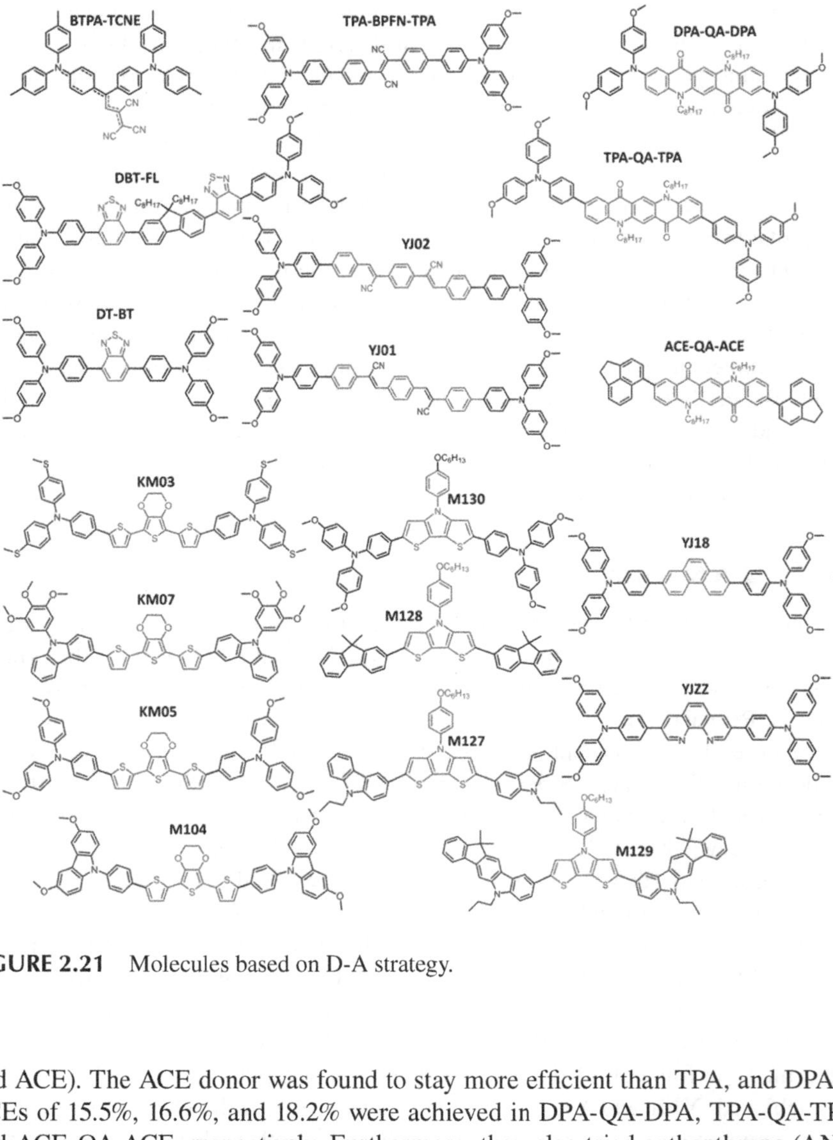

FIGURE 2.21 Molecules based on D-A strategy.

and ACE). The ACE donor was found to stay more efficient than TPA, and DPA.[227] PCEs of 15.5%, 16.6%, and 18.2% were achieved in DPA-QA-DPA, TPA-QA-TPA, and ACE-QA-ACE, respectively. Furthermore, they also tried anthanthrone (ANT) with TPA and ACE and achieved PCEs of 13.1% and 17.5%, respectively.[229] Through these different comparisons, we can analyze that the design of both paired blocks (D and A) affects the push-pull property and performs the electron delocalization differently. All the TPA blocks used by Pham et al. possess OMe as the functionality of TPA, which has a good effect on hydrophobicity and defect passivation. The ketonic group was also included in MPA-BTI and MPA-BTTI D-A based molecule, where MBA-BTTI was made planar with two extra thiophene units than the MPA-BTI.[22] High long-term stability and PCE (21.17 %) were reported in MPA-BTTI, while MPA-BTI-based devices showed an efficiency of 17.5%.

In another work, different types of moieties (SMe and OMe) at donor were judged by synthesizing a series of D-A based SM-HTMs, i.e., KM03, KM07, KM05, and M104, where 3,4-ethylene dioxythiophene (EDOT) was taken as acceptor unit.[245] KM03 was with SMe moieties and provided the lowest PCE of 7%, while KM05 and KM07 delivered PCEs of 14.44% and 13.3%, respectively. M104 incorporation as SM-HTM delivered a champion PCE of 15.37%, which can be attributed to the donor unit's extra bond insertion, which can fix the molecule to a rigid and planar position. M104 was not based on a TPA donor. Furthermore, it is also claimed that OMe based TPA reduces the T_g and possesses low oxidation potential. Wang et al. designed a series of four D-A-based two donors, M127, M128, M129, and M130, with three different acceptors, OMe-TPA, dimethyl fluorine, carbazole, 11, 11-dimethyl-5-propyl-5, 11-dihydroindeno [1, 2-b] carbazole donors respectively.[246] The PCEs are reported in order, M128 (5.75%) < M127 (18.27%) < M130 (18.59 %) < M129 (20.35%). The long-term stability was also recorded higher in M129. YZ18 and YZ22 are introduced in which the YZ22 possesses two extra nitrogen atoms at the acceptor unit to provide defect passivation and Pb-N interaction.[228] Therefore, YZ22 achieved a PCE of 22.4% higher than YZ18 (18.1%). In 2023, Ji et al. synthesized a series of SMs, namely XF1, XF2, and XF3 (Figure 2.22).[247] Side chains of benzo [1,2-b:4,5-b] dithiophene (BDT) unit were introduced to spread the 2D-conjugated system. Out of these three molecules, XF2 and XF3 adopted face-on orientation with impressive hole mobility. Furthermore, these molecules were also decorated with abundant Lewis base groups. In addition, the morphological stabilities were improved due to the rigid backbone of the donor block. As a result, PCEs of 18.84%, 20.59%, and 21.44% were reported in XF1, XF2, and XF3, respectively.

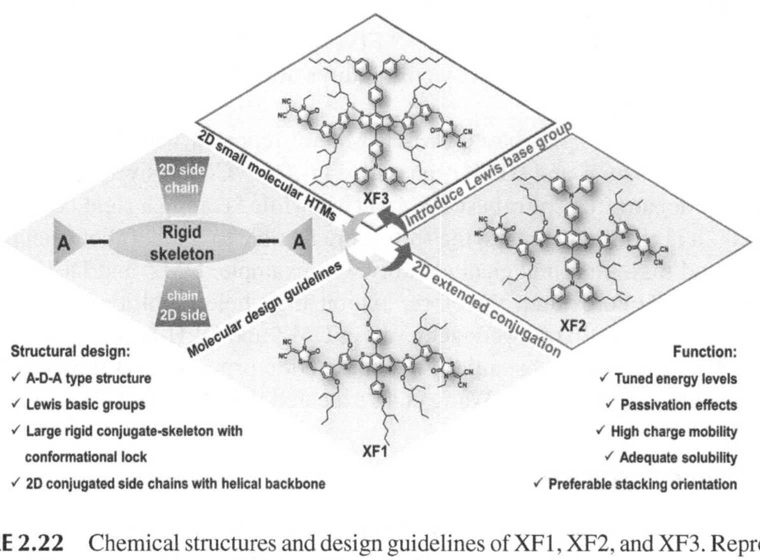

FIGURE 2.22 Chemical structures and design guidelines of XF1, XF2, and XF3. Reproduced with permission. Copyright 2023, Wiley.[247]

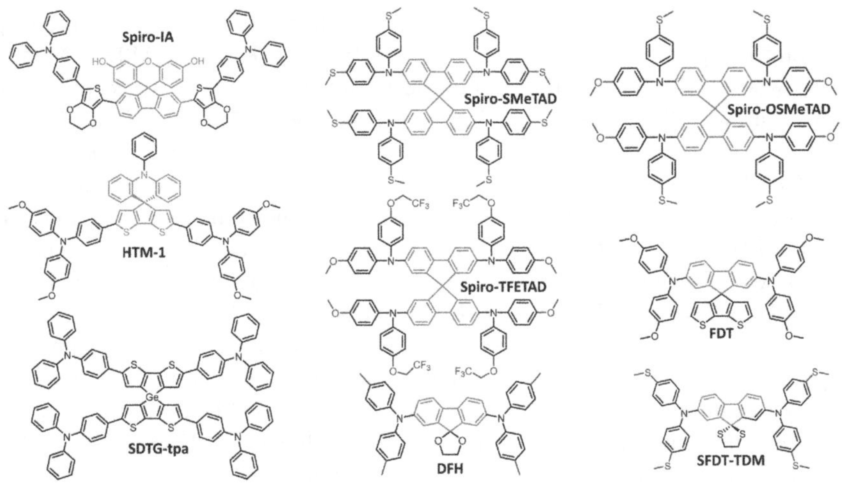

FIGURE 2.23 Molecules based on spiro core.

2.4.3 Spiro-Based SMHTMs

In spiro-based SMHTMs, the rigid 3D core establishes low redox reorganization energies and high transition temperatures (Figure 2.23). Therefore, the morphological stabilities of the layers based on spiro SM-HTMs are high. Nevertheless, the spiro core enhances the intermolecular distances, and hole mobility is reduced, which is one of the reasons that dopants are required in Spiro-OMeTAD for high PCE. Wang et al. introduced SAF-OMe and reported PCEs of 12.39% and 16.73% without and with dopants, respectively.[248] Other spiro-based HTMs were proposed to replace Spiro-OMeTAD and comparatively high PCEs were recorded, such as HTM-1 with 21%,[249] Spiro-4TFETAD with 21.1%,[250] FDT with 20.2%,[251] spiro-OSMeTAD with 18.24%,[252] etc. However, dopants were required to achieve high efficiency in these spiro-cored SM-HTMs.

Some dopant-free spiro-based SM-HTMs are reported with low PCEs, such as Spiro-IA with 15.66% and s-DTG-tpa with 14.67%.[232] Conclusively, enhancing hole mobility in dopant-free spiro-based SM-HTMs while keeping a rigid core is a strenuous task. It is required to make the spiro-core flexible to tradeoff between high hole mobility and high morphological stability. For example, DMZ and DFH molecules have flexible 3D cores and, therefore, exhibit high hole mobilities.[233] As a result, PCEs of 18.61% and 20.6% were achieved in DMZ and DFH-based devices, respectively.[215] By adding the substantial Pb-S interaction property in a flexible 3D cored molecule named SFDT-TDM, Wang et al. executed the PCE of 21.7%.

2.5 CONCLUSIONS

Categorically, there are three kinds of HTMs for PSCs, i.e., inorganic, small molecule, and polymeric HTMs. Among inorganic HTMs, NiO_X is reported regularly with different modifications. NiO_X is adoptable by industrially available methods due to its high chemical stability. Nevertheless, it makes a buried interface after joining with the perovskite layer, and in large-area PSCs, this interface area enhances, and the defects do too. On the other hand, polymeric HTMs are efficient and designable with optimized energy levels, defect passivation capabilities, smooth film formation, etc. On top of that, polymeric HTMs are also adoptable by different industrially available solution printing methods. D-A-type polymers can hit higher hole mobility than non-D-A-type polymers because of the dipole–dipole interactions derived from the D-A structure. Adding the π-bridges between D and A units can add up the value of hole mobility and is advantageous for getting rid of unstable dopants. It's worth noticing that the non-D-A type polymers are generally applied to the n-i-p structure in the form of thin films, which is helpful in obtaining high PCE even at low hole mobility. Nonetheless, the batch-to-batch molecular weight variation and high cost are two main reasons researchers reported almost a thousand small molecules. Small molecules are relatively cheaper, easily designable, and can be synthesized in a facile way. The alteration between the different molecular blocks by changing linkage positions, altering twisting angles, locking the molecule at a planar position, or enhancing the push-pull electronic flow between D and A part of D-A-based SMs can help in high hole mobility. Further properties such as defect passivation, hydrophobicity, solubility, etc. can easily be obtained by making minute changes (Figure 2.24).

Nevertheless, there is a big question mark, i.e., are they industrially adoptable? This is because the film formation in small molecules is different from the polymers. Small molecules follow intermolecular interactions to interact, while polymers, due to their spreadable structure, are easily entangled and form more dense and smooth films than small molecules. It can be the reason that small molecule-based film depositions through industrial methods are scarcely reported.

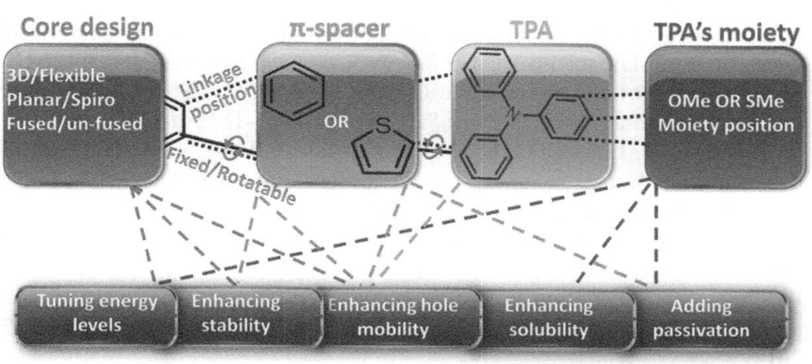

FIGURE 2.24 An empirical some golden law developed by Xu and coworkers; Reproduced with permission. Copyright 2023, Wiley.[23]

REFERENCES

1. H. V. Cooper, *et al.* Greenhouse gas emissions resulting from conversion of peat swamp forest to oil palm plantation. *Nat Commun* **2020**, *11*, 407.
2. T. D. Lee, A. U. Ebong, A review of thin film solar cell technologies and challenges. *Renewable Sustainable Energy Rev* **2017**, *70,* 1286–1297.
3. National Renewable Energy Laboratory Best Research-Cell Efficiency Chart. https://www.nrel.gov/pv/assets/pdfs/best-research-cell-efficiencies.pdf (Accessed: April 4, 2024).
4. B. Gopal Krishna, G. S. Rathore, N. Shukla, S. Tiwari, 18—Perovskite solar cells: A review of architecture, processing methods, and future prospects. In *Woodhead Publishing Series in Composites Science and Engineering* (eds. Khan, I. et al.) 375–412 (Woodhead Publishing, 2021). https://doi.org/10.1016/B978-0-12-819977-0.00018-4.
5. L. Duan, A. Uddin, Defects and stability of perovskite solar cells: a critical analysis. *Mater Chem Front* **2022**, *6*, 400–417.
6. B. Park, S. I. Seok, Intrinsic instability of inorganic–organic hybrid halide perovskite materials. *Adv Mater* **2019**, *31*, 1805337.
7. L. Meng, J. You, Y. Yang, Addressing the stability issue of perovskite solar cells for commercial applications. *Nat Commun* **2018**, *9*, 5265.
8. M. M. Byranvand, *et al.* Recent progress in mixed A-site cation halide perovskite thin-films and nanocrystals for solar cells and light-emitting diodes. *Adv Opt Mater* **2022**, *10*, 2200423.
9. J. A. S. Syed, D. Khan, W. Ahmad, G. S. Soomro, A. A. Dar, Formamidinium post-dripping on methylammonium lead iodide to achieve stable and efficient perovskite solar cells. *Int J Energy Res 46*, 5306–5314 **2022**.
10. S. Sajid, *et al.* Antisolvent-fumigated grain growth of active layer for efficient perovskite solar cells. *Solar Energy 225*, 1001–1008 **2021**.
11. H. Chen, *et al.* Advances to high-performance black-phase $FAPbI_3$ perovskite for efficient and stable photovoltaics. *Small Struct 2*, 2000130 **2021**.
12. Y. Tang, A. Lesage, P. Schall, CsPbI3 nanocrystal films: Towards higher stability and efficiency. *J Mater Chem C 8*, 17139–17156 **2020**.
13. F. Xu, T. Zhang, G. Li, Y. Zhao, Mixed cation hybrid lead halide perovskites with enhanced performance and stability. *J Mater Chem A 5*, 11450–11461 **2017**.
14. M. Mateen, *et al.* Engineering the intermediate adduct phase to control the crystallization of perovskites for efficient and stable perovskite solar cells. *Mater Chem Front 7*, 4080–4091 **2023**.
15. S. Chen, *et al.* Defect passivation of $CsPbBr_3$ with AgBr for high-performance all-inorganic perovskite solar cells. *Adv Energy Sustainability Res 2*, 2000099 **2021**.
16. Y. Xu, *et al.* Flexible, all-inorganic $CsPbBr_3$ perovskite solar cells tailored by heat-resistant muscovite substrates. *ChemSusChem 14*, 1512–1516 **2021**.
17. W. Zhang, *et al.* Enhanced efficiency of air-stable $CsPbBr_3$ perovskite solar cells by defect dual passivation and grain size enlargement with a multifunctional additive. *ACS Appl Mater Interfaces 12*, 36092–36101 **2020**.
18. W. Ke, M. G. Kanatzidis, Prospects for low-toxicity lead-free perovskite solar cells. *Nat Commun 10*, 965 **2019**.
19. R. Singh, P. K. Singh, B. Bhattacharya, H.-W. Rhee, Review of current progress in inorganic hole-transport materials for perovskite solar cells. *Appl Mater Today 14*, 175–200 **2019**.
20. X. Jing, Z. Zhang, T. Chen, J. Luo, Review of inorganic hole transport materials for perovskite solar cells. *Energy Technol 11*, 2201005 **2023**.
21. L. Calió, S. Kazim, M. Grätzel, S. Ahmad, Hole-transport materials for perovskite solar cells. *Angew Chem Int Ed 55*, 14522–14545 **2016**.

22. Y. Wang, *et al.* Dopant-free small-molecule hole-transporting material for inverted perovskite solar cells with efficiency exceeding 21%. *Adv Mater 31*, 1902781 **2019**.
23. D. Khan, *et al.* Nexuses between the chemical design and performance of small molecule dopant-free hole transporting materials in perovskite solar cells. *Small 19*, 2205926 **2023**.
24. G. Qu, *et al.* Reformation of thiophene-functionalized phthalocyanine isomers for defect passivation to achieve stable and efficient perovskite solar cells. *J Energy Chem 67*, 263–275 **2022**.
25. R. Swartwout, M. T. Hoerantner, V. Bulović, Scalable deposition methods for large-area production of perovskite thin films. *Energy Environ Mater 2*, 119–145 **2019**.
26. N.-G., Park, K. Zhu, Scalable fabrication and coating methods for perovskite solar cells and solar modules. *Nat Rev Mater 5*, 333–350 **2020**.
27. P. Liu, G. Tang, F. Yan, Strategies for large-scale fabrication of perovskite films for solar cells. *Solar RRL 6*, 2100683 **2022**.
28. T. H. Schloemer, J. A. Christians, J. M. Luther, A. Sellinger, Doping strategies for small molecule organic hole-transport materials: Impacts on perovskite solar cell performance and stability. *Chem Sci 10*, 1904–1935 **2019**.
29. B. L. Watson, N. Rolston, K. A. Bush, L. Taleghani, R. H. Dauskardt, Synthesis and use of a hyper-connecting cross-linking agent in the hole-transporting layer of perovskite solar cells. *J Mater Chem A 5*, 19267–19279 **2017**.
30. N. J. Jeon, *et al.* A fluorene-terminated hole-transporting material for highly efficient and stable perovskite solar cells. *Nat Energy 3*, 682–689 **2018**.
31. Y. Yue, *et al.* Enhanced stability of perovskite solar cells through corrosion-free pyridine derivatives in hole-transporting materials. *Adv Mater 28*, 10738–10743 **2016**.
32. T. Malinauskas, *et al.* Enhancing thermal stability and lifetime of solid-state dye-sensitized solar cells via molecular engineering of the hole-transporting material Spiro-OMeTAD. *ACS Appl Mater Interfaces 7*, 11107–11116 **2015**.
33. W. Li, *et al.* Montmorillonite as bifunctional buffer layer material for hybrid perovskite solar cells with protection from corrosion and retarding recombination. *J Mater Chem A 2*, 13587–13592 **2014**.
34. S. Wang, *et al.* Unveiling the role of tBP–LiTFSI complexes in perovskite solar cells. *J Am Chem Soc 140*, 16720–16730 **2018**.
35. M. Shahinuzzaman, *et al.* Roles of inorganic oxide based HTMs towards highly efficient and long-term stable PSC—a review. *Nanomaterials 12*, 3003 **2022**.
36. Z. Yu, L. Sun, Inorganic hole-transporting materials for perovskite solar cells. *Small Methods 2*, 1700280 **2018**.
37. F. Ma, *et al.* Nickel oxide for inverted structure perovskite solar cells. *J Energy Chem 52*, 393–411 **2021**.
38. K. G. Brooks, M. K. Nazeeruddin, Laser processing methods for perovskite solar cells and modules. *Adv Energy Mater 11*, 2101149 **2021**.
39. D. Li, *et al.* A review on scaling up perovskite solar cells. *Adv Funct Mater 31*, 2008621 **2021**.
40. S. Seo, S. Jeong, H. Park, H. Shin, N.-G. Park, Atomic layer deposition for efficient and stable perovskite solar cells. *Chem Comm 55*, 2403–2416 **2019**.
41. W. Chen, *et al.* Efficient and stable large-area perovskite solar cells with inorganic charge extraction layers. *Science 350*, 944–948 **2015**.
42. H. Rao, *et al.* A 19.0% efficiency achieved in CuOx-based inverted CH3NH3PbI3-xClx solar cells by an effective Cl doping method. *Nano Energy 27*, 51–57 **2016**.
43. Q. He, *et al.* Room-temperature and solution-processable Cu-doped nickel oxide nanoparticles for efficient hole-transport layers of flexible large-area perovskite solar cells. *ACS Appl Mater Interfaces 9*, 41887–41897 **2017**.

44. N. Wijeyasinghe, *et al.* Copper(I) thiocyanate (CuSCN) hole-transport layers processed from aqueous precursor solutions and their application in thin-film transistors and highly efficient organic and organometal halide perovskite solar cells. *Adv Funct Mater 27*, 1701818 **2017**.
45. H. Zhang, H. Wang, W. Chen, A. K. Y. Jen, $CuGaO_2$: A promising inorganic hole-transporting material for highly efficient and stable perovskite solar cells. *Adv Mater 29*, 1604984 **2017**.
46. Z. K. Yu, *et al.* Solution-processed CuOx as an efficient hole-extraction layer for inverted planar heterojunction perovskite solar cells. *Chin Chem Lett 28*, 13–18 **2017**.
47. S. Yue, *et al.* Efficacious engineering on charge extraction for realizing highly efficient perovskite solar cells. *Energy Environ Sci 10*, 2570–2578 **2017**.
48. W. Chen, *et al.* Cesium doped NiOx as an efficient hole extraction layer for inverted planar perovskite solar cells. *Adv Energy Mater 7*, 1700722 **2017**.
49. C. Hu, *et al.* Tuning the A-site cation composition of FA perovskites for efficient and stable NiO-based p-i-n perovskite solar cells. *J Mater Chem A 5*, 21858–21865 **2017**.
50. G. Li, *et al.* Overcoming the limitations of sputtered nickel oxide for high-efficiency and large-area perovskite solar cells. *Advanced Science 4*, 1700463 **2017**.
51. K. Yao, *et al.* A copper-doped nickel oxide bilayer for enhancing efficiency and stability of hysteresis-free inverted mesoporous perovskite solar cells. *Nano Energy 40*, 155–162 **2017**.
52. N. Arora, *et al.* Perovskite solar cells with CuSCN hole extraction layers yield stabilized efficiencies greater than 20%. *Science 358*, 768–771 **2017**.
53. T. Wang, *et al.* High-performance inverted perovskite solar cells with mesoporous NiO x hole transport layer by electrochemical deposition. *ACS Omega 3*, 18434–18443 **2018**.
54. D. Ouyang, *et al.* Strategic synthesis of ultrasmall $NiCo_2O_4$ NPs as hole transport layer for highly efficient perovskite solar cells. *Adv Energy Mater 8*, 1702722 **2018**.
55. H. Zhang, *et al.* Low-temperature solution-processed $CuCrO_2$ hole-transporting layer for efficient and photostable perovskite solar cells. *Adv Energy Mater 8*, 1702762 **2018**.
56. Z. Liu, *et al.* High-performance planar perovskite solar cells using low temperature, solution–combustion-based nickel oxide hole transporting layer with efficiency exceeding 20%. *Adv Energy Mater 8*, 1703432 **2018**.
57. S. Seo, S. Jeong, C. Bae, N. G. Park, H. Shin, Perovskite solar cells with inorganic electron- and hole-transport layers exhibiting long-term (≈500 h) stability at 85 °C under continuous 1 sun illumination in ambient air. *Adv Mater 30*, 1801010 **2018**.
58. M. Najafi, *et al.* Highly efficient and stable flexible perovskite solar cells with metal oxides nanoparticle charge extraction layers. *Small 14*, 1702775 **2018**.
59. X. Xia, *et al.* Lithium and silver co-doped nickel oxide hole-transporting layer boosting the efficiency and stability of inverted planar perovskite solar cells. *ACS Appl Mater Interfaces 10*, 44501–44510 **2018**.
60. X. Wan, *et al.* Zinc as a new dopant for NiOx-based planar perovskite solar cells with stable efficiency near 20%. *ACS Appl Energy Mater 1*, 3947–3954 **2018**.
61. J. Zhang, *et al.* Solution-processed Sr-doped NiOx as hole transport layer for efficient and stable perovskite solar cells. *Solar Energy 174*, 1133–1141 **2018**.
62. W. Chen, *et al.* Understanding the doping effect on NiO: Toward high-performance inverted perovskite solar cells. *Adv Energy Mater 8*, 1703519 **2018**.
63. X. Yin, *et al.* A Cu_3PS_4 nanoparticle hole selective layer for efficient inverted perovskite solar cells. *J Mater Chem A 7*, 4604–4610 **2019**.
64. X. Li, *et al.* Perovskite solar cells employing an eco-friendly and low-cost inorganic hole transport layer for enhanced photovoltaic performance and operational stability. *J Mater Chem A 7*, 7065–7073 **2019**.
65. C. Liu, *et al.* Hydrophobic Cu_2O quantum dots enabled by surfactant modification as top hole-transport materials for efficient perovskite solar cells. *Adv Sci 6*, 1801169 **2019**.

66. P. S. Chandrasekhar, Y. H. Seo, Y. J. Noh, S. I. Na, Room temperature solution-processed Fe doped NiOx as a novel hole transport layer for high efficient perovskite solar cells. *Appl Surf Sci 481*, 588–596 **2019**.
67. J. H. Lee, Y. W. Noh, I. S. Jin, S. H. Park, J. W. Jung, A solution-processed cobalt-doped nickel oxide for high efficiency inverted type perovskite solar cells. *J Power Sources 412*, 425–432 **2019**.
68. P. C. Chen, S. H. Yang, Potassium-doped nickel oxide as the hole transport layer for efficient and stable inverted perovskite solar cells. *ACS Appl Energy Mater 2*, 6705–6713 **2019**.
69. Q. Fu, S. Xiao, X. Tang, T. Hu, High-performance inverted planar perovskite solar cells based on solution-processed rubidium-doped nickel oxide hole-transporting layer. *Org Electron 69*, 34–41 **2019**.
70. S. Wang, *et al.* Achieving high performance and stable inverted planar perovskite solar cells using lithium and cobalt co-doped nickel oxide as hole transport layers. *J Mater Chem C 7*, 9270–9277 **2019**.
71. S. Xiao, *et al.* An ultra-low concentration of gold nanoparticles embedded in the NiO hole transport layer boosts the performance of p-i-n perovskite solar cells. *Solar RRL 3*, 1800278 **2019**.
72. D. Koushik, *et al.* Plasma-assisted atomic layer deposition of nickel oxide as hole transport layer for hybrid perovskite solar cells. *J Mater Chem C 7*, 12532–12543 **2019**.
73. Q. Han, *et al.* Low-temperature processed inorganic hole transport layer for efficient and stable mixed Pb-Sn low-bandgap perovskite solar cells. *Sci Bull 64*, 1399–1401 **2019**.
74. T. Abzieher, *et al.* Electron-beam-evaporated nickel oxide hole transport layers for perovskite-based photovoltaics. *Adv Energy Mater 9*, 1802995 **2019**.
75. W. Chen, *et al.* Alkali chlorides for the suppression of the interfacial recombination in inverted planar perovskite solar cells. *Adv Energy Mater 9*, 1803872 **2019**.
76. Y. Cheng, *et al.* Impact of surface dipole in NiOx on the crystallization and photovoltaic performance of organometal halide perovskite solar cells. *Nano Energy 61*, 496–504 **2019**.
77. T. Wang, *et al.* Efficient inverted planar perovskite solar cells using ultraviolet/ozone-treated NiOx as the hole transport layer. *Solar RRL 3*, 1900045 **2019**.
78. S. Pang, *et al.* Efficient NiOx hole transporting layer obtained by the oxidation of metal nickel film for perovskite solar cells. *ACS Appl Energy Mater 2*, 4700–4707 **2019**.
79. D. Hou, *et al.* Pb and Li co-doped NiOx for efficient inverted planar perovskite solar cells. *J Colloid Interface Sci 559*, 29–38 **2020**.
80. U. K. Thakur, *et al.* Consistently high Voc values in p-i-n type perovskite solar cells using Ni3+-doped NiO nanomesh as the hole transporting layer. *ACS Appl Mater Interfaces 12*, 11467–11478 **2020**.
81. Y. Wu, *et al.* Unraveling the dual-functional mechanism of light absorption and hole transport of $Cu_2Cd_XZn_{1-X}SnS_4$ for achieving efficient and stable perovskite solar cells. *ACS Appl Mater Interfaces 12*, 17509–17518 **2020**.
82. M. Feng, *et al.* High-efficiency and stable inverted planar perovskite solar cells with pulsed laser deposited Cu-doped NiOx hole-transport layers. *ACS Appl Mater Interfaces 12*, 50684–50691 **2020**.
83. A. M. Elseman, L. Luo, Q. L. Song, Self-doping synthesis of trivalent Ni_2O_3 as a hole transport layer for high fill factor and efficient inverted perovskite solar cells. *Dalton Trans 49*, 14243–14250 **2020**.
84. X. Zheng, *et al.* Interface modification of sputtered NiOx as the hole-transporting layer for efficient inverted planar perovskite solar cells. *J Mater Chem C 8*, 1972–1980 **2020**.
85. C. C. Boyd, *et al.* Overcoming redox reactions at perovskite-nickel oxide interfaces to boost voltages in perovskite solar cells. *Joule 4*, 1759–1775 **2020**.

86. P. Lee, *et al.* Solar energy materials and solar cells high-efficiency perovskite solar cell using cobalt doped nickel oxide hole transport layer fabricated by NIR process. *Sol Energy Mater Sol Cells 208*, 110352 **2020**.
87. H. Eggers, *et al.* Inkjet-printed micrometer-thick perovskite solar cells with large columnar grains. *Adv Energy Mater 10*, 1903184 **2020**.
88. H. Bao, *et al.* Samarium-doped nickel oxide for superior inverted perovskite solar cells: Insight into doping effect for electronic applications. *Adv Funct Mater 31*, 2102452 **2021**.
89. D. Ouyang, J. Zheng, Z. Huang, L. Zhu, W. C. H. Choy, An efficacious multifunction codoping strategy on a room-temperature solution-processed hole transport layer for realizing high-performance perovskite solar cells. *J Mater Chem A 9*, 371–379 **2021**.
90. X. Yuan, *et al.* Enhanced p-type conductivity of NiOx films with divalent Cd ion doping for Efficient inverted perovskite solar cells. *ACS Appl Mater Interfaces 14*, 17434–17443 **2022**.
91. M. Du, *et al.* Surface redox engineering of vacuum-deposited NiOx for top-performance perovskite solar cells and modules. *Joule 6*, 1931–1943 **2022**.
92. N. J. Jeon, *et al.* Solvent engineering for high-performance inorganic–organic hybrid perovskite solar cells. *Nat Mater 13*, 897–903 **2014**.
93. J. A. Christians, R. C. M. Fung, P. V. Kamat, An inorganic hole conductor for organo-lead halide perovskite solar cells: Improved hole conductivity with copper iodide. *J Am Chem Soc 136*, 758–764 **2014**.
94. G. A. Sepalage, *et al.* Copper(I) iodide as hole-conductor in planar perovskite solar cells: Probing the origin of J-V hysteresis. *Adv Funct Mater 25*, 5650–5661 **2015**.
95. W.-Y. Chen, *et al.* Low-cost solution-processed copper iodide as an alternative to PEDOT:PSS hole transport layer for efficient and stable inverted planar heterojunction perovskite solar cells. *J Mater Chem A 3*, 19353–19359 **2015**.
96. W. Sun, *et al.* Room-temperature and solution-processed copper iodide as the hole transport layer for inverted planar perovskite solar cells. *Nanoscale 8*, 15954–15960 **2016**.
97. H. Wang, *et al.* Efficient and stable inverted planar perovskite solar cells employing CuI as hole-transporting layer prepared by solid–gas transformation. *Energy Technol 5*, 1836–1843 **2017**.
98. B. O'Regan, F. Lenzmann, R. Muis, J. Wienke, A solid-state dye-sensitized solar cell fabricated with pressure-treated P25–TiO_2 and CuSCN: Analysis of pore filling and IV characteristics. *Chem Mater 14*, 5023–5029 **2002**.
99. B. C. O'Regan, S. Scully, A. C. Mayer, E. Palomares, J. Durrant, The effect of Al_2O_3 barrier layers in TiO_2/Dye/CuSCN photovoltaic cells explored by recombination and DOS characterization using transient photovoltage measurements. *J Phys Chem B 109*, 4616–4623 **2005**.
100. E. V. A. Premalal, *et al.* Tuning chemistry of CuSCN to enhance the performance of TiO_2/N719/CuSCN all-solid-state dye-sensitized solar cell. *Chem Comm 46*, 3360–3362 **2010**.
101. A. Perumal, *et al.* High-efficiency, solution-processed, multilayer phosphorescent organic light-emitting diodes with a copper thiocyanate hole-injection/hole-transport layer. *Adv Mater 27*, 93–100 **2015**.
102. N. Yaacobi-Gross, *et al.* High-efficiency organic photovoltaic cells based on the solution-processable hole transporting interlayer copper thiocyanate (CuSCN) as a replacement for PEDOT:PSS. *Adv Energy Mater 5*, 1401529 **2015**.
103. S. Ito, *et al.* Carbon-double-bond-free printed solar cells from $TiO_2/CH_3NH_3PbI_3$/CuSCN/Au: Structural control and photoaging effects. *ChemPhysChem 15*, 1194–1200 **2014**.
104. P. Qin, *et al.* Inorganic hole conductor-based lead halide perovskite solar cells with 12.4% conversion efficiency. *Nat Commun 5*, 3834 **2014**.

105. V. E. Madhavan, *et al.* Copper thiocyanate inorganic hole-transporting material for high-efficiency perovskite solar cells. *ACS Energy Lett 1*, 1112–1117 **2016**.
106. M. Jung, *et al.* Thermal stability of CuSCN hole conductor-based perovskite solar cells. *ChemSusChem 9*, 2592–2596 **2016**.
107. K. Zhao, *et al.* Solution-processed inorganic copper(I) thiocyanate (CuSCN) hole transporting layers for efficient p–i–n perovskite solar cells. *J Mater Chem A 3*, 20554–20559 **2015**.
108. A. S. Subbiah, *et al.* Inorganic hole conducting layers for perovskite-based solar cells. *J Phys Chem Lett 5*, 1748–1753 **2014**.
109. H. Rao, *et al.* Solution-processed CuS NPs as an inorganic hole-selective contact material for inverted planar perovskite solar cells. *ACS Appl Mater Interfaces 8*, 7800–7805 **2016**.
110. S. Chatterjee, A. J. Pal, Introducing Cu_2O thin films as a hole-transport layer in efficient planar perovskite solar cell structures. *J Phys Chem C 120*, 1428–1437 **2016**.
111. K. Akimoto, *et al.* Thin film deposition of Cu_2O and application for solar cells. *Solar Energy 80*, 715–722 **2006**.
112. W. L. Yu, *et al.* Ultrathin Cu_2O as an efficient inorganic hole transporting material for perovskite solar cells. *Nanoscale 8*, 6173–6179 **2016**.
113. W. H. Sun, *et al.* High-performance inverted planar heterojunction perovskite solar cells based on a solution-processed CuOx, hole transport layer. *Nanoscale 8*, 10806–10813 **2016**.
114. R. Islam, *et al.* Investigation of the changes in electronic properties of nickel oxide (NiOx) due to UV/Ozone treatment. *ACS Appl Mater Interfaces 9*, 17201–17207 **2017**.
115. J. W. Jung, C.-C. Chueh, A. K.-Y. Jen, A low-temperature, solution-processable, Cu-doped nickel oxide hole-transporting layer via the combustion method for high-performance thin-film perovskite solar cells. *Adv Mater 27*, 7874–7880 **2015**.
116. H. Zhang, *et al.* Pinhole-free and surface-nanostructured NiOx film by room-temperature solution process for high-performance flexible perovskite solar cells with good stability and reproducibility. *ACS Nano 10*, 1503–1511 **2016**.
117. J. H. Heo, *et al.* Efficient inorganic–organic hybrid heterojunction solar cells containing perovskite compound and polymeric hole conductors. *Nat Photonics 7*, 486–491 **2013**.
118. Q. Sun, *et al.* Bright, multicoloured light-emitting diodes based on quantum dots. *Nat Photonics 1*, 717–722 **2007**.
119. J. Wang, *et al.* Reducing surface recombination velocities at the electrical contacts will improve perovskite photovoltaics. *ACS Energy Lett 4*, 222–227 **2019**.
120. G. Qu, *et al.* Spontaneous decoration of ionic compounds at perovskite interfaces to achieve 23.38% efficiency with 85% fill factor in NiO -based perovskite solar cells. *J Energy Chem 85*, 39–48 **2023**.
121. J. Zhang, *et al.* Efficient and ultraviolet durable planar perovskite solar cells via a ferrocenecarboxylic acid modified nickel oxide hole transport layer. *Nanoscale 10*, 5617–5625 **2018**.
122. D. D. Girolamo, *et al.* Stability and dark hysteresis correlate in NiO-based perovskite solar cells. *Adv Energy Mater 9*, 1901642 **2019**.
123. S. Lee, *et al.* Defect-passivating organic/inorganic bicomponent hole-transport layer for high efficiency metal-halide perovskite device. *ACS Appl Mater Interfaces 12*, 40310–40317 **2020**.
124. X. Lian, J. Chen, S. Shan, G. Wu, H. Chen, Polymer modification on the NiOx hole transport layer boosts open-circuit voltage to 1.19 V for perovskite solar cells. *ACS Appl Mater Interfaces 12*, 46340–46347 **2020**.
125. P. Ru, *et al.* High electron affinity enables fast hole extraction for efficient flexible inverted perovskite solar cells. *Adv Energy Mater 10*, 1903487 **2020**.

126. Y. Z. Wu, *et al.* Thermally stable $MAPbI_3$ perovskite solar cells with efficiency of 19.19% and area over 1 cm2 achieved by additive engineering. *Adv Mater 29*, 1701073 **2017**.
127. J. H. Park, *et al.* Efficient $CH_3NH_3PbI_3$ perovskite solar cells employing nanostructured p-type NiO electrode formed by a pulsed laser deposition. *Adv Mater 27*, 4013–4019 **2015**.
128. S. Seo, *et al.* An ultra-thin, un-doped NiO hole transporting layer of highly efficient (16.4%) organic–inorganic hybrid perovskite solar cells. *Nanoscale 8*, 11403–11412 **2016**.
129. P. Docampo, J. M. Ball, M. Darwich, G. E. Eperon, H. J. Snaith, Efficient organometal trihalide perovskite planar-heterojunction solar cells on flexible polymer substrates. *Nat Commun 4*, 2761 **2013**.
130. M. Xiao, *et al.* Efficient perovskite solar cells employing inorganic interlayers. *ChemNanoMat 2*, 182–188 **2016**.
131. X. Sun, X. Yu, Z. Li, Recent advances of dopant-free polymer hole-transporting materials for perovskite solar cells. *ACS Appl Energy Mater 3*, 10282–10302 **2020**.
132. D. H. Sin, *et al.* Decoupling charge transfer and transport at polymeric hole transport layer in perovskite solar cells. *ACS Appl Mater Interfaces 8*, 6546–6553 **2016**.
133. D. Liu, *et al.* Improved performance of inverted planar perovskite solar cells with F4-TCNQ doped PEDOT:PSS hole transport layers. *J Mater Chem A 5*, 5701–5708 **2017**.
134. W. Hu, *et al.* High open-circuit voltage of 1.134 V for inverted planar perovskite solar cells with sodium citrate-doped PEDOT:PSS as a hole transport layer. *ACS Appl Mater Interfaces 11*, 22021–22027 **2019**.
135. Y.-C. Chin, M. Daboczi, C. Henderson, J. Luke, J.-S. Kim, Suppressing PEDOT:PSS doping-induced interfacial recombination loss in perovskite solar cells. *ACS Energy Lett 7*, 560–568 **2022**.
136. P. Li, M. I. Omer Mohamed, C. Xu, X. Wang, X. Tang Electrical property modified hole transport layer (PEDOT:PSS) enhance the efficiency of perovskite solar cells: Hybrid co-solvent post-treatment. *Org Electron 78*, 105582 **2020**.
137. K. M. Reza, *et al.* Tailored PEDOT:PSS hole transport layer for higher performance in perovskite solar cells: Enhancement of electrical and optical properties with improved morphology. *J Energy Chem 44*, 41–50 **2020**.
138. M. Zhang, D. Chi, J. Wang, F. Wu, S. Huang, Improved performance of lead-tin mixed perovskite solar cells with PEDOT:PSS treated by hydroquinone. *Solar Energy 201*, 589–595 **2020**.
139. H. Zhang, *et al.* Tuning the phase separation of PEDOT:PSS affords efficient lead-free perovskite solar cells. *Chem Comm 59*, 2251–2254 **2023**.
140. W. Zhang, *et al.* Systematic improvement in charge carrier mobility of air stable triarylamine copolymers. *J Am Chem Soc 131*, 10814–10815 **2009**.
141. S. Ryu, *et al.* Voltage output of efficient perovskite solar cells with high open-circuit voltage and fill factor. *Energy Environ Sci 7*, 2614–2618 **2014**.
142. S. Tsarev, *et al.* A new polytriarylamine derivative for dopant-free high-efficiency perovskite solar cells. *Sustain Energy Fuels 3*, 2627–2632 **2019**.
143. Y. Kim, *et al.* Sequentially fluorinated PTAA polymers for enhancing V_{OC} of high-performance perovskite solar cells. *Adv Energy Mater 8*, 1801668 **2018**.
144. Y. Kim, *et al.* Methoxy-functionalized triarylamine-based hole-transporting polymers for highly efficient and stable perovskite solar cells. *ACS Energy Lett 5*, 3304–3313 **2020**.
145. J. Luo, *et al.* Toward high-efficiency, hysteresis-less, stable perovskite solar cells: Unusual doping of a hole-transporting material using a fluorine-containing hydrophobic Lewis acid. *Energy Environ Sci 11*, 2035–2045 **2018**.
146. J. Luo, *et al.* An efficient and hydrophobic molecular doping in perovskite solar cells. *Nano Energy 82*, 105751 **2021**.
147. G. Zhang, *et al.* Renewed prospects for organic photovoltaics. *Chem Rev 122*, 14180–14274 **2022**.

148. D. Khan, Z. Ali, D. Asif, M. Kumar Panjwani, I. Khan, Incorporation of carbon nanotubes in photoactive layer of organic solar cells. *Ain Shams Eng J 12*, 897–900 **2021**.
149. X. Guo, *et al.* Highly efficient perovskite solar cells based on a dopant-free conjugated DPP polymer hole transport layer: influence of solvent vapor annealing. *Sustainable Energy Fuels 2*, 2154–2159 **2018**.
150. C.-H. Tsai, *et al.* Efficient and UV-stable perovskite solar cells enabled by side chain-engineered polymeric hole-transporting layers. *J Mater Chem A 6*, 12999–13004 **2018**.
151. X. Sun, X. Yu, Z. Li Recent advances of dopant-free polymer hole-transporting materials for perovskite solar cells. *ACS Appl Energy Mater 3*, 10282–10302 **2020**.
152. F. Cai, *et al.* Molecular engineering of conjugated polymers for efficient hole transport and defect passivation in perovskite solar cells. *Nano Energy 45*, 28–36 **2018**.
153. M. Kyeong, J. Lee, K. Lee, S. Hong, BODIPY-based conjugated polymers for use as dopant-free hole transporting materials for durable perovskite solar cells: Selective tuning of HOMO/LUMO levels. *ACS Appl Mater Interfaces 10*, 23254–23262 **2018**.
154. G.-W. Kim, *et al.* Dopant-free polymeric hole transport materials for highly efficient and stable perovskite solar cells. *Energy Environ Sci 9*, 2326–2333 **2016**.
155. F. Zhang, *et al.* Polymeric, cost-effective, dopant-free hole transport materials for efficient and stable perovskite solar cells. *J Am Chem Soc 141*, 19700–19707 **2019**.
156. H.-C. Liao, *et al.* Dopant-free hole transporting polymers for high efficiency, environmentally stable perovskite solar cells. *Adv Energy Mater 6*, 1600502 **2016**.
157. J. Lee, *et al.* Green-solvent-processable, dopant-free hole-transporting materials for robust and efficient perovskite solar cells. *J Am Chem Soc 139*, 12175–12181 **2017**.
158. Q. Fu, *et al.* Multifunctional two-dimensional conjugated materials for dopant-free perovskite solar cells with efficiency exceeding 22%. *ACS Energy Lett 6*, 1521–1532 **2021**.
159. G.-W. Kim, J. Lee, G. Kang, T. Kim, T. Park, Donor–acceptor type dopant-free, polymeric hole transport material for planar perovskite solar cells (19.8%). *Adv Energy Mater 8*, 1701935 **2018**.
160. J. Lee, G.-W. Kim, M. Kim, S. A. Park, T. Park, Nonaromatic green-solvent-processable, dopant-free, and lead-capturable hole transport polymers in perovskite solar cells with high efficiency. *Adv Energy Mater 10*, 1902662 **2020**.
161. K. Kranthiraja, *et al.* Accomplishment of multifunctional π-conjugated polymers by regulating the degree of side-chain fluorination for efficient dopant-free ambient-stable perovskite solar cells and organic solar cells. *ACS Appl Mater Interfaces 9*, 36053–36060 **2017**.
162. K. Kranthiraja, *et al.* High-performance long-term-stable dopant-free perovskite solar cells and additive-free organic solar cells by employing newly designed multirole π--conjugated polymers. *Adv Mater 29*, 1700183 **2017**.
163. F. Qi, *et al.* A dopant-free polymeric hole-transporting material enabled high fill factor over 81% for highly efficient perovskite solar cells. *Adv Energy Mater 9*, 1902600 **2019**.
164. P. Wang, *et al.* Dopant-free polymeric hole transport materials for efficient CsPbI2Br perovskite cells with a fill factor exceeding 84%. *J Mater Chem C 8*, 8507–8514 **2020**.
165. G. You, *et al.* Donor–acceptor type polymer bearing carbazole side chain for efficient dopant-free perovskite solar cells. *Adv Energy Mater 12*, 2102697 **2022**.
166. G. You, *et al.* Dopant-free, donor–acceptor-type polymeric hole-transporting materials for the perovskite solar cells with power conversion efficiencies over 20%. *Adv Energy Mater 10*, 1903146 **2020**.
167. G.-W. Kim, *et al.* Dopant-free polymeric hole transport materials for highly efficient and stable perovskite solar cells. *Energy Environ Sci 9*, 2326–2333 **2016**.
168. S. Öz, *et al.* Lead(II) propionate additive and a dopant-free polymer hole transport material for $CsPbI_2Br$ perovskite solar cells. *ACS Energy Lett 5*, 1292–1299 **2020**.
169. L. Liang, *et al.* Face-on oriented hydrophobic conjugated polymers as dopant-free hole-transport materials for efficient and stable perovskite solar cells with a fill factor approaching 85%. *J Mater Chem A 10*, 3409–3417 **2022**.
170. Q. Fu, *et al.* Tunable molecular packing of dopant-free hole-transport polymers for perovskite solar cells. *ACS Energy Lett* 2878–2885 **2023**.

171. Z. Xie, *et al.* Thienothiophene-assisted property optimization for dopant-free π-conjugation polymeric hole transport material achieving over 23% efficiency in perovskite solar cells. *Adv Energy Mater 13*, 2202680 **2023**.
172. Z. Li, C. -C. Chueh, A. K.-Y. Jen, Recent advances in molecular design of functional conjugated polymers for high-performance polymer solar cells. *Prog Polym Sci 99*, 101175 **2019**.
173. M.-H. Li, *et al.* High-efficiency $CsPbI_2Br$ perovskite solar cells with dopant-free poly(3-hexylthiophene) hole transporting layers. *Adv Energy Mater 10*, 2000501 **2020**.
174. E. H. Jung, *et al.* Efficient, stable and scalable perovskite solar cells using poly(3-hexylthiophene). *Nature 567*, 511–515 **2019**.
175. X. Li, *et al.* Improving efficiency and reproducibility of perovskite solar cells through aggregation control in polyelectrolytes hole transport layer. *ACS Appl Mater Interfaces 9*, 31357–31361 **2017**.
176. W. Zhang, *et al.* A dopant-free polyelectrolyte hole-transport layer for high efficiency and stable planar perovskite solar cells. *J Mater Chem A 7*, 18898–18905 **2019**.
177. L. Wan, *et al.* Achieving over 21% efficiency in inverted perovskite solar cells by fluorinating a dopant-free hole transporting material. *J Mater Chem A 8*, 6517–6523 **2020**.
178. L. Zhang, *et al.* Intensive exposure of functional rings of a polymeric hole-transporting material enables efficient perovskite solar cells. *Adv Mater 30*, 1804028 **2018**.
179. L. Zhang, *et al.* Side-chain engineering on dopant-free hole-transporting polymers toward highly efficient perovskite solar cells (20.19%). *Adv Funct Mater 29*, 1904856 **2019**.
180. L. Zhang, *et al.* Conjugated polyelectrolyte with potassium cations enables inverted perovskite solar cells with an efficiency over 20%. *J Mater Chem A 8*, 8238–8243 **2020**.
181. L. Zhang, *et al.* Ladder-like conjugated polymers used as hole-transporting materials for high-efficiency perovskite solar cells. *J Mater Chem A 7*, 14473–14477 **2019**.
182. X. Kong, *et al.* Dopant-free F-substituted benzodithiophene copolymer hole-transporting materials for efficient and stable perovskite solar cells. *J Mater Chem A 8*, 1858–1864 **2020**.
183. F. Zhang, *et al.* Polymeric, cost-effective, dopant-free hole transport materials for efficient and stable perovskite solar cells. *J Am Chem Soc 141*, 19700–19707 **2019**.
184. Y. Xie, *et al.* Dopant-free hole-transporting polycarbazoles with tailored backbones for efficient inverted perovskite solar cells. *Macromolecules 52*, 4757–4764 **2019**.
185. Y. Guo, *et al.* A phenanthrocarbazole-based dopant-free hole-transport polymer with noncovalent conformational locking for efficient perovskite solar cells. *Angew Chem Int Ed 61*, e202114341 **2022**.
186. Z. Yao, *et al.* Pyrene-based dopant-free hole-transport polymers with fluorine-induced favorable molecular stacking enable efficient perovskite solar cells. *Angew Chem Int Ed 61*, e202201847 **2022**.
187. Y. Xu, *et al.* Non-conjugated polymer as an efficient dopant-free hole-transporting material for perovskite solar cells. *ChemSusChem 10*, 2578–2584 **2017**.
188. J. Gu, *et al.* Non-conjugated polymer based on polyethylene backbone as dopant-free hole-transporting material for efficient and stable inverted quasi-2D perovskite solar cells. *Solar RRL 4*, 2000184 **2020**.
189. J. Wu, *et al.* Polystyrene with a methoxytriphenylamine-conjugated-thiophene moiety side-chain as a dopant-free hole-transporting material for perovskite solar cells. *J Mater Chem A 6*, 13123–13132 **2018**.
190. J. Wu, *et al.* Side-chain polymers as dopant-free hole-transporting materials for perovskite solar cells—the impact of substituents' positions in carbazole on device performance. *ACS Appl Mater Interfaces 11*, 26928–26937 **2019**.
191. L. Lan, X. Deng, J. Zhang, J. Luo, A. K.-Y. Jen, Synthesis of a side-chain hole transporting polymer through Mitsunobu post-functionalization for efficient inverted perovskite solar cells. *Polym Chem 11*, 2883–2888 **2020**.

192. Q. Xiao, *et al.* A pseudo-two-dimensional conjugated polysquaraine: An efficient p-type polymer semiconductor for organic photovoltaics and perovskite solar cells. *J Mater Chem A 6*, 13644–13651 **2018**.
193. Q. Xiao, *et al.* Dopant-free squaraine-based polymeric hole-transporting materials with comprehensive passivation effects for efficient all-inorganic perovskite solar cells. *Angew Chem Int Ed 58*, 17724–17730 **2019**.
194. R. Chen, *et al.* Robust hole transport material with interface anchors enhances the efficiency and stability of inverted formamidinium–cesium perovskite solar cells with a certified efficiency of 22.3%. *Energy Environ Sci 15*, 2567–2580 **2022**.
195. Q. Fu, *et al.* Ionic dopant-free polymer alloy hole transport materials for high-performance perovskite solar cells. *J Am Chem Soc 144*, 9500–9509 **2022**.
196. Z. Li, Z. Zhu, C.-C. Chueh, J. Luo, A. K.-Y. Jen, Facile thiol-ene thermal crosslinking reaction facilitated hole-transporting layer for highly efficient and stable perovskite solar cells. *Adv Energy Mater 6*, 1601165 **2016**.
197. Z. Zhu, *et al.* Highly efficient and stable perovskite solar cells enabled by all-crosslinked charge-transporting layers. *Joule 2*, 168–183 **2018**.
198. W. J. Nimens, *et al.* Understanding hydrogen bonding interactions in crosslinked methylammonium lead iodide crystals: Towards reducing moisture and light degradation pathways. *Angew Chem Int Ed 58*, 13912–13921 **2019**.
199. D. Vaitukaitytė, *et al.* Enamine-based cross-linkable hole-transporting materials for perovskite solar cells. *Solar RRL 5*, 2000597 **2021**.
200. N. Y. Kwon, *et al.* Improved stability of all-polymer solar cells using crosslinkable donor and acceptor polymers bearing vinyl moieties in the side-chains. *ACS Appl Mater Interfaces 13*, 16754–16765 **2021**.
201. L. Zhang, X. Zhou, C. Liu, X. Wang, B. Xu, A review on solution-processable dopant-free small molecules as hole-transporting materials for efficient perovskite solar cells. *Small Methods 4*, 2000254 **2020**.
202. W. Wang, J. Zhou, W. Tang, Design of dopant-free small molecular hole transport materials for perovskite solar cells: A viewpoint from defect passivation. *J Mater Chem A 10*, 1150–1178 **2022**.
203. X. Yin, *et al.* Dithieno[3,2-b:2′,3′-d]pyrrol-cored hole transport material enabling over 21% efficiency dopant-free perovskite solar cells. *Adv Funct Mater 29*, 1904300 **2019**.
204. H. Chen, *et al.* Molecular Engineered Hole-Extraction Materials to Enable Dopant-Free, Efficient p-i-n Perovskite Solar Cells. *Adv Energy Mater 7*, 1700012 **2017**.
205. S. J. Park, *et al.* Inverted planar perovskite solar cells with dopant free hole transporting material: Lewis base-assisted passivation and reduced charge recombination. *J Mater Chem A 5*, 13220–13227 **2017**.
206. X. Yu, *et al.* Dopant-free dicyanofluoranthene-based hole transporting material with low cost enables efficient flexible perovskite solar cells. *Nano Energy 82*, 105701 **2021**.
207. C. Shen, *et al.* Semi-locked tetrathienylethene as a building block for hole-transporting materials: Toward efficient and stable perovskite solar cells. *Angew Chem Int Ed 58*, 3784–3789 **2019**.
208. Z. Chang, *et al.* Central-core engineering of dopant-free hole transport materials for efficient n-i-p structured perovskite solar cells. *Solar RRL 5*, 2100184 **2021**.
209. H. Guo, *et al.* A coplanar π-extended quinoxaline based hole-transporting material enabling over 21 % efficiency for dopant-free perovskite solar cells. *Angew Chem Int Ed 60*, 2674–2679 **2021**.
210. K.-M. Lee, *et al.* A star-shaped cyclopentadithiophene-based dopant-free hole-transport material for high-performance perovskite solar cells. *Chem Commun 57*, 6444–6447 **2021**.
211. J. Qiu, H. Liu, X. Li, S. Wang, Position effect of arylamine branches on pyrene-based dopant-free hole transport materials for efficient and stable perovskite solar cells. *Chem Eng J 387*, 123965 **2020**.

212. H. D. Pham, *et al.* Boosting inverted perovskite solar cell performance by using 9,9-bis(4-diphenylaminophenyl)fluorene functionalized with triphenylamine as a dopant-free hole transporting material. *J Mater Chem A 7*, 12507–12517 **2019**.
213. S. B. Akula, *et al.* Novel thieno-imidazole salt-based hole transport material for dopant-free, efficient inverted perovskite solar cell applications. *J Power Sources 483*, 229177 **2021**.
214. X. Liu, *et al.* Facile synthesis of a dopant-free hole transporting material with a phenothiazine core for planar perovskite solar cells. *RSC Adv 7*, 53604–53610 **2017**.
215. X. Lai, *et al.* High-performance inverted planar perovskite solar cells enhanced by thickness tuning of new dopant-free hole transporting layer. *Small 15*, 1904715 **2019**.
216. W. Chen, *et al.* Molecular tailor-making of zinc phthalocyanines as dopant-free hole-transporting materials for efficient and stable perovskite solar cells. *J Power Sources 505*, 230095 **2021**.
217. Z. Yu, *et al.* Intramolecular electric field construction in metal phthalocyanine as dopant-free hole transporting material for stable perovskite solar cells with >21% efficiency. *Angew Chem Int Ed 60*, 6294–6299 **2021**.
218. Y. Feng, *et al.* High-performance and stable perovskite solar cells based on dopant-free arylamine-substituted copper(II) phthalocyanine hole-transporting materials. *Adv Energy Mater 9*, 1901019 **2019**.
219. G. Qu, *et al.* Dopant-free phthalocyanine hole conductor with thermal-induced holistic passivation for stable perovskite solar cells with 23% efficiency. *Adv Funct Mater 32*, 2206585 **2022**.
220. L. Rui, *et al.* Hydrogen-bonded dopant-free hole transport material enables efficient and stable inverted perovskite solar cells. *CCS Chem 4*, 3084–3094 **2021**.
221. H. D. Pham, *et al.* All-rounder low-cost dopant-free D-A-D hole-transporting materials for efficient indoor and outdoor performance of perovskite solar cells. *Adv Electron Mater 6*, 1900884 **2020**.
222. Y. K. Wang, *et al.* Fluorinating dopant-free small-molecule hole-transport material to enhance the photovoltaic property. *ACS Appl Mater Interfaces 13*, 7705–7713 **2021**.
223. Y. Wang, *et al.* Annealing- and doping-free hole transport material for p-i-n perovskite solar cells with efficiency achieving over 21%. *Chem Eng J 433*, 133265 **2022**.
224. Q. Cheng, *et al.* Molecular self-assembly regulated dopant-free hole transport materials for efficient and stable n-i-p perovskite solar cells and scalable modules. *Angew Chem Int Ed 61*, e202210613 **2022**.
225. J. Yuan, Y. Chen, X. Liu, S. Xue, Dopant-free hole-transporting materials for CH3NH3PbI3 inverted perovskite solar cells with an approximate efficiency of 20%. *ACS Appl Energy Mater 4*, 5756–5766 **2021**.
226. T. Niu, *et al.* D-A-π-A-D-type dopant-free hole transport material for low-cost, efficient, and stable perovskite solar cells. *Joule 5*, 249–269 **2021**.
227. H. D. Pham, *et al.* Dopant-free novel hole-transporting materials based on quinacridone dye for high-performance and humidity-stable mesoporous perovskite solar cells. *J Mater Chem A 7*, 5315–5323 **2019**.
228. B. X. Zhao, *et al.* A hole-transport material that also passivates perovskite surface defects for solar cells with improved efficiency and stability. *Energy Environ Sci 13*, 4334–4343 **2020**.
229. H. D. Pham, *et al.* Molecular engineering using an anthanthrone dye for low-cost hole transport materials: A strategy for dopant-free, high-efficiency, and stable perovskite solar cells. *Adv Energy Mater 8*, 1703007 **2018**.
230. K.-M. Lee, *et al.* High-performance perovskite solar cells based on dopant-free hole-transporting material fabricated by a thermal-assisted blade-coating method with efficiency exceeding 21%. *Chem Eng J 427*, 131609 **2022**.
231. Y. Tian, *et al.* Facile synthesized fluorine substituted benzothiadiazole based dopant-free hole transport material for high efficiency perovskite solar cell. *Dyes Pigments 184*, 108786 **2021**.

232. J. Ohshita, K. Kondo, Y. Adachi, M. Song, S.-H. Jin, Synthesis of spirodithienogermole with triphenylamine units as a dopant-free hole-transporting material for perovskite solar cells. *J Mater Chem C 9*, 2001–2007 **2021**.
233. Y. Cao, *et al.* Dopant-free molecular hole transport material that mediates a 20% power conversion efficiency in a perovskite solar cell. *Energy Environ Sci 12*, 3502–3507 **2019**.
234. J. Wang, *et al.* Dopant-free hole-transporting material with enhanced intermolecular interaction for efficient and stable n-i-p perovskite solar cells. *Adv Energy Mater 11*, 2100967 **2021**.
235. K. Jiang, *et al.* Dopant-free organic hole-transporting material for efficient and stable inverted all-inorganic and hybrid perovskite solar cells. *Adv Mater 32*, 1908011 **2020**.
236. X. Yang, *et al.* A dopant-free twisted organic small-molecule hole transport material for inverted planar perovskite solar cells with enhanced efficiency and operational stability. *Nano Energy 64*, 103946 **2019**.
237. Z. Li, *et al.* Fluorinated triphenylamine-based dopant-free hole-transporting material for high-performance inverted perovskite solar cells. *Chem Eng J 402*, 125923 **2020**.
238. C. Huang, *et al.* Dopant-free hole-transporting material with a C3h symmetrical truxene core for highly efficient perovskite solar cells. *J Am Chem Soc 138*, 2528–2531 **2016**.
239. A. A. Sutanto, *et al.* Isomeric carbazole-based hole-transporting materials: Role of the linkage position on the photovoltaic performance of perovskite solar cells. *Chem Mater 33*, 3286–3296 **2021**.
240. H. D. Pham, *et al.* Boosting inverted perovskite solar cell performance by using 9,9-bis(4-diphenylaminophenyl)fluorene functionalized with triphenylamine as a dopant-free hole transporting material. *J Mater Chem A 7*, 12507–12517 **2019**.
241. E. Rezaee, *et al.* Phthalocyanine in perovskite solar cells: A review. *Mater Chem Front 7*, 1704–1736 **2023**.
242. K. Yang, *et al.* Intramolecular noncovalent interaction-enabled dopant-free hole-transporting materials for high-performance inverted perovskite solar cells. *Angew Chem Int Ed 61*, e202113749 **2022**.
243. X. Liu, *et al.* Tetra-propyl-substituted copper (II) phthalocyanine as dopant-free hole transporting material for planar perovskite solar cells. *Solar RRL 2*, 1800050 **2018**.
244. Z. Li, *et al.* Rational design of dipolar chromophore as an efficient dopant-free hole-transporting material for perovskite solar cells. *J Am Chem Soc 138*, 11833–11839 **2016**.
245. K.-M. Lu, K.-M. Lee, C.-H. Lai, C.-C. Ting, C.-Y. Liu, One-pot synthesis of D–π–D–π–D type hole-transporting materials for perovskite solar cells by sequential C–H (hetero)arylations. *Chem Commun 54*, 11495–11498 **2018**.
246. J. Wang, *et al.* Indeno[1,2-b]carbazole as methoxy-free donor group: Constructing efficient and stable hole-transporting materials for perovskite solar cells. *Angew Chem Int Ed 58*, 15721–15725 **2019**.
247. X. Ji, *et al.* Dopant-free two-dimensional hole transport small molecules enable efficient perovskite solar cells. *Adv Energy Mater 13*, 2203756 **2023**.
248. Y.-K. Wang, *et al.* Dopant-free spiro-triphenylamine/fluorene as hole-transporting material for perovskite solar cells with enhanced efficiency and stability. *Adv Funct Mater 26*, 1375–1381 **2016**.
249. S. Akin, *et al.* Cyclopentadithiophene-based hole-transporting material for highly stable perovskite solar cells with stabilized efficiencies approaching 21%. *ACS Appl Energy Mater 3*, 7456–7463 **2020**.
250. Z. Zhang, *et al.* A trifluoroethoxyl functionalized spiro-based hole-transporting material for highly efficient and stable perovskite solar cells. *Solar RRL 6*, 2100944 **2022**.
251. M. Saliba, *et al.* A molecularly engineered hole-transporting material for efficient perovskite solar cells. *Nat Energy 1*, 15017 **2016**.
252. X. Zhang, *et al.* Heteroatom engineering on spiro-type hole transporting materials for perovskite solar cells. *J Energy Chem 67*, 19–26 **2022**.

3 Electrode Materials for Perovskite Solar Cells

Guozheng Du, Li Yang, and Jinbao Zhang

3.1 INTRODUCTION

Photovoltaic technologies represent one of the most clean and environmental-friendly methods to mitigate the environmental problems such as climate change and resources shortage. However, the traditional solar cells (such as silicon, CdTe, CIGS) have many disadvantages including high material costs, complex fabrication process as well as application limitation.[1] Based on these challenges, solar cells with new absorbing materials like hybrid perovskites have emerged as one of the most promising candidates for next generation photovoltaic technologies. The perovskite materials have a similar structure to traditional inorganic mineral $CaTiO_3$, which is well-known as the low-cost fabrication,[2] high absorption coefficient,[3] superior defect tolerance,[4] and long charge carrier lifetime.[5] After tremendous optimization of interfacial materials,[6], [7] perovskite compositions,[8] and devices' architecture,[9], [10] the certified power conversion efficiency (PCE) of more than 26% has been achieved in single-junction perovskite solar cells.[11]

A typical PSC is composed of different function layers including perovskite absorbing layer, electron transport layer (ETL), hole transport layer (HTL), and two contact electrode layers. The contact electrodes usually divide into transparent conducting glasses (such as FTO, ITO) and counter electrodes (such as Au and Ag). According to the different position of HTL and ETL, PSCs are typically fabricated in the n-i-p (substrate/cathode/ETL/Perovskite/HTL/anode) or p-i-n (substrate/anode/HTL/Perovskite/ETL/cathode) heterojunction configuration.[12] In a solid-state PSC, perovskite layer absorbs light within a certain wavelength range, generating electron-hole pairs (exciton), and then the electrons and holes are separated and collected by selective charge transport layers (ETL and HTL) and finally transfer to the corresponding electrodes. The built-in potential and work function at different interfaces are the driving force for charge transport and collection.[13] Therefore, the properties of formed contacts about the electrodes play an important role in device performance. In this part, we provide a comprehensive overview about counter electrodes in perovskite solar cells.

3.2 FUNDAMENTAL ROLES OF COUNTER ELECTRODES

According to the traditional PSC sandwiched structures, the counter electrode is usually located above the light-absorbing layer, regardless of whether it is HTL (such as

DOI: 10.1201/9781003400493-3

Spiro-OMeTAD, CuPc, PTAA, P3HT) or ETL (such as PCBM) between the perovskite and the counter electrode layer. Therefore, in both n-i-p and p-i-n structured devices, the counter electrodes are crucial to the PSC stability. The counter electrodes not only effectively segregate the charge transport layer as well as the light-absorbing layer from the external environment but also could prevent the leakage of harmful ingredients during long-term operations.[14] High atoms' or molecules' arranged density in the electrode is essential for maintaining devices' stability.[15] Moreover, the counter materials should be chemically stable and should not react with charge transport layers and perovskite layers.

During PSC operation, the main role of counter electrode is to extract and transport, transfer and collect the light-induced charge carriers and ultimately form a closed loop. Therefore, the film conductivity of the counter electrode is a crucial parameter for PSCs to achieve high-performance. Besides, the counter electrodes also should have proper energy-level alignment between the charge transport layers and the perovskite layers. For example, in n-i-p structured devices, the work function of counter electrode should be higher than that of HTLs and perovskite. In addition to the previous requirements, the price of counter electrode materials should be as low as possible to reduce the preparation cost of devices.[16] The metals, e.g. Au and Ag, are the most frequently used counter electrodes. However, the high cost and relatively low mechanical properties limit their application in large-area modules and flexible PSCs. In order to solve the problem, some other electrodes have been applied to PSCs such as carbon-based electrodes, polymer-based electrodes and transparent electrodes. In the following text, we will discuss the details about different counter electrodes.

3.3 METAL ELECTRODES

Metallic materials, such as Al, Ag, Au, et al. with proper energy-level alignment as well as conductivity, have been reported to be used as the counter electrode in the current high-efficiency PSCs. The metal thin films are usually fabricated as the top electrode by thermal evaporation, which are also called back-contact electrodes in some cases. The basic physical properties of commonly used metal electrodes in PSCs are shown in Table 3.1.

In conclusion, the choice of counter electrodes is very important in perovskite solar cells. The following contents are exhibiting to provide the research summaries of state-of-the-art perovskite devices with metal electrodes.

3.3.1 GOLD (AU)

Gold as a luxury metal has been one of the most common electrode materials for high-efficiency PSCs since the first application in 2012 because the energy level of Au matches common HTLs such as Spiro-OMeTAD, CuPc, and NiO_x.[17] At present, the PSCs with an Au electrode still maintain the highest efficiency record.[11] Contrary to the conventional electrode thickness (around 100 nm), Au layers must be fabricated extremely thin to ensure the optical transparency, while the film conductivity

TABLE 3.1
The Basic Physical Properties of Commonly Used Metal Electrodes in PCSs (Date of price collection: September 16, 2023)

Metal	Work function (eV)	Resistance (μΩ cm)	General thickness (nm)	Price ($/oz)
Au	−5.1	2.35	80 ~ 100	1,952.00
Ag	−4.2	1.59	80 ~ 100	23.23
Al	−4.2	2.83	80 ~ 100	0.06
Cu	−4.8	1.72	80 ~ 100	0.27
Ni	−4.9	6.84	~300	1.64

would be threatened. Roldan-Carmona and coworkers fabricated the inverted perovskite devices with the structure of ITO/PEDOT:PSS/polyTPD/perovskite/PCBM/Au, the thicknesses of perovskite and Au electrode were around 100 nm and 6 nm, respectively.[18] Finally, the semitransparent devices exhibited a PCE of more than 6% with 30% light transmission. Fan et al.[19] assembled four-terminal tandem solar cells by combining n-i-p structured PSCs with the structure of FTO/TiO_2/perovskite/Spiro-OMeTAD/Au and silicon heterojunction bottom cells, ultimately achieving an optimal PCE of 14.8%. Vijayaraghavan et al.[20] proposed a facile sputtered technique to deposit Au nanoparticles to decorate carbon electrode. The sputtered Au layer effectively enhanced the hydrophobicity and decreased the sheet resistivity of the electrode. Finally, the devices with the optimal electrode thickness (40 nm) achieved a maximum PCE of 16.87% with a high V_{oc} of 1.13 V.

Although Au material exhibits excellent chemical stability in various application scenarios, it has been evidenced that Au atoms could migrate into the perovskite layer through the charge transporting layers. Some factors such as moisture in atmosphere, light illumination and heat would accelerate the metal diffusion, resulting in the severe degradation of photovoltaic performance of the devices. It is recently found that high temperature (more than 70 °C) would cause an irreversible metal diffusion towards the perovskite layer.[21] According to ToF-SIMS results, high element intensity was observed at the perovskite/HTL interface, directly proving the serious migration occurred in perovskite devices. Shlenskaya et al.[22] proposed that the degradation mechanism of perovskite devices mainly originated from the reaction between the gold electrode and iodide-containing products. The intensive UV irradiation would cause the decomposition of the perovskite layer, resulting in the formation of highly active polyiodide melts MAI-nI_2 because of the volatile I_2 and MAI, which then react with Au at room temperature causing the production of $[AuI_2]$- and $[AuI_4]^-$. Cacovich et al.[23] also investigated the combined effects of both light and heat on the metal diffusion; various testing results signified the formation of clusters caused by the reaction between Au and perovskite.

In order to alleviate the metal electrode diffusion in PSCs, some intermediate layers were introduced to ensure the long-term stability of devices. Zai et al.[7] introduced a molecular interlayer F4TCNQ in both perovskite/HTL and HTL/Au interfaces to construct a sandwiched structure. The Au atoms diffusion was effectively alleviated according to the TOF-SIMS results because of the strong attractive interaction between the molecule and the metal surface. Shi et al.[24] deposited a Ti metal layer between perovskite and Au electrode to fabricate HTL-free perovskite devices. The Ti interlayer achieved better contact and less sheet resistance at the interface and prevented the direct contact between perovskite and Au electrode. Hence, the devices with Ti-Au electrodes achieved the highest PCE of around 13% and maintained more than 70% of initial PCE after storing for 300 h in an ambient condition. The Au layer also has been introduced as an interlayer under the Ag or Al electrode to inhibit chemical reaction between perovskite and halogens.[25] The chemically stable layer effectively hindered the diffusion of perovskite halide ions and blocked the inward diffusion of electrode metals, leading to the improvement of device stability.

However, the single-time utilization of Au electrode materials severely wastes resources and would cause environment pollution. Kim et al.[26] applied a two-step selective dissolution technique to reuse the Au electrode and HTL. PSCs were first immersed into the chlorobenzene solution. Spiro-OMeTAD molecules were dissolved in the solution. Due to the absence of attachment, the Au layer was then fell off the devices and reused. Yang et al.[27] developed a dry-transfer process to deposit a nonporous Au layer. Compared with traditional thermal evaporation technique, this dry-transfer method made the Au electrode easy to peel off and recycle. More importantly, the Au electrodes can be recycled more than 12 times, further reducing the consumption of Au and the expenditure on perovskite solar cells.

3.3.2 Silver (Ag)

Sliver (Ag) is also an alternative metal commonly used as counter electrodes in PSCs due to its suitable work function and high conductivity; both of them are the prerequisites for efficiently extracting, collecting, and transferring charge carriers in the devices. As an electrode, Ag is a cheaper material compared with Au and its conductivity is also higher than that of Au. However, most of the highest performing devices still used Au as counter electrodes since Au has a lower work function. With the increasing work function, the electric field between charge transport layer and metal electrode decreases, leading to lower charge collection yield and V_{oc}.[28] However, there is no denying that Ag is still a potential counter electrode material for achieving high power conversion efficiency of PSCs. In 2012, Snaith et al.[29] first used Ag as the counter electrode in PSCs and achieved a PCE of 10.9% with the device structure of FTO/TiO_2/Al_2O_3/MAPb(I, Cl)$_3$/Spiro-OMeTAD/Ag. Up to now, the PCEs of Ag-based PSCs have improved dramatically and exceeded 24%.[30]

Besides PCEs, the stability is also a key parameter for PSCs. The stability issue is a vital challenge for PSCs to further commercialization due to the vulnerability of perovskite to oxygen, moisture, and light illumination. Especially for the PSCs with silver electrodes, there is concern regarding the long-term stability as the corrosion of sliver electrode. Cappel et al.[31] revealed the interaction between Ag electrode and perovskite at the interface through various characteristic techniques. Severe Ag diffusion occurred in all devices with different compositional perovskites after being stored for 7 d, and the XPS results indicated that the Fermi Level of perovskites moved to the valance band, leading to the decrease of devices' efficiency. Lu et al.[32] explored the long-term stability of PSCs with an ultra-thin sliver electrode and they pointed that the corrosion resulted from the oxidation of I^- to I_2 at the interface between perovskite and the transporting layer. Wang et al.[33] evaluated the thermal degradation process induced by sliver electrode and found that I- and MA^+ diffused through the PCBM ETL and accumulated at the Ag inner surface under 85°C. In addition, silver diffusion through the charge transport material into the perovskite layer can also induce the damage of devices. Du and coworkers[33] used first-principle calculation to investigate the formation energetics and diffusion dynamics of different metal impurities in $MAPbI_3$, and the results uncovered that Ag_i has low formation energy and the diffusion barrier of Ag_i^+ is lower than that of Cu_i^+, Au_i^+ and so on; all of the results indicated Ag can efficiently diffuse into $MAPbI_3$.

To expedite the practical application of PSCs, the stability of the Ag-based devices should be paid more attentions and more efforts are needed to tackle the corresponding issues. To prevent the corrosion of sliver electrode, some interfacial layers were designedly inserted between the charge transport layer and the Ag electrode. Yang et al.[34] eliminated the electron contact barrier through inserting an interlayer of Bphen:Cs_2CO_3:BCP between $PC_{61}BM$ and the Ag electrode, resulting in PCE improvement from 17.57% to 18.29%. H_xMoO_3 was introduced into the HTL/Ag interface. Given the better energy-level bending orientation, the charge transport was highly promoted, finally achieving a PCE of 23.16%. McLachlan and coworkers inserted a thin Cu layer below the Ag electrode. The Cu layer effectively hindered the interfacial interactions between perovskite and Ag, hence improving the operation stability of the devices.[35] For semitransparent PSCs, the counter electrode based on an ultrathin metal film tends to be degraded by the corrosive decomposition products of the perovskite. To protect both perovskite and the ultrathin silver electrode, T. Riedl[36] added a SnO_x film between perovskite and Ag. The corresponding devices revealed outstanding stability over more than 4,500 h regardless if they were exposed to ambient atmosphere or to elevated temperatures.

3.3.3 Aluminum (Al)

Due to the significant reduction in material cost, aluminum exhibits significant advantages as an electrode material for low-cost perovskite PVs. Al has been typically applied in inverted perovskite solar cells because of its low work-function. Jeng et al.[37] fabricated the inverted devices with the structure of ITO/PEDOT:PSS/Perovskite/PCBM/BCP/Al, which showed a maximum PCE of 3.9%. And then they substituted PEDOT:PSS with NiO_x HTL to further increase the PCE to 7.8% with a higher V_{oc} and FF.[38] Considering the influence of precursor solution on perovskite crystalline quality, Wang et al.[39] fabricated Al-based PSCs with a PCE of 12.2%, in which a low-temperature solution deposition technique was introduced to deposit the $MAPbI_3$ layer. Seo and coworkers[40] inserted a LiF interlayer between the PCBM layer and Al, which improved the charge carrier transportation and reduced the non-radiative recombination at the interface, resulting in dramatic improvement in FF and J_{sc}, and finally the devices realized the maximum efficiency of 14.1%. You et al.[41] prepared inverted PSCs with configuration of ITO/NiO_x/Perovskite/ZnO/Al.

By introducing solution-processed p-typed NiO_x and n-typed ZnO as the HTL and ETL, respectively, the devices exhibited an PCE up to 16.1%. After being stored in air conditions for 60 d, the Al-based devices still maintained more than 90% of the initial PCE.

Kim et al.[10] fabricated fully vacuum-processed devices by employing MoO_3/NPB as HTMs and Al as counter electrode material. Due to the excellent hole extraction at the interface, they achieved a high V_{oc} of 1.12V, resulting in the maximum PCE of 13.7%. Nouri et al.[42] introduced graphene oxide as ETL in inverted Al-based PSCs and a maximum PCE of 10.2% was reported. Todorov et al.[43] introduced semi-transparent Al with a thickness of 20–25 nm as the counter electrode in CZTS-perovskite tandem solar cells; the reported PCE went up to 4.6%, and the PSCs with the structure of ITO/PEDOT:PSS/$MAPbI_3$/PCBM/Al showed the maximum PCE more than 12%. Sanehira et al.[44] studied the effect of different metal electrode configuration. They found that the devices with the structure of the configuration of FTO/TiO_2/Perovskite/MoO_3/Al exhibited more stable performance compared to traditional Au-based or Al-based devices.

Al could also cause some stability problems. Zhao et al.[45] found that a fast decrease of devices' PCE and the degradation of perovskite layers occurred in Al-based devices. By several in-situ techniques, the immanent cause was proved to be the redox reaction between perovskite and Al, and this redox reaction would have happened even without external environmental factors such as heat, moisture, and light. In addition, the in-situ synchrotron radiation photoemission spectroscopy and XPS results also indicated the reaction occurred at the Al/perovskite interface,[46] resulting in the formation of AlI_3, which acted as an energy barrier due to its improper energy-level alignment, leading to the insufficient charge extraction at the interface.

3.3.4 Copper (Cu)

Copper is also an attractive electrode material due to its high conductivity, abundant resources, and low cost. The Cu electrode is usually applied as both anode and cathode due to the similar work-function to ITO with a value of 4.8 eV. Shao et al.[47] introduced commercially available Cu tape as the counter electrode; the device structure is ITO/PEDOT:PSS/Perovskite/PCBM/Cu, showing the highest efficiency of 12.7% and maintaining more than 90% of initial PCE after being stored in ambient conditions for 250 h. Deng and coworkers[48] achieved a PCE of 18.3% in inverted Cu-based PSCs. Kaltenbrunner et al.[49] improved the Cu-based device stability by introducing a Cr_2O_3/Cr interlayer to suppress the formation of CuI. The devices with the structure of ITO/PEDOT:PSS/Perovskite/Cr_2O_3/Cr/Cu displayed a maximum PCE of 12.5%, which was comparable with the devices with Au electrode. Surprisingly, the Cu-based devices showed much better stability than Au-based devices under long-term light illumination processes, with more than 80% of initial PCE maintained after 8 h. Zhao et al.[50] achieved the maximum PCE of 20.7% in the devices with the structure of ITO/PTAA/Perovskite/C60/BCP/Cu. The unencapsulated devices sustained 98% of initial PCE after storing in ambient conditions for 800 h, and the encapsulated devices were projected to be stable for around 170 years and 22 years at room temperature and 40 °C, respectively. Flexible devices with Cu counter electrode were achieved by Nejand et al.[51] as well. They introduced Cu foil as the substrate and prepared devices with the structure of Cu/CuI/Perovskite/ZnO/Ag, which exhibited the highest PCE of 12.8% and maintained a stable efficiency output under continuous light illumination. In addition, more than 70% of initial PCE was retained after bending 1,000 cycles, representing excellent bending stability.

The drawback of Cu material applied in perovskite devices as counter electrode mainly lies in the unmatched energy level alignment (-4.8 eV), which creates a large Schottky barrier during the process of charge transfer. Xu et al.[52] demonstrated the mechanism of limited factors in Cu-based n-i-p structured devices. The mismatched Fermi level between HTL and Cu electrode inhibited the charge carrier transportation process; only a few HTLs have the matched Fermi level with Cu electrode. Furthermore, the easy oxidation of Cu poses a fatal threat to the long-term stability of devices. In order to solve the problems, Cheng et al.[53] introduced mercaptopyridine (MP-CF3)-based molecules at the Spiro-OMeTAD/Cu interface to regulate the Fermi level and the oxidation potential of Cu. The devices with a MP-CF3 interlayer achieved a maximum PCE of 22.90% and 18.55% in small and large area perovskite devices, respectively.

3.3.5 Nickel (Ni)

Low material cost of Nickel is attractive for PSC electrodes, and its high work function similar to Au also ensures the efficient charge collection. Jiang et al.[54] deposited Ni electrode for devices with the structure of FTO/TiO_2/Perovskite/Spiro-OMeTAD/Ni. Compared to Au-based devices (PCE = 11.6%), the Ni-based devices achieved the highest PCE of 10.4%. The main factor for limiting the device performance is the high series resistance of Ni layer, which could be resolved by decreasing the electrode thickness. Ku et al.[55] employed printable mesoporous Ni thin films as counter electrodes to achieve high efficiency perovskite solar cells. The film with a thickness of 2 um exhibited the best conductivity and, based on this, the devices realized the highest PCE of 13.6%. Notably, the devices fabricated by this method promised the reuse of the porous substrates, and the reused-based devices achieved the highest PCE of 12.1%. Wang and his group[56] introduced a Ni-mesh electrode in flexible photovoltaic devices. The Ni-based electrode exhibited both high conductivity as well as high transmittance and mechanical durability. Finally, a PCE of 17.3% was attained, which was the highest PCE based on a flexible metal electrode at that time.

3.3.6 Metal Alloy

Alloying the metals is an appropriate method to modify the material properties such as conductivity, work function, and chemical sensitivity. Jiang et al.[57] introduced Ag-Al alloy as the counter electrode to improve the photovoltaic performance and stability of PSCs. Al atoms preferred to migrate away from the Ag-Al layer, resulting in the formation of AlO_x at the PCBM/alloy electrode interface. Such interlayer prevented the ion migration of Ag atoms and moisture invading from atmosphere into the perovskite layer. Wang et al.[58] fabricated the inverted devices by introducing the semiconductor material Bi_2Te_3 as the electrode, with the optimal thickness around 200 nm. This alloy-based device exhibited a higher PCE of 18.2% than traditional Ag-based devices (15.9%) because the Be_2Ti_3-based cells exhibited higher IPCE performance, compensating the charge collection efficiency of high thickness. Han et al.[59] designed Cu-Ni alloy as the counter electrode in normal devices with the structure of FTO/SnO_2/Perovskite/Spiro-OMeTAD/EVA/CNG. The work function of Cu can be effectively tuned through adding different Ni contents. Besides, they introduced high quality graphene on both sides of the alloy electrode to improve the electrochemical as well as environmental stability. The resulting perovskite devices showed a maximum PCE of 24.34% (0.09 cm^2) and 20.76% (1.02 cm^2). In addition, the devices maintained more than 95% of initial PCE in double-85 testing conditions for 1,440 h and at MPP tracking for 5,000 h. Han et al.[60] combined a low-melting point Sn-Bi alloy film with a Ag layer to fabricate the inverted perovskite solar cells. This strategy improved the physical contact between electrode and ETL and reduced the charge transfer resistance from perovskite layers. Ultimately, the devices based on $MAPbI_3$ perovskite exhibited a maximum PCE of 13.52%.

3.4 CARBON-BASED ELECTRODES

Carbon-based electrodes have been extensively used for dye-sensitized solar cells (DSSCs), in which a high temperature annealing process is necessary to dissolve the polymers.[61] However, due to the all solid-state device structure of PSCs, the thermal annealing process can be avoided. In terms of electrical properties, carbon-based electrodes usually exhibit a proper energy-level alignment (around -5.0 eV), which is comparable with Au electrode.[62] In addition, carbon-based electrodes present remarkable protection effects due to its inherent chemical stability.

Generally, carbon-based PSCs have two forms in device structure; one of them is monolithic layered structure and the other is conventional layered structure. The former is fabricated by depositing a mesoporous insulating layer such as Al_2O_3 sandwiched by mesoporous TiO_2 and porous carbon electrode, and then the perovskite layer infiltrates into the mesoporous structures. This structure is simple to achieve and can realize large-scale fabrication through printing techniques. However, a high annealing temperature is required for the deposition of metal oxide layers, which hinders the commercialization process of this structure. The poor interface between perovskite and carbon-based electrode is another challenge. Therefore, the carbon-based devices usually exhibit inferior photovoltaic performance compared to metal electrode-based PSCs. The latter is conventional layer structure in which the carbon electrode locates on the top of perovskite layer or HTL. In this structure, the interlayer layer such as Al_2O_3 is not required. Due to its high chemical stability, the carbon layer usually acts as HTL to reduce the cost of devices. The interfaces between carbon-based electrode and perovskite or HTL layer play an important role in charge transportation and collection. Various efforts have been attempted to optimizing the carbon-based electrodes; the relevant details can be found in previous reviews. Here, we summarize some representative work on carbon-based electrodes such as materials composition, devices structure, and scalable fabrication.

Many efforts have been made in PSCs based on carbon electrodes to optimize the photovoltaic performance. In 2013, Han and his group[63] introduced carbon electrode (around 9 um) in HTL-free structured monolithic devices. Compared with traditional metal electrodes, the thickness of the carbon electrode is much thicker in these cells, which led to high film resistance as well as long charge carrier diffusion distance. In order to regulate the photovoltaic performance of carbon-based PSCs, Han et al.[64] introduced heteroatom doping in carbon films to improve conductivity and charge collection and finally obtained a PCE with a value of 13.6%. Han's group[65] optimized the perovskite composition through adding 5-AVA into the perovskite precursor solution, a certified PCE of 12.8% was achieved, and the devices exhibited excellent stability in ambient air and under one sun condition. Carbon-based electrode was also applied to HTL-free structured PSCs, such as TiO_2/Al_2O_3/NiO/perovskite/carbon, and a high PCE of 15% was achieved.[66] Yang et al.[67] added a PbI_2 interlayer to improve the electrical contact between perovskite and carbon electrode, finally gaining a high efficient HTL-free carbon-based PSC with a PCE over 11%. Tao et al.[68] introduced polystyrene spheres as poor forming agent to regulate the permeability of carbon electrode in printable PSCs, and a PCE over 22% was obtained.

Various types of other carbon electrodes such as graphite, graphene, and carbon nanotube have also been applied in PSCs. Zhang et al.[69] fabricated a commercial carbon paste layer through the doctor-blading technique. In order to improve the physical contact, a nanorod-liked CuPc interlayer was deposited between perovskite and carbon electrode via thermal evaporation process. The state-of-the-art devices exhibited comparable PCEs with traditional Au-based PSCs. However, the carbon-based devices maintained more than 90% of initial PCE after 500 h, which is superior to Au-based devices. Anders and his group[70] proposed a new technique, hot pressure transfer method, to deposit carbon cloth on carbon paste and obtained PSCs with the highest PCE of 19.11%. This technique provided the possibility of large-area film fabrication. Yang et al.[71] deposited conformable perovskite/carbon bilayers through an inkjet printing process. This method not only passivated the defects in grain boundaries of perovskite but also promoted grain growth. According to cross-sectional SEM results, the pinholes and voids in control devices were diminished after applying carbon films and subsequent $MAPbI_3$ bath, which ascribed to the in-situ chemical transformation of perovskite at the time of carbon film formation. In addition, carbon black interlayer was introduced to improve the physical contact at the interface. In order to enhance the interfacial charge collection, Zhang et al.[72] added CBC and CMNS in traditional carbon paste. The device PCE increased significantly after the addition of 0.1% CBC or CMNS, which realized better interface contact between the carbon layer and the perovskite.

Graphene is a universal material applied in photovoltaic devices as well due to its superior optoelectronic properties such as mechanical flexibility, proper energy-level alignment, and excellent optical transmittance. Yang et al.[73] investigated the optimal thickness of graphene in PSCs. Compared with single layer graphene electrode, multilayered graphene exhibited better hole extraction, although both multilayered graphene and single-layered graphene showed ohmic contact at the perovskite/carbon electrode interface. Comparatively, the devices based on multilayered graphene displayed superior performance mainly due to the long carrier lifetime, long diffusion length, and high transport coefficient. Guo et al.[74] improved the PSC PCE up to 12.39% through decreasing the work function of graphene. They developed Ni-doped graphene in perovskite devices; the N atoms can coordinate with Ni atoms and carbon atoms, increasing hole mobility of the film. B-doping is also an effective way to enhance the photovoltaic performance of carbon-based electrode. B-doping can decrease the work function as well as increase the film conductivity and carrier concentration of carbon electrode. Wei and coworkers[75] synthesized three-dimensional honeycomb carbon electrodes through the reaction between potassium and CO_2. The highest PCE of more than 10% was achieved because of the high conductivity and large surface area of the electrode. Lee et al.[76] also enhanced the photovoltaic performance of carbon-based electrode through combining graphene and carbon nanotube; the perovskite devices performed the highest PCE of 15.36% and superior stability by maintaining more than 80% of initial PCE after 500 h in atmosphere. In addition to the application in n-i-p structured devices and HTL-free devices, graphene materials were also introduced as the top electrode in inverted PSCs. Yan et al.[77] first applied the graphene electrodes in inverted devices and obtained ultrathin and flexible devices with high power output with a value of 5W/g. Trans et al.[78] deposited graphene films directly on the polymer substrates through plasma assisted chemical vapor deposition which provided the possibility of making large area and transfer-free graphene counter electrodes. Finally, the highest PCE of 14.2% and excellent mechanical flexibility were achieved in flexible devices.

Carbon nanotube has been widely used as the counter electrode in semi-transparent PSCs due to its good transparency at a certain thickness. However, the main challenge of this material is the poor physical contact between the light-absorbing layer and carbon nanotube. In order to enhance the devices' performance, various efforts have been devoted to optimize the interface contact. Mhaisalkar et al.[79] directly transferred carbon nanotube films on perovskite surface and fabricated HTL-free structured PSCs. The as-prepared carbon nanotube can be easily peeled off and transferred to different substrates, but the carbon nanotube films showed relatively high resistance (around 2–5 kΩ). The highest PCE of around 7% was achieved due to the enhanced hole injection. In order to improve the perovskite/carbon electrode interface, Yang et al.[80] developed PEI/carbon nanotube hybrid films as the counter electrode which effectively decreased the defects at the perovskite surface, resulting in the reduced interfacial recombination. The device PCE increased from 7.4% to 10.55%. Huang et al.[81] introduced carbon nanotube and carbon fiber in PSCs as the counter electrode. The enhanced hole mobility and reduced non-radiative recombination effectively enhanced the device PCE from 8.9% to 11.8%. Choi and coworkers[82] mixed the polymer HTMs with carbon nanotubes to enhance the stability of devices. This process reduced the material cost significantly, and the resulting devices maintained more than 80% of initial PCE after light stability test in ambient conditions for 2,200 h. More information about the utilization of carbon nanotube in perovskite solar cells can be found in the reviews.[83]

3.5 MULTILAYER ELECTRODES

Multilayer electrodes have been demonstrated to be a promising method to regulate the optoelectrical properties such as film conductivity and work function. In order to replace the Au electrode, Xu et al.[84] combined Ni-doped natural graphite layer with a

Bi-In alloy as the counter electrode in vacuum-free devices with the structure of FTO/SnO_2/Perovskite/Spiro-OMeTAD/Ni-Graphite/Bi-In alloy. The graphite on Spiro-OMeTAD layer effectively regulated the work function at the interface while the alloy electrode improved the conductivity. Finally, they achieved a maximum PCE of 21.0% in $FAPbI_3$-based devices. Domanski et al.[21] deposited a Cr layer (10 nm) between Au and HTL to form the Cr/Au bilayer electrode. The Cr effectively worked as a buffer layer to reduce the metal diffusion to perovskite. The bilayer-electrode-based devices maintained more than 80% of initial PCE in a thermal annealing MPP condition, slightly lower than that of perovskite devices based on Au-electrode. In addition, Mei et al.[25] applied Au/Ag bilayers as the counter electrode. After 200 h, the single Ag-based devices only maintained 40% of initial PCE, but the 80% of initial PCE was kept for the bilayer-based devices because the Au layer would inhibit the reaction between halogen and metal electrode. Wu et al.[85] demonstrated a facile strategy by introducing a Bi interlayer between HTL and Ag electrode, which highly prevented the halide ion migration from perovskite to the Ag layer. According to XPS results, Bi-based samples did not exhibit metallic Pb signals. Under the optimal thickness (20 nm), the devices showed a maximum PCE of more than 18%, and maintained more than 85% of initial PCE in air conditions for 6,000 h. Lin et al.[35] introduced Cu and Ag as the bilayer electrode in inverted devices to enhance the device stability. After storing in air conditions for 80 h, the devices based on single Cu or Ag electrode maintained around 60% and 10% of initial performance, respectively. In contrast, the bilayer-electrode-based devices maintained more than 85% of initial PCE. According to TOF-SIMs and other testing results, a large amount of halide ions such as I- accumulated at the interface, resulting in the formation of AgI_{2-} in single Ag-based devices and CuO in single Cu-based devices. In the control devices, the bilayer effectively prevented the oxidation reaction between Cu electrode and O_2, and the Cu electrode layer blocked the halide ion migration to the Ag layer. The synergistic effects of different metal electrodes significantly improved the device stability.

Considering the suitable work function of Ti (-4.3 eV), Chen et al.[86] introduced a Ti interlayer as cathode between perovskite layer and metal electrode. The devices with the structure of ITO/PTAA/$MAPbI_3$/Ti/Au exhibited a maximum PCE of over 18% with a high FF of more than 78%. Liu et al.[87] deposited a thin MoO_x interlayer by thermal evaporation technique. MoO_x effectively modified the work function, which enhanced the charge carrier extraction at the interface. The devices performed excellent thermal stability and maintained more than 88% and 75% of initial PCE after storing in air conditions and nitrogen atmosphere with thermal annealing for 90 min, respectively. Similarly, Sanehira et al.[44] also found that the MoO_x interlayer could improve the light stability as well as the long-term stability of devices. Compared with single Al-based devices, MoO_x probably reacted with Al electrode, resulting in the formation of Al_2O_3 or other nonstoichiometric aluminum oxides, finally preventing the erosion of moisture and Al atoms migration.

Organic-based barriers have also been introduced in perovskite devices as the multilayer electrode to enhance the energy level alignment and conductivity. Bi et al.[88] applied nanostructured carbon quantum dost (CQDs) in inverted PSCs to regulate the ion migration. CQDs interlayer effectively enhanced charge conductivity and modified the ohmic contact at the CQDs/Ag interface. Ultimately, the devices displayed a highest PCE of 15.6% for centimeter-squared cells. Further, they deposited polydimethylsiloxane (PDMS) as an interlayer in the same structured devices.[89] According to TOF-SIMs results, a 2D barrier composed by tri-s-triazine-based graphitic carbon nitride (g-C_3N_4) was formed, leading to the prevention of halide ions and reduction in charge carrier non-radiative recombination. Finally, the solar modules with an area of 36 cm^2 maintained more than 95% of initial PCE after thermal annealing at 85 °C for 1,000 h. Li et al.[90] introduced a benzotriazole (BTA) anticorrosion layer in Cu-based perovskite devices. The devices based on Cu/BTA multilayer electrode exhibited a strong chemical coordination of BTA-Cu, resulting in the enhancement of device stability. Under 1,000 h MPP tracking conditions, BTA-Cu based devices still maintained more than 88% of initial PCE. Kim et al.[91] investigated the number of amine groups at the end of BTA molecules. They speculated that the bonding strength was increased with the increasing number of amine groups, leading to the inhibition effect of halide ion migration.

Introducing dielectric/metal/dielectric multilayer is a promising method to prevent the reaction between perovskite layer and metal electrode. A variety of multilayers based on this structure has been reported, such as ZTO/Ag/ZTO,[92] ZnS/Ag/WO_3,[93] and SnO_x/Ag/SnO_x.[36] Xie and coworkers[94] deposited MoO_3/Ag/MoO_3 as the counter electrode in PSCs. Compared with traditional MoO_x electrode, a very high conductivity of the film was gained. The PCE of devices based on different bandgap perovskites showed high PCEs from 13.6% to 15.33%. Back et al.[95] deposited amine-mediated TiO_x as the chemical inhibition layer to mitigate the ion migration between the Ag electrode and perovskite layer. The TiO_x layer effectively reduced the sheet resistance, and the amine group acted as a barrier to prevent the halide ion migration from perovskite to Ag anodes. Eventually, the devices displayed the best PCE of more than 16%. Furthermore, the anime-mediated TiOx-Ag based devices maintained more than 90% of initial PCE after storing in ambient conditions for 1,000 h. Haliegnaw et al.[96] deposited TiO_x as an interlayer between ETL and Al electrode as well, which effectively decreased the interface resistance from around 380 Ω to less than 20 Ω, leading to the PCE up to 14.6%. Wang et al.[97] deposited nano-ZnMgO as the interlayer in inverted structured perovskite solar cells. When the stoichiometric ratio was $Zn_{0.85}Mg_{0.15}O$, the energy-level alignment was effectively modified, resulting in the significant improvement in charge carrier collection. Finally, the ZnMgO-based devices exhibited a maximum PCE of 15.61%. The multilayer composed of ZnO nanorods and BCP was also applied in perovskite devices.[98] When the concentration of ZnO nanorods was 70%, the inverted devices exhibited a maximum PCE of more than 18%, much higher than that of ZnO nanorods-based and BCP-based devices.

3.6 POLYMER-BASED ELECTRODES AND TRANSPARENT ELECTRODES

Conducting polymers can achieve good electrodes for perovskite solar cells with versatile fabrication processes. PEDOT:PSS is widely used as a flexible or transparent electrode because this polymer shows proper optoelectronic properties such as conductivity, work function, and optical transmittance. However, the commercial

PEDOT:PSS is often dispersed in water solution, hindering the direct application on perovskite surface. Zhou et al. deposited PEDOT:PSS as the counter electrode through a transfer-lamination technique. The PEDOT:PSS layer was first deposited on a PDMS substrate and then transferred to the perovskite top surface to act as the counter electrode. The PEDOT:PSS-based devices showed a maximum PCE over 11%. Further, they proposed the light management technique through the optimization of PEDOT:PSS film thickness, achieving the modulation of color and antireflection confined at a specific region and wavelength. Ultimately, they achieved the maximum PCE with varied values from 11% to 15% in different-color-based devices (such as red, green, orange, blue, and purple).

Many conductive oxides have also been applied as counter electrodes in PSCs. Giuliano et al.[99] achieved a device PCE of 12.5% based on a semitransparent counter electrode composed of MoO_x/Au/Cu/MoO_x. The dielectric/metal/dielectric multilayer structure effectively inhibited the reaction at the interface and reduced the total cost. The Au seed layer reduced the sheet resistance from 1 MΩ sq^{-1} to 10 Ω sq^{-1} and also effectively prevented the self-aggregation of Cu and the diffusion to MoO_x. ITO is a very common type of transparent conductive oxide due to its high conductivity and high transmittance. However, the high energy during sputtering process of ITO poses a threat on the stability of perovskite layers. In order to mitigate the damaging effect, Bush et al.[100] introduced ZnO nanoparticles as the interlayer between perovskite and electrode. Compared with traditional devices, the ITO-ZnO-based devices exhibited an improved PCE, maintained more than 80% of initial PCE after thermal annealing at 100 °C for 124 h. Fu et al.[101] deposited In_2O_3: H as the electrode with high conductivity and high transmittance in near-infrared region. More importantly, the deposition of In_2O_3: H layer did not need high temperature, reducing the damage to the perovskite during magnetron sputtering process. Given this strategy, they achieved the maximum PCE of 20.5% in four-terminal tandem solar cells. Werner et al.[102] deposited IZO as a transparent layer to achieve the light management in the range of 400–1,200 nm. Due to the high conductivity and low optical transmission loss, the author achieved a maximum PCE of 9.7% and 10% in Spiro-OMeTAD-based devices and Spiro-OMeTAD-MoO_x-based devices, respectively.

In addition to single-junction perovskite devices, the semitransparent electrodes with high conductivity and proper energy level alignment were also applied in tandem solar cells. The most commonly used semitransparent electrodes are the metal oxides such as ITO, aluminum-doped zinc oxide (AZO), and zinc doped indium oxide (IZO).[103] For example, Herlina et al.[104] deposited a transparent ITO layer as the top electrode for both single-junction perovskite devices and tandem solar cells by DC sputtering technique. After optimizing the composition of perovskite, the maximum PCE of more than 17% was achieved in $Cs_{0.05}(MA_{0.17}FA_{0.83})_{0.95}Pb(I_{0.83}Br_{0.17})_3$-based single-junction devices. By combining the perovskite devices (bandgap = 1.58eV) with silicon solar cells, a four-terminal (4T) tandem devices with a maximum PCE of 25.5% was obtained. Lukas et al.[105] applied the NIR-transparent, ZnO/ZnO: Al, as the semitransparent electrode in perovskite-CIGS tandem solar cells. The improvement in light transmittance between the range of 800–100 nm resulted in the photovoltaic enhancement in both single-junction perovskite devices and tandem solar cells, achieving the champion efficiency of 12.1% and 19.5%, respectively.

Colin and his coworkers[106] applied Ag nanowires as the electrode to achieve transparent perovskite devices. They dissolved Ag nanowires in methanol and then sprayed it with a pneumatic airbrush-type method on to the PET films to improve the uniformity and repeatability on large area. After the evaporation of solvent, the Ag nanowires were transferred from PET layer to the top surface of device by applying a vertical force (around 500 g). Finally, the semi-transparent tandem devices based on CIGS and Si solar cells exhibited the highest PCE of more than 25%. Li et al.[107] also applied ultrathin Ag films with the optimal thickness of lower than 6 nm in perovskite-Si tandem solar cells. Due to the improved light transmittance, the 4T perovskite-silicon tandem solar cell exhibited the maximum PCE of over 28%. In addition to the Ag metal electrode, Cu has also been applied as the semitransparent electrode in tandem solar cells.[108] Given the high transmittance of the Cu electrode, the top cells showed high transmittance in 750–1,100 nm wavelength range, ensuring the potential applied in 4T perovskite-silicon tandem solar cells. Finally, the 4T tandem devices with large area achieved the maximum PCE of 7.31%, much higher than that of single-junction perovskite devices (with the maximum PCE of 5.07%). Yang et al.[109] introduced a Cr seed layer under the ultrathin Au film to fabricate top semitransparent electrode. Due to the large surface energy of Cr, the resulting ultrathin Au layer exhibited a continuous morphology, with significant improvement in both transmittance and conductivity. After the optimization of film thickness (Cr 1 nm/Au 7 nm), the semitransparent electrodes performed average transmittance of 78.6% between visible and near-infrared region. Finally, the semitransparent single-junction perovskite devices and 4T perovskite-silicon tandem solar cells achieved the highest PCE of 19.8% and 28.3%, respectively.

3.7 PROSPECTS AND CONCLUSION

Here, we summarized the recent development of counter electrodes in perovskite solar cells, including the metal electrodes, carbon electrodes, polymer-based electrodes, and multilayer electrodes. In addition, we also summed up the problems of stability about common metal electrodes. Although various methods have been applied in traditional metal-based devices to enhance the device stability, most of those processes and techniques still cannot meet the needs of commercialization. The commonly used noble metal electrodes such as Au and Ag suffer from severe issues about stability. Cu is an excellent candidate for commercialization due to its high conductivity, low cost, and relatively high stability, while the deposition technique should be optimized to meet the large-scale application. Metal alloys are also excellent candidates for achieving high-efficiency PSCs, but investigations have barely focused on this field. Therefore, much effort should be paid on low-cost, stable, and uniform metal counter electrodes. The advantages and weakness of carbon-based counter electrodes including carbon nanotubes, graphene, and graphite were also discussed in this chapter. The poor physical and electrical contact between electrodes and charge transport layers is the main factor for the limitation of device efficiency. Hence, more work should focus on the interfacial contact to enhance charge carrier transport and reduce non-radiative recombination. This chapter provides fundamental comments about the roles and limitations of counter electrodes in PSCs, which would act as a reference of future research for the improvement of stability and photovoltaic performance of perovskite solar cells.

REFERENCES

1. I. Massiot, A. Cattoni, S. Collin, Progress and prospects for ultrathin solar cells. *Nature Energy* **2020**, *5*(12), 959–972.
2. C. Ma, N.-G. Park, A realistic methodology for 30% efficient perovskite solar cells. *Chem* **2020**, *6*(6), 1254–1264.
3. A. W. Faridi, M. Imran, G. H. Tariq, S. Ullah, S. F. Noor, S. Ansar, F. Sher, Synthesis and characterization of high-efficiency halide perovskite nanomaterials for light-absorbing applications. *Industrial & Engineering Chemistry Research* **2022**, *62*(11), 4494–4502.
4. J. Kang, L.-W. Wang, High defect tolerance in lead halide perovskite CsPbBr3. *The Journal of Physical Chemistry Letters* **2017**, *8*(2), 489–493.
5. C.-J. Tong, L. Li, L.-M. Liu, O. V. Prezhdo, Long carrier lifetimes in PbI_2-rich perovskites rationalized by Ab initio nonadiabatic molecular dynamics. *ACS Energy Letters* **2018**, *3*(8), 1868–1874.
6. F. Cheng, R. He, S. Nie, C. Zhang, J. Yin, J. Li, N. Zheng, B. Wu, Perovskite quantum dots as multifunctional interlayers in perovskite solar cells with dopant-free organic hole transporting layers. *Journal of American Chemical Society* **2021**, *143*(15), 5855–5866.
7. H. Zai, *et al.*, Sandwiched electrode buffer for efficient and stable perovskite solar cells with dual back surface fields. *Joule* **2021**, *5*(8), 2148–2163.
8. R. Ji, *et al.*, Perovskite phase heterojunction solar cells. *Nature Energy* **2022**, *7*(12), 1170–1179.
9. T. Niu, *et al.*, D-A-π-A-D-type dopant-free hole transport material for low-cost, efficient, and stable perovskite solar cells. *Joule* **2021**, *5*(1), 249–269.
10. B.-S. Kim, T.-M. Kim, M.-S. Choi, H.-S. Shim, J.-J. Kim, Fully vacuum-processed perovskite solar cells with high open circuit voltage using MoO_3/NPB as hole extraction layers. *Organic Electronics* **2015**, *17*, 102–106.
11. J. Park, *et al.*, Controlled growth of perovskite layers with volatile alkylammonium chlorides. *Nature* **2023**, *616*, 724–730.
12. A. K. Jena, A. Kulkarni, T. Miyasaka, Halide perovskite photovoltaics: Background, status, and future prospects. *Chemical Reviews* **2019**, *119*(5), 3036–3103.
13. B. Wang, *et al.*, The charge carrier dynamics, efficiency and stability of two-dimensional material-based perovskite solar cells. *Chemical Society Reviews* **2019**, *48*(18), 4854–4891.
14. B. Lyu, L. Yang, Y. Luo, X. Zhang, J. Zhang, Counter electrodes for perovskite solar cells: Materials, interfaces and device stability. *Journal of Materials Chemistry C* **2022**, *10*(30), 10775–10798.
15. G. Du, L. Yang, C. Zhang, X. Zhang, N. Rolston, Z. Luo, J. Zhang, Evaporated undoped spiro-OMeTAD enables stable perovskite solar cells exceeding 20% efficiency. *Advanced Energy Materials* **2022**, *12*(22), 2103966.
16. X. Zhao, C. Xu, X. Wang. J. Guo, M. Wu, Construction of multilevel network structured carbon nanofiber counter electrode and back interface engineering in all inorganic HTL-free perovskite solar cells. *Colloids and Surfaces A: Physicochemical and Engineering Aspects* **2022**, *648*(5), 129420.
17. J. H. Heo, *et al.*, Efficient inorganic-organic hybrid heterojunction solar cells containing perovskite compound and polymeric hole conductors. *Nature Photonics* **2013**, *7*(6), 486–491.
18. C. Roldán-Carmona, *et al.*, High efficiency single-junction semitransparent perovskite solar cells. *Energy Environmental Science* **2014**, *7*(9), 2968–2973.
19. L. Fan, *et al.*, Perovskite/silicon-based heterojunction tandem solar cells with 14.8% conversion efficiency via adopting ultrathin Au contact. *Journal of Semiconductors* **2017**, *38*(1), 014003.
20. S. N. Vijayaraghavan, *et al.*, Carbon electrode with sputtered Au coating for efficient and stable perovskite solar cells. *ACS Applied Materials & Interfaces* **2023**, *15*(12), 15290–15297.

21. K. Domanski, *et al.*, Not all that glitters is gold: Metal-migration-induced degradation in perovskite solar cells. *ACS Nano* **2016**, *10*(6), 6306–6314.
22. N. N. Shlenskaya, N. A. Belich, M. Grätzel, E. A. Goodilin, A. B. Tarasov, Light-induced reactivity of gold and hybrid perovskite as a new possible degradation mechanism in perovskite solar cells. *Journal of Materials Chemistry A* **2018**, *6*(4), 1780–1786.
23. S. Cacovich, *et al.*, Gold and iodine diffusion in large area perovskite solar cells under illumination. *Nanoscale* **2017**, *9*(14), 4700–4706.
24. T. Shi, J. Chen, J. Zheng, X. Li, B. Zhou, H. Cao, Y. Wang, Ti/Au cathode for electronic transport material-free organic-inorganic hybrid perovskite solar cells. *Scientific Reports* **2016**, *6*(1), 39132.
25. S. Mei, Z. Yin, P. Gu, H.-Q. Wang, J. Wang, W. Song, Improved operational stability of perovskite solar cells via Au barrier layer incorporation. *ACS Applied Energy Materials* **2021**, *4*(10), 11062–11068.
26. H. J. Kim, O. Y. Gong, Y. J. Kim, G. W. Yoon, G. S. Han, H. Shin, H. S. Jung, Environmentally viable solvent management in perovskite solar cell recycling process. *ACS Energy Letters* **2023**, *8*, 4330–4337.
27. F. Yang, *et al.*, Recycled utilization of a nanoporous Au electrode for reduced fabrication cost of perovskite solar cells. *Advanced Science* **2020**, *7*(6), 201902474.
28. X. Rong, H. Tian, W. Bi, H. Jin, T. Zhang, D. Guo, K. Zhao, Impact of metal electrode work function of $CH_3NH_3PbI_3$/p-Si planar heterojunction perovskite solar cells. *Solar Energy* **2017**, *158*, 424–431.
29. M. Lee, J. Teuscher, T. Miyasaka, T. Murakami, H. Snaith, Efficient hybrid solar cells based on meso-superstructured organometal halide perovskites. *Science* **2012**, *338*(6107), 643–647.
30. R. Azmi, *et al.*, Damp heat-stable perovskite solar cells with tailored-dimensionality 2D/3D heterojunctions. *Science* **2022**, *376*(6588), 73–77.
31. S. Svanström, T. J. Jacobsson, G. Boschloo, E. M. J. Johansson, H. Rensmo, U. B. Cappel, Degradation mechanism of silver metal deposited on lead halide perovskites. *ACS Applied Materials & Interfaces* **2020**, *12*(6), 7212–7221.
32. H. Li, *et al.*, Probing the stability of perovskite solar cell under working condition through an ultra-thin silver electrode: Beyond the halide ion diffusion and metal diffusion. *Chemical Engineering Journal* **2023**, *458*, 141405.
33. W. Ming, D. Yang, T. Li, L. Zhang, M. H. Du, Formation and diffusion of metal impurities in perovskite solar cell material $CH_3NH_3PbI_3$: Implications on solar cell degradation and choice of electrode. *Advanced Science* **2017**, *5*(2), 1700662.
34. J. Tao, *et al.*, Enhanced efficiency in perovskite solar cells by eliminating the electron contact barrier between the metal electrode and electron transport layer. *Journal of Materials Chemistry A* **2019**, *7*(3), 1349–1355.
35. C.-T. Lin, *et al.*, Enhancing the operational stability of unencapsulated perovskite solar cells through Cu-Ag bilayer electrode incorporation. *Journal of Materials Chemistry A* **2020**, *8*(17), 8684–8691.
36. J. Zhao, *et al.*, Self-encapsulating thermostable and air-resilient semitransparent perovskite solar cells. *Advanced Energy Materials* **2017**, *7*(14), 1602599.
37. J.-Y. Jeng, *et al.*, $CH_3NH_3PbI_3$ Perovskite/fullerene planar-heterojunction hybrid solar cells. *Advanced Materials* **2013**, *25*(27), 3727–3732.
38. J.-Y. Jeng, *et al.*, Nickel oxide electrode interlayer in $CH_3NH_3PbI_3$Perovskite/PCBM planar-heterojunction hybrid solar cells. *Advanced Materials* **2014**, *26*(24), 4107–4113.
39. Q. Wang, Y. Shao, Q. Dong, Z. Xiao, Y. Yuan, J. Huang, Large fill-factor bilayer iodine perovskite solar cells fabricated by a low-temperature solution-process. *Energy Environmental Science* **2014**, *7*(7), 2359–2365.
40. J. Seo, S. Park, Y. Chan Kim, N. J. Jeon, J. H. Noh, S. C. Yoon, S. I. Seok, Benefits of very thin PCBM and LiF layers for solution-processed p–i–n perovskite solar cells. *Energy Environmental Science* **2014**, *7*(8), 2642–2646.

41. J. You, *et al.*, Improved air stability of perovskite solar cells via solution-processed metal oxide transport layers. *Nature Nanotechnology* **2015**, *11*(1), 75–81.
42. E. Nouri, M. R. Mohammadi, P. Lianos, Inverted perovskite solar cells based on lithium-functionalized graphene oxide as an electron-transporting layer. *Chemical Communications* **2017**, *53*(10), 1630–1633.
43. T. Todorov, T. Gershon, O. Gunawan, C. Sturdevant, S. Guha, Perovskite-kesterite monolithic tandem solar cells with high open-circuit voltage. *Applied Physics Letters* **2014**, *105*(17).
44. E. M. Sanehira, *et al.*, Influence of electrode interfaces on the stability of perovskite solar cells: Reduced degradation using MoOx/Al for hole collection. *ACS Energy Letters* **2016**, *1*(1), 38–45.
45. L. Zhao, R. A. Kerner, Z. Xiao, Y. L. Lin, K. M. Lee, J. Schwartz, B. P. Rand, Redox chemistry dominates the degradation and decomposition of metal halide perovskite optoelectronic devices. *ACS Energy Letters* **2016**, *1*(3), 595–602.
46. H. Ding, *et al.*, In situ investigations of Al/perovskite interfacial structures. *ACS Applied Materials & Interfaces* **2020**, *12*(25), 28861–28868.
47. D. Ebner, M. Bauch, T. Dimopoulos, High performance and low cost transparent electrodes based on ultrathin Cu layer. *Optics Express* **2017**, *25*(8), 282897.
48. Y. Deng, Q. Dong, C. Bi, Y. Yuan, J. Huang, Air-stable, efficient mixed-cation perovskite solar cells with Cu electrode by scalable fabrication of active layer. *Advanced Energy Materials* **2016**, *6*(11), 201600372.
49. M. Kaltenbrunner, *et al.*, Flexible high power-per-weight perovskite solar cells with chromium oxide-metal contacts for improved stability in air. *Nature Materials* **2015**, *14*(10), 1032–1039.
50. J. Zhao, *et al.*, Is Cu a stable electrode material in hybrid perovskite solar cells for a 30-year lifetime? *Energy & Environmental Science* **2016**, *9*(12), 3650–3656.
51. B. Abdollahi Nejand, P. Nazari, S. Gharibzadeh, V. Ahmadi, A. Moshaii, All-inorganic large-area low-cost and durable flexible perovskite solar cells using copper foil as a substrate. *Chemical Communications* **2017**, *53*(4), 747–750.
52. Z. Xu, N. Li, X. Niu, H. Liu, G. Liu, Q. Chen, H. Zhou, Balancing energy-level difference for efficient n-i-p perovskite solar cells with Cu electrode. *Energy Material Advances* **2022**, *2022*, 9781073.
53. F. Cheng, *et al.*, Interfacial property tuning enables copper electrodes in high-performance n-i-p perovskite solar cells. *Journal of the American Chemical Society* **2023**, *15*(36), 20081–20087.
54. Q. Jiang, X. Sheng, B. Shi, X. Feng, T. Xu, Nickel-cathoded perovskite solar cells. *The Journal of Physical Chemistry C* **2014**, *118*(45), 25878–25883.
55. Z. Ku, X. Xia, H. Shen, N. H. Tiep, H. J. Fan, A mesoporous nickel counter electrode for printable and reusable perovskite solar cells. *Nanoscale* **2015**, *7*(32), 13363–13368.
56. M. Li, *et al.*, Embedded nickel-mesh transparent electrodes for highly efficient and mechanically stable flexible perovskite photovoltaics: Toward a portable mobile energy source. *Advanced Materials* **2020**, *32*(38), 202003422.
57. Z. Jiang, *et al.*, Amazing stable open-circuit voltage in perovskite solar cells using AgAl alloy electrode. *Solar Energy Materials and Solar Cells* **2016**, *146*, 35–43.
58. M. Wang, Q. Fu, L. Yan, W. Pi, G. Wang, Z. Zheng, W. Luo, A Bi_2Te_3 Topological insulator as a new and outstanding counter electrode material for high-efficiency and endurable flexible perovskite solar cells. *ACS Applied Materials & Interfaces* **2019**, *11*(51), 47868–47877.
59. X. Lin, *et al.*, In situ growth of graphene on both sides of a Cu-Ni alloy electrode for perovskite solar cells with improved stability. *Nature Energy* **2022**, *7*(6), 520–527.
60. S. Han, F. Wu, W. Qin, H. Cao, L. Yang, S. Yin, Perovskite solar cell based on double-layer Ag/SnBi alloy as cathode. *Journal of Alloys and Compounds* **2021**, *888*, 161455.

61. M. Younas, T. N. Baroud, M. A. Gondal, M. A. Dastageer, E. P. Giannelis, Highly efficient, cost-effective counter electrodes for dye-sensitized solar cells (DSSCs) augmented by highly mesoporous carbons. *Journal of Power Sources* **2020**, *468*, 228359.
62. X. Meng, *et al.*., Scrutinizing defects and defect density of selenium-doped graphene for high-efficiency triiodide reduction in dye-sensitized solar cells. *Angewandte Chemie International Edition* **2018**, *57*(17), 4682–4686.
63. Z. Ku, Y. Rong, M. Xu, T. Liu, H. Han, Full printable processed mesoscopic $CH_3NH_3PbI_3/TiO_2$ heterojunction solar cells with carbon counter electrode. *Scientific Reports* **2013**, *3*(1), 3132.
64. M. Duan, *et al.*, Boron-doped graphite for high work function carbon electrode in printable hole-conductor-free mesoscopic perovskite solar cells. *ACS Applied Materials & Interfaces* **2017**, *9*(37), 31721–31727.
65. J. Chen, *et al.*, Hole-conductor-free fully printable mesoscopic solar cell with mixed-anion perovskite $CH_3NH_3PbI_{(3-x)}(BF_4)_x$. *Advanced Energy Materials* **2016**, *6*(5), 201502099.
66. K. Cao, *et al.*, Efficient screen printed perovskite solar cells based on mesoscopic TiO_2/Al_2O_3/NiO/carbon architecture. *Nano Energy* **2015**, *17*, 171–179.
67. Z. Wei, *et al.*, Cost-efficient clamping solar cells using candle soot for hole extraction from ambipolar perovskites. *Energy Environmental Science* **2014**, *7*(10), 3326–3333.
68. H. Tao, *et al,*, High permeable microporous structured carbon counter electrode assisted by polystyrene sphere for fully printable perovskite solar cells. *Solid State Communications* **2018**, *271*, 71–75.
69. F. Zhang, X. Yang, M. Cheng, W. Wang, L. Sun, Boosting the efficiency and the stability of low cost perovskite solar cells by using CuPc nanorods as hole transport material and carbon as counter electrode. *Nano Energy* **2016**, *20*, 108–116.
70. S. Gholipour, *et al.*, Highly efficient and stable perovskite solar cells based on a low-cost carbon cloth. *Advanced Energy Materials* **2016**, *6*(20), 201601116.
71. Z. Wei, H. Chen, K. Yan, S. Yang, Inkjet printing and instant chemical transformation of a CH3NH3PbI3/nanocarbon electrode and interface for planar perovskite solar cells. *Angewandte Chemie International Edition* **2014**, *53*(48), 13239–13243.
72. L. Jiang, *et al.*, A counter electrode modified with renewable carbonized biomass for an all-inorganic $CsPbBr_3$ perovskite solar cell. *Journal of Alloys and Compounds* **2022**, *902*(5), 163725.
73. K. Yan, *et al.*, High-performance graphene-based hole conductor-free perovskite solar cells: Schottky junction enhanced hole extraction and electron blocking. *Small* **2015**, *11*(19), 2269–2274.
74. M. Guo, *et al.*, Composite electrode based on single-atom Ni doped graphene for planar carbon-based perovskite solar cells. *Materials & Design* **2021**, *209*(1), 109972.
75. W. Wei, *et al.*, Potassium-chemical synthesis of 3D graphene from CO_2 and its excellent performance in HTM-free perovskite solar cells. *Journal of Materials Chemistry A* **2017**, *5*(17), 7749–7752.
76. M. Tian, *et al.*, Printable free-standing hybrid graphene/dry-spun carbon nanotube films as multifunctional electrodes for highly stable perovskite solar cells. *ACS Applied Materials & Interfaces* **2020**, *12*(49), 54806–54814.
77. Z. Liu, P. You, C. Xie, G. Tang, F. Yan, Ultrathin and flexible perovskite solar cells with graphene transparent electrodes. *Nano Energy* **2016**, *28*, 151–157.
78. V.-D. Tran, S. V. N. Pammi, B.-J. Park, Y. Han, C. Jeon, S.-G. Yoon, Transfer-free graphene electrodes for super-flexible and semi-transparent perovskite solar cells fabricated under ambient air. *Nano Energy* **2019**, *65*, 104018.
79. Z. Li, *et al.*, Laminated carbon nanotube networks for metal electrode-free efficient perovskite solar cells. *ACS Nano* **2014**, *8*(7), 6797–6804.
80. Y. Yang, H. Chen, C. Hu, S. Yang, Polyethyleneimine-functionalized carbon nanotubes as an interlayer to bridge perovskite/carbon for all inorganic carbon-based perovskite solar cells. *Journal of Materials Chemistry A* **2019**, *7*(38), 22005–22011.

81. M. Guo, J. Liu, Y. Yuan, Z. Zhang, S. Yin, J. Leng, N. Huang, CNTs/Cf based counter electrode for highly efficient hole-transport-material-free perovskite solar cells. *Journal of Photochemistry and Photobiology A: Chemistry* **2020**, *403*, 112843.
82. N. Ahn, I. Jeon, J. Yoon, E. I. Kauppinen, Y. Matsuo, S. Maruyama, M. Choi, Carbon-sandwiched perovskite solar cell. *Journal of Materials Chemistry A* **2018**, *6*(4), 1382–1389.
83. Q. Luo, *et al.*, Recent advances in carbon nanotube utilizations in perovskite solar cells. *Advanced Functional Materials* **2020**, *31*(6), 202004765.
84. M. Li, *et al.*, Nickel-doped graphite and fusible alloy bilayer back electrode for vacuum-free perovskite solar cells. *ACS Energy Letters* **2023**, *8*(7), 2940–2945.
85. S. Wu, *et al.*, A chemically inert bismuth interlayer enhances long-term stability of inverted perovskite solar cells. *Nature Communications* **2019**, *10*(1), 1161.
86. T. Chen, *et al.*, Efficient perovskite solar cells with titanium cathode interlayer. *Solar RRL* **2018**, *2*(11), 201800167.
87. C. Liu, W. Li, J. Chen, J. Fan, Y. Mai, R. E. I. Schropp, Ultra-thin MoOx as cathode buffer layer for the improvement of all-inorganic $CsPbIBr_2$ perovskite solar cells. *Nano Energy* **2017**, *41*, 75–83.
88. E. Bi, *et al.*, Diffusion engineering of ions and charge carriers for stable efficient perovskite solar cells. *Nature Communications* **2017**, *8*(1), 15530.
89. E. Bi, *et al.*, Efficient perovskite solar cell modules with high stability enabled by iodide diffusion barriers. *Joule* **2019**, *3*(11), 2748–2760.
90. X. Li, S. Fu, W. Zhang, S. Ke, W. Song, J. Fang, Chemical anti-corrosion strategy for stable inverted perovskite solar cells. *Science Advances* **2020**, *6*(51), abd1580.
91. Y. R. Kim, *et al.*, Inner encapsulating approach for moisture-stable perovskite solar cells. *Solar RRL* **2021**, *5*(9), 202100351.
92. T. Winkler, *et al.*, Efficient large area semitransparent organic solar cells based on highly transparent and conductive ZTO/Ag/ZTO multilayer top electrodes. *Organic Electronics* **2011**, *12*(10), 1612–1618.
93. S. Han, S. Lim, H. Kim, H. Cho, S. Yoo, Versatile multilayer transparent electrodes for ITO-free and flexible organic solar cells. *IEEE Journal of Selected Topics in Quantum Electronics* **2010**, *16*(6), 1656–1664.
94. X. Xie, C. Wu, S. Sun, X. Xu, W. Xu, G. Qin, L. Xiao, Semitransparent perovskite solar cells with dielectric/metal/dielectric top electrodes. *Energy Technology* **2019**, *8*(4), 201900868.
95. H. Back, *et al.*, Achieving long-term stable perovskite solar cells via ion neutralization. *Energy & Environmental Science* **2016**, *9*(4), 1258–1263.
96. B. Hailegnaw, G. Adam, H. Heilbrunner, D. H. Apaydin, C. Ulbricht, N. S. Sariciftci, M. C. Scharber, Inverted (p-i-n) perovskite solar cells using a low temperature processed TiO_x interlayer. *RSC Advances* **2018**, *8*(44), 24836–24846.
97. Z. Wang, *et al.*, Solution processed nano-ZnMgO interfacial layer for highly efficient inverted perovskite solar cells. *Journal of Energy Chemistry* **2019**, *28*, 107–110.
98. M. S. Selim, A. M. Elseman, Z. Hao, ZnO nanorods: An advanced cathode buffer layer for inverted perovskite solar cells. *ACS Applied Energy Materials* **2020**, *3*(12), 11781–11791.
99. G. Giuliano, S. Cataldo, M. Scopelliti, F. Principato, D. Chillura Martino, T. Fiore, B. Pignataro, Nonprecious copper-based transparent top electrode via seed layer–assisted thermal evaporation for high-performance semitransparent n-i-p perovskite solar cells. *Advanced Materials Technologies* **2019**, *4*(5), 201800688.
100. K. A. Bush, *et al.*, Thermal and environmental stability of semi-transparent perovskite solar cells for tandems enabled by a solution-processed nanoparticle buffer layer and sputtered ITO electrode. *Advanced Materials* **2016**, *28*(20), 3937–3943.

101. F. Fu, *et al.*, Low-temperature-processed efficient semi-transparent planar perovskite solar cells for bifacial and tandem applications. *Nature Communications* **2015**, *6*(1), 8932.
102. J. Werner, *et al.*, Sputtered rear electrode with broadband transparency for perovskite solar cells. *Solar Energy Materials and Solar Cells* **2015**, *141*, 407–413.
103. Y. Cheng, L. Ding, Perovskite/Si tandem solar cells: Fundamentals, advances, challenges, and novel applications. *SusMat* **2021**, *1*(3), 324–344.
104. H. A. Dewi, *et al.*, Highly efficient semitransparent perovskite solar cells for four terminal perovskite-silicon tandems. *ACS Applied Materials & Interfaces* **2019**, *11*(37), 34178–34187.
105. L. Kranz, *et al.*, High-efficiency polycrystalline thin film tandem solar cells. *The Journal of Physical Chemistry Letters* **2015**, *6*(14), 2676–2681.
106. C. D. Bailie, *et al.*, Semi-transparent perovskite solar cells for tandems with silicon and CIGS. *Energy & Environmental Science* **2015**, *8*(3), 956–963.
107. Z. Li, *et al.*, Semitransparent perovskite solar cells with ultrathin silver electrodes for tandem solar cells. *Solar Energy* **2020**, *206*, 294–300.
108. M. R. Samantaray, D. S. Ghosh, N. Chander, Four-terminal perovskite/silicon tandem solar cells based on large-area perovskite solar cells utilizing low-cost copper semi-transparent electrode. *Applied Physics A* **2022**, *128*(2), 111.
109. D. Yang, *et al.*, 28.3%-efficiency perovskite/silicon tandem solar cell by optimal transparent electrode for high efficient semitransparent top cell. *Nano Energy* **2021**, *84*, 105934.

4 Single-Crystal Perovskite Solar Cells

Zemin Zhang, Linchuan Ma, and Yuelong Li

4.1 INTRODUCTION

Since the discovery of solar cells with perovskite (PVK) materials as the active layer, polycrystalline PVK devices have been deeply studied.[1–5] Many performance optimization strategies, such as solvent engineering, additive engineering, and interface engineering, have been used to increase the power conversion efficiency (PCE) and operation stability of polycrystalline PVK thin-film solar cells. However, there are still many problems that need to deal with in polycrystalline PVK thin-film solar cells. To achieve commercial preparation, effective solutions are required to eliminate pinholes and ensure uniformity of the polycrystalline thin film. Grain boundaries and traps in polycrystalline thin films lead to the nonequilibrium carrier transport, ion migration, material degradation, and other problems. Compared to polycrystalline PVK materials, monocrystalline PVK materials provide a platform to study the intrinsic properties of PVK materials because of their inherent advantages, such as long-ordered lattice structure, no grain boundaries, fewer defects, and longer carrier diffusion length and life.[6–10] To fully understand and study the underlying mechanisms in PVK materials, high-quality monocrystalline PVK materials are required to be synthesized. Simultaneously, owing to the excellent optoelectronic properties, monocrystalline PVK materials become ideal candidates for photovoltaic applications. However, there were fewer reports of highly efficient monocrystalline perovskite solar cells (PSCs) compared with polycrystalline PSCs.

To develop the potential of monocrystalline PSCs, researchers still need to solve the following key problems. 1. Optimizing the growth methods of crystals: monocrystalline PVK materials prepared by the traditional growth method always have large sizes and poor crystal quality. Using the improved growth method could improve the crystal quality and prepare large-area independent monocrystalline PVK wafers for optoelectronic devices. 2. Controlling the thickness of monocrystalline PVK: the thickness of monocrystalline PVK needs to be in the micro or sub-micro meters scale to match the carrier diffusion distance. The suitable crystal thickness will significantly reduce the loss of photogenerated carriers. 3. Suppression of surface defects: the light absorption coefficient of monocrystalline PVK is large, which determines the carrier generated near the surface of monocrystalline PVK in the light incident side. However, surface recombination caused by surface defects results in carrier loss and serious charge accumulation, which reduces the efficiency of carrier collection and degrades device performance.

 DOI: 10.1201/9781003400493-4

In the past decade, monocrystalline PSCs have made great progress. Researchers have conducted in-depth research on the aforementioned issues. Monocrystalline PSCs are gradually becoming an important development direction of PVK materials in photovoltaic devices. Here, we summarize development processes in monocrystalline PSCs. The basic properties of the material, such as crystal structure, optical properties, electrical properties, defects, and material stability, are introduced. The problems encountered in the development of monocrystalline PVK from materials to devices are discussed in detail. We think that large-scale production of the high-quality, controllable-thickness, and low-surface-defects monocrystalline PVK thin film will promote the development of monocrystalline PSCs. Finally, the potential solutions and prospects of monocrystalline PSCs are pointed to make guiding suggestions for the future developments.

4.2 PROPERTIES OF MONOCRYSTALLINE PVK MATERIALS

4.2.1 Crystal Structure

The components of monocrystalline PVK materials are usually ABX_3; A is a monovalent organic or inorganic cation, B is a divalent metal cation, and X is a monovalent halogen ion.[11] The BX_6 octahedron and the disordered A-site cations filled in the octahedral gap formed a three-dimensional (3D) network structure.[12] When large-sized organic cations are introduced into the A-site, the dimension of the PVK crystal structure changes to a two-dimensional (2D) structure (Figure 4.1a).[13] Furthermore, the same components still have phase transitions within a certain temperature range, which significantly affects the optoelectronic properties of PVK materials.[14] Common PVK phases include the cubic, orthorhombic, and tetragonal phases (Figure 4.1b). The stability of PVK crystal structure is determined by elemental composition and ionic radius. The tolerance factor (t) and octahedral factor (μ) are used to evaluate the stability of PVK crystal structures. In the range of $0.813 \leq t \leq 1.107$ and $0.442 \leq \mu \leq 0.895$, the structure of PVK materials can be ensured to good stability of crystal structure (Figure 4.1c). Otherwise, lattice distortion will occur and form a low-symmetry, unstable PVK crystal structure. The photographs of several monocrystalline PVK crystals of different components are shown in Figure 4.1d. Through the component engineering, multiple A-site cations and mixed halogen monocrystalline PVK can be formed, which can effectively release lattice strain and form a stable PVK phase with high photoelectric properties (Figure 4.1e). This is an effective strategy for designing and synthesizing monocrystalline PVK materials with great properties.[15]

Another research hot topic is the anisotropic photoelectric properties of the monocrystalline PVK, which are dependent on the different crystal structures.[16] Under different growth conditions, it is possible to grow into different crystals with different exposed surfaces.[17] The termination atoms on each exposed surface are different, resulting in different material properties, such as ion migration, carrier transport, and defect types. The structure-dependent optoelectronic properties of monocrystalline PVK and the underlying anisotropy mechanism should be

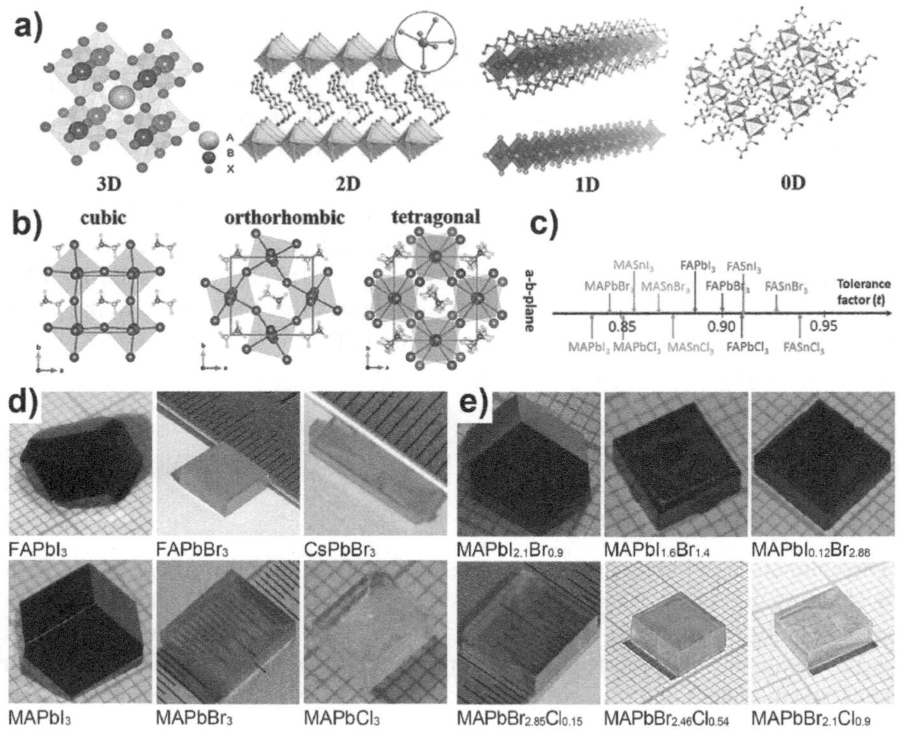

FIGURE 4.1 a) The typical 0–3D PVK structure; b) the comparison of cubic, orthorhombic, and tetragonal phases for $MAPbI_3$; c) the *t* values of a series of metal halide PVK materials; d) $FAPbX_3$ (X = I, Br), $CsPbBr_3$, and $MAPbX_3$ (X = I, Br, Cl) single crystals. Reproduced with permission from Ref.[12] Copyright 2023, WILEY-VCH. e) A series of mixed-halide $MAPbI_xBr_{3-x}$ and $MAPbBr_xCl_{3-x}$ single crystals ($0 \leq x \leq 3$). Reproduced with permission from Ref.[18] Copyright 2021, Springer Nature.

investigated deeply. It is necessary to design the crystal growth experiment to prepare the oriented-grown monocrystalline PVK with the specific crystal structure. Furthermore, simulating the crystal structure of monocrystalline PVK is also helpful to calculate surface-dependent defect formation energy and study the optoelectronic properties of specific crystal structures.

4.2.2 Optical Properties

The optical properties of monocrystalline PVK materials include absorption range, absorption coefficient, photoluminescence intensity and lifetime, etc.[19–21] In 3D PVK materials, the conduction band is mainly dominated by the 6p orbit of Pb, while the p orbit of halide is mainly composed of the valence band (Figure 4.2a).[22] The introduction or complete replacement of Sn ions can reduce the band gap and enable the solar cell to achieve the best absorption effect.[23] The bandgap values of monocrystalline PVK materials could be obtained by ultraviolet-visible-near infrared (UV-vis–NIR)

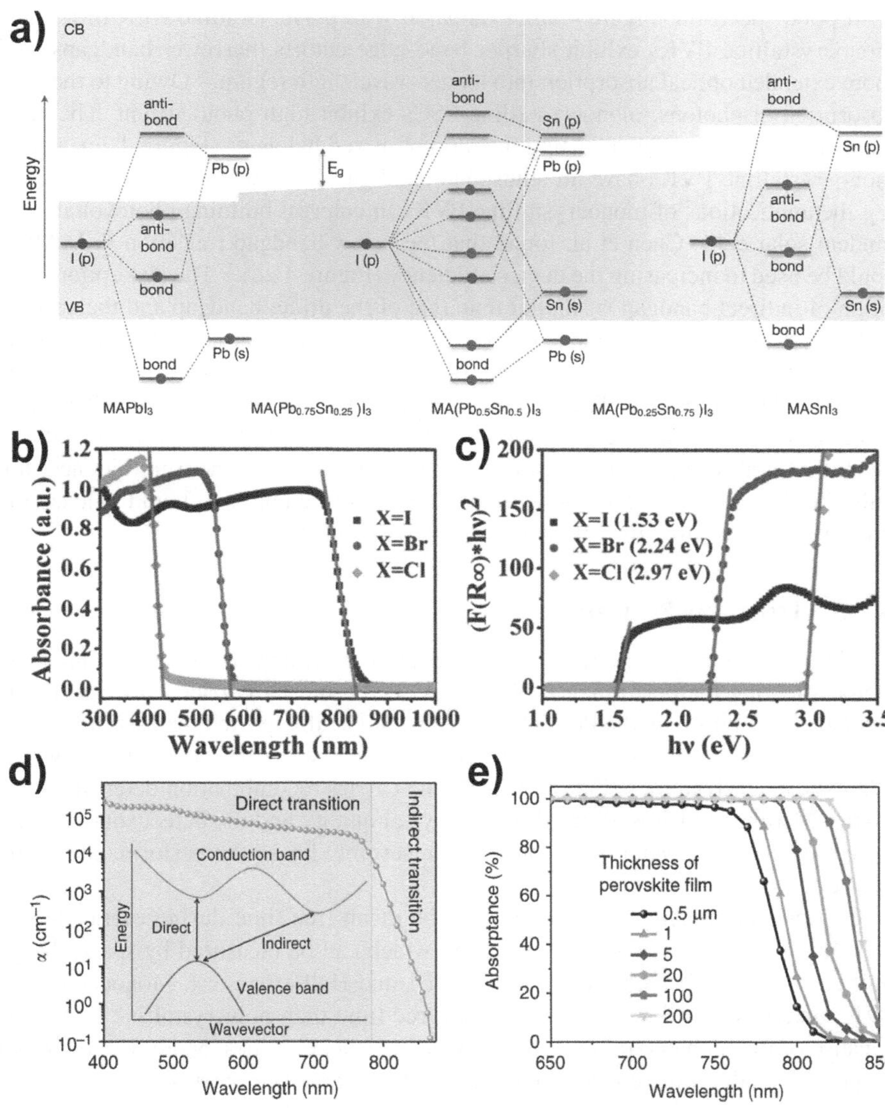

FIGURE 4.2 a) Schematic summarizing the origin of the band gap bowing in $MA(Pb_{1-x}Sn_x)I_3$. Shaded regions represent the valence and conduction bands with thick lines showing the molecular orbital picture of the formation of electronic bands in the alloy. Reproduced with permission from Ref.[22] Copyright 2018, American Chemical Society. b) UV-vis–NIR absorption spectrum and c) bandgap determination properties of $MAPbX_3$ (X=Cl, Br, I). Reproduced with permission from Ref.[24] Copyright 2015, WILEY-VCH d) Schematic illustration of the direct, below-bandgap transitions and absorption coefficient of $MAPbI_3$ from a polycrystalline film. e) The calculated absorption of $MAPbI_3$ films with different thicknesses. Reproduced with permission from Ref.[25] Copyright 2017, Springer Nature

absorption spectrum (Figure 4.2b). Compared with polycrystalline PVK thin films, monocrystalline PVKs exhibit sharper band edge cutoffs, narrower bandgaps, and more extended optical absorption into longer-wavelength region.[24] Owing to the high absorption of photons, monocrystalline PSCs exhibit high photocurrent. The band gap could be changed by adjusting the ratio of mixed halogens (Figure 4.2c), so that monocrystalline PVKs have an adjustable band gap from narrow to wide, expanding the applications of monocrystalline PVKs in colorful building photovoltaic and tandem solar cells. Chen et al. found that the below-bandgap transition in $MAPbI_3$ could be used to increasing the device efficiency (Figure 4.2d).[25] The absorption transition of indirect bandgap is smaller than that of the direct bandgap and the absorption spectrum become broadened. The absorption coefficient corresponding to the below-bandgap transition is two to four orders of magnitude weaker than that of the direct bandgap transition. The calculated results indicated that the thick PVK layer has a wide absorption spectrum (Figure 4.2e). Therefore, to fully utilize the advantage of light absorption properties of monocrystalline PVKs, the film thickness not only should be increased to a few micrometers or more but also should balance the film thickness and the limitation of carrier diffusion length.

4.2.3 Electrical Properties

Carrier lifetime means the time from production to disappearance is an important factor affecting the performance of devices. The minority carrier with long life is conducive to improving the PCE of solar cells. Shi et al. reported the carrier lifetime (τ) of $MAPbBr_3$ single crystal was evaluated to the fast ($\tau \approx 74$ ns) and the slow ($\tau \approx 978$ ns) carrier dynamics (Figure 4.3a).[26] The carrier recombination determines the carrier lifetime, which is affected by the crystal quality and the defect state density. Defect-induced non-radiative recombination severely limits the performance of the device.

Carrier mobility (μ) is determined by the mean free time during carrier movement and the effective mass of the carrier, which can be measured by space charge limited current (SCLC), time of flight (TOF), and Hall effect test. Monocrystalline PVK materials have longer carriers mean free time than polycrystalline thin films because of the high crystal mass and low defect state densities. Moreover, owing to no grain boundary, there is less scattering in the charge transport process, resulting in large carrier mobility of monocrystalline PVK materials. As shown in Figure 4.3b, monocrystalline PVK exhibited high carrier mobility ($\mu = 2.5$ cm^2 V^{-1} s^{-1}) and the value of the conductivity is 1×10^{-8} (ohm·cm)$^{-1}$.

Carrier diffusion length is determined by carrier mobility and carrier lifetime. High mobility and long lifetime result in long carrier diffusion length.[27, 28] Dong et al. show that the carrier diffusion length in monocrystalline PVK materials is longer than that in polycrystalline PVK materials.[29] However, the surface charge recombination significantly decreases the carrier diffusion length. The low crystal quality indicates high bulk trap density and short carrier diffusion length, which affected the efficiency and operational stability of monocrystalline PSCs. The carrier transport length in high-quality $FA_{0.55}MA_{0.45}PbI_3$ thin single crystals was discovered

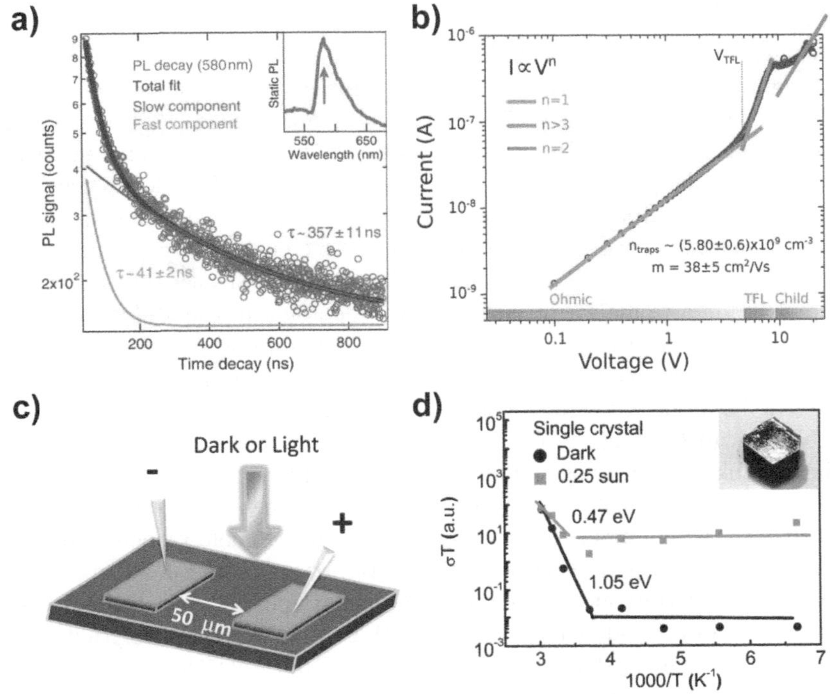

FIGURE 4.3 a) PL time decay trace on a $MAPbBr_3$ crystal; b) Characteristic I-V trace (purple markers) showing three different regimes for $MAPbBr_3$ (at 300 K). Reproduced with permission from Ref.[26] Copyright 2015, AAAS. c) A scheme of the conductivity measurement setup. d) The temperature-dependent conductivity of a single crystal. The inset pictures of d) is a photograph of the test sample. Reproduced with permission from Ref.[31] Copyright 2016, The Royal Society of Chemistry.

to be beyond 200 μm. The monocrystalline PVK with high crystal quality, low surface defects density, and long carrier diffusion length could improve the collection efficiency of photogenerated carriers and reduce the probability of recombination for enhancing the PCE of monocrystalline PSCs.

Ion migration phenomenon exists in the PVK films, which results in the current hysteresis of PSCs.[30] Xing et al. fabricated a lateral-structure device that consists of $MAPbI_3$ single crystal and two Au electrodes on it for the ion migration test (Figure 4.3c).[31] The activation energy of ion migration was extracted from the dependence of the conductivity of the $MAPbI_3$ films on temperatures. The measurement result showed that the activation energies of ion migration in single crystals are 1.05 eV in the dark and 0.47 eV under illumination (Figure 4.3d). The results confirmed that light illumination reduced the ion conducting energy barrier of single crystal surfaces and made the migration easier.

4.2.4 Defects

The defects in the interior and on the surface of crystals form electron defects in the band gap, which could trap free carriers. The non-radiative recombination of trapped carriers reduces the fill factor (FF) and open circuit voltage (V_{OC}) of solar cells. Although the interior defects of monocrystalline PVK thin film are significantly reduced due to the periodic order of the lattice, the high surface defect density still exists in monocrystalline PVK thin film.[32] Ni et al. characterized spatial distributions of carrier and trap densities in PVK materials by the drive-level capacitance profiling (DLCP) method (Figure 4.4a). The DLCP test on the Si solar cell was shown in Figure 4.4b to prove the accuracy of the carrier density measured. Figure 4.4c and 4.4d are the dependence of the trap density on the profiling distance of a $MAPbI_3$ bulk and thin single crystal, respectively. The defect density sharply varies from the surface to the inside of the film. Trap density of states (tDOS) spectra showed three trap bands with E_w values of 0.27 eV, 0.35 eV, and greater than 0.40 eV. Previous studies speculated that the deep trap states were mainly related to the surface defects of the perovskite and that shallower trap states were more likely from inside the perovskite.[33] These surface defects trap carriers and promote recombination, thus slowing carrier transport. They may also cause ion migration problems and seriously affect the performance of monocrystalline PSCs. Using the polish process and passivation with the wide-bandgap oxysalts (Figure 4.4e) significantly reduced the trap density of monocrystalline PVK thin films (Figure 4.4f) Therefore, the surface engineering should be further developed to reduce the surface defects of monocrystalline PVK thin films, such as optimizing the preparation method, adding surface passivation, and hydrophobic post-treatment.

4.2.5 Stability

PVK materials are sensitive to water, oxygen, light, high temperature, and other environmental factors.[35], [36] The decomposition of PVK films generally originates from surface defects and grain boundaries.[37–39] They act as channels for ion migration and attacking sites for water and oxygen. More surface defects and grain boundaries cause more severe degradation, which make the material unstable. For example, water molecules form hydration reactions with PVK molecules which generate hydrate products leading to the structure collapse (Figure 4.5a).[40] Wang et al. reported a scaling behavior of moisture-induced grain degradation in polycrystalline films and compared the stability of the monocrystalline and polycrystalline film (Figure 4.5b).[41] They hypothesized the amorphous and defective region at grain boundaries facilitates the quick permeation of moisture along the grain boundaries (Figure 4.5c). Grancini et al. designed the stability tests of monocrystalline PVK under different atmospheres, such as vacuum, low humidity, and high humidity.[42] The photoluminescence (PL) spectra, PL dynamics, and Ramen spectra (Figure 4.5d) showed the undesirable effects of water on the surface of monocrystalline PVK. Likewise, oxygen molecules preferentially attack defect and grain boundary locations resulting in the formation of PbI_2 and MA. Under the light illumination, small organic cations and halogen ions in PVK materials can overcome the migration energy barrier and migrate in defects

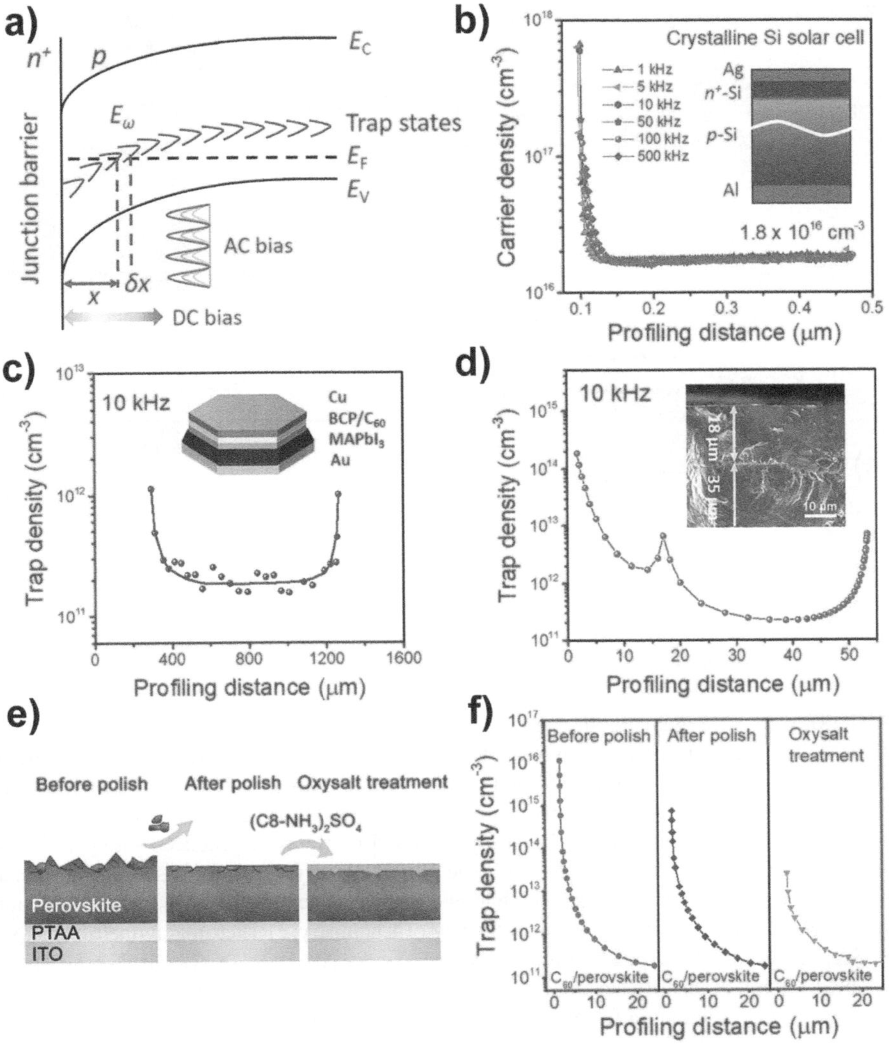

FIGURE 4.4 a) Schematic of band bending of a p-type semiconductor with deep trap states in an n$^+$-p junction. E_C, E_V, and E_F indicate the conduction band edge, valence band edge, and Fermi level, respectively; b) dependence of the carrier density on the profiling distance of a-Si solar cell at different ac frequencies measured by DLCP. The inset shows the schematic of the device structure; c) dependence of the trap density on the profiling distance of a $MAPbI_3$ single crystal measured by DLCP. The inset shows the device structure; d) dependence of the trap density on the profiling distance of a double-layer $MAPbI_3$ thin single crystal. The inset shows the cross-sectional SEM image of the double-layer $MAPbI_3$ thin single crystal; e) schematics of a $MAPbI_3$ thin single crystal on a PTAA/ITO substrate before mechanical polish, after mechanical polish, and after oxysalt treatment; f) trap density near the junction barrier of a $MAPbI_3$ thin single crystal before mechanical polish, after mechanical polish, and after oxysalt treatment. Reproduced with permission from Ref.[34] Copyright 2020, AAAS.

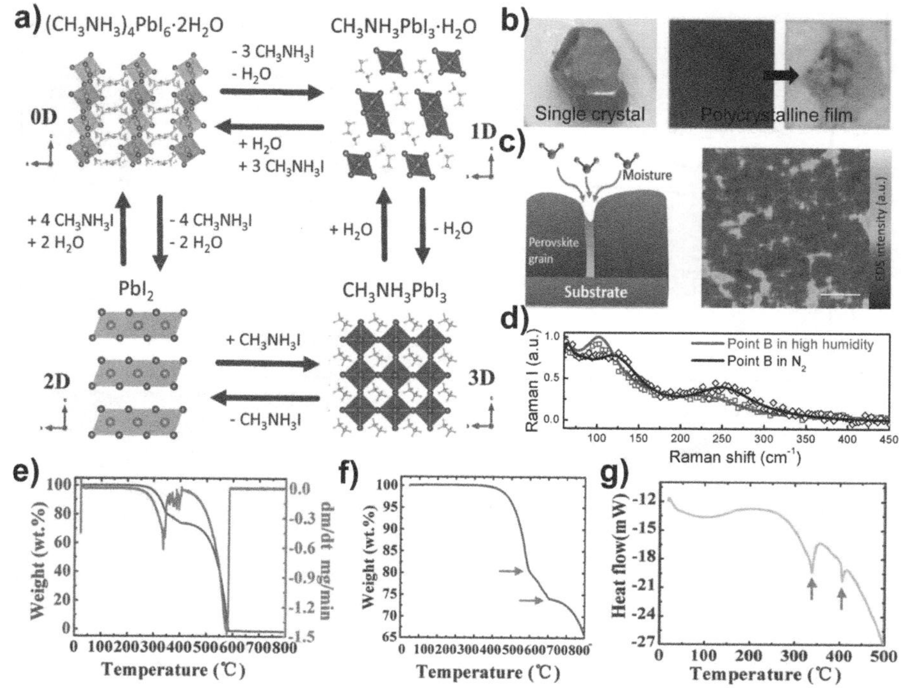

FIGURE 4.5 a) Schematic diagram of the phase equilibria in the PbI_2-MAI-H_2O system. Reproduced with permission from Ref.[40] Copyright 2016, WILEY-VCH. B) The photograph of degradation of single crystal (left) and polycrystalline films (right). C) Schematic showing the interaction of moisture with a bare perovskite film (left). Oxygen element mapping image showing the distributions of elemental oxygen in the film (right). Reproduced with permission from Ref.[41] Copyright 2017, The Royal Society of Chemistry. D) Comparison of the Raman spectra, under dry N_2 versus under a high humidity environment. Reproduced with permission from Ref.[42] Copyright 2015, The Royal Society of Chemistry. e) Thermogravimetric analysis (TGA) curves and corresponding first derivatives for the $MAPbX_3$ (X=Cl, Br, I) powders; f) zoomed version to identify accurate temperatures; g) differential scanning calorimetry (DSC) heating curve of the region for organic component. Reproduced with permission from Ref.[24] Copyright 2015, WILEY-VCH.

and grain boundaries leading to vacancy defects and ion accumulation areas. Shao et al. reported that ion migration at grain boundaries is faster than that within grains in polycrystalline PVK films.[43] PSCs with large grain sizes exhibited less hysteresis due to the reduction of grain boundaries. The monocrystalline PVK has a low defect density and no grain boundary; the ion migration phenomenon would be suppressed. The absence of grain boundaries in single crystals gave the largest energy barrier for ion migration. The performance of monocrystalline PVK devices would be improved significantly without ion migrations. Temperature also causes the phase transition of PVK materials. High-temperature conditions will accelerate the decomposition of perovskite and significantly reduce the performance and stability of the device.

Because of its few defects and no grain boundary, monocrystalline PVK shows a good stability. In addition, monocrystalline PVK materials have a high temperature tolerance. Liu et al. showed that monocrystalline PVK materials decomposed at higher temperatures than the PVK polycrystalline thin films (Figure 4.5e–g).[24] Introducing additional cations could increase the symmetry and rigidity of the crystal structure, increase defect formation energy, and inhibit lattice vibration, which leads to enhanced thermal stability and reduced phase transitions.[44], [45] Accurately controlling the formation of multi-component single crystals could reduce the formation of defects and enhance the stability of crystal structure.

4.3 CRYSTALLIZATION METHODS OF MONOCRYSTALLINE PVK MATERIALS

High quality monocrystalline material is the key to the preparation of monocrystalline PSCs. We will introduce the most basic methods of synthesizing monocrystalline materials. Understanding the growth mechanism of monocrystalline crystals could improve the crystal quality, defect density, and environmental stability of monocrystalline PSCs cells towards high efficiency.

4.3.1 Solution Temperature-Lowering (STL) Method

Dang et al. reported that the solubility of $MAPbI_3$ in HI solution increased with the rising temperatures.[46] When the temperature reached 65 °C, the $MAPbI_3$ precursor solution became saturated. The solution was preheated at 70 °C for 24 h to ensure that the solute was dissolved. In the cooling stage from 65 °C to 40 °C, the solute reached saturation and gradually separated crystals. After one month, a monocrystalline PVK bulk crystal with a size of 10 mm×10 mm×8 mm finished the growth stage. Dong et al. used the pre-obtained small $MAPbI_3$ single crystals as raw materials to grow the monocrystalline PVK materials (Figure 4.6a).[29] Small crystals were placed at the bottom to maintain the saturation of the solution. Nucleation occurs on the silicon wafer above the container due to air cooling. Owning to the solution convection caused by the temperature difference, the materials were transported to the large $MAPbI_3$ single crystal to rapidly promote growth. Lian et al. developed the bottom-seed assisted monocrystalline PVK growth method.[47] Platinum wires were used to support the seed crystals to avoid the growth of many small crystals. After cooling for 15 days, monocrystalline PVK materials were obtained. To increase the rate of crystal growth, Lian et al. added the Cl element into the precursor solution for preparing the $MAPbI_3$ single crystals.[48] Figure 4.6b showed the photograph and microscopic image of the $MAPbI_3(Cl)$ bulk single crystal. The Cl element changed the solubility of $MAPbI_3$ and impacted the growth kinetics. This method reduced the growth time of monocrystalline PVK materials. Furthermore, Fang et al. reported the preparation of monocrystalline PVK materials with mixed halogen components.[49] The supersaturated solution of perovskite was prepared at 100 °C from MA, a mixture of hydrochloric acid with different halide ratios, and lead acetate. The mixed-halogen components of monocrystalline PVK grown at low temperatures achieved a wide range of adjustable bandgaps.

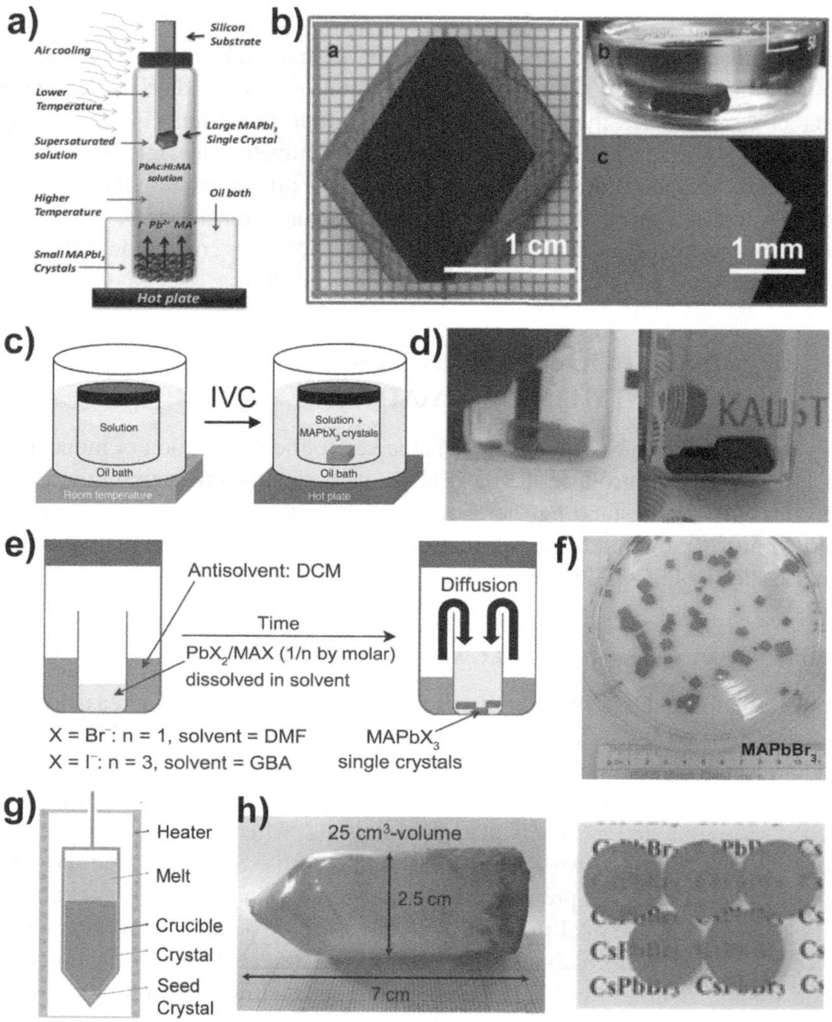

FIGURE 4.6 a) The top-seeded solution growth (TSSG) method set up for large single crystal growth. Reproduced with permission from Ref.[29] Copyright 2015, AAAS. b) Photograph of seed-induced-grown $MAPbI_3(Cl)$ bulk single crystal. Photograph and microscopic image of $MAPbI_3(Cl)$ bulk single crystal. Reproduced with permission from Ref.[48] Copyright 2016, American Chemical Society. c) Schematic representation of the ITC apparatus in which the crystallization vial is immersed within a heating bath. d) The photograph of shape-controlled crystals of $MAPbBr_3$ (orange) and $MAPbI_3$ (black). Reproduced with permission from Ref.[51] Copyright 2015, Springer Nature. e) Schematic diagram of the AVC process. f) Photograph of a batch of the as-grown $MAPbBr_3$ single crystals obtained within one week. Reproduced with permission from Ref.[26] Copyright 2015, AAAS. g) Schematic illustration of the crystal growth via the BSC method. h) Photograph of the initial $CsPbBr_3$ single-crystal ingot and wafers. Reproduced with permission from Ref.[62] Copyright 2017, WILEY-VCH.

4.3.2 Inverse Temperature Crystallization (ITC) Method

Kadro et al. reported an abnormal solubility curve of monocrystalline PVK precursor solution.[50] When the solvent of $MAPbI_3$ precursor solution was γ-butyrolactone (γ-GBL), the monocrystalline PVK began to grow under gradually increasing temperatures in a short time. This work provided a fast and simple preparation method for monocrystalline PVK growth. Saidaminov et al. proposed that the key to inverse temperature crystallization (ITC) is the selection of suitable solvents before the growth of different monocrystalline PVK materials.[51] For example, *N, N*-dimethylformamide (DMF) is a suitable solvent for $MAPbBr_3$ single crystals. Dimethylsulphoxide (DMSO) is suitable for dissolving $MAPbCl_3$ materials, and $MAPbI_3$ materials have an ITC phenomenon in γ-GBL solvent. The schematic diagram of the ITC method and the photograph of monocrystalline PVK were shown in Figure 4.6c and d. To grow inch-scale monocrystalline PVK materials in a short time, Liu et al. combined the seed crystal method with the ITC method to prepare two-inch $MAPbI_3$ single crystals.[24] They proposed that seed crystals provide a nucleation site and an additional growth layer on the surface to promote crystal growth. By optimizing the temperature and concentration conditions, monocrystalline PVK materials with low trap density, high carrier mobility, and high stability were obtained. Zhang et al. successfully prepared double halide perovskites by mixing precursors with the ITC method.[52] The prepared monocrystalline PVK had an adjustable band gap. With a further understanding of this method, researchers prepared FA-based, Sn-based, and three-cation mixtures monocrystalline PVK materials with high crystal quality.[53–56]

4.3.3 Anti-Solvent Vapor-Assisted Crystallization (AVC) Method

Shi et al. used dichloromethane (DCM) as the antisolvent to prepare monocrystalline PVK materials (Figure 4.6e).[26] As an antisolvent of PVK precursor, DCM cannot form hydrogen bonds with MAX and PbX_2 leading to the reducing asymmetric ions interaction. The photograph of $MAPbBr_3$ single crystals were shown in Figure 4.6f. This work provided a rapid preparation method for monocrystalline PVK materials. Liao et al. prepared $MAPbBr_3$ single-crystal microplates on a glass substrate via the AVC method.[57] DCM vapor led to nucleation and growth. High-quality $MAPbX_3$ single crystals were obtained by controlling the evaporation rate of the solvent. Furthermore, researchers successfully used this fast and efficient AVC method to synthesize a variety of single crystals, such as 2D and lead-free PVK single crystals.[58–61]

4.3.4 Bridgeman-Stockbarger Crystallization (BSC) Method

The Bridgeman-Stockbarger Crystallization (BSC) is a method that gradually cooled the melt in a crucible to crystallize the single-crystal ingot. It is often used to prepare large single-crystal ingots for producing wafers. Song et al. first prepared a large $CsPbBr_3$ single-crystal ingot using an improved BSC method.[62] The setup of the BSC method and the photographs of $CsPbBr_3$ single-crystal ingot and wafers were shown in Figure 4.6g and h. The main steps of wafer preparation are: (i) synthesizing $CsPbBr_3$ seed crystals with an equal molar mass of CsBr and $PbBr_2$, (ii) increasing the furnace temperature for crystal growth, and (iii) slicing the ingot along the growth direction and polishing the surface of wafers. By carefully controlling the heating and cooling process, Zhang et al. successfully grew large $CsPbBr_3$ single-crystal ingots and fabricated translucent and oriented $CsPbBr_3$ single-crystal wafers with reduced internal impurities.[63] Furthermore, Li et al. used the same method to prepare $Cs_3Bi_2Br_9$ single-crystal ingots.[64] This lead-free PVK ingots had high crystal quality, high carrier mobility, high thermal stability, and environmental friendliness, which were great candidate materials for X-ray applications.

4.4 FABRICATION OF MONOCRYSTALLINE THIN FILMS FOR PSCs

4.4.1 Towards Controlled Thickness

Chen et al. developed a space-confined growth technique based on two parallel substrates immersed in the solution.[65] The thickness of monocrystalline PVK film was controlled by adjusting the distance between the substrates from the nanometer to the micron scale. Owing to the temperature difference, bottom-up solute convection provides the raw material for the continuous growth of PVK single crystals. Most importantly, monocrystalline PVK films prepared via this method can be grown on different substrates. Liu et al. designed a geometry-defined dynamic-flow reaction system for preparing ultrathin monocrystalline PVK wafers with adjustable thickness, high crystal quality, and large size (Figure 4.7a and b).[66] The spacer between the two substrates determined the wafer thickness. Continuous crystal growth was achieved by constantly pumping the solution. This technique provided an effective strategy for preparing low-cost and thickness-suitable monocrystalline PVK wafers. Chen et al. covered the hydrophobic PTAA hole transport layer on two ITO glasses as the confined space.[25] It was found that the hydrophobic substrates accelerated the ion transportation, which made the monocrystalline PVK film grow continuously and rapidly. Yang et al. improved the space-confined and ITC methods to grow monocrystalline PVK films with nanometer thickness, high crystal quality, and large area.[67] The innovative point of this work was the introduction of the anisotropic growth environment composed of different hydrophilic substrates (Figure 4.7c). The mechanical hot-pressing setup provided the different pressure on the gap, which controlled the ultra-thin thickness of the monocrystalline PVK film. The relationship between pressure and thickness and the optical microscope image of monocrystalline PVK thin film were shown in Figure 4.7d and e, respectively. Tang et al. studied the deciding factors in the growth of monocrystalline PVK films on the hole transport layer. The experiment results showed that low surface energy, slow heating rate, and medium concentration of precursor are the best conditions for preparing high-quality monocrystalline PVK thin films. Song et al. developed a low surface energy perfluorinated-gel modified substrate for traditional space-confined growth methods.[68] The flexible soft contact layer on the substrate eliminated the stress caused by lattice mismatch during crystal growth and significantly reduced the surface and bulk defects of monocrystalline PVK thin films. The monocrystalline PVK films had suitable thickness, high crystal quality, and excellent carrier transport properties.

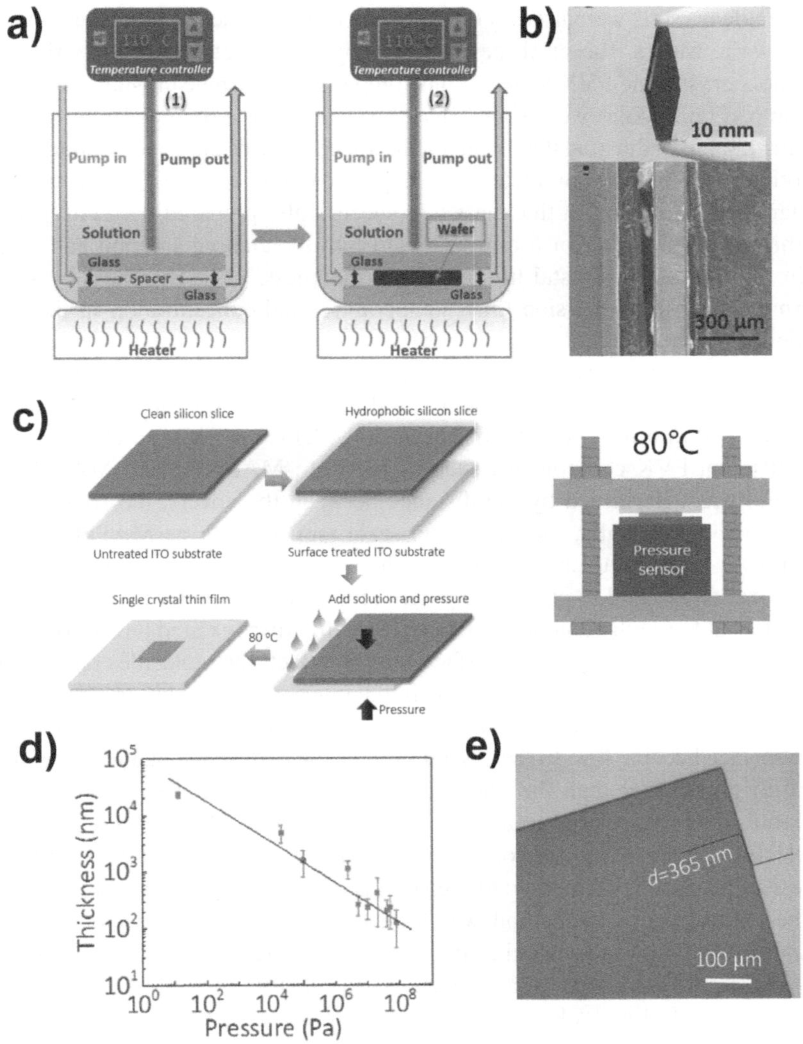

FIGURE 4.7 a) A schematic illustration for the ultrathin monocrystalline PVK wafer preparation. Two flat glass slides are separated apart by two spacers to form a microreactor in which the monocrystalline PVK wafer is formed. b) Photos of the monocrystalline PVK wafers and cross-sectional view of monocrystalline PVK wafers showing ≈150 μm thicknesses. Reproduced with permission from Ref.[66] Copyright 2016, WILEY-VCH. c) The schematic diagram of growth process and pressure measurement device. d) Thickness of the film under different pressure growth conditions. e) The optical image of a monocrystalline PVK thin film with a thickness of 365 nm. Reproduced with permission from Ref.[67] Copyright 2018, WILEY-VCH.

4.4.2 Towards Large Area

The preparation method of single crystal was further improved and large-size bulk PVK single crystals were used as the raw material to fabricate large-size monocrystalline PVK wafers. Liu et al. developed a top-down method for preparing large-area monocrystalline PVK wafers.[69] The prepared bulk single crystals with 120 mm in length (Figure 4.8a) were sliced into thin wafers by a diamond line microtome (Figure 4.8b). By adjusting the linear slicing speed, wire diameter, crystal feed speed, and other experimental parameters, the large-area monocrystalline PVK wafer with regular shape and 100 μm thickness was successfully prepared. Lv et al. put a 200 μm-thick $MAPbI_3$ wafer of 6 mm in length into the etching solution and obtained a thin $MAPbI_3$ single-crystal film of 15 μm thickness.[70] The thickness of wafers is determined by the immersion time, temperature, and concentration in the etching solution.

To obtain the monocrystalline PVK thin films on a centimeter scale, Zhumekenov et al. developed the surface tension-assisted method to grow large-area monocrystalline PVK thin films with a high aspect ratio.[71] This method also extended to prepare different PVK components, such as $MAPbI_3$, $MAPbBr_3$, and $MASnBr_3$. The thickness was determined by the distance between the covers, the solution height and temperature. Finally, the centimeter-scale monocrystalline $MAPbI_3$ thin films were grown in 30 min. Liu et al. used a thick-controllable surface tension-assisted crystallization method to prepare large-area monocrystalline PVK thin films.[72] The nucleation and growth of monocrystalline PVK thin films existed on the solution surface (Figure 4.8c). Owing to the surface-tension effect, the molecules on the surface had a larger distance compared to those in the interior of the solution. Therefore, the interaction between molecules on the surface was reduced and the nucleation barrier at the surface became low. In the crystal growth stage, the edge of a monocrystalline thin film grew faster than the bottom. As shown in Figure 4.8d, monocrystalline PVK thin films with a large area of 1 cm^2 and a thickness of 5–10 μm were obtained by selecting a suitable mixture of solvent and precursor concentration. Furthermore, the researchers used this method to prepare 2D monocrystalline PVK thin films.[73], [74].

The epitaxial growth method was used to prepare large-area monocrystalline PVK thin films through matching the lattice parameter between growth materials and substrate. To achieve large-area monocrystalline PVK thin films, Chen et al. prepared monocrystalline PVK thin films with 10 mm of length and 5 mm of width on $SrTiO_3$ (STO) (100) substrates using the vapor phase epitaxial growth method.[75] STO substrates are commonly used for the epitaxial growth of PVK materials. Its crystal structure was the same as that of $CsPbBr_3$ but different in lattice constants. However, it was interesting that the two units of $CsPbBr_3$ lattice matched three units of STO lattice, which were suited for successful epitaxial growth. Furthermore, other substrate materials for epitaxial growth were metal alkali halides that had similar chemical properties and lattice parameters to perovskite materials. Wang et al. found that $CsPbBr_3$ has a good lattice match with NaCl (100) and prepared large-area monocrystalline PVK thin films on a NaCl substrate.[76] The results showed that high-temperature epitaxy growth was helpful to prepare high-quality monocrystalline PVK thin films. The orthographic monocrystalline $CsPbBr_3$(100) thin films with high crystal

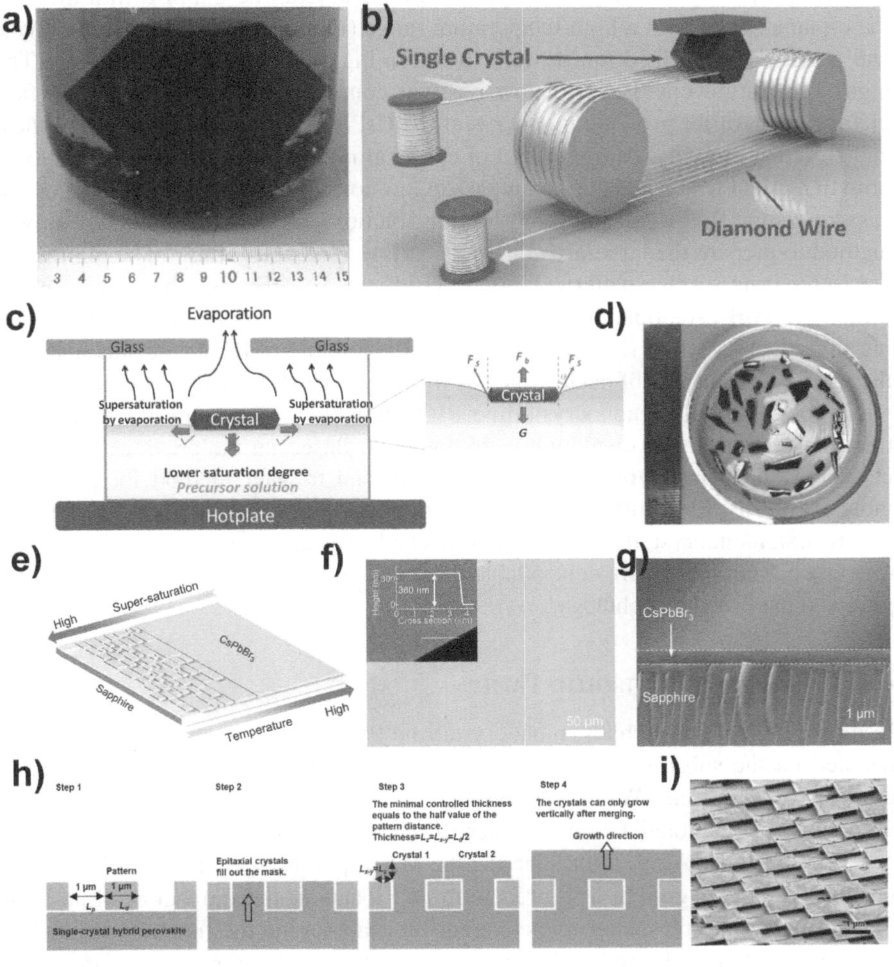

FIGURE 4.8 a) Photograph taken from a $MAPbI_3$ crystal in growth solution; b) the schematic illustration of the wafer slicing process. Reproduced with permission from Ref.[69] Copyright 2017, Springer Nature. c) Scheme of a fast and scalable bottom-up technique for crystal growth; d) photograph of pieces of monocrystalline $MAPbI_3$ wafers grown from 5 cm diameter container. Reproduced with permission from Ref.[72] Copyright 2019, WILEY-VCH. e) Schematic illustration of nucleation process of monocrystalline $CsPbBr_3$ thin films by vapor phase epitaxy method on c-plane sapphire substrate. f) Optical and AFM (inset) images of monocrystalline $CsPbBr_3$ thin films. The line profile shows the thickness of the film. The scale bar of the optical image is 50 μm. g) Cross-sectional SEM images of the monocrystalline $CsPbBr_3$ thin films showing high uniformity. Reproduced with permission from Ref.[69] Copyright 2020, American Chemical Society. h) Detailed schematic growth steps of photoresist-assisted epitaxy and transfer method. i) Titled SEM images of monocrystalline thin films grown under low temperatures and precursor concentrations. Reproduced with permission from Ref.[80] Copyright 2020, Springer Nature.

quality and controllable morphology were grown on the ZnSe substrate.[77] Zhong et al. prepared large-scale (5 mm × 10 mm) monocrystalline $CsPbBr_3$ thin films on the c-plane sapphire at a high-temperature range (Figure 4.8e).[78] The optical image and cross-sectional SEM image were shown in Figure 4.8f and g, respectively. The high temperature could control low reactant concentration and sufficient surface diffusion length for growing monocrystalline $CsPbBr_3$ thin films with good optical and structure stability. Another kind of epitaxial method was the solution epitaxial growth method which had the advantages of low cost, low-temperature treatment and easy operation. Lei et al. first reported the solution-process homoepitaxial growth method to prepare the large-area monocrystalline PVK thin films.[79] The main steps were preparing surface-patterned monocrystalline $MAPbBr_3$ and the growth in saturated $MAPbBr_3$ solution. Solvent evaporation was slow at room temperature resulting in a low nucleation rate. However, the diffusion rate of the solute was high and the interlayer transport of nucleation was inhibited, which contributed to the growth on the plane. Finally, monocrystalline PVK thin films with controllable distribution, morphology, and orientation were obtained. Based on the previous work, Lei et al. developed a photoresist-assisted epitaxy and transfer method for preparing monocrystalline PVK thin films in the solution process (Figure 4.8h).[80] As shown in Figure 4.8i, monocrystalline PVK thin films had large area (continuous thin films up to about 5.5 cm × 5.5 cm), controllable thickness (from about 600 nm to about 100 μm), and patterned morphology.

4.4.3 Towards Controlled Pattern

In most preparation methods, monocrystalline PVK thin films were randomly distributed on the substrates, which were difficult for preparing advanced integrated optoelectronic devices. Wang et al. pre-deposited PbI_2 microplates on the substrates with surface functionalization treatment.[81] These seed crystals were converted to monocrystalline $MAPbI_3$ after injecting the MA vapor process. By controlling crystal nucleation and growth at a predetermined location, the monocrystalline PVK grew directly onto pre-patterning electrodes for fabricating the functional optoelectrical device. Gu et al. used inkjet-printing seed crystals to prepare large-area, controllable-thickness, and high-yield monocrystalline PVK films (Figure 4.9a).[82] The advantages of this method were the control of nucleation density and the preparation on different substrates. The monocrystalline PVK films arrays were achieved via inhibiting random nucleation and using space-confined growth method. Xu et al. reported a photolithographic pattern-assisted method for preparing monocrystalline PVK arrays (Figure 4.9b).[83] The PVK precursor solution was scraping-coated on a pre-patterned array and the crystallization of monocrystalline PVK microplate happened through Ostwald ripping. This process was compatible with most optoelectronic integrated devices. Lee et al. exhibited a geometrically confined method for preparing patterned wafer-scale monocrystalline PVK thin films via the roller coating technology (Figure 4.9c).[84] The patterned polydimethylsiloxane (PDMS) mold with a periodic groove wrapped the metal roller. The PVK ink contacted the preheated PDMS template and crystallized in the channel under the geometric confined space. Finally, the patterned monocrystalline PVK thin films with controlled

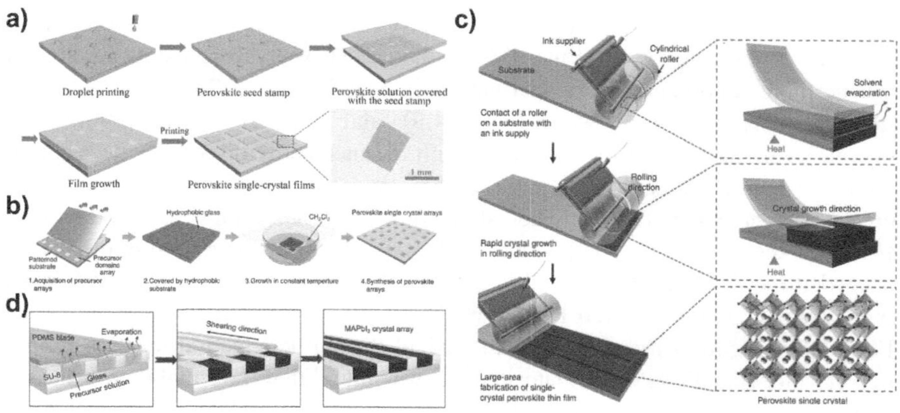

FIGURE 4.9 a) Schematic illustration of the scalable growth of monocrystalline PVK films. Scale bar, 1 mm. Reproduced with permission from Ref.[82] Copyright 2018, AAAS. b) Schematic illustration for the synthesis process of monocrystalline PVK array. Reproduced with permission from Ref.[83] Copyright 2023, Springer Nature. c) Schematic of the manufacturing procedure for monocrystalline PVK thin films using geometrically confined lateral crystal growth with a rolling mold. Reproduced with permission from Ref.[84] Copyright 2017, Springer Nature. d) Schematic illustration of the growth process of monocrystalline PVK within the microchannels. Reproduced with permission from Ref.[85] Copyright 2022, WILEY-VCH.

thickness and high crystal quality were obtained. Sun et al. proposed a novel 3D limited crystallization strategy.[85] The high-quality monocrystalline PVK array was prepared in a pre-made channel by scraping coating method, which achieved centimeter-sized crystal growth (Figure 4.9d). The results of hydrodynamic simulation showed that the narrow microchannel effectively inhibited the random flow of fluid and increased the flow rate, inducing unidirectional material transport in the microchannel and promoting the growth of crystals in a single direction. In addition, the manifold and flow rate in the microchannel were related to the channel width but not to the depth. Therefore, the thickness of the monocrystalline PVK array could be regulated within a certain range.

4.5 ARCHITECTURE AND PERFORMANCE OF MONOCRYSTALLINE PSCs

4.5.1 Vertical-Structure Devices

In monocrystalline thin films PSCs, the carriers could pass through the perovskite layer uniformly owing to the long carrier lifetime and long carrier diffusion length. However, the bulk single crystal is too thick to effectively separate and transport photogenerated carriers. Furthermore, the previous preparation methods of monocrystalline PVK for fabricating the vertical-structure solar cells are unsuitable, because the independent monocrystalline PVK with a small aspect ratio was grown without the TCO-coated substrate. Peng et al. reported a cavitation-triggered asymmetric crystallization method for preparing monocrystalline PVK thin films on various substrates.[86] The monocrystalline PSCs without HTL was fabricated as an attempt. Furthermore, another monocrystalline PSC with FTO/TiO_2/$MAPbBr_3$/Au structure was fabricated. As shown in Figure 4.10a and b, the TiO_2 ETL optimized the energy level matching of the interface. With the increasing thickness, the recombination probability of carrier increases, and V_{OC} was significantly reduced, but short-circuit current density (J_{SC}) only had slight variation from 1 μm to 12 μm thickness. However, the J_{SC} was seriously reduced when the thickness increased to 60 μm and exceeded the diffusion length and loss width (Figure 4.10c). The champion device with 1 μm thickness showed PCE = 6.53%, V_{OC} = 1.360 V, J_{SC} = 6.96 mA cm^{-2}, FF = 0.69 (Figure 4.10d). Chen et al. studied the sub-bandgap absorption properties of monocrystalline PVK materials.[25] It was found that the suitable thickness of monocrystalline PVK films provided sufficient light absorption and reduced carrier recombination. The calculated data showed that the thickness of $MAPbI_3$ film increased from 500 nm to 200 μm and the absorption spectrum redshifted to 850 nm. The value of PCE tended to increase first and then decrease with the increasing monocrystalline PVK thin-film thickness. The vertical structure of the ITO/PTAA/$MAPbI_3$/PCBM/C60/BCP/Cu solar cell was fabricated. The PCBM ETL was spin-coated on the PVK layer and the C60/BCP layer was prepared via the vapor deposition method to avoid short circuits. The monocrystalline thin-film PSC with a thickness of 10 μm exhibited a J_{SC} of 20.5 mA cm^{-2}, a V_{OC} of 1.060 V, an FF of 0.74, and a

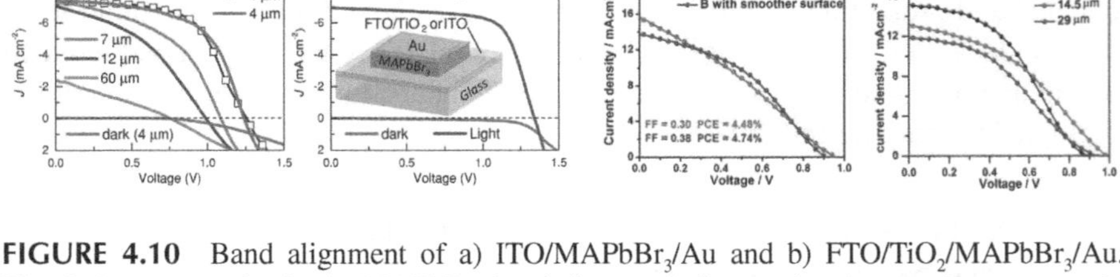

FIGURE 4.10 Band alignment of a) ITO/$MAPbBr_3$/Au and b) FTO/TiO_2/$MAPbBr_3$/Au. The dark orange color in the $MAPbBr_3$ band diagram indicates the photocarriers are mostly generated near the interface of ITO/$MAPbBr_3$ (TiO_2/$MAPbBr_3$) due to the large absorption coefficient. The direction of the built-in field E_{bi} is also marked by the solid arrow. Dark and illuminated J–V curves of c) ITO-based and d) FTO/TiO_2-based monocrystalline solar cells, respectively. Reproduced with permission from Ref.[86] Copyright 2016, WILEY-VCH. e) The proposed schematic diagram and the J–V curves of ITO/NiO_x/$(FAPbI_3)_{0.85}(MAPbBr_3)_{0.15}$/$TiO_2$/Ag devices: f) with TiO_2 layer of 60 nm thick deposited at 100 °C (upper line) and without TiO_2 layer (nether line); g) single crystal A and B have almost the same thickness but different roughness; h) monocrystalline PVKs have the different thickness. Reproduced with permission from Ref.[89] Copyright 2018, WILEY-VCH.

PCE of 16.10%. Although the bulk defects of monocrystalline PVK films were low, the surface defects cannot be negligible. The formation of these charge traps might be caused by the loss of MAI molecules on the surface during preparation. MAI molecule as a passivation layer was spin-coated on the single-crystal surface. Under optimized conditions, the PCE of the champion device was increased to 17.8%, J_{SC} is 21.00 mA cm^{-2}, V_{OC} is 1.08V, and FF is 0.78. This work provided a new direction for further improving the efficiency and stability of monocrystalline PSCs. It laid a good foundation for the development of vertical-structure monocrystalline PSCs with high performance. There are still great challenges for the fabrication of monocrystalline thin-film PSCs with controllable thickness and large areas. Rao et al. proposed a flowing-solution method to achieve continuous growth of monocrystalline PVK thin films.[87] The monocrystalline $MAPbBr_3$ film with a thickness of 16 μm and size of 6 mm × 8 mm was successfully prepared on FTO glass in a polytetrafluoroethylene (PTFE) box. Owing to low defect density, high mobility, and high crystallinity, the champion device showed a PCE of 7.11%, a V_{OC} of 1.310 V, a J_{SC} of 8.77 mA cm^{-2}, an FF of 0.62, and held a 93% initial efficiency after operating for 1,000 h. After comparing the carrier dynamics of different solar cell structures with and without the transport layer, the results showed that the introductions of ETL and HTL could reduce interface recombination and enhance carrier separation and extraction leading to the improved efficiency of monocrystalline PSCs. Zhao et al. prepared monocrystalline PVK thin films on two parallel TiO_2/FTO substrates which were immersed into the precursor solution.[88] The capillary force and temperature gradient provided the driving force for the solution to enter the confined space. The monocrystalline PSC with FTO/TiO_2/$MAPbI_3$/Spiro-OMeTAD/Ag structure had a Voc of 0.660 V, a Jsc of 22.28 mA cm^2, an FF of 0.59, and a PCE of 8.78%. The mixed-cation PVK solar cells had high PCE and stability. In addition to the conventional n-i-p structured solar cell, the p-i-n structured monocrystalline PSCs had been successfully prepared. Huang et al. prepared PSCs based on ITO/NiO_x/$(FAPbI_3)_{0.85}(MAPbBr_3)_{0.15}$/$TiO_2$/Ag structure via the PDMS-assisted space-confined crystallization.[89] The carrier dynamics of the p-i-n structured device were studied in detail (Figure 4.10e). The smooth surface of the monocrystalline PVK thin films and TiO_2 ETL interface were beneficial for electron and hole extraction (Figure 4.10f and g). Finally, the performance of the p-i-n device was significantly improved to a PCE of 12.18%, a J_{SC} of 23.14 mA cm^{-2}, a V_{OC} of 1.030, and an FF of 0.51. This work provided some suggestions for improving device performance through interface engineering.

4.5.2 Lateral-Structure Devices

In lateral-structure monocrystalline PSCs, the electrodes are deposited on the monocrystalline PVK surface, which could absorb enough light and further promote the absorption efficiency. It would contribute to the enhanced photocurrent and efficiency of monocrystalline PSCs. The lateral-structure does not require expensive transparent electrodes leading to a low cost. The carrier diffusion length of about 150 μm was found in the monocrystalline PVK with few grain boundaries and defects, which was suitable for fabricating the lateral-structure solar cells. Dong et al. found a strong piezoelectric effect on monocrystalline PVK under a non-uniform electric field.[90] The grain boundaries generated by piezoelectric polarization became channels for ion migration. The lateral-structure monocrystalline PSC was fabricated via deposing two parallel gold electrodes with a 50 μm separation distance on the single crystal surface (Figure 4.11a). The PCE was 1.88% at room temperature and showed a slight hysteresis effect. When the temperature was as low as 170 K, the photocurrent increased significantly and the maximum PCE reached 5.36% because of the enhanced carrier mobility and decreased charge recombination. The PCE was improved via optimizing the matching of work function and realizing the level alignment. This work opened a new view for low-cost and solution-process of lateral-structure monocrystalline PSCs. The sizes of monocrystalline PVK bulk materials were difficult to apply in the lightweight and highly integrated optoelectronic devices. Fabricating thin-film-based lateral-structure monocrystalline PSCs became a potential solution. Lee et al. used the roll coating method to prepare patterned monocrystalline PVK thin films.[84] The 200-nm-thick monocrystalline PVK film deposited two electrodes and obtained the carrier mobility of 45.64 $cm^2\ v^{-1}\ s^{-1}$ via the SCLC test. In addition, the lateral structure PSCs with gold electrode, PCBM layer, and silver electrode were prepared. The monocrystalline PSC with an active area of 30 × 22.2 μm^2 showed a PCE of 4.83%, a J_{SC} of 18.33 mA cm^{-2}, a V_{OC} of 0.800 V, and an FF of 0.33 via inducing polarization effect in the monocrystalline PVK thin films. Lin et al. further improved the PCE of the lateral structure monocrystalline PSCs based on the high aspect ratio monocrystalline PVK thin films that grew at the interface between the precursor solution and air.[72] The monocrystalline wafer had a large area, an adjustable thickness, and high crystal quality. In previous works, the polarization process damaged the crystals and limited the performance of devices. Therefore, the introduction of the transport layer became a solution for the manufacture of non-polarized devices. The monocrystalline PSC with Au/$MAPbI_3$/C60/BCP/Au structure was prepared and the final device had an active area of 0.0364 mm^2. Under 0.25 sunlight, the champion device had a J_{SC} of 5.06 mA cm^{-2}, a V_{OC} of 0.660 V, an FF of 0.44, and a PCE of 5.80%. The surface defects on monocrystalline PVK thin films caused the carriers recombination, which seriously limited the device performance. Song et al. optimized the anode contact of lateral-structure monocrystalline PSCs via the surface passivation.[91] The fabrication and image of lateral-structure monocrystalline PSCs were shown in Figure 4.11g and h. The V_{OC} and FF of devices were significantly increased after surface passivation. The experiment results showed that the surface defect density decreased and the carrier lifetime increased after MAI treatment. Finally, a champion device with a PCE of 11.52%, a J_{SC} of 22.49 mA cm^{-2}, a V_{OC} of 0.930 V, and an FF of 0.55 was obtained under one sun. There was no hysteresis effect in the devices and the operation stability was maintained for 200 h. This work had a great inspiration for studies of lateral-structure monocrystalline PSCs. With the development of material science and industry techniques, lateral-structure monocrystalline PSCs prepared via solution process without expensive transparent electrodes had broad application prospects.

4.5.3 Optimized Device Performance and Stability

In 2019, Chen et al. improved the PCE of $MAPbI_3$ single-crystal solar cells by over 21%.[92] The ITO/PTAA/$MAPbI_3$/C60/BCP/Cu structured device was prepared

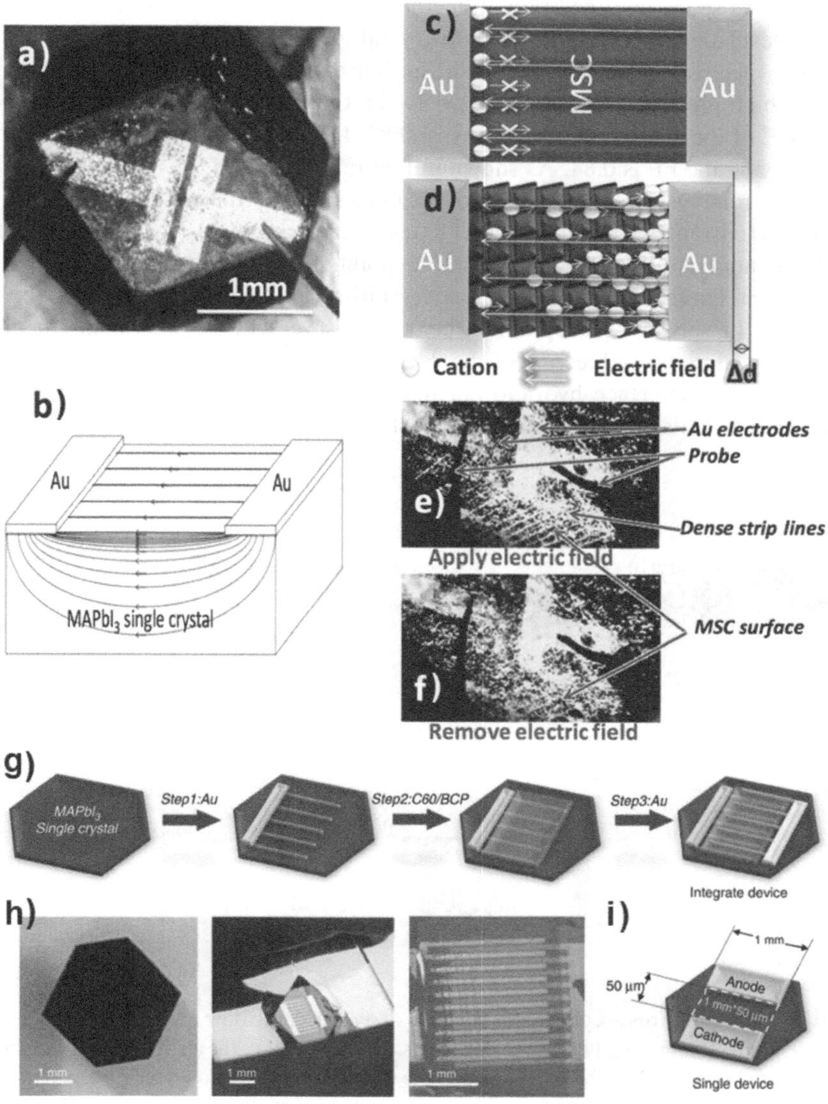

FIGURE 4.11 a) An image of lateral-structure monocrystalline $MAPbI_3$ device. b) Illustration of the poling electric field distribution in top surface and cross-section of a lateral monocrystalline device. c, d) Scheme of electromechanical strain generated grain boundaries and iron migration in $MAPbI_3$ single crystals. e, f) Images of lateral-structure monocrystalline $MAPbI_3$ devices with and without applied electric field. Reproduced with permission from Ref.[90] Copyright 2016, WILEY-VCH. g) Schematic diagram of preparation process of large-area lateral-structure monocrystalline PSCs. h) Image of the $MAPbI_3$ single crystal and photographs of the lateral-structure monocrystalline PSCs. i) Structure of a regular single device with the area of 50 μm × 1 mm. The large-area device was integrated by 19 single devices (65 μm× 1.5 mm) with a total device area of 1.85 mm^2. Reproduced with permission from Ref.[91] Copyright 2020, Springer Nature.

via ITC and space-confined growth methods. The cross-sectional SEM image and energy-level diagram of $MAPbI_3$ single-crystal solar cells was shown in Figure 4.12a and b, respectively. The smooth crystal surface could be completely covered by the ETL, which prevented direct contact between the metal electrode and the crystal. The PCE of the champion device was 21.09%, the J_{SC} is 23.46 mA cm^{-2}, the V_{OC} is 1.070 V, and the FF is 0.84. As shown in Figure 4.12c and d, this device had a negligible current hysteresis and the J_{SC} of the device was verified and corrected via the integrated current calculated by the EQE test. Furthermore, to avoid the degradation of the monocrystalline PVK material, two points should be paid attention to in the experiment. On the one hand, the monocrystalline PVK thin films could not be dissolved by the solvent when spinning the transfer layer. On the other hand, the manufacturing, storage, and testing of the devices should be conducted in the N_2 glove box to prevent interface hydration or degradation of monocrystalline PVK thin-film devices. This work increased the PCE of $MAPbI_3$-based single-crystal solar

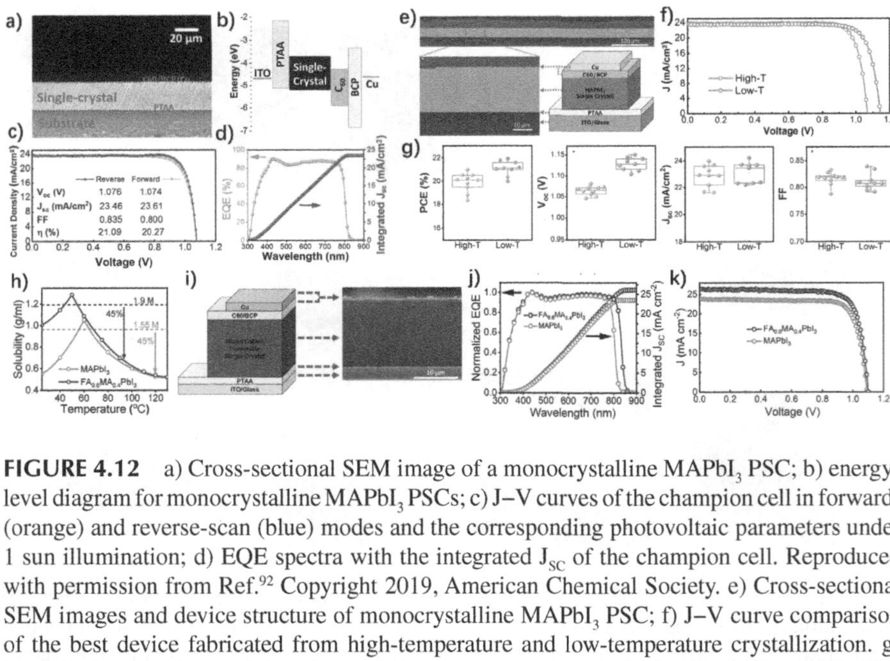

FIGURE 4.12 a) Cross-sectional SEM image of a monocrystalline $MAPbI_3$ PSC; b) energy-level diagram for monocrystalline $MAPbI_3$ PSCs; c) J–V curves of the champion cell in forward-(orange) and reverse-scan (blue) modes and the corresponding photovoltaic parameters under 1 sun illumination; d) EQE spectra with the integrated J_{SC} of the champion cell. Reproduced with permission from Ref.[92] Copyright 2019, American Chemical Society. e) Cross-sectional SEM images and device structure of monocrystalline $MAPbI_3$ PSC; f) J–V curve comparison of the best device fabricated from high-temperature and low-temperature crystallization. g) Statistical comparison of the photovoltaic parameters for ten devices including PCE, V_{OC}, J_{SC}, and FF. All data were measured under 1 sun illumination. Reproduced with permission from Ref.[93] Copyright 2020, American Chemical Society. h) Temperature-dependent solubility of $FA_{0.6}MA_{0.4}PbI_3$ and $MAPbI_3$ in γ-GBL. The horizontal dashed lines represent the starting concentration of the precursor solution. i) Device architecture and cross-sectional SEM image of the monocrystalline PSC; j) EQE spectra and k) J–V curve comparison between $MAPbI_3$ and $FA_{0.6}MA_{0.4}PbI_3$ devices. All measurements were conducted under 1 sun illumination. Reproduced with permission from Ref.[94] Copyright 2021, The Royal Society of Chemistry.

cells and the high performance of devices proved the great development potential of monocrystalline PSCs. Alsalloum et al. proposed a method of low-temperature growth of monocrystalline PVK films to improve the PCE of $MAPbI_3$ single-crystal solar cells.[93] Introducing polycarbonate (PC) molecules into the precursor solution could decrease the nucleation and crystallization temperatures. Therefore, the thermal decomposition of the PVK material during the high-temperature preparation process was avoided, which increased the surface defects and significantly reduced the V_{OC} of devices. The cross-sectional SEM image and device structure of monocrystalline $MAPbI_3$ PSCs was shown in Figure 4.12e. The monocrystalline PVK thin films prepared in low-temperature solution had high crystal quality, long carrier lifetime, and low surface defect density. The champion device showed a J_{SC} of 23.68 mA cm^{-2}, a V_{OC} of 1.140, an FF of 0.810, and a PCE of 21.93% (Figure 4.12f and g). Although the space-confined methods were used to prepare high-quality monocrystalline PVK films, there was still an urgent need to study the industry standard technique for preparing large-area monocrystalline PVK films in a scalable process.

The interface engineering strategy was also the key to improving the performance of monocrystalline PSCs. In 2021, Alsalloum et al. increased the PCE of monocrystalline PSCs to 22.80%.[94] They used a mixed cationic monocrystalline PVK absorption layer ($FA_{0.6}MA_{0.4}PbI_3$), which achieved the redshift of the EQE band edge. Owning to the introduction of FA ions, the solute concentration, solubility, and crystallization temperature in the precursor solution were changed (Figure 4.12h). The mixed cationic monocrystalline PVK thin films with high yield and large area could be successfully prepared via optimizing the solution ratio. The cross-sectional SEM image and device structure of monocrystalline $FA_{0.6}MA_{0.4}PbI_3$ PSCs was shown in Figure 4.12i. The thick single-crystal film also expanded the spectral absorption of the device. Therefore, the PCE of monocrystalline $FA_{0.6}MA_{0.4}PbI_3$ PSC reached 22.80% (Figure 4.12k). Compared with single-component single-crystal devices, mixed cationic single-crystal devices had better thermal stability and lower surface defect density. Xie et al. focused on the photostability problem related to charge defect density and ion migration of monocrystalline PSCs.[95] Thin single-crystal wafers had lower crystal quality, higher defect state density, and more ion migration channels than bulk monocrystalline materials. The monocrystalline PVK films with improved crystal quality grew under rapid diffusion of ions in a confined space. The 20-μm-thick monocrystalline $FA_{0.55}MA_{0.45}PbI_3$ solar cell showed a J_{SC} of 22.75 mA cm^{-2}, a V_{oc} of 1.060 V, an FF of 0.760, and a PCE of 18.41%. The 200-μm-thick devices had a lower PCE of 16.47%. However, the thick devices operated stably after 330 h under V_{mpp} conditions with no loss of PCE. If the quality of 20-μm monocrystalline PVK could be optimized, its stability would be significantly improved. Yang et al. studied the surface orientation engineering of monocrystalline PVK thin films.[96] After adjusting the ratio of PbI_2 and MAI in the precursor solution, monocrystalline PSCs with different surface orientations were obtained and their structure was shown in Figure 4.13a and e. These two kinds of (001) and (100) monocrystalline films exhibited different photoelectric properties because of the different terminating atoms of different surface orientations. PL spectrum (Figure 4.13b

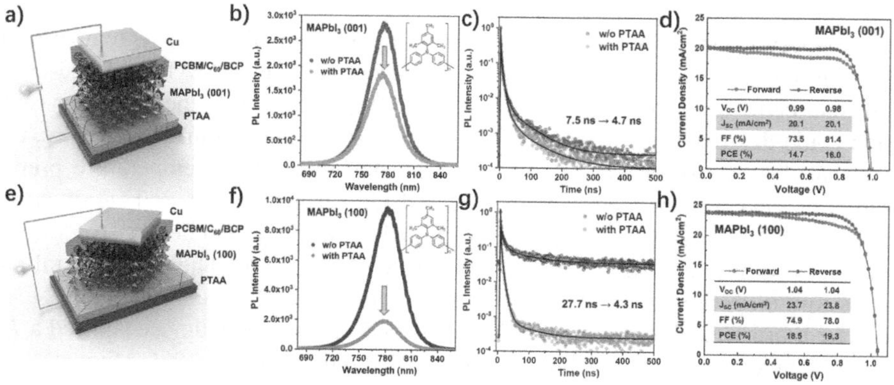

FIGURE 4.13 a, e) Schematic diagram of the $MAPbI_3$ (001) and $MAPbI_3$ (100) single-crystal film solar cells. b, c) Photoluminescence (PL) spectra and time-resolved PL decays for $MAPbI_3$ (001) single crystals without and with the deposition of a PTAA film; f, g) PL spectra and time-resolved PL decays for $MAPbI_3$ (100) single crystals without and with the deposition of a PTAA film; J–V curves of the d) $MAPbI_3$ (001) and h) $MAPbI_3$ (100) solar cells obtained in forward-scan (red) and reverse-scan (green) modes measured in the nitrogen-filled glovebox. The corresponding maximum photovoltaic parameters under 1 sun illumination are given as insets. Reproduced with permission from Ref.[96] Copyright 2022, American Chemical Society

and f) and time-resolved PL decay tests (Figure 4.13c and g) showed that PTAA HTL promoted the charge extraction of devices. The reasonable energy level matching between $MAPbI_3$ terminating surface and carrier transport layer allowed effective charge transfer between interfaces. Studying the relationship between the surface orientation and the interface carrier dynamics had great significance for improving the performance of monocrystalline PSCs.

Previous work demonstrated that high-efficiency monocrystalline thin-film PSCs could be obtained via preparing single crystals on the hydrophobic PTAA HTL. Li et al. introduced hydrophobic P3HT molecules into the PTAA HTL for interacting with uncoordinated Pb^{2+} ions which caused the formation of deep-level defects (Figure 4.14a).[97] Owing to the low surface defects and improved crystal quality, the monocrystalline PSC exhibited the increasing interface carrier lifetime, reducing non-radiative recombination, and high device performance (Figure 4.14b). Furthermore, P3HT:PTAA HTL with a large contact angle decreased the interaction between substrate and solvent ions, which was conducive to rapid ion diffusion, low nucleation density, and fast crystal growth (Figure 4.14c–f). The monocrystalline PSC devices P3HT:PTAA HTL showed a PCE of 22.10%, a V_{OC} of 1.130 V, a J_{SC} of 23.88 mA cm^{-2}, and an FF of 81.80%. Guo et al. found that the bifunctional molecule 3-mercaptopropyl(dimethoxy)methylsilane (MDMS) could passivate surface defects and prevent the hydration on the surface of monocrystalline PVK thin films (Figure 4.14g).[98] The molecule interacted with the uncoordinated Pb^{2+}

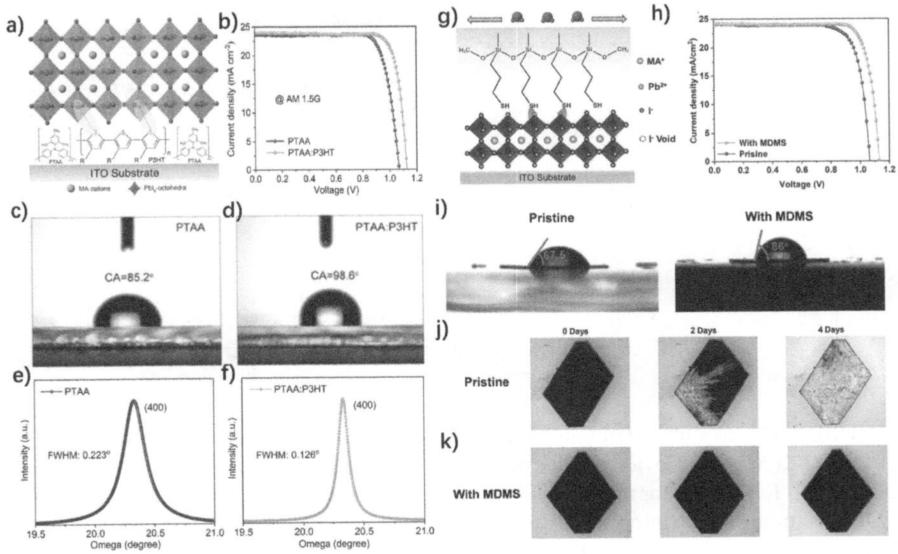

FIGURE 4.14 a) Schematic illustration of the interaction between P3HT molecule and perovskite single crystal; b) J–V curves of the devices with PTAA and PTAA:P3HT HTLs under simulated AM 1.5G illumination. The surface contact angles of c) PTAA and d) PTAA:P3HT HTLs. X-ray rocking curves of monocrystalline $MAPbI_3$ grown on e) PTAA and f) PTAA:P3HT HTLs. Reproduced with permission from Ref.[97] Copyright 2022, WILEY-VCH. g) Schematic illustration of the interaction between MDMS molecules and $MAPbI_3$ crystal surface. h) J–V curves of single-crystal solar cells with and without MDMS modification under AM 1.5 light illumination. i) Contact angles of deionized water on the surface of $MAPbI_3$ thin single crystals with and without MDMS modification. Optical microscope images of the j) pristine and k) MDMS-modified $MAPbI_3$ thin single crystals after aging in high-humidity air for 4 d (T = 25 ± 2 °C, RH = 85 ± 5%). Reproduced with permission from Ref.[98] Copyright 2023, WILEY-VCH.

on the surface of $MAPbI_3$ single crystal by sulfur atoms, reducing the surface trap density and inhibiting the non-radiative recombination. The performance of the MDMS-treatment device reached a J_{SC} of 24.10 mA cm^{-2}, a V_{OC} of 1.130 V, an FF of 81.50%, and a PCE of 22.20%. This was the highest PCE for PSCs based on monocrystalline $MAPbI_3$ (Figure 4.14h). The cross-linking reaction of silane molecules formed a covering layer on the crystal surface, which increased the contact angle and improved the water stability of materials (Figure 4.14i). The buried interface was closely related to the crystal growth and the carrier transport. Almasabi et al. used hydrophilic self-assembled MeO-2PACz molecules as the HTL for fabricating the monocrystalline PSCs.[99] The MeO-2PACz layer had a strong adhesion with the monocrystalline PVK surface, which enhanced the stability of the single-crystal film. The DFT results showed that MeO-2PACz molecules interact strongly with the surface of perovskite (110) (Figure 4.15a). The device

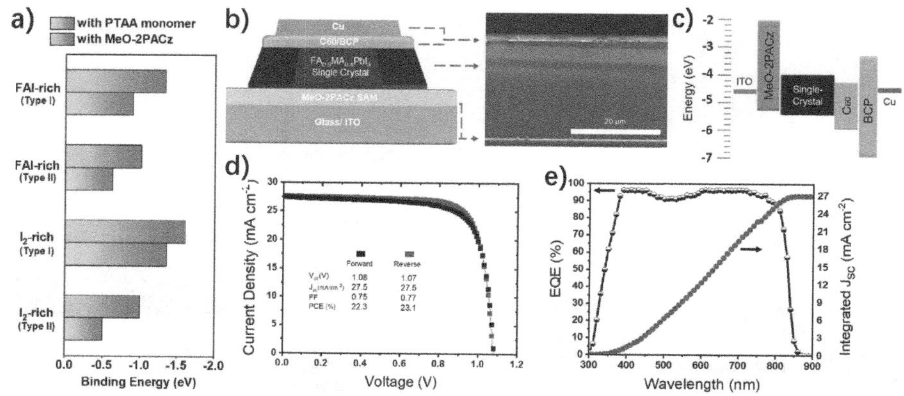

FIGURE 4.15 a) Calculated binding energies of $FAPbI_3$ (110) slabs adsorbed with PTAA and MeO-2PACz. b) Device architecture, cross-sectional SEM image, and c) energy-level diagram. d) J–V curves for the champion monocrystalline PSC under 1 sun illumination and e) EQE spectrum with the integrated *J*sc of the champion cell. Reproduced with permission from Ref.[99] Copyright 2023, American Chemical Society.

structure, cross-sectional SEM image, and energy-level diagram of monocrystalline PSCs was shown in Figure 4.15b and c. The PSCs showed a J_{SC} of 27.50 mA cm^{-2}, a V_{OC} of 1.070 V, an FF of 0.77, and a PCE of 23.10% (Figure 4.15d). After maximum power point tracking (MPPT) measurements, approximately 25% of initial performance was lost after 730 h. Surprisingly, the device recovered about 9% of its performance after 21 days in the dark. One of the potential reasons was the redistribution of ions and the self-healing of defects. Another potential reason was the strong interaction between monocrystalline PVK film with the SAM layer, which reduced the damage to the device in long-term operation. This work opened a new path for monocrystalline PSCs based on the SAMs HTL. Yeddu et al. found that a slow spontaneous efficiency enhancement (SEE) in the PSCs with ITO/PTAA/$MAPbI_3$/C60/BCP/Cu structure.[100] The device showed the SEE phenomenon and the PCE of the device from 22.91% to 34.39% after 14 d and then decreased again. Under white light (the power of 48 mW cm^{-2}) illumination, the *J-V* characteristic of the device exhibited that the V_{OC} is 1.080 V, the J_{SC} is 19.78 mA cm^{-2}, the FF is 77.26%, and the PCE is 34.39%. The test results proved that γ-GBL existed in $MAPbI_3$ single crystals and the amount of γ-GBL captured dramatically decreased with increasing aging time. During the long aging time, the γ-GBL trapped in the PTAA/$MAPbI_3$ interface and the top surface of single crystals escaped, which improved carrier collection at interfaces and increased the J_{SC}. When the monocrystalline PVK device was heated at a low temperature of 40 °C for 12 h, the residual γ-GBL could be effectively removed, resulting in the improving performance of the devices. Finally, the performance of monocrystalline PSCs and the tendency chart of best efficiencies were summarized in Table 4.1 and Figure 4.16, respectively.

TABLE 4.1
Performance of Monocrystalline PSCs

Device architecture	Thickness	J_{SC} (mA cm^{-2})	V_{OC} (V)	FF (%)	PCE (%)	Year	Refs.
FTO/TiO_2/$MAPbBr_3$/Au	1 μm	6.69	1.360	69.0	6.53	2016	86
ITO/PTAA/$MAPbI_3$/PCBM/C_{60}/BCP/Cu	10 μm	21.0	1.080	78.6	17.80	2017	25
Au/$MAPbI_3$/PCBM/Ag	200 nm	18.33	0.800	32.9	4.83	2017	84
FTO/TiO_2/$MAPbBr_3$/Spiro-OMeTAD/Au	16 μm	8.77	1.310	62.0	7.11	2017	87
FTO/TiO_2/$MAPbI_3$/Spiro-OMeTAD/Ag	~3 μm	22.28	0.680	59.0	8.78	2017	88
ITO/NiO_x/$(FAPbI_3)_{0.85}(MAPbBr_3)_{0.15}$/$TiO_2$/Ag	24.5 μm	23.14	1.030	51.0	12.18	2018	89
ITO/PEDOT:PSS/$MAPbI_3$/PCBM/Ag	50 μm	22.15	0.750	27.0	4.40	2018	101
ITO/PTAA/$MAPbI_3$/C_{60}/BCP/Cu	20 μm	23.46	1.076	83.5	21.09	2019	92
Au/$MAPbI_3$/C_{60}/BCP/Au	38 μm	5.06	0.660	44.0	5.90	2019	72
ITO/PTAA/$MAPbI_3$/C_{60}/BCP/Cu	~20 μm	23.68	1.144	81.0	21.93	2020	93
ITO/PEDOT:PSS/$MAPbI_3$/PCBM/BCP/Ag	300 nm	22.60	1.080	82.5	20.10	2020	102
PET/ITO/SnO_2/$MAPb_{0.5+x}Sn_{0.5x}I_3$/Spiro-OMeTAD/Au/PDMS/Su 8	2 μm	24.0	1.050	79.5	20.04	2020	103
Au/$MAPbI_3$/C_{60}/BCP/Au	–	22.49	0.930	55.1	11.52	2020	91
ITO/PTAA/$FA_{0.6}MA_{0.4}PbI_3$/C_{60}/BCP/Cu	15 μm	26.2	1.100	79.0	22.80	2021	94
ITO/PTAA:P3HT/$MAPbI_3$/C_{60}/BCP/Cu	20 μm	23.88	1.130	81.8	22.10	2022	97
ITO/PTAA/$FA_{0.55}MA_{0.45}PbI_3$/C_{60}/BCP/Cu	20 μm	22.75	1.062	76.2	18.41	2022	95
ITO/PTAA/$MAPbI_3$(100)/PCBM/C_{60}/BCP/Cu	20–25 μm	23.80	1.040	78.0	19.30	2022	96
ITO/PTAA/$MAPbI_3$(001)/PCBM/C_{60}/BCP/Cu	20–25 μm	20.10	0.980	81.4	16.00	2022	96
ITO/MeO-2PACz/$FA_{0.6}MA_{0.4}PbI_3$/PCBM/C_{60}/BCP/Cu	~20 μm	27.50	1.070	77.0	23.10	2023	99
ITO/PTAA/$MAPbI_3$/MDMS/C_{60}/BCP/Cu	40 μm	24.1	1.130	81.5	22.20	2023	98

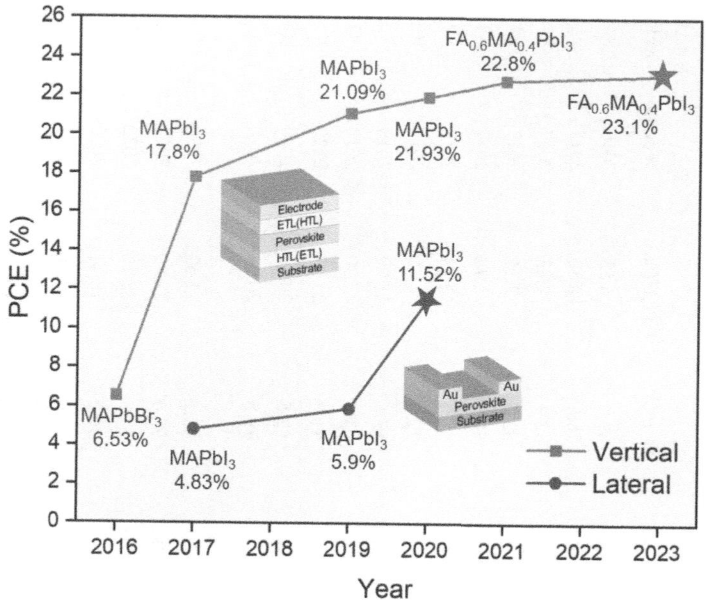

FIGURE 4.16 Best research efficiencies of monocrystalline PSCs from 2016 to 2023 (the line with a square point represents the vertical-structure devices and the line with a circle point represents the lateral-structure devices).

4.6 CONCLUSION AND OUTLOOK

In previous sections, we reviewed the material properties, crystallization process, and growth methods of monocrystalline PVK materials, which determined the solar cell performance as the material basis. The preparation methods of monocrystalline PVK thin films with controllable thickness, large area, and ordered distribution for fabricating the monocrystalline PSCs were summarized. The developments of monocrystalline PSCs with lateral structure and vertical structure were concluded in chronological order. The mechanism of device structure design, surface passivation engineering, and interface engineering for improving the device performance were introduced. In following paragraphs, we divided it into five parts to summarize the problems which existed in monocrystalline PSCs and give promising solutions and guiding suggestions (Figure 4.17).

1. Growth of monocrystalline PVK and fabrication of monocrystalline PSCs: optimizing crystal growth, extending the active area, and reducing crystal thickness are promising solutions to improve the performance of monocrystalline PSCs.[104–107] First, there are still many challenges to the growth of multi-component monocrystalline PVK, such as different crystal thermodynamics and dynamics of different components. The reasonable designs of precursor solution components, ligand-induced nucleation, and solution stabilization strategies are beneficial to the growth of multi-component

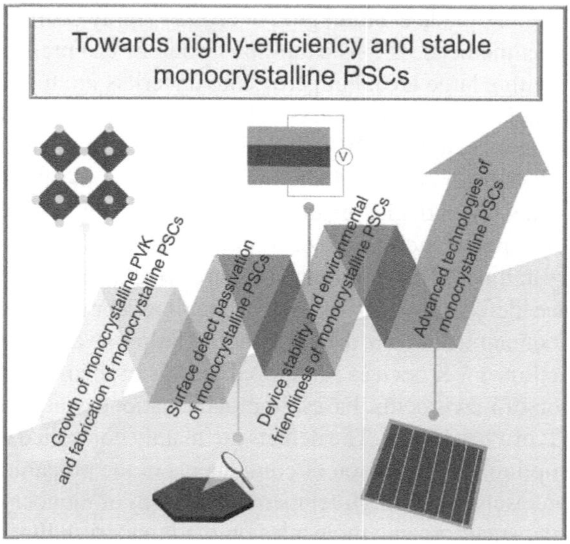

FIGURE 4.17 The methods towards high-efficiency and stable monocrystalline PSCs.

monocrystalline PVK. Second, the common growth methods of STL, ITC, AVC, and BSC are all for growing the independent bulk single crystals.[108] These methods should be combined with the thickness-controlled methods for preparing the single-crystal thin films, such as ITC combined with the space-confined method. A stable supply of saturated solution is needed to ensure the continuous growth of small crystals into large-area single-crystal films. The monocrystalline PVK films with a high aspect ratio are also expected for fabricating large-size monocrystalline PSCs. Third, the thickness of the monocrystalline film should match the carrier diffusion length of materials because the conflict between large thickness and short carrier length results in poor performance of large-area monocrystalline PSCs. The large thickness of monocrystalline film (more than 10 μm) compared to that of polycrystalline films (400–500 nm) reduces the defects density and the driving force of ion migration, which is conducive to the stable operation of the device.[109] However, the thickness of monocrystalline film still needs to be precisely controlled. If the thickness of monocrystalline film is too large, the carriers need to travel a longer distance before being collected and recombination probably occurs. The advanced confined growth methods and vapor deposition methods could effectually reduce the thickness of monocrystalline film.

2. Large-scale fabrication of monocrystalline PSCs: the feasible improvements for large-scale fabricating monocrystalline PVK films by the solution-process method are providing a low temperature, stable solution supersaturation, and protective atmosphere. It is also necessary to improve the utilization of materials and develop scalable solution-process preparation

methods, such as blade coating, groove coating, spray coating, inkjet printing, and other methods.[110,111] The top-down strategy can prepare single-crystal wafers with a large area, but perovskite materials are too fragile to slice the crystal below 50 μm thickness. The technical problems for preparing large-area monocrystalline PVK film by the top-down method are growing perfect bulk monocrystalline PVK ingots, reducing the thickness of bulk single crystal without damaging the surface, and polishing the wafer surface.

3. Surface defect passivation of monocrystalline PSCs: reducing surface defects of monocrystalline PVK is important for improving device performance.[109,112] The surface defects are derived from dangling bonds, dislocations, and chemical contamination on the surface.[113] The passivation engineering of monocrystalline PVK devices is required for this problem. For example, the introduction of Lewis acids, bases, or organic cations could passivate uncoordinated I^- or Pb^{2+} defects. The defects are mainly controlled by the Pb^0 and PbI_2. The optimized proportion of components in the preparation should be adjusted precisely. In the high-temperature growth of monocrystalline PVK thin film, the small organic molecules on the monocrystalline PVK surface are easy to be decomposed via high-temperature heating, resulting in the absence of MAI on the surface. The loss of cations could be prevented by decreasing the growth temperature. Another potential method is washing the residue on the single crystal surface with fresh supersaturated solution. Furthermore, during the cooling process after crystal growth, most monocrystalline PVK materials undergo a temperature-dependent phase transition process, which may introduce defects and strains in the single crystals. How to avoid the phase-transition-caused defects is also an existing problem in the preparation of monocrystalline PVK materials. Furthermore, the non-ideal contact between the monocrystalline perovskite and charge transport layer caused by peel-off process and lattice mismatch also significantly affects device efficiency. This problem could be improved by methods of lattice matching, in-situ growth, and interface passivation. The performance of monocrystalline PSCs could be further improved via reducing the defects density. Studies of defect formation and passivation mechanism, theoretical calculation of surface defect formation energy, and deep understanding of surface defect dynamics are contributed to fabricating the high-efficiency monocrystalline PSCs.
4. Device stability of monocrystalline PSCs: device stability is a great challenge that cannot be ignored in the industrialization and commercialization of monocrystalline PSCs.[114] Optimized stability under high temperature, light, humidity, and oxygen ambient conditions of monocrystalline PSCs should be achieved. In addition, the mechanical properties of the monocrystalline PVK thin film also need to be improved. The thin films grow directly to the conductive substrate. It is easily destroyed during the stripping process, which reduces the interface carrier injection rate and reduces EQE and J_{SC}. How to achieve monocrystalline PVK thin films with good mechanical properties is a problem that needs to be solved. To

improve the stability of monocrystalline PSCs, it is necessary to study improving the crystal quality of monocrystalline PVK materials and optimization of the device structure.

5. Environmental friendliness of monocrystalline PSCs: to produce environmentally friendly monocrystalline PSCs, solvent pollution and lead leakage in the manufacturing and recycling process must be addressed.[115] On the one hand, the synthesis of monocrystalline PVK materials require plenty of organic solvents which all have certain toxicity, which is bad for researchers and the environment. Therefore, it is urgent to develop green solvents to prepare monocrystalline PVK materials and use few solvents to produce high-yield monocrystalline PVK materials. On the other hand, the main devices are the lead-based monocrystalline PSCs presently whose PCE is higher than that of other metal-based PVK solar cells. However, the toxicity of lead could not be ignored, which can cause harm to the human body and the natural environment. Monocrystalline PVK degradation caused by ion migration and phase separation is difficult to prevent by encapsulation. The potential solutions are developing low-toxicity Sn-based or non-toxic monocrystalline PVK materials and stabilizing the degradation and loss of lead elements.
6. Advanced technologies of monocrystalline PSCs: some of the following advanced techniques for improving device performance and stability could be applied in monocrystalline PSCs. The optical management structure could be introduced on the monocrystalline PVK surface to further improve the J_{SC} of the device.[116] The efficiency of the device can be further improved by interconnecting a wide-bandgap monocrystalline PSC with a narrow-bandgap silicon or PSC to achieve a tandem solar cell.[117] The monocrystalline PVK with heterojunction structure exhibits improved carrier transport kinematics and matched interfacial energy level, resulting in optimized performance of the devices.

To achieve practical application of monocrystalline PSCs, the biggest challenge is the growth of high-quality and large-area monocrystalline PVK film. The high surface and interface defect density caused by the surface without passivation and poor interface contact are the main factors for limiting carrier transmission in monocrystalline devices, which seriously affects the device efficiency of monocrystalline PSCs. Therefore, more deep studies are required to fully understand the crystallization and growth of single crystals and the operation and performance attenuation of monocrystalline PSCs, providing the basis for the leap in device performance. Monocrystalline PSCs are a promising development direction at present, which exhibit low-cost, high-efficiency, and long-stability characteristics. Overcoming the technical challenges of monocrystalline PVK thin-film or wafer growth and device design, crystal quality improvement, and surface defects will provide opportunities to approach or even break the Shockley-Queisser limit of solar cells. We believe that monocrystalline PSCs will occupy a place in the photovoltaic market in the future.

REFERENCES

1. T. Baikie, Y. N. Fang, J. M. Kadro, M. Schreyer, F. X. Wei, S. G. Mhaisalkar, M. Graetzel, and T. J. White, Synthesis and crystal chemistry of the hybrid perovskite (CH_3NH_3)PbI_3 for solid-state sensitised solar cell applications. *J. Mater. Chem. A* **2013**, *1*, 5628.
2. T. Leijtens, G. E. Eperon, S. Pathak, A. Abate, M. M. Lee, and H. J. Snaith, Overcoming ultraviolet light instability of sensitized TiO_2 with meso-superstructured organometal trihalide perovskite solar cells. *Nat. Commun.* **2013**, *4*, 2885.
3. M. Z. Liu, M. B. Johnston, and H. J. Snaith, Efficient planar heterojunction perovskite solar cells by vapour deposition. *Nature* **2013**, *501*, 395.
4. J. Park, J. Kim, H.-S. Yun, M. J. Paik, E. Noh, H. J. Mun, M. G. Kim, T. J. Shin, and S. I. Seok, Controlled growth of perovskite layers with volatile alkylammonium chlorides. *Nature* **2023**, *616*, 724.
5. S. Tan *et al.*, Stability-limiting heterointerfaces of perovskite photovoltaics. *Nature* **2022**, *605*, 268.
6. J. Ding and Q. Yan, Progress in organic-inorganic hybrid halide perovskite single crystal: growth techniques and applications. *Sci. China Mater.* **2017**, *60*, 1063.
7. J. Li, Y. Gu, Z. Han, J. Liu, Y. Zou, and X. Xu, Further advancement of perovskite single crystals. *J. Phys. Chem. Lett.* **2022**, *13*, 274.
8. X. Zhao, S. M. Wang, X. Y. Shan, G. Meng, and X. D. Fang, Fabrications of halide perovskite single-crystal slices and their applications in solar cells, photodetectors, and LEDs. *Cryst. Growth Des.* **2021**, *21*, 5983.
9. Y. Li and L. Ding, Single-crystal perovskite devices. *Sci. Bull* **2021**, *66*, 214.
10. J. Li, Z. Han, Y. Gu, D. Yu, J. Liu, D. Hu, X. Xu, and H. Zeng, Perovskite single crystals: synthesis, optoelectronic properties, and application. *Adv. Funct. Mater.* **2021**, *31*, 2008684.
11. T. Oku, Crystal structures of perovskite halide compounds used for solar cells. *Rev. Adv. Mater. Sci.* **2020**, *59*, 264.
12. H. Li, N. Shen, S. Chen, F. Guo, and B. Xu, Recent progress on synthesis, intrinsic properties and optoelectronic applications of perovskite single crystals. *Adv. Funct. Mater.* **2023**, *33*, 2214339.
13. W. Peng *et al.*, Engineering of $CH_3NH_3PbI_3$ perovskite crystals by alloying large organic cations for enhanced thermal stability and transport properties. *Angew. Chem. Int. Ed.* **2016**, *55*, 10686.
14. E. C. Schueller, G. Laurita, D. H. Fabini, C. C. Stoumpos, M. G. Kanatzidis, and R. Seshadri, Crystal structure evolution and notable thermal expansion in hybrid perovskites formamidinium tin Iodide and formamidinium lead bromide. *Inorg. Chem.* **2018**, *57*, 695.
15. Y. Zhang, Y. Liu, and S. Liu, Composition engineering of perovskite single crystals for high-performance optoelectronics. *Adv. Funct. Mater.* **2022**, *33*, 2210335.
16. J. X. Ding, L. Jing, X. H. Cheng, Y. Zhao, S. J. Du, X. Y. Zhan, and H. Z. Cui, Design growth of $MAPbI_3$ single crystal with (220) facets exposed and its superior optoelectronic properties. *J. Phys. Chem. Lett.* **2018**, *9*, 216.
17. M. H. Jao, C. F. Lu, P. Y. Tai, and W. F. Su, Precise facet engineering of perovskite single crystals by ligand-mediated strategy. *Cryst. Growth Des.* **2017**, *17*, 5945.
18. L. Ma *et al.*, A polymer controlled nucleation route towards the generalized growth of organic-inorganic perovskite single crystals. *Nat. Commun.* **2021**, *12*, 2023.
19. Y. J. Fang, H. T. Wei, Q. F. Dong, and J. S. Huang, Quantification of re-absorption and re-emission processes to determine photon recycling efficiency in perovskite single crystals. *Nat. Commun.* **2017**, *8*, 14417.
20. E. M. Hutter, M. C. Gelvez-Rueda, A. Osherov, V. Bulovic, F. C. Grozema, S. D. Stranks, and T. J. Savenije, Direct–indirect character of the bandgap in methylammonium lead iodide perovskite. *Nat. Mater.* **2017**, *16*, 115.

21. T. Wang, B. Daiber, J. M. Frost, S. A. Mann, E. C. Garnett, A. Walsh, and B. Ehrler, Indirect to direct bandgap transition in methylammonium lead halide perovskite. *Energy Environ. Sci.* **2017**, *10*, 509.
22. A. Goyal, S. McKechnie, D. Pashov, W. Tumas, M. van Schilfgaarde, and V. Stevanovic, Origin of pronounced nonlinear band gap behavior in lead–tin hybrid perovskite alloys. *Chem. Mater.* **2018**, *30*, 3920.
23. Y. Ogomi *et al.*, $CH_3NH_3Sn_xPb_{(1-x)}I_3$ perovskite solar cells covering up to 1060 nm. *J. Phys. Chem. Lett.* **2014**, *5*, 1004.
24. Y. Liu *et al.*, Two-inch-sized perovskite $CH_3NH_3PbX_3$ (X=Cl,Br,I) crystals: growth and characterization. *Adv. Mater.* **2015**, *27*, 5176.
25. Z. Chen *et al.*, Thin single crystal perovskite solar cells to harvest below-bandgap light absorption. *Nat. Commun.* **2017**, *8*, 1890.
26. D. Shi *et al.*, Low trap-state density and long carrier diffusion in organolead trihalide perovskite single crystals. *Science* **2015**, *347*, 519.
27. A. A. Zhumekenov *et al.*, Formamidinium lead halide perovskite crystals with unprecedented long carrier dynamics and diffusion length. *ACS Energy Lett.* **2016**, *1*, 32.
28. B. Turedi *et al.*, Single-crystal perovskite solar cells exhibit close to half a millimeter electron-diffusion length. *Adv. Mater.* **2022**, *34*, e2202390.
29. Q. Dong, Y. Fang, Y. Shao, P. Mulligan, J. Qiu, L. Cao, and J. Huang, Electron-hole diffusion lengths > 175 μm in solution-grown $CH_3NH_3PbI_3$ single crystals. *Science* **2015**, *347*, 967.
30. C. Eames, J. M. Frost, P. R. F. Barnes, B. C. O'Regan, A. Walsh, and M. S. Islam, Ionic transport in hybrid lead iodide perovskite solar cells. *Nat. Commun.* **2015**, *6*, 7497.
31. J. Xing, Q. Wang, Q. F. Dong, Y. B. Yuan, Y. J. Fanga, and J. S. Huang, Ultrafast ion migration in hybrid perovskite polycrystalline thin films under light and suppression in single crystals. *Phys. Chem. Chem. Phys* **2016**, *18*, 30484.
32. Y. Yang, Y. Yan, M. J. Yang, S. Choi, K. Zhu, J. M. Luther, and M. C. Beard, Low surface recombination velocity in solution-grown $CH_3NH_3PbBr_3$ perovskite single crystal. *Nat. Commun.* **2015**, *6*, 7961.
33. Y. Shao, Z. Xiao, C. Bi, Y. Yuan, and J. Huang, Origin and elimination of photocurrent hysteresis by fullerene passivation in $CH_3NH_3PbI_3$ planar heterojunction solar cells. *Nat. Commun.* **2014**, *5*, 5784.
34. Z. Ni *et al.*, Resolving spatial and energetic distributions of trap states in metal halide perovskite solar cells. *Science* **2020**, *367*, 1352.
35. C. C. Wang, B. R. Ecker, H. T. Wei, J. S. Huang, and Y. L. Gao, Environmental surface stability of the $MAPbBr_3$ single crystal. *J. Phys. Chem. C* **2018**, *122*, 3513.
36. D. Bryant, N. Aristidou, S. Pont, I. Sanchez-Molina, T. Chotchunangatchaval, S. Wheeler, J. R. Durrant, and S. A. Haque, Light and oxygen induced degradation limits the operational stability of methylammonium lead triiodide perovskite solar cells. *Energy Environ. Sci.* **2016**, *9*, 1655.
37. B. P. Dhamaniya, P. Chhillar, B. Roose, V. Dutta, and S. K. Pathak, Unraveling the effect of crystal structure on degradation of methylammonium lead halide perovskite. *ACS Appl. Mater. Interfaces* **2019**, *11*, 22228.
38. C. Ma, M.-C. Kang, S.-H. Lee, S. J. Kwon, H.-W. Cha, C.-W. Yang, and N.-G. Park, Photovoltaically top-performing perovskite crystal facets. *Joule* **2022**, *6*, 2626.
39. C. Ma, F. T. Eickemeyer, S.-H. Lee, D.-H. Kang, S. J. Kwon, M. Grätzel, and N.-G. Park, Unveiling facet-dependent degradation and facet engineering for stable perovskite solar cells. *Science* **2023**, *379*, 173.
40. Z. Song, A. Abate, S. C. Watthage, G. K. Liyanage, A. B. Phillips, U. Steiner, M. Graetzel, and M. J. Heben, Perovskite solar cell stability in humid air: partially reversible phase transitions in the PbI_2-CH_3NH_3I-H_2O system. *Adv. Energy Mater.* *6*, 1600846 **2016**.

41. Q. Wang, B. Chen, Y. Liu, Y. H. Deng, Y. Bai, Q. F. Dong, and J. S. Huang, Scaling behavior of moisture-induced grain degradation in polycrystalline hybrid perovskite thin films. *Energy Environ. Sci.* **2017**, *10*, 516.
42. G. Grancini *et al.*, $CH_3NH_3PbI_3$ perovskite single crystals: surface photophysics and their interaction with the environment. *Chem. Sci. 6*, 7305 **2015.**
43. Y. C. Shao *et al.*, Grain boundary dominated ion migration in polycrystalline organic–inorganic halide perovskite films. *Energy Environ. Sci.* **2016**, *9*, 1752.
44. L. Q. Xie, L. Chen, Z. A. Nan, H. X. Lin, T. Wang, D. P. Zhan, J. W. Yan, B. W. Mao, and Z. Q. Tian, Understanding the cubic phase stabilization and crystallization kinetics in mixed cations and halides perovskite single crystals. *J. Am. Chem. Soc.* **2017**, *139*, 3320.
45. L. Chen *et al.*, Toward long-term stability: single-crystal alloys of cesium-containing mixed cation and mixed halide perovskite. *J. Am. Chem. Soc.* **2019**, *141*, 1665.
46. Y. Dang, Y. Liu, Y. Sun, D. Yuan, X. Liu, W. Lu, G. Liu, H. Xia, and X. Tao, Bulk crystal growth of hybrid perovskite material $CH_3NH_3PbI_3$. *CrystEngComm.* **2015**, *17*, 665.
47. Z. Lian *et al.*, High-performance planar-type photodetector on (100) facet of $MAPbI_3$ single crystal. *Sci. Rep. 5*, 16563 **2015.**
48. Z. P. Lian, Q. F. Yan, T. T. Gao, J. Ding, Q. R. Lv, C. G. Ning, Q. Li, and J. L. Sun, Perovskite $CH_3NH_3PbI_3$(Cl) single crystals: rapid solution growth, unparalleled crystalline 1uality, and low trap density toward 10^8 cm^{-3}. *J. Am. Chem. Soc.* **2016**, *138*, 9409.
49. Y. J. Fang, Q. F. Dong, Y. C. Shao, Y. B. Yuan, and J. S. Huang, Highly narrowband perovskite single-crystal photodetectors enabled by surface-charge recombination. *Nat. Photon.* **2015**, *9*, 679.
50. J. M. Kadro, K. Nonomura, D. Gachet, M. Graetzel, and A. Hagfeldt, Facile route to freestanding $CH_3NH_3PbI_3$ crystals using inverse solubility. *Sci. Rep. 5*, 11654 **2015.**
51. M. I. Saidaminov *et al.*, High-quality bulk hybrid perovskite single crystals within minutes by inverse temperature crystallization. *Nat. Commun.* **2015**, *6*, 7586.
52. Y. X. Zhang, Y. C. Liu, Y. J. Li, Z. Yang, and S. Z. Liu, Perovskite $CH_3NH_3Pb(Br_xI_{1-x})_3$ single crystals with controlled composition for fine-tuned bandgap towards optimized optoelectronic applications. *J. Mater. Chem. C* **2016**, *4*, 9172.
53. W. F. Wang, J. Su, L. Zhang, Y. Lei, D. Wang, D. Lu, and Y. Bai, Growth of mixed-halide perovskite single crystals. *CrystEngComm.* **2018**, *20*, 1635.
54. M. I. Saidaminov, A. L. Abdelhady, G. Maculan, and O. M. Bakr, Retrograde solubility of formamidinium and methylammonium lead halide perovskites enabling rapid single crystal growth. *Chem. Commun.* **2015**, *51*, 17658.
55. Z. H. Yuan *et al.*, Growing $MASnI_3$ perovskite single-crystal films by inverse temperature crystallization. *J. Phys. Conf. Ser.* **2022**, *34*, 144009.
56. R. Jiang *et al.*, Low-cost fabrication of efficient perovskite photovoltaics using low-purity lead lodide via powder engineering. *ACS Appl. Mater. Interfaces* **2023**, *15*, 20958.
57. Q. Liao, K. Hu, H. H. Zhang, X. D. Wang, J. N. Yao, and H. B. Fu, Perovskite microdisk microlasers self-assembled from solution. *Adv. Mater.* **2015**, *27*, 3405.
58. J. X. Ding, S. J. Du, Z. Y. Zuo, Y. Zhao, H. Z. Cui, and X. Y. Zhan, High detectivity and rapid response in perovskite $CsPbBr_3$ single-crystal photodetector. *J. Phys. Chem. C* **2017**, *121*, 4917.
59. F. Ledee, G. Trippe-Allard, H. Diab, P. Audebert, D. Garrot, J. S. Lauret, and E. Deleporte, Fast growth of monocrystalline thin films of 2D layered hybrid perovskite. *CrystEngComm.* **2017**, *19*, 2598.
60. H. Tian, L. F. Zhao, X. F. Wang, Y. W. Yeh, N. Yao, B. P. Rand, and T. L. Ren, Extremely low operating current resistive memory based on exfoliated 2D perovskite single crystals for neuromorphic computing. *ACS Nano* **2017**, *11*, 12247.
61. C. T. Zuo and L. M. Ding, Lead-free perovskite materials $(NH_4)_3Sb_2I_xBr_{9-x}$. *Angew. Chem. Int. Ed.* **2017**, *56*, 6528.

62. J. Song *et al.*, Ultralarge all-Inorganic perovskite bulk single crystal for high-performance visible–infrared dual-modal photodetectors. *Adv. Opt. Mater.* **2017**, *5*, 1700157.
63. P. Zhang, Q. Sun, Y. Xu, X. Li, L. Liu, G. Zhang, and X. Tao, Enhancing carrier transport properties of melt-grown $CsPbBr_3$ single crystals by eliminating inclusions. *Cryst. Growth Des.* **2020**, *20*, 2424.
64. X. Li *et al.*, Ultralow detection limit and robust hard X-ray imaging detector based on inch-sized lead-free perovskite $Cs_3Bi_2Br_9$ single crystals. *ACS Appl. Mater. Interfaces* **2022**, *14*, 9340.
65. Y.-X. Chen *et al.*, General space-confined on-substrate fabrication of thickness-adjustable hybrid perovskite single-crystalline thin films. *J. Am. Chem. Soc.* **2016**, *138*, 16196.
66. Y. Liu, Y. Zhang, Z. Yang, D. Yang, X. Ren, L. Pang, and S. Liu, Thinness- and shape-controlled growth for ultrathin single-crystalline perovskite wafers for mass production of superior photoelectronic devices. *Adv. Mater.* **2016**, *28*, 9204.
67. Z. Yang *et al.*, High-performance single-crystalline perovskite thin-film photodetector. *Adv. Mater.* **2018**, *30*, 1704333.
68. Y. Song *et al.*, Detector-grade perovskite single-crystal wafers via stress-free gel-confined solution growth targeting high-resolution ionizing radiation detection. *Light Sci. Appl.* **2023**, *12*, 85.
69. Y. C. Liu *et al.*, 120 mm single-crystalline perovskite and wafers: towards viable applications. *Sci. China Chem.* **2017**, *60*, 1367.
70. Q. R. Lv, Z. P. Lian, W. H. He, J. L. Sun, Q. Li, and Q. F. Yan, A universal top-down approach toward thickness-controllable perovskite single-crystalline thin films. *J. Mater. Chem. C* **2018**, *6*, 4464.
71. A. A. Zhumekenov *et al.*, The role of surface tension in the crystallization of metal halide perovskites. *ACS Energy Lett.* **2017**, *2*, 1782.
72. Y. Liu, Q. Dong, Y. Fang, Y. Lin, Y. Deng, and J. Huang, Fast growth of thin $MAPbI_3$ crystal wafers on aqueous solution surface for efficient lateral-structure perovskite solar cells. *Adv. Funct. Mater.* **2019**, *29*, 1807707.
73. K. Wang, C. C. Wu, D. Yang, Y. Y. Jiang, and S. Priya, Quasi-two-dimensional halide perovskite single crystal photodetector. *ACS Nano* **2018**, *12*, 4919.
74. Y. Liu *et al.*, Surface-tension-controlled crystallization for high-quality 2D perovskite single crystals for ultrahigh photodetection. *Matter* **2019**, *1*, 465.
75. J. Chen *et al.*, Single-crystal thin films of cesium lead bromide perovskite epitaxially grown on metal oxide perovskite ($SrTiO_3$). *J. Am. Chem. Soc.* *139*, 13525 **2017**.
76. Y. Wang, X. Sun, Z. Chen, Y.-Y. Sun, S. Zhang, T.-M. Lu, E. Wertz, and J. Shi, High-temperature Ionic epitaxy of halide perovskite thin film and the hidden carrier dynamics. *Adv. Mater.* **2017**, *29*, 1702643.
77. Y. Wang, F. Yang, X. Li, F. Ru, P. Liu, L. Wang, W. Ji, J. Xia, and X. Meng, Epitaxial growth of large-scale orthorhombic $CsPbBr_3$ perovskite thin films with anisotropic photoresponse property. *Adv. Funct. Mater.* **2019**, *29*, 1904913.
78. Y. G. Zhong *et al.*, Large-scale thin $CsPbBr_3$ single-crystal film grown on sapphire via chemical vapor deposition: toward laser array application. *ACS Nano* **2020**, *14*, 15605.
79. Y. Lei *et al.*, Controlled homoepitaxial growth of hybrid perovskites. *Adv. Mater.* **2018**, *30*, 1705992.
80. Y. S. Lei *et al.*, A fabrication process for flexible single-crystal perovskite devices. *Nature* **2020**, *583*, 790.
81. G. Wang *et al.*, Wafer-scale growth of large arrays of perovskite microplate crystals for functional electronics and optoelectronics. *Sci. Adv.* **2015**, *1*, e1500613.
82. Z. Gu, Z. Huang, C. Li, M. Li, and Y. Song, A general printing approach for scalable growth of perovskite single-crystal films. *Sci. Adv.* **2018**, *4*, eaat2390.

83. Z. Xu *et al.*, Controlled on-chip fabrication of large-scale perovskite single crystal arrays for high-performance laser and photodetector integration. *Light Sci. Appl.* **2023**, *12*, 67.
84. L. Lee, J. Baek, K. S. Park, Y.-E. Lee, N. K. Shrestha, and M. M. Sung, Wafer-scale single-crystal perovskite patterned thin films based on geometrically-confined lateral crystal growth. *Nat. Commun.* **2017**, *8*, 15882.
85. Y. Y. Sun, X. Y. Liu, W. Deng, X. J. Zhang, Z. J. Lu, Z. Z. Chang, X. C. Fang, D. Wu, and J. S. Jie, A three-dimensional confined crystallization strategy toward controllable growth of high-quality and large-area perovskite single crystals. *Adv. Funct. Mater.* **2022**, *32*, 2112758.
86. W. Peng *et al.*, Solution-grown monocrystalline hybrid perovskite films for hole-transporter-free solar cells. *Adv. Mater.* **2016**, *28*, 3383.
87. H.-S. Rao, B.-X. Chen, X.-D. Wang, D.-B. Kuang, and C.-Y. Su, A micron-scale laminar $MAPbBr_3$ single crystal for an efficient and stable perovskite solar cell. *Chem. Commun.* **2017**, *53*, 5163.
88. J. J. Zhao *et al.*, Single crystalline $CH_3NH_3PbI_3$ self-grown on FTO/TiO_2 substrate for high efficiency perovskite solar cells. *Sci. Bull.* **2017**, *62*, 1173.
89. Y. Huang, Y. Zhang, J. Sun, X. Wang, J. Sun, Q. Chen, C. Pan, and H. Zhou, The exploration of carrier behavior in the inverted mixed perovskite single-crystal solar cells. *Adv. Mater. Interfaces* **2018**, *5*, Unsp 1800224.
90. Q. F. Dong, J. F. Song, Y. J. Fang, Y. C. Shao, S. Ducharme, and J. S. Huang, Lateral-structure single-crystal hybrid perovskite solar cells via piezoelectric poling. *Adv. Mater.* **2016**, *28*, 2816.
91. Y. L. Song, W. H. Bi, A. R. Wang, X. T. Liu, Y. F. Kang, and Q. F. Dong, Efficient lateral-structure perovskite single crystal solar cells with high operational stability. *Nat. Commun.* **2020**, *11*, 274.
92. Z. L. Chen, B. Turedi, A. Y. Alsalloum, C. Yang, X. P. Zheng, I. Gereige, A. AlSaggaf, O. F. Mohammed, and O. M. Bakr, Single-crystal $MAPbI_3$ perovskite solar cells exceeding 21% power conversion efficiency. *ACS Energy Lett.* **2019**, *4*, 1258.
93. A. Y. Alsalloum *et al.*, Low-temperature crystallization enables 21.9% efficient single-crystal $MAPbI_3$ inverted perovskite solar cells. *ACS Energy Lett.* **2020**, *5*, 657.
94. A. Y. Alsalloum, B. Turedi, K. Almasabi, X. P. Zheng, R. Naphade, S. D. Stranks, O. F. Mohammed, and O. M. Bakr, 22.8%-Efficient single-crystal mixed-cation inverted perovskite solar cells with a near-optimal bandgap. *Energy Environ. Sci.* **2021**, *14*, 2263.
95. S. D. Xie *et al.*, Bulk defect suppression of micrometer-thick perovskite single crystals enables stable photovoltaics. *ACS Materials Lett.* **2022**, 4, 7, 1332–1340.
96. C. Yang, J. Yin, H. Li, K. Almasabi, L. Gutierrez-Arzaluz, I. Gereige, J.-L. Bredas, O. M. Bakr, and O. F. Mohammed, Engineering surface orientations for efficient and stable hybrid perovskite single-crystal solar cells. *ACS Energy Lett.* **2022**, *7*, 1544.
97. N. Li *et al.*, Engineering the hole extraction interface enables single-crystal $MAPbI_3$ perovskite solar cells with efficiency exceeding 22% and superior indoor response. *Adv. Energy Mater.* **2022**, *12*, 2103241.
98. X. Guo *et al.*, Mitigating surface deficiencies of perovskite single crystals enables efficient solar cells with enhanced moisture and reverse-bias stability. *Adv. Funct. Mater.* **2023**, 2213995.
99. K. Almasabi *et al.*, Hole-transporting self-Assembled monolayer enables efficient single-crystal perovskite solar cells with enhanced stability. *ACS Energy Lett.* **2023**, *8*, 950.
100. V. Yeddu *et al.*, Slow spontaneous efficiency enhancement of single-crystal perovskite solar cells due to trapped solvent. *ACS Appl. Energy Mater* **2023**, *6*, 2257.
101. H.-L. Yue, H.-H. Sung, and F.-C. Chen, Seeded space-limited crystallization of $CH_3NH_3PbI_3$ single-crystal plates for perovskite solar cells. *Adv. Electron. Mater.* **2018**, *4*, 1700655.

102. W. C. Kong *et al.*, Ultrathin perovskite monocrystals boost the solar cell performance. *Adv. Energy Mater.* **2020**, *10*, 2000453.
103. Y. S. Lei *et al.*, A fabrication process for flexible single-crystal perovskite devices. *Nature* **2020**, *583*, 790.
104. X. M. Jiang, X. W. Fu, D. X. Ju, S. Yang, Z. L. Chen, and X. T. Tao, Designing large-area single-crystal perovskite solar cells. *ACS Energy Lett.* **2020**, *5*, 1797.
105. Y. S. Lei, Y. M. Chen, and S. Xu, Single-crystal halide perovskites: opportunities and challenges. *Matter.* **2021**, *4*, 2266.
106. W. X. Gao, Z. L. Zhang, R. K. Xu, E. M. Chan, G. L. Yuan, and J. M. Liu, Development and prospects of halide perovskite single crystal films. *Adv. Electron. Mater.* **2022**, *8*, 2100980.
107. X.-D. Wang, W.-G. Li, J.-F. Liao, and D.-B. Kuang, Recent advances in halide perovskite single-crystal thin films: fabrication methods and optoelectronic applications. *Sol. RRL* **2019**, *3*, 1800294.
108. Y. Liu, Z. Yang, and S. Liu, Recent progress in single-crystalline perovskite research Iincluding crystal preparation, property evaluation, and applications. *Adv. Sci.* **2018**, *5*, 1700471.
109. X. Cheng, S. Yang, B. Q. Cao, X. T. Tao, and Z. L. Chen, Single crystal perovskite solar cells: development and perspectives. *Adv. Funct. Mater.* **2020**, *30*, 1905021.
110. Y. H. Deng, X. P. Zheng, Y. Bai, Q. Wang, J. J. Zhao, and J. S. Huang, Surfactant-controlled ink drying enables high-speed deposition of perovskite films for efficient photovoltaic modules. *Nat. Energy* **2018**, *3*, 560.
111. M. A. Haque, J. Troughton, and D. Baran, Processing-performance evolution of perovskite solar cells: from large grain polycrystalline films to single crystals. *Adv. Energy Mater.* *10*, 1902762 **2020**.
112. K. Wang, D. Yang, C. Wu, J. Shapter, and S. Priya, Mono-crystalline perovskite photovoltaics toward ultrahigh efficiency? *Joule* **2019**, *3*, 311.
113. M. Ghasemi, S. Yuan, J. Fan, B. Jia, and X. Wen, Challenges in the development of metal-halide perovskite single crystal solar cells. *J. Mater. Chem. A* **2023**, *11*, 3822.
114. Y. Rong, Y. Hu, A. Mei, H. Tan, M. I. Saidaminov, S. I. Seok, M. D. McGehee, E. H. Sargent, and H. Han, Challenges for commercializing perovskite solar cells. *Science* **2018**, *361*, eaat8235.
115. M. G. Ju, M. Chen, Y. Y. Zhou, J. Dai, L. Ma, N. P. Padture, and X. C. Zeng, Toward eco-friendly and stable perovskite materials for photovoltaics. *Joule* **2018**, *2*, 1231.
116. C. Q. Ma and N. G. Park, A realistic methodology for 30% efficient perovskite solar cells. *Chem* **2020**, *6*, 1254.
117. Q. Fu and A. K. Y. Jen, Perovskite solar cell developments, what's next? *Next Energy* **2023**, *1*, 100004.

5 Flexible Perovskite Solar Cells

Zhu Ma, Yanlin Li, Shanyue Hou, Zhangfeng Huang, and Zhuowei Du

5.1 FLEXIBLE SUBSTRATE

Flexible substrates play a critical role in the realization of flexible electronic devices, particularly in the field of flexible perovskite solar cells (FPSCs). These substrates offer a range of benefits such as flexibility, lightweight properties, and portability, which are essential for FPSCs.[1,2] To ensure optimal photovoltaic performance in flexible perovskite solar cells, flexible substrates must satisfy several requirements: (i) high optical transparency: an elevated level of optical transparency is crucial for maximizing light absorption by the perovskite layer. Additionally, substrates with high haze can promote further light utilization. (ii) Mechanical durability and flexibility: the substrate material must possess sufficient strength to endure subsequent fabrication processes and normal wear and tear during usage. Furthermore, it should demonstrate enough flexibility to withstand deformation when being bent or flexed. (iii) Chemical stability: the flexible substrate should exhibit inertness towards solvents commonly employed in perovskite solar cells, ensuring compatibility and preventing chemical reactions that can degrade the device. (4) High heat resistance: to avoid shape deformation and potential damage, the coefficient of thermal expansion of the substrate should match closely with that of adjacent layers, mitigating the risk of mechanical stress and cracking during temperature fluctuations. (5) Environmental friendliness: there is a growing urgency to develop renewable and biodegradable materials for flexible substrates to minimize environmental pollution.[3] Additionally, excellent oxygen and water barrier properties, along with low surface roughness, are vital for facilitating the growth of high-quality perovskite films and achieving stable device performance.

Flexible substrates for FPSCs are typically categorized into four main types: polymer-based substrates,[4,5] metal foils,[6] biodegradable substrates,[7] and flexible glass (Figure 5.1).[8] Table 5.1 summarizes the device performance of lightweight F-PSCs based on these flexible substrates.

5.1.1 Polymer

Currently, most reported high-weight power lightweight F-PSCs typically rely on conductive polymer substrates, such as polyethylene terephthalate (PET),[9] polyethylene 2,6-naphthalate (PEN),[5] and polyimide (PI).[10,11] These polymer materials possess

 DOI: 10.1201/9781003400493-5

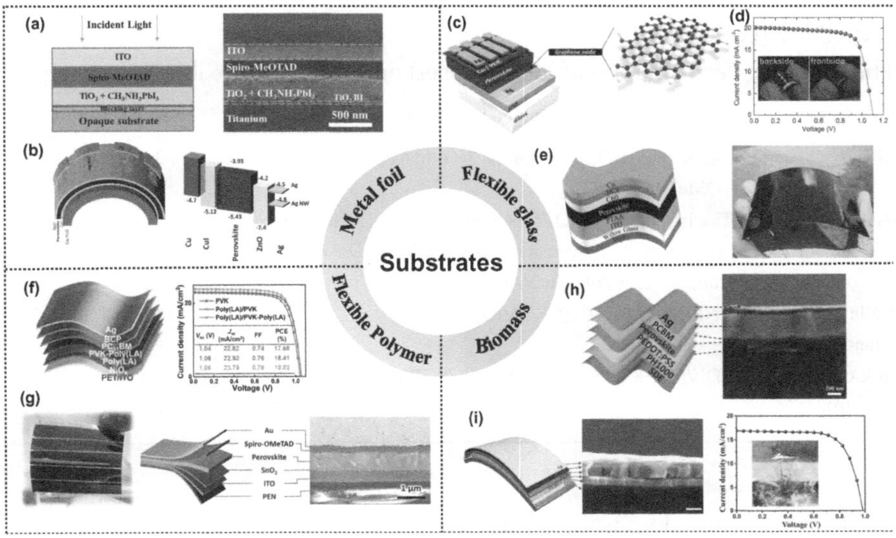

FIGURE 5.1 Different flexible substrates for FPSCs. (a) Schematic image and cross-sectional image of PSCs using a Ti substrate.[32] Copyright (2015), Royal Society of Chemistry; (b) Schematic illustration of the device architecture and energy-level diagram of Cu/CuI/Perovskite/ZnO/Ag.[22] Copyright (2017), Royal Society of Chemistry; (c) Schematic structure of a typical perovskite solar cell with a GO HTL.[8] Copyright (2021), American Chemical Society; (d) J–V characteristics of flexible PSCs on PET substrates with an optimized APM1 electrode.[31] Copyright (2019), Wiley; (e) Schematic illustration of a FPSCs structure and photograph of a flexible perovskite module.[30] Copyright (2020), Wiley-VCH; (f) Structure diagram of flexible device and J–V curves of flexible devices.[27] Copyright (2023), Wiley-VCH; (g) Photograph of a flexible PSM with a series connection of five sub-cells; structure diagram and cross-sectional SEM image of the flexible PSC.[28] Copyright (2023), Springer; (h) Device architecture and cross-sectional SEM image of flexible PSCs.[33] Copyright (2021), Wiley-VCH; (i) The schematic diagram of the flexible PSCs structure; the cross-section SEM image of the perovskite films based on the b-CNF electrode and J–V curve of the b-CNF based PSCs.[7] Copyright (2019), Wiley-VCH.

TABLE 5.1
Photovoltaic Performance of FPSCs Based on Different Flexible Substrates

Substrate	Device structure	PCE (%)	Ref.
PET	PET/ITO/NiOX/Poly (LA)/PVK-Poly (LA)/$PC_{61}BM$/BCP/Ag	19.03%	27
PEN	PEN/ITO/SnO_2/PVK/Spiro-OMeTAD/Au	21.8%	28
PET	PET/ITO/SnO_2/PVK/Spiro-OMeTAD/Au	23.94%	29
Glass	MgF_2/Willow glass/ITO/PTAA/$MAPbI_3$/C_{60}/BCP/Cu	19.72	30
Glass	Glass/APM/ZnO/PVK/Spiro-OMeTAD/MoO_3/Ag	17.06%	31

(Continued)

TABLE 5.1 *(Continued)*
Photovoltaic Performance of FPSCs Based on Different Flexible Substrates

Substrate	Device structure	PCE (%)	Ref.
Ti foil	Ti/compact TiO_2/mesoporous TiO_2/$MAPbI_3$/Spiro-OMeTAD/Ag/ITO	11.01%	32
Ti	Ti wire/compact TiO_2/mesoporous TiO_2/CH3NH_3PbI_3-xClx (vapor-assisted deposition)/Spiro-OMeTAD/Au/Au wire	10.79%	6
Cu foil	Cu/CuI/Perovskite/ZnO/Ag	12.80	22
Cellulose nanofibril	CNF/IZO/PEDOT:PSS/PVK/PCBM/Ag	11.68%	7
Silk-derived electrodes	SDE/PH1000/PEDOT:PSS/PVK/PCBM/Ag	10.4%	33

excellent flexibility, low weight, high corrosion resistance, high optical transparency, good mechanical properties, and commercial availability.[12,13] PET is widely regarded as an outstanding flexible substrate due to its main chain containing methylene groups, aromatic rings, and esters, with conjugation formed between the ester and aromatic rings, imparting strong rigidity to PET macromolecules. Furthermore, high tensile strength, stiffness, hardness, and good wear resistance enable PET to maintain excellent mechanical performance over a wide temperature range.

However, the drawbacks of polymer substrates include poor water and oxygen barrier properties and relatively low thermal resistance. The glass transition temperatures of PET, PEN, and PI are 78, 120, and 200 °C, respectively.[14] When PET and PEN have poor thermal resistance, deformation occurs at high processing temperatures. The resistance of ITO deposited on PET substrates significantly increases when annealing temperature exceeds 200 °C. Therefore, the entire device fabrication process must be carried out at low temperatures. Considering that the processing temperature required for perovskite thin films is typically below 150 °C, efforts have been made to prepare high-quality charge transport layers at low temperatures to enhance the photovoltaic performance of F-PSCs. PI can withstand temperatures as high as 400 °C and as low as -200 °C, making it suitable for applications in the fields of optoelectronic devices, electrical systems, architecture.[15] However, transparent PI is significantly more expensive. In addition, polycarbonate (PC) is also a good substrate material with high transmittance and excellent impact resistance, commonly used in industries.[16]

5.1.2 Flexible Glass

In addition to flexible polymers coated with ITO, ultra-thin flexible glass (FG) substrates have also been utilized in FPSC devices operated under indoor conditions.[17] Rigid perovskite solar cell substrates typically consist of fluorine-doped tin oxide (FTO) glass and indium tin oxide (ITO) glass, featuring high transparency, significant conductivity, remarkable water resistance, and good chemical stability. When the thickness is reduced to below a few hundred micrometers, glass substrates can achieve mechanical flexibility while maintaining high transmittance, high-temperature compatibility, good dimensional stability, significant water resistance, gas barrier performance, and compatibility with most chemical solvents. Willow glass, being a perfect sealing barrier, can protect sensitive materials from moisture and oxygen, facilitating the fabrication of FPSCs while achieving good recyclability. Tavakoli and his team first proposed the use of 50-μm-thick willow glass as a flexible substrate for FPSC fabrication, achieving an efficiency of 12.06%.[18] Importantly, FPSCs exhibit good flexibility, with 96% efficiency retained after 200 bending cycles. Furthermore, flexible glass substrates can also be utilized for roll-to-roll (R2R) manufacturing, making the production of large-area FPSCs feasible due to the smooth and non-deformable surface at high temperatures.

However, while easily bendable, willow glass can also be susceptible to fracture if bent excessively. Flexible glass meets many desirable criteria for flexible substrates, particularly thermal stability, but its mechanical stability is poor and difficult to improve. Therefore, there is a need to explore flexible matrices with better recyclability.

5.1.3 Metal Foil

One strategy to overcome temperature-related issues is the utilization of metal foils with high heat resistance, such as titanium (Ti) and copper (Cu).[19] Metal foils, known for their high flexibility, thermal conductivity, and low thermal resistance, are considered promising flexible substrates.[20] They can serve as both flexible substrates and bottom electrodes, simplifying the manufacturing process.

Among metal substrates, titanium foil stands out due to its smoother surface compared to other materials, which aids in improving the uniformity of the perovskite layer and minimizing shunt currents. Additionally, these metal foils can withstand high-temperature processes such as chemical vapor deposition (CVD) and high-temperature sterilization.[12] Thermally oxidizing the surface of the titanium foil to generate a titanium dioxide (TiO_2) electron transport layer (ETL) can streamline the process of device fabrication and result in cost savings on materials. Han et al. reported that reducing the concentration of oxygen vacancies in TiO_2 can significantly enhance the efficiency of electron collection. FPSCs incorporating a top transparent electrode composed of a 7-nm-thick layer of gold (Au) and a 1-nm-thick layer of copper (Cu) achieved PCE of 14.9%. Furthermore, because of the high crystalline quality of the TiO_2 layer, these FPSCs maintained their initial PCE of 100% even after undergoing 1,000 bending cycles with a bending radius of 4 mm.[21] Besides titanium foil, other metal substrates such as copper foil and stainless foil have also been applied by some researchers. Nejand et al.[22] chose copper foil as a flexible and conductive substrate due to its similar working functionality. Furthermore, copper iodide (CuI) formed on copper substrates can serve as a hole transport layer (HTL). FPSCs with a Cu/CuI/perovskite/ZnO/Ag nanowire inverted structure exhibited a PCE of 12.80%.

Despite the advantages of metal foils as flexible substrates, there are limitations. Thicker metal foils are not transparent, while thinner metal films have lower conductivity.[23] The lack of suitable top transparent electrodes with both high conductivity and flexibility hinders the development of FPSCs fabricated on metal foils.

5.1.4 Biomass Substrates

Considering the environmental pollution caused by plastic substrates and traditional petroleum-based polymers, various biodegradable, renewable, and biocompatible biomass substrates have been developed for the field of FPSCs due to their excellent performance. Biomass substrates exhibit characteristics such as high thermal stability (exceeding 200 °C), foldability, and low production costs, demonstrating great potential for manufacturing flexible green electronic products.

Several paper-based substrates, including opaque cellulose paper and transparent nanocellulose paper, have been utilized in perovskite solar cells,[24], [25] with most of them already commercialized. Specifically, cellulose paper offers advantages such as flexibility, lightweight, low cost, and ease of accessibility, making it a suitable flexible substrate. Environmentally friendly and biodegradable substrates based on nano fibrillated cellulose have been explored for eco-friendly flexible energy devices due to their excellent mechanical properties, renewability, and availability of abundant raw materials. For example, nanocellulose paper (NCP) extracted from wood, cotton, and other cellulose-rich materials exhibits exceptional characteristics such as high transparency, adequate mechanical strength, thermal stability, and chemical stability.[26] By combining the "green," biodegradable, and low-cost features, this alternative approach addresses current energy and environmental pollution issues, providing a promising solution for recyclable flexible photovoltaic devices.[23] Furthermore, devices based on paper substrates can be easily cut and disposed of through incineration with minimal environmental pollution caused by the combustion residue primarily composed of carbon materials. However, during the incineration process, polyethylene terephthalate (PET) tends to produce solid ash residue, airborne particulate matter, and toxic heavy metals.

While biomass substrates have many advantages in the fabrication of flexible devices, the efficiency of the produced FPSCs still requires improvement due to their high surface roughness. Therefore, it is urgently necessary to design and synthesize high-performance FPSCs with high transparency, low surface roughness, high conductivity, and low cost. Biomass substrates can also be applied to other electronic devices, contributing to the development of novel eco-friendly flexible electronic products.

5.2 FLEXIBLE ELECTRODE

5.2.1 Transparent Conductive Oxides

The commonly used electrode materials for transparent conductive oxides (TCO) include indium tin oxide (ITO), fluorine-doped tin oxide (FTO), indium zinc oxide (IZO),[34] and aluminum-doped zinc oxide (AZO).[35] Among them, FTO is typically deposited at substrate temperatures above 350°C,[36] IZO requires a deposition temperature of 100°C,[37] while ITO, AZO, and IWO can be prepared at room temperature.[38] However, AZO can react chemically with perovskite films, which negatively affects the performance of Flexible Perovskite Solar Cells (FPSC). ITO, on the other hand, has been widely used in FPSC due to its low-temperature manufacturing process, chemical stability with perovskite materials, high optical transparency, and good energy level alignment. Nevertheless, ITO electrodes still have some issues. Cracks and increased sheet resistance during bending cycles can lead to performance degradation of flexible devices. Furthermore, the surface roughness of ITO thin films caused by sputtering effects can result in leakage currents and poor flexibility. In addition, the scarcity of indium and the expensive vacuum deposition techniques (such as sputtering, evaporation, and electroplating) make ITO the most costly component in PSC (accounting for about 50–60% of the total material cost), limiting its potential for large-scale production.[39]

Reducing the thickness of ITO optimizes its optical, electrical, and mechanical properties, making it more flexible with fewer fractures during bending tests. As the ITO thickness decreases from around 400 nm to 160 nm, Liu et al. found that the increase in ITO layer thickness resulted in reduced transmittance, while the conductivity increased significantly as expected. By coating ITO with a thickness of 160 nm, a power conversion efficiency (PCE) of 16.64% was achieved in the fabrication of F-PSC, while maintaining good mechanical durability even at a bending radius as small as 4 mm.[40] Another study utilized ion-assisted arc ion plating for ITO film deposition (Figure 5.2a and b). The ITO film showed low sheet resistance of 15.75 ohms per square, high average optical transmittance of 85.88%, and a small external bending radius of 5 mm. Additionally, due to improved crystallinity, adhesion, and

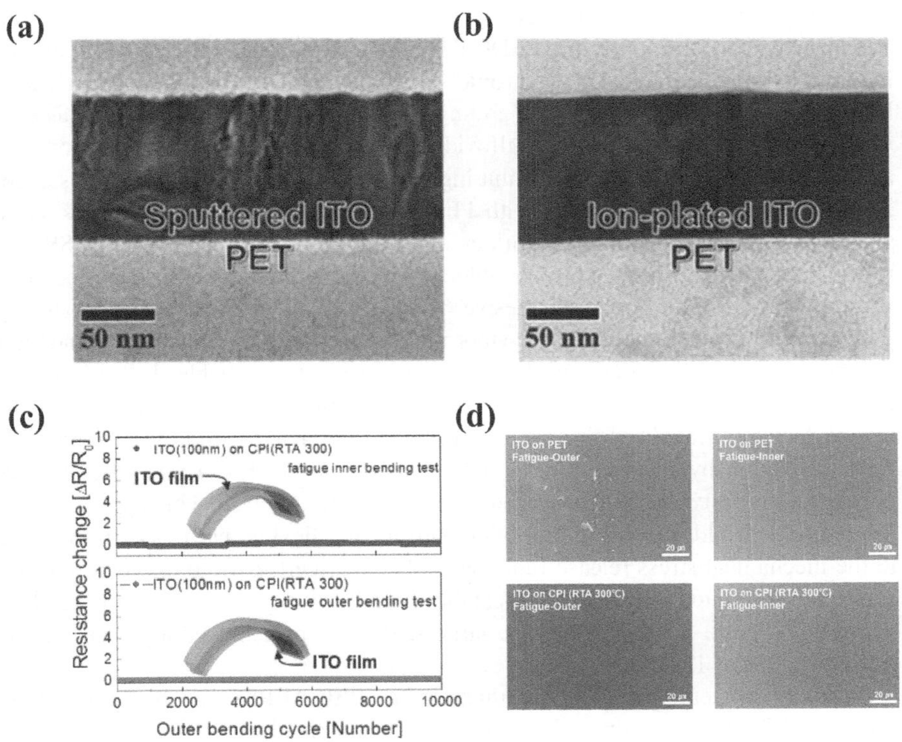

FIGURE 5.2 Cross-section images about the (a) sputtered ITO film. Copyright (2017), Wiley-VCH and (b) ion-plated ITO film. Copyright (2017), Wiley-VCH; (c) dynamic outer and inner bending fatigue test results of annealed ITO/CPI substrates. Copyright (2018), Royal Society of Chemistry; (d) surface FESEM images acquired after 10,000 cycles of outer and inner bending. Copyright (2018), Royal Society of Chemistry.

a flat smooth surface, flexible perovskite solar cells manufactured on ion-plated ITO electrodes exhibited a PCE of 16.8%.[41] Furthermore, high-temperature annealing can also have an effect on ITO films (Figure 5.2c and d). The ITO/Clear Polyimide (CPI) sample subjected to rapid thermal annealing at 300°C exhibited lower sheet resistance, higher transmittance, and good mechanical flexibility.[42]

ITO possesses excellent optical and electrical properties, making it the most popular flexible transparent electrode in FPSC, especially as a bottom electrode. However, the application of ITO as a top electrode is often challenging due to the sputtering process and high-temperature environment during TCO film fabrication, which can have adverse effects on the perovskite layers and organic charge transport layers.

5.2.2 Polymer

Despite the significant advantages of ITO as an electrode, the fracture of ITO electrodes during bending with a curvature radius smaller than 10 mm always leads to device failure. In contrast, the conductive polymer PEDOT:PSS can effectively compensate for the drawbacks of ITO. PEDOT:PSS, as an electrode, possesses several notable advantages, including solution processability, high flexibility, and high transparency.[43] It is suitable for roll-to-roll processes, which attract large-scale manufacturing due to low production costs.

Furthermore, by modifying PEDOT:PSS, its work function can be adjusted, making it more suitable for electronic collection and enhancing conductivity.[44] Researchers have achieved this by adding high-boiling-point solvents such as ethylene glycol (EG) and dimethyl sulfoxide (DMSO) for modification (Figure 5.3a and b). The optimized samples exhibit high transmittance and low sheet resistance. Additionally, flexible solar cells with PEDOT anodes on flexible substrates show superior mechanical flexibility.[45] Park et al. used DMSO-modified PEDOT:PSS as a transparent electrode and an electron-blocking layer, applying it to a shape-memory substrate with Norland optical adhesive 63 (NOA 63). The flexible shape-memory polymer with a super-smooth surface morphology ensures good consistency and high flexibility between layers (Figure 5.3c). The constructed stretchable electrode allows devices to have high mechanical stability without compromising photovoltaic performance during repeated bending cycles.[43] Hu et al. significantly improved the device's light utilization by incorporating a hole transport layer (HTL) into PEDOT:PSS doped with polystyrene in nano-sized pores (NC-PEDOT:PSS). This approach also acted as a scaffold, providing highly crystalline and flexible perovskite films. Due to the mechanical stress release function of the nano-cell scaffold, these perovskite solar cells exhibited remarkable repeatability and stability, with excellent bending durability (Figure 5.3d). The flexible large-scale devices achieved a power conversion efficiency of 12.32%.[46]

Furthermore, enhancing the conductivity of PEDOT:PSS can be achieved by removing PSS insulation through strong acid immersion[47]. However, strong acids like sulfuric acid may damage the polymer substrate, adversely affecting device performance. Sun et al. reduced this adverse effect by treating PEDOT:PSS with methanesulfonic acid (MSA). The MSA-treated PEDOT:PSS film exhibited a conductivity greater than 2000 S cm^{-1} and high transmittance in the visible range. Flexible PSCs

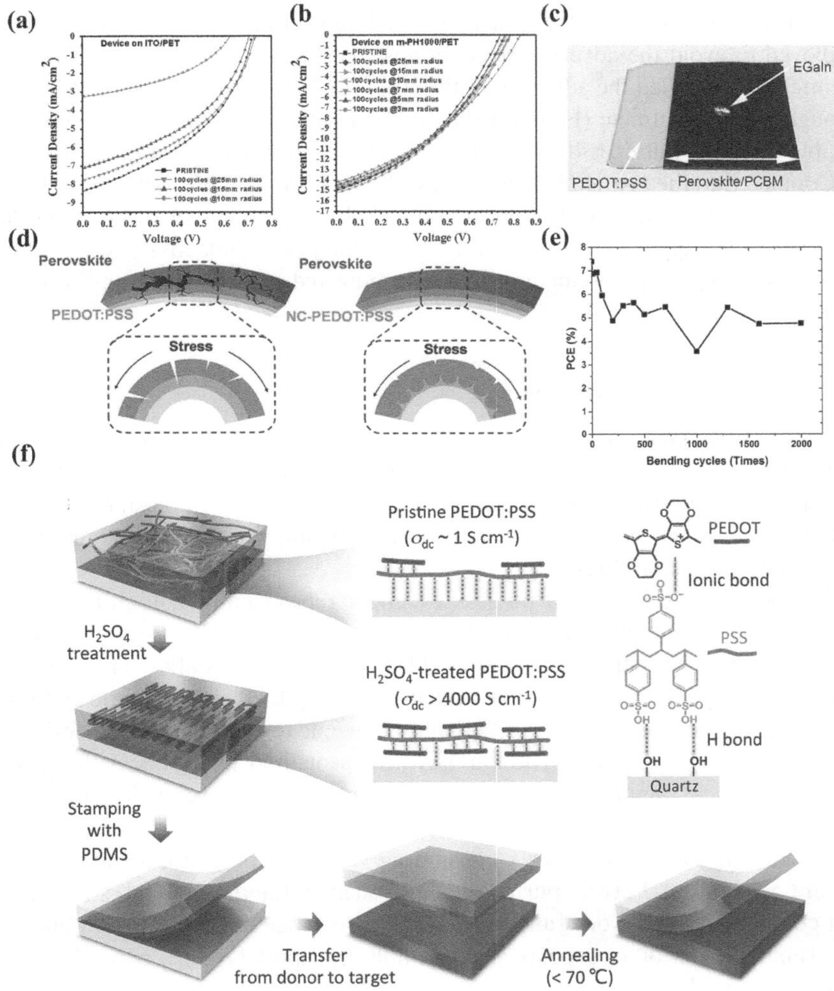

FIGURE 5.3 Changes in J–V characteristics for two solar cells on flexible PET substrates during repeated 100 bending cycles at different radius: (a) Organo metal perovskite device on PET/ITO. Copyright (2015), Elsevier; (b) Solar cell on plastic with modified PH1000 anode. Copyright (2015), Elsevier; (c) Photographs of as-prepared flexible devices. Copyright (2015), American Chemical society; (d) Scheme of stresses release by the NC-PEDOT:PSS structure. Copyright (2017), Wiley-VCH; (e) Variation of the PCE of a flexible PSC with the bending cycle. Copyright (2015), Wiley-VCH; (f) Schematic illustration of the procedure for the transfer printing of H_2SO_4-treated PEDOT:PSS film from a donor quartz substrate to a target substrate. The mechanism of transfer printing is based on the chemically controlled adhesion of the PEDOT:PSS to the substrate via H_2SO_4 treatment. Copyright (2015), Wiley-VCH.

endured small bending radii of 2 mm and retained over 50% of the initial PCE after 2,000 repeated bending cycles in the air (Figure 5.3e). Moreover, MSA treatment reduced the work function of PEDOT:PSS from 5.0 to 4.8 eV, favoring charge carrier transfer and collection.[48] The acid-assisted transfer printing strategy can also be employed to avoid the adverse effects of acid on the polymer substrate. Kim et al. chemically controlled the adhesion of PEDOT:PSS film to the donor/target substrate through H_2SO_4 treatment (Figure 5.3f), achieving highly transparent ($T > 90\%$) and flexible (bending radius < 400 μm) electrodes using carrier-assisted imprinting.[49]

Compared to the brittle ITO electrodes, PEDOT:PSS is more suitable for flexible device applications due to its solution processability and higher mechanical flexibility. In FPSC applications, the main issue is the poor conductivity, which can be improved through the aforementioned acid doping and acid-assisted transfer printing strategies.

5.2.3 Carbon Nanomaterials

Due to its abundant carbon composition, ultra-flexibility, excellent optoelectronic performance, and high air stability,[50] carbon nanomaterials (such as CNTs and graphene) are advantageous flexible electrodes in FPSCs. With outstanding optical and electrical properties, carbon nanomaterials exhibit performance comparable to ITO counterparts in FPSCs.[51]

Carbon nanotubes (CNTs) possess excellent reproducibility, high-temperature resistance, good optoelectronic performance, and exceptional air stability, making them effective alternatives to high-cost ITO. CNTs are typically transferred onto substrates using a thin-film transfer strategy, securing the base with tape and utilizing a few drops of toluene, ethanol, or chlorobenzene to ensure robust adhesion and improved electrical contact. The main synthesis method for CNTs is the floating catalyst chemical vapor deposition method. Some studies have explored manufacturing by pressing CNT network film layers onto $CH_3NH_3PbI_3$ substrates as hole collectors, eliminating the energy-consuming vacuum process for metal deposition (Figure 5.4a and b). Further purification and chemical doping of CNTs to enhance film conductivity and work function can improve efficiency.[52] CNTs can also play important roles in other aspects, such as hole transport layers,[53] perovskite additives,[54] and interlayers.[55]

CNTs can be categorized based on the number of graphene cylinders as single-wall carbon nanotubes (SWNTs), double-wall carbon nanotubes (DWNTs), and multiwall carbon nanotubes (MWNTs). SWNTs, the simplest type composed of a single graphene cylinder,[56] exhibit excellent mechanical durability, ease of manufacturing, and hydrophobicity, making them widely used in FPSCs as electrodes.[57] However, the hydrophobic nature of SWNTs can be detrimental to achieving uniform films. Additionally, due to the relationship between the high thin-layer resistance and conductivity of SWNT films, different methods have been proposed to effectively dope SWNT transparent electrodes, improving their optoelectronic properties without adversely affecting cell performance.[58] Jeon et al. studied the effects of various dopants on SWNTs, finding that modifying PEDOT:PSS with isopropanol (IPA) was

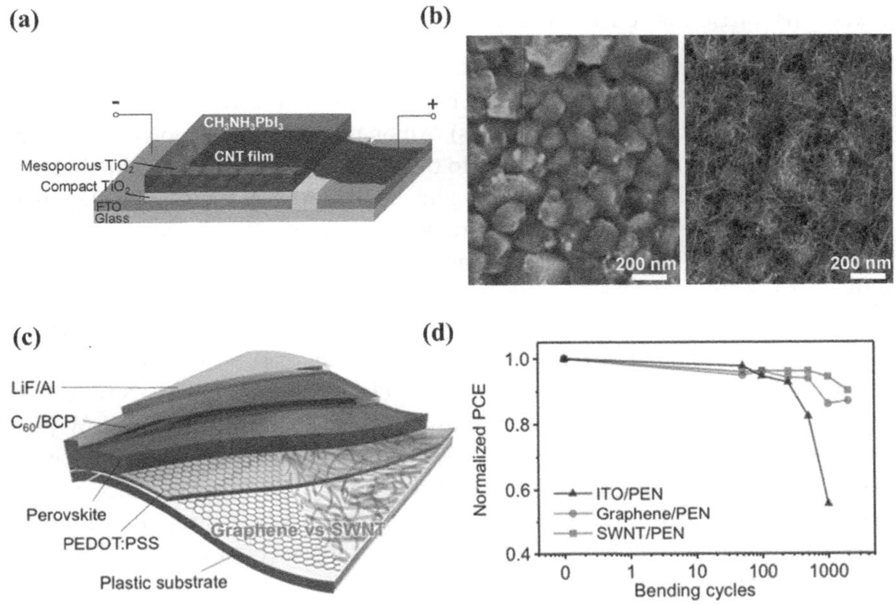

FIGURE 5.4 (a) Schematic of $CH_3NH_3PbI_3$ perovskite solar cell with CNT film electrode. Copyright (2014), American Chemical society; (b) Top view SEM images of $CH_3NH_3PbI_3$ perovskite substrate before (left) and after (right) CNT transfer. Copyright (2014), American Chemical society; (c) Schematic illustration of the flexible PSCs using graphene and SWNT on a PEN substrate. Copyright (2017), *The Journal of Physical Chemistry Letters*; (d) Cyclic flexural test results for flexible PSCs using ITO/PEN, graphene/PEN, and SWNT/PEN. Copyright (2017), *The Journal of Physical Chemistry Letters*.

more effective than surfactant-modified PEDOT:PSS. However, both showed low shunt resistance and high thin-layer resistance, resulting in overall low performance. Doping SWNTs with MoO_3, despite energy level mismatch, led to lower efficiency. Even in its diluted form, HNO_3-doped SWNTs exhibited the best performance at up to 35 v/v%, but the energy level mismatch between the Fermi level of perovskite and SWNT electrode affected hole mobility.[58]

Compared to CNTs, graphene with single-atomic thickness is smoother, more conductive, and more transparent in a broad wavelength range. The transparency of graphene films can even surpass ITO, especially in the near-infrared range, making graphene an attractive candidate for transparent electrodes in FPSCs (Figure 5.4c and d). Overall, despite higher performance, graphene transfer steps have lower reproducibility. Graphene-based cells are also more susceptible to strain compared to CNT-based cells.[59] Both CNTs and graphene exhibit excellent chemical stability and mechanical flexibility, facilitating the realization of low-cost, high-stability, and durable FPSCs. In comparison, CNT films are more robust during bending, while graphene films provide a smoother surface to ensure the quality of subsequent layers.

5.2.4 Metallic Nanostructures

Due to their excellent conductivity, metal nanostructures, including metal nanowires and metal meshes, have significant potential as alternatives to ITO electrodes in flexible perovskite solar cells (FPSCs). Among them, silver nanowires (AgNW) are considered the most promising due to their low-cost fabrication techniques and excellent optoelectronic performance. Highly flexible AgNW transparent electrodes can achieve low sheet resistance comparable to ITO at 80–90% transmittance. The high haze ratio caused by light scattering from AgNWs is advantageous for light collection in solar cells.[60] Key issues with AgNWs include poor stability, low coverage, and high resistance.[61] Additionally, a chemical reaction between silver and halides forms insulating silver halides, leading to increased thin-layer resistance and degradation of AgNW-based transparent electrodes.[62] While AgNWs exhibit high conductivity along the nanowires, high thin-layer resistance at AgNW junctions reduces the film conductivity, adversely impacting FPSCs. Relevant research has deposited fluorine-doped ZnO (FZO) above AgNWs layers to enhance conductivity and mechanical stability.[63] Chen et al. adopted a hybrid structure composed of AgNWs and relatively low-cost commercial graphene (Figure 5.5a and b). They achieved 22 Ω/cm sheet resistance at 88% optical transparency (stabilized to 13 Ω/cm after four months) and demonstrated excellent stability under atmospheric exposure, mechanical pressure, and bending.[64] Dong et al. encapsulated silver nanowire (AgNWs) networks between a monolayer graphene and a polymer film as a sandwich structure, creating a typical thin and fully flexible hybrid electrode. This composite electrode featured an ultra-smooth surface (RMS = 2.58 nm), ultra-thin thickness (1.8 μm; Figure 5.5d), and lightweight construction. Furthermore, the composite electrode exhibited excellent mechanical strength and corrosion resistance.[62]

Metal mesh is a promising contender that competes with widely used TCO electrodes in flexible devices. Metal mesh, including Ag, Au, Cu, and Ni mesh, is often combined with other transparent conductive materials to form efficient composite electrodes. Jin et al. developed a high-durability and highly flexible transparent electrode (FTE) by applying a composite of thin metal grids and doped conductive polymers onto a NOA 63-coated substrate (Figure 5.5c). The prepared FTE demonstrated good mechanical stability under 10,000 cycles of bending deformation with a bending radius less than 1 mm, and the conductivity did not decrease. Li et al. investigated FPSCs prepared with low-cost Cu electrodes (Figure 5.5e). Both transparent bottom electrodes and opaque top electrodes were manufactured with Cu. FPSCs made with these Cu electrodes achieved an efficiency of 13.58%, maintaining over 90% after 1,000 cycles of bending with a small curvature radius of 5 mm.[65]

This section summarizes the key issues of using AgNWs as transparent electrodes, including poor chemical stability of perovskite materials, surface roughness, and high junction resistance. These weaknesses can be improved by coating additional materials or adopting new manufacturing processes, resulting in efficient and stable FPSCs. Metal meshes, such as Au, Ag, Cu, and Ni, also present attractive alternatives to ITO electrodes.

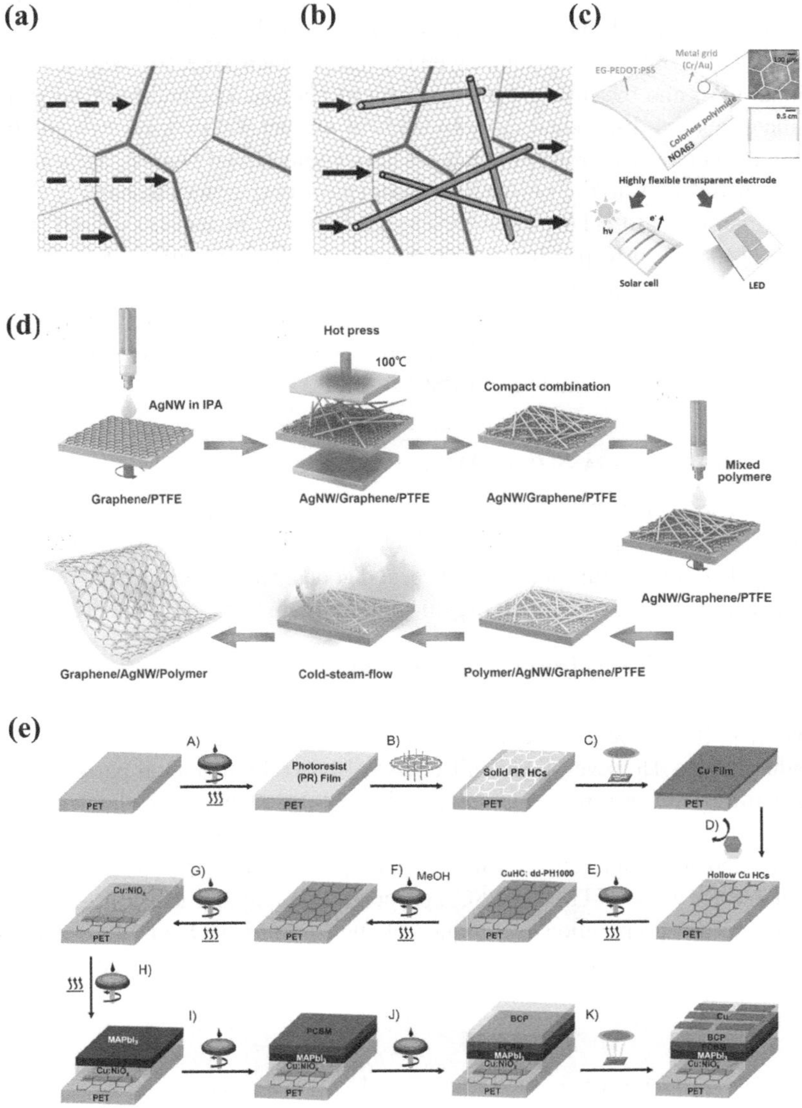

FIGURE 5.5 The illustration of the co-percolating conduction for AgNWs/graphene composite electrode (b) compared to the single layer graphene which has high-resistance grain boundaries (a). Copyright (2013), Wiley-VCH; (c) Schematic illustration of the fabricated FTE for perovskite optoelectronic devices (solar cells and LEDs). Copyright (2018), American Chemical society; (d) Schematic illustration of the procedures in fabricating composite PET/graphene/AgNWs electrode. Copyright (2016), American Chemical Society; (e) Fabrication process of the hybrid transparent electrode of PET/Cu grid. Copyright (2020), American Chemical Society.

5.3 CHARGE CARRIERS TRANSPORT MATERIALS

Charge transport materials, mainly electron transport materials (ETM) and hole transport materials (HTM), are designed to transport charge carriers. There are two structural types of FPSCs, depending on the light incident direction, such as the conventional n-i-p and the inverted p-i-n structures (Figure 5.6a and b).[66]Charge transport materials are distributed on both sides of the perovskite layer, which absorbs light and generates electrons and holes. These photoexcited charges are transferred from the perovskites to their respective charge transport materials, namely electrons to ETM and holes to HTM. Then the charge transport materials on both sides are connected to the corresponding electrodes to complete the circuit. For the charge transport material used in FPSCs, it should have the ability to be fabricated at low temperature, excellent carrier extraction ability, good interface contact, low carrier transfer resistance, suitable energy level and high recyclability. These characteristics will determine the performance of FPSCs.

5.3.1 Electronic Transport Materials

In high-efficiency FPSCs, ETM, as a key component, is required to be highly transparent across the entire solar spectrum, while providing good electron extraction capability, low carrier transport resistance, and good energy levels.[67] Charge transport materials are mainly divided into organic and inorganic charge transport materials.

5.3.1.1 Inorganic Electronic Transport Materials

Many inorganic ETMs in rigid PSCs exhibit excellent properties, and most of these ETMs are still suitable for FPSCs.[68] For example, TiO_2, ZnO, SnO_2, WO_X, Zn_2SnO_4 and other oxides. However, ETM in FPSCs is limited by technical parameters such as annealing temperature, roughness, and flexibility. Generally, the widely used TiO_2 requires high temperature annealing (~500 °C), and flexible substrates cannot withstand such high temperatures, so mesoporous TiO_2 prepared by sintering is not suitable for use as ETM in FPSCs, especially on polymer substrates. Therefore, low-temperature preparation techniques of mesoporous TiO_2 or the use of other excellent inorganic ETM are necessary for FPSCs. Kim et al. developed a method for

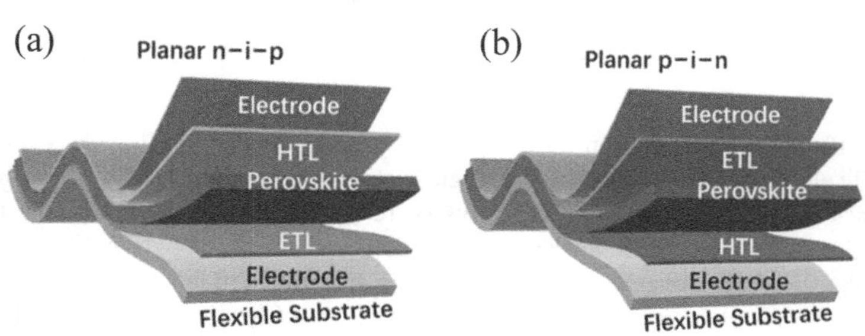

FIGURE 5.6 (a) and (b) show the n-i-p and inverted p-i-n structures of FPSCs.[66] Copyright (2023), Nano-Micro Letters.

the preparation of mesoporous TiO_2 at low temperature. Mesoporous TiO_2 was spin-coated on compact TiO_2 and layered with an organic binder to form a paste at 80 °C. Then, mesoporous TiO_2 was treated by reactive ion etching (RIE) to remove organic residues. The FPSCs with the structure of PEN/ITO/compact TiO_2/mesoporous TiO_2/perovskite/spiro-OMeTAD/Au realized the highest PCE of 17.29% (Figure 5.7a).[69]

At present, the low-temperature preparation technology of TiO_2 in FPSCs has been relatively mature, such as atomic layer deposition (ALD),[70] magnetron sputtering,[71] electron beam evaporation, and low-temperature pulsed laser deposition.[72] Among them, the devices fabricated by magnetron sputtering and their properties are shown in Figure 5.7b. However, the PCE of FPSCs based on cryogenic TiO_2 is still low (< 20%). Therefore, the use of other ETMs is critical for the performance of FPSCs. Zinc oxide (ZnO), a commonly used ETM, has been introduced into FPSCs due to its suitable energy level matching with perovskite and mature low-temperature preparation process. Despite these advantages, ZnO-based devices are less stable when ZnO is directly in contact with the perovskite, which will accelerate the degradation of the perovskite and in turn lead to rapid device deterioration.[73]

In recent years, SnO_2 as an ETM can be well matched with the energy levels of perovskites due to its low processing temperature, high electron mobility, strong photostability, and wide optical band gap, making it ideal for FPSCs applications. In addition, the PCE of PSCs using SnO_2 as ETM has reached 25.6%.[74] At the same time, the PCE of FPSCs using SnO_2 as ETM also increased rapidly. In 2019, Wang et al. used nanocrystalline SnO_2 prepared by UV ozone pretreatment as the ETM for constructing F-PSCs. The PCE of FPSCs based on SnO_2 ETL was 17.5%.[65] The high-performance FPSCs based on SnO_2 can be further realized by introducing additives. Yang et al. used ethylenediamine tetraacetic acid (EDTA) to modify the ETL of SnO_2 treated by low temperature, and the PCE of FPSCs reached 18.28%.[75] Zheng et al. used the organic additive ammonium formate ($HCOONH_4$) to improve SnO_2. This embedded additive reduces surface oxygen vacancies and improves the electron extraction layer of SnO_2. The residual strain and microstrain in the perovskite films are reduced, and the defect density within the ETM and perovskite films is reduced. The PCE of F-PSCs increased from 20.2% to 22.37%. After 4,000 bending cycles with a bending radius of 7 mm, the efficiency remained above 90% of the initial value (Figure 5.7c).[76] Other halogen salts, bases, and organic additives are also being used to achieve high-performance FPSCs. Given that the efficiency gap between FPSCs and rigid PSCs is still large, the recombination mechanism between low-temperature fabricated SnO_2 films and perovskite layers should be investigated for future SnO_2-based FPSCs to further reduce the energy loss at the SnO_2/perovskite interface and achieve high-performance SnO_2-based FPSCs.

5.3.1.2 Organic Electronic Transport Materials

Organic ETMs are also often used for FPSCs because they are easy to process at low temperatures and compatible with flexible substrates. Organic ETMs are mainly fullerenes and their derivatives, such as C_{60} and PCBM. C_{60} is a hollow cage structure composed of 12 five-element rings and 20 six-element rings of carbon atoms. Because the five-element ring is a tensile ring and the six-element ring is non-tensile, the C_{60} caged structure is stronger than all known particles and more durable than

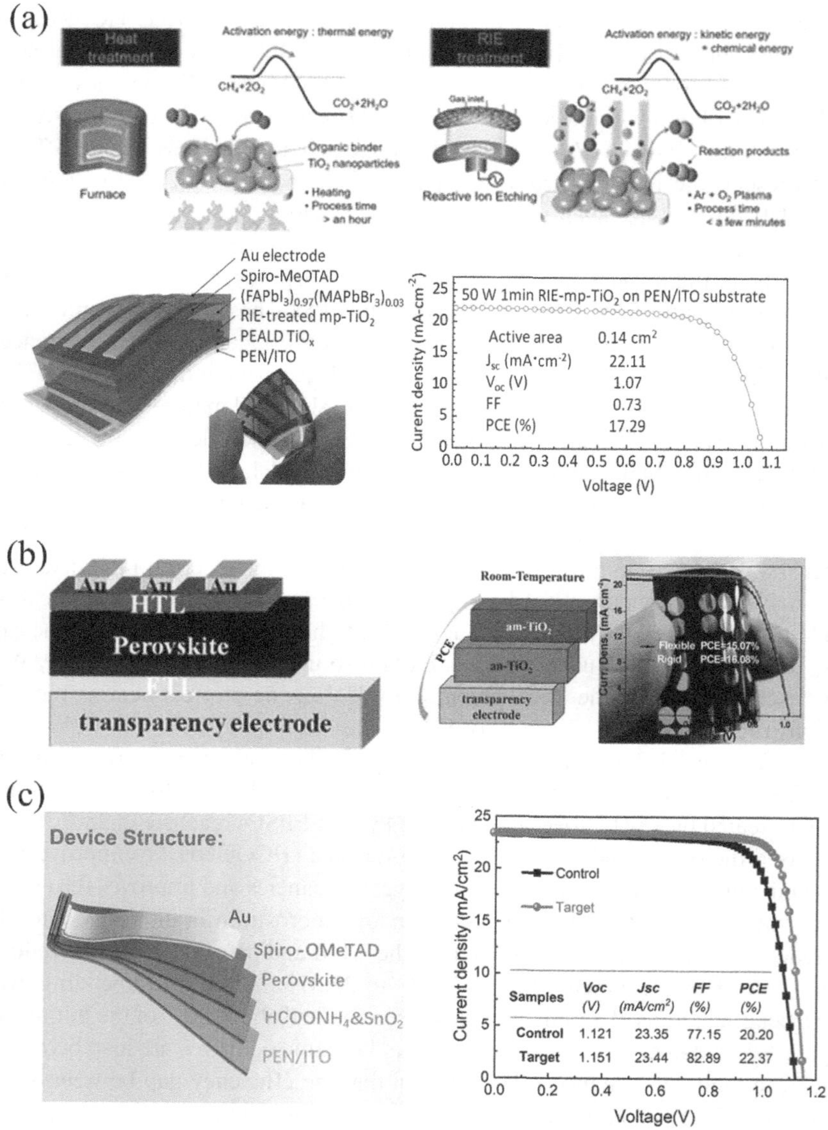

FIGURE 5.7 (a) Diagram of the preparation process and chemical reaction of mesoporous TiO_2 treated by RIE and pictures of the FPSC and its J–V curve. Copyright (2020), ACS Applied Materials Interfaces; (b) Structure and performance diagram of the device fabricated by magnetron sputtering. Copyright (2015), Energy Environmental Science; (c) Structure and performance of FPSCs with SnO_2 as ETM. Copyright (2022) Advanced Materials.

diamond. Compared with PCBM, C_{60} has better electron mobility and conductivity. However, the solubility of C_{60} is low in common solvents, making it difficult to fabricate a dense C_{60} layer. In this regard, the C_{60} layer is usually deposited by thermal evaporation. Yoon's group successfully deposited C_{60} onto the substrate by thermal deposition with a PCE of 14% for its device. PCE can maintain 5% of its initial value after 100 bends within a radius of 95 mm (Figure 5.8a).[77] PCBM is an essential

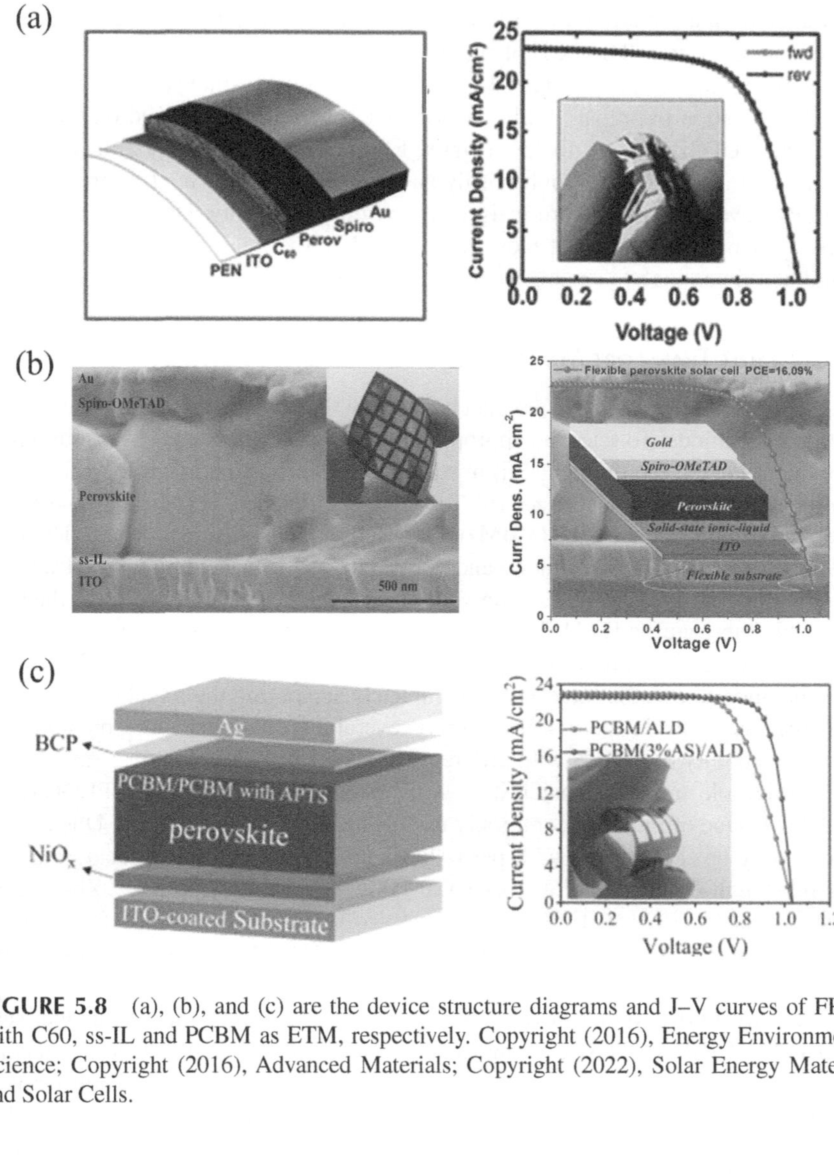

FIGURE 5.8 (a), (b), and (c) are the device structure diagrams and J–V curves of FPSCs with C60, ss-IL and PCBM as ETM, respectively. Copyright (2016), Energy Environmental Science; Copyright (2016), Advanced Materials; Copyright (2022), Solar Energy Materials and Solar Cells.

ETM in FPSCs due to its advantages of simple fabrication process, low temperature preparation, high conductivity, and electronic extractability. But the molecules aggregate, and inferior adhesion to the perovskite layer may lead to the rough morphology of PCBM. ETM is often optimized to produce high-performance FPSCs. Yang et al. used ss-IL (1-benzyl-3-methylimidazole chloride) as ETM to obtain efficient FPSCs with good reproducibility by low-temperature treatment. The ss-IL helps to reduce the density of electron trap states on the surface of the perovskite absorbent. ss-IL ETM has excellent optoelectronic properties, including wide band gap, anti-reflection, high electron mobility, and suitable work function (WF), which eliminates hysteresis while reducing the density of electron trap states on the surface of perovskite absorbents. The PCE of FPSCs was 16.09% (Figure 5.8b).[78] In 2015,

Konrad et al. first prepared n-i-p FPSCs without hysteresis effects using C_{60} as ETM on flexible substrates. The PCE of the device is 10.4%, which shows good stability and recoverability.[79] Li et al. added 3-aminopropyl triethoxysilane (APTS) molecules to PCBM solution to prepare high-performance inverted FPSC based on optimized ETM, and the PCE reached 18.62% (Figure 5.8c).[80] Compared to inorganic ETM, the properties of organic ETM can be easily customized by modifying the structure and generally have higher moisture resistance and high flexibility. However, due to the relatively poor conductivity of most organic ETMs, the thickness of organic ETMs should be thin enough to ensure efficient electron transport.

5.3.2 Hole Transport Materials

The role of HTM in FPSCs is to extract holes from the perovskite layer and deliver the holes to the cathode. To select an appropriate HTM in FPSCs, the HTM is required to meet the following requirements: (i) in addition to transporting holes, the HTM must also block the influx of electrons. (ii) To improve efficiency, HTMs are expected to have a maximum valence band (VBM or highest occupied molecular orbital (HOMO)) that matches well with the cathode and perovskite and to have a high hole transport capacity. (iii) Unlike rigid PSC, water vapor has a high permeability through the plastic flexible substrate, so HTM should have better environmental stability to avoid corrosion or degradation caused by oxygen, steam, metal electrodes, etc. (iv) Reduce the recombination rate at the interface by completely separating the metal electrode and the perovskite active layer. (v) It has good recoverability. The structures, photographs, and recoverability of FPSCS with different HTMs are shown in Figure 5.9.

Organic hole transport materials mainly include spiro-OMeTAD, PEDOT:PSS and PTAA. One of the most commonly used HTMS is spiro-OMeTAD. Due to their good energy level matching with perovskite and low temperature preparation process, most high-efficiency FPSCs use spiro-OMeTAD as HTM. In 2022, Zheng et al. reported that the PCE of FPSC-based on spiro-OMeTAD was 22.37%.[76] Li et al. reported a PCE of 23.6% for FPSC based on spiro-OMeTAD. This indicates that the application of spiro-OMeTAD in efficient FPSCs is very important.[1]

Pure spiro-OMeTAD has low conductivity, so some dopant must be added, such as Li-TFSI, 4-tBP. Li-TFSI can improve the hole mobility and reduce the recombination rate of hole-electron pairs. However, Li-TFSI is hygroscopic and tends to become liquid after hygroscopic. The tBP was used to further improve the solubility of Li-TFSI in spiro-OMeTAD solution and prevent the accumulation of Li-TFSI. However, tBP can dissolve PbI_2 and thus decompose the perovskite layer. This also makes the FPSCs unstable.

PEDOT:PSS has become a commonly used HTM due to its good hole conductivity, suitable HOMO level, low temperature fabrication process, and low cost. However, it also has the disadvantages of hygroscopicity and acidity (pH < 1). When the PEDOT:PSS layer comes into contact with the perovskite, acidity rapidly degrades the perovskite layer, which leads to the instability of FPSCs. Optimization has been performed to improve the performance of PEDOT:PSS. Yang et al. proposed the additive $HAuCl_4{\cdot}_3H_2O$ and the interface modifying N, N'-bis-(1-naphthalenyl)-N, N'-bis-phenyl-(1,1';-biphenyl)-4,4';-diamine (NPB) was used to improve the conductivity

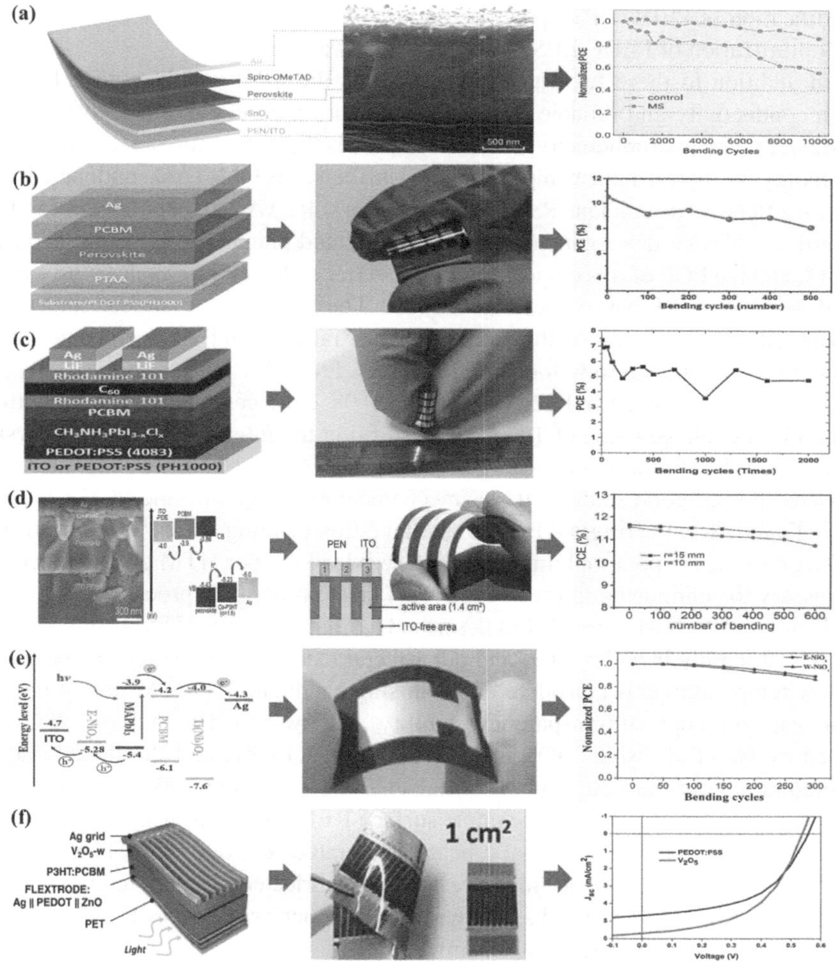

FIGURE 5.9 FPSCs with different HTM. (a) Schematic of device structure, cross-section SEM image of MS-FPSC, and bending stability of FPSCs (R = 6 mm). (b) Schematic of inverted oxidized PSC and photographs of flexible device with PTAA as HTM, and degradation of FPSCs in PCE at different stages of bendability testing. (c) Schematic of the structure of PSCs and photographs of FPSCS with MSA-PEDOT:PSS/PET. The PSC was manually deformed with a bending radius of 2 mm. PCE of FPSC as a function of bending cycle. (d) Cross-section SEM images and energy maps of PSC in each layer of the device ITO, PCBM, perovskite, Co-P3HT, and Au. A photograph of a flexible device with Co-P3HT as HTL. The inset shows the device structure. PCE of flexible devices after repeated bending for 600 bending cycles at $r = 15$ mm and $r = 10$ mm. (e) Energy level diagram of the material used in the flexible device using NiOx as HTL. Photos of the equipment. The normalized PCE of the E-NiO_X-based flexible device as a function of a bending cycle with a radius of 5 mm. (f) Schematic representation of an organic solar cell with V_2O_5 and PEDOT:PSS as HTL. A photograph of FPSCs. IV profile of an inverted organic solar cell. Copyright (2023), Angewandte Chemie International Edition.

and hole mobility of PEDOT:PSS. The PCE of FPSCs was 14.04%.[81] Although the modification of PEDOT:PSS plays an important role, the efficiency is still low, and the optimization of PEDOT:PSS still has a long way to go.

In addition to these two materials, PTAA is the third widely used HTM. The high conductivity and reliable robustness make it an excellent alternative to spiro-OMeTAD. The hydrophobicity of PTAA improves the perovskite crystallization and promotes the performance improvement. Unlike spiro-OMeTAD, undoped PTAA is still effective in efficient PSC. In recent years, PTAA has been at the forefront of efficient device development. Xu et al. fabricated planar FPSCs using PTAA as HTM, and the PCE of this device was 17.5%. After eight bending cycles with a radius of 100 mm, the PCE was reduced by only 5%. The main cause of degradation can be attributed to microcracks on the perovskite layer rather than PTAA, since no significant cracks or damage were found on the PTAA layer.[82]

Gao et al. achieved a PCE of 23.68% (0.08 cm^2, certified: 23.35%) with an open-circuit voltage (V_{OC}) of 1.17 V by modifying the interface between perovskite and PTAA by ammonium pentanoacetate (PenAAc). In addition, the fabricated devices showed good stability under mechanical bending, with the PCE remaining more than 91% of the original PCE even after 5,000 bending cycles.[83] However, since PTAA coverage is poor and thinner PTAA will lead to poor HTM performance, it is necessary to continue to optimize the PTAA layer preparation process.

In addition to the organic HTM described earlier, inorganic HTM is often cited in FPSCs. It mainly includes NiO_X and other metal oxides. NiO_X, which can be treated at low temperatures, is considered a promising candidate for HTM due to its wide band gap, good optical transparency, stability, and low cost. The PCE of FPSCs prepared by He et al. using NiO_X as HTM was up to 15.9%, and the flexible device maintained more than 80% of the initial PCE after bending 300 times at a radius of 5 mm.[84] Du et al. proposed a simple surface REDOX engineering (SRE) method to adjust the surface properties of e-beam evaporated Ni oxide films. The efficiency reached 23.4% and 21.3% on small area devices fabricated on rigid and flexible substrates, respectively.[85] NiOx has shown excellent performance in FPSCs and is a promising HTM for future FPSCs commercialization.

Many other metal oxides have also shown excellent performance in FPSCs, such as some cu-based materials (CuI, CuO, Cu_2O and CuCrO, etc.); these materials have the characteristics of wide bandwidth gap, high conductivity, low cost, and relatively simple synthesis methods. Harsh process conditions, such as high post-heating temperature and high vacuum, largely limit the application of these candidate materials on polymer substrates. So far, only a few reports have applied copper derivatives on flexible substrates. Qin et al. used CuCrO at low temperature and obtained 15.6% PCE for the prepared F-PSCs.[86] This reveals the importance of the inorganic binary metal oxide HTL, providing a promising metal system for the preparation of high performance and more stable FPSCs.

Molybdenum oxide (MoO_X) has also attracted much interest due to its high transmittance and proper band alignment with perovskites. Yoon et al. combined heat-deposited MoO_3 with a single-layer graphene electrode, MoO3 modified graphene, to improve conductivity. With a bending radius of 2 mm, the initial efficiency of the flexible device remained above 85% after 5,000 bending cycles, and its PCE reached 16.8%.[59]

5.4 PREPARATION METHOD OF FLEXIBLE SOLAR CELL

There are numerous methods currently used for the preparation of flexible perovskite solar cells. Here, we summarize several of the most common methods used in both laboratory-scale and large-scale industrial applications.

5.4.1 Spin Coating

Spin-coating is currently the mainstream method for laboratory-scale preparation of small-area flexible perovskite solar cells. This method is simple to operate and allows for precise control of film thickness, enabling the production of uniform, pinhole-free high-quality films. There are generally two methods for spin-coating flexible perovskite films: the one-step method and the two-step method. The one-step method typically involves the use of an anti-solvent, and the choice of anti-solvent for different perovskite systems is a crucial factor in forming high-quality perovskite films. Cheng's group introduced chlorobenzene (CB) as an effective anti-solvent during spin-coating of perovskite in N,N-dimethylformamide (DMF), resulting in highly uniform perovskite layers.[87] Seok's group used toluene (TB) as an anti-solvent to fabricate perovskite in a mixed solvent of dimethyl sulfoxide (DMSO) and gamma-butyrolactone (GBL).[88] However, the one-step method requires precise control of the dropping time and volume of the anti-solvent, with a narrow processing window that is not suitable for large-area flexible perovskite film preparation. In contrast, the two-step method seems to not have similar issues. The solvent-free process is more favorable for controlling interface effects, but edge effects may occur and become more severe with larger areas. Additionally, the two-step method suffers from incomplete conversion of PbI_2, severely limiting experimental reproducibility and the preparation of large-area films. In conclusion, due to the increased likelihood of defects, pinholes, and uncovered regions when scaling up the spin-coating method, there is a significant efficiency loss in device performance. Moreover, the inherent drawbacks of the spin-coating process result in substantial waste of precursor materials, making it currently widely used only in the laboratory and not suitable for industrial production.

5.4.2 Blade Coating

Blade coating is an effective method for early preparation of DSSCs. In the field of perovskite solar cells, blade coating is also used for depositing charge transport layers and electrode materials (Figure 5.10c). This method has a similar level of complexity to spin-coating and comparable equipment costs. To some extent, blade coating is similar to the one-step spin-coating method but consumes relatively less precursor material, typically only around 20% of the precursor solution used in spin-coating. It has been reported that Yang et al. were able to prepare all layers except the Ag electrode using blade coating, achieving a PCE of 7.1% with the configuration of PET/ITO/PEDOT:PSS/$CH_3NH_3PbI_xCl_{3}I_x$/$PC_{61}BM$/Bis-C60/Ag.[89] Later, Li et al. achieved a PCE of 17.5% using ITO/PEI:PCBM/$CH_3NH_3PbI_3$/PTAA/MoO_3/Ag prepared by blade coating, with perovskite grain sizes larger than those obtained from spin-coating.[90] The high adaptability of blade coating to different functional layers

FIGURE 5.10 Large-scale perovskite fabrication techniques. (a) Spray coating; (b) inkjet printing; (c) blade coating; (d) slot-die coating.

makes its application more extensive compared to spin-coating. Additionally, blade coating enables the nucleation and crystallization process of perovskite films at relatively low environments without requiring high-temperature annealing, thus expanding the selection of flexible substrates. Recently, a PCE of 22% was achieved using blade coating on rigid substrates. Therefore, we believe that blade coating holds great potential as a method for future commercial large-area flexible perovskite solar cells.

5.4.3 Slot-Die Coating

Another process used for the preparation of flexible perovskite solar cells is called slot-die coating. The slot-die coating system consists mainly of a moving substrate, a slot die head, pumps, and a pressure system (Figure 5.10d). Different equipment responsible for various functions can effectively control the quality of the film and achieve precise thickness control. By adjusting the flow rate, substrate speed, and printing width, we can achieve the fabrication of perovskite films with different thicknesses and sizes. Sears et al. achieved a PCE of 11% by using slot-die coating to prepare the HTM while the perovskite layer was still formed by spin-coating.[91] Remeika et al. used the slot-die method to prepare perovskite layers in an ambient atmosphere with the PET/ITO/ZnO/$MAPbI_3$/P_3HT/Au structure, achieving an average PCE of 3.6% with an active area of 1 cm^2.[92] In addition, slot-die coating has good compatibility with other functional layers of perovskite. Currently, except for electrode thermal evaporation, which cannot be completed on slot-die equipment, the preparation of all other functional layers can be achieved through slot-die coating. If the thermally evaporated electrode is replaced with a printable carbon electrode, then the entire process of perovskite solar cell fabrication can be realized using slot-die coating.

5.4.4 Roll-to-Roll

The roll-to-roll (R2R) printing technique is currently a mature and industrially oriented method for the fabrication of flexible perovskite solar cells. In recent years, microgravity techniques suitable for depositing films with thicknesses in the range of a few micrometers have emerged. This method often utilizes pre-made gravure plates to enable the fabrication of specific patterned components. In the large-scale industrial production of flexible perovskite solar cells, avoiding the impact of moisture and oxygen in the air is a common challenge that researchers are trying to address. Additionally, wet films often struggle to form high-quality perovskite films during the flow and slow drying process, which is another issue that needs to be resolved. In early 2015, Hwang et al. achieved R2R printing of F-PSCs and modules through sequential deposition.[93] Gong et al. reported R2R-printed F-PSCs manufactured in air (40% relative humidity) using a two-step process (Figure 5.11).[94] They performed R2R microgravure printing of SnO_2 diluted with IPA and water, followed by an annealing process. Then, a precursor solution of doped PbI_2/CsI was also R2R microgravure printed onto SnO_2 and annealed at 70 °C (Figure 5.11a). A mixed solution of cation halide salts was slot-die coated, reacted with PbI_2 under N_2 gas blowing. The resulting perovskite film was annealed to form perovskite crystals (Figure 5.11b). With the optimized structure of ITO/SnO_2/perovskite/spiro-OMeTAD/Ag, the F-PSCs achieved a PCE of 10.57%. Jin et al. developed an R2R gravure printing strategy for the manufacture of efficient F-PSCs.[95]

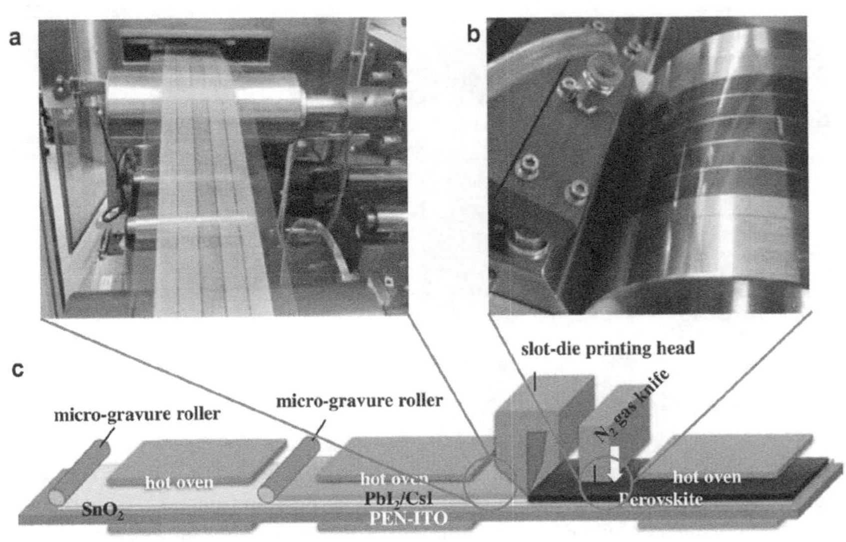

FIGURE.5.11 (a, b) Photograph of as-printed PbI2/CsI (a) and perovskite (b) on PET substrate. (c) Schematic process of R2R printing of f-PSCs in air. Copyright (2022), American Chemical Society.

The equipment is illustrated in Figure 5.12a. R2R printing can also enable the patterned fabrication of perovskite films. By placing a pre-designed gravure plate pattern into the R2R printing device and introducing the perovskite precursor solution into the engraving drum, the precursor solution is fixed onto the patterned substrate using the rotating printing drum and pressure roller. Excess perovskite solution is removed by a doctor blade on the side, and the patterned perovskite is transferred onto the ITO/PET substrate under the pressure from the embossing roll. This way, we can design and fabricate flexible perovskite solar cells with different specifications and patterns. Figure 5.12b shows F-PSCs fabricated through full-scale R2R gravure printing experiments. After printing the precursor solution of

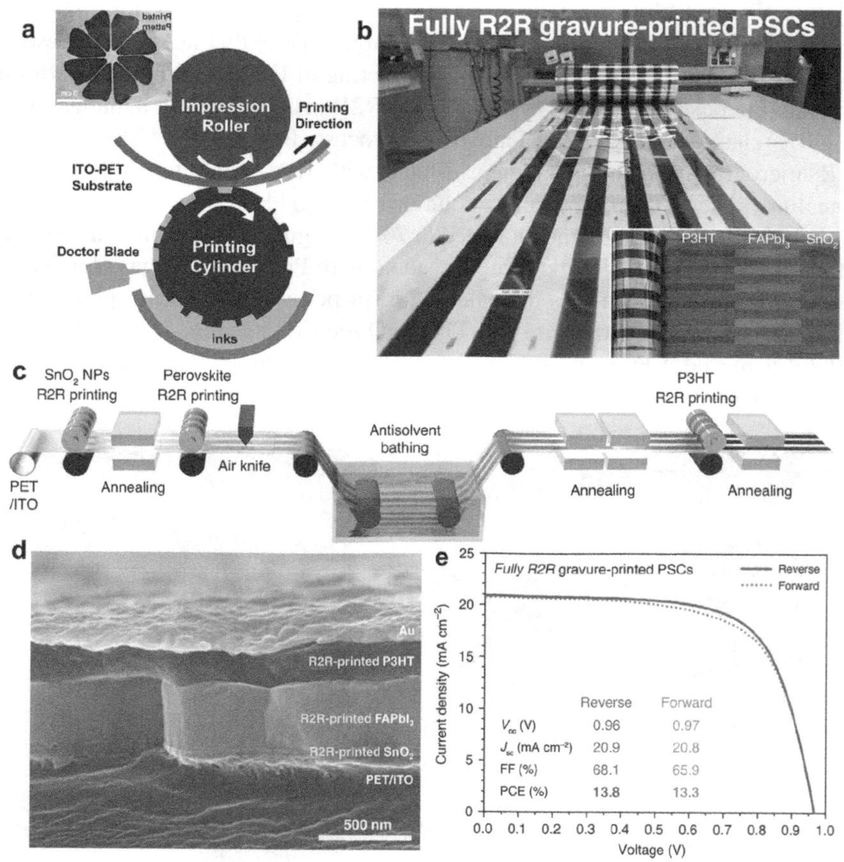

FIGURE 5.12 (a) Schematic diagram of gravure printing process. The inset is a photograph of a gravure-printed perovskite pattern. Adapted with permission from ref.[94] Copyright 2019, Wiley. (b) Photograph of fully R2R gravure-printed f-PSCs. The inset shows every constituent layer of R2R gravure-printed f-PSCs by removing the corresponding layers. (c) Detailed steps of R2R gravure printing process to produce f-PSCs. (d) Cross-sectional SEM characterization of R2R gravure-printed f-PSCs. (e) Forward and reverse scans of J–V characteristics of fully R2R gravure-printed f-PSCs. Adapted with permission from ref.[95] Copyright 2020, Springer Nature.

$(FAPbI_3)_{0.95}(MAPbBr_3)_{0.05}$, the perovskite starts to crystallize during annealing using an eco-friendly tBuOH: EA anti-solvent bath (Figure 5.12c). A gravure-printed and annealed P_3HT HTL was also used. The full-scale R2R gravure printed F-PSCs exhibit good morphology as shown in the cross-sectional SEM image and achieve a PCE of 13.8% (Figure 5.12 d and e).

5.4.5 Thermal Evaporation

Another method for fabricating flexible perovskite solar cells is thermal evaporation, which involves the use of vacuum equipment. This method can produce finer perovskite grains, resulting in pinhole-free films with precise control over film thickness. Thermal evaporation is a process where solid-state perovskite precursor materials are heated to their sublimation temperature inside a vacuum chamber, and the vapor molecules diffuse from bottom to top onto the substrate in a high vacuum environment. This method can be classified into two categories based on the number of evaporative sources: single-source and dual-source (or multi-source) evaporation. As the name suggests, single-source evaporation involves only one evaporative source (depositing one material or a mixture), while the remaining steps are formed through other means. Recently, Liang et al. have demonstrated the direct evaporation of perovskite powder to produce perovskite films.[96] For dual-source (or multi-source) evaporation, both inorganic and organic materials are evaporated simultaneously. The first reported example of dual-source vapor deposition was illustrated by Snaith et al.[97] One major advantage of this method is that it does not involve high-temperature processing during deposition, making it feasible for manufacturing flexible PSCs. However, the additional vacuum processes increase the infrastructure costs and complexity, which are incompatible with the mass production of F-PSCs. Compared to the dual-source method, the single-source method is easier to follow and more energy-efficient. After a comprehensive cost assessment (including precursor, equipment, and process costs, etc.), thermal evaporation may still be an indispensable component of F-PSC fabrication. Li et al. used single-source thermal evaporation to manufacture perovskite layers on PEN/ITO substrates, achieving a PCE of 13.8%.[98] In their work, a smooth PbI_2 layer was first deposited by thermal evaporation, followed by spin-coating MAI on top to form $MAPbI_3$. In the past, thermal evaporation methods were challenging to ensure uniformity and continuity in large-area film fabrication. However, recent improvements and optimizations in the equipment have enabled thermal evaporation to be used for large-scale industrial production of flexible perovskite solar cells.

5.4.6 Inkjet Printing

Recently, inkjet printing technology, initially used for organic solar cells, has shown great promise in the field of perovskite solar cells. Inkjet printing not only allows for significant savings in precursor materials and cost reduction but also enables printing without masks and with specific patterns through integration with computer programming. However, due to some inherent limitations of inkjet printing technology, such as the dispersion of ink on the substrate, the use of non-solvents is not feasible. To achieve high-quality films with uniformity and no pinholes, optimization of substrate and ink temperatures, as well as the incorporation of additional techniques like vacuum annealing, are typically required. So far, many research groups have successfully utilized this technique to fabricate high-performance flexible perovskite films. For example, Gheno et al. employed inkjet printing to fabricate WoO_X, $MAPbI_xCl_{3-X}$, and Spiro-OMeTAD on an ITO substrate. The entire printing process was conducted under ambient conditions and low temperature (< 90 °C), resulting in a PCE of 10%.[99]

5.4.7 Spray Coating

Spray coating is a method that transforms a solution into fine droplets through atomization equipment and then sprays it onto a substrate under the blow of an inert gas. The quality of perovskite films prepared by this method mainly depends on several parameters, including the composition of the precursor solution, substrate temperature, inert gas flow rate, and droplet size. Due to the high temperature required to evaporate the precursor solution into a gas form, spray coating is largely limited in its application on polymer substrates. However, in recent years, with the maturity of atomization technology, many groups have developed low-temperature processes to fabricate organic or inorganic films, greatly expanding the range of substrate materials for flexible perovskite solar cells. Huang et al. coated TiO_2 on a PEI substrate and built $MAPbI_3$/Spiro-OMeTAD/Ag on top, achieving a PCE of 8.5%.[100] Das et al. produced F-PSCs by spray coating and achieved a PCE of 11%, with the architecture being PET/ITO/TiO_2/$MAPbI_XCl_{3-X}$/Spiro-OMeTAD/Au.[101] The F-PSCs produced through spray coating demonstrated impressive durability, with minimal PCE loss after bending for 1,000 times. Their work provides a possible pathway for manufacturing low-cost and high-performance F-PSCs. Following this work, they also demonstrated 25 cm^2 pinhole-free perovskite films through spray coating. In their work, a mixture of CH_3NH_3I and $PbCl_2$ precursor solutions was ultrasonically sprayed onto various substrates, followed by a 1-hour thermal annealing at 100°C in air to drive chemical reactions and crystallization. With significant breakthroughs achieved by various groups in the direction of low-temperature spray coating, the fabrication of large-area and industrially viable flexible perovskite solar cells through this method shows great potential.

5.5 DEDICATION

Zhu Ma wrote the manuscript and proofread the article; Yanlin Li mainly wrote the part of Flexible substrate; Shanyue Hou mainly wrote the part of preparation method of flexible solar cell; Zhangfeng Huang mainly wrote the part of flexible electrode; Zhuowei Du mainly wrote the part of carrier transport materials.

ACKNOWLEDGMENTS

This work was supported by the Open Fund of Sichuan Science and Technology Program (No. 2022NSFSC0226), Photovoltaic Industry Production-Education Integration Comprehensive Demonstration Base of Sichuan Province (Sichuan Financial Education [2022] No.106), Project of Southwest Petroleum University (2021JBGS08), and Sichuan Science and Technology Program (2022YFSY0040).

REFERENCES

1. J. Liu, T. Ye, D. Yu, S. F. Liu and D. Yang, Recoverable flexible perovskite solar cells for next-generation portable power sources. *Angewandte Chemie International Edition*, **2023**, *62*, e202307225.
2. Y. Wu, G. Xu, J. Xi, Y. Shen, X. Wu, X. Tang, J. Ding, H. Yang, Q. Cheng, Z. Chen, Y. Li and Y. Li, In situ crosslinking-assisted perovskite grain growth for mechanically robust flexible perovskite solar cells with 23.4% efficiency. *Joule*, **2023**, *7*, 398–415.
3. Y. Xu, Z. Lin, J. Zhang, Y. Hao, J. Ouyang, S. Liu and J. Chang, Flexible perovskite solar cells: material selection and structure design. *Applied Physics Reviews*, **2022**, *9*.
4. H. Cheng, C. Liu, J. Zhuang, J. Cao, T. Wang, W. Y. Wong and F. Yan, KBF_4 additive for alleviating microstrain, improving crystallinity, and passivating defects in inverted perovskite solar cells. *Advanced Functional Materials*, **2022**, *32*.
5. C. Ge, X. Liu, Z. Yang, H. Li, W. Niu, X. Liu and Q. Dong, Thermal dynamic self-healing supramolecular dopant towards efficient and stable flexible perovskite solar cells. *Angewandte Chemie International Edition*, **2022**, *61*, e202116602.
6. B. Dong, J. Hu, X. Xiao, S. Tang, X. Gao, Z. Peng and D. Zou, High-efficiency fiber – shaped perovskite solar cell by vapor – assisted deposition with a record efficiency of 10.79%. *Advanced Materials Technologies*, **2019**, *4*.
7. K. Zhu, Z. Lu, S. Cong, G. Cheng, P. Ma, Y. Lou, J. Ding, N. Yuan, M. H. Rummeli and G. Zou, Ultraflexible and lightweight bamboo-derived transparent electrodes for perovskite solar cells. *Small*, **2019**, *15*, e1902878.
8. M.-A. Park, S. J. Sung, Y. J. Ahn, I. Hong, I. J. Park, C. R. Park and J. Y. Kim, Bifunctional graphene oxide hole-transporting and barrier layers for transparent bifacial flexible perovskite solar cells. *ACS Applied Energy Materials*, **2021**, *4*, 8824–8831.
9. S. Wu, Z. Li, J. Zhang, X. Wu, X. Deng, Y. Liu, J. Zhou, C. Zhi, X. Yu, W. C. H. Choy, Z. Zhu and A. K. Jen, Low-bandgap organic bulk-heterojunction enabled efficient and flexible perovskite solar cells. *Advanced Materials*, **2021**, *33*, e2105539.
10. G. Jeong, D. Koo, J. Seo, S. Jung, Y. Choi, J. Lee and H. Park, Suppressed interdiffusion and degradation in flexible and transparent metal electrode-based perovskite solar cells with a graphene interlayer. *Nano Letters*, **2020**, *20*, 3718–3727.
11. J. Yoon, U. Kim, Y. Yoo, J. Byeon, S. K. Lee, J. S. Nam, K. Kim, Q. Zhang, E. I. Kauppinen, S. Maruyama, P. Lee and I. Jeon, Foldable perovskite solar cells using carbon nanotube-embedded ultrathin polyimide conductor. *Advanced Science (Weinheim)*, **2021**, *8*, 2004092.
12. J. Zhang, W. Zhang, H.-M. Cheng and S. R. P. Silva, Critical review of recent progress of flexible perovskite solar cells. *Materials Today*, **2020**, *39*, 66–88.
13. M. Soldera, Q. Wang, F. Soldera, V. Lang, A. Abate and A. F. Lasagni, Toward high-throughput texturing of polymer foils for enhanced light trapping in flexible perovskite solar cells using roll-to-roll hot embossing. *Advanced Engineering Materials*, **2020**, *22*.
14. H. S. Jung, G. S. Han, N.-G. Park and M. J. Ko, Flexible perovskite solar cells. *Joule*, **2019**, *3*, 1850–1880.
15. J.-I. Park, J. H. Heo, S.-H. Park, K. I. Hong, H. G. Jeong, S. H. Im and H.-K. Kim, Highly flexible InSnO electrodes on thin colourless polyimide substrate for high-performance flexible $CH_3NH_3PbI_3$ perovskite solar cells. *Journal of Power Sources*, **2017**, *341*, 340–347.
16. Y. Gao, K. Huang, C. Long, Y. Ding, J. Chang, D. Zhang, L. Etgar, M. Liu, J. Zhang and J. Yang, Flexible perovskite solar cells: from materials and device architectures to applications. *ACS Energy Letters*, **2022**, *7*, 1412–1445.
17. G. Tang and F. Yan, Recent progress of flexible perovskite solar cells. *Nano Today*, **2021**, *39*.
18. M. M. Tavakoli, K. H. Tsui, Q. Zhang, J. He, Y. Yao, D. Li and Z. Fan, Highly efficient flexible perovskite solar cells with antireflection and self-cleaning nanostructures. *ACS Nano*, **2015**, *9*, 10287–10295.

19. A. Fuad, A. A. Fibriyanti, N. Mufti, A. Taufiq, S. Maryam and N. Hidayat, Growth of $CH_3NH_3PbI_3$ perovskite on stainless steel substrate layered by ZnO nanoparticles using one-step spin coating route. *Journal of Physics: Conference Series*, **2018**, *1011*.
20. Z. Xu, Q. Zhuang, Y. Zhou, S. Lu, X. Wang, W. Cai and Z. Zang, Functional layers of inverted flexible perovskite solar cells and effective technologies for device commercialization. *Small Structures*, **2023**, *4*.
21. G. S. Han, S. Lee, M. L. Duff, F. Qin and J. K. Lee, Highly bendable flexible perovskite solar cells on a nanoscale surface oxide layer of titanium metal plates. *ACS Appl Mater Interfaces*, **2018**, *10*, 4697–4704.
22. B. Abdollahi Nejand, P. Nazari, S. Gharibzadeh, V. Ahmadi and A. Moshaii, All-inorganic large-area low-cost and durable flexible perovskite solar cells using copper foil as a substrate. *Chemical Communications (Cambridge)*, **2017**, *53*, 747–750.
23. Y. Hu, T. Niu, Y. Liu, Y. Zhou, Y. Xia, C. Ran, Z. Wu, L. Song, P. Müller-Buschbaum, Y. Chen and W. Huang, Flexible perovskite solar cells with high power-per-weight: progress, application, and perspectives. *ACS Energy Letters*, **2021**, *6*, 2917–2943.
24. J. Zhang, N. Luo, X. Zhang, L. Xu, J. Wu, J. Yu, J. He and J. Zhang, All-cellulose nanocomposites reinforced with in situ retained cellulose nanocrystals during selective dissolution of cellulose in an ionic liquid. *ACS Sustainable Chemistry & Engineering*, **2016**, *4*, 4417–4423.
25. P. T. Gomathi, P. Sahatiya and S. Badhulika, Large-area, flexible broadband photodetector based on ZnS-MoS_2 hybrid on paper substrate. *Advanced Functional Materials*, **2017**, *27*.
26. L. Gao, L. Chao, M. Hou, J. Liang, Y. Chen, H.-D. Yu and W. Huang, Flexible, transparent nanocellulose paper-based perovskite solar cells. *NPJ Flexible Electronics*, **2019**, *3*.
27. J. Yang, W. Sheng, X. Li, Y. Zhong, Y. Su, L. Tan and Y. Chen, Synergistic toughening and self-healing strategy for highly efficient and stable flexible perovskite solar cells. *Advanced Functional Materials*, **2023**, *33*.
28. Z. Su, J. Li, R. Jiang, S. Zhang, C. Jin, F. Ye, B. Ke, M. Zhou, J. Tong, H. Park, F. Huang, Y.-B. Cheng and T. Bu, Efficient flexible perovskite solar cells and modules using a stable SnO_2-nanocrystal isopropanol dispersion. *Nano Research*, **2024**, *17*, 2704–2711,.
29. X. Wu, G. Xu, F. Yang, W. Chen, H. Yang, Y. Shen, Y. Wu, H. Chen, J. Xi, X. Tang, Q. Cheng, Y. Chen, X.-m. Ou, Y. Li and Y. Li, Realizing 23.9% flexible perovskite solar cells via alleviating the residual strain induced by delayed heat transfer. *ACS Energy Letters*, **2023**, *8*, 3750–3759.
30. X. Dai, Y. Deng, C. H. Van Brackle, S. Chen, P. N. Rudd, X. Xiao, Y. Lin, B. Chen and J. Huang, Scalable fabrication of efficient perovskite solar modules on flexible glass substrates. *Advanced Energy Materials*, **2019**, *10*.
31. Q. Sun, J. D. Chen, J. W. Zheng, T. Y. Qu, T. Y. Jin, Y. Q. Li and J. X. Tang, Surface plasmon-assisted transparent conductive electrode for flexible perovskite solar cells. *Advanced Optical Materials*, **2019**, *7*.
32. M. Lee, Y. Jo, D. S. Kim, H. Y. Jeong and Y. Jun, Efficient, durable and flexible perovskite photovoltaic devices with Ag-embedded ITO as the top electrode on a metal substrate. *Journal of Materials Chemistry A*, **2015**, *3*, 14592–14597.
33. P. Ma, Y. Lou, S. Cong, Z. Lu, K. Zhu, J. Zhao and G. Zou, Malleability and pliability of silk-derived electrodes for efficient deformable perovskite solar cells. *Advanced Energy Materials*, **2020**, *10*.
34. B. Dou, E. M. Miller, J. A. Christians, E. M. Sanehira, T. R. Klein, F. S. Barnes, S. E. Shaheen, S. M. Garner, S. Ghosh, A. Mallick, D. Basak and M. F. A. M. van Hest, High-performance flexible perovskite solar cells on ultrathin glass: Implications of the TCO. *The Journal of Physical Chemistry Letters*, **2017**, *8*, 4960–4966.
35. S. Pisoni, F. Fu, R. Widmer, R. Carron, T. Moser, O. Groening, A. N. Tiwari and S. Buecheler, Impact of interlayer application on band bending for improved electron extraction for efficient flexible perovskite mini-modules. *Nano Energy*, **2018**, *49*, 300–307.

36. J. W. Bae, S. W. Lee and G. Y. Yeom, Doped-fluorine on electrical and optical properties of tin oxide films grown by ozone-assisted thermal CVD. *Journal of The Electrochemical Society*, **2007**, *154*, D34.
37. M. P. Taylor, D. W. Readey, M. F. A. M. van Hest, C. W. Teplin, J. L. Alleman, M. S. Dabney, L. M. Gedvilas, B. M. Keyes, B. To, J. D. Perkins and D. S. Ginley, The remarkable thermal stability of amorphous In-Zn-O transparent conductors. *Advanced Functional Materials*, **2008**, *18*, 3169–3178.
38. S. H. Kang, M. J. Jeong, Y. K. Eom, I. T. Choi, S. M. Kwon, Y. Yoo, J. Kim, J. Kwon, J. H. Park and H. K. Kim, Porphyrin sensitizers with donor structural engineering for superior performance dye-sensitized solar cells and tandem solar cells for water splitting applications. *Advanced Energy Materials*, **2017**, *7*, 1602117.
39. M. Cai, Y. Wu, H. Chen, X. Yang, Y. Qiang and L. Han, Cost-performance analysis of perovskite solar modules. *Advanced Science*, **2017**, *4*, 1600269.
40. J. Feng, X. Zhu, Z. Yang, X. Zhang, J. Niu, Z. Wang, S. Zuo, S. Priya, S. Liu and D. Yang, Record efficiency stable flexible perovskite solar cell using effective additive assistant strategy. *Advanced Materials*, **2018**, *30*, 1801418.
41. J.-H. Kim, H.-J. Seok, H.-J. Seo, T.-Y. Seong, J. H. Heo, S.-H. Lim, K.-J. Ahn and H.-K. Kim, Flexible ITO films with atomically flat surfaces for high performance flexible perovskite solar cells. *Nanoscale*, **2018**, *10*, 20587–20598.
42. Y. Zhu, L. Shu, Q. Zhang, Y. Zhu, S. Poddar, C. Wang, Z. He and Z. Fan, Moth eye-inspired highly efficient, robust, and neutral-colored semitransparent perovskite solar cells for building-integrated photovoltaics. *EcoMat*, **2021**, *3*, e12117.
43. M. Park, H. J. Kim, I. Jeong, J. Lee, H. Lee, H. J. Son, D.-E. Kim and M. J. Ko, Mechanically recoverable and highly efficient perovskite solar cells: investigation of intrinsic flexibility of organic–inorganic perovskite. *Advanced Energy Materials*, **2015**, *5*, 1501406.
44. M. Vosgueritchian, D. J. Lipomi and Z. Bao, Highly conductive and transparent PEDOT:PSS films with a fluorosurfactant for stretchable and flexible transparent electrodes. *Advanced Functional Materials*, **2012**, *22*, 421–428.
45. M. Dianetti, F. Di Giacomo, G. Polino, C. Ciceroni, A. Liscio, A. D'Epifanio, S. Licoccia, T. M. Brown, A. Di Carlo and F. Brunetti, TCO-free flexible organo metal trihalide perovskite planar-heterojunction solar cells. *Solar Energy Materials and Solar Cells*, **2015**, *140*, 150–157.
46. X. Hu, Z. Huang, X. Zhou, P. Li, Y. Wang, Z. Huang, M. Su, W. Ren, F. Li, M. Li, Y. Chen and Y. Song, Wearable large-scale perovskite solar-power source via nanocellular scaffold. *Advanced Materials*, **2017**, *29*, 1703236.
47. X. Fan, W. Nie, H. Tsai, N. Wang, H. Huang, Y. Cheng, R. Wen, L. Ma, F. Yan and Y. Xia, PEDOT:PSS for Flexible and stretchable electronics: modifications, strategies, and applications. *Advanced Science*, **2019**, *6*, 1900813.
48. K. Sun, P. Li, Y. Xia, J. Chang and J. Ouyang, Transparent conductive oxide-free perovskite solar cells with PEDOT:PSS as transparent electrode. *ACS Applied Materials & Interfaces*, **2015**, *7*, 15314–15320.
49. N. Kim, H. Kang, J.-H. Lee, S. Kee, S. H. Lee and K. Lee, Highly conductive PEDOT:PSS nanofibrils induced by solution-processed crystallization. *Advanced Materials*, **2015**, *27*, 2317–2323.
50. M. Xu, J. Feng, Z.-J. Fan, X.-L. Ou, Z.-Y. Zhang, H.-Y. Wang and H.-B. Sun, Flexible perovskite solar cells with ultrathin Au anode and vapour-deposited perovskite film. *Solar Energy Materials and Solar Cells*, **2017**, *169*, 8–12.
51. J. H. Heo, D. H. Shin, D. H. Song, D. H. Kim, S. J. Lee and S. H. Im, Super-flexible bis(trifluoromethanesulfonyl)-amide doped graphene transparent conductive electrodes for photo-stable perovskite solar cells. *Journal of Materials Chemistry A*, **2018**, *6*, 8251–8258.
52. Z. Li, S. A. Kulkarni, P. P. Boix, E. Shi, A. Cao, K. Fu, S. K. Batabyal, J. Zhang, Q. Xiong, L. H. Wong, N. Mathews and S. G. Mhaisalkar, Laminated carbon nanotube networks for metal electrode-free efficient perovskite solar cells. *ACS Nano*, **2014**, *8*, 6797–6804.

53. T. J. Macdonald, M. Batmunkh, C.-T. Lin, J. Kim, D. D. Tune, F. Ambroz, X. Li, S. Xu, C. Sol, I. Papakonstantinou, M. A. McLachlan, I. P. Parkin, J. G. Shapter and J. R. Durrant, Origin of performance enhancement in TiO_2-carbon nanotube composite perovskite solar cells. *Small Methods* 2019, *3*, 1900164.
54. V. T. Tiong, N. D. Pham, T. Wang, T. Zhu, X. Zhao, Y. Zhang, Q. Shen, J. Bell, L. Hu, S. Dai and H. Wang, Octadecylamine-functionalized single-walled carbon nanotubes for facilitating the formation of a monolithic perovskite layer and stable solar cells. *Advanced Functional Materials*, **2018**, *28*, 1705545.
55. P. Schulz, A.-M. Dowgiallo, M. Yang, K. Zhu, J. L. Blackburn and J. J. Berry, Charge transfer dynamics between carbon nanotubes and hybrid organic metal halide perovskite films. *The Journal of Physical Chemistry Letters*, **2016**, *7*, 418–425.
56. R. A. Hatton, A. J. Miller and S. R. P. Silva, Carbon nanotubes: A multi-functional material for organic optoelectronics. *Journal of Materials Chemistry*, **2008**, *18*, 1183–1192.
57. S. N. Habisreutinger, T. Leijtens, G. E. Eperon, S. D. Stranks, R. J. Nicholas and H. J. Snaith, Carbon nanotube/polymer composites as a highly stable hole collection layer in perovskite solar cells. *Nano Letters*, **2014**, *5*, 4207–4212.
58. I. Jeon, T. Chiba, C. Delacou, Y. Guo, A. Kaskela, O. Reynaud, E. I. Kauppinen, S. Maruyama and Y. Matsuo, Single-walled carbon nanotube film as electrode in indium-free planar heterojunction perovskite solar cells: investigation of electron-blocking layers and dopants. *Nano Letters*, **2015**, *15*, 6665–6671.
59. J. Yoon, H. Sung, G. Lee, W. Cho, N. Ahn, H. S. Jung and M. Choi, Superflexible, high-efficiency perovskite solar cells utilizing graphene electrodes: Towards future foldable power sources. *Energy & Environmental Science*, **2017**, *10*, 337–345.
60. W. Shin, W. Cho and S. J. Baik, Silver nanowires network encapsulated by low temperature sol–gel ZnO for transparent flexible electrodes with ambient stability. *Materials Research Express*, **2018**, *5*, 015050.
61. X. Chen, G. Xu, G. Zeng, H. Gu, H. Chen, H. Xu, H. Yao, Y. Li, J. Hou and Y. Li, Realizing ultrahigh mechanical flexibility and >15% efficiency of flexible organic solar cells via a "welding" flexible transparent electrode. *Advanced Materials*, **2020**, *32*, 1908478.
62. H. Dong, Z. Wu, Y. Jiang, W. Liu, X. Li, B. Jiao, W. Abbas and X. Hou, Thin graphene/silver nanowires/polymer hybrid transparent electrode for optoelectronic devices. *ACS Applied Materials & Interfaces*, **2016**, *8*, 31212–31221.
63. J. Han, S. Yuan, L. Liu, X. Qiu, H. Gong, X. Yang, C. Li, Y. Hao and B. Cao, Fully indium-free flexible Ag nanowires/ZnO:F composite transparent conductive electrodes with high haze. *Journal of Materials Chemistry A*, **2015**, *3*, 5375–5384.
64. R. Chen, S. R. Das, C. Jeong, M. R. Khan, D. B. Janes and M. A. Alam, Co-percolating graphene-wrapped silver nanowire network for high performance, highly stable, transparent conducting electrodes. *Advanced Functional Materials*, **2013**, *23*, 5150–5158.
65. A. Pang, J. Li, X. F. Wei, Z. W. Ruan, M. Yang and Z. N. Chen, UV-O_3 Treated annealing-free cerium oxide as electron transport layers in flexible planar perovskite solar cells. *Nanoscale Advances*, **2020**, *2*, 4062–4069.
66. X. Li, H. Yu, Z. Liu, J. Huang, X. Ma, Y. Liu, Q. Sun, L. Dai, S. Ahmad, Y. Shen and M. Wang, Progress and challenges toward effective flexible perovskite solar cells, *Nanomicro Letters*, **2023**, *15*, 206.
67. K. Mahmood, S. Sarwar and M. T. Mehran, Current status of electron transport layers in perovskite solar cells: Materials and properties, *RSC Advances*, **2017**, *7*, 17044–17062.
68. J. H. Heo, M. H. Lee, H. J. Han, B. R. Patil, J. S. Yu and S. H. Im, Highly efficient low temperature solution processable planar type $CH_3NH_3PbI_3$ perovskite flexible solar cells. *Journal of Materials Chemistry A*, **2016**, *4*, 1572–1578.
69. B. J. Kim, S. L. Kwon, M. C. Kim, Y. U. Jin, D. G. Lee, J. B. Jeon, Y. Yun, M. Choi, G. Boschloo, S. Lee and H. S. Jung, High-efficiency flexible perovskite solar cells enabled by an ultrafast room-temperature reactive ion etching process, *ACS Applied Materials & Interfaces*, **2020**, *12*, 7125–7134.

70. F. Di Giacomo, V. Zardetto, A. D'Epifanio, S. Pescetelli, F. Matteocci, S. Razza, A. Di Carlo, S. Licoccia, W. M. M. Kessels, M. Creatore and T. M. Brown, Flexible perovskite photovoltaic modules and solar cells based on atomic layer deposited compact layers and UV-irradiated TiO_2 scaffolds on plastic substrates, *Advanced Energy Materials*, **2015**, *5*.
71. D. Yang, R. Yang, J. Zhang, Z. Yang, S. Liu and C. Li, High efficiency flexible perovskite solar cells using superior low temperature TiO_2, *Energy & Environmental Science*, **2015**, *8*, 3208–3214.
72. Y. Li, J. K. Cooper, W. Liu, C. M. Sutter-Fella, M. Amani, J. W. Beeman, A. Javey, J. W. Ager, Y. Liu, F. M. Toma and I. D. Sharp, Defective TiO_2 with high photoconductive gain for efficient and stable planar heterojunction perovskite solar cells, *Nature Communications*, **2016**, *7*, 12446.
73. Y. Cheng, Q. D. Yang, J. Xiao, Q. Xue, H. W. Li, Z. Guan, H. L. Yip and S. W. Tsang, Decomposition of organometal halide perovskite films on zinc oxide nanoparticles, *ACS Applied Materials & Interfaces*, **2015**, *7*, 19986–19993.
74. Y. Zhao, F. Ma, Z. H. Qu, S. Q. Yu, T. Shen, H. X. Deng, X. B. Chu, X. X. Peng, Y. B. Yuan, X. W. Zhang and J. B. You, Inactive $(PbI_2)_2RbCl$ stabilizes perovskite films for efficient solar cells, *Science*, **2022**, *377*, 531–534.
75. D. Yang, R. Yang, K. Wang, C. Wu, X. Zhu, J. Feng, X. Ren, G. Fang, S. Priya and S. F. Liu, High efficiency planar-type perovskite solar cells with negligible hysteresis using EDTA-complexed SnO_2, *Nature Communications*, **2018**, *9*, 3239.
76. Z. Zheng, F. Li, J. Gong, Y. Ma, J. Gu, X. Liu, S. Chen and M. Liu, Pre-buried additive for cross-layer modification in flexible perovskite solar cells with efficiency exceeding 22%, *Advanced Materials*, **2022**, *34*, e2109879.
77. H. Yoon, S. M. Kang, J.-K. Lee and M. Choi, Hysteresis-free low-temperature-processed planar perovskite solar cells with 19.1% efficiency, *Energy & Environmental Science*, **2016**, *9*, 2262–2266.
78. D. Yang, R. Yang, X. Ren, X. Zhu, Z. Yang, C. Li and S. F. Liu, Hysteresis-suppressed high-efficiency flexible perovskite solar cells using solid-state ionic-liquids for effective electron transport, *Advanced Materials*, **2016**, *28*, 5206–5213.
79. K. Wojciechowski, T. Leijtens, S. Siprova, C. Schlueter, M. T. Horantner, J. T. Wang, C. Z. Li, A. K. Jen, T. L. Lee and H. J. Snaith, C60 as an Efficient n-type compact layer in perovskite solar cells, *The Journal of Physical Chemistry Letters*, **2015**, *6*, 2399–2405.
80. W. Wang, Z. Yang, J. Ding, J. Kong and X. Li, Improving water-resistance of inverted flexible perovskite solar cells via tailoring the top electron-selective layers, *Solar Energy Materials and Solar Cells*, **2022**, *238*.
81. Y. Sun, S. Yang, Z. Pang, S. Chi, X. Sun, L. Fan, F. Wang, X. Liu, M. Wei, L. Yang and J. Yang, Hydrogen tetrachloroaurate-modulated pedot:pss film assembled with conductive NPB buffer layer for high-performance planar perovskite solar cells, *Chemical Engineering Journal*, **2022**, *432*.
82. C. Xu, Z. Liu and E.-C. Lee, High-performance metal oxide-free inverted perovskite solar cells using poly(Bis(4-Phenyl)(2,4,6-Trimethylphenyl)Amine) as the hole transport layer, *Journal of Materials Chemistry C*, **2018**, *6*, 6975–6981.
83. D. Gao, B. Li, Z. Li, X. Wu, S. Zhang, D. Zhao, X. Jiang, C. Zhang, Y. Wang, Z. Li, N. Li, S. Xiao, W. C. H. Choy, A. K. Jen, S. Yang and Z. Zhu, Highly efficient flexible perovskite solar cells through pentylammonium acetate modification with certified efficiency of 23.35%, *Advanced Materials*, **2023**, *35*, e2206387.
84. J. He, E. Bi, W. Tang, Y. Wang, Z. Zhou, X. Yang, H. Chen and L. Han, Ligand-free, highly dispersed NiOx nanocrystal for efficient, stable, low-temperature processable perovskite solar cells, *Solar RRL*, **2018**, *2*.
85. M. Du, S. Zhao, L. Duan, Y. Cao, H. Wang, Y. Sun, L. Wang, X. Zhu, J. Feng, L. Liu, X. Jiang, Q. Dong, Y. Shi, K. Wang and S. Liu, Surface redox engineering of vacuum-deposited NiOx for top-performance perovskite solar cells and modules, *Joule*, **2022**, *6*, 1931–1943.

86. P.-L. Qin, Q. He, C. Chen, X.-L. Zheng, G. Yang, H. Tao, L.-B. Xiong, L. Xiong, G. Li and G.-J. Fang, High-performance rigid and flexible perovskite solar cells with low-temperature solution-processable binary metal oxide hole-transporting materials, *Solar RRL*, **2017**, *1*.
87. Z. Yang, C. C. Chueh, F. Zuo, J. H. Kim, P. W. Liang and A. K. Y. Jen, High-performance fully printable perovskite solar cells via blade-coating technique under the ambient condition, *Advanced Energy Materials*, **2015**, *5*.
88. M. Xiao, F. Huang, W. Huang, Y. Dkhissi, Y. Zhu, J. Etheridge, A. Gray-Weale, U. Bach, Y. B. Cheng and L. Spiccia, A fast deposition-crystallization procedure for highly efficient lead iodide perovskite thin-film solar cells, *Angewandte Chemie International Edition in English*, **2014**, *53*, 9898–9903.
89. N. J. Jeon, J. H. Noh, Y. C. Kim, W. S. Yang, S. Ryu and S. I. Seok, Solvent engineering for high-performance inorganic-organic hybrid perovskite solar cells, *Nature Materials*, **2014**, *13*, 897–903.
90. C. Feng, B. Zhao and Y. Bi, Constructing large-size and ultrathin Nicop nanosheets on an Fe_2O_3 photoanode toward efficient solar water splitting, *Journal of Materials Chemistry A*, **2022**, *10*, 12811–12816.
91. K. K. Sears, M. Fievez, M. Gao, H. C. Weerasinghe, C. D. Easton and D. Vak, ITO-free flexible perovskite solar cells based on roll-to-roll, slot-die coated silver nanowire electrodes, *Solar RRL*, **2017**, *1*.
92. M. Remeika, L. K. Ono, M. Maeda, Z. Hu and Y. Qi, High-throughput surface preparation for flexible slot die coated perovskite solar cells, *Organic Electronics*, **2018**, *54*, 72–79.
93. C. Gong, S. Tong, K. Huang, H. Li, H. Huang, J. Zhang and J. Yang, Flexible planar heterojunction perovskite solar cells fabricated via sequential roll-to-roll microgravure printing and slot-die coating deposition, *Solar RRL*, **2019**, *4*.
94. K. Hwang, Y. S. Jung, Y. J. Heo, F. H. Scholes, S. E. Watkins, J. Subbiah, D. J. Jones, D. Y. Kim and D. Vak, Toward large scale roll-to-roll production of fully printed perovskite solar cells, *Advanced Materials*, **2015**, *27*, 1241–1247.
95. Y. Y. Kim, T. Y. Yang, R. Suhonen, M. Valimaki, T. Maaninen, A. Kemppainen, N. J. Jeon and J. Seo, Gravure-printed flexible perovskite solar cells: toward roll-to-roll manufacturing, *Advanced Science (Weinheim)*, **2019**, *6*, 1802094.
96. Z. H. Zheng, H. B. Lan, Z. H. Su, H. X. Peng, J. T. Luo, G. X. Liang and P. Fan, Single source thermal evaporation of two-dimensional perovskite thin films for photovoltaic applications, *Scientific Reports*, **2019**, *9*, 17422.
97. M. Liu, M. B. Johnston and H. J. Snaith, Efficient planar heterojunction perovskite solar cells by vapour deposition, *Nature*, **2013**, *501*, 395–398.
98. T. Maeda, K. Kaneko, T. Namba, Y. Koshino, Y. Sato, R. Teranishi and Y. Aruga, Structural and compositional study of precipitates in under-aged Cu-added Al-Mg-Si alloy, *Scientific Reports*, **2018**, *8*, 16629.
99. A. Gheno, Y. Huang, J. Bouclé, B. Ratier, A. Rolland, J. Even and S. Vedraine, Ligand-free, highly dispersed NiOx nanocrystal for efficient, stable, low-temperature processable perovskite solar cells, *Solar RRL*, **2018**, *2*.
100. A. Huang, J. Zhu, Y. Zhou, Y. Yu, Y. Liu, S. Yang, S. Ji, L. Lei and P. Jin, One step spray-coated TiO_2 electron-transport layers for decent perovskite solar cells on large and flexible substrates, *Nanotechnology*, **2017**, *28*, 01LT02.
101. Y. Zhang, X. Chen, B. Cai, H. Luan, Q. Zhang and M. Gu, Photonics empowered passive radiative cooling, *Advanced Photonics Research*, **2021**, *2*.

6 Perovskite Tandem Solar Cells

Long Zhou and Chunfu Zhang

6.1 INTRODUCTION

The increasing demand for renewable energy sources has led to the development of various solar cell technologies, shown in Figure 6.1.[1–4] Among these, silicon (Si) solar cells are the most widely used due to their established manufacturing infrastructure and long-term stability.[5] However, their efficiency has plateaued, and it is imperative to explore alternative approaches to further improve their performance. Perovskite solar cells have recently emerged as a promising candidate due to their high power conversion efficiency (PCE) and low-cost potential.[6] However, their long-term stability remains a concern, and their commercialization has been hindered by the lack of established manufacturing infrastructure.[7–9] Perovskite/silicon tandem solar cells offer a solution to overcome these limitations by combining the advantages of both technologies. Perovskite/silicon tandem solar cells represent a promising technology for increasing the efficiency and reducing the cost of solar energy. Perovskite/silicon tandem solar cells can overcome the limitations of each technology by integrating them into a single device. In the tandem cells, the perovskite top cell absorbs the high-energy photons, while the silicon bottom cell absorbs the low-energy photons. This configuration allows for a higher overall efficiency than either cell alone, and the established manufacturing infrastructure of the silicon cell can be used to produce the tandem cell.[10]

Perovskite/silicon tandem solar cells can be divided into three types based on their different cell structures, including two-terminal structure (integrated structure, 2-T), triple-terminal (spectrally split structure, 3-T), and four-terminal structure (mechanically stacked structure, 4-T) on the basis of device structures.[11] Both 4-T tandems and 2-T tandems are typical configurations in tandem solar cells, shown in Figure 6.2. The 4-T tandems exhibited mechanically stacked configuration and spectrally split configurations. One of attractive advantages for 4-T tandems is unconstrained current-matching and each subcell can be fabricated or optimized independently.[12] However, the complexity of fabrication and assembly as well as the higher material cost due to the multiple substrates usually result in higher overall cost. Monolithic two-terminal tandems are electrically connected through a recombination layer or tunnel junction, which required only one transparent electrode and showed tremendous potential for integrating into a photovoltaic system.[13] One of great challenges for 2-T tandems is that current matching between perovskites and silicon subcell for minimal power loss. The main advantage of the two-terminal tandem structure is

DOI: 10.1201/9781003400493-6

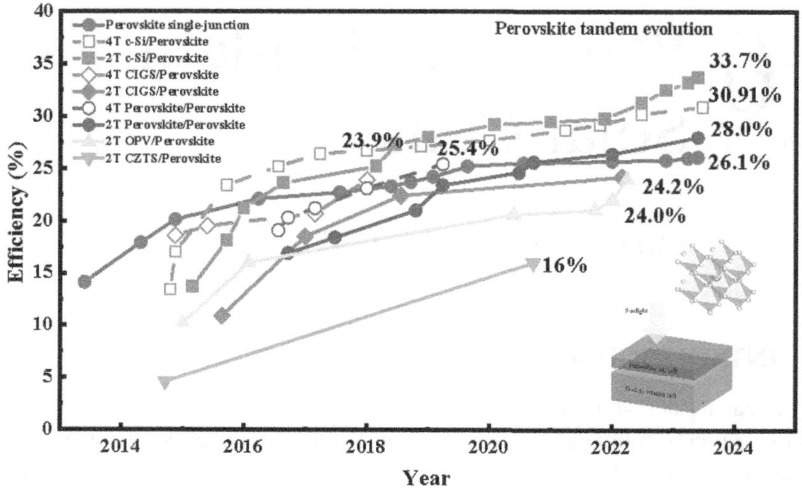

FIGURE 6.1 Efficiency evolution of different perovskite tandem solar cells.

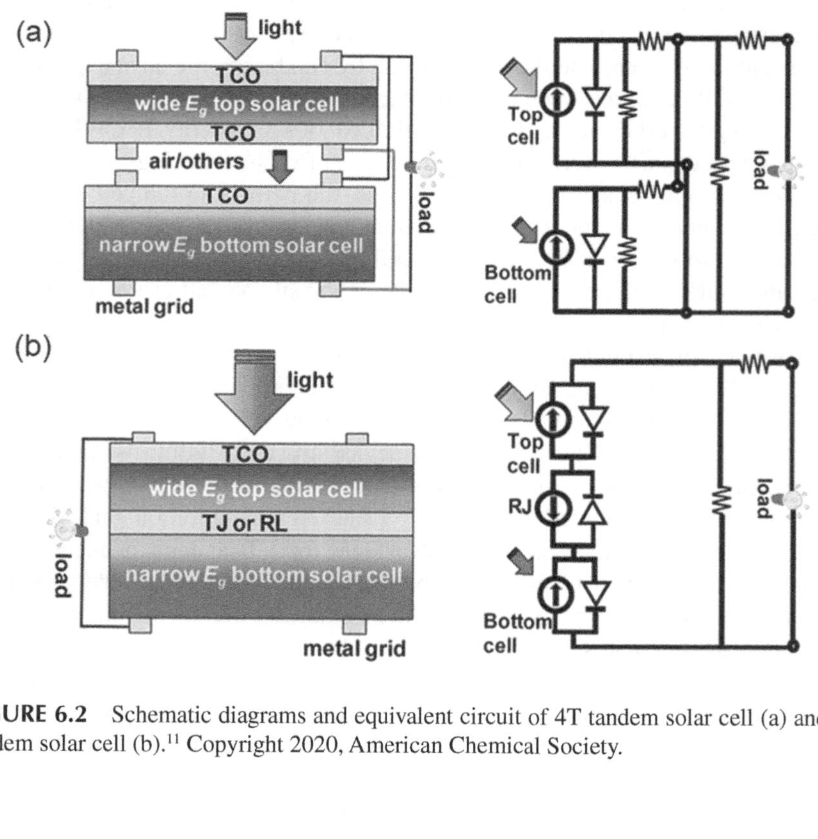

FIGURE 6.2 Schematic diagrams and equivalent circuit of 4T tandem solar cell (a) and 4T tandem solar cell (b).[11] Copyright 2020, American Chemical Society.

its simplicity, which can lead to reduced manufacturing complexity and potentially lower costs.

In this chapter, we summarized the recent progress of perovskite tandem solar cells and reviewed the novel strategies for improving the performance of perovskite tandem solar cells. We systematically discussed the semi-transparent perovskite solar cells, perovskite/Si tandem solar cells, and other perovskite tandem solar cells. This chapter provides a novel perspective on the development of perovskite tandem solar cells and emerging photovoltaic technologies.

6.2 SEMI-TRANSPARENT PEROVSKITE SOLAR CELLS

6.2.1 Transparent Conducting Electrodes

Organic-inorganic hybrid perovskites are good light absorbers as top cells in silicon/perovskites solar cells due to their tunable band gaps and high efficiency.[14] As is known, Si is the most favorable absorber material for bottom cell due to its optimal band gap (1.2 eV). The ideal band gap of perovskites is regions of 1.67 ~ 1.75 eV for top cell when considering the long-term energy yield. It is most important that the perovskite solar cells as top cell should exhibit excellent photovoltaic performance and outstanding transparency to allow as many photons in the long wavelength region as possible to reach the Si cell. Thus, semi-transparent perovskite solar cells are necessary to develop for achieving high efficiency perovskite/silicon tandem solar cells. The conventional opaque metal electrodes need to be replaced by transparent electrodes such as Sn-doped In_2O_3 (ITO), Al-doped ZnO (AZO), B-doped ZnO (BZO), and zinc oxide-doped indium oxide (IZO).[15–17] Among them, ITO is the most widely used in tandem solar cells due to its highest available transmittance and lowest electrical resistivity. ITO was used as transparent electrode to obtain high fill factor (FF) and well-matched band alignment. However, ITO sputtering process will damage the organic transport layer or perovskite absorbers due to its high energy, resulting in poor performances and serious deteriorated junction. To reduce the effect of sputtering process, MoO_3 and SnO_2 were widely used as buffer layer to avoid the damage and provided an efficient charge extraction and transportation. Philipp et al.[18] fabricated a semi-transparent perovskite solar cell based on the methyl ammonium lead triiodide ($CH_3NH_3PbI_3$), which showed broad-band transparency owing to the large transmittance in the near-infrared region. They introduced MoO_x layer as hole collecting buffer layer and prepared ITO layer as a semi-transparent electrode with sputter deposition to achieve a PCE of 6.2%, shown in Figure 6.3a and b. Even though the PCE of a semi-transparent solar cell is low, it provides a reference for developing perovskites in the tandem solar cells. Yan et al.[19] used MoO_x/ITO as transparent electrodes to achieve a high transmittance of 70% in the wavelength range beyond 700 nm. Han et al.[20] fabricated ITO transparent electrode in combination with MgF_2, yielding a higher transmittance of 80%. The high optical transparency of semi-transparent solar cells allows most of the long-wavelength photons absorbed by the bottom cell to achieve high efficiency for tandem solar cells, shown in Figure 6.3c and d. They used ZnO nanoparticles as buffer layer to increase interface contact and electron transport efficiency. As a result, they achieved a wide band gap perovskite semi-transparent solar cell with an impressive PCE of 15.7%.

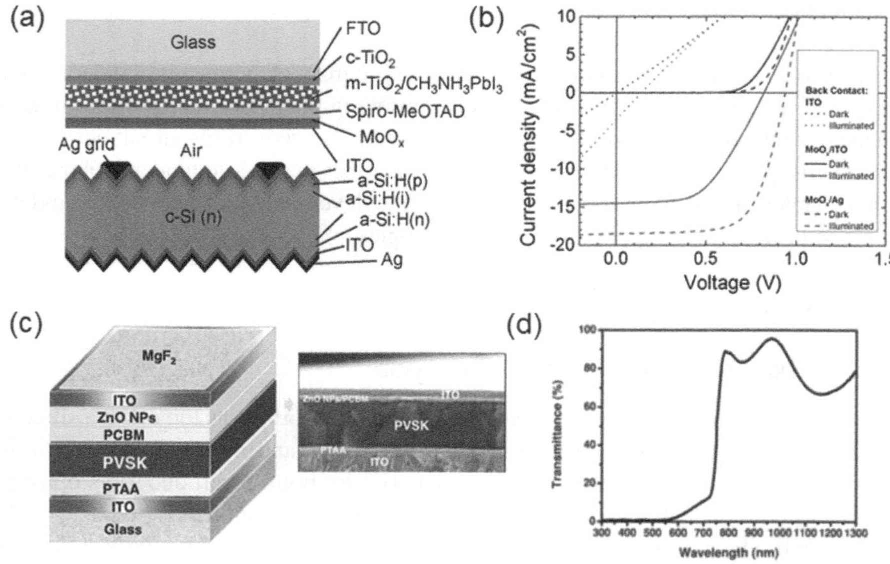

FIGURE 6.3 (a) MoO_x layer as hole collecting buffer layer. (b) Device performance of semi-transparent perovskite solar cell.[18] Copyright 2015, the Owner Societies. (c) ITO transparent electrode combined with MgF_2. (d) Transmittance spectrum through the entire device stack.[20] Copyright 2018, American Association for the Advancement of Science.

Although ITO has low resistivity and high transmittance, it suffers from the high reflectivity and free-carrier absorption. The high conductivity of ITO is attributed to the high doping concentration, which causes high free-carrier absorption. It is necessary to achieve a balance between conductivity and free carrier absorption. There are more strategies to minimize the free-carrier absorption with the reduced doping concentration and increased charger carrier mobility. Zhang et al.[21] optimized the materials properties and light management of ITO to achieve a record near-infrared (NIR) transmittance of 92% and a small free carrier absorption. They further fabricated a planar semi-transparent solar cell with an efficiency of 15.7%, shown in Figure 6.4a and b. As is known, ITO is In_2O_3 with Sn dopant, and the In_2O_3 exhibits wide band gap, high electron mobility, and high NIR transmittance. It is noted that Sn of ITO replaced with Zr or Zn can significantly increase the carrier mobility and conductivity. Erkan Aydin[22] et al. presented Zr-doped indium oxide as transparent electrode to achieve high electron mobility and low sheet resistance (Figure 6.4c). The infrared transparent film was applied as transparent electrode in semi-transparent perovskite solar cell, resulting in a good PCE of 15.6%. Hydrogenated indium oxide (IO:H) exhibited a hall mobility of 140 $cm^2V^{-1}s^{-1}$, which is much higher than that of ITO, shown in Figure 6.4d. It has been demonstrated that the introduction of IO:H between buffer layer and ITO layer can greatly increase the transparency of ITO. Werner et al.[23] first presented a low-temperature process for semi-transparent perovskite solar cell with IO:H/ITO transparent electrodes. The IO:H/ITO as transparent electrodes exhibited a transparence of 70% in the NIR region, resulting in a PCE of 14.5% for the semi-transparent solar cells.

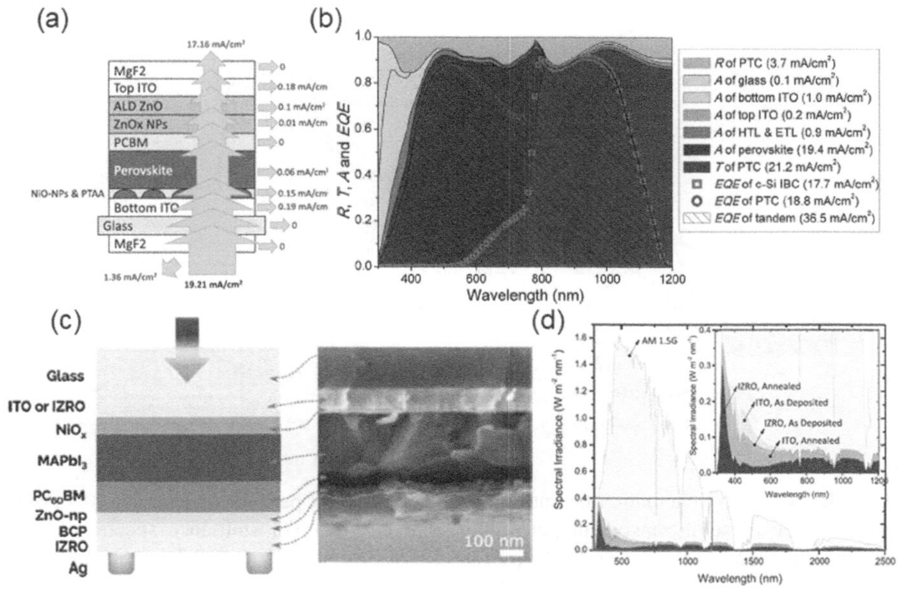

FIGURE 6.4 (a) Optical loss analysis of semi-transparent perovskite solar cell with optimum ITO. (b) Simulated optical spectra of semi-transparent perovskite solar cell.[21] Copyright 2018, Elsevier B.V. (c) Schematics of semi-transparent perovskite solar cell with IZRO. (d) Solar-weighted absorbance spectra of the sputtered IZRO and ITO films measured with integrating sphere.[22] Copyright 2019, WILEY-VCH Verlag GmbH and Co.

Indium zinc oxide is widely used in semi-transparent solar cell due to its excellent electrical and optical properties. Werner et al.[24] demonstrated a semi-transparent perovskite solar cell with the sputtered amorphous indium zinc oxide (120 nm) as transparent electrode, shown in Figure 6.5a. The semi-transparent cells achieved a PCE of 10.6% with inserting a thin molybdenum oxide buffer layer and showed more than 60% transmittance in NIR region, shown in Figure 6.5b. The thicknesses and deposition conditions of IZO layer need to be further optimized towards higher performance and less sputter damages, shown in Figure 6.5c. Shen et al.[17] presented a semi-transparent wide band gap perovskite solar cell with 40 nm IZO layer, shown in Figure 6.5d. The thinner IZO layer could effectively increase the transmittance and reduce sheet resistance, which is beneficial to achieve better charge collection and a high PCE of 18.1% (Figure 6.5e and f). The electrical and optical properties of transparent electrodes is necessary to achieve high performance for semi-transparent perovskite solar cells.

Al-doped zinc oxide exhibited essential potential in transparent electrodes due to its excellent optical transparency in visible and NIR regions.[25,26] Jiang et al.[27] fabricated a semi-transparent solar cell with a PCE of 10.3%, which used the sputtered AZO layer combined with a ZnO nano parties buffer layer as transparent electrodes, shown in Figure 6.6a–c. Compared with ITO and IZO transparent electrodes, the AZO showed advantages in high carrier collection efficiency, low cost, and deposition process with low temperatures. The AZO layer could be deposited

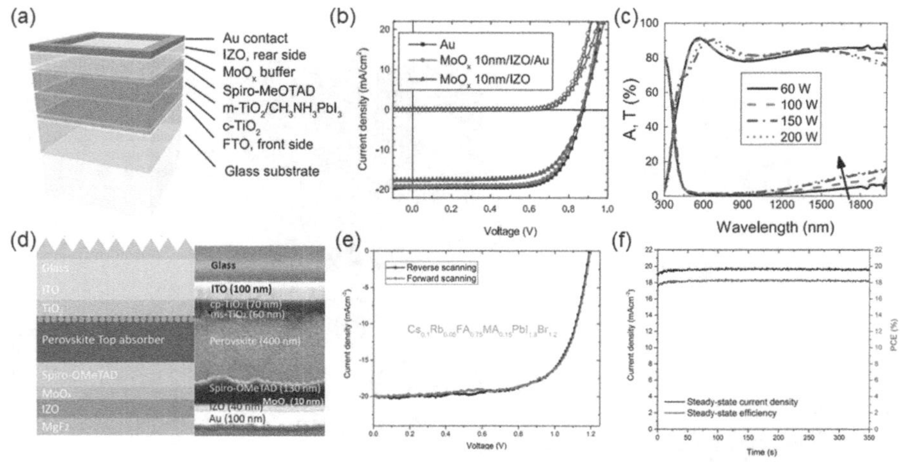

FIGURE 6.5 (a) Schematic illustration of semi-transparent perovskite solar cell. (b) Device performance of semi-transparent perovskite solar cell. (c) Transmittance spectra of semi-transparent perovskite solar cell.[24] Copyright 2015, Elsevier B.V. (d) Schematic illustration of semi-transparent perovskite solar cell with IZO layer. (e-f) Device performance of semi-transparent perovskite solar cell.[17] Copyright 2018, The Royal Society of Chemistry.

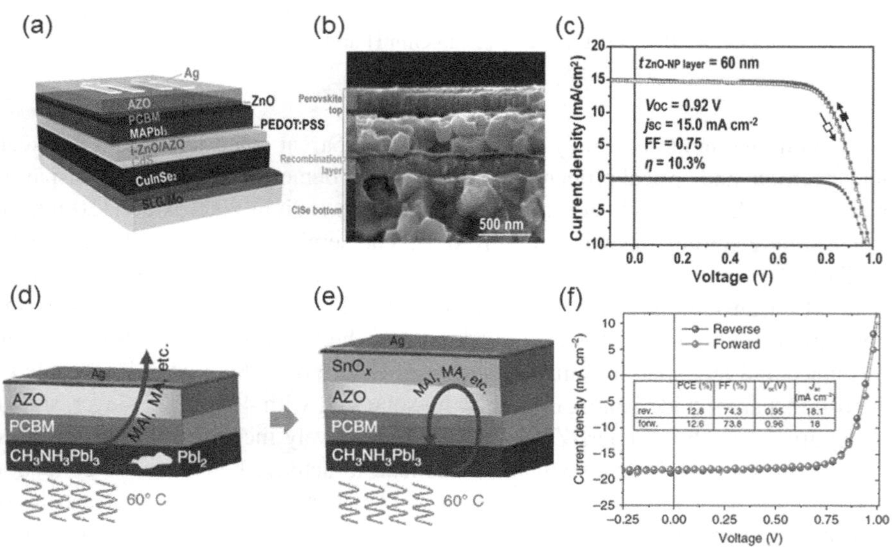

FIGURE 6.6 (a) Schematic of perovskite tandem solar cell with an AZO layer. (b) Cross-sectional SEM image of device. (c) J-V curves of semi-transparent perovskite solar cell.[27] Copyright 2017, The Royal Society of Chemistry. (d-e) Schemes of ageing and decomposition of the perovskite in case of the AZO. (f) Device performance of semi-transparent perovskite solar cell.[28] Copyright 2017, Springer Nature.

via metal-organic chemical vapor deposition (MOCVD) at a deposition temperature of 200°C. T. Riedl et al.[28] introduced the AZO layer to fabricate bilayer AZO/SnO_x, which could afford perovskite solar cell with a remarkable resilience against moisture and heat (Figure 6.6d and e). The optimized semi-transparent solar cell achieved a champion PCE of 12.8% based on the bilayers and excellent stability with the suppressed out-diffusion of decomposition, shown in Figure 6.6f.

6.2.2 Metal Nanowires Transparent Electrodes

Metal nanowires (Figure 6.7a) is a good candidate as transparent electrode for semi-transparent solar cells due to its better transparency, excellent electrical conductivity, low sheet resistance, and good compatibility.[29–32] Jonathan et al.[33] fabricated silver nanowire (AgNW) with the sprayed method and transferred it onto the spiro-OMeTAD layer as transparent electrode. The AgNWs exhibited a low sheet resistance of 9 © / *sq* and an excellent transmission of 89.5%. Similar method has been applied to achieve the semi-transparent solar cell (Figure 6.7b), leading to a PCE of 12.7%.[32] One of the main issues is the stability of semi-transparent solar cell with AgNWs as transparent electrode. The polar solvents in AgNWs could dissolve the perovskite layer and damage the device stability.[34] To overcome the stability issue, a thin metal layer is introduced to cover metal nanowires with thermal evaporation method. Guo et al.[35] demonstrated an efficient semi-transparent perovskite solar cell

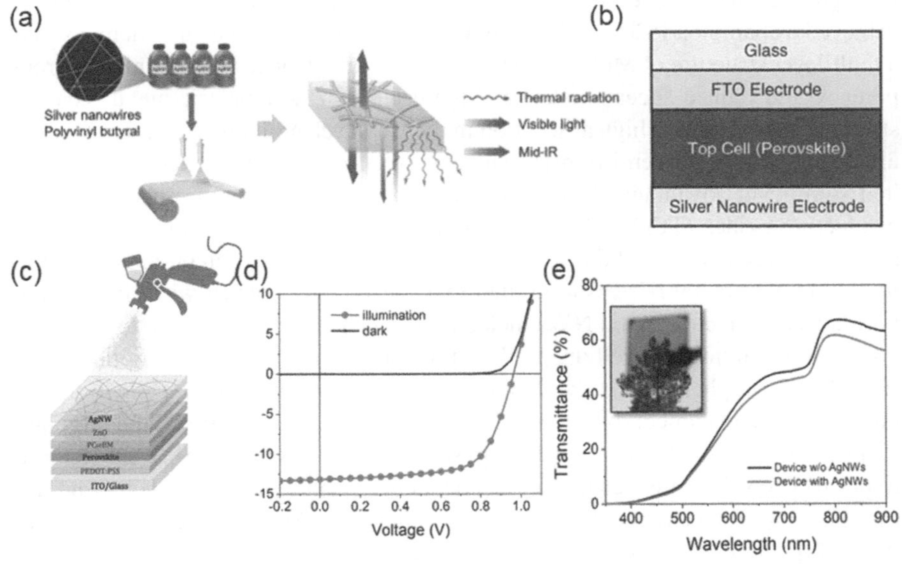

FIGURE 6.7 (a) Schematic illustrating of the AgNWs fabricated methods.[30] Copyright 2019, Elsevier B.V. (b–c) Semi-transparent perovskite solar cell with AgNWs.[32] Copyright 2015, The Royal Society of Chemistry. (d) J-V curves of semi-transparent perovskite solar cell. (e) Transmittance spectra of the device with AgNWs top electrode.[35] Copyright 2015, The Royal Society of Chemistry.

with using solution-processed AgNWs. They introduced a thin layer of ZnO nanoparticles to obtain ohmic contact and protect the perovskite film from solvent damage, shown in Figure 6.7c–e. Huang et al.[36] demonstrated a semi-transparent metal electrode with Au/Cu thin metal layer which exhibited good electrical conductivity and optical transparency. It is noted that 1 nm Cu seed layer was introduced to obtain continuous and ultrathin Au layer due to its higher surface energy. The growth of the Au layer changed from Volmer-Weber Island growth mode to Frank van der Merve layer by layer growth mode.[37,38] As a result, the semi-transparent solar cell with Cu/Au nanowires (NWs) achieved a PCE of 16.5%, which could provide a 95% photons absorption in the 600–800 nm region. Similar methods have been applied to obtain Au/Ag NWs layer as transparent electrodes. Metal nanowires provide an effective strategy to obtain semi-transparent solar cell via the thermal deposited method.

A novel strategy is that the dielectric-metal NWs-dielectric layer deposited with thermal evaporation is used as transparent electrodes in the semi-transparent solar cell. The bottom dielectric layer could help metal NWs to form continuous layer film. And the top dielectric layer was used to improve the transparency of the semi-transparent solar cell. Yang et al.[39] first presented a multilayer transparent electrode based on MoO_x/Au/Ag/MoO_x structure for a semi-transparent solar cell, which delivers an efficiency of 11.5% (Figure 6.8a and b). An ultrathin gold seeding layer was introduced to improve the conductivity and the top MoO_x layer was applied to increase the optical transparency and match the absorption spectra of perovskites layer. Ren et al.[40] presented an optical engineering with using the transparent electrode MoO_x/Au/MoO_x to obtain high transmission at long wavelengths and maximize the light passing. The semi-transparent perovskite solar cell with MoO_x/Au/MoO_x achieved a champion PCE of 17.9%, shown in Figure 6.8c. Yan et al. [41] demonstrated a multilayer structure of MoO_x/Au/MoO_x as transparent electrodes to enhance transparency and reduce sheet resistance (Figure 6.8d). The dielectric-metal dielectric structure still exhibited high absorption in the IR wavelength range, leading to a high efficiency of 18% for semi-transparent solar cells. Although the metal NWs exhibited significant advantages in term of flexibility compared with other transparent conducting oxides (TCOs), the spray methods limited its further developments for semi-transparent solar cell due to the polar solvents in fabrication process.[35] And the morphology of metal NWs is too rough to achieve better contact with charge transport layers. Dielectric-metal NWs-dielectric layers are still a challenge to be used in a semi-transparent solar cell due to their thickness limitation and absorption losses. In addition, graphene is also widely used as TCOs owing to its excellent electron mobility and mechanical robustness.[42] Liang et al.[43] first demonstrated the large-area graphene grown by chemical vapor deposition (CVD) as semi-transparent electrode in perovskite solar cells, leading to identical charge collection efficiencies and high transparency. The better contact between graphene and Au electrode resulted in high internal quantum efficiency and increased transmittance. This work provided a way to outreach the application of graphene in perovskites solar cells.

6.2.3 Optimization of Perovskites Component

Semi-transparent solar cells are more important for a tandem solar cell, which is widely employed as the top cell. The band gap of perovskites depends on the tandem

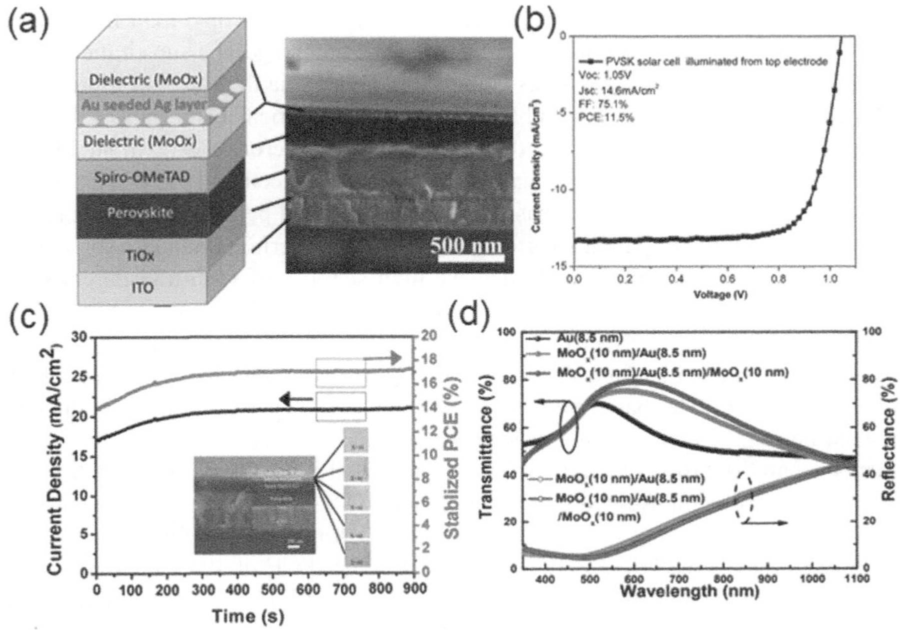

FIGURE 6.8 (a) Schematic structure of semi-transparent perovskite solar cell. (b) J-V curve of semi-transparent perovskite solar cell.[39] Copyright 2019 American Chemical Society. (c) Device performance of semi-transparent perovskite solar cell.[40] Copyright 2019, Elsevier B.V. (d) Optical spectra of MoO_x/Au/MoO_x transparent electrode.[41] Copyright 2017, Springer Nature.

solar cell structure. The perovskites top cell of 4T configuration exhibited larger band gap region compared with 2T configuration. The optimal band gap region of 1.67 eV–1.75 eV is necessary for 2T configuration according to the light management. For instance, $MAPbI_3$ is the first one to use as semi-transparent solar cell and combined with copper indium gallium selenide (CIGS) bottom cell to fabricate 4T tandem solar cells.[44] However, $MAPbI_3$ showed worse stability due to the suffering from air. The optimization of perovskites components is necessary to achieve better stability and high performance. The band gap of perovskite increased with the dopant of Br content, resulting in the region of 1.55–2.30 eV, shown in Figure 6.9a. The $MAPb(I_{1-x}Br_x)_3$ showed photo-induced phase segregation and thermally instability, leading to the larger deficit in voltage for semi-transparent solar cell.[45] Notably, the substitution of Cs in perovskites was used to overcome the phase segregation and photo-instability problems. Jen et al.[46] proposed the $Cs_{0.1}MA_{0.9}Pb(I_{0.6}Br_{0.4})_3$ with the band gap of 1.82 eV is well matched the optical management as the top cell, resulting in a PCE of 12.5% with a larger V_{oc} deficit of 0.62. Sahli et al. further increased the ratio of Cs component as semi-transparent solar cell with a PCE of 13.75%. The substitution of MA with FA has been considered as an effective strategy to achieve high stability for the top cell, which is mainly contributed to the strong interaction between FA and I. Zhao et al.[41] fabricated semi-transparent solar cells based on $FA_{0.3}MA_{0.7}PbI_3$

with a PCE of 18%. It is noted that the FAI is more hygroscopic than MAI, leading to worse moisture stability. To circumvent the challenge, the substitution of Cs is used to optimize the perovskites component. Thus, Snaith et al.[47] first presented MA-free perovskites $Cs_{0.17}FA_{0.83}Pb(I_{0.6}Br_{0.4})_3$ with an optical band gap of 1.74 eV (Figure 6.9b). They demonstrated that the CsFA-based perovskites exhibited better light stability and no significant phase segregation. An excellent PCE of 17% with a large V_{oc} of 1.2 V was achieved based on the CsFA-based perovskites, which is further used as semi-transparent solar cell to apply in four terminal tandem solar cells. This work provides a reference to achieve high stability of semi-transparent solar cell. Tan et al.[48] demonstrated that the incorporation of dipolar MA cation in CsFA-based perovskites could effectively heal deep trap defects and reduce the trap-assisted non-radiative recombination, shown in Figure 6.9c. They achieved a stabilized PCE of 19.1% for CsFAMA-based perovskites with a high V_{oc} of 1.25 V. Thus, multicomponent perovskites have been considered as a good candidate to achieve high performance and stability. Qiu et al.[49] further fabricated four types of CsFAMA-based perovskites and demonstrated that the presence of Cs could effectively promote the crystal growth and suppress defect formation. As a result, the thin film exhibited a lower trap density of 1.13×10^{16} cm^{-3} and a higher PCE of 19.84% based on the $Cs_{0.12}FA_{0.5}MA_{0.38}Pb(I_{0.68}Br_{0.32})_3$ perovskites component. Rb plays an important role in perovskites film crystallinity, carrier mobility and defect migration due to the smaller ionic size of Rb. Duong et al.[50] fabricated RbCsFAMA-based perovskites with a E_g

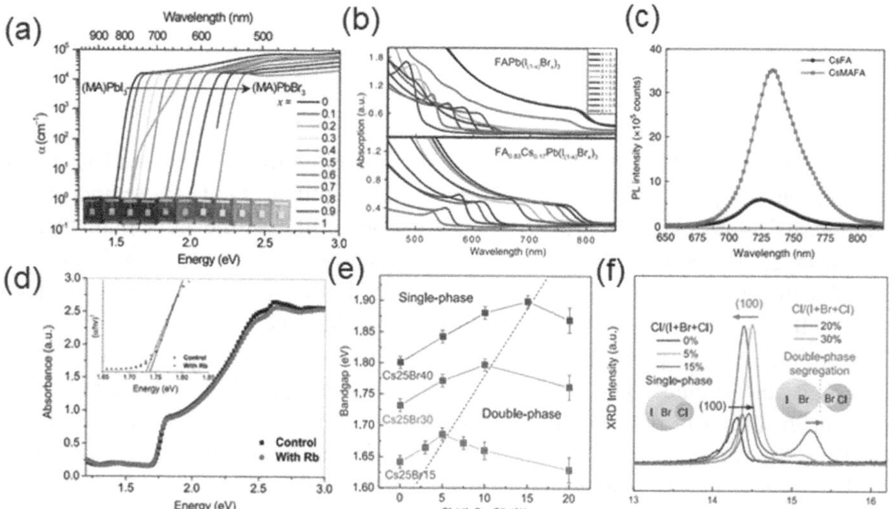

FIGURE 6.9 (a) Absorption spectra of perovskite $MAPb(I_{1-x}Br_x)_3$.[45] Copyright 2015, The Royal Society of Chemistry and (b) MA-free perovskites.[47] Copyright 2016, American Association for the Advancement of Science. (c) PL spectra of CsMAFA-based perovskite.[48] (d) Absorption of Rb-based perovskites.[50] Copyright 2017 WILEY-VCH Verlag GmbH and Co. KGaA. (e) Band gap of wide band gap perovskites. (f) XRD spectra of perovskites with Cl incorporation.[51] Copyright 2020, American Association for the Advancement of Science.

of 1.73 eV (Figure 6.9d) and achieve a negligible hysteresis solar cell with better light stability. Wu et al. used less Rb dopant to achieve good uniformity and crystallinity for perovskite film. They obtained a semi-transparent solar cell with a PCE of 15.9%, which is further applied in tandem solar cell. Moreover, the I/Br ratio is mainly used to optimize the band gap and phase segregation. It is noted that the Cl addition could affect the crystallized process, suppress the phase segregation, and increase the electrical properties of perovskites. Xu et al.[51] demonstrated that the Cl addition could greatly suppress the phase segregation of $Cs_{0.22}FA_{0.78}Pb(I_{0.85}Br_{0.15})_3$ and enhance carrier mobility from 29 $cm^2V^{-1}s^{-1}$ to 53 $cm^2V^{-1}s^{-1}$, shown in Figure 6.9e and f.

One of the smart strategies is the long-chain organic salt as additive to form low dimension materials. For instance, phenethylammonium iodide (PEAI) is widely used to reduce film trap density and improve the electrical quality of perovskites. Meanwhile, the formation of low-dimension perovskites could increase the band gap of perovskite film and suppress the non-radioactive recombination. The wide band gap perovskites exhibited larger V_{oc} and lower voltage deficits, which could well contribute to obtain high voltage for semi-transparent solar cell and tandem solar cell. Yue Yu et al.[52] used 4-fluoro-phenylethylammonium iodide (FPEAI) to form 2D perovskite on the grain boundary and surface of 3D crystals, which could passivate defects and protect the perovskite film from moisture erosion as well as suppressed ion migration (Figure 6.10a). The formation of 2D perovskites could induce a better energy level alignment and mitigate the interfacial carrier recombination. As a result, the modified cells achieved an ultra-high V_{oc} of 1.3 V with a FF of 77.8%, leading to

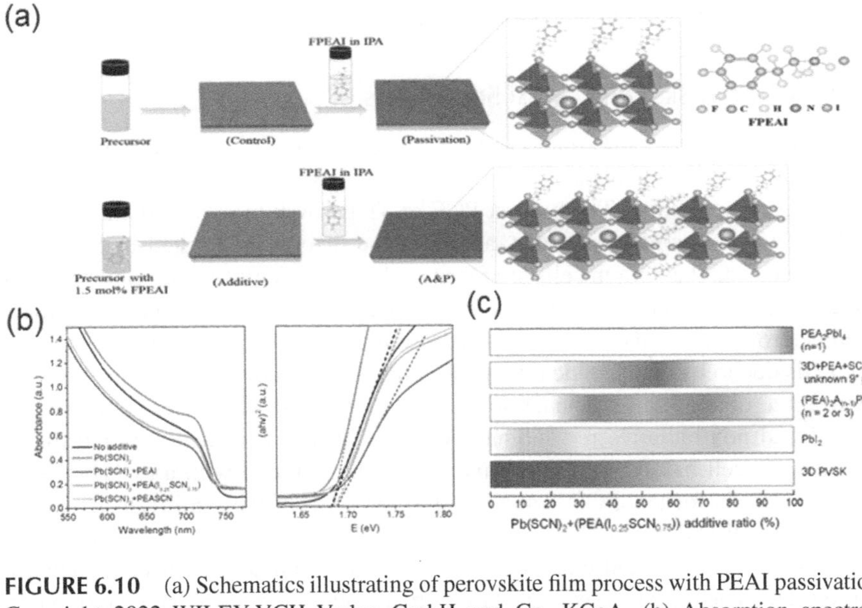

FIGURE 6.10 (a) Schematics illustrating of perovskite film process with PEAI passivation.[52] Copyright 2022 WILEY-VCH Verlag GmbH and Co. KGaA. (b) Absorption spectra of perovskites film with PEAI passivation. (c) Color-coded visualization of different phases formed.[53] Copyright 2020, American Association for the Advancement of Science.

a remarkable PCE of 19.1% for wide-band gap solar cell. Daehan Kim et al.[53] demonstrated an anion engineering with 2D perovskites passivation on PbI_2-framework to obtain a stable perovskite solar cell with a band gap of 1.7 eV (Figure 6.10b and c), retaining over 80% of its initial PCE under continuous light soaking for 1,000 h. A high PCE of 20.7% was achieved based on the 2D perovskites passivation located at the grain boundaries, which validated the use of perovskites in semi-transparent solar cell. Similar works have been reported and provide a reference to fabricated high efficiency semi-transparent solar cell.

6.3 PEROVSKITE/SI TANDEM SOLAR CELLS

The PCE of silicon solar cells have reached 26.8%, approaching their theoretical limiting PCE of 29.1%.[6] It is a great challenge to further increase the PCE and decrease the cost of silicon solar cell. The PCE of single-junction perovskite solar cells has exceeded 26%, comparable with that of silicon solar cells. Although the efficiency of perovskite solar cells has made a significant breakthrough, it is extremely difficult to replace the Si solar cell considering the stability and market share. Furthermore, the development of perovskite solar cells (PSCs) has brought a new opportunity to achieve lower cost and higher PCE. The tandem solar cells based on perovskites combined with silicon have been considered as the most ideal candidates for photovoltaic energy field. The progress of perovskite/Si tandem solar cells has developed quickly due to its unique advantages. The record PCE of perovskite/Si tandem solar cells has increased from 13.4%[18] in 2015 to 33.7%[6] in 2023, which is higher than that of Si and perovskite single solar cells. The quick development of perovskite/Si tandem solar cell benefits from the optimized perovskites component, the design of tandem configurations, and the modification of interface engineering.

6.3.1 4T Perovskite/Si Tandem Solar Cells

The main advantages of 4T perovskite/Si tandem solar cells are mechanical connection and no current management, leading to low technical barriers and high processing compatibility. The highest PCE of 4T perovskite solar cells has reached 30.91%,[54] which is still a large distance from the theoretical limited PCE of 46%. The top cells and the bottom cells are fabricated independently, which can well mind texture, Si polarity, fabricated processing, interface stress, and temperature. Silicon based heterojunction (SHJ) solar cells are wildly used as bottom cells in 4T tandem solar cells due to its double-side texture structure and excellent NIR absorption. It is important to obtain high efficiency for perovskite top cell with lower voltage deficit and reduced non-radiative recombination. Loper et al.[18] first demonstrated the 4T tandem solar cell based on $MAPbI_3$ and SHJ Si bottom cell with a PCE of 13.4%. The main reason for low PCE is attributed to low PCE of the semi-transparent perovskite top cell and silicon bottom cell. Ren et al.[40] reported an efficient strategy to improve the optical and electrical properties of 4T tandem solar cell, shown in Figure 6.11a. They used oxygen annealing treatment to further reduce the trap density of perovskites film and used optimized optical transparent electrode to obtain high transmission. A high PCE of 22.4% for 4T tandem solar cell is obtained based on the c-Si bottom cell (Figure 6.11b). Huang et al.[36] further achieved a PCE of 23.0% based on the better semi-transparent $MAPbI_3$ top cell and double-side texture Si bottom cell (Figure 6.11d). Furthermore, Snaith et al.[47] fabricated the stable $Cs_{0.17}FA_{0.83}Pb(I_{0.6}Br_{0.4})_3$ perovskite solar cell as the top cell and achieved a stabilized PCE of 25.2% based on the SHJ solar cell with 19% efficiency for 4T tandem solar cell, shown in Figure 6.11e. Erkan Aydin et al.[22] fabricated a 4T tandem solar cell based on the $MAPbI_3$ top cell and SHJ bottom cell, leading to a high PCE of 26.2% with better light harvesting (Figure 6.11f). Even with these results, the cost of SHJ solar cell limited its development and application.

Various Si solar cells have been used as the bottom cells, such as multicrystalline (mc)-Si solar cells, aluminum back surface field (Al-BSF) Si solar cells, and Si interdigitated back contact (IBC) solar cells. Bailie et al.[55] used low quality mc-Si bottom cells as bottom cell in a mechanical stacked tandem solar cell (Figure 6.12a) to obtain a certified PCE of 17.3%, which provides a reference to achieve high-efficiency tandem solar cells with low cost. Ramos et al.[56] demonstrated that the perovskite/Si tandem solar cell based on the Al-BSF Si bottom cell exhibited a PCE of 19.52%, which showed the great potential of silicon in tandem solar cells, shown in Figure 6.12b and c. Compared with SHJ and Al-BSF Si solar cell, IBC solar cells eliminate shading losses by putting both contacts on the rear sides of the cells, which eliminates front contact shading losses and reduces series resistance. Zhang et al.[16] used IBC c-Si as a

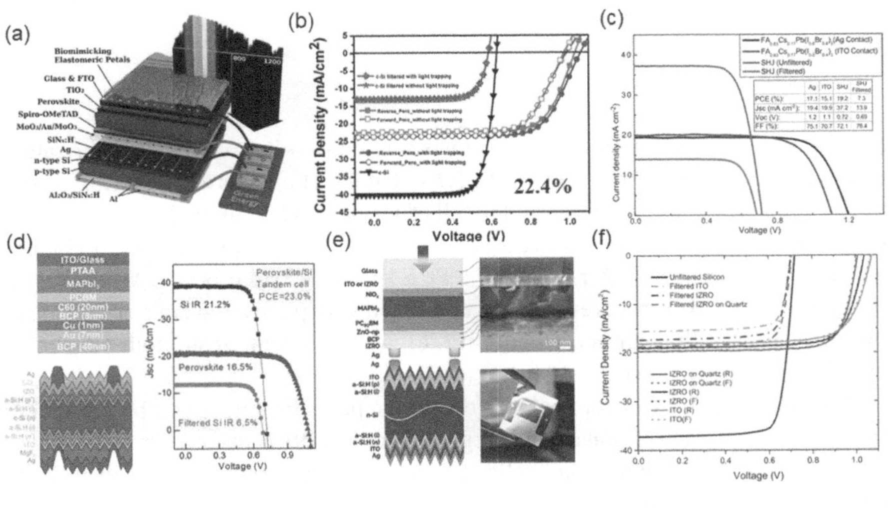

FIGURE 6.11 (a) Device architecture of the four-terminal perovskite/Si tandem solar cell. (b) The J-V characteristics of the c-Si solar cell, perovskite/Si tandem solar cell.[40] Copyright 2018, Elsevier B.V. (c) Forward bias to short-circuit J-V curves of the best perovskite devices, SHJ cell and tandem solar cell.[47] Copyright 2016, American Association for the Advancement of Science. (d) Device architecture and performance of perovskite/Si tandem solar cell.[36] Copyright 2016 WILEY-VCH Verlag GmbH and Co. KGaA. (e) Device structure and images of perovskite/Si tandem solar cell. (f) J-V curves of perovskite/Si tandem solar cell.[22] Copyright 2019 WILEY-VCH Verlag GmbH and Co. KGaA.

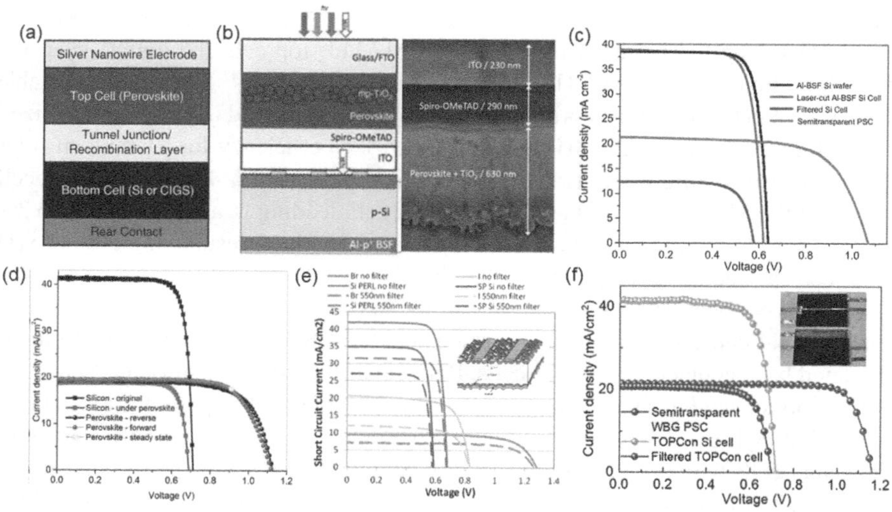

FIGURE 6.12 (a) Schematic of a monolithic polycrystalline tandem solar cell.[55] Copyright 2015, The Royal Society of Chemistry. (b) Scheme of the 4-terminal tandem device based on Al-BSF Si solar cell. (c) I-V curves of Al-BSF Si solar cell and tandem solar cells.[56] Copyright 2018, Springer Nature. (d) J-V curves of the silicon cell and tandem solar cell.[50] Copyright 2017 WILEY-VCH Verlag GmbH & Co. KGaA. (e) J-V curves of passivated emitter rear cell (PERC) Si cell and tandem solar cell.[57] Copyright 2015 American Chemical Society. (f) J-V curves of TOPcon Si cell and tandem solar cell.[54] Copyright 2023, American Association for the Advancement of Science.

bottom cell combined with the semi-transparent perovskite solar cell to fabricate 4T tandem solar cell, resulting a PCE of 25.7%. Duong et al.[50] introduced RbI passivation to improve semi-transparent solar cell efficiency and combined with silicone gel on IBC c-Si bottom cell to achieve a PCE of 26.4%, shown in Figure 6.12d. Sheng et al.[57] reported a tandem solar cell with a PCE of 23.4% based on the $MAPbBr_3$ perovskite as the semi-transparent top cell combined with passivated emitter rear locally diffused (PERL) Si solar cell (Figure 6.12e), which benefits from high voltage and excellent spectral response.

The Si cell also suffers optical loss due to the parasitic absorption by the spectral splitter after the cut-off wavelength region. The low NIR absorption is the biggest challenge for 4T tandem solar cells, leading to large optical losses in c-Si bottom cells and eventually reduce the performances of the whole tandem solar cells (Figure 6.12f). In addition, the utility of three layers transparent conductive electrodes (TCEs) in the front and bottom solar cells further induces NIR parasitic absorption and reflective optical losses coming from layers with different refractive indexes. The low NIR absorption is the biggest challenge for 4T tandem solar cells (TSCs), leading to large optical losses in c-Si bottom cells and eventually reducing the performances of the whole TSCs. In addition, the utility of three layers TCEs in the front and bottom solar cells further induces NIR parasitic absorption and reflective optical losses coming from layers with different refractive indexes. Depending on the thickness and doping concentration, hole transport layers (HTLs) and electron transport layers (ETLs) in PSCs also induce parasitic absorption losses.

6.3.2 2T Perovskite/Si Tandem Solar Cells

Compared with the 4T tandem solar cell, the 2T tandem solar cell has been considered the most promising for industrialization. The theoretical limited PCE of 2T tandem solar cells is up to 45.7%, which is slightly lower than that of 4T tandem solar cells.[58] The main advantages of 2T tandem solar cells are lower cost and great potential in industrialization due to their less functional layers and excellent optical natures. It is extremely challenging to achieve high efficiency for 2T tandem solar cells due to the requirement of current-matching and processing compatibility. Many works focus on the high efficiency and stability of 2T tandem solar cells via various strategies. The Si bottom cells play an important role for achieving high efficiency of 2T tandem solar cells and the continuity of perovskite films, shown in Figure 6.13. SHJ solar cells are widely used as bottom cells due to their high PCE low voltage deficit, low temperature coefficient, and excellent NIR absorption ability. The double texture structure of SHJ solar cells can greatly reduce the optical loss, which is beneficial for 2T tandem solar cells.

The ITO layers are commonly introduced to increase the contact of top cells and bottom cells.[59,60] The main challenge of the 2T tandem solar cell is how to fabricate high-quality perovskite film on the top of textured Si solar cells. The shunting problems and accumulation of perovskite films will limit the development and the PCE of 2T tandem solar cells. As a result, a double-side-polished SHJ solar cell has been used as the bottom cell to avoid the accumulation of perovskite films and the shunting problems. Werner et al.[23] first presented a low-temperature process to fabricate perovskite films on double-side-polished SHJ bottom cell, resulting a monolithic tandem solar cell with a PCE of 21.4% (Figure 6.14a). The planar tandem configuration based on double-side-polished SHJ solar cell could result in strong optical

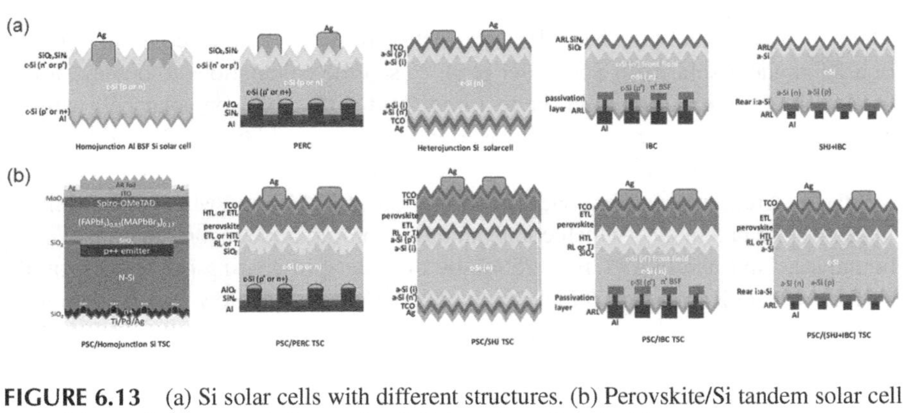

FIGURE 6.13 (a) Si solar cells with different structures. (b) Perovskite/Si tandem solar cell with different Si structures.[11] Copyright 2020 American Chemical Society.

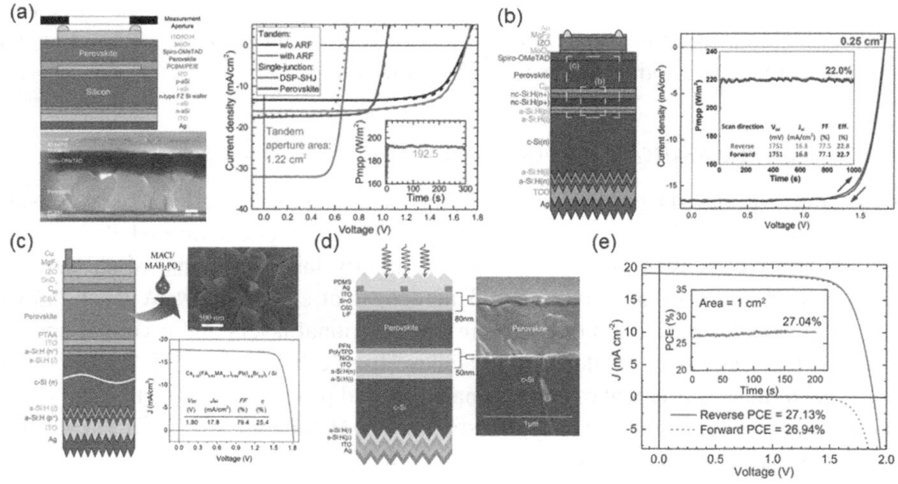

FIGURE 6.14 (a) Schematic drawing of the planar monolithic perovskite/silicon heterojunction tandem cell layer and J-V curves of perovskite/SHJ monolithic tandem solar cell.[23] Copyright 2015 American Chemical Society. (b) Schematic view and J-V curves of the perovskite/SHJ monolithic tandem cell with a nc-Si:H recombination junction.[61] Copyright 2018 WILEY-VCH Verlag GmbH and Co. KGaA (c) Schematic structures and device performance of perovskite/silicon monolithic tandem solar cell.[63] Copyright 2019, Elsevier Inc. (d) Schematic of the two-terminal monolithic tandem structure with a-Si:H denotes hydrogenated amorphous silicon. (e) Light and dark J-V curves and maximum power point (MPP) tracking (inset) of the champion tandem solar cells.[64] Copyright 2020, American Association for the Advancement of Science.

interferences, larger reflection loss, and poor matched currents. The rear side texture and single-side polished Si solar cell are used to enhance optical absorption. Sahli et al.[61] employed a rear-side textured SHJ as the bottom cell with a front emitter configuration (Figure 6.14b), leading to optical losses and reduced shunt resistance. The tandem solar cell achieved a PCE of 22.7% combined with a Cs-based perovskite top cell, which provided a new way to obtain a high-efficiency 2T tandem solar cell. The rear texture and polished SHJ solar cells are widely employed as bottom cell in a 2T tandem solar cell. Schneider et al.[62] used the calculations to achieve a theoretical PCE of 35% based on the textured front surface silicon solar cell with the combination of an optimized single layer anti reflection and brag reflector. Huang et al.[63] fabricated a 2T tandem solar cell based on the single side mirror polished Si and conducted grain engineering with incorporation of MACl and MAH_2PO_2 additives in perovskite precursors, which could suppress non-radiative recombination at grain boundaries and obtain a PCE of 25.4% by directly spin-coating perovskites on the surface of ITO RL (Figure 6.14c). Xu et al.[64] demonstrated a low-cost tandem technology to obtain a PCE of 27% based on the single-side mirror polished Si bottom cell, shown in Figure 6.14d and e. They used the addition of $MAPbCl_3$ in perovskite precursors to shrink the lattice parameter and stabilize phase segregation. It is found that the introduction of Cl is beneficial for increasing lifetime and charger carrier mobility. The thicker perovskite films and high quality of films could effectively avoid the shunting problem. Kim et al.[53] fabricated a 2T tandem solar cell using the rear-textured SHJ solar cell as the bottom cell and combined with wide band gap perovskites (1.7 eV) as the top cell, resulting a stabilized PCE of 26.7% and 26.2% for areas of 0.188 and 1.001 cm^2, respectively. They improved the electrical quality and phase stability with the modification of 2D perovskite layers formed by PEA addition. The Si solar cells without front-side texture are widely used in 2T tandem solar cell due to the improved quality of perovskite films and high efficiency.

Compared with the fully textured SHJ solar cells, the single rear texture Si cell exhibited lower current density due to the suppression of the strong reflection. Sahli et al.[65] developed a deposition process to obtain high quality of perovskites film on the micrometer-size pyramids of double side textured silicon (Figure 6.15a), resulting a certified steady-state PCE of 25.2% and exploiting the high photocurrent potential of monolithic tandem solar cell, shown in Figure 6.15b. Shen et al.[66] have presented the improved configuration perovskite/c-Si monolithic tandem solar cell with addition of pyramids structure in front surface, which can effectively increase the PCE to 29.05%. This work suggests that the commercial SHJ solar cells with double texture structure as bottom cells could achieve a high PCE for 2T monolithic tandem solar cells. Huang et al.[67] proposed a new tandem configuration that enables perovskite films blade-coated on the textured silicon solar cell with pyramids less than 1 μm

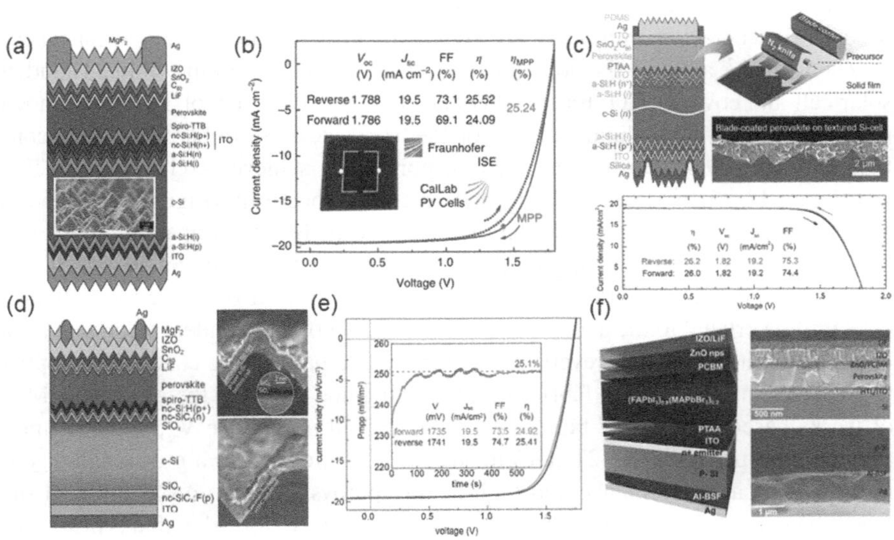

FIGURE 6.15 (a) Schematic view of a fully textured monolithic perovskite/SHJ tandem solar cell. (b) Corresponding certified J-V curves of tandem solar cell.[65] Copyright 2018, Springer Nature. (c) Schematic structure and device performance of perovskite/Si tandem solar cell.[67] Copyright 2020, Elsevier Inc. (d) Schematic view of the perovskite/p-type Si bottom tandem solar cell. (e) J-V properties and maximum power point tracking of tandem solar cell.[69] Copyright 2019, American Chemical Society. (f) Schematic and scanning electron microscopy cross-sectional images of a fully planar monolithic perovskite/Si tandem device.[70] Copyright 2019, Elsevier Inc.

in height, which obtained a PCE of 26.0% for 2T perovskite/SHJ tandem solar cells, shown in Figure 6.15c. This work opens a promising way to obtain high quality perovskite films with blade-coating method for perovskite/silicon 2T tandem solar cell. Even though n-Si based SHJ solar cells showed high efficiency, they are still limited by its temperature tolerance due to low thermal stability of i-a-Si. Thus, it is necessary to improve the temperature tolerance of the Si cell and restrict the processing temperature of perovskites subcells. The highest PCE of homojunction Si solar cell is 25.7%, slightly lower than that of IBC-SHJ solar cell, which showed great potential in 2T perovskite/silicon tandem solar cells. Mailoa et al.[33] pioneered work for using a homojunction Si solar cell as the bottom cell, obtaining a lower PCE of 13.7% without any optimization. Wu et al.[68] presented an innovative design to feature a homojunction Si bottom cell combined with the mesoscopic perovskite subcell. This structure allows high temperature processed compact and mesoporous TiO_2 electron transport layer in perovskite top cell, maintaining a high V_{oc} for c-Si subcell. Nogay et al.[69] fabricated a 2T tandem solar cell based on featuring a p-type Si bottom cell with a flat rear side and a textured front side, and introduced an SiO_x layer to increase contact grown by UV-O_3 exposure (Figure 6.15d and e). Kim et al.[70] used homojunction p-type Si with Al_2O_3/SiN_x passivation to better match the refractive index and suppress the reflection loss, shown in Figure 6.15f. The Al_2O_3/SiN_x passivated layer in homojunction Si solar cell could reduce reflectivity and increase the interface resistance. Thus, the replacement or removement of dielectric layer and contact openings are used to overcome larger interface resistance. The metal stacks and ITO film are used for the contact openings as a recombination layer to reduce contact resistance.

Zheng et al.[71] adopted the homojunction PERC Si as the bottom cell in 2T tandem solar cell and covered SiO_2 for metal contacts via a metal stack of Ti/Pd/Ag, shown in Figure 6.16a. The p-n homojunction on the front side of Si could reduce the recombination due to the improved carrier transport. They achieved a champion PCE of 21.0% and 17.6% for areas of 4 cm^2 and 16 cm^2, respectively (Figure 6.16b and c). Wu et al.[72] used numerical modeling to rigorously analyze the impact of doped c-Si bottom cell and showed the higher PCE of low resistivity p-type Si cell on 2T tandem solar cell (Figure 6.16f). The implementation of such optimized c-Si bottom cells will be crucial to the achievement of over 30% efficient tandem solar cells. The localized emitter rear and reversed tunneling oxide passivating contact are applied to further simplify tandem fabricated processing and improve the efficiency. Most of works based on the SHJ bottom cell achieved an impressive V_{oc}, which is more promising in perovskite/Si 2T tandem solar cell. However, it is a great challenge to improve optical performance and high-quality perovskites film on textured Si surface. Shen et al.[73] used the poly-Si/SiO_x heterojunction combined with a typical n-i-p perovskite solar cell to achieve a tandem PCE of 24.5%, shown in Figure 6.16d and e. High doping poly Si technology showed great potential in 2T tandem solar cell with excellent contact. For the homojunction Si bottom cell, it is important to achieve high passivation quality and low optical loss for 2T tandem solar cell.

In addition, the photocurrent matching is necessary to achieve high current and PCE via modification of the perovskite films thickness and band gap. The current density of perovskite subcell is related with the film quality, trap density, and phase segregation of perovskites.

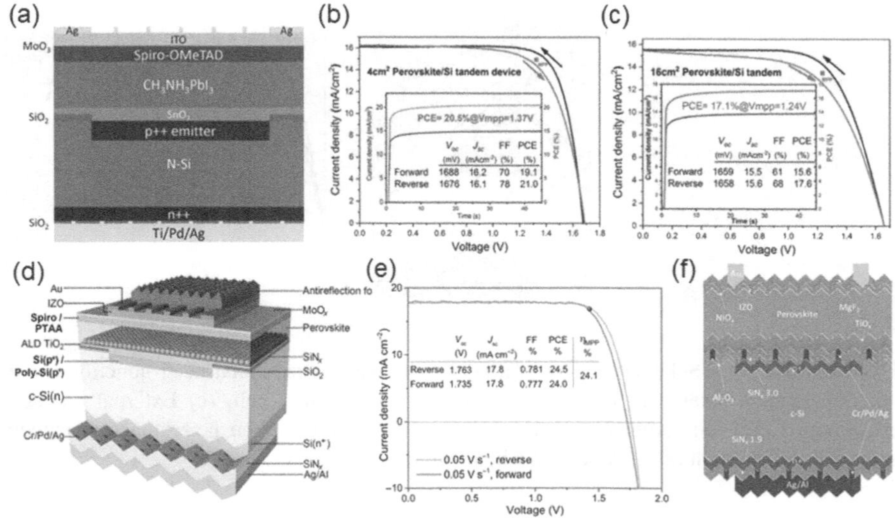

FIGURE 6.16 (a) Schematic device design of interface-layer-free perovskite/silicon-homojunction solar cells. (b–c) J–V curves of perovskite/Si tandem cell with large areas.[71] Copyright 2018, The Royal Society of Chemistry. (d) Schematic of the interlayer-free monolithic perovskite/crystalline-silicon tandem solar cell. (e) J-V behavior of the tandem device with both reverse and forward scanning.[73] Copyright 2018, American Association for the Advancement of Science (f) Schematic drawing for monolithic perovskite/c-Si tandem with double-sided textures.[72] Copyright 2018 WILEY-VCH Verlag GmbH and Co. KGaA.

6.3.3 Triple-Junction Perovskite/Si Tandem Solar Cells

Compared with 2T perovskite/Si tandem solar cell, the triple-junction perovskite/Si tandem solar cell is composed of two perovskites layers and a Si solar cell, which can achieve a trajectory PCE of 38.8%. Werner et al.[74] fabricated the triple-junction perovskite/Si tandem solar cell based on double-side textured Si bottom cell and used the sequential deposition method to obtain the perovskite middle and top cell, which provided a novel alternative way to achieve a low-cost and high-efficiency device. In this work, the authors used evaporated method combined with spin-coating method to regulate the growth of perovskite films on the micrometer-sized pyramids of the Si bottom cell, shown in Figure 6.17. For the perovskite middle cell, the Spiro-TTB is deposited as HTL on the p-type nc-Si:H formed with plasma-enhanced chemical vapor deposition (PECVD). The obtained perovskite middle cell exhibited a band gap of 1.53–1.55 eV and the band gap of perovskite top cell is around 1.77–1.80 eV. However, this work provided a proof-of-concept for the triple-junction perovskite/Si tandem solar cell, resulting in a low V_{oc} of 2.69 V. The main reason is that the thin thickness of perovskite top cell is applied to match current between subcells. The major challenge for triple-junction perovskite/Si tandem solar cell is the fabrication of the middle and top perovskites solar cells.

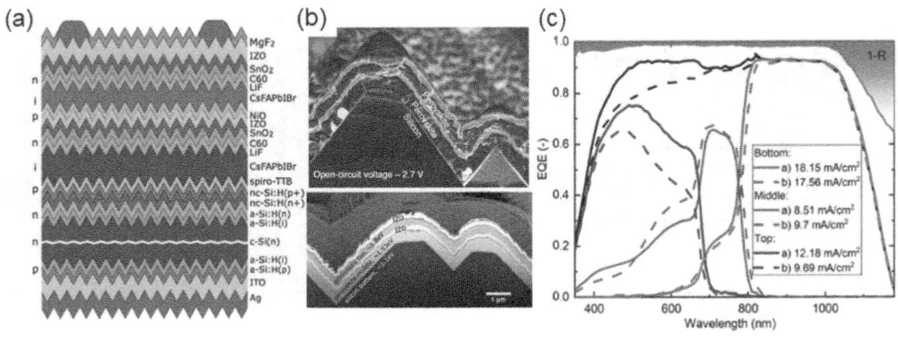

FIGURE 6.17 (a) Schematic view of a perovskite/perovskite/silicon heterojunction triple-junction tandem solar cell. (b) SEM images of tandem solar cell. (c) External quantum efficiency (EQE) measurements of the perovskite/perovskite/silicon heterojunction tandem solar cell.[74] Copyright 2018, American Chemical Society.

6.4 OTHER PEROVSKITE TANDEM SOLAR CELLS

6.4.1 All Perovskite Tandem Solar Cells

Perovskite materials have been considered as good candidates due to their tunable band gaps from 1.2–2.3 eV, which offers great promise to all perovskite tandem solar cells.[74] The wide band gap perovskite top cells combined with narrow band gap perovskite bottom cells have been widely developed due to their excellent performance and low cost. The highest PCE of all perovskite tandem solar cells has reached 29%,[6] exceeding the record PCE of single-junction perovskite solar cell and Si solar cells. In 2015, Sheng et al.[7] presented the first all-perovskite tandem solar cell based on $MAPbBr_3/MAPbI_3$ 4T configuration and obtained a tandem PCE of 13.4%. Even though the tandem PCE is lower, this work provides a proof-of-concept for all perovskite tandem solar cells. Zhou et al.[75] demonstrated the first solution-processed, 2T, all-perovskite tandem solar cell using a novel charge recombination layer and obtained a PCE of 7%. The progress of all-perovskite tandem solar cells has been developed quickly in recent years. Kai Zhu et al.[76] reported a record PCE of 25.4% for 4T all-perovskite tandem solar cells, attributed to the high carrier lifetimes of 1 μs. Hairen Tan et al.[6] demonstrated a record PCE of 29.0% for 2T all perovskite tandem solar cells. All-perovskite tandem solar cells are composed of wide band gap perovskite solar cells and narrow band gap perovskite solar cells. For wide band gap perovskite solar cells, the band gap of perovskite increases with the bromine ratio optimization. However, perovskites with increased bromide ration suffer from the light-induced phase segregation and film defects, leading to a decrease in band gap and absorption. For narrow band gap perovskite solar cells, it is essential to balance light distribution and the utilization of low energy photons. The substitution of Pb with Sn can effectively reduce band gap, comparable to the Si solar cell. However, the stability is an essential challenge for narrow band gap perovskite solar cells. Yang et al.[77] employed Pb-Sn perovskites with a band gap of 1.33 eV and combined with $MAPbI_3$ as the top cell to fabricate 4T all-perovskite tandem solar cells

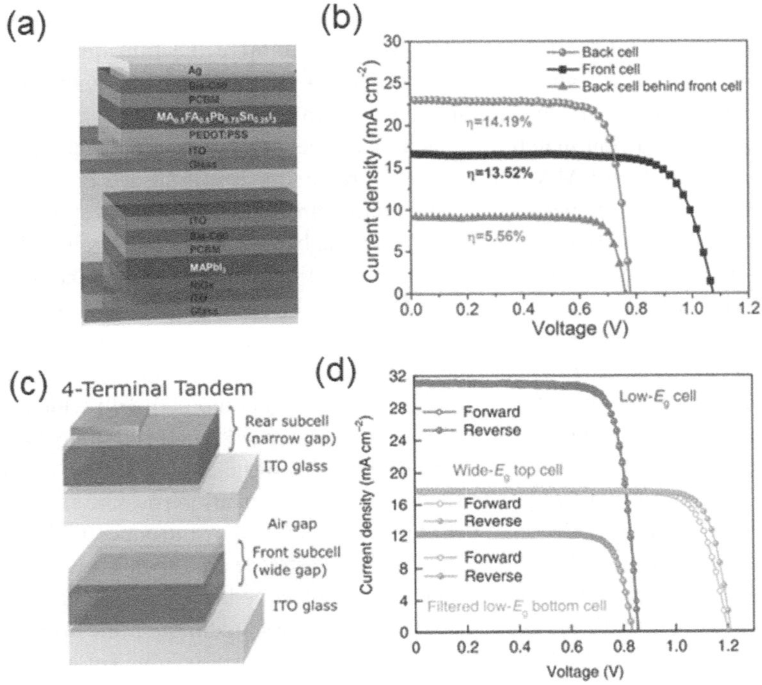

FIGURE 6.18 (a-c) Schematic view of 4T all-perovskite tandem solar cell.[78] Copyright 2016, American Association for the Advancement of Science. (b-d) I-V curves of 4T all-perovskite tandem solar cell.[79] Copyright 2022, Springer Nature.

(Figure 6.18a), resulting in a tandem PCE of 19.08%, shown in Figure 6.18b. This work provides a reference for narrow band gap perovskites with Sn as infrared light absorber. The composition and additive engineering play an important role to achieve high quality of perovskite film for all perovskite tandem solar cells. Snaith et al.[78] made mechanically stacked four-terminal tandem solar cells with a champion PCE of 20.3%, which is composed of the 1.6 eV wide band gap perovskite and 1.2 eV narrow band gap perovskite. Wang et al.[79] fabricated the 4T all-perovskite tandem solar cell based on Pb-Sn perovskites with band gap of 1.25 eV and wide band gap perovskites with band gap of 1.75 eV in Figure 6.18c. They presented a universal close-space annealing strategy to improve the film crystallinity and reduce the film trap density. A higher PCE of 25.15% was achieved with the enlarged grains and longed carrier lifetimes, shown in Figure 6.18d.

Compared with 4T all-perovskite tandem solar cells, 2T all-perovskite tandem solar cells are connected in series with interconnected layers and show lower parasitic light absorption loss and reflection loss. The theoretical limited PCE of all perovskite tandem solar cells is predicted to yield 40%.[80] Eperon et al.[78] fabricated the 2T all-perovskite tandem solar cell via employing the wide band gap of 1.8 eV and narrow band gap of 1.2 eV. They used ITO as the recombination layer and achieved a tandem PCE of 16.9%; the lower PCE is mainly attributed to the unsatisfactory film quality and the current mismatch due to the thickness of narrow band gap perovskites. Tong et al.[76] introduced GuaSCN to optimize the film quality with the thickness up to 1 μm and achieve level carrier lifetime and millimeter diffusion lengths. This work provides a template to obtain thick narrow band gap perovskite film and better current match for tandem solar cell. Berry et al.[81] developed a novel strategy to introduce DMA in perovskites precursors, resulting in a tandem PCE of 23% with the improved perovskites film quality and high energy utilization, shown in Figure 6.19a and b. Zhou et al.[82] optimized wide band gap perovskites with MASCN additive to suppress interface charge recombination and obtained a tandem PCE of 23.65% (Figure 6.19c and d). Tan et al.[83] reported a strategy to reduce Sn vacancies in Pb-Sn perovskites via using metallic tin in perovskite precursors, avoiding the oxidation of Sn^{2+}. The narrow band gap perovskites with the metallic tin optimization achieved longer carrier diffusion length of 3 μm and combined with wide band gap perovskites to form monolithic all-perovskite tandem solar cell (Figure 6.19e) with a certified PCE of 24.8%. After that, they reported a steric engineering to obtain high quality and photostable wide band gap perovskites for all perovskite tandem cells, resulting in a tandem PCE of 26.0% (Figure 6.19f).[84] They introduced DMA and Cl in wide band gap perovskites to reduce the lattice strain and trap density. Zhao et al.[85] developed a self-assembled monolayer as hole selective layer for wide band gap perovskite solar cells with a high V_{oc} of 1.31 eV and fabricated monolithic all-perovskite tandem solar cells with a higher PCE of 27.0%. Furthermore, Zhao et al.[86] designed a donor-acceptor type molecule as HTL for wide band gap perovskite and narrow

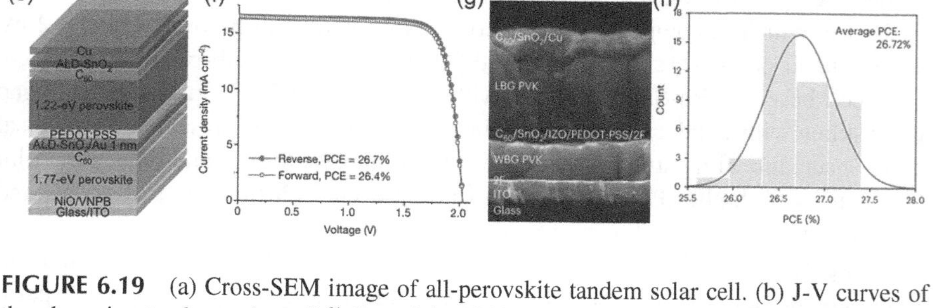

FIGURE 6.19 (a) Cross-SEM image of all-perovskite tandem solar cell. (b) J-V curves of the champion tandem solar cell.[81] Copyright 2019, Elsevier Inc. (c) Schematic view of all-perovskite tandem solar cell. (d) Stabilized PCE of the champion tandem solar cell.[82] Copyright 2023, American Chemical Society. (e) Schematic view of all-perovskite tandem solar cell. (f) I-V curves of the best-performing tandem solar cell with reverse and forward scanning.[84] Copyright 2022, Springer Nature. (g) Cross-SEM image of all-perovskite tandem solar cell. (h) The average PCE of tandem solar cell.[86] Copyright 2023, Springer Nature.

band gap perovskite subcells (Figure 6.19g), leading to a tandem PCE of 27.22% and improved operational stability, shown in Figure 19h.

The large-area all-perovskite tandem solar module has been considered the most promising technology to achieve perovskites industrialization. Dai et al.[87] developed a hot gas-assisted strategy to prepare narrow band gap perovskite films with a high quality and increased carrier recombination lifetime, which was further applied to fabricate all perovskite tandem solar modules with a champion PCE of 21.6% for 14.3 cm^2, shown in Figure 6.20a–c. Xiao et al.[88] optimized Cs ratio of perovskites to improve the homogeneity of crystallization and fabricated all-perovskite tandem solar modules with a certified PCE of 21.7% for 20 cm^2, shown in Figure 6.20d–f.

6.4.2 Perovskite/CIGS Tandem Solar Cell

Compared with the Si solar cell, CIGS is an excellent candidate as bottom cell to a fabricated tandem solar cell due to its suitable optimal band gap of 1.08–1.15 eV. In addition, the band gap of CIGS could be tuned with the region of 1.00–1.70 eV by optimizing the component ratios, which provided a high capability to achieve better current match.[88] CIGS thin-film solar cells are widely investigated due to their direct band gap, high absorption coefficient, low temperature coefficient, high stability, and low cost. Another advantage of the CIGS solar cell is that the flexible CIGS cells have great potential applications in building integrated photovoltaics (BIPV), flexible wearable devices, and space applications.[44,55] With the rapid progress of perovskite

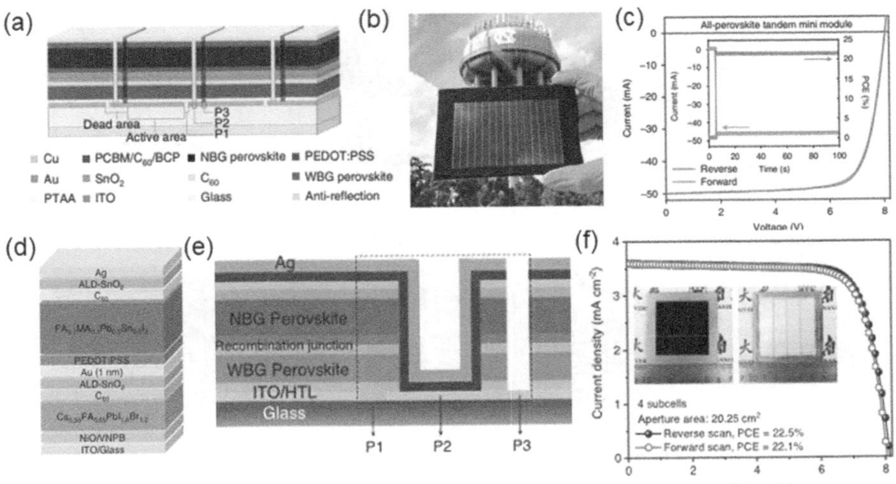

FIGURE 6.20 (a) Schematic structure of the all-perovskite tandem mini module. (b) Picture of an all-perovskite tandem mini module. (c) I–V curve of an all-perovskite tandem mini module.[87] Copyright 2022, Springer Nature. (d) Configuration schematic of an all-perovskite tandem device. (e) Schematic diagram of the structure of the series-connected all-perovskite tandem module. (f) I-V curves of the champion all-perovskite tandem module.[88] Copyright 2022, American Association for the Advancement of Science.

solar cells, the current record PCE of a perovskite/CIGS tandem solar cell is 25.9% and 23.2% for 4T configuration and 2T configuration, respectively.[11] Pisoni et al.[25] reported that the flexible perovskite solar cell grown on a typical flexible CIGS solar cell resulted in a proof-of-concept 18.2% efficiency, shown in Figure 6.21a and b. Fan Fu et al.[89] reported a semi-transparent planar perovskite solar cell (Figure 6.21c) with a high V_{oc} of 1.116 V and directly combined with CIGS as bottom cell to achieve an efficiency of 22.1%, shown in Figure 6.21d. They innovatively used ZnO NPs and doped poly [bis (4-phenyl) (2,4,6-triMethylphenyl) aMine] (PTAA) to increase interface carrier transport and improve the PCE of perovskite solar cells. This work provided a new idea to achieve high efficiency based on perovskite solar cells. Imec scientists[11] achieved a PCE of 24.6% for perovskite/CIGS tandem solar cell. It is found that the most suitable band gap is optimized to 1.72 eV, allowing the perovskite solar cell to absorb visible light. Bailie et al.[55] presented the semi-transparent $MAPbIBr_2$ as a top cell and combined with CIGS bottom cell, resulting a tandem PCE of 18.6%. Shen et al.[17] fabricated semi-transparent perovskite solar cells on CIGS cell to achieve a tandem PCE of 23.9% for 4T configuration (Figure 6.21e and f). They found that the perovskite solar cell has a strong absorption at the short wavelength region for the tandem solar cell, avoiding the light loss from CdS in tandem solar cell. Zhu et al.[90] presented a bimolecular additive engineering approach to obtain high quality perovskite films, achieving a tandem PCE of 25.9% based on a 2D/3D perovskite solar cell combined with the CIGS solar cell, shown in Figure 6.21g and h.

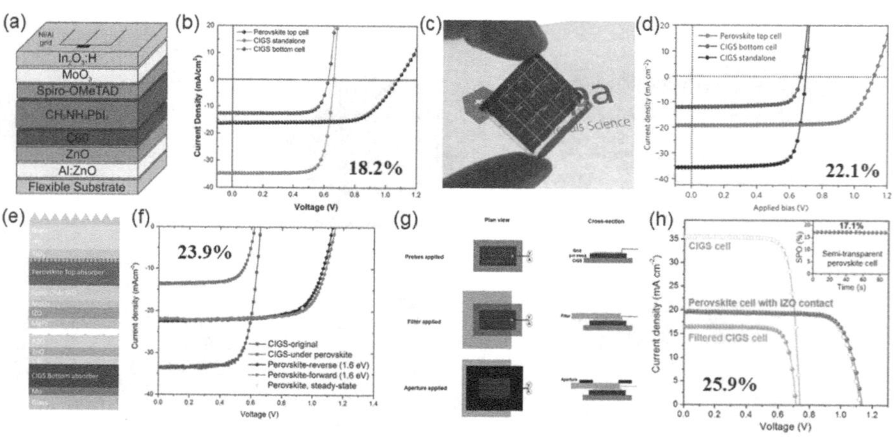

FIGURE 6.21 (a) Schematic representation of the superstrate configuration flexible NIR-transparent perovskite solar cell. (b) J-V curves of the flexible CIGS stand- alone cell.[25] Copyright 2017, The Royal Society of Chemistry. (c) Photograph of the solar cell grown on glass. (d) J-V curves of the perovskite/CIGS in four-terminal tandem configuration.[89] Copyright 2019, Springer Nature. (e) Schematic of the 4-T perovskite/CIGS tandem solar cell. (f) J-V curves of the perovskite/CIGS tandem solar cell.[17] Copyright 2018, The Royal Society of Chemistry. (g) Schematic illustration of the arrangement of tandem solar cell. (h) Curves of the PSC with a transparent (IZO) top contact and the perovskite/CIGS tandem solar cell.[90] Copyright 2019, Elsevier Inc.

In addition, 2T perovskite/CIGS tandem solar cells have been widely investigated via modification of subcell and design of configuration. CIGS solar cell with high efficiency are usually fabricated via the sputtered method and co-evaporation, leading to large roughness for CIGS films.[90]–[92] It is a challenge to fabricate the perovskite films on the top of GIGS cells with larger roughness. Jost et al.[93] introduced a 10 nm NiO_x layer as HTL on the rough CIGS cell to avoid the shunting problem, shown in Figure 6.22a. The device performance is further improved via introducing polymer PTAA at the NiO/perovskite interface, showing a stabilized PCE of 21.6% for tandem solar cells (Figure 6.22b). The tandem solar cell achieved a larger V_{oc} of 1.58 and an FF of 76.0%, indicating good interface contact and suppressed carrier non-radioactive recombination. This work provides an effective way to achieve high efficiency for perovskite/CIGS tandem solar cell. One of the effective strategies to achieve the high quality of perovskite films on CIGS cells is directly polishing surface or solution fabricated method to reduce the roughness of bottom cell. Todorov et al.[94] fabricated CIGS film from hydrazine solutions and obtained a record performance of 15.9% with the optimization of perovskite top cell. They used two-step-method to fabricate perovskite films with adjusting temperature, pressure,

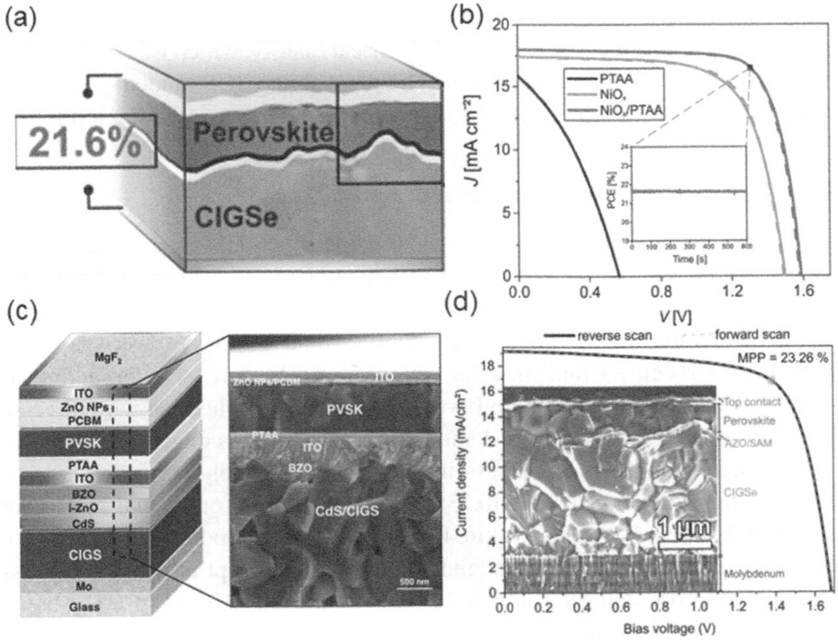

FIGURE 6.22 (a) Schematic view of 2T perovskite/CIGS tandem solar cell. (b) J-V characteristics under illumination of the monolithic perovskite/CIGSe tandem solar cell.[93] Copyright 2019 American Chemical Society. (c) Schematic and cross-sectional SEM image of the monolithic perovskite/CIGS tandem solar cell.[96] Copyright 2018, American Association for the Advancement of Science. (d) J-V curve of a monolithic CIGS/perovskite tandem solar cell.[95] Copyright 2019, The Royal Society of Chemistry.

and precursor optimization. The temperature of fabrication process is lower than the CIGS degradation temperature. Uhl et al.[95] synthesized CIGS film by using a novel molecular ink optimization to decrease the roughness of CIGS from 132 nm to 77 nm. They used *N,N'*-dimethylformamide (DMF) solvent to improve the ink stability and accelerate the deposition process, leading to a record PCE of 23.26% for perovskite/CIGS tandem solar cell (Figure 6.22d). Another smart strategy is the increased perovskite thickness to avoid the formation of shunt paths. The tandem PCE is mainly limited by lower V_{oc} and FF due to its poor morphology and short carrier diffusion length of perovskite film on CIGS cell. Compared with the solution fabricated method for CIGS, the vacuum methods can form high crystalline quality and grade composition. Han et al.[96] fabricated a CIGS cell with a sputtered method and introduced a 300 nm ITO layer to reduce roughness of subcell surface. The chemical mechanical polishing process with a commercialized SiO_2 slurry is widely used to obtain a smooth surface for the CIGS cell, shown in Figure 6.22c. The chemical mechanical polishing process could well retain the CIGS cell structure without any damage. The perovskite film is directly deposited on the CIGS bottom cell with the smooth surface, resulting in the enhanced the FF and V_{oc}. This work provides a new strategy to achieve a balance between band gap and thickness for perovskite/CIGS tandem solar cell. Compared with the perovskite/Si tandem solar cell, the progress of perovskite/CIGS tandem solar cells is limited by the rough surface of the bottom cell and complex compositions. The Ag alloying and alkali metal ions co-doping are applied to improve film crystallinity and electrical quality, which is beneficial for the interface transport and band alignment. The perovskite/CIGS tandem solar cell has great potential due to its ionization hardness, low temperature, and flexibility.

6.4.3 Perovskite/OPV Tandem Solar Cell

Similar with narrow band gap perovskites, organic solar cells are considered a good candidate for tandem solar cells due to its narrow band gap, high absorption coefficient, low temperature process, and flexible solution process.[97] The hydrophobic property of organic film showed better stability than Pb-Sn perovskites, which is suitable to as top cell for tandem solar cell. Meanwhile, perovskite/organic tandem solar cells with solution fabrication have great potential application with large-scale roll-to-roll process. Yang et al.[98] demonstrated a proof-of-concept for perovskite/organic tandem solar cells by integrating perovskite and polymer:fullerene as subcell, obtaining a tandem PCE of 10.23%, shown in Figure 6.23a and b. The lower PCE is mainly attributed to perovskite film quality and components. Liu et al.[99] presented a novel perovskite/polymer tandem solar cell and optimized the perovskite thickness of the top cell and obtained a higher tandem PCE of 16.0%, shown in Figure 6.23c and d. Li et al.[100] first employed the all-inorganic perovskite $CsPbBr_3$ as bottom cell and fabricated perovskite/organic tandem solar cell (Figure 6.23e) with a PCE of 14.03%. Zeng et al.[101] used all-inorganic perovskite $CsPbI_2Br$ as a bottom cell and combined it with PTB7-Th:COi8DFIC:$PC_{71}BM$ active layer to obtain a tandem PCE of 15.04%. Tan et al.[102] developed all-inorganic perovskite/organic tandem solar cell with $CsPbI_2Br$ and PTB7-Th:IEICO-4F active layers, resulting in a tandem PCE of 17.24% with the optimization of organic active layer thickness. Jang et al. [103]

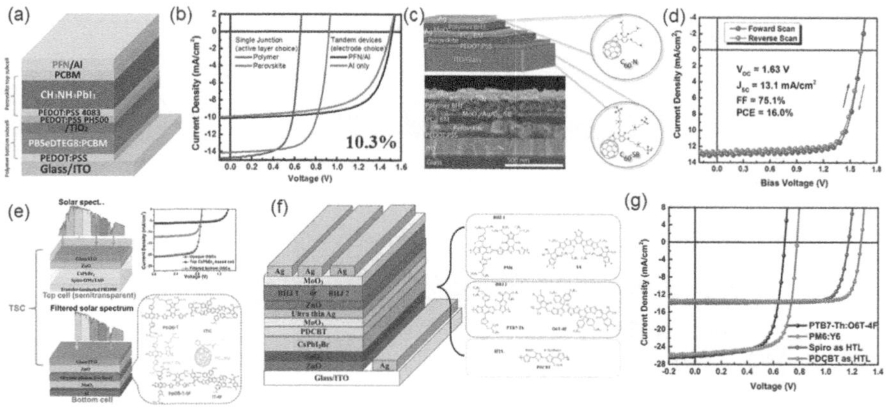

FIGURE 6.23 (a) Device architecture of the hybrid tandem solar cell composed of perovskite and polymer absorbers. (b) J-V curves of the hybrid tandem device.[98] Copyright 2015, The Royal Society of Chemistry. (c) Device structure of tandem perovskite/polymer solar cells and the fullerene interlayer materials. (d) J-V curves and device metrics of the optimal polymer/perovskite hybrid tandem solar cell.[99] Copyright 2016 American Chemical Society. (e) Device structure and performance of four-terminal perovskite/organic tandem solar cell.[100] Copyright 2018, WILEY-VCH Verlag GmbH and Co. KGaA. (f) Monolithic perovskite/organic tandem cell structure. (g) J-V curves of the perovskite/organic tandem solar cell.[104] Copyright 2020, Elsevier B.V.

fabricated a 2T hybrid tandem solar cell based on inorganic perovskites and organic bulk heterojunction to achieve the current match. The inorganic perovskite $CsPbI_2Br$ as top cell showed a wide band gap of 1.85 eV, which was integrated with organic BHJ as the bottom cell (1.25 eV) to achieve a stabilized tandem PCE of 18.04% with a V_{oc} of 1.73 eV. Xie et al.[104] developed perovskite/organic tandem solar cells (Figure 6.23f) with a PM6:Y6-based or PTB7-Th:O6T-4F-based bulk-heterojunction organic top cell, resulting in tandem PCEs of 18.38% and 18.06% (Figure 6.23g), respectively. Chen et al.[105] demonstrated a successful monolithic perovskite/organic tandem solar cell with a landmark PCE of 20.6% via the gas pumping method. Qin et al.[106] further fabricated a perovskite/organic tandem solar cell with the passivation method to obtain a high tandem efficiency of 22.0%. Brinkmann et al.[106] demonstrated perovskite/organic tandem solar cells with an efficiency of 24.0% and a high V_{oc} of 2.15 V, which is the best PCE among reported works. Even though the PCE of the perovskite/organic tandem solar cell is lower than another tandem solar cell, it has great potential application in the photovoltaic energy field.

6.4.4 Perovskite/CdTe Tandem Solar Cell

CdTe solar cells have been considered as the commercially thin-film solar cells due to their low cost, low energy manufacturing, and short payback time.[107] The CdTe solar cell exhibited a record PCE of 22.1% and has been considered as the most competitive candidate compared with Si or CIGS cells.[6] The CdTe solar cell with

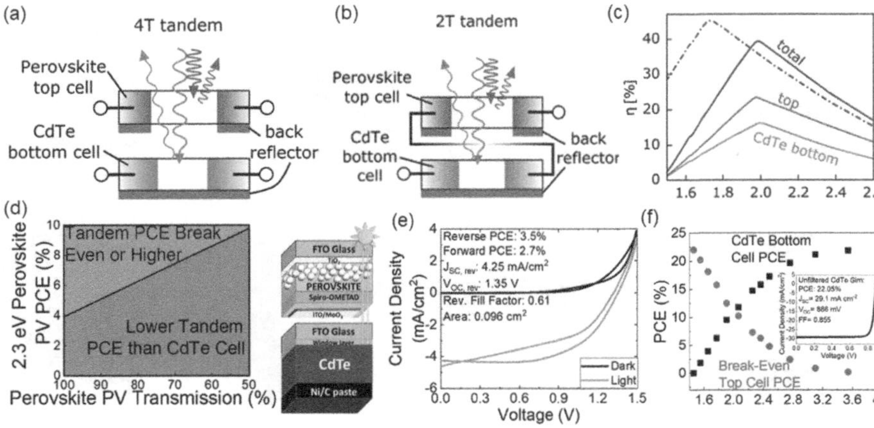

FIGURE 6.24 (a–b) Schematic view of the perovskite/CdTe tandem solar cell. (c) Simulated efficiency of tandem solar cell.[109] Copyright 2017, WILEY-VCH Verlag GmbH & Co. KGaA. (d) Architecture of the perovskite/CdTe tandem solar cell. (e) I-V curves of a semi-transparent perovskite solar cell. (f) Simulated efficiency of perovskite/CdTe tandem solar cell.[110] Copyright 2019, Elsevier B.V.

selenium doping could effectively reduce band gap (1.36 eV) and achieve high PCE, which is suitable to fabricate perovskite/CdTe tandem solar cells. Zhou et al.[108] used the theoretical methods to design a novel perovskite/CdTe tandem solar cell and obtained a tandem PCE of 20.6%. The perovskite $MAPb(I_{1-x}Br_x)_3$ is applied as the bottom cell to achieve the current match. Ruhle et al.[109] investigated the relationship between band gap of subcell materials and device performance via the calculated methods, shown in Figure 6.24a and b. When the top and bottom band gap is 2.11 eV and 1.47 eV, respectively, the highest PCE of perovskite/CdTe tandem solar cell could reach 39.5% (Figure 6.24c). They found that the PCE of perovskite/CdTe tandem solar cell is above 37% based on the top and bottom band gap of 1.7–2.6 eV and 1.47 eV, respectively. Siegler et al.[110] used the calculated simulator method to design a perovskite/CdTe tandem solar cells and achieved a tandem PCE of 40%. They fabricated perovskite/CdTe tandem solar cell based on $MAPbBr_3$ perovskites and obtained a tandem PCE of 3.5%, which is a proof-of-concept for tandem solar cells, shown in Figure 6.24d-f. The PCE of the perovskite/CdTe tandem solar cell could be further improved via the optimization of perovskites and structure design.

6.5 CHALLENGES AND OUTLOOK

The progress of perovskite materials has been developed rapidly in the past decade, which is widely used in solar cells, LEDs, detectors, and transistors. Perovskite tandem solar cells have been considered as the most promising way to achieve photovoltaic industrialization. Compared with other typical photovoltaic materials, perovskites have a long way to go to achieve practical application due to their stability and high-quality film fabricated process. Among the perovskite tandem solar cells, the perovskite/Si tandem solar cell and all-perovskite tandem solar cells play an important role in photovoltaic field, which has been considered as the new star of photovoltaic industrialization. All-perovskite tandem solar cells exhibited an all-solution fabricated process and flexible application in BIPV. However, the stability of narrow band gap perovskites and phase separation of wide band gap perovskites will be the greatest challenge for the development of all-perovskite tandem solar cells. Meanwhile, the Si cells possess a large market share (more than 90% PV market) and mature process line. It presents more challenges for current photovoltaic (PV) technologies to replace the Si solar cells in PV market, while it is an effective strategy to achieve high efficiency based on the perovskite/Si tandem solar cells. Despite these results and advanced developments, the perovskite tandem solar cells still need to be further improved. The development of the perovskite/Si tandem solar cell needs to achieve high quality perovskite film on textured Si bottom cells and a large scale up for matching current Si cells. The perovskite film fabricated process needs to be developed with novel technologies to achieve high-quality perovskite film with micrometer level thickness, which could effectively avoid the shunting problems. In addition, the large area perovskite film will be the key factor for the development of perovskite/Si tandem solar cells. The current typical fabricated processes such as blade-coating and slot die coating have been widely used in the field of up-scalable perovskite films. Despite the aforementioned challenges, more promising strategies are expected to overcome these challenges and fulfill the broad implementation of the perovskite/Si tandem solar cells. The large market and potential of perovskite/silicon tandem solar cells will further accelerate their developments.

REFERENCES

1. C. J. Brabec, Organic photovoltaics: Technology and market. *Sol. Energy Mater. Sol. Cells* **2004**, *83*(2–3), 273–292. https://doi.org/10.1016/j.solmat.2004.02.030
2. R. F. Service, Is It Time to Shoot for the Sun? *Science* **2005**, *309*(5734), 548–551. https://doi.org/10.1126/science.309.5734.548
3. C. Yu, Q. Zou, Q. Wang, Y. Zhao, X. Ran et al., Silicon Solar Cell with Undoped tin Oxide Transparent Electrode. *Nat. Energy* **2023**, *8*(10), 1119–1125. https://doi.org/10.1038/s41560-023-01331-7
4. H. J. Snaith, Perovskites: The Emergence of a New Era for Low-Cost, High-Efficiency Solar Cells. *J. Phys. Chem. Lett.* **2013**, *4*(21), 3623–3630. https://doi.org/10.1021/jz4020162
5. Z. J. Yu, J. V. Carpenter, Z. C. Holman, Techno-Economic Viability of Silicon-Based Tandem Photovoltaic Modules in the United States. *Nat. Energy* **2018**, *3*(9), 747–753. https://doi.org/10.1038/s41560-018-0201-5
6. M. A. Green, E. D. Dunlop, M. Yoshita, N. Kopidakis, K. Bothe et al., Solar Cell Efficiency Tables (Version 62). *Prog. Photovoltaics Res. Appl.* 651–663 (May 2023). https://doi.org/10.1002/pip.3726
7. H. Shen, D. Walter, Y. Wu, K. C. Fong, D. A. Jacobs et al., Monolithic Perovskite/Si Tandem Solar Cells: Pathways to Over 30% Efficiency. *Adv. Energy Mater.* **2020**, *10*(13), 1902840. https://doi.org/10.1002/aenm.201902840
8. M. H. Futscher, B. Ehrler, Efficiency Limit of Perovskite/Si Tandem Solar Cells. *ACS Energy Lett.* **2016**, *1*(4), 863–868. https://doi.org/10.1021/acsenergylett.6b00405

9. Y. Cheng, L. Ding, Perovskite/Si Tandem Solar Cells: Fundamentals, Advances, Challenges, and Novel Applications. *SusMat.* **2021**, *1*(3), 324–344. https://doi.org/10.1002/sus2.25
10. J. H. Park, S. G. Ji, Y. S. Yoon, M. Park, S. H. Lee et al., Monolithic Perovskite/Si Tandem Solar Cells—Silicon Bottom Cell Types and Characterization Methods. *Adv. Mater. Technol.* **2023**, *2201006*, 1–14. https://doi.org/10.1002/admt.202201006
11. H. Li, W. Zhang, Perovskite Tandem Solar Cells: From Fundamentals to Commercial Deployment. *Chem. Rev.* **2020**, *120*(18), 9835–9950. https://doi.org/10.1021/acs.chemrev.9b00780
12. E. Raza, Z. Ahmad, Review on Two-Terminal and Four-Terminal Crystalline-Silicon/Perovskite Tandem Solar Cells; Progress, Challenges, and Future Perspectives. *Energy Rep.* **2022**, *8*, 5820–5851. https://doi.org/10.1016/j.egyr.2022.04.028
13. F. Jiang, T. Liu, B. Luo, J. Tong, F. Qin et al., A Two-Terminal Perovskite/Perovskite Tandem Solar Cell. *J. Mater. Chem. A* **2016**, *4*(4), 1208–1213. https://doi.org/10.1039/C5TA08744A
14. G. Yang, Z. Ni, Z. J. Yu, B. W. Larson, Z. Yu et al., Defect Engineering in Wide-Bandgap Perovskites for Efficient Perovskite–Silicon Tandem Solar Cells. *Nat. Photonics* **2022**, *16*(8), 588–594. https://doi.org/10.1038/s41566-022-01033-8
15. M. Ledinsky, S. Nicolay, J. Bailat, J. Yum, S. De Wolf, Silicon Four-Terminal Tandem Solar Cells †. *Phys. Chem. Chem. Phys.* **2015**, 1619–1629. https://doi.org/10.1039/c4cp03788j
16. D. Zhang, M. Najafi, V. Zardetto, M. Dörenkämper, X. Zhou et al., High Efficiency 4-Terminal Perovskite/C-Si Tandem Cells. *Sol. Energy Mater. Sol. Cells 188*, 1–5 (July 2018). https://doi.org/10.1016/j.solmat.2018.07.032
17. H. Shen, T. Duong, J. Peng, D. Jacobs, N. Wu et al., Mechanically-Stacked Perovskite/Cigs Tandem Solar Cells with Efficiency of 23.9% and Reduced Oxygen Sensitivity. *Energy Environ. Sci.* **2018**, *11*(2), 394–406. https://doi.org/10.1039/c7ee02627g
18. P. Löper, S.-J. Moon, S. Martín de Nicolas, B. Niesen, M. Ledinsky et al., Organic–Inorganic Halide Perovskite/Crystalline Silicon Four-Terminal Tandem Solar Cells. *Phys. Chem. Chem. Phys.* **2015**, *17*(3), 1619–1629. https://doi.org/10.1039/C4CP03788J
19. D. Zhao, C. Wang, Z. Song, Y. Yu, C. Chen et al., Four-Terminal All-Perovskite Tandem Solar Cells Achieving Power Conversion Efficiencies Exceeding 23%. *ACS Energy Lett.* **2018**, *3*(2), 305–306. https://doi.org/10.1021/acsenergylett.7b01287
20. Q. Han, Y.-T. Hsieh, L. Meng, J.-L. Wu, P. Sun et al., High-Performance Perovskite/Cu(In,Ga)Se 2 Monolithic Tandem Solar Cells. *Science.* **2018**, *361*(6405), 904–908. https://doi.org/10.1126/science.aat5055
21. D. Zhang, M. Najafi, V. Zardetto, M. Dörenkämper, X. Zhou et al., High Efficiency 4-Terminal Perovskite/C-Si Tandem Cells. *Sol. Energy Mater. Sol. Cells.* **2018**, *188*, 1–5. https://doi.org/10.1016/j.solmat.2018.07.032
22. E. Aydin, M. De Bastiani, X. Yang, M. Sajjad, F. Aljamaan et al., Zr-Doped Indium Oxide (IZRO) Transparent Electrodes for Perovskite-Based Tandem Solar Cells. *Adv. Funct. Mater.* **2019**, *29*(25), 1901741. https://doi.org/10.1002/adfm.201901741
23. J. Werner, C.-H. Weng, A. Walter, L. Fesquet, J. P. Seif et al., Efficient Monolithic Perovskite/Silicon Tandem Solar Cell with Cell Area >1 cm^2. *J. Phys. Chem. Lett.* **2016**, *7*(1), 161–166. https://doi.org/10.1021/acs.jpclett.5b02686
24. J. Werner, G. Dubuis, A. Walter, P. Löper, S.-J. Moon et al., Sputtered Rear Electrode with Broadband Transparency for Perovskite Solar Cells. *Sol. Energy Mater. Sol* **2015**, . *Cells 141*, 407–413. https://doi.org/10.1016/j.solmat.2015.06.024
25. S. Pisoni, F. Fu, T. Feurer, M. Makha, B. Bissig et al., Flexible Nir-Transparent Perovskite Solar Cells for All-Thin-Film Tandem Photovoltaic Devices. *J. Mater. Chem. A* **2017**, *5*(26), 13639–13647. https://doi.org/10.1039/C7TA04225F

26. H. Kim, C. M. Gilmore, J. S. Horwitz, A. Piqué, H. Murata et al., Transparent Conducting Aluminum-Doped Zinc Oxide Thin Films for Organic Light-Emitting Devices. *Appl. Phys. Lett.* **2000**, *76*(3), 259–261. https://doi.org/10.1063/1.125740
27. Y. H. Jang, J. M. Lee, J. W. Seo, I. Kim, D. Lee, Monolithic Tandem Solar Cells Comprising Electrodeposited Cuinse 2 and Perovskite Solar Cells with a Nanoparticulate Zno Buffer Layer. *J. Mater. Chem. A* **2017**, *5*(36), 19439–19446. https://doi.org/10.1039/C7TA06163C
28. K. O. Brinkmann, J. Zhao, N. Pourdavoud, T. Becker, T. Hu et al., Suppressed Decomposition of Organometal Halide Perovskites by Impermeable Electron-Extraction Layers in Inverted Solar Cells. *Nat. Commun.* **2017**, *8*(1), 13938. https://doi.org/10.1038/ncomms13938
29. S. Lin, X. Bai, H. Wang, H. Wang, J. Song et al., Roll-to-Roll Production of Transparent Silver-Nanofiber-Network Electrodes for Flexible Electrochromic Smart Windows. *Adv. Mater.* **2017**, *29*(41), 1703238. https://doi.org/10.1002/adma.201703238
30. S. Lin, H. Wang, X. Zhang, D. Wang, D. Zu et al., Direct Spray-Coating of Highly Robust and Transparent Ag Nanowires for Energy Saving Windows. *Nano Energy* **2019**, *62*, 111–116. https://doi.org/10.1016/j.nanoen.2019.04.071
31. S. Lin, H. Wang, F. Wu, Q. Wang, X. Bai et al., Room-Temperature Production of Silver-Nanofiber Film for Large-Area, Transparent and Flexible Surface Electromagnetic Interference Shielding. *NPJ Flex. Electron.* **2019**, *3*(1), 6. https://doi.org/10.1038/s41528-019-0050-8
32. C. D. Bailie, M. G. Christoforo, J. P. Mailoa, A. R. Bowring, E. L. Unger et al., Semi-Transparent Perovskite Solar Cells for Tandems with Silicon and Cigs. *Energy Environ. Sci.* **2015**, *8*(3), 956–963. https://doi.org/10.1039/C4EE03322A
33. J. P. Mailoa, C. D. Bailie, E. C. Johlin, E. T. Hoke, A. J. Akey et al., A 2-Terminal Perovskite/Silicon Multijunction Solar Cell Enabled by a Silicon Tunnel Junction. *Appl. Phys.Lett.* **2015**, *106*(12),. https://doi.org/10.1063/1.4914179
34. Z. Hawash, L. K. Ono, S. R. Raga, M. V Lee, Y. Qi, Air-Exposure Induced Dopant Redistribution and Energy Level Shifts in Spin-Coated Spiro-MeOTAD Films. *Chem. Mater.* **2015**, *27*(2), 562–569. https://doi.org/10.1021/cm504022q
35. F. Guo, H. Azimi, Y. Hou, T. Przybilla, M. Hu et al., High-Performance Semitransparent Perovskite Solar Cells with Solution-Processed Silver Nanowires as Top Electrodes. *Nanoscale* **2015**, *7*(5), 1642–1649. https://doi.org/10.1039/C4NR06033D
36. B. Chen, Y. Bai, Z. Yu, T. Li, X. Zheng et al., Efficient Semitransparent Perovskite Solar Cells for 23.0%-Efficiency Perovskite/Silicon Four-Terminal Tandem Cells. *Adv. Energy Mater.* **2016**, *6*(19), 1601128. https://doi.org/10.1002/aenm.201601128
37. C. Nagl, E. Platzgummer, M. Schmid, P. Varga, S. Speller et al., Direct Observation of a New Growth Mode: Subsurface Island Growth of Cu on Pb(111). *Phys. Rev. Lett.* **1995**, *75*(16), 2976–2979. https://doi.org/10.1103/PhysRevLett.75.2976
38. H. Park, I. J. Park, M. G. Lee, K. C. Kwon, S. P. Hong et al., Water Splitting Exceeding 17% Solar-to-Hydrogen Conversion Efficiency Using Solution-Processed Ni-Based Electrocatalysts and Perovskite/Si Tandem Solar Cell. *ACS Appl. Mater. Interfaces* **2019**, *11*(37), 33835–33843. https://doi.org/10.1021/acsami.9b09344
39. Y. (Michael) Yang, Q. Chen, Y.-T. Hsieh, T.-B. Song, N. De Marco et al., Multilayer Transparent Top Electrode for Solution Processed Perovskite/Cu(In,Ga)(Se,S) 2 Four Terminal Tandem Solar Cells. *ACS Nano* **2015**, *9*(7), 7714–7721. https://doi.org/10.1021/acsnano.5b03189
40. Z. Ren, J. Zhou, Y. Zhang, A. Ng, Q. Shen et al., Strategies for High Performance Perovskite/Crystalline Silicon Four-Terminal Tandem Solar Cells. *Sol. Energy Mater. Sol. Cells* **2018**, *179*, 36–44. https://doi.org/10.1016/j.solmat.2018.01.004
41. D. Zhao, Y. Yu, C. Wang, W. Liao, N. Shrestha et al., Low-Bandgap Mixed Tin–Lead Iodide Perovskite Absorbers with Long Carrier Lifetimes for All-Perovskite Tandem Solar Cells. *Nat. Energy* **2017**, *2*(4), 17018. https://doi.org/10.1038/nenergy.2017.18

42. L. Lin, B. Deng, J. Sun, H. Peng, Z. Liu, Bridging the Gap between Reality and Ideal in Chemical Vapor Deposition Growth of Graphene. *Chem. Rev.* **2018**, *118*(18), 9281–9343. https://doi.org/10.1021/acs.chemrev.8b00325
43. F. Lang, M. A. Gluba, S. Albrecht, J. Rappich, L. Korte et al., Perovskite Solar Cells with Large-Area CVD-Graphene for Tandem Solar Cells. *J. Phys. Chem. Lett.* **2015**, *6*(14), 2745–2750. https://doi.org/10.1021/acs.jpclett.5b01177
44. F. Fu, T. Feurer, T. Jäger, E. Avancini, B. Bissig et al., Low-Temperature-Processed Efficient Semi-Transparent Planar Perovskite Solar Cells for Bifacial and Tandem Applications. *Nat. Commun.* **2015**, *6*, 8932. https://doi.org/10.1038/ncomms9932
45. E. T. Hoke, D. J. Slotcavage, E. R. Dohner, A. R. Bowring, H. I. Karunadasa et al., Reversible Photo-Induced Trap Formation in Mixed-Halide Hybrid Perovskites for Photovoltaics. *Chem. Sci.* **2015**, *6*(1), 613–617. https://doi.org/10.1039/C4SC03141E
46. A. Rajagopal, Z. Yang, S. B. Jo, I. L. Braly, P. Liang et al., Highly Efficient Perovskite–Perovskite Tandem Solar Cells Reaching 80% of the Theoretical Limit in Photovoltage. *Adv. Mater.* **2017**, *29*(34), 1–10. https://doi.org/10.1002/adma.201702140
47. D. P. McMeekin, G. Sadoughi, W. Rehman, G. E. Eperon, M. Saliba et al., A Mixed-Cation Lead Mixed-Halide Perovskite Absorber For Tandem Solar Cells. *Science* **2016**, *351*(6269), 151–155. https://doi.org/10.1126/science.aad5845
48. H. Tan, F. Che, M. Wei, Y. Zhao, M. I. Saidaminov et al., Dipolar Cations Confer Defect Tolerance in Wide-Bandgap Metal Halide Perovskites. *Nat. Commun.* **2018**, *9*(1), 3100. https://doi.org/10.1038/s41467-018-05531-8
49. Z. Qiu, Z. Xu, N. Li, N. Zhou, Y. Chen et al., Monolithic Perovskite/Si Tandem Solar Cells Exceeding 22% Efficiency Via Optimizing Top Cell Absorber. *Nano Energy*. **2018**, *53*, 798–807. https://doi.org/10.1016/j.nanoen.2018.09.052
50. T. Duong, Y. Wu, H. Shen, J. Peng, X. Fu et al., Rubidium Multication Perovskite with Optimized Bandgap for Perovskite-Silicon Tandem with over 26% Efficiency. *Adv. Energy Mater.* **2017**, *7*(14), 1700228. https://doi.org/10.1002/aenm.201700228
51. J. Xu, C. C. Boyd, Z. J. Yu, A. F. Palmstrom, D. J. Witter et al., Triple-Halide Wide–Band Gap Perovskites with Suppressed Phase Segregation for Efficient Tandems. *Science* **2020**, *367*(6482), 1097–1104. https://doi.org/10.1126/science.aaz5074
52. Y. Yu, R. Liu, C. Liu, X. Shi, H. Yu et al., Synergetic Regulation of Oriented Crystallization and Interfacial Passivation Enables 19.1% Efficient Wide-Bandgap Perovskite Solar Cells. *Adv. Energy Mater.* **2022**, *12*(33), 2201509. https://doi.org/10.1002/aenm.202201509
53. D. Kim, H. J. Jung, I. J. Park, B. W. Larson, S. P. Dunfield et al., Efficient, Stable Silicon Tandem Cells Enabled by Anion-Engineered Wide-Bandgap Perovskites. *Science* **2020**, *368*(6487), 155–160. https://doi.org/10.1126/science.aba3433
54. W. Chai, L. Li, W. Zhu, D. Chen, L. Zhou et al., Graded Heterojunction Improves Wide-Bandgap Perovskite for Highly Efficient 4-Terminal Perovskite/Silicon Tandem Solar Cells. *Research* **2023**, *6*, 1–10. https://doi.org/10.34133/research.0196
55. C. D. Bailie, M. G. Christoforo, J. P. Mailoa, A. R. Bowring, E. L. Unger et al., Semi-Transparent Perovskite Solar Cells for Tandems with Silicon and Cigs. *Energy Environ. Sci.* **2015**, *8*(3), 956–963. https://doi.org/10.1039/c4ee03322a
56. F. J. Ramos, S. Jutteau, J. Posada, A. Bercegol, A. Rebai et al., Highly Efficient Moox-Free Semitransparent Perovskite Cell for 4T Tandem Application Improving the Efficiency of Commercially-Available Al-BSF Silicon. *Sci. Rep.* **2018**, *8*(1), 16139. https://doi.org/10.1038/s41598-018-34432-5
57. R. Sheng, A. W. Y. Ho-Baillie, S. Huang, M. Keevers, X. Hao et al., Four-Terminal Tandem Solar Cells Using $CH_3NH_3PbBr_3$ by Spectrum Splitting. *J. Phys. Chem. Lett.* **2015**, *6*(19), 3931–3934. https://doi.org/10.1021/acs.jpclett.5b01608
58. T. Leijtens, K. A. Bush, R. Prasanna, M. D. McGehee, Opportunities and Challenges for Tandem Solar Cells using Metal Halide Perovskite Semiconductors. *Nat. Energy* **2018**, *3*(10), 828–838. https://doi.org/10.1038/s41560-018-0190-4

59. A. Louwen, W. van Sark, R. Schropp, A. Faaij, A Cost Roadmap for Silicon Heterojunction Solar Cells. *Sol. Energy Mater. Sol Cells* **2016**, *147*, 295–314. https://doi.org/10.1016/j.solmat.2015.12.026
60. J. Haschke, O. Dupré, M. Boccard, C. Ballif, Silicon Heterojunction Solar Cells: Recent Technological Development and Practical Aspects—from Lab to Industry. *Sol. Energy Mater. Sol. Cells.* **2018**, *187*, 140–153. https://doi.org/10.1016/j.solmat.2018.07.018
61. F. Sahli, B. A. Kamino, J. Werner, M. Bräuninger, B. Paviet-Salomon et al., Improved Optics in Monolithic Perovskite/Silicon Tandem Solar Cells with a Nanocrystalline Silicon Recombination Junction. *Adv. Energy Mater.* **2018**, *8*(6), 1701609. https://doi.org/10.1002/aenm.201701609
62. B. W. Schneider, N. N. Lal, S. Baker-Finch, T. P. White, Pyramidal Surface Textures for Light Trapping and Antireflection in Perovskite-on-Silicon Tandem Solar Cells. *Opt Express* **2014**, *22*(S6), A1422. https://doi.org/10.1364/OE.22.0A1422
63. B. Chen, Z. Yu, K. Liu, X. Zheng, Y. Liu et al., Grain Engineering for Perovskite/Silicon Monolithic Tandem Solar Cells with Efficiency of 25.4%. *Joule* **2019**, *3*(1), 177–190. https://doi.org/10.1016/j.joule.2018.10.003
64. N. Kataoka, Y. Shima, K. Nakajima, K. Nakamura, A Central Master Driver of Psychosocial Stress Responses in the Rat. *Science* **2020**, *367*(6482), 1105–1112. https://doi.org/10.1126/science.aaz4639
65. F. Sahli, J. Werner, B. A. Kamino, M. Bräuninger, R. Monnard et al., Fully Textured Monolithic Perovskite/Silicon Tandem Solar Cells with 25.2% Power Conversion Efficiency. *Nat. Mater.* **2018**, *17*(9), 820–826. https://doi.org/10.1038/s41563-018-0115-4
66. L. Ba, H. Liu, W. Shen, Perovskite/c-Si tandem Solar Cells with Realistic Inverted Architecture: Achieving High Efficiency by Optical Optimization. *Prog. Photovoltaics Res. Appl.* **2018**, *26*(11), 924–933. https://doi.org/10.1002/pip.3037
67. B. Chen, Z. J. Yu, S. Manzoor, S. Wang, W. Weigand et al., Blade-Coated Perovskites on Textured Silicon for 26%-Efficient Monolithic Perovskite/Silicon Tandem Solar Cells. *Joule* **2020**, *4*(4), 850–864. https://doi.org/10.1016/j.joule.2020.01.008
68. Y. Wu, D. Yan, J. Peng, T. Duong, Y. Wan et al., Monolithic Perovskite/Silicon-Homojunction Tandem Solar Cell with over 22% Efficiency. *Energy Environ. Sci.* **2017**, *10*(11), 2472–2479. https://doi.org/10.1039/C7EE02288C
69. G. Nogay, F. Sahli, J. Werner, R. Monnard, M. Boccard et al., 25.1%-Efficient Monolithic Perovskite/Silicon Tandem Solar Cell Based on a p -type Monocrystalline Textured Silicon Wafer and High-Temperature Passivating Contacts. *ACS Energy Lett.* **2019**, *4*(4), 844–845. https://doi.org/10.1021/acsenergylett.9b00377
70. C. U. Kim, J. C. Yu, E. D. Jung, I. Y. Choi, W. Park et al., Optimization of Device Design for Low Cost and High Efficiency Planar Monolithic Perovskite/Silicon Tandem Solar Cells. *Nano Energy.* **2019**, *60*, 213–221. https://doi.org/10.1016/j.nanoen.2019.03.056
71. J. Zheng, C. F. J. Lau, H. Mehrvarz, F.-J. Ma, Y. Jiang et al., Large Area Efficient Interface Layer Free Monolithic Perovskite/Homo-Junction-Silicon Tandem Solar Cell with Over 20% Efficiency. *Energy Environ. Sci.* **2018**, *11*(9), 2432–2443. https://doi.org/10.1039/C8EE00689J
72. Y. Wu, A. Fell, K. J. Weber, A Step-by-Step Optimization of the c-Si Bottom Cell in Monolithic Perovskite/c-Si Tandem Devices. *Sol RRL* **2018**, *2*(11), 1800193. https://doi.org/10.1002/solr.201800193
73. H. Shen, S. T. Omelchenko, D. A. Jacobs, S. Yalamanchili, Y. Wan et al., In Situ Recombination Junction Between p-Si and TiO_2 Enables High-Efficiency Monolithic Perovskite/Si Tandem Cells. *Sci Adv.* **2018**, *4*(12). https://doi.org/10.1126/sciadv.aau9711
74. J. Werner, F. Sahli, F. Fu, J. J. Diaz Leon, A. Walter et al., Perovskite/Perovskite/Silicon Monolithic Triple-Junction Solar Cells with a Fully Textured Design. *ACS Energy Lett.* **2018**, *3*(9), 2052–2058. https://doi.org/10.1021/acsenergylett.8b01165

75. F. Jiang, T. Liu, B. Luo, J. Tong, F. Qin et al., A Two-Terminal Perovskite/Perovskite Tandem Solar Cell. *J. Mater. Chem. A* **2016**, *4*(4), 1208–1213. https://doi.org/10.1039/C5TA08744A
76. J. Tong, Z. Song, D. H. Kim, X. Chen, C. Chen et al., Carrier Lifetimes of ≫1 Ms in Sn-Pb Perovskites Enable Efficient All-Perovskite Tandem Solar Cells. *Science* **2019**, *364*(6439), 475–479. https://doi.org/10.1126/science.aav7911
77. Z. Yang, A. Rajagopal, C.-C. Chueh, S. B. Jo, B. Liu et al., Stable Low-Bandgap Pb-Sn Binary Perovskites for Tandem Solar Cells. *Adv. Mater.* **2016**, *28*(40), 8990–8997. https://doi.org/10.1002/adma.201602696
78. G. E. Eperon, T. Leijtens, K. A. Bush, R. Prasanna, T. Green et al., Perovskite-Perovskite Tandem Photovoltaics with Optimized Band Gaps. *Science* **2016**, *354*(6314), 861–865. https://doi.org/10.1126/science.aaf9717
79. C. Wang, Y. Zhao, T. Ma, Y. An, R. He et al., A Universal Close-Space Annealing Strategy Towards High-Quality Perovskite Absorbers Enabling Efficient All-Perovskite Tandem Solar Cells. *Nat. Energy* **2022**, *7*(8), 744–753. https://doi.org/10.1038/s41560-022-01076-9
80. L. Mao, H. Tsai, W. Nie, L. Ma, J. Im et al., Role of Organic Counterion in Lead- and Tin-Based Two-Dimensional Semiconducting Iodide Perovskites and Application in Planar Solar Cells. *Chem. Mater.* **2016**, *28*(21), 7781–7792. https://doi.org/10.1021/acs.chemmater.6b03054
81. A. F. Palmstrom, G. E. Eperon, T. Leijtens, R. Prasanna, S. N. Habisreutinger et al., Enabling Flexible All-Perovskite Tandem Solar Cells. *Joule* **2019**, *3*(9), 2193–2204. https://doi.org/10.1016/j.joule.2019.05.009
82. X. Zhou, H. Lai, T. Huang, C. Chen, Z. Xu et al., Suppressing Nonradiative Losses in Wide-Band-Gap Perovskites Affords Efficient and Printable All-Perovskite Tandem Solar Cells with a Metal-Free Charge Recombination Layer. *ACS Energy Lett.* **2023**, *8*(1), 502–512. https://doi.org/10.1021/acsenergylett.2c02156
83. R. Lin, K. Xiao, Z. Qin, Q. Han, C. Zhang et al., Monolithic All-Perovskite Tandem Solar Cells with 24.8% Efficiency Exploiting Comproportionation to Suppress Sn(Ii) Oxidation In Precursor Ink. *Nat. Energy* **2019**, *4*(10), 864–873. https://doi.org/10.1038/s41560-019-0466-3
84. R. Lin, J. Xu, M. Wei, Y. Wang, Z. Qin et al., All-Perovskite Tandem Solar Cells with Improved Grain Surface Passivation. *Nature.* **2022**, *306*, 302–306. https://doi.org/10.1038/s41586-021-04372-8
85. R. He, W. Wang, Z. Yi, F. Lang, C. Chen et al., Improving Interface Quality for 1-cm2 All-Perovskite Tandem Solar Cells. *Nature* **2023**, *618*(7963), 80–86. https://doi.org/10.1038/s41586-023-05992-y
86. J. Zhu, Y. Luo, R. He, C. Chen, Y. Wang et al., A Donor–Acceptor-Type Hole-Selective Contact Reducing Non-Radiative Recombination Losses in Both Subcells Towards Efficient All-Perovskite Tandems. *Nat. Energy* **2023**, *8*(7), 714–724. https://doi.org/10.1038/s41560-023-01274-z
87. X. Dai, S. Chen, H. Jiao, L. Zhao, K. Wang et al., Efficient Monolithic All-Perovskite Tandem Solar Modules with Small Cell-to-Module Derate. *Nat. Energy* **2022**, *7*(10), 923–931. https://doi.org/10.1038/s41560-022-01102-w
88. K. Xiao, Y.-H. Lin, M. Zhang, R. D. J. Oliver, X. Wang et al., Scalable Processing for Realizing 21.7%-Efficient All-Perovskite Tandem Solar Modules. *Science* **2022**, *376*(6594), 762–767. https://doi.org/10.1126/science.abn7696
89. F. Fu, T. Feurer, T. P. Weiss, S. Pisoni, E. Avancini et al., High-Efficiency Inverted Semi-Transparent Planar Perovskite Solar Cells In Substrate Configuration. *Nat Energy* **2016**, *2*(1), 16190. https://doi.org/10.1038/nenergy.2016.190
90. D. H. Kim, C. P. Muzzillo, J. Tong, A. F. Palmstrom, B. W. Larson et al., Bimolecular Additives Improve Wide-Band-Gap Perovskites for Efficient Tandem Solar Cells with CIGS. *Joule* **2019**, *3*(7), 1734–1745. https://doi.org/10.1016/j.joule.2019.04.012

91. S. Ishizuka, J. Nishinaga, M. Iioka, H. Higuchi, Y. Kamikawa et al., Si-Doped Cu(In,Ga) Se 2 Photovoltaic Devices with Energy Conversion Efficiencies Exceeding 16.5% without a Buffer Layer. *Adv. Energy Mater.* **2018**, *8*(11), 1702391. https://doi.org/10.1002/aenm.201702391
92. M. Powalla, S. Paetel, E. Ahlswede, R. Wuerz, C. D. Wessendorf et al., Thin-Film Solar Cells Exceeding 22% Solar Cell Efficiency: An Overview On Cdte-, Cu(In,Ga)Se2-, and Perovskite-Based Materials. *Appl. Phys Rev.* **2018**, *5*(4). https://doi.org/10.1063/1.5061809
93. M. Jošt, T. Bertram, D. Koushik, J. A. Marquez, M. A. Verheijen et al., 21.6%-Efficient Monolithic Perovskite/Cu(In,Ga)Se 2 Tandem Solar Cells with Thin Conformal Hole Transport Layers for Integration on Rough Bottom Cell Surfaces. *ACS Energy Lett.* **2019**, *4*(2), 583–590. https://doi.org/10.1021/acsenergylett.9b00135
94. T. Todorov, T. Gershon, O. Gunawan, Y. S. Lee, C. Sturdevant et al., Monolithic Perovskite-CIGS Tandem Solar Cells via in Situ Band Gap Engineering. *Adv. Energy Mater.* **2015**, *5*(23), 1500799. https://doi.org/10.1002/aenm.201500799
95. A. Al-Ashouri, A. Magomedov, M. Roß, M. Jošt, M. Talaikis et al., Conformal Monolayer Contacts with Lossless Interfaces for Perovskite Single Junction and Monolithic Tandem Solar Cells. *Energy Environ. Sci.* **2019**, *12*(11), 3356–3369. https://doi.org/10.1039/C9EE02268F
96. Q. Han, Y.-T. Hsieh, L. Meng, J.-L. Wu, P. Sun et al., High-Performance Perovskite/ $Cu(In,Ga)Se_2$ Monolithic Tandem Solar Cells. *Science.* **2018**, *361*(6405), 904–908. https://doi.org/10.1126/science.aat5055
97. A. Polman, M. Knight, E. C. Garnett, B. Ehrler, W. C. Sinke, Photovoltaic materials: Present Efficiencies and Future Challenges. *Science* **2016**, *352*(6283). https://doi.org/10.1126/science.aad4424
98. C.-C. Chen, S.-H. Bae, W.-H. Chang, Z. Hong, G. Li et al., Perovskite/Polymer Monolithic Hybrid Tandem Solar Cells Utilizing a Low-Temperature, Full Solution Process. *Mater. Horizons* **2015**, *2*(2), 203–211. https://doi.org/10.1039/C4MH00237G
99. Y. Liu, L. A. Renna, M. Bag, Z. A. Page, P. Kim et al., High Efficiency Tandem Thin-Perovskite/Polymer Solar Cells with a Graded Recombination Layer. *ACS Appl. Mater. Interfaces* **2016**, *8*(11), 7070–7076. https://doi.org/10.1021/acsami.5b12740
100. W. Chen, J. Zhang, G. Xu, R. Xue, Y. Li et al., A Semitransparent Inorganic Perovskite Film for Overcoming Ultraviolet Light Instability of Organic Solar Cells and Achieving 14.03% Efficiency. *Adv. Mater.* **2018**, *30*(21), 1–10. https://doi.org/10.1002/adma.201800855
101. Q. Zeng, L. Liu, Z. Xiao, F. Liu, Y. Hua et al., A Two-Terminal All-Inorganic Perovskite/Organic Tandem Solar Cell. *Sci. Bull.* **2019**, *64*(13), 885–887. https://doi.org/10.1016/j.scib.2019.05.015
102. K. Lang, Q. Guo, Z. He, Y. Bai, J. Yao et al., High Performance Tandem Solar Cells with Inorganic Perovskite and Organic Conjugated Molecules to Realize Complementary Absorption. *J. Phys. Chem. Lett.* **2020**, *11*(22), 9596–9604. https://doi.org/10.1021/acs.jpclett.0c02794
103. H. Aqoma, I. F. Imran, F. T. A. Wibowo, N. V. Krishna, W. Lee et al., High-Efficiency Solution-Processed Two-Terminal Hybrid Tandem Solar Cells Using Spectrally Matched Inorganic and Organic Photoactive Materials. *Adv. Energy Mater.* **2020**, *10*(37), 2001188. https://doi.org/10.1002/aenm.202001188
104. S. Xie, R. Xia, Z. Chen, J. Tian, L. Yan et al., Efficient Monolithic Perovskite/Organic Tandem Solar Cells and Their Efficiency Potential. *Nano Energy* **2020**, *78*(August), 105238. https://doi.org/10.1016/j.nanoen.2020.105238
105. X. Chen, Z. Jia, Z. Chen, T. Jiang, L. Bai et al., Efficient and Reproducible Monolithic Perovskite/Organic Tandem Solar Cells with Low-Loss Interconnecting Layers. *Joule* **2020**, *4*(7), 1594–1606. https://doi.org/10.1016/j.joule.2020.06.006
106. S. Qin, C. Lu, Z. Jia, Y. Wang, S. Li et al., Constructing Monolithic Perovskite/Organic Tandem Solar Cell with Efficiency of 22.0% via Reduced Open-Circuit Voltage Loss and Broadened Absorption Spectra. *Adv. Mater.* **2022**, *34*(11), 2108829. https://doi.org/10.1002/adma.202108829

107. H. Li, P. Harrison, J. Wu, H. Yao, Probing the Correlation of Twin Boundaries and Charge Transport of CdTe Solar Cells Using Electron Backscattering Diffraction and Conductive Atomic Force Microscopy. *ACS Appl. Energy Mater.* **2018**, *1*(8), 3646–3653. https://doi.org/10.1021/acsaem.8b00382
108. T. Tang, H. Zhang, X. Du, Y. Liu, H. Zhou, Theoretical Design of Perovskite/CdTe Four-Terminal Tandem Solar Cells. In *2017 IEEE 44th Photovoltaic Specialist Conference (PVSC)* 2558–2561 (June 2017). IEEE. https://doi.org/10.1109/PVSC.2017.8366478
109. S. Rühle, The Detailed Balance Limit of Perovskite/Silicon and Perovskite/CdTe Tandem Solar Cells. *Phys. Status Solidi.* **2017**, *214*(5), 1600955. https://doi.org/10.1002/pssa.201600955
110. T. D. Siegler, T. M. Shimpi, W. S. Sampath, B. A. Korgel, Development of Wide Bandgap Perovskites for Next-Generation Low-Cost CdTe Tandem Solar Cells. *Chem. Eng. Sci.* **2019**, *199*, 388–397. https://doi.org/10.1016/j.ces.2019.01.003

7 Lead-Free Perovskites for Solar Cells and Beyond

Shurong Wang, Huanhuan Yao, and Feng Hao

7.1 INTRODUCTION

Hybrid lead halide perovskite ($APbX_3$) solar cell has had a rapid rise in power conversion efficiency (PCE) from the original 3.8% to a phenomenal 26.1% in recent years.[1,2] These semiconductors have unique characteristics such as high defect tolerance, small exciton binding energy, high absorption coefficient, adjustable bandgap, long carrier diffusion length, ambipolar charge-carrier transport, and so on.[3–5] Despite their excellent photoelectric performance, they still face unprecedented challenges due to the toxicity of Pb for both the environment and organisms.[6,7] There are some encapsulation methods to prevent Pb leakage, such as employing Pb-absorbing materials, functionalized metal-organic frameworks, epoxy resin with great self-healing ability, cation-exchange resins, etc.[8–11] However, the development of metal halide perovskites has aroused the exploration of lead-free counterparts in photovoltaic applications and other related optoelectronic devices. These lead-free counterparts are extensively crucial for clean and green photovoltaic applications and the future commercialization of optoelectronic devices.

Inspired by the remarkable characteristics of Pb perovskites, other perovskite-like lattice structures have been investigated on the isovalent substitutions (e.g. Sn and Ge) and heterovalent substitutions (e.g. Bi and Sb). So far, the tin (Sn)-based PSCs reached a PCE of over 15% from accredited laboratories, which showed great potential for practical photovoltaics (PV) development.[12] The urgent efforts focus on the inferior photovoltaic performance compared to the Pb-PSCs. Therefore, we first analyzed the main risk of Pb analogs on environmental and toxicological implications. Then, a series of eco-friendly candidates (Figure 7.1) were summarized from their structural, electronic, and optical properties. Furthermore, we discussed the photovoltaic development in Sn-based PSCs including MA-, FA-, and Cs-based Sn halide perovskites as well as low/mixed-dimensional Sn halide perovskites. Then, the application in LEDs, thermoelectric devices (TEs), photodetectors, thin-film transistors (TFTs), photocatalysis, and memristors was also presented. In the end, future perspectives and the development of lead-free PSCs were discussed.

DOI: 10.1201/9781003400493-7

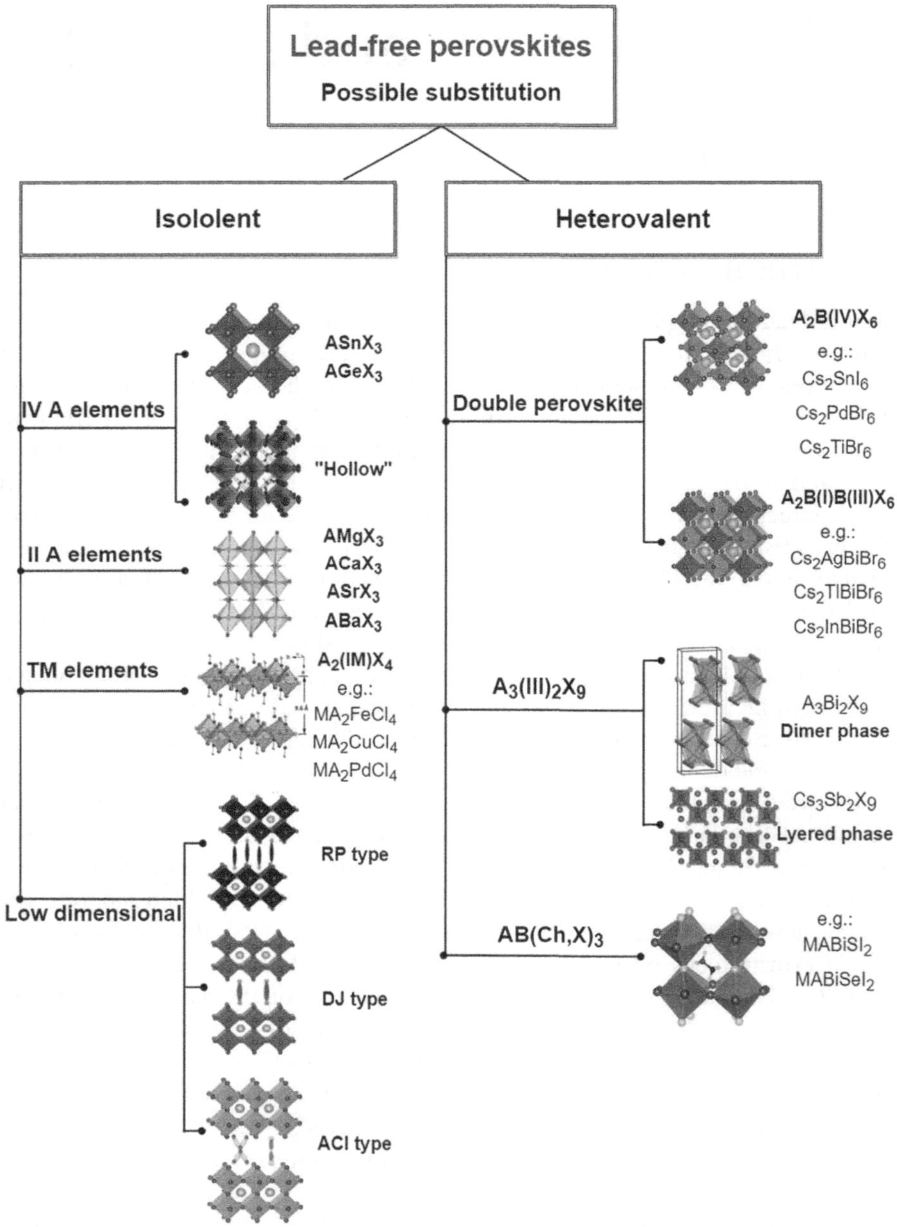

FIGURE 7.1 Summary of recently reported lead-free perovskites.

7.2 ENVIRONMENTAL AND TOXICOLOGICAL IMPLICATIONS OF LEAD PEROVSKITES

Lead (Pb) has raised a big concern in environmental and toxicological implications so far. Though the current concentration of Pb in PSCs is well below the limits of the employment of heavy metals in electronics,[13–15] it accumulates in organisms (such as bones, liver, brain, and kidneys) and moves up into the food chain and therefore affects our body systems, for instance, neurological, gastrointestinal, renal, and cardiovascular. Only a few micrograms of Pb in the human body, in particular for pregnant women and children, can cause a catastrophic impact. Therefore, the toxic problem of Pb-based perovskites is a Gordian knot and may prejudice the implementation of PSC applications in the photovoltaic field. In 2003 and 2011, the European Union promulgated "RoHS1" and "RoHS2" as lead-free directives and restricted to 0.1 wt% as the maximum concentration in perovskites.[13] The positive bivalent Pb^{2+} is stable and can intervene in the human body at a molecular lever like other divalent elements like Zn^{2+}, Fe^{2+}, and Ca^{2+}. Moreover, the International Agency for Research on Cancer (IARC) considers Pb to be carcinogenic to humans. Beyond the toxicity, Pb has a long half-life of around 20~30 years and long-term preservation in the environment.

For the fabrication of Pb-based PSCs, the main risk is the perovskite precursor compounds. The solubility of these compounds is determined by the anionic species. $Pb(NO_3)_2$, $Pb(CH_3COO)_2$, and $Pb(OH)_2$ are more soluble in both H_2O and fats. PbI_2, $PbBr_2$, and $PbCl_2$ are less soluble in water. Therein, high-polar solvents such as N, N-dimethylformamide (DMF) and dimethyl sulfoxide (DMSO) were used in the precursor solutions, which certainly increased the risk of toxicity issues. When leaking into the water, the halide perovskites dissolve into the original compounds and then into methylamine and hydroiodic acid (HI).[15]

In the case of Sn halide perovskite, the SnI_2 and oxidative product SnI_4 can easily hydrolyze to the almost insoluble SnO and SnO_2, respectively.[16] Moreover, the inorganic Sn halide perovskites appear slightly soluble in H_2O and poor absorption in tissues. Their toxicity is very low in inorganic compounds. Therefore, $ASnX_3$ perovskites show less toxicity as compared with Pb analog. Abate et al. provided a simple way to monitor the metal uptake by capturing the Sn and Pb elements in mint.[9] The $MASnI_3$ precursor showed more reduced bioavailability than the $MAPbI_3$ precursor and offered lower bioaccumulation. It is worth noting that the toxicity of Sn in $ASnX_3$ perovskites is complex. On the one hand, the organic Sn compounds can easily enter the bloodstream because of their liposoluble property. On the other hand, the SnI_2 as well as SnI_4 hydrolysis produce HI, resulting in the acidification of H_2O.[17]

$$8SnI_2 + 4H_2O + O_2 \rightarrow 2SnI_4 + 6Sn(OH)I + 2HI \tag{7.1}$$

$$SnI_4 + 2H_2O \rightarrow SnO_2 + 4HI \tag{7.2}$$

The acidification of the soil is the key role of the toxic effect of Sn perovskite compounds. Generally, Sn is less available in organisms than Pb due to the fast formation of inert oxides when considering the oxidation from Sn^{2+} to Sn^{4+}. Therefore, Sn perovskites are still considered a safe alternative with much lower bioavailability compared with Pb counterparts.

7.3 LEAD-FREE HALIDE PEROVSKITES

7.3.1 Isovalent Elements Substitutions

7.3.1.1 IVA Elements

$ASnX_3$ Tin halide perovskites (THPs) is the most potential candidate due to their similar electronic structures. Given the appropriate ion radius of Sn(II) (112 pm for Sn(II) and 119 pm for Pb(II)), the THPs present a similar crystal structure to $APbI_3$. The combinations of eight inorganic frameworks of $[SnX_6]^{4-}$ octahedra along with the center A-site cations construct a stable 3D perovskite network. $MASnI_3$, $FASnI_3$, and $CsSnI_3$ were widely adopted in Sn-based 3D perovskites (Figure 7.2a).[18–20]

The crystal structure for these three 3D tin perovskites depends on temperature, the orientation of $[SnX_6]^{4-}$ octahedra, and the length and orientation of the Sn-X-Sn bond.[21] The highest symmetric crystal structure is the cubic phase, where the Sn-X-Sn bond angle of corner-shared octahedra is 180°. With a decreased temperature, the cubic phase could transit to a tetragonal or orthorhombic phase with a low symmetricity. The Sn-X-Sn bonds of tetragonal and orthorhombic phases deviated

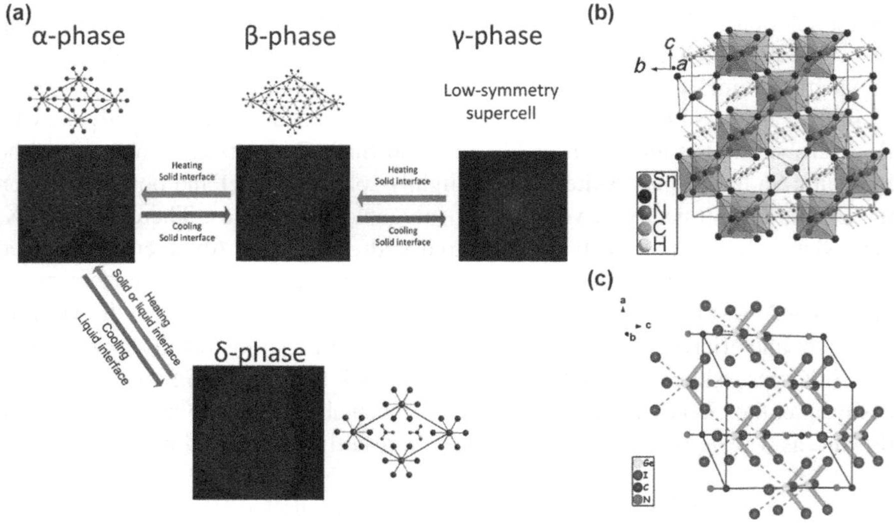

FIGURE 7.2 (a) The phase transition when changing the temperature for Sn hybrid perovskites. Reproduced with permission.[30] Copyright 2013, American Chemical Society. (b) Supercell of hollow perovskite ($(FA)_{16}Sn_{14}I_{44}$) with two SnI_2 vacancies. Reproduced with permission.[38] Copyright 2017, American Association for the Advancement of Science. (c) The unit cell of $MAGeI_3$. Reproduced with permission.[44] Copyright 2015, American Chemical Society.

from the ideal 180°. For $MASnI_3$, a phase transition from the cubic phase at 295 K (*Pm3m*) to the tetragonal phase around 275 K (*I4/mcm*) is observed (Figure 7.2a).[22] When the temperature is further lowered to 108~114 K, the phase could further transform to orthorhombic (*Pbn21* or *Pbnm*).[22,23] The theoretical calculations revealed that $MASnI_3$ has a direct bandgap of 1.1 eV, which is close to the experimental value of 1.3 eV.[24–26] The increase of octahedral distortions or the size of the A-site cation would result in a wider bandgap. For instance, the $MASnI_3$ with an orthorhombic structure had a higher bandgap than that of a tetragonal structure,[27] while increasing the size of the X site inversely leads to a smaller bandgap. The Pb-based perovskites have similar trends.[28]

The original discovery of $FASnI_3$ crystal dates back to 1997; therein a high-temperature cubic phase is observed.[29] Kanatzidis et al. determined that $FASnI_3$ had three phases with temperature: pseudocubic (*Amm2*) phase at 340 K, orthorhombic (*Imm2*) phase below 180 K, and a primitive monoclinic structure close to 100 K.[30] Other reports by Schueller et al. elucidated that the phase transition from cubic (*Pm3m*) to tetragonal (*P4/mbm*) occurred at 225 K to 250 K, then to orthorhombic (*pnma*) structure at 125–150 K.[31] $FASnI_3$ also has a direct bandgap.[32,33] Different from $MASnI_3$, $FASnI_3$ has a longer Sn-I bond due to the larger size of FA^+ than MA^+, therefore resulting in weakened antibonding interaction between Sn 5*s* and I 5*p* and wider bandgap.[33]

$CsSnI_3$ has four polymorphs: the black cubic (*Pm3m*) phase at 500 K, the black tetragonal (*P4/mbm*) phase at 380 K, the black orthorhombic (*Pnma*) phase at 300 K, and the yellow orthorhombic (*Pnma*) phase at 300 K.[34] The A-site cations show indirect influence in bandgap when considering the distortion of $[SnI_6]^{4-}$ octahedra.[35,36] $CsSnI_3$ has a lower direct bandgap (1.25 eV) than $FASnI_3$ and $MASnI_3$. Meanwhile, the increased tendency of bandgap is determined when the phase transiting from cubic to orthorhombic structure with the increased tilting of the $[SnI_6]^{4-}$ network.[37] Moreover, $CsSnI_3$ shows strong PL ability in the near-infrared region (around 950 nm). The high hole mobility (~585 $cm^2\ V^{-1}\ s^{-1}$) and carrier concentration (~$10^{17}\ cm^{-3}$) have been observed.[34] Noteworthily, the electrical and optical properties of $CsSnI_3$ depend on their Sn-related defects and crystal transition. This is the origin of metallic for $CsSnI_3$ though it is a semiconductor.

Hollow perovskites: Kanatzidis et al. reported a new class of 3D Sn perovskites named as hollow structure (Figure 7.2b), which replaces the A-site cations with ethylenediammonium (EN), trimethylenediammonium (TN), or propylenediammonium (PN) with 3D structure retention.[38–41] When new cations incorporate, it causes large lattice discontinuities and disrupts the perovskite structures. The hollow vacancy framework results in a blue-shifted bandgap and has an increased tendency when loading more A-site cations.[42] The perovskites with hollow structures have significantly different electronic properties than the 3D counterparts. When increasing EN cations to 0.4 M, an increased bandgap of around 40% was observed compared with the original perovskites.[42] The wider bandgap of hollow perovskites is due to the perovskite lattice instead of its thermal expansion.[43] The orbital overlap of Sn/I decreased because of the I^- and Sn^{2+}-related vacancies, thus leading to a narrow bandwidth of CB and VB. Meanwhile, the presence of defects may act as a nonradiative recombination center and affect the photoluminescence intensity.

$AGeX_3$ Similar to Sn and Pb, geranium (Ge) also belongs to IVA elements and is deemed to be another low-toxic alternative to Pb elements.[45] Stoumpos et al. synthesized a series of $AGeI_3$ (A = MA^+, formamidinium (FA), acetamidinium (MFO), GA^+, etc.) and observed a bandgap ranging from 1.63 to 2.8 eV depending on the A cations.[44] It is noted that the $AGeX_3$ perovskites show a much larger bandgap than Sn and Pb analogs. One of the origins is the deep Ge 4*s* states, which leads to weak hybridization with halide *p* states and further endows shallow VBM for Ge analogs.[46] The structural distortion of $[GeX_6]^{4-}$ octahedra is due to the smaller ionic radius of Ge^{2+} (0.73 Å), where Ge^{2+} is off the center of $[GeX_6]^{4-}$ octahedra might be another reason (Figure 7.2c). Meanwhile, $AGeX_3$ perovskites have poor solubility in most organic polar solvents, thus affecting the film morphology and device performance.[45]

7.3.1.2 IIA Elements Substitutions

According to the Goldschmidt rules, Mg^{2+}, Ca^{2+}, Sr^{2+}, and Ba^{2+} have nearly identical ionic radii and are suitable for perovskite structure.[47,48] The DFT calculation implies that the $FAMgI_3$, $MAMgI_3$, and $CsMgI_3$ have a direct band gap of 0.9 eV, 1.5 eV, and 1.7 eV, respectively.[49] However, it is worth noting that $CsMgI_3$ only exists in a $CsNiCl_3$-type structure and has a very wide bandgap (~4.8 eV).[50] Similar DFT calculations of $MACaI_3$ perovskites also indicate a bandgap of about 2.95 eV.[47] The experimental value of bandgap in $MACaI_3$ perovskites is 3.78 eV with a pseudocubic structure. The bandgap of $MASrI_3$ is also too large (3.6 eV) to be a suitable absorber in PSCs though it can form a perovskite scaffold.[51] The bandgap decreases monotonically for Ca-, Sr-, and Ba-based perovskites for the alkaline-earth metal series (Figure 7.3a and b).[47] $MABaI_3$ compound is also susceptible under ambient conditions like $MASrI_3$, which impedes the process of material synthesis and characterization.[47]

7.3.1.3 Divalent Transition Metal (TM) Elements Substitutions

The divalent transition metal (TM) cations such as Cu^{2+}, Fe^{2+}, and Pd^{2+} are also regarded as potential candidates for lead-free perovskites.[52–57] Due to their small ionic radius, the TM cations typically form 2D layered perovskites. Meanwhile, the TM cations have multiple oxidation states, which usually lead to deep trap states and instability against oxidation similar to Sn and Ge perovskites. Meanwhile, Hg and Cd are highly toxic.

The general chemical formula for this kind of perovskite is $MA_2(TM)X_4$, similar to the feature of RP-type perovskites of MA_2FeCl_4, $MA_2CuCl_xBr_{4-x}$, and MA_2PdCl_4.[52,54,58,59] However, among these $MA_2(TM)X_4$ perovskites, only Cu^{2+}-based components have been investigated in PSCs. $MA_2CuCl_xBr_{4-x}$ has a monoclinic or orthorhombic crystal structure as determined by the ratio of Cl/Br.[52] The bandgap is 2.48 eV, 2.12 eV, 1.90 eV, and 1.80 eV for MA_2CuCl_4, $MA_2CuCl_2Br_2$, $MA_2CuClBr_3$, and $MA_2CuCl_{0.5}Br_{3.5}$, respectively. It is worth noting that their high density of trap states, heavy carrier effective masses, and low absorption are identified as the main limitations of high-efficiency Cu^{2+}-based PSCs (Figure 7.3c). For instance, the $MA_2CuCl_xBr_{4-x}$ PSC device only has a PCE below 0.2%.[52]

Based on the DFT calculation, a rising family of ABX_3 structure has emerged, chalcogenide perovskites, consisting of alkaline earth metals (e.g. Ca, Sr, and Ba)

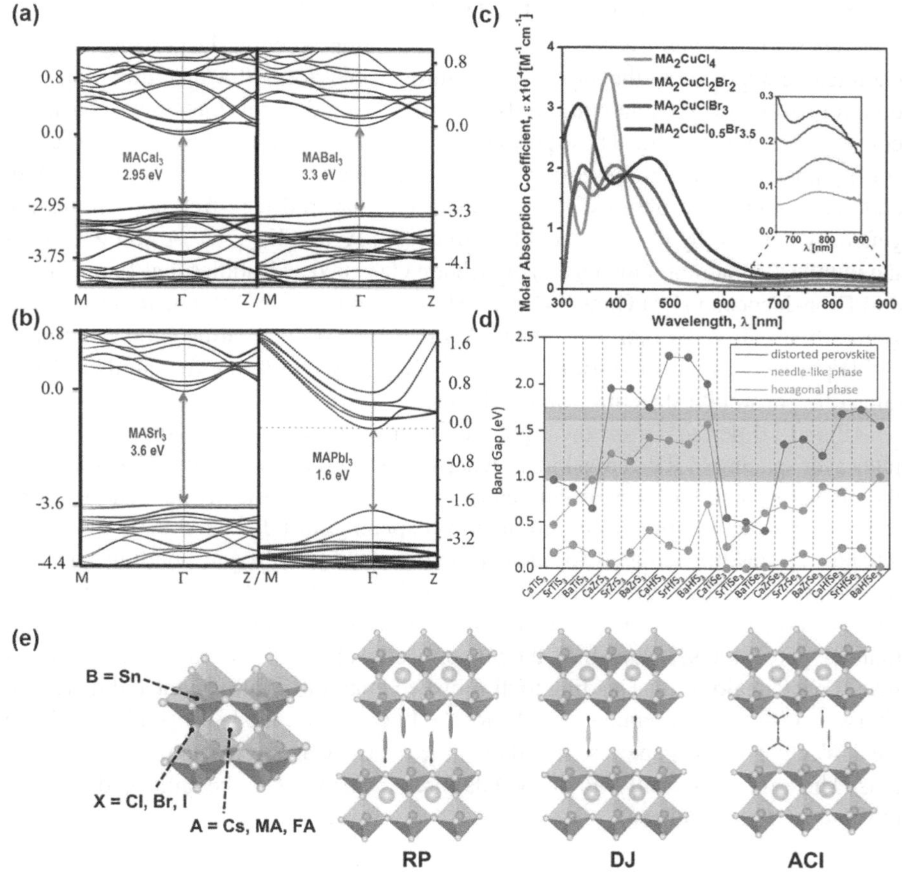

FIGURE 7.3 Calculated band structure of (a) $MACaI_3$, $MABaI_3$, (b) $MASrI_3$, and $MAPbI_3$. Reproduced with permission.[47] Copyright 2016, American Physical Society. (c) The absorption coefficient for perovskites of the series $MA_2CuCl_xBr_{4-x}$. Reproduced with permission.[52] Copyright 2016, American Chemical Society. (d) Band gap of 18 ABX_3 compounds in the distorted perovskite, needle-like, and hexagonal phases. Reproduced with permission.[60] Copyright 2014, American Chemical Society. (e) Crystal structure of ABX_3, RP, DJ, and ACI perovskites. Reproduced with permission.[65] Copyright 2022, Elsevier.

at A site, TM elements (e.g. Ti, Zr, and Hf) at B site, and S^{2-} or Se^{2-} at X site.[60–62] $BaTiS_3$ has a hexagonal phase with $P6_3/mmc$ space group.[63] Other needle-like phases like $SrZrS_3$, $SrZrSe_3$, and $BaZrSe_3$ have a *pnma* space group.[61,64] For ABX_3 chalcogenides, the bandgap of the hexagonal phase is too small to apply to solar cells, and the needle-like phases have a pseudodirect bandgap and are not suitable for PV application. Fortunately, $CaTiS_3$, $CaZrSe_3$, $CaHfSe_3$, and $BaZrS_3$ with distorted perovskite structures can be used as promising light absorption materials due to their suitable bandgap and high absorption by theoretical calculation (Figure 7.3d).[60]

7.3.1.4 Low-Dimensional Layered Halide Perovskites

Layered lead-free perovskites gained great attention recently. Depending on their chemical composition, the layered structures could form 0D quantum dots, 1D quantum rods, or 2D quantum wells. The 2D layered perovskites also consist of corner-sharing octahedra. These perovskites slice parent 3D structure along a specific (hkl) plane. There are three common categories: the (100)-oriented, (110)-oriented, and (111)-oriented structures.[66] Low-dimensional layered lead-free perovskites are generally denoted by two chemical formulas: $(A')_2A_{n-1}Sn_nX_{3n+1}$, $(A')A_{n-1}Sn_nX_{3n+1}$ (*n* is the number of the inorganic layers), which are classified as Ruddlesden–Popper (RP) and Dion-Jacobson (DJ) Quasi-2D perovskites (Figure 7.3e).[65–68] The $n = 1$ phase is generally an ideal 2D inorganic sheet. Another type of alternating cation interlayer (ACI) perovskites has the formula of $A'A_nB_nX_{3n+1}$. By increasing the number of *n*, a quasi-2D perovskite structure can be obtained.

RP and DJ perovskites show a great difference in the interlayer spacing.[69,70] The interlayer spacing of RP perovskite is generally larger than that of the DJ phase due to the bilayer structure of monovalent cation pairs. Owing to the divalent cation pairs' monolayer structure, the minimum interlayer spacing for DJ perovskite is 2 Å.[70] The RP perovskites are tethered to adjacent octahedra by hydrogen bonding and Van der Waals, while DJ perovskites are connected by hydrogen bonding.[71,72] Due to the enhanced interaction of inorganic layers, DJ perovskites tend to show better stability than RP perovskites. Meanwhile, the DJ perovskites have shorter interlayer distances than the RP analogs, thus leading to diminished quantum confinement and increased charge transfer of inorganic slabs. As for ACI perovskite, the hydrogen bonds and a weak Van der Waals gap presented due to the alternate arrangement of A and A' cations. Thus, the weakened Van der Waals force contributes to the enhanced stability of these ACI perovskites.[73]

7.3.2 Heterovalent Elements Substitutions

7.3.2.1 Double Perovskites

$A_2B(IV)X_6$ (0D): implementing tetravalent B^{4+} cations as the heterovalent substitution usually adopts by removing half of B^{2+} cations.[74] Considering the significant difference between B-site vacancies and the remained B-site cations, the perovskites structure is arranged in a rock-salt way, thus calling them vacancy-ordered lead-free double perovskites with the formula of $A_2B(IV)X_6$. It is exactly because of the lack between $[BX_6]^{4-}$ octahedra that these $A_2B(IV)X_6$ perovskites (e.g. Cs_2SnI_6, Cs_2PdBr_6, and Cs_2TiBr_6) are regarded as 0D non-perovskites (Figure 7.4a).

Due to the isolated $[BX_6]^{4-}$ octahedra, the optoelectronic properties of $A_2B(IV)X_6$ compounds are different from $APbX_3$, $ASnX_3$, and $AGeX_3$. Their VBM is decided by nonbonding X *p* orbitals instead of the long pair *s*-X *p* antibonding states in $APbX_3$, $ASnX_3$, and $AGeX_3$, which leads to rather large m_h.[75–77] The CBM is decided by the *s* or *d* antibonding states of cations and X *p* orbitals.[76,78] Meanwhile, the CBM is relatively dispersive because of charge redistribution of B-site cations, therefore m_e is generally smaller than m_h in $A_2B(IV)X_6$ compounds. It is also the origin of intrinsic *n*-type conductivity characteristic of these counterparts.

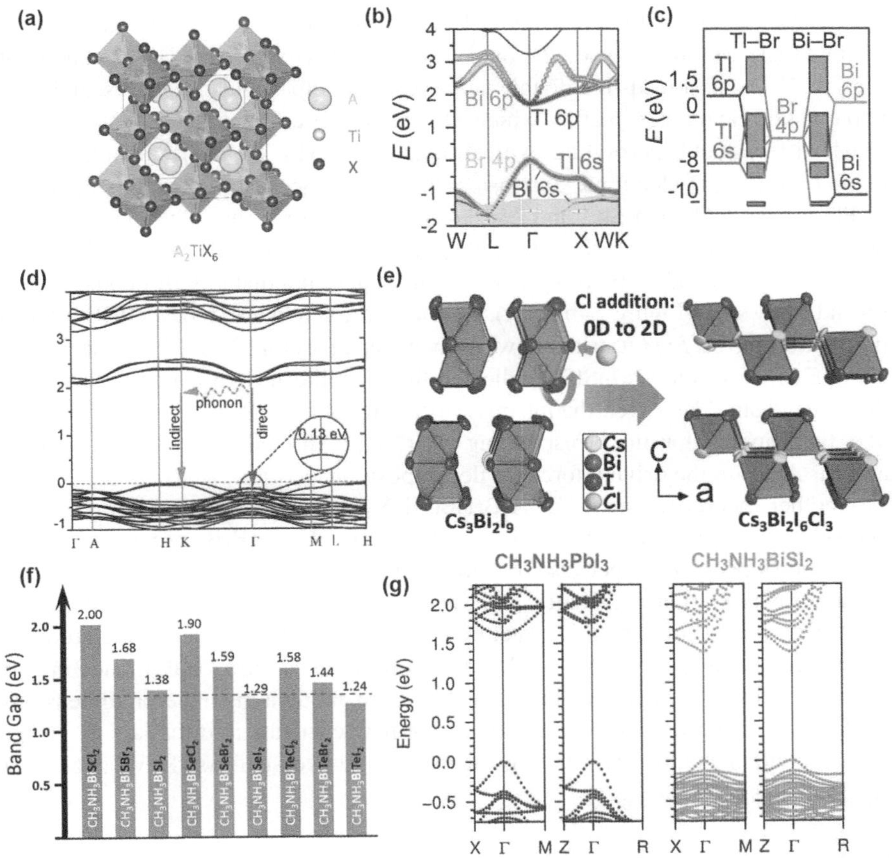

FIGURE 7.4 (a) Schematic crystal structure of OD A_2TiX_6 perovskites. Reproduced with permission.[77] Copyright 2018, American Chemical Society. (b) Band structures and (c) schematic energy diagrams for $Cs_2TlBiBr_6$ perovskites. Reproduced with permission.[90] Copyright 2018, American Chemical Society. (d) Band structure of $Cs_3Bi_2I_9$ perovskites. Reproduced with permission.[95] Copyright 2017, American Chemical Society. (e) Structure comparison of 0D $Cs_3Bi_2I_9$ and 2D $Cs_3Bi_2I_6Cl_3$ viewed down the b-axis. Reproduced with permission.[96] Copyright 2019, American Chemical Society. (f) Calculated bandgap of $CH_3NH_3BiXY_2$ compounds (with X = S, Se, or Te and Y = Cl, Br, or I). (g) Comparison of band structures of $MAPbI_3$ and MA $BiSI_2$. Reproduced with permission.[97] Copyright 2016, Royal Society of Chemistry.

Notably, the bandgap of Cs_2SnI_6 is direct at the Γ point, while the bandgap values of Cs_2PdBr_6 and Cs_2TiBr_6 are indirect with CBM at the X point and VBM at the Γ point.[75,77,79] For defect properties, using TM element Ti in Cs_2TiX_6 perovskites also reveals similar defect chemistry to vacancy-ordered Cs_2SnI_6 counterparts.[77,80] The bandgap of Cs_2SnI_6, Cs_2PdBr_6, and Cs_2TiBr_6 is 1.5 eV, 1.6 eV, and 1.8 eV, respectiv ely.[76,80,81] Other reports indicated that the $Cs_2TiI_2Br_4$ and Cs_2TiBr_6 possess a bandgap of around 1.38 eV and 1.78 eV, respectively, which show great possibility towards single-junction PSCs and tandem photovoltaic application.[77]

$A_2B(I)B(III)X_6$ (3D): double perovskites are also perovskite structures with alternating B-site cations of B^+ or B^{3+} oxidation states instead of just B^{2+} as center cations.[79,82,83] The first lead-free double perovskites or "elpasolites" were researched as photovoltaic candidates in 2016, which were named after the mineral K_2NaAlF_6.[84–86] The lead-free double perovskites with all-lone-pair B^+ (In, Tl) and B^{3+} cations (Sb, Bi) exhibit similar 3D electronic dimensionality and optoelectronic features compared to Pb analog.[87] Though Sb^{3+} and Bi^{3+} are much less toxic than Pb^{2+}, Tl^+ as a toxic heavy metal is not suitable for solar cell applications.[88–90] The VBM of $Cs_2TlBiBr_6$ is decided by Tl 6*s* and Br 6*s*-Br 4*p* antibonding states, and that of CBM consists of Bi 6*p* and Tl 6*p* state (Figure 7.4b and c).[90] Besides, the synthesized $MA_2TlBiBr_6$ double perovskites have a *Fm3m* structure with a bandgap of around 2.0 eV.[91,92] Other reports about Cs_2InBiX_6 and Cs_2InSbX_6 indicated their small effective masses and suitable direct bandgap.[93] However, the intrinsic instability of Cs_2InBiX_6 and Cs_2InSbX_6 hinders their application in corresponding solar cells. The reduction-oxidation for In^+ and Bi^{3+} leads to the driving force for decomposition.

Another theoretical study for $A_2B(I)B(III)X_6$ (B^+ = Cu, Au, and Ag, B^{3+} = Sb, and Bi) shows a bandgap range of 0~2.7 eV.[86] So far, $Cs_2AgBiBr_6$ stands out among these double perovskites because of similar optoelectronic properties to its Pb counterpart (e.g. 3D structure, small carrier effective mass, and long carrier lifetime).[84] $Cs_2AgBiBr_6$ has a similar cubic structure with *Fm3m* group at room temperature and an indirect bandgap (1.95 eV), while phase transitions to tetragonal, monoclinic, or triclinic could occur when decreasing temperature, modulating material dimensions, treating with high pressure, or incorporating new elements into the lattice.[94] Their defects contain shallow defects (Ag vacancies) and deep defects (Bi vacancies and Ag_{Bi} antisites).

7.3.2.2 $A_3B(III)_2X_9$ Perovskites

Dimer phase (0D): using Bi^{3+} and Sb^{3+}, the formation of $A_3B(III)_2X_9$ low-dimensional perovskites is achievable, which derives from the ABX_3 structure by removing 1/3 of B-site cations.[98] There are two different polymorphs of $A_3B(III)_2X_9$: 0D non-perovskites (dimer phase) and 2D vacancy-ordered layered perovskite (layer phase). 0D $A_3B(III)_2X_9$ counterpart (*P6$_3$/mmc* structure) presents undesired optoelectronic properties because of the low-dimensional structure and thus electronic dimensionality. The bandgap of $Cs_3Bi_2I_9$ shows an indirect value of 2.10 eV.[95] Herein CBM consists of Bi 6*p* anti-bonding states and I 5*p* orbitals, and VBM is derived from mainly I 5*p* and Bi 6*s* orbitals (Figure 7.4d). The CBM and VBM are both localized due to the isolation of $[Bi_2I_9]^{3-}$octahedra, thus resulting in large carrier effective masses.[95] The dimmer structure of $Cs_3Bi_2I_9$ also causes large exciton binding energies.[95,99,100] The defects in $Cs_3Bi_2I_9$ lie in the deep-level states and act as recombination centers.

Layer phase (2D): the layered $A_3B(III)_2X_9$ perovskites such as $Cs_3Sb_2I_9$ (*P*3*m*1 structure) are derived from the hypothetical perovskite ABX_3 (e.g. $Cs_3Sb_3I_9$) by removing every third B layer along ⟨111⟩ to maintain charge balance.[101] The layered perovskites have higher structural/electronic dimensionality than the dimer counterparts. Thus, the layered perovskites deliver enhanced optoelectronic properties when compared to their dimer counterpart. The layered $Cs_3Sb_2I_9$ has a nearly direct bandgap of 2.06 eV and a high-level absorption compared to pure $MAPbI_3$.[101] The dimer modification of $Cs_3Sb_2I_9$ is also simulated with an indirect bandgap of 2.4 eV, making it less promising than the layered $Cs_3Sb_2I_9$. Meanwhile, the VBM and CBM of layered $Cs_3Sb_2I_9$ are more dispersive than the dimer modification, thus rendering better carrier transport properties. For defects, only Cs vacancies and Cs interstitials are shallow acceptors and donors; all other defects act as deep-level states in the bandgap. Besides, the halide mixing in $Cs_3Bi_2X_9$ perovskites (e.g. $Cs_3Bi_2I_6Cl_3$, $Cs_3Bi_2I_6Br_3$) are regarded as a new number of 2D layered perovskites with ⟨111⟩-oriented defects (Figure 7.4e).[96,102]

7.3.2.3 Mixed-Anion Perovskites

AB(Ch,X)$_3$: the AB(Ch,X)$_3$ perovskites introduce dual anions (e.g. chalcogen (Se, S, and Te) and halogen anions) and replace Pb with heterovalent elements to keep the charge neutrality.[97,103] Herein, $MABiSI_2$ and $MABiSeI_2$ yield the optimal bandgap (1.38 eV and 1.29 eV, respectively) for solar absorbers based on the Shockley–Queisser theory (Figure 7.4f).[97,104] Similar to the band structure of Pb perovskites, the CBM of $MABiSI_2$ is derived from Bi 6*p* states, and VBM consists of mainly Bi 6*s* anti-bonding states and I 5*p*/S 3*p* orbitals (Figure 7.4g). Meanwhile, the $MABiSI_2$ perovskites also yield high optical absorption coefficients in the visible range. Thus it can be seen that the AB(Ch,X)$_3$ perovskites particularly $MABiSI_2$ are potential photovoltaic absorber candidates. Notably, their thermodynamic instability should be a major consideration.

7.4 DEVELOPMENTS OF TIN HALIDE PEROVSKITE SOLAR CELLS

7.4.1 Methylammonium Tin Halides

Sn^{2+} stoichiometry: the pioneering works of $MASnI_3$ delivered a PCE ranging 5~6% reported by Kanatzidis and Snaith et al.[24,25] Kanatzidis et al. reported $CH_3NH_3SnI_{3-x}Br_x$ (x = 0, 1, 2, 3) perovskite as Sn absorber, which can tune bandgap from 1.30 eV to 2.15 eV (Figure 7.5a).[25] These approaches represent a critical step for Pb-free solid-state solar cells. In the earlier research, SnF_2 is a remarkable additive for suppressing the degree of oxidation from Sn^{2+} to Sn^{4+} and reducing Sn-related defects such as Sn vacancies.[105–107] The photocarrier lifetime of Sn perovskite films increased by an order of magnitude after adding suitable SnF_2. Besides the easy oxidation of Sn^{2+}, their structural stability showed great relevance to pressure treatment.[108] A phase transition from tetragonal (*P4mm*, phase I) to orthorhombic (*Pnma*, phase II) was observed with variations of lattice parameters as well as cell volume (Figure 7.5b and c).[108] When increasing pressure, orthorhombic $MASnI_3$ had a large anisotropy behavior of compressibility. The *c*-axial parameter decreased about twice more rapidly than *an* axial and *b* axial, which is attributed to different interactions between dumbbell-shape MA^+ and $[SnI_6]^{4-}$ octahedra along in-phase and out-of-phase directions.

Reducing agents: besides, Wakamiya et al. realized the in-situ growth of Sn^0 nanoparticles by a selective reaction between dihydropyridine and SnF_2 additives (Figure 7.5d), thus delivering a high PCE of 11.5% with a voltage output up to 0.76 V.[109] The initial attempt at vapor-assisted technology was achieved by Kanatzidis et al.[110] By evaporating MAI powder at 150°C, the MA gas reacts with SnI_2 films as well as

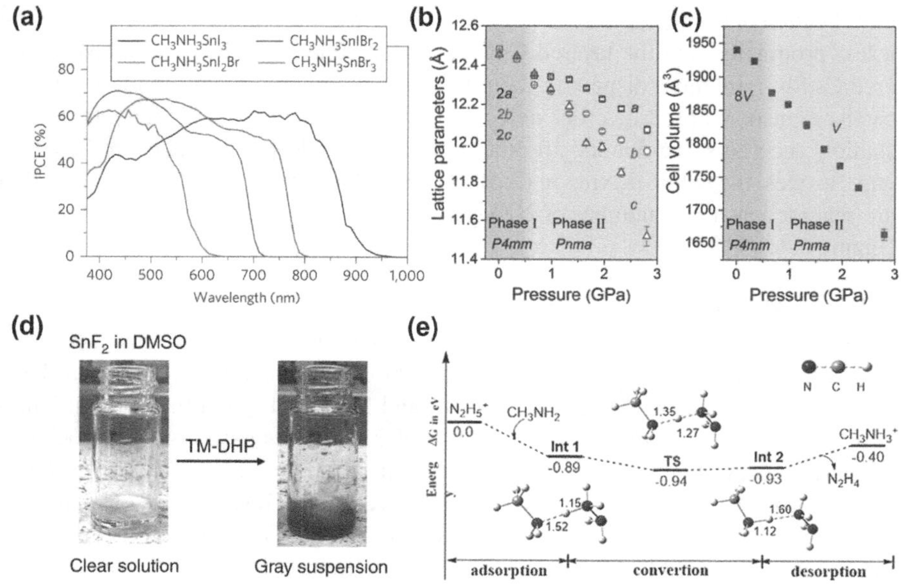

FIGURE 7.5 (a) IPCE characteristics based on $MASnI_{3-x}Br_x$ ($x = 0, 1, 2, 3$) perovskites. Reproduced with permission.[25] Copyright 2014, Springer Nature. Variations of crystal parameters of (b) lattice constant and (c) cell volume evolution with pressure. Reproduced with permission.[108] Copyright 2016, Wiley-VCH. (d) Patterns of the SnF_2 solution before (left) and after (right) the addition of TM-DHP. Reproduced with permission.[109] Copyright 2020, Springer Nature. (e) Free energy patterns for chemical conversion. Reproduced with permission.[113] Copyright 2019, Wiley-VCH.

the invasion of H_2O. Therefore, the final films contained not only $MASnI_3$ compounds but also SnO and $Sn(OH)_2$ in a Sn^{2+}-rich environment.[110] These excess Sn^{2+} analogs are beneficial to low hole-doping behavior in films. A similar strategy was also reported by Song et al., which was developed by insertion reactions/ion exchange between MAI gas and SnF_2.[111] These synergetic approaches realized pinhole-free, uniform, and continuous perovskite films. Another reliable strategy such as hot anti-solvent treatment to obtain high-density nucleation sites before annealing and high-quality films with average crystallite size after annealing was also discovered by Wakamiya et al.[112] As for the ion exchange strategy, Li et al. devoted cation exchange by hydrazinium iodide (HAI) in the precursor.[113] First, the $HASnI_3$ perovskite was deposited by HAI and SnI_2 compounds and then transformed into $MASnI_3$ in an MA atmosphere. As shown in Figure 7.5e, MA molecules diffused inside $HASnI_3$ perovskite films as Int 1 process and then transformed to $NH_2NH_2{\cdot\cdot}H{\cdot\cdot}CH_3NH_2$ states. In the end, the generated N_2H_4 gas was released and $MASnI_3$ films formed. The changes in free energy are favorable for the cation exchange process. As a result, the original $HASnI_3$ with 1D $[SnI_3]_n^{n-}$ chains were reconstructed and an ideal $MASnI_3$ was formed.

Hollow $MASnI_3$-based perovskites: moreover, hollow (en) $MASnI_3$ perovskites as absorbers and corresponding PSCs were also reported by Kanatzidis et al.[39,42] The (en) $MASnI_3$ is still a 3D structure with larger bandgap than 3D $MASnI_3$ (from ~1.25

eV to ~1.40 eV). These uncompact structures could significantly enhance carrier lifetime and reduce electron trap-state density, thus showing high device efficiency and stability.

7.4.2 Formamidinium Tin Halides

Reducing agents: compared to the poor stability of $MASnI_3$, $FASnI_3$ is a stable perovskite structure featuring the high formation energy of Sn vacancies. This enhanced stability in $FASnI_3$ is attributed to the weaker anti-bonding coupling between Sn 5*s* and I 5*p* orbitals than that of $MASnI_3$ analogs.[114] Meanwhile, $FASnI_3$ has enhanced thermal stability due to the strong hydrogen bonding between FA^+ and the inorganic matrix.[115] Employing FA^+ in Sn perovskites is deemed to be the most popular A-site cations in the recent research of Sn-halide devices. Han et al. controlled the surface energy by tailoring pentafluorophen-oxyethylammonium iodide (FOEI) additive, which achieved a high-oriented and smooth $FASnI_3$ film with fewer defects.[116] They also reported a templated growth with the assistance of n-propylammonium iodide (PAI) and reconstructed the intermediate phase to induce a high-oriented $FASnI_3$ film along (100) plane, which delivered a stable PCE of 11.22% in TPSCs.[117] Abate et al. confirmed that n-butylammonium acetate (BAAc) had strong coordination interaction by forming specific N-H···X hydrogen bonding and O···Sn chelating bonding.[118] During the annealing progress as monitored by the GIWAXS method (Figure 7.6a and b), the blurred scattering halos in the early 45 s reflected poor crystallinity for the control film, while the BAAc film was significantly enhanced in the early 45 s. The existence of hydrogen bonding and chelating bonding prolonged the transfer process to $[SnX_6]^{4-}$ octahedra, therefore delaying crystallization. Hao et al. incorporated formamidine acetate (FAAc) and achieved a high PCE of 12.43%.[119] The color change in Figure 7.6c is clear evidence for retard oxidation of Sn^{2+}.

To maintain the stability of Sn halide perovskites, antioxidant additives or reducing agents such as gallic acid (GA), pyrazine, hydrazinium chloride (N_2H_5Cl), phenylhydrazine hydrochloride (PHCl), hydroquinone sulfonic acid (KHQSA), etc. were employed.[120–124] The GA as a secondary additive can restrict $SnCl_2$ aggregation via the formation of $SnCl_2$-GA compounds that enveloped the perovskite grain surface.[120] As a result, GA could scavenge O_2 continuously and protected Sn halide perovskites against oxidative damage. The TPSCs with N_2H_5Cl showed increased shelf life up to 1,000 h.[125] The hydrazine dissociated into the solution and released electrons, thus suppressing the oxidation reaction of Sn^{2+}. The reducing hydroxybenzene group of KHQSA endowed the antioxidant activity to prevent Sn^{2+} oxidation, and the sulfonate group (SO_3^-) could interact with perovskites via coordination interaction and electrostatic attraction. Seok et al. elucidated the coordination chemistry of the SnF_2-pyrazine compound and the corresponding device displayed superior stability over 100 d without encapsulation.[122] By the O-H···I^- hydrogen bonding effect, the addition of poly(vinyl alcohol) (PVA) increased nucleation sites, retarded crystal growth, and suppressed the ion migration of I^-.[126]

Green solvent: besides the additives in the precursor, green anti-solvent acetic acid (HAc) was first used in Sn halide perovskites.[127] Compared to the common anti-solvent chlorobenzene (CB), HAc had stronger polarity, which can form a hydrogen

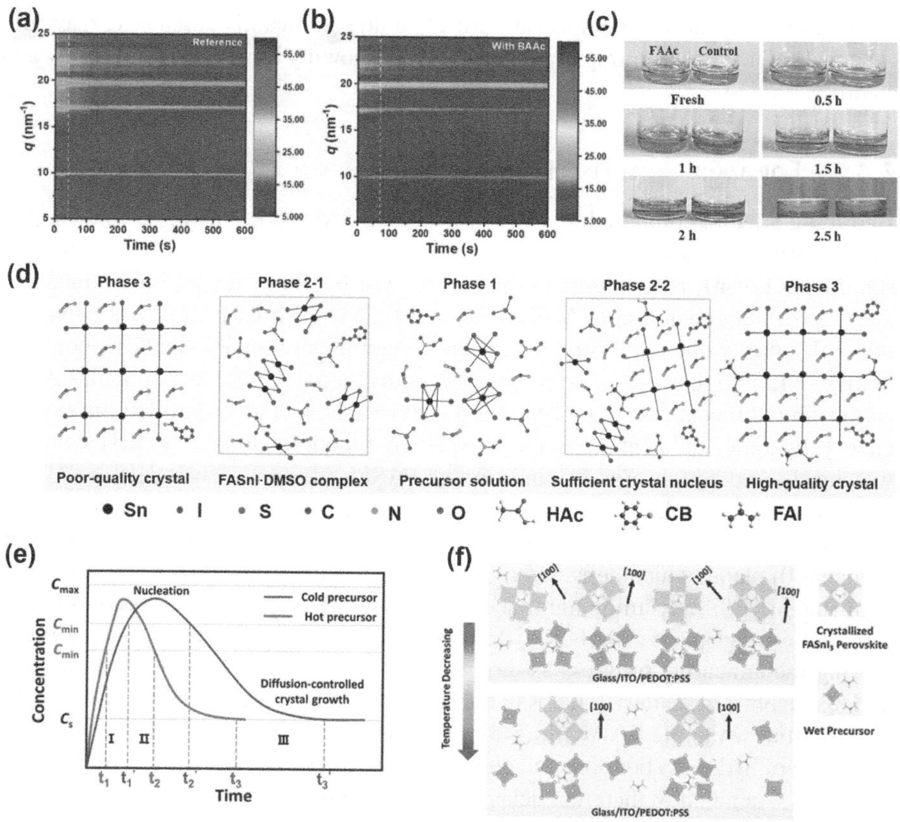

FIGURE 7.6 Time-resolved azimuthally integrated intensity patterns along the outside plane direction (azimuth at 90°) for (a) control and (b) BAAc films. Reproduced with permission.[118] Copyright 2021, Wiley-VCH. (c) The color changes of SnI_2 solution with or without FAAc. Reproduced with permission.[119] Copyright 2023, Wiley-VCH. (d) The composition of perovskite films with CB or HAc as an anti-solvent before and after each step. Reproduced with permission.[127] Copyright 2023, Royal Society of Chemistry. (e) Lamer diagram of the overall nucleation and crystal growth for thin films with different precursor solution temperatures. (f) Schematic illustration of nuclei orientation and crystal growth. Reproduced with permission.[128] Copyright 2021, Wiley-VCH.

bond with O=S bond in DMSO, thus decreasing the solvation degree of SnI_2-DMSO and promoting rapid nucleation (phase 2–2) as demonstrated in Figure 7.6d. The CB treatment just formed relatively insufficient crystal nuclei (phase 2–1). During the annealing process, the volatilization rate of HAc has slowed due to the increased potential barrier of HAc evaporation by hydrogen bonding with organic ammonium salt (e.g. PEAI), which is conducive to forming a high-quality perovskite film (phase 3) with the defect passivation effect. Han et al. employed a cold precursor solution in $FASnI_3$ and also favored ordered (h00) crystalline orientation (Figure 7.6e and f).[128] By Lamer analysis in Figure 7.6e, the low temperature of the precursor can restrict the rate of solute diffusion and retard crystal growth during Stage III. The subsequent Ostwald ripening via solvent mediation also plays a significant role in obtaining large grains. In this case, a cold solution could slow the evaporation rate of solvents (stage III), resulting in more residual solvents in films than that of hot precursor films, thus prolonging the Ostwald ripening process. Moreover, this strategy inhibited renucleation because of the lower nucleation concentration than the critical concentration of nucleation.

Surface modification: for $FASnI_3$, as we know, the VBM is shallow (around -4.7 eV), while the work function (WF) of PEDOT:PSS is around -5.1 eV, which showed a large energy-level difference in the interface.[129] Han et al. developed polyethylene glycol (PEG) in PEDOT:PSS by minimizing the interfacial energy level mismatch, tuning the WF of PEDOT:PSS to -4.79 eV.[130] To simultaneously fulfill great stability and reliability, Chen et al. deposited a GeO_2 native layer between the NiO_x and $FASn_{0.9}Ge_{0.1}I_3$ interface to enhance the interfacial mechanical bonding.[131] Via bias voltage treatment in Kelvin probe force microscopy (KPFM) as displayed in Figure 7.7a and b, the deduced electric field between NiO_x and pristine perovskite film showed a small peak, which was attributed to a leakage current in this interface. However, the target PSCs with dielectric GeO_2 native layer had no apparent leakage current in the same interface, which is expected to reduce interfacial recombination sites and high V_{oc} as well as FF in devices. Different from the metallic oxide, Yan

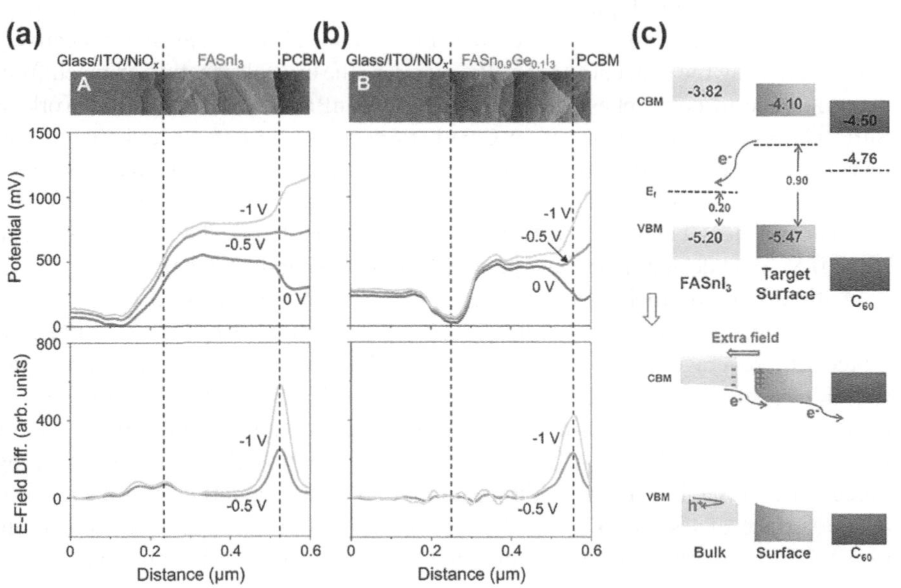

FIGURE 7.7 Special characterization of Sn halide devices based on (a) $FASnI_3$ and (b) $FASn_{0.9}Ge_{0.1}I_3$ for AFM images, potential profiles, and calculated electric-field-difference profiles, respectively. Reproduced with permission.[131] Copyright 2022, American Chemical Society. (c) Energy level patterns based on UPS spectra (top) and back-surface field formation at the perovskite surface with surface treatment (bottom). Reproduced with permission.[134] Copyright 2022, American Chemical Society.

et al. used 2D transition-metal dichalcogenides (WSe_2) as a growth template and achieved a epitaxial growth of thin films along (100) plane by van der Waal interaction.[132] The optimization of interface energy could promote cascade hole extraction and charge transfer in relevant devices. Moreover, the surface modification of films with improved hydrophobicity and oxidation resistance is also considered important in Sn halide perovskites. The surface passivation strategy was also investigated by using 3-(trifluoromethyl) phenethylamine hydroiodide (CF_3PEAI).[133] Hao et al. certified that the formed 2D layer $(CF_3PEA)_2SnI_4$ further reduced the interface energy loss. Another surface-modulation strategy was realized by 6-maleimidohexanehydrazide trifluoroacetate salt (MHATFA) reported by Yin et al.[134] Encouragingly, the MHATFA treatment favored a transform of the perovskite surface from p-type to n-type (Figure 7.7c), which delivered an extra built-in electrical field, thus suppressing the charge carrier recombination in the perovskite/ETL interface.

7.4.3 Cesium Tin Halides

Sn^{2+} stoichiometry: though the performance of $CsSnI_3$ PSCs is not as efficient as those of FA or MA base due to the high p-type metallic behavior; they have no volatile organic cations and are more stable in different environmental conditions. In 2012, Chen et al. explored the ratio of Br/I without the addition of SnF_2, which reflected that Br^- could reduce the Sn vacancies known as the dominant defects in $CsSnI_3$.[135] As drawn in Figure 7.8a, the bandgap transformed from 1.27 eV for pure $CsSnI_3$ to 1.37 eV ($CsSnI_2Br$), 1.65 eV ($CsSnIBr_2$), and 1.75 eV ($CsSnBr_3$) respectively when increasing the concentration of Br^-, whereas the crystal structure changed from orthorhombic to cubic phase from $CsSnI_3$ to $CsSnBr_3$. Later, the similar work of SnF_2 addition was also explored in $CsSnI_3$ to decrease Sn defects and free carrier densities.[136,137] This work plays an important role in the development and subsequent design of other additives. Hatton et al. proved the orthorhombic $CsSnI_3$ films with low defect density based on CuI/$CsSnI_3$/fullerene structure.[138] The V_{oc} of $CsSnI_3$ can reach 0.55 V by designing the energetics at the perovskite/fullerene interface.

Antioxidant channels: Kanatzidis et al. compared $MASnI_3$, $CsSnI_3$, and $CsSnBr_3$ under a reducing vapor atmosphere.[139] The high volatility of hydrazine vapor promoted full coverage on top of the ETL layer. The Sn^{4+}/Sn^{2+} ratio in $MASnI_3$ and $CsSnI_3$ are markedly reduced compared to $CsSnBr_3$, therefore presenting much enhanced efficiency after being exposed to hydrazine vapor. Taking these factors into account, Marshall et al. achieved $CsSnI_3$ devices with high FF via co-depositing precursor and $SnCl_2$.[20] The Cl^- did not incorporate into the lattice and just existed at the surface of $CsSnI_3$. The $SnCl_2$ layer at the surface features a stable hydrate ($SnCl_2 \cdot 2H_2O$) and the oxidized product SnO_2, which could sacrificially oxidize and reduce the oxidation of $CsSnI_3$. By passivator-assisted sequential vapor deposition strategy, Yin et al. used thiosemicarbazide to interact with perovskite by S=C-N···Sn bonding via Lewis acid-base interaction.[140] When performed by photo-thermal induced resonance (PTIR) combining AFM techniques, the surface topography and corresponding PTIR image recording C=S stretching intensity under 120°C annealing was shown in Figure 7.8b and c. The passivator aggregated at the grain boundary (GB) and did not volatilize from perovskite film, which is crucial to crystal regulation

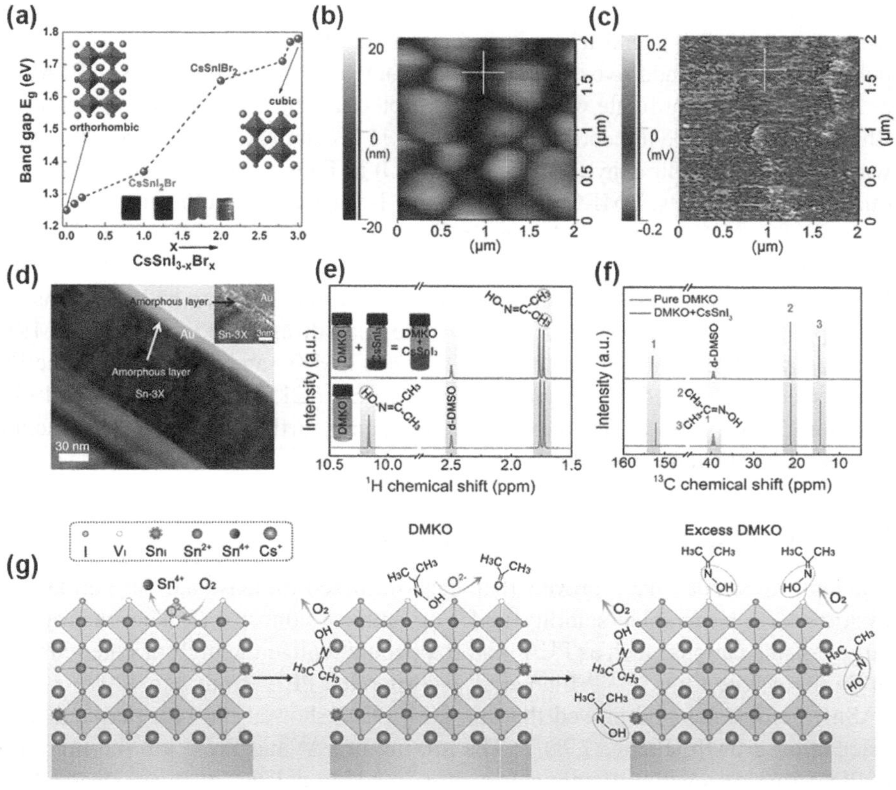

FIGURE 7.8 (a) Bandgap variation with respect to Br– concentration. Reproduced with permission.[135] Copyright 2015, American Chemical Society. (b) AFM images and (c) corresponding photo-thermal induced resonance patterns of $CsSnI_3$ films with TSC. Reproduced with permission.[140] Copyright 2021, Wiley-VCH. (d) TEM images of Sn-3X film. Reproduced with permission.[144] Copyright 2020, Springer Nature. (e) ^{1}H NMR and (f) ^{13}C NMR spectra of pure DMKO and DMKO-$CsSnI_3$ samples in deuterated DMSO-d_6 solution. (g) Schematic illustration of antioxidation and defect coordination at grain surfaces of $CsSnI_3$ films enabled by DMKO. Reproduced with permission.[143] Copyright 2022, Wiley-VCH.

and passivation at GB. As we can see, the vapor deposition strategy highlights a novel pathway to fulfilling high-efficiency, lead-free, Cs-based PSCs. Han et al. formed a stable amorphous-polycrystalline structure with outstanding resistance to H_2O, O_2, and ion diffusion (Figure 7.8d).[140] The Sn-3X film was composed of $CsFASnI_3$ with 20 mol% $SnCl_2$ and 10 mol% SnF_2, which formed a layered structure as shown in the TEM image. Similar to previous reports, the Cl^- and F^- in the surface were higher than that in the solution, proving that the amorphous layer contained Sn triple halide (I^-, Br^-, Cl^-), not only the $SnCl_2$ layer.

Reducing agents: Yan et al. obtained a certified PCE of 10.9% for durable $CsSnI_3$ PSCs adopted by a bifunctional carbazide (CBZ) additive.[141] The incorporation of CBZ can slow the crystallization process during the phase transition towards black orthorhombic (B-γ) $CsSnI_3$ film. Similarly, Ye et al. stabilized (B-γ) $CsSnI_3$ perovskites via the special trigeminal coordination bonding with Sn^{2+} cations with phthalimide (PTM) additive.[142] The long pairs of two C=O and N-H units in PTM act as active sites to coordinate with Sn^{2+}, thus protecting the Sn^{2+} oxidation. Moreover, Zhang et al. delivered a record PCE as high as 11.2% in the $CsSnI_3$-xBrx component with the deoxidant dimethyl ketoxime (DMKO) in Figure 7.8e–g.[143] Different from other used deoxidants, DMKO could passivate I– vacancies and Sn_I anti-site defects by neutralizing O_2 and then reducing Sn^{4+} (Figure 7.8g). To verify the hypothesized mechanism, the ^1H NMR spectra is used (Figure 7.8e and f) for pure DMKO and $CsSnI_3$-DMKO samples. The characteristic peak of =NOH group had a response at 10.16 ppm while the corresponding resonance peak disappeared in $CsSnI_3$-DMKO samples, which indicated the interaction between DMKO and Sn^{2+}. Besides, the ^{13}C NMR spectra displayed three signals of 152.2 ppm for C1, 21.4 ppm for C2, and 14.5 ppm for C3, respectively. The upshift of these peaks further verified the interaction between DMKO and perovskites.

7.4.4 Low/Mixed-Dimensional Tin Halide Perovskites

Previous researches have proved that the low/mixed-dimensional Sn halide perovskites have remarkable stability under an ambient atmosphere. The inclusion of large organic cations, such as PEA^+ and BA^+, could realize the high-oriented growth of Sn perovskite films.[66,67,145] Liao et al. incorporated PEA^+ from 0 to 100% in bulk $FASnI_3$ perovskites and proved that the 20% PEA^+ showed the best perpendicular orientation growth (Figure 7.9a).[146] The mixing of FA^+ and PEA^+ are the mixtures of 2D perovskites with different values of *n*. In addition, Chen et al. achieved a low-dimensional interlayer by dropping PEABr solution at the $FASnI_3$ films resulting in a champion PCE of 7.05% with negligible *J-V* hysteresis based on p-i-n scaffold.[147] Later, Wu et al. employed 2-fluoro-phenethylammonium iodide (2-F-PEAI) to realize a robust structure benefiting the hydrogen bonding effects, resulting in a high PCE of over 10% with remarkable stability under light exposure over 1,600 h.[148] In 2021, He et al. delivered a 2D/3D heterogeneous film where 2D phases embracing 3D grains and located at GB and the surface by adding 4-fluoro-phenethylammonium bromide (FPEABr), which achieved a new record PCE of 14.81% based on p-i-n devices with ICBA ETL.[149] Recently, Hao et al. incorporated PEASCN containing pseudohalide SCN– anions and Sn^{2+} cations into FA-based perovskites, which promoted a preferred alignment of perovskites with high *n* values and high-quality crystal growth.[67]

Though the PEA^+ can stabilize the Sn perovskites, it is insulating and leads to poor charge transport. From this, Wu et al. introduced 3-phenyl-2-propen-1-amine (PPA) into perovskite precursor to form $PPA_xFA_{1-x}SnI_3$.[150] The more polarizable and conjugated backbones than PEA^+ lead to enhanced *J-V* performance especially J_{sc} due to the promoted photoexcited charge transport of PSCs. To probe the orientation with PPA, the GIXRD measurement was carried out in Figure 7.9b and c. The pure $FASnI_3$ showed a random orientation with the diffraction ring at q around 1.0 Å^{-1}, 2.2 Å^{-1}, 1.7 Å^{-1}, 1.9 Å^{-1}, and 2.2 Å^{-1}, which corresponded to (100), (120), (200), and (122) planes, respectively. However, the film with PPA demonstrated the high intensity of diffraction rings in (100) and (200) planes and decreased intensity in (122)

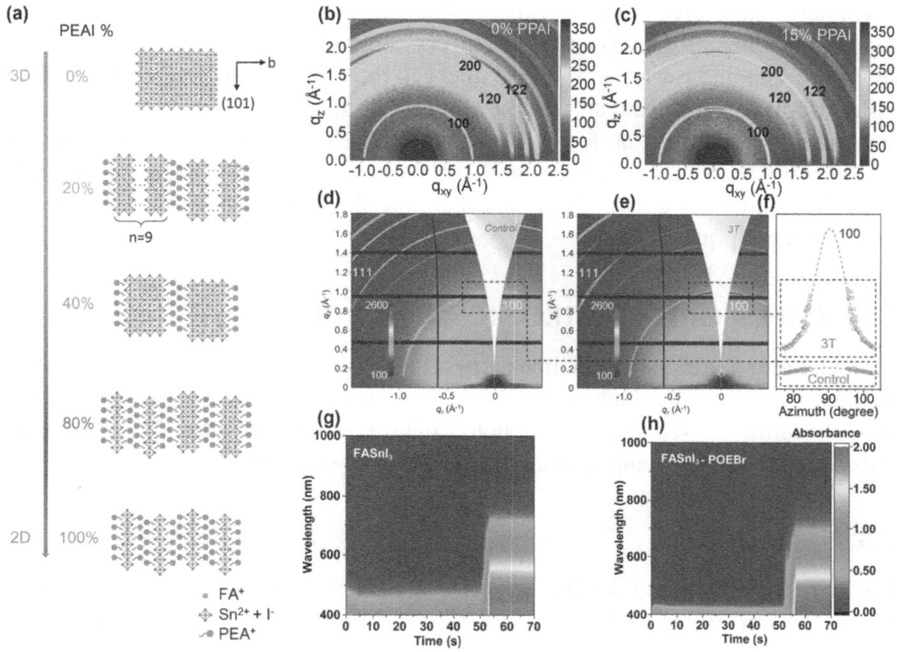

FIGURE 7.9 (a) Structural patterns of $FASnI_3$, PEA_2SnI_4, and mixed FA-PEA Sn halide perovskites. Reproduced with permission.[146] Copyright 2017, American Chemical Society. 2D-GIXRD profiles of (b) 0% PPAI and (c) 15% PPAI films. Reproduced with permission.[150] Copyright 2019, Elsevier. GIXRD profiles of (d) control and (e) 3T-treated films. (f) Gaussian peak fitting of intensity curves along the 100 rings in panels (d) and (e). Reproduced with permission.[152] Copyright 2022, American Chemical Society. In-situ UV-vis absorption spectra during the spin-coating process for (g) $FASnI_3$ and (h) $FASnI_3$-FOEBr films. Reproduced with permission.[153] Copyright 2023, American Chemical Society.

planes with no Bragg spots of 2D analogs, implying crystalline orientation along (h00) planes.

To grow the Quasi-2D layer, Diau et al. deposited organic space layers by hexafluoro-2-propanol (HFP) solution instead of isopropanol (IPA) without destroying the $GA_{0.2}FA_{0.8}SnI_{3}$_1% $EDAI_2$ perovskites.[151] Meanwhile, the anilinium (AN) has a self-healing ability via surface passivation combining thermal stress and light-soaking conditions. Benefiting from the properties of the H-bond donor and Lewis base, Mi et al. employed trimethylthiourea (3T) and extended the PL lifetime to 123 ns from the original value of 18 ns for $FASnI_3$ film.[152] As shown in Figure 7.9d–f, the film with 3T resulted in converged Bragg rings especially in (100) and (111) planes. The azimuthal Gaussian width of 8° in the (100) peak appeared especially in 3T-regulated film (Figure 6f). As a result, the efficiency with 3T PSCs obtained a certified PCE of 14% based on ICBA ETL. Recently, Meng et al. revealed the crystal growth mechanism by a new crystal growth mechanism called oriented attachment (OA) with the assistance of 2-phenoxyethylamine bromide (POEBr).[153] The asymmetric structure of POEBr with π–π stacking of the benzene group was expected to bonding with perovskites and facilitate crystal growth. The in-situ UV-vis spectra in Figure 7.9g and h were introduced to check the crystal growth rate with POEBr. After dropping antisolvent at 50 s, the absorption intensity increased and broadened immediately, implying the formation of perovskite crystallites. Different from the $FASnI_3$ film, the POEBr-$FASnI_3$ film showed a slowly absorption intensity within 4 s (50 s-54 s) under OA growth dynamics, therefore slowing the crystal growth rate and enabling a high-quality film.

7.5 OTHER OPTOELECTRONIC APPLICATIONS

Though the application of Sn halide perovskite in PSCs was discussed in the previous section, they are still widely applied in LEDs, thermoelectric devices, photodetectors, thin-film transistors (TFTs), photocatalytic, and memristors.[154,155] Herein, we introduced their clever application in other fields beyond PSCs.

7.5.1 LEDs

Compared to Pb-based LEDs, lead-free LEDs display better environmentally friendly characteristics and broader emission regions from the violet to near-infrared wavelength.[15] The Sn-based LEDs can approach to a near-infrared emission to 950 nm owing to the tunable bandgap.[156] Lin et al. prepared TEA_2SnI_4 with the addition of pentanoic acid, which delivered a record PLQY up to 21% with superior photostability.[157] The absorption and photoluminescence results revealed that the degree of redshift had a great influence on different organic cations. Lai et al. achieved a near-infrared LED with a radiance of 3.4 W sr^{-1} by varying the ratio of Br/I of $MASn(I_xBr_{1-x})_3$.[158] Later, Haque et al. first reported visible electroluminescence with the 2D $(PEA)_2SnI_xBr_{4-x}$ devices.[159] More importantly, the 2D layered $(PEA)_2SnI_4$ counterpart showed higher PL quantum efficiency and slower radiative recombination dynamics than the 3D $MASnI_3$.[159] Inspired by the high emissive of 2D layered Sn halide perovskites, Wang et al. synthesized the $(OCTAm)_2SnX_4$ (OCTAm is octylammonium) in an acidic aqueous solution, which appeared ultra-high PL emission with a long lifetime (3.3 μs), large Stokes shifts (250 nm), zero-overlap between the emission and absorption measurements, and high absolute PLQY near unity in the solid states.[160] Based on the CIE color in Figure 7.10a, the light intensity of fabricated white LED steadily increased from 3 V to 12 V without energy transfer among these three phosphors.

By supramolecular dynamic coordination strategy, Ning et al. united crown and sodium trifluoroacetate to coordinate with A- and B-site ions, respectively.[161] Though the coordination ability retarded nucleation and growth of low-dimensional structure at room temperature, the transformation of supramolecular structure enabled the secondary step growth by releasing both A- and B-site ions. The formed insular nanocrystal was favorable of effective carrier injection and high PLQY. The devices with a structure of ITO/NiO_x/Poly-TPD/Perovskite/Perovskites/TPBi/LiF/Al demonstrated a narrow linewidth (~20 nm) electroluminescence (Figure 7.10b and c). Zhang et al. also certified that the SnF_2-derived Cs_4SnBr_6 had a high PLQY from 2.8% to

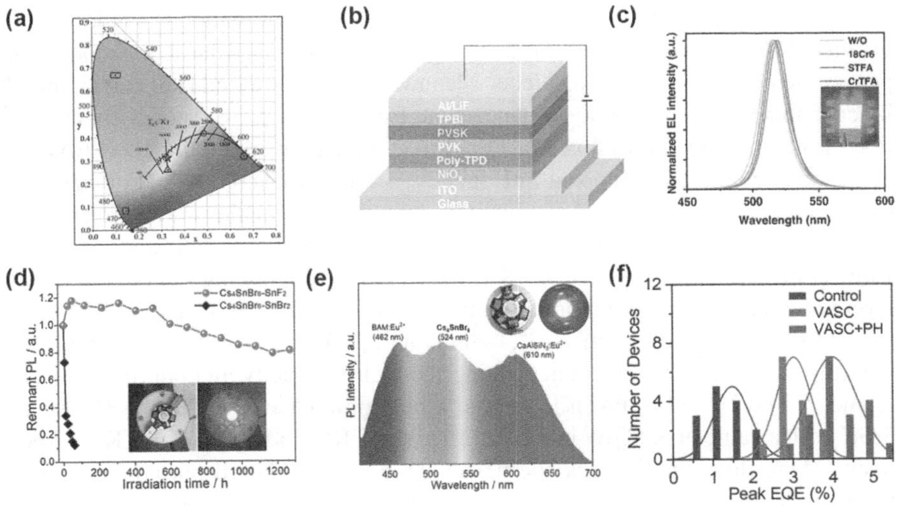

FIGURE 7.10 (a) Chromaticity coordinates of different ratios of the phosphor mixture plotted on the CIE1931 chromaticity chart. Reproduced with permission.[160] Copyright 2019, The Royal Society of Chemistry. (b) Device structure and (c) EL spectra of perovskite LED. Reproduced with permission.[161] Copyright 2023, Springer Nature. (d) Photostability of Cs_4SnBr_6-$SnBr_2$ and Cs_4SnBr_6-SnF_2 under illumination. (e) The emission spectrum of the LED device. Reproduced with permission.[162] Copyright 2022, Wiley VCH. (f) EQE histogram of devices with different incorporation. Reproduced with permission.[163] Copyright 2023, Wiley-VCH.

62.8% and superior stability against H_2O, O_2, and light radiation of corresponding devices.[162] The thermal and moisture resistance has also been proven by monitoring the PL intensity (Figure 7.10d); the Cs_4SnBr_6-SnF_2 perovskites retained over 70% of the pristine value. Meanwhile, the LED chip (Figure 7.10e) with UV-curing resins still maintained 80% of pristine PL intensity over 1,200 h. To further improve the film morphology, Zhang et al. developed a vapor-assisted spin-coating strategy to fabricate the high-efficiency LEDs, which resulted in 4-fold enhancement of PL quantum efficiency (~30%) compared to the control method.[163] Notably, a resulted EQE of 5.3% was obtained, which is the highest value among the reported works in Sn LEDs.

7.5.2 Thermoelectric Devices (TEs)

The thermoelectric devices (TEs) are contributed to energy converters from heat energy to that of electricity. The unique organic-inorganic hybrid crystal structure in lead-free TE materials endows them with electron-inorganic phonon-glass behavior. The inorganic frameworks of lead-free perovskites conduct charge carriers, and the organic molecules scatter phonons via a cluster-rattling mechanism.[164] Maximizing dimensionless figures of merit (ZT) is the primary goal for practical TEs. Qian et al. synthesized $CsGe_xSn_{1-x}I_3$ (x = 0.1, 0.2, and 0.3) alloy bulk crystals by solid-state sintering method.[165] The optimized $CsGe_{0.2}Sn_{0.8}I_3$ delivered a ZT value of 0.123 at 473

K (Figure 7.11a), which is strikingly better than the $MAPbI_3$ halide perovskites.[166,167] Further, the Hall carrier concentration in Figure 7.11b increased due to the more stable perovskite phase with reduced defects. Meanwhile, the decreased Hall carrier mobility contributed to conductivity enhancement. Encouraged by the successful application of 3D Sn halide perovskites in TEs, Liu et al. demonstrated the 2D $(PEA)_2SnI_4$ perovskites with 5 mol% dopant SnI_4 in TEs.[168] The 2D layered Sn halide materials possessed even lower thermal conductivity than their 3D analogs. The different TE parameters in Figure 7.11c with the varying ratio of dopant SnI_4 indicated the highest value of power factor (3.92 μW m^{-1} K^{-2}) with 5 mol% SnI_4 doping, suggesting the 2D layered Sn halide perovskites are attractive for efficient TEs. Hsu et al. also reported the 2D $(4Tm)_2FASn_2I_7$ (4Tm is 3‴,4′-dimethyl[2,2′:5′,2″:5″,2‴-quaterthiophen]-5-yl) ethan-1-ammonium) in TEs beneficiating the hydrophobicity and bulkiness of 4Tm ligand.[169] The 2D $(4Tm)_2FASn_2I_7$ showed a high power factor at 293 K, 313 K, and 333K with a value of 3.03 μW m^{-1} K^{-2}, 3.15 μW m^{-1} K^{-2}, and 7.07 μW m^{-1} K^{-2}, respectively. Meanwhile, the Seebeck coefficient (*S*) in Figure 7.11d displayed a maximum value on the carrier concentration (σ) of 3×10^{18} cm^{-3}. Yang et al. optimized the phase control of RP-type $(PEA)_2(MA)_{n-1}Sn_nI_{3n+1}$ perovskites via the degree of MACl doping.[170] To reduce the phase separation of $(PEA)_2(MA)_{n-1}Sn_nI_{3n+1}$, they established the optimal combination between the ratio of PEAI/SnI_2 (*x*) and DMSO solution: 10% DMSO in $x = 2$, 30% DMSO in $x = 1$, 2/3, and 1/2, respectively. After that, the TE

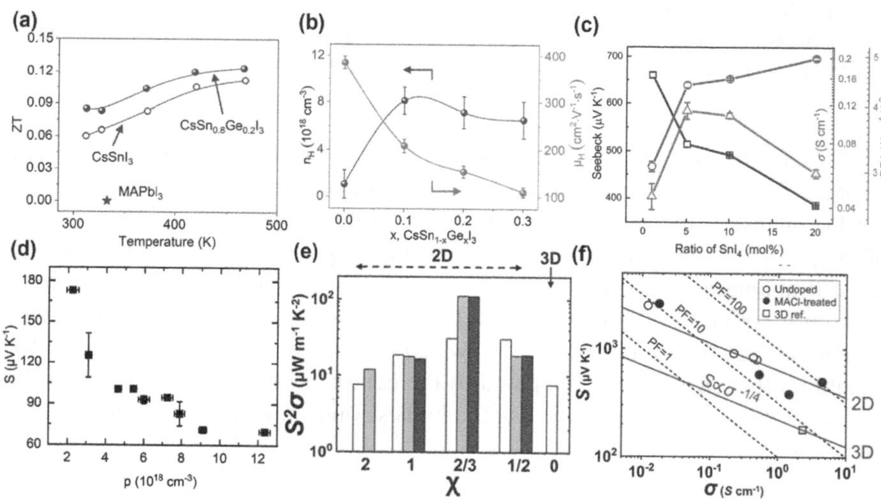

FIGURE 7.11 (a) Variation of thermal conductivities of $CsSnI_3$ and $CsSn_{0.8}Ge_{0.2}I_3$ as a function of temperature. (b) Variation of the Hall carrier concentration and Hall carrier mobility as a function of *x* of $CsSn_{1-x}Ge_xI_3$ alloys at room temperature. Reproduced with permission.[165] Copyright 2020, American Chemical Society. (c) Seebeck coefficients, electrical conductivities, and power factors of PEA_2SnI_4 films with different doping ratios. Reproduced with permission.[168] Copyright 2022, Elsevier. (d) Room temperature electrical conductivity of $(4Tm)_2FASn_2I_7$ thin film. Reproduced with permission.[169] Copyright 2021, American Chemical Society. (e) Power factor with respect to different ratios of PEAI/SnI_2 (*x*). (f) The fitted line of different films according to the slope of the trade-off relation. Reproduced with permission.[170] Copyright 2021, Wiley-VCH.

properties with the effect of MACl doping and x were demonstrated in Figure 7.11e and f. $PEA_2MA_2Sn_3I_{10}$ with 20 mg mL^{-1} MACl ($x = 2/3$) showed a high $S^2\sigma$ value (111 μW m^{-1} K^{-2}). The S–σ relation in Figure 7.11f further proved the S–σ point for 2D Sn halide films was higher than the 3D $MASnI_3$.

7.5.3 Photodetectors

Photodetectors based on Sn halide materials are expected to combine their merits such as tunable light absorption, low toxicity, long charge-carrier lifetime, and large charge-carrier mobility. Waleed et al. first fabricated the 3D $MASnI_3$ nanowire arrays and addressed their integration and stability issues via the vapor phase chemical reaction approach (Figure 7.12a).[171] Specifically, the $MASnI_3$ nanowire arrays grow

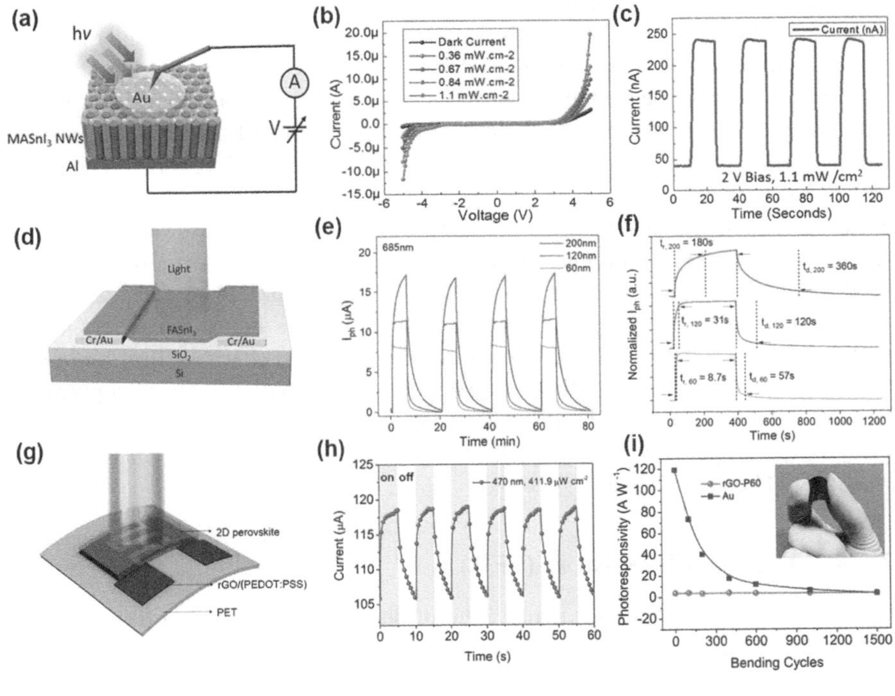

FIGURE 7.12 (a) Device schematic for Sn perovskite nanowires. (b) *I-V* curves for Sn perovskite nanowires with different light intensities under 5 V bias voltage. (c) *I-t* curves of Sn perovskite photodetector device with the light intensity of 1.1 mW/cm² under a bias voltage of 2 V. Reproduced with permission.[171] Copyright 2017, American Chemical Society. (d) Structure device for $FASnI_3$ photodetectors. (e) Transient response under four on–off illumination cycles for the photodetectors with various perovskite thicknesses. (f) Normalized photocurrent as a function of time for perovskite photodetectors with different thicknesses under one on–off illumination cycle. Reproduced with permission.[172] Copyright 2019, Wiley-VCH. (g) The structure of a PEA_2SnI_4 flexible photoconductor. (h) The time dependence of current for the flexible device at 5 V bias. (i) Plots of photoresponsivity against the bending cycle number for the flexible device with the Au electrode or the rGO-P60 electrode. Reproduced with permission.[175] Copyright 2018, The Royal Society of Chemistry.

in high tensity (4×10^8/cm^2) and are tightly embedded in a robust porous alumina template. The current-voltage (*I-V*) patterns in Figure 7.12b observed the increased current when increasing the light intensity because of asymmetric metal contact. The response speed of the $MASnI_3$ device (Figure 7.12c) suggested that the reduced length of nanowires could reduce corresponding resistance, thus consequently reducing carrier transit time. Meanwhile, the photodetectors with $MASnI_3$ nanowires showed 500 times slower light current decays than the thin-film one.

Besides the $MASnI_3$ photodetector, Yan et al. reported a low-voltage $FASnI_3$ photodetector with a high responsivity (10^5 A W^{-1}) benefiting from the p-doping behavior and high-hole mobility (Figure 7.12d–f).[172] By monitoring the responsivity of photodetector with different perovskite thicknesses from 60 to 200 nm, it increased with the increase of thin film thickness even if under different light intensities (Figure 7.12e). Therefore, the device with a thin film of 60 nm exhibited a rising time of around 8.7 s, which was much lower than that with the thickest film (200 nm). Moreover, the $FASnI_3$ perovskite materials were also applied in the flexible substrate and the related device demonstrated robust bending stability.

Moreover, low-dimensional Sn perovskites photodetectors also have proven good detectivity and response time.[172–175] Qian et al. used PEA_2SnI_4 as a photodetector on PET substrate with high stability and reproducibility (Figure 7.12g–i).[175] The related device had a rise time of 0.63 s and a long relaxation time of 3.6 s (Figure 7.12h). Meanwhile, the devices with rGO/(PEDOT:PSS) electrodes showed more better mechanical flexibility than the Au devices (Figure 7.12i). Furthermore, Zhou et al. used $Cs_2SnCl_{6-x}Br_x$ crystals as narrowband single-crystal photodetectors, which displayed a high detectivity (2.71×10^{10} Jones) and ion diffusion barriers.[174] This narrowband photodetection contributed to the suppressed charge collection in short wavelength excitation due to the strong charge recombination-induced suppression. Recently, He et al. fulfilled a self-powdered low-dimensional photodetector based on $FA_{0.8}PEA_{0.2}SnI_3$ perovskites.[176] The flexible device also possessed ultrafast responses with 10,000 bending cycles.

7.5.4 Thin-Film Transistors (TFTs)

Generally, the phototransistor is one kind of photodetector, wherein thin-film transistors (TFTs) appear suppressed noise and enlarged photocurrent signals as well as superior photoelectric detection performance. Liu et al. optimized TFTs based on 3D $CsSnI_3$ semiconducting channels under CsI-rich conditions. Therefore, the p-channel TFTs exhibited a field-effect mobility (μ_{FE}) over 50 cm^2 V-1 s^{-1} with high operational stability (Figure 7.13a).[177] Moreover, the efficient and hysteresis-free TFTs with 3D $MASnI_3$ (Figure 7.13b) via halogen engineering enabled a μ_{FE} of 20 cm^2 V-1 s^{-1} and a current ratio of I_{on}/I_{off} over 10^7.[178] Benefiting the co-substituting of Br^- and Cl^- in precursor, the perovskite films enabled reduced vacancy defects and suppressed ion migration.

Compared to 3D perovskite counterparts, their 2D analogs expect to suppress severe ion migration and appear intrinsic hole-transport properties, which cater to the realization of complementary circuits with n-type metal oxide semiconductors. Based on the mechanism, Zhu et al. developed a self-passivation strategy for GBs via

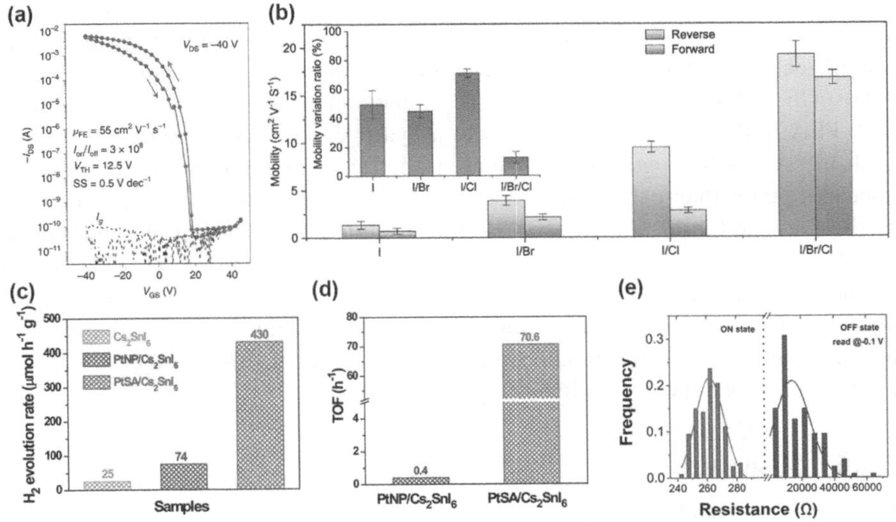

FIGURE 7.13 (a) Transfer characteristics of the optimized TFTs ($CsI/SnI_2 = 1.25$). Reproduced with permission.[177] Copyright 2022, Springer Nature. (b) Histogram of the extracted mobilities from the transfer characteristics under different scan directions. Reproduced with permission.[178] Copyright 2022, Springer Nature. (c) The rate of photocatalytic H_2 evolution over PtSA/Cs_2SnI_6, PtNP/Cs_2SnI_6, and Cs_2SnI_6 catalysts. (d) TOF of PtSA/Cs_2SnI_6 and PtNP/Cs_2SnI_6 catalysts. Reproduced with permission.[184] Copyright 2021, Springer Nature. (e) The histogram and Gaussian fitting of on/off resistance distribution. Reproduced with permission.[187] Copyright 2019, Wiley-VCH.

excess PEAI addition in PEA_2SnI_4 perovskites, which contributed to carrier transport medium with decreased resistance.[179] They also focused on grain crystallization engineering by Lewis base adducts to boost the performance of TFTs. Further, Kim et al. achieved a 2D-3D core-shell structure by quasi-2D $PEA_2FASn_2I_7$ in TFT application, which obtained a μ_{FE} of 25 cm^2 V-1 s^{-1} and I_{on}/I_{off} of 10^6.[180] The corresponding inverter showed a high voltage gain with ideal output characteristics. Further, Gao et al. achieved a 10-fold enhanced charge-carrier mobility in TFTs by $(4Tm)_2SnI_4$ featuring π-conjugated oligothiophene ligand 4Tm compared to the PEA_2SnI_4 one.[181] The $(4Tm)_2SnI_4$ possesses more negative formation energy and better stability than PEA_2SnI_4. More recently, Liu et al. explored the feasibility of vacancy-ordered double perovskites Cs_2SnI_6 in TFTs.[182] With 3 mol% Mn^{2+} doping in Cs_2SnI_6, the Mn^{2+} could weaken n-type conductivity of Cs_2SnI_6 due to the p-type dopant of Mn^{2+}, therefore realizing the first all-perovskite complementary inverter.

7.5.5 Photocatalysis

In the field of photocatalysis, the substitution of Pb divides into two main groups: Sn or Ge for equivalents and Bi or Sb for heterovalent elements.[183] In Sn-based perovskites, only Cs_2SnI_6, PEA_2SnBr_4, and $DMASnBr_3$ were reported as photocatalysis.

Zhou et al. found that Cs_2SnI_6 dispersing in Pt-I_3 species (PtSA) had superior tolerance towards HI aqueous solution (Figure 7.13c and d).[184] Remarkably, the PtSA/Cs_2SnI_6 catalyst presented accelerated production of H_2 with a turnover frequency (TOF) OF 70.6 H^{-1} per Pt (Figure 7.13d). Romani et al. used a co-catalytic system with PEA_2SnBr_4/g-C_3N_4 composites with a hydrogen evolution rate of 1,600 μmol g^{-1} h^{-1}.[185] As for hydrogen photogeneration, the PEA_2SnBr_4/g-C_3N_4 composites also appeared better than carbon nitride. Particularly, the composites still decomposed methylene blue for 40 m even after 90 m of irradiation. Romani et al. reported the superior water stability of $DMASnBr_3$ (DMA is dimethylammonium), which could promote photocatalysis in aqueous medium, especially the $DMASnBr_3$/g-C_3N_4 composites.[186] The hydrogen evolution rate reached over 1,700 μmol g^{-1} h^{-1} with this composites. The further verification in water containing glucose also resulted a 100-fold improvement than the g-C_3N_4 in hydrogen generation.

7.5.6 Memristors

Memristors are rarely reported in Sn-based materials. In 2019, Qian et al. reported the $MASnBr_3$ Memristors with the structure of ITO/Perovskite/Au and showed a well-separated on/off state with a resistance ratio ranging in 10^2–10^3 (Figure 7.13e).[187] Moreover, it can distinguish four resistive states with a retention time of around 10^4 s for repeat written/rea/erased. Another work in memristor was successfully fabricated by Liu et al. featuring $MASnI_3$ with PEACl as the key perovskite material.[188]

7.6 CONCLUSIONS AND PERSPECTIVE

Addressing the toxicity of Pb elements, we have given a systematic discussion of the structural, electronic, and optical properties of lead-free perovskites in this chapter. Especially, the unique crystal structure of Sn-based perovskites guarantees more attention and research than the others. Numerous applications in solar cells, LEDs, thermoelectric devices (TEs), photodetectors, thin-film transistors (TFTs), photocatalysis, and memristors were presented. Though Sn halide perovskite has been investigated in these areas, its performance is still far behind the Pb analogs. Many breakthroughs should be further explored in some respects.

First, eliminating the fast Sn^{2+} oxidation and heavy p-doping in PSCs is highly important. The majority of efforts were focused on decreasing the p-doped hole densities, while the deep-level defects in Sn halide perovskites still deserve great attention. A deep understanding of defect tolerance and defect physics is the key to realizing high-quality Sn-based thin films. Devoting intensive research activities in reducing atmosphere, reducing agents, suitable solvents or anti-solvents as well as large organic spacers incorporations is necessary. At this point, figuring out what types of defects (vacancies, anti-site substitutions, or interstitial atoms) are suitable to defer Sn^{2+} oxidation should be better considered.

Second, regulating the nucleation and crystallization is beneficial to retard fast crystallization rate and leads to smooth and pinhole-free Sn halide films. With the assistance of a stable intermediate phase, ion exchange strategies, templated growth, or Ostwald ripening efforts, their nucleation and crystallization process can be easy controlled. Herein, choosing green and efficient anti-solvents to replace toxic anti-solvents such as chlorobenzene or methylbenzene can contribute to rapid nucleation and slow crystallization with decreased defect state densities as well as less toxicity of solvent systems.

Third, considering the low/mixed-dimensional Sn halide perovskites is a crucial effort for improving PCE and stability for application. The quasi-2D Sn halide perovskites with RP type are relatively complete in current research. However, the van der Waal effect and poor carrier transport for RP perovskites significantly affect their performance. Therefore, it is urgent to find new organic cations to enhance the interaction between octahedral and large cations.

Finally, since there are rare reports of Sn halide perovskites in photocatalysis and memristors, researchers could draw on corresponding Pb-related achievements to experiment with them. Meanwhile, exploring novel synthesis methods for crystal perovskites achieves high hydrogen generation in photocatalysis and low-power Sn halide perovskite memristors.

REFERENCES

1. www.nrel.gov/pv/cell-efficiency.html (September 2023).
2. A. Kojima; K. Teshima; Y. Shirai; T. Miyasaka, *J. Am. Chem. Soc.* Organometal halide perovskites as visible-light sensitizers for photovoltaic cells. **2009**, *131*, 6050–6051.
3. S. Wang; A. Wang; F. Hao, Toward stable lead halide perovskite solar cells: A knob on the A/X sites components. *iScience* **2022,** *25*, 103599.
4. S. Wang; C. Wu; H. Yao; L. Ding; F. Hao, The nonhalides in perovskite solar cells. *Mater. Chem. Front.* **2023,** *7*, 789–805.
5. M. A. Green; A. Ho-Baillie; H. J. Snaith, The emergence of perovskite solar cells. *Nat. Photonics* **2014,** *8*, 506–514.
6. K. Dedecker; G. Grancini, Dealing with lead in hybrid perovskite: a challenge to tackle for a bright future of this technology? *Adv. Energy Mater.* **2020,** *10*, 2001471.
7. A. Abate, Perovskite solar cells go lead free. *Joule* **2017,** *1*, 659–664.
8. S. Wu; Z. Li; M.-Q. Li; Y. Diao; F. Lin; T. Liu; J. Zhang; P. Tieu; W. Gao; F. Qi, 2D metal -organic framework for stable perovskite solar cells with minimized lead leakage. *Nat. Nanotechnol.* **2020,** *15*, 934–940.
9. J. Li; H.-L. Cao; W.-B. Jiao; Q. Wang; M. Wei; I. Cantone; J. Lü; A. Abate, Biological impact of lead from halide perovskites reveals the risk of introducing a safe threshold. *Nat. Commun.* **2020,** *11*, 310.
10. Y. Jiang; L. Qiu; E. J. Juarez-Perez; L. K. Ono; Z. Hu; Z. Liu; Z. Wu; L. Meng; Q. Wang; Y. Qi, Reduction of lead leakage from damaged lead halide perovskite solar modules using self-healing polymer-based encapsulation. *Nat. Energy* **2019,** *4*, 585–593.
11. X. Li; F. Zhang; H. He; J. J. Berry; K. Zhu; T. Xu, On-device lead sequestration for perovskite solar cells. *Nature* **2020,** *578*, 555–558.
12. B. B. Yu; Z. Chen; Y. Zhu; Y. Wang; B. Han; G. Chen; X. Zhang; Z. Du; Z. He, Heterogeneous 2D/3D tin-halides perovskite solar cells with certified conversion efficiency breaking 14%. *Adv. Mater.* **2021,** *33*, 2102055.
13. M. Saliba; J.-P. Correa-Baena; M. Grätzel; A. Hagfeldt; A. Abate, Perovskite solar cells: from the atomic level to film quality and device performance. *Angew. Chem. Int. Ed.* **2018,** *57*, 2554–2569.
14. Q. Zhang; F. Hao; J. Li; Y. Zhou; Y. Wei; H. Lin, Perovskite solar cells: must lead be replaced–and can it be done?. *Sci. Technol. Adv. Mat.* **2018,** *19*, 425–442.

15. A. Babayigit; A. Ethirajan; M. Muller; B. Conings, Toxicity of organometal halide perovskite solar cells. *Nat. Mater.* **2016,** *15*, 247–251.
16. D. Rai; M. Yui; H. T. Schaef; A. Kitamura, Thermodynamic Model for SnO_2(cr) and SnO_2(am) Solubility in the Aqueous Na^+–H^+–OH^-–Cl^-–H_2O System. *J. Solution Chem.* **2011,** *40*, 1155–1172.
17. A. Babayigit; D. Duy Thanh; A. Ethirajan; J. Manca; M. Muller; H.-G. Boyen; B. Conings, Assessing the toxicity of Pb- and Sn-based perovskite solar cells in model organism Danio rerio. *Sci. Rep.* **2016,** *6*, 18721.
18. F. Hao; C. C. Stoumpos; P. Guo; N. Zhou; T. J. Marks; R. P. Chang; M. G. Kanatzidis, Solvent-Mediated Crystallization of $CH_3NH_3SnI_3$ Films for Heterojunction Depleted Perovskite Solar Cells. *J. Am. Chem. Soc.* **2015,** *137*, 11445–11452.
19. W. Ke; C. C. Stoumpos; J. L. Logsdon; M. R. Wasielewski; Y. Yan; G. Fang; M. G. Kanatzidis, TiO_2–ZnS Cascade Electron Transport Layer for Efficient Formamidinium Tin Iodide Perovskite Solar Cells. *J. Am. Chem. Soc.* **2016,** *138*, 14998–15003.
20. K. Marshall; M. Walker; R. Walton; R. Hatton, Enhanced stability and efficiency in hole-transport-layer-free $CsSnI_3$ perovskite photovoltaics. *Nat. Energy* **2016,** *1*, 1–9.
21. W. Travis; E. Glover; H. Bronstein; D. Scanlon; R. Palgrave, On the application of the tolerance factor to inorganic and hybrid halide perovskites: a revised system. *Chem. Sci.* **2016,** *7*, 4548–4556.
22. Y. Takahashi; R. Obara; Z.-Z. Lin; Y. Takahashi; T. Naito; T. Inabe; S. Ishibashi; K. Terakura, Charge-transport in tin-iodide perovskite $CH_3NH_3SnI_3$: origin of high conductivity. *Dalton Trans.* **2011,** *40*, 5563–5568.
23. G. Laurita; D. H. Fabini; C. C. Stoumpos; M. G. Kanatzidis; R. Seshadri, Chemical tuning of dynamic cation off-centering in the cubic phases of hybrid tin and lead halide perovskites. *Chem. Sci.* **2017,** *8*, 5628–5635.
24. N. K. Noel; S. D. Stranks; A. Abate; C. Wehrenfennig; S. Guarnera; A.-A. Haghighirad; A. Sadhanala; G. E. Eperon; S. K. Pathak; M. B. Johnston, Lead-free organic-inorganic tin halide perovskites for photovoltaic applications. *Energy Environ. Sci.* **2014,** *7*, 3061–3068.
25. F. Hao; C. C. Stoumpos; D. H. Cao; R. P. Chang; M. G. Kanatzidis, Lead-free solid-state organic-inorganic halide perovskite solar cells. *Nat. Photonics* **2014,** *8*, 489–494.
26. C. Bernal; K. Yang, First-Principles Hybrid Functional Study of the Organic-Inorganic Perovskites $CH_3NH_3SnBr_3$ and $CH_3NH_3SnI_3$. *J. Phys. Chem. C* **2014,** *118*, 24383–24388.
27. C. C. Stoumpos; C. D. Malliakas; M. G. Kanatzidis, Semiconducting tin and lead iodide perovskites with organic cations: phase transitions, high mobilities, and near-infrared photoluminescent properties. *Inorg. Chem.* **2013,** *52*, 9019–9038.
28. T. J. Jacobsson; J.-P. Correa-Baena; M. Pazoki; M. Saliba; K. Schenk; M. Grätzel; A. Hagfeldt, Exploration of the compositional space for mixed lead halogen perovskites for high efficiency solar cells. *Energy Environ. Sci.* **2016,** *9*, 1706–1724.
29. D. B. Mitzi; K. Liang, Synthesis, resistivity, and thermal properties of the cubic perovskite $NH_2CH = NH_2SnI_3$ and related systems. *J. Solid State Chem.* **1997,** *134*, 376–381.
30. C. C. Stoumpos; C. D. Malliakas; M. G. Kanatzidis, Semiconducting tin and lead iodide perovskites with organic cations: phase transitions, high mobilities, and near-infrared photoluminescent properties. *Inorg. Chem.* **2013,** *52*, 9019–9038.
31. E. C. Schueller; G. Laurita; D. H. Fabini; C. C. Stoumpos; M. G. Kanatzidis; R. Seshadri, *Inorg.* Crystal structure evolution and notable thermal expansion in hybrid perovskites formamidinium tin iodide and formamidinium lead bromide. *Chem.* **2018,** *57*, 695–701.
32. Z.-Q. Ma; H. Pan; P. K. Wong, A first-principles study on the structural and electronic properties of Sn-based organic-inorganic halide perovskites. *J. Electron. Mater.* **2016,** *45*, 5956–5966.

33. S. Kahmann; O. Nazarenko; S. Shao; O. Hordiichuk; M. Kepenekian; J. Even; M. V. Kovalenko; G. R. Blake; M. A. Loi, Negative Thermal Quenching in $FASnI_3$ Perovskite Single Crystals and Thin Films. *ACS Energy Lett.* **2020,** *5*, 2512–2519.
34. I. Chung; J.-H. Song; J. Im; J. Androulakis; C. D. Malliakas; H. Li; A. J. Freeman; J. T. Kenney; M. G. Kanatzidis, $CsSnI_3$: Semiconductor or Metal? High Electrical Conductivity and Strong Near-Infrared Photoluminescence from a Single Material. High Hole Mobility and Phase-Transitions. *J. Am. Chem. Soc.* **2012,** *134*, 8579–8587.
35. L. M. Herz, Charge-Carrier Mobilities in Metal Halide Perovskites: Fundamental Mechanisms and Limits. *ACS Energy Lett.* **2017,** *2*, 1539–1548.
36. L. Ma; F. Hao; C. C. Stoumpos; B. T. Phelan; M. R. Wasielewski; M. G. Kanatzidis, Carrier Diffusion Lengths of over 500 nm in Lead-Free Perovskite $CH_3NH_3SnI_3$ Films. *J. Am. Chem. Soc.* **2016,** *138*, 14750–14755.
37. R. Prasanna; A. Gold-Parker; T. Leijtens; B. Conings; A. Babayigit; H.-G. Boyen; M. F. Toney; M. D. McGehee, Band gap tuning via lattice contraction and octahedral tilting in perovskite materials for photovoltaics. *J. Am. Chem. Soc.* **2017,** *139*, 11117–11124.
38. W. Ke; C. C. Stoumpos; M. Zhu; L. Mao; I. Spanopoulos; J. Liu; O. Y. Kontsevoi; M. Chen; D. Sarma; Y. Zhang; M. R. Wasielewski; M. G. Kanatzidis, Enhanced photovoltaic performance and stability with a new type of hollow 3D perovskite {en}$FASnI_3$. *Sci. Adv.* **2017,** *3*, e1701293.
39. W. Ke; C. C. Stoumpos; I. Spanopoulos; L. Mao; M. Chen; M. R. Wasielewski; M. G. Kanatzidis, Efficient Lead-Free Solar Cells Based on Hollow {en}$MASnI_3$ Perovskites. *J. Am. Chem. Soc.* **2017,** *139*, 14800–14806.
40. W. Ke; C. C. Stoumpos; I. Spanopoulos; M. Chen; M. R. Wasielewski; M. G. Kanatzidis, Diammonium Cations in the $FASnI_3$ Perovskite Structure Lead to Lower Dark Currents and More Efficient Solar Cells. *ACS Energy Lett.* **2018,** *3*, 1470–1476.
41. I. Spanopoulos; I. Hadar; W. Ke; P. Guo; E. M. Mozur; E. Morgan; S. Wang; D. Zheng; S. Padgaonkar; G. Manjunatha Reddy, Tunable broad light emission from 3D "hollow" bromide perovskites through defect engineering. *J. Am. Chem. Soc.* **2021,** *143*, 7069–7080.
42. I. Spanopoulos; W. Ke; C. C. Stoumpos; E. C. Schueller; O. Y. Kontsevoi; R. Seshadri; M. G. Kanatzidis, Unraveling the chemical nature of the 3D "hollow" hybrid halide perovskites. *J. Am. Chem. Soc.* **2018,** *140*, 5728–5742.
43. D. H. Fabini; G. Laurita; J. S. Bechtel; C. C. Stoumpos; H. A. Evans; A. G. Kontos; Y. S. Raptis; P. Falaras; A. Van der Ven; M. G. Kanatzidis; R. Seshadri, Dynamic Stereochemical Activity of the Sn_2+ Lone Pair in Perovskite $CsSnBr_3$. *J. Am. Chem. Soc.* **2016,** *138*, 11820–11832.
44. C. C. Stoumpos; L. Frazer; D. J. Clark; Y. S. Kim; S. H. Rhim; A. J. Freeman; J. B. Ketterson; J. I. Jang; M. G. Kanatzidis, Hybrid germanium iodide perovskite semiconductors: active lone pairs, structural distortions, direct and indirect energy gaps, and strong nonlinear optical properties. *J. Am. Chem. Soc.* **2015,** *137*, 6804–6819.
45. R. Chiara; M. Morana; L. Malavasi, Germanium-based halide perovskites: materials, properties, and applications. *ChemPlusChem* **2021,** *86*, 879–888.
46. T. Krishnamoorthy; H. Ding; C. Yan; W. L. Leong; T. Baikie; Z. Zhang; M. Sherburne; S. Li; M. Asta; N. Mathews, Lead-free germanium iodide perovskite materials for photovoltaic application. *J. Mater. Chem. A* **2015,** *3*, 23829–23832.
47. M. Pazoki; T. J. Jacobsson; A. Hagfeldt; G. Boschloo; T. Edvinsson, Effect of metal cation replacement on the electronic structure of metalorganic halide perovskites: Replacement of lead with alkaline-earth metals. *Phys. Rev. B* **2016,** *93*, 144105.
48. J. I. Uribe; D. Ramirez; J. M. Osorio-Guillén; J. Osorio; F. Jaramillo, $CH_3NH_3CaI_3$ Perovskite: Synthesis, Characterization, and First-Principles Studies. *J. Phys. Chem. C* **2016,** *120*, 16393–16398.

49. M. R. Filip; F. Giustino, Computational screening of homovalent lead substitution in organic–inorganic halide perovskites. *J. Phys. Chem. C* **2016,** *120*, 166–173.
50. G. L. McPherson; K. Talluto, Luminescent centers in doped crystals of $CsMgI_3$: Energy storage and thermoluminescence. *Solid State Commun.* **1982,** *43*, 331–334.
51. T. J. Jacobsson; M. Pazoki; A. Hagfeldt; T. Edvinsson, Goldschmidt's Rules and Strontium Replacement in Lead Halogen Perovskite Solar Cells: Theory and Preliminary Experiments on $CH_3NH_3SrI_3$. *J. Phys. Chem. C* **2015,** *119*, 25673–25683.
52. D. Cortecchia; H. A. Dewi; J. Yin; A. Bruno; S. Chen; T. Baikie; P. P. Boix; M. Grätzel; S. Mhaisalkar; C. Soci, Lead-Free $MA_2CuCl_xBr_{4-x}$ Hybrid Perovskites *Inorg. Chem.* **2016,** *55*, 1044–1052.
53. M. F. Mostafa; R. D. Willett, Magnetic properties of ferrous chloride complexes with two-dimensional structures. *Phys. Rev. B* **1971,** *4*, 2213.
54. T. J. Huang; Z. X. Thiang; X. Yin; C. Tang; G. Qi; H. Gong, $(CH_3NH_3)_2PdCl_4$: A Compound with Two-Dimensional Organic–Inorganic Layered Perovskite Structure. *Chem Eur. J.* **2016,** *22*, 2146–2152.
55. M. Pazoki; T. Edvinsson, Metal replacement in perovskite solar cell materials: chemical bonding effects and optoelectronic properties. *Sustain. Energy & Fuels* **2018,** *2*, 1430–1445.
56. J. Gebhardt; A. M. Rappe, Transition metal inverse-hybrid perovskites. *J. Mater. Chem. A* **2018,** *6*, 14560–14565.
57. D. Sen; T. Saha-Dasgupta, Rational Design of Magnetic Hybrid Transition Metal Double Perovskites in Polyatomic Anionic Systems. *Chem. Mater.* **2022,** *34*, 1821–1828.
58. H. Tsai; W. Nie; J.-C. Blancon; C. C. Stoumpos; R. Asadpour; B. Harutyunyan; A. J. Neukirch; R. Verduzco; J. J. Crochet; S. Tretiak, High-efficiency two-dimensional Ruddlesden–Popper perovskite solar cells. *Nature* **2016,** *536*, 312–316.
59. J. Schurter; R. Barnes; R. Willett. *Hyperfine fields and critical point exponents of some two-dimensionally layered Fe (2+) salts*, AIP Conference Proceedings, American Institute of Physics: **1975**; pp 307–308.
60. Y.-Y. Sun; M. L. Agiorgousis; P. Zhang; S. Zhang, Chalcogenide perovskites for photovoltaics. *Nano Lett.* **2015,** *15*, 581–585.
61. C.-S. Lee; K. M. Kleinke; H. Kleinke, Synthesis, structure, and electronic and physical properties of the two $SrZrS_3$ modifications. *Solid State Sci.* **2005,** *7*, 1049–1054.
62. J. W. Bennett; I. Grinberg; A. M. Rappe, Effect of substituting of S for O: The sulfide perovskite $BaZrS_3$ investigated with density functional theory. *Phys. Rev. B* **2009,** *79*, 235115.
63. A. Clearfield, *Acta Crystallogr.* The synthesis and crystal structures of some alkaline earth titanium and zirconium sulfides. **1963,** *16*, 135–142.
64. L. J. Tranchitella; B.-H. Chen; J. C. Fettinger; B. W. Eichhorn, Structural Evolutions in the $Sr_{1-x}Ba_xZrSe_3$ Series. *J. Solid State Chem.* **1997,** *130*, 20–27.
65. H. Yao; S. Wang; Z. Jin; L. Ding; F. Hao, Dimensionality regulation in tin halide perovskite solar cells: Toward high performance and stability. *J. Energy Chem.* **2022,** *77*, 144–156.
66. S. Wang; C. Wu; H. Yao; F. Hao, Dimensional tailoring endows tin halide perovskite solar cells with high efficiency and stability. *Mater. Lab* **2022,** *2*, 220047.
67. S. Wang; C. Wu; L. Xie; L. Ding; F. Hao, Pseudohalide-modulated crystallization for efficient quasi-2D tin perovskite solar cells with minimized voltage deficit. *ACS Materials Lett.* **2023,** *5*, 936–943.
68. S. Wang; L. Yan; W. Zhu; Z. Cao; L. Zhou; L. Ding; F. Hao, Suppressing the formation of tin vacancy yields efficient lead-free perovskite solar cells. *Nano Energy* **2022,** *99*, 107416.
69. P. Cheng; Z. Xu; J. Li; Y. Liu; Y. Fan; L. Yu; D.-M. Smilgies; C. Müller; K. Zhao; S. F. Liu, Highly Efficient Ruddlesden–Popper Halide Perovskite $PA_2MA_4Pb_5I_{16}$ Solar Cells *ACS Energy Lett.* **2018,** *3*, 1975–1982.

70. C. Ma; D. Shen; T. W. Ng; M. F. Lo; C. S. Lee, 2D perovskites with short interlayer distance for high-performance solar cell application. *Adv. Mater.* **2018,** *30*, 1800710.
71. J. Fu; Q. Xu; I. Abdelwahab; R. Cai; B. Febriansyah; T. Yin; K. P. Loh; N. Mathews; H. Sun; T. C. Sum, Strain propagation in layered two-dimensional halide perovskites. *Sci. Adv.* **2022,** *8*, eabq1971.
72. A. V. Bondi, van der Waals volumes and radii. *J. Phys. Chem. C* **1964,** *68*, 441–451.
73. W. Ke; L. Mao; C. C. Stoumpos; J. Hoffman; I. Spanopoulos; A. D. Mohite; M. G. Kanatzidis, Compositional and solvent engineering in Dion–Jacobson 2D perovskites boosts solar cell efficiency and stability. *Adv. Energy Mater.* **2019,** *9*, 1803384.
74. M. Brik; I. Kityk, Modeling of lattice constant and their relations with ionic radii and electronegativity of constituting ions of A_2XY_6 cubic crystals (A=K, Cs, Rb, Tl; X=tetravalent cation, Y=F, Cl, Br, I). *J. Phys. Chem. Solids* **2011,** *72*, 1256–1260.
75. Z. Xiao; H. Lei; X. Zhang; Y. Zhou; H. Hosono; T. Kamiya, Ligand-Hole in $[SnI_6]$ Unit and Origin of Band Gap in Photovoltaic Perovskite Variant Cs_2SnI_6. *B Chem. Soc. Jpn.* **2015,** *88*, 1250–1255.
76. N. Sakai; A. A. Haghighirad; M. R. Filip; P. K. Nayak; S. Nayak; A. Ramadan; Z. Wang; F. Giustino; H. J. Snaith, Solution-Processed Cesium Hexabromopalladate(IV), Cs_2PdBr_6, for Optoelectronic Applications. *J. Am. Chem. Soc.* **2017,** *139*, 6030–6033.
77. M.-G. Ju; M. Chen; Y. Zhou; H. F. Garces; J. Dai; L. Ma; N. P. Padture; X. C. Zeng, Earth-abundant nontoxic titanium (IV)-based vacancy-ordered double perovskite halides with tunable 1.0 to 1.8 eV bandgaps for photovoltaic applications. *ACS Energy Lett.* **2018,** *3*, 297–304.
78. G. M. Dalpian; Q. Liu; C. C. Stoumpos; A. P. Douvalis; M. Balasubramanian; M. G. Kanatzidis; A. Zunger, Changes in charge density vs changes in formal oxidation states: The case of Sn halide perovskites and their ordered vacancy analogues. *Phys. Rev. Mater.* **2017,** *1*, 025401.
79. L. Chu; W. Ahmad; W. Liu; J. Yang; R. Zhang; Y. Sun; J. Yang; X. A. Li, Lead-free halide double perovskite materials: a new superstar toward green and stable optoelectronic applications. *Nano-Micro Lett.* **2019,** *11*, 16.
80. X. Qiu; Y. Jiang; H. Zhang; Z. Qiu; S. Yuan; P. Wang; B. Cao, Lead-free mesoscopic Cs_2SnI_6 perovskite solar cells using different nanostructured ZnO nanorods as electron transport layers. *Phys Status Solidi-R* **2016,** *10*, 587–591.
81. M. Chen; M.-G. Ju; A. D. Carl; Y. Zong; R. L. Grimm; J. Gu; X. C. Zeng; Y. Zhou; N. P. Padture, Cesium titanium (IV) bromide thin films based stable lead-free perovskite solar cells. *Joule* **2018,** *2*, 558–570.
82. F. Giustino; H. J. Snaith, Toward lead-free perovskite solar cells. *ACS Energy Lett.* **2016,** *1*, 1233–1240.
83. H. Lei; D. Hardy; F. Gao, Lead-Free Double Perovskite $Cs_2AgBiBr_6$: Fundamentals, Applications, and Perspectives. *Adv. Funct. Mater.* **2021,** *31*, 2105898.
84. A. H. Slavney; T. Hu; A. M. Lindenberg; H. I. Karunadasa, A Bismuth-Halide Double Perovskite with Long Carrier Recombination Lifetime for Photovoltaic Applications. *J. Am. Chem. Soc.* **2016,** *138*, 2138–2141.
85. E. T. McClure; M. R. Ball; W. Windl; P. M. Woodward, Cs_2AgBiX_6 (X = Br, Cl): New Visible Light Absorbing, Lead-Free Halide Perovskite Semiconductors. *Chem. Mater.* **2016,** *28*, 1348–1354.
86. G. Volonakis; M. R. Filip; A. A. Haghighirad; N. Sakai; B. Wenger; H. J. Snaith; F. Giustino, Lead-free halide double perovskites via heterovalent substitution of noble metals. *J. Phys. Chem. Lett.* **2016,** *7*, 1254–1259.
87. W. Meng; X. Wang; Z. Xiao; J. Wang; D. B. Mitzi; Y. Yan, Parity-forbidden transitions and their impact on the optical absorption properties of lead-free metal halide perovskites and double perovskites. *J. Phys. Chem. Lett.* **2017,** *8*, 2999–3007.

88. W. Serfontein; R. Mekel, *Res.* Bismuth toxicity in man II. Review of bismuth blood and urine levels in patients after administration of therapeutic bismuth formulations in relation to the problem of bismuth toxicity in man. *Commun. Chem. Pathol. Pharmacol.* **1979,** *26*, 391–411.
89. A. Slikkerveer; F. A. de Wolff, Pharmacokinetics and toxicity of bismuth compounds. *Med. Toxicol. Adverse Drug Experience* **1989,** *4*, 303–323.
90. Z. Xiao; Y. Yan; H. Hosono; T. Kamiya, Roles of Pseudo-Closed s_2 Orbitals for Different Intrinsic Hole Generation between Tl–Bi and In–Bi Bromide Double Perovskites. *J. Phys. Chem. Lett.* **2018,** *9*, 258–262.
91. G. García-Espejo; D. Rodríguez-Padrón; R. Luque; L. Camacho; G. de Miguel, Mechanochemical synthesis of three double perovskites: $Cs_2AgBiBr_6$, $(CH_3NH_3)_2TlBiBr_6$ and $Cs_2AgSbBr_6$. *Nanoscale* **2019,** *11*, 16650–16657.
92. Z. Deng; F. Wei; S. Sun; G. Kieslich; A. K. Cheetham; P. D. Bristowe, Exploring the properties of lead-free hybrid double perovskites using a combined computational-experimental approach. *J. Mater. Chem. A* **2016,** *4*, 12025–12029.
93. Z. Xiao; K.-Z. Du; W. Meng; J. Wang; D. B. Mitzi; Y. Yan, Intrinsic Instability of $Cs_2In(I)M(III)X_6$ (M = Bi, Sb; X = Halogen) Double Perovskites: A Combined Density Functional Theory and Experimental Study. *J. Am. Chem. Soc.* **2017,** *139*, 6054–6057.
94. F. Ji; J. Klarbring; F. Wang; W. Ning; L. Wang; C. Yin; J. S. M. Figueroa; C. K. Christensen; M. Etter; T. Ederth, Lead-Free Halide Double Perovskite $Cs_2AgBiBr_6$ with Decreased Band Gap. *Angew. Chem. Int. Ed.* **2020,** *132*, 15303–15306.
95. Y. Zhang; J. Yin; M. R. Parida; G. H. Ahmed; J. Pan; O. M. Bakr; J.-L. Brédas; O. F. Mohammed, Direct-indirect nature of the bandgap in lead-free perovskite nanocrystals. *J. Phys. Chem. Lett.* **2017,** *8*, 3173–3177.
96. K. M. McCall; C. C. Stoumpos; O. Y. Kontsevoi; G. C. Alexander; B. W. Wessels; M. G. Kanatzidis, From 0D $Cs_3Bi_2I_9$ to 2D $Cs_3Bi_2I_6Cl_3$: Dimensional Expansion Induces a Direct Band Gap but Enhances Electron–Phonon Coupling. *Chem. Mater.* **2019,** *31*, 2644–2650.
97. Y.-Y. Sun; J. Shi; J. Lian; W. Gao; M. L. Agiorgousis; P. Zhang; S. Zhang, Discovering lead-free perovskite solar materials with a split-anion approach. *Nanoscale* **2016,** *8*, 6284–6289.
98. J.-H. Chang; T. Doert; M. Ruck, Structural Variety of Defect Perovskite Variants $M_3E_2X_9$ (M = Rb, Tl, E = Bi, Sb, X = Br, I). *Z. fur Anorg. Allg. Chem.* **2016,** *642*, 736–748.
99. K. M. McCall; C. C. Stoumpos; S. S. Kostina; M. G. Kanatzidis; B. W. Wessels, Strong Electron–Phonon Coupling and Self-Trapped Excitons in the Defect Halide Perovskites $A_3M_2I_9$ (A = Cs, Rb; M = Bi, Sb). *Chem. Mater.* **2017,** *29*, 4129–4145.
100. V. Machulin; F. Motsnyi; O. Smolanka; G. Svechnikov; E. Y. Peresh, Effect of temperature variation on shift and broadening of the exciton band in $Cs_3Bi_2I_9$ layered crystals. *Low Temp. Phys.* **2004,** *30*, 964–967.
101. B. Saparov; F. Hong; J.-P. Sun; H.-S. Duan; W. Meng; S. Cameron; I. G. Hill; Y. Yan; D. B. Mitzi, Anionic order and band gap engineering in vacancy ordered triple perovskites. *Chem. Mater.* **2015,** *27*, 5622–5632.
102. T. L. Hodgkins; C. N. Savory; K. K. Bass; B. L. Seckman; D. O. Scanlon; P. I. Djurovich; M. E. Thompson; B. C. Melot, Thin-Film Preparation and Characterization of $Cs_3Sb_2I_9$: A Lead-Free Layered Perovskite Semiconductor. *Chem. Commun.* **2019,** *55*, 3164–3167.
103. F. Hong; B. Saparov; W. Meng; Z. Xiao; D. B. Mitzi; Y. Yan, Viability of lead-free perovskites with mixed chalcogen and halogen anions for photovoltaic applications. *J. Phys. Chem. C* **2016,** *120*, 6435–6441.
104. W. Shockley; H. J. Queisser, Detailed balance limit of efficiency of p-n junction solar cells. *J. Appl. Phys.* **1961,** *32*, 510–519.

105. W. Liao; D. Zhao; Y. Yu; C. R. Grice; C. Wang; A. J. Cimaroli; P. Schulz; W. Meng; K. Zhu; R. G. Xiong, Lead-free inverted planar formamidinium tin triiodide perovskite solar cells achieving power conversion efficiencies up to 6.22%. *Adv. Mater.* **2016,** *28*, 9333–9340.
106. T. M. Koh; T. Krishnamoorthy; N. Yantara; C. Shi; W. L. Leong; P. P. Boix; A. C. Grimsdale; S. G. Mhaisalkar; N. Mathews, Formamidinium tin-based perovskite with low E_g for photovoltaic applications. *J. Mater. Chem. A* **2015,** *3*, 14996–15000.
107. T. Handa; T. Yamada; H. Kubota; S. Ise; Y. Miyamoto; Y. Kanemitsu, Photocarrier Recombination and Injection Dynamics in Long-Term Stable Lead-Free $CH_3NH_3SnI_3$ Perovskite Thin Films and Solar Cells. *J. Phys. Chem. C* **2017,** *121*, 16158–16165.
108. X. Lü; Y. Wang; C. C. Stoumpos; Q. Hu; X. Guo; H. Chen; L. Yang; J. S. Smith; W. Yang; Y. Zhao, Enhanced Structural Stability and Photo Responsiveness of $CH_3NH_3SnI_3$ Perovskite via Pressure-Induced Amorphization and Recrystallization. *Adv. Mater.* **2016,** *28*, 8663–8668.
109. T. Nakamura; S. Yakumaru; M. A. Truong; K. Kim; J. Liu; S. Hu; K. Otsuka; R. Hashimoto; R. Murdey; T. Sasamori, Sn (IV)-free tin perovskite films realized by in situ Sn (0) nanoparticle treatment of the precursor solution. *Nat. Commun.* **2020,** *11*, 3008.
110. T. Yokoyama; T.-B. Song; D. H. Cao; C. C. Stoumpos; S. Aramaki; M. G. Kanatzidis, The origin of lower hole carrier concentration in methylammonium tin halide films grown by a vapor-assisted solution process. *ACS Energy Lett.* **2017,** 2, 22–28.
111. P. Wang; F. Li; K. J. Jiang; Y. Zhang; H. Fan; Y. Zhang; Y. Miao; J. H. Huang; C. Gao; X. Zhou, Ion exchange/insertion reactions for fabrication of efficient methylammonium tin iodide perovskite solar cells. *Adv. Sci.* **2020,** *7*, 1903047.
112. J. Liu; M. Ozaki; S. Yakumaru; T. Handa; R. Nishikubo; Y. Kanemitsu; A. Saeki; Y. Murata; R. Murdey; A. Wakamiya, Lead-free solar cells based on tin halide perovskite films with high coverage and improved aggregation. *Angew. Chem. Int. Ed.* **2018,** *130*, 13405–13409.
113. F. Li; C. Zhang; J. H. Huang; H. Fan; H. Wang; P. Wang; C. Zhan; C. M. Liu; X. Li; L. M. Yang, A cation-exchange approach for the fabrication of efficient methylammonium tin iodide perovskite solar cells. *Angew. Chem. Int. Ed.* **2019,** *58*, 6688–6692.
114. T. Shi; H.-S. Zhang; W. Meng; Q. Teng; M. Liu; X. Yang; Y. Yan; H.-L. Yip; Y.-J. Zhao, Effects of organic cations on the defect physics of tin halide perovskites. *J. Mater. Chem. A* **2017,** *5*, 15124–15129.
115. A. Amat; E. Mosconi; E. Ronca; C. Quarti; P. Umari; M. K. Nazeeruddin; M. Gratzel; F. De Angelis, Cation-induced band-gap tuning in organohalide perovskites: interplay of spin–orbit coupling and octahedra tilting. *Nano Lett.* **2014,** *14*, 3608–3616.
116. X. Meng; Y. Wang; J. Lin; X. Liu; X. He; J. Barbaud; T. Wu; T. Noda; X. Yang; L. Han, Surface-controlled oriented growth of $FASnI_3$ crystals for efficient lead-free perovskite solar cells. *Joule* **2020,** *4*, 902–912.
117. X. Liu; T. Wu; J.-Y. Chen; X. Meng; X. He; T. Noda; H. Chen; X. Yang; H. Segawa; Y. Wang; L. Han, Templated growth of $FASnI_3$ crystals for efficient tin perovskite solar cells. *Energy Environ. Sci.* **2020,** *13*, 2896–2902.
118. G. Li; Z. Su; M. Li; F. Yang; M. H. Aldamasy; J. Pascual; F. Yang; H. Liu; W. Zuo; D. Di Girolamo, Ionic liquid stabilizing high-efficiency tin halide perovskite solar cells. *Adv. Energy Mater.* **2021,** *11*, 2101539.
119. S. Wang; H. Yao; W. Zhu; C. Wu; Z. Tang; J. Liu; L. Ding; F. Hao, Stabilization of Perovskite Lattice and Suppression of Sn^{2+}/Sn^{4+} Oxidation via Formamidine Acetate for High Efficiency Tin Perovskite Solar Cells. *Adv. Funct. Mater.* **2023**, 2215041.
120. T. Wang; Q. Tai; X. Guo; J. Cao; C.-K. Liu; N. Wang; D. Shen; Y. Zhu; C.-S. Lee; F. Yan, Highly air-stable tin-based perovskite solar cells through grain-surface protection by gallic acid. *ACS Energy Lett.* **2020,** *5*, 1741–1749.

121. M. E. Kayesh; T. H. Chowdhury; K. Matsuishi; R. Kaneko; S. Kazaoui; J.-J. Lee; T. Noda; A. Islam, Enhanced Photovoltaic Performance of $FASnI_3$-Based Perovskite Solar Cells with Hydrazinium Chloride Coadditive. *ACS Energy Lett.* **2018,** *3*, 1584–1589.
122. S. J. Lee; S. S. Shin; Y. C. Kim; D. Kim; T. K. Ahn; J. H. Noh; J. Seo; S. I. Seok, Fabrication of Efficient Formamidinium Tin Iodide Perovskite Solar Cells through SnF_2–Pyrazine Complex. *J. Am. Chem. Soc.* **2016,** *138*, 3974–3977.
123. Q. Tai; X. Guo; G. Tang; P. You; T. W. Ng; D. Shen; J. Cao; C. K. Liu; N. Wang; Y. Zhu, Antioxidant grain passivation for air-stable tin-based perovskite solar cells. *Angew. Chem. Int. Ed.* **2019,** *58*, 806–810.
124. C. Wang; F. Gu; Z. Zhao; H. Rao; Y. Qiu; Z. Cai; G. Zhan; X. Li; B. Sun; X. Yu, Self-repairing tin-based perovskite solar cells with a breakthrough efficiency over 11%. *Adv. Mater.* **2020,** *32*, 1907623.
125. W. Ke; C. C. Stoumpos; M. Zhu; L. Mao; I. Spanopoulos; J. Liu; O. Y. Kontsevoi; M. Chen; D. Sarma; Y. Zhang, Enhanced photovoltaic performance and stability with a new type of hollow 3D perovskite {en}$FASnI_3$. *Sci. Adv.* **2017,** *3*, e1701293.
126. X. Meng; J. Lin; X. Liu; X. He; L. Han, Highly Stable and Efficient $FASnI_3$-Based Perovskite Solar Cells by Introducing Hydrogen Bonding. *Adv. Mater.* **2019,** *31*, 1903721.
127. Y. Su; J. Yang; H. Rao; Y. Zhong; W. Sheng; L. Tan; Y. Chen, Environmentally friendly anti-solvent engineering for high-efficiency tin-based perovskite solar cells. *Energy Environ. Sci.* **2023,** *16*, 2177–2186.
128. D. Cui; X. Liu; T. Wu; X. Lin; X. Luo; Y. Wu; H. Segawa; X. Yang; Y. Zhang; Y. Wang, Making Room for Growing Oriented $FASnI_3$ with Large Grains via Cold Precursor Solution. *Adv. Funct. Mater.* **2021,** *31*, 2100931.
129. X. Liu; Z. Yang; C.-C. Chueh; A. Rajagopal; S. T. Williams; Y. Sun; A. K.-Y. Jen, Improved efficiency and stability of Pb–Sn binary perovskite solar cells by Cs substitution. *J. Mater. Chem. A* **2016,** *4*, 17939–17945.
130. X. Liu; Y. Wang; F. Xie; X. Yang; L. Han, Improving the performance of inverted formamidinium tin iodide perovskite solar cells by reducing the energy-level mismatch. *ACS Energy Lett.* **2018,** *3*, 1116–1121.
131. M. Chen; Q. Dong; C. Xiao; X. Zheng; Z. Dai; Y. Shi; J. M. Luther; N. P. Padture, Lead-free flexible perovskite solar cells with interfacial native oxide have > 10% efficiency and simultaneously enhanced stability and reliability. *ACS Energy Lett.* **2022,** *7*, 2256–2264.
132. T. Wang; F. Zheng; G. Tang; J. Cao; P. You; J. Zhao; F. Yan, 2D WSe_2 Flakes for Synergistic Modulation of Grain Growth and Charge Transfer in Tin-Based Perovskite Solar Cells. *Adv. Sci.* **2021,** *8*, 2004315.
133. B. Chen; S. Wang; X. Zhang; W. Zhu; Z. Cao; F. Hao, Reducing the interfacial voltage loss in tin halides perovskite solar cells. *Chem. Eng. J.* **2022,** *445*, 136769.
134. H. Li; B. Chang; L. Wang; Z. Wang; L. Pan; Y. Wu; Z. Liu; L. Yin, Surface reconstruction for tin-based perovskite solar cells. *ACS Energy Lett.* **2022,** *7*, 3889–3899.
135. D. Sabba; H. K. Mulmudi; R. R. Prabhakar; T. Krishnamoorthy; T. Baikie; P. P. Boix; S. Mhaisalkar; N. Mathews, Impact of Anionic Br^- Substitution on Open Circuit Voltage in Lead Free Perovskite ($CsSnI_{3-x}Br_x$) Solar Cells. *J. Phys. Chem. C* **2015,** *119*, 1763–1767.
136. S. Gupta; D. Cahen; G. Hodes, How SnF_2 Impacts the Material Properties of Lead-Free Tin Perovskites. *J. Phys. Chem. C* **2018,** *122*, 13926–13936.
137. M. H. Kumar; S. Dharani; W. L. Leong; P. P. Boix; R. R. Prabhakar; T. Baikie; C. Shi; H. Ding; R. Ramesh; M. Asta, Lead-free halide perovskite solar cells with high photocurrents realized through vacancy modulation. *Adv. Mater.* **2014,** *26*, 7122–7127.
138. K. P. Marshall; R. I. Walton; R. A. Hatton, Tin perovskite/fullerene planar layer photovoltaics: improving the efficiency and stability of lead-free devices. *J. Mater. Chem. A* **2015,** *3*, 11631–11640.

139. T.-B. Song; T. Yokoyama; C. C. Stoumpos; J. Logsdon; D. H. Cao; M. R. Wasielewski; S. Aramaki; M. G. Kanatzidis, Importance of reducing vapor atmosphere in the fabrication of tin-based perovskite solar cells. *J. Am. Chem. Soc.* **2017,** *139*, 836–842.
140. B. Li; H. Di; B. Chang; R. Yin; L. Fu; Y.-N. Zhang; L. Yin, Efficient Passivation Strategy on Sn Related Defects for High Performance All-Inorganic $CsSnI_3$ Perovskite Solar Cells. *Adv. Funct. Mater.* **2021,** *31*, 2007447.
141. C. Duan; F. Zou; Q. Wen; M. Qin; J. Li; C. Chen; X. Lu; L. Ding; K. Yan, A Bifunctional Carbazide Additive For Durable $CsSnI_3$ Perovskite Solar Cells. *Adv. Mater.*, 2300503.
142. T. Ye; X. Wang; K. Wang; S. Ma; D. Yang; Y. Hou; J. Yoon; K. Wang; S. Priya, Localized Electron Density Engineering for Stabilized B-γ $CsSnI_3$-Based Perovskite Solar Cells with Efficiencies >10%. *ACS Energy Lett.* **2021,** *6*, 1480–1489.
143. W. Zhang; Y. Cai; H. Liu; Y. Xia; J. Cui; Y. Shi; R. Chen; T. Shi; H. L. Wang, Organic-free and lead-free perovskite solar cells with efficiency over 11%. *Adv. Energy Mater.* **2022,** *12*, 2202491.
144. X. Liu; Y. Wang; T. Wu; X. He; X. Meng; J. Barbaud; H. Chen; H. Segawa; X. Yang; L. Han, Efficient and stable tin perovskite solar cells enabled by amorphous-polycrystalline structure. *Nat. Commun.* **2020,** *11*, 2678.
145. J. Liu; H. Yao; S. Wang; C. Wu; L. Ding; F. Hao, Origins and suppression of Sn (II)/Sn (IV) oxidation in tin halide perovskite solar cells. *Adv. Energy Mater.* **2023,** 2300696.
146. Y. Liao; H. Liu; W. Zhou; D. Yang; Y. Shang; Z. Shi; B. Li; X. Jiang; L. Zhang; L. N. Quan, Highly oriented low-dimensional tin halide perovskites with enhanced stability and photovoltaic performance. *J. Am. Chem. Soc.* **2017,** *139*, 6693–6699.
147. K. Chen; P. Wu; W. Yang; R. Su; D. Luo; X. Yang; Y. Tu; R. Zhu; Q. Gong, Low-dimensional perovskite interlayer for highly efficient lead-free formamidinium tin iodide perovskite solar cells. *Nano Energy* **2018,** *49*, 411–418.
148. P. Li; H. Dong; J. Xu; J. Chen; B. Jiao; X. Hou; J. Li; Z. Wu, Ligand orientation-induced lattice robustness for highly efficient and stable tin-based perovskite solar cells. *ACS Energy Lett.* **2020,** *5*, 2327–2334.
149. B. B. Yu; Z. Chen; Y. Zhu; Y. Wang; B. Han; G. Chen; X. Zhang; Z. Du; Z. He, Heterogeneous 2D/3D Tin-Halides Perovskite Solar Cells with Certified Conversion Efficiency Breaking 14%. *Adv. Mater.* **2021,** *33*, 2102055.
150. C. Ran; W. Gao; J. Li; J. Xi; L. Li; J. Dai; Y. Yang; X. Gao; H. Dong; B. Jiao; I. Spanopoulos; C. D. Malliakas; X. Hou; M. G. Kanatzidis; Z. Wu, Conjugated organic cations enable efficient self-healing $FASnI_3$ solar cells. *Joule* **2019,** *3*, 3072–3087.
151. E. Jokar; P.-Y. Cheng; C.-Y. Lin; S. Narra; S. Shahbazi; E. Wei-Guang Diau, Enhanced performance and stability of 3D/2D tin perovskite solar cells fabricated with a sequential solution deposition. *ACS Energy Lett.* **2021,** *6*, 485–492.
152. Z. Zhu; X. Jiang; D. Yu; N. Yu; Z. Ning; Q. Mi, Smooth and Compact $FASnI_3$ Films for Lead-Free Perovskite Solar Cells with over 14% Efficiency. *ACS Energy Lett.* **2022,** *7*, 2079–2083.
153. J. Wang; C. Yang; H. Chen; M. Lv; T. Liu; H. Chen; D.-J. Xue; J.-S. Hu; L. Han; S. Yang; X. Meng, Oriented attachment of tin halide perovskites for photovoltaic applications. *ACS Energy Lett.* **2023,** *8*, 1590–1596.
154. X. Li; X. Gao; X. Zhang; X. Shen; M. Lu; J. Wu; Z. Shi; V. L. Colvin; J. Hu; X. Bai, Lead-free halide perovskites for light emission: recent advances and perspectives. *Adv. Sci.* **2021,** *8*, 2003334.
155. Y. Zhang; Y. Ma; Y. Wang; X. Zhang; C. Zuo; L. Shen; L. Ding, Lead-free perovskite photodetectors: progress, challenges, and opportunities. *Adv. Mater.* **2021,** *33*, 2006691.
156. J. Lu; C. Yan; W. Feng; X. Guan; K. Lin; Z. Wei, Lead-free metal halide perovskites for light-emitting diodes. *EcoMat* **2021,** *3*, e12082.

157. J.-T. Lin; C.-C. Liao; C.-S. Hsu; D.-G. Chen; H.-M. Chen; M.-K. Tsai; P.-T. Chou; C.-W. Chiu, Harnessing dielectric confinement on tin perovskites to achieve emission quantum yield up to 21%. *J. Am. Chem. Soc.* **2019,** *141*, 10324–10330.
158. M. L. Lai; T. Y. Tay; A. Sadhanala; S. E. Dutton; G. Li; R. H. Friend; Z.-K. Tan, Tunable Near-Infrared Luminescence in Tin Halide Perovskite Devices. *J. Phys. Chem. Lett.* **2016,** *7*, 2653–2658.
159. L. Lanzetta; J. M. Marin-Beloqui; I. Sanchez-Molina; D. Ding; S. A. Haque, Two-dimensional organic tin halide perovskites with tunable visible emission and their use in light-emitting devices. *ACS Energy Lett.* **2017,** *2*, 1662–1668.
160. A. Wang; Y. Guo; Z. Zhou; X. Niu; Y. Wang; F. Muhammad; H. Li; T. Zhang; J. Wang; S. Nie, Aqueous acid-based synthesis of lead-free tin halide perovskites with near-unity photoluminescence quantum efficiency. *Chem. Sci.* **2019,** *10*, 4573–4579.
161. H. Wang; W. Xu; Q. Wei; S. Peng; Y. Shang; X. Jiang; D. Yu; K. Wang; R. Pu; C. Zhao; Z. Zang; H. Li; Y. Zhang; T. Pan; Z. Peng; X. Shen; S. Ling; W. Liu; F. Gao; Z. Ning, In-situ growth of low-dimensional perovskite-based insular nanocrystals for highly efficient light emitting diodes. *Light-Sci. Appl.* **2023,** *12*, 62.
162. Q. Zhang; S. Liu; M. He; W. Zheng; Q. Wan; M. Liu; X. Liao; W. Zhan; C. Yuan; J. Liu; H. Xie; X. Guo; L. Kong; L. Li, Stable Lead-Free Tin Halide Perovskite with Operational Stability > 1200 h by Suppressing Tin (II) Oxidation. *Angew. Chem. Int. Ed.* **2022,** *61*, e202205463.
163. F. Zhang; H. Min; Y. Zhang; Z. Kuang; J. Wang; Z. Feng; K. Wen; L. Xu; C. Yang; H. Shi, Vapor-Assisted In Situ Recrystallization for Efficient Tin-Based Perovskite Light-Emitting Diodes. *Adv. Mater.* **2022,** *34*, 2203180.
164. A. Giri; A. Z. Chen; A. Mattoni; K. Aryana; D. Zhang; X. Hu; S.-H. Lee; J. J. Choi; P. E. Hopkins, Ultralow thermal conductivity of two-dimensional metal halide perovskites. *Nano Lett.* **2020,** *20*, 3331–3337.
165. F. Qian; M. Hu; J. Gong; C. Ge; Y. Zhou; J. Guo; M. Chen; Z. Ge; N. P. Padture; Y. Zhou; J. Feng, Enhanced Thermoelectric Performance in Lead-Free Inorganic $CsSn_{1-x}Ge_xI_3$ Perovskite Semiconductors. *J. Phys. Chem. C* **2020,** *124*, 11749–11753.
166. Y. Xiong; L. Xu; P. Wu; L. Sun; G. Xie; B. Hu, Bismuth Doping–Induced Stable Seebeck Effect Based on $MAPbI_3$ Polycrystalline Thin Films. *Adv. Funct. Mater.* **2019,** *29*, 1900615.
167. S. Zhou; Y. Ma; G. Zhou; X. Xu; M. Qin; Y. Li; Y.-J. Hsu; H. Hu; G. Li; N. Zhao; J. Xu; X. Lu, Ag-doped halide perovskite nanocrystals for tunable band structure and efficient charge transport. *ACS Energy Lett.* **2019,** *4*, 534–541.
168. Y. Liu; P.-A. Chen; X. Qiu; J. Guo; J. Xia; H. Wei; H. Xie; S. Hou; M. He; X. Wang; Z. Zeng; L. Jiang; L. Liao; Y. Hu, Doping of Sn-based two-dimensional perovskite semiconductor for high-performance field-effect transistors and thermoelectric devices. *iScience* **2022,** *25*, 104109.
169. S.-N. Hsu; W. Zhao; Y. Gao; Akriti; M. Segovia; X. Xu; B. W. Boudouris; L. Dou, Thermoelectric performance of lead-free two-dimensional halide perovskites featuring conjugated ligands. *Nano Lett.* **2021,** *21*, 7839–7844.
170. S. J. Yang; D. Kim; J. Choi; S. H. Kim; K. Park; S. Ryu; K. Cho, Enhancing thermoelectric power factor of 2D organometal halide perovskites by suppressing 2D/3D phase separation. *Adv. Mater.* **2021,** *33*, 2102797.
171. A. Waleed; M. M. Tavakoli; L. Gu; Z. Wang; D. Zhang; A. Manikandan; Q. Zhang; R. Zhang; Y.-L. Chueh; Z. Fan, Lead-Free Perovskite Nanowire Array Photodetectors with Drastically Improved Stability in Nanoengineering Templates. *Nano Lett.* **2017,** *17*, 523–530.
172. C. K. Liu; Q. Tai; N. Wang; G. Tang; H. L. Loi; F. Yan, Sn-based perovskite for highly sensitive photodetectors. *Adv. Sci.* **2019,** *6*, 1900751.

173. Y. Yang; H. Zhang; S. Hou; T. Wang; W. Chen; S. Xian; Z. Zhang; Y. Mao, Sn-based quasi-two-dimensional organic–inorganic hybrid halide perovskite for high-performance photodetectors. *Appl. Phys. Lett.* **2021,** *119*, 161106.
174. J. Zhou; J. Luo; X. Rong; P. Wei; M. S. Molokeev; Y. Huang; J. Zhao; Q. Liu; X. Zhang; J. Tang, Lead-Free Perovskite Derivative $Cs_2SnCl_{6-x}Br_x$ Single Crystals for Narrowband Photodetectors. *Adv. Opt. Mater.* **2019,** *7*, 1900139.
175. L. Qian; Y. Sun; M. Wu; C. Li; D. Xie; L. Ding; G. Shi, A lead-free two-dimensional perovskite for a high-performance flexible photoconductor and a light-stimulated synaptic device. *Nanoscale* **2018,** *10*, 6837–6843.
176. M. He; Z. Xu; C. Zhao; Y. Gao; K. Ke; N. Liu; X. Yao; F. Kang; Y. Shen; L. Lin, Sn-Based Self-Powered Ultrafast Perovskite Photodetectors with Highly Crystalline Order for Flexible Imaging Applications. *Adv. Funct. Mater.* **2023**, 2300282.
177. A. Liu; H. Zhu; S. Bai; Y. Reo; T. Zou; M.-G. Kim; Y.-Y. Noh, High-performance inorganic metal halide perovskite transistors. *Nat. Electron.* **2022,** *5*, 78–83.
178. H. Zhu; A. Liu; K. I. Shim; H. Jung; T. Zou; Y. Reo; H. Kim; J. W. Han; Y. Chen; H. Y. Chu, High-performance hysteresis-free perovskite transistors through anion engineering. *Nat. Commun.* **2022,** *13*, 1741.
179. H. Zhu; A. Liu; K. I. Shim; J. Hong; J. W. Han; Y. Y. Noh, High-performance and reliable lead-free layered-perovskite transistors. *Adv. Mater.* **2020,** *32*, 2002717.
180. J. Kim; Y. S. Shiah; K. Sim; S. Iimura; K. Abe; M. Tsuji; M. Sasase; H. Hosono, High-Performance P-Channel Tin Halide Perovskite Thin Film Transistor Utilizing a 2D–3D Core–Shell Structure. *Adv. Sci.* **2022,** *9*, 2104993.
181. Y. Gao; Z. Wei; P. Yoo; E. Shi; M. Zeller; C. Zhu; P. Liao; L. Dou, Highly stable lead-free perovskite field-effect transistors incorporating linear π-conjugated organic ligands. *J. Am. Chem. Soc.* **2019,** *141*, 15577–15585.
182. A. Liu; H. Zhu; Y. Reo; M.-G. Kim; H. Y. Chu; J. H. Lim; H.-J. Kim; W. Ning; S. Bai; Y.-Y. Noh, Modulation of vacancy-ordered double perovskite Cs_2SnI_6 for air-stable thin-film transistors. *Cell Rep. Phys. Sci.* **2022,** *3*, 100812.
183. P. Cheng; K. Han; J. Chen, Recent advances in lead-free halide perovskites for photocatalysis. *ACS Mater. Lett.* **2023,** *5*, 60–78.
184. P. Zhou; H. Chen; Y. Chao; Q. Zhang; W. Zhang; F. Lv; L. Gu; Q. Zhao; N. Wang; J. Wang, Single-atom Pt-I_3 sites on all-inorganic Cs_2SnI_6 perovskite for efficient photocatalytic hydrogen production. *Nat. Commun.* **2021,** *12*, 4412.
185. L. Romani; A. Bala; V. Kumar; A. Speltini; A. Milella; F. Fracassi; A. Listorti; A. Profumo; L. Malavasi, PEA_2SnBr_4: a water-stable lead-free two-dimensional perovskite and demonstration of its use as a co-catalyst in hydrogen photogeneration and organic-dye degradation. *J. Mater. Chem. C.* **2020,** *8*, 9189–9194.
186. L. Romani; A. Speltini; F. Ambrosio; E. Mosconi; A. Profumo; M. Marelli; S. Margadonna; A. Milella; F. Fracassi; A. Listorti, Water-Stable $DMASnBr_3$ Lead-Free Perovskite for Effective Solar-Driven Photocatalysis. *Angew. Chem. Int. Ed.* **2021,** *133*, 3655–3662.
187. W. H. Qian; X. F. Cheng; J. Zhou; J. H. He; H. Li; Q. F. Xu; N. J. Li; D. Y. Chen; Z. G. Yao; J. M. Lu, Lead-free perovskite $MASnBr_3$-based memristor for quaternary information storage. *InfoMat* **2020,** *2*, 743–751.
188. Z. Liu; H. Tang; P. Cheng; R. Kang; J. Zhou; X. Zhao; J. Zhao; Z. Zuo, Inorganic Lead-Free and Bismuth-Based Perovskite Nanoscale-Thick Films for Memristors and Artificial Synapse Applications. *Adv. Elect. Mater.* **2023,** *9*, 2201005.

8 Inorganic Perovskite Solar Cells

Kang Wang and Shengzhong Liu

8.1 INTRODUCTION

Organic-inorganic perovskite solar cells have indeed achieved impressive photovoltaic performance, with a certified power conversion efficiency (PCE) of up to 26.1% according to NREL. However, one of the challenges with organic-inorganic perovskites is the instability of the organic components, such as methylamine (MA^+) and formamidine (FA^+), especially under moisture or thermal stress conditions.

To address the instability issues associated with organic components, researchers have turned to replacing them with inorganic metal cesium ions (Cs^+) to form $CsBX_3$ perovskites, where B can be elements like lead (Pb), germanium (Ge), bismuth (Bi), tin (Sn), antimony (Sb), or mixtures thereof, and X represents halide ions such as iodide (I), bromide (Br), chloride (Cl), or mixtures thereof. Among the $CsBX_3$ perovskites, $CsPbX_3$ and $CsBiX_3$ have shown rapid development in solar cells and detectors.

Specifically, $CsPbX_3$ perovskites, including $CsPbI_3$, $CsPbI_2Br$, $CsPbIBr_2$, and $CsPbBr_3$, have attracted significant attention due to their promising photovoltaic performance. These perovskites have achieved efficiencies close to the Shockley-Queisser (S-Q) limit, with $CsPbI_3$ reaching an efficiency of 28.9%, $CsPbI_2Br$ achieving 24.4%, $CsPbIBr_2$ reaching 22.1%, and $CsPbBr_3$ achieving 15.9%. These high efficiencies suggest a strong potential for commercialization.

On the other hand, $CsSnX_3$, $CsGeX_3$, and $CsSbX_3$ perovskites have faced challenges due to their susceptibility to oxidation, which has limited their development in solar cells. However, ongoing research is focused on addressing these challenges to improve their stability and performance. In addition to the aforementioned perovskite compositions, researchers have also fabricated derived structures of $CsBX_3$, such as Cs_2BX_6 (e.g., Cs_2SnI_6, Cs_2PbBr_6, Cs_2PbI_6) and $Cs_3B_2X_9$ (e.g., $Cs_3Sb_2I_9$, $Rb_3Sb_2I_9$, $Cs_3Bi_2I_9$). Furthermore, $Cs_2B^{1+}B^{3+}X_6$ (e.g., $Cs_2BiAgCl_6$, $Cs_2BiAgBr_6$) perovskites have also been explored.

8.2. CRYSTAL STRUCTURE

In inorganic $CsBX_3$ perovskite materials, the B atom is situated at the center of a regular octahedron formed by X atoms. The octahedron structure is obtained by connecting the X atoms at their vertices, creating a 3D network. The Cs^+ cation then occupies the spaces within this octahedral arrangement. This arrangement leads to

DOI: 10.1201/9781003400493-8

a stable crystal lattice structure and is crucial for the unique properties exhibited by inorganic $CsBX_3$ perovskite materials, such as their excellent optoelectronic properties and photoconversion efficiencies. The stabilized typical inorganic $CsBX_3$ can be related to two parameters: the Goldschmidt tolerance factor t (Equation 1) and the octahedral factor μ (Equation 2)

$$\tau = (r_{Cs} + r_x)/\sqrt{2}(r_B + r_x) \quad \text{(Equation 1)}$$

$$\mu = r_B/r_x \quad \text{(Equation 2)}$$

where r_{Cs}, r_B, and r_X are the radii of the corresponding ions. The Goldschmidt tolerance factor provides insight into the structural compatibility between the constituent atoms in the perovskite lattice. When τ is close to 1, it indicates that the cation and anion sizes are well balanced, resulting in a stable perovskite structure. A value of μ close to 1 indicates minimal distortion and suggests a more stable perovskite structure. By considering these two parameters, researchers can assess the stability and feasibility of different $CsBX_3$ perovskite compositions, guiding the design and synthesis of efficient and stable inorganic perovskite materials for various applications, including solar cells.

8.2.1 $CsPbX_3$

In $CsPbX_3$ perovskite materials, where X represents halogen atoms such as Br or I, there are different compositions and crystal structures can exist. And it has four types of crystal polymorphs, including cubic structure (α-, *Pm3m*), tetragonal structure (β-, *4/mbm*), orthorhombic structure (γ-, *Pbnm*), and non-perovskite structure (δ-, *Pnma*). The interconversion of these crystal structures is possible by manipulating the Pb-X-Pb bond angle within the $[PbX_6]^{4-}$ octahedral unit. The bending of this angle can induce structural transitions between the different polymorphs, leading to changes in the material's properties and performance.

8.2.1.1 $CsPbI_3$

$CsPbI_3$ perovskite was first discovered in 1890 by Russian scientist *Alexei Fyodorovich Ioffe* and it was not until 1958 that *Howard C. Dunlap* and *F. Wooten* identified $CsPbI_3$ as a photoconductive material while studying its electrical and optical properties. The non-perovskite structure of $CsPbI_3$, also referred to as the δ-phase or *Pnma* phase, can be obtained through a synthesis method using an aqueous solution under ambient conditions. This low-temperature phase is stable and can exist at room temperature. However, it was found that the α-$CsPbI_3$ phase only stabilized at high temperatures above 300 °C. In the process of cooling down, a spontaneous transformation occurred, leading to the formation of lower symmetry phases, as per Figure 8.1. The β-phase of $CsPbI_3$, with the space group *P4/mbm*, was observed with a bandgap of 1.68 eV at 260 °C. The crystal structure of the β-phase expands along the b-axis. At a lower temperature of 175 °C, the γ-phase of $CsPbI_3$ emerged, characterized by the *Pbnm* space group and a bandgap of 1.73 eV. In this phase, the crystal structure grows along the c-axis. It's interesting to note that each phase of $CsPbI_3$ is

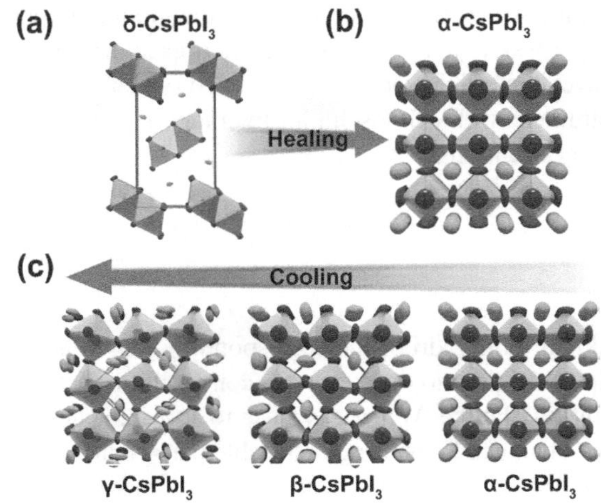

FIGURE 8.1 Structural phase transitions in $CsPbI_3$ versus temperature. (a) Initial yellow perovskitoid phase (δ-$CsPbI_3$, NH_4CdCl_3-type) converts to (b) black perovskite phase (α-$CsPbI_3$, $CaTiO_3$-type) as the temperature exceeds the transition temperature. (c) Upon cooling, the black perovskite phase is retained and can be undercooled below the transition temperature, where the typical perovskite distortions (tetragonal β-$CsPbI_3$, orthorhombic γ-$CsPbI_3$) can be observed to room temperature. These phases are metastable and transform to the thermodynamic δ-$CsPbI_3$ upon standing.

associated with a different Pb-I-Pb angle. The Pb-I-Pb angle for the α-phase is 180°, while it decreases to 163.6° for the β-phase and further to 153.2° for the γ-phase.[1]

8.2.1.2 $CsPbBr_3$

$CsPbBr_3$, similar to $CsPbI_3$, exhibits a perovskite-like crystal structure. At room temperature, $CsPbBr_3$ crystallizes in the orthorhombic (*Pnma*) space group. However, upon heating, $CsPbBr_3$ undergoes phase transitions to different crystal structures. The first phase transition occurs at around 88 °C, where $CsPbBr_3$ transforms into a tetragonal structure with the space group *P4/mbm*. This phase is characterized by an elongation in one crystallographic axis while maintaining tetragonal symmetry. Further heating leads to a second phase transition at approximately 130 °C, resulting in the cubic (*Pm-3m*) phase. This cubic phase has a higher symmetry compared to the previous phases.

Apart from these transitions, $CsPbBr_3$ also exhibits two other derivative phases, namely $CsPb_2Br_5$ and Cs_4PbBr_6. $CsPb_2Br_5$ has a 2D-layered tetragonal structure. In this structure, Cs^+ ions are located between two layers of Pb-Br-coordinated polyhedra. The coordination numbers for Cs^+ and Pb^{2+} in this tetragonal structure are 10 and 8, respectively.

The formation of the $CsPb_2Br_5$ phase is induced by the excessive $PbBr_2$ on the following reaction mechanism[2,3]:

$$CsBr + PbBr_2 = CsPbBr_3$$

$$CsPbBr_3 + PbBr_2 = CsPb_2Br_5$$

The possible processes can be described as follows: the connections of the $PbBr_6$ octahedrons were broken along the c-axis direction to form an expanded interlayer. In the a-b axis plane, the $PbBr_6$ octahedrons rearranged and the polyhedrons changed from $PbBr_6$ octahedrons to $PbBr_8$ capped-triangular prisms. The phase transformation from $CsPbBr_3$ to $CsPb_2Br_5$ appears to be between 150 and 250 °C. The reverse reaction ($CsPb_2Br_5 \rightarrow PbBr_2 + CsPbBr_3$) takes place above 300 °C. After annealing at 400 °C, the tetragonal $CsPb_2Br_5$ fully disappears.

Cs_4PbBr_6 indeed has a different crystal structure compared to $CsPbBr_3$ and $CsPb_2Br_5$. It is a 0-dimensional (0D) structure, meaning it does not have extended layers or continuous connectivity in the crystal lattice. In Cs_4PbBr_6, the $[PbBr_6]^{4-}$ octahedra are separated from each other by CsBr bridges. These bridges composed of Cs^+ and Br- ions connect the $[PbBr_6]^{4-}$ octahedra in a way that creates a distinct 0D structure. The excessive CsBr leads to the formation of Cs_4PbBr_6 that can be explained by the following reaction[4]:

$$4CsBr + PbBr_2 \rightarrow Cs_4PbBr_6$$

$$3CsBr + CsPbBr_3 \rightarrow Cs_4PbBr_6$$

8.2.1.3 $CsPbI_xBr_{3-x}$

This section contains two perovskite compounds: $CsPbI_2Br$ and $CsPbIBr_2$. Similar to $CsPbI_3$, $CsPbI_2Br$ also has multiple phases, specifically the α-, β-, and γ-phases. However, only the α-phase of $CsPbI_2Br$ has been observed experimentally thus far. According to theoretical calculations, the optimized lattice parameters for α-$CsPbI_2Br$ are indeed a = 6.40 Å, b = 6.40 Å, and c = 5.97 Å. And for β-$CsPbI_2Br$, the optimized lattice parameters are a = 8.95 Å, b = 8.95 Å, and c = 6.03 Å. In both the α- and β-phases of $CsPbI_2Br$, the lattice parameter c is slightly smaller than the lattice parameters a and b. This arrangement creates a pseudo-cubic structure in the α-phase and a pseudo-tetragonal structure in the β-phase. These terms refer to the crystal structures being close to a perfect cubic or tetragonal symmetry but with slight distortions as a result of the differing lattice parameters. Additionally, the bond length between Pb and Br (Pb-Br) in $CsPbI_2Br$ (measuring 2.98 Å) is slightly smaller than the bond length between Pb and I (Pb-I; measuring 3.21 Å). This difference can be attributed to the distinct ionic radii of Br and I ions, as Br ions are smaller in size compared to I ions.[5]

$CsPbIBr_2$ exhibits two distinct phases: the room-temperature non-perovskite phase (referred to as the δ-perovskite phase) and the high-temperature perovskite phase (known as the α-perovskite phase). The δ-perovskite phase of $CsPbIBr_2$ indeed belongs to the space group *Pmnb*, with lattice parameters of a = 4.797 Å, b = 9.982 Å, c = 17.184 Å. On the other hand, the α-perovskite phase of $CsPbIBr_2$ has a lattice parameter of a = 5.926 Å and belongs to the space group *Pm-3m*.[6]

8.2.2 Sn-Based Inorganic Perovskite

8.2.2.1 $CsSnI_3$

$CsSnI_3$ is known to exhibit four different phases, including three black phases and one yellow phase. The black cubic phase (α) is formed at above 425 K, which converts to tetragonal phase (β) and orthorhombic phase (γ) at below 426 K and 351 K, respectively. The yellow phase is a non-perovskite phase and can coexist with the γ orthorhombic phase at room temperature. This phase is characterized by the orthorhombic *Pnma* space group.[7,8] According to theoretical calculations, the introduction of Br and Rb into the $CsSnI_3$ framework can potentially suppress the formation of the yellow phase. This suppression occurs due to the reduced oxidation of Sn^{2+} into Sn^{4+} within the perovskite structure. As a result, the γ-$Rb_yCs_{1-y}SnI_3$ perovskite phase becomes more stable.[9] The oxidation instability in $CsSnI_3$ results in a phase transition when exposed to water vapor. The initial photoactive black orthorhombic (γ) phase transforms into a photoinactive 2D yellow phase. This spontaneous transformation of the yellow phase leads to the formation of a weakly absorbing one-dimensional Cs_2SnI_6 structure.[10]

8.2.2.2 $CsSnBr_3$ Perovskite

$CsSnBr_3$ undergoes three successive phase transitions as the temperature decreases[11]:

1. Above 292 K: $CsSnBr_3$ adopts a cubic perovskite structure.
2. Above 247 K: the cubic phase transitions into a tetragonal phase.
3. Below 247 K: the tetragonal phase further transforms into a monoclinic phase.

8.2.2.3 Cs_2SnX_6 Perovskite

At room temperature, Cs_2SnX_6 (where X is Cl, Br, or I) exhibits a cubic structure with the space group *Fm-3m*. Experimental synthesis has been successful for Cs_2SnX_6 and $Cs_2SnI_4X_2$ (where X is Cl or Br). The lattice constants for Cs_2SnCl_6, Cs_2SnBr_6, and Cs_2SnI_6 are 7.59 Å, 7.96 Å, and 8.52 Å, respectively. The calculated lattice constants align well with the experimental values, indicating good agreement. When two iodine (I) atoms in Cs_2SnI_6 are substituted with chlorine (Cl) or bromine (Br), compounds such as $Cs_2SnI_4Cl_2$, Cs_2SnI_4ClBr, and $Cs_2SnI_4Br_2$ can be formed. The decomposition energy for these three mixed halide compounds is 0.12 eV/f.u., 0.08 eV/f.u., and −0.04 eV/f.u., respectively. So three mixed halide compounds can stabilize at low temperature.[12]

8.2.3 $CsGeX_3$ Perovskite

$CsGeX_3$, where X represents the halogen atoms, exhibits a phase transition from the rhombohedral phase to the cubic phase as the temperature increases. At room temperature, $CsGeX_3$ adopts a rhombohedral structure with specific lattice parameters of $a = 5.98$ Å and $\alpha = 88.6°$. The rhombohedral structure at room temperature is very close to the cubic structure, which is observed at high temperatures. The cubic phase is characterized by the following parameters of $a = 5.99$ Å and $\alpha = 90°$.[13] The difference in the structural behavior of Ge-based halide perovskites, such as $CsGeX_3$, compared to Sn-based ($CsSnX_3$) and Pb-based ($CsPbX_3$) perovskites, can be attributed to the stronger tendency of Ge to form sp^3 hybrid orbitals. This allows both the s and p electrons of Ge to participate in bonding. In $CsGeX_3$, the Ge atom can adopt an off-center position along the diagonal within its octahedral coordination environment. This off-center displacement enables stronger bonding between Ge and three of its halogen neighbors compared to the other three halogen neighbors. Essentially, Ge prefers to form fewer but stronger bonds.[14] In the low-temperature structures of Ge-based halide perovskites, this off-center displacement leads to the elongation of three Ge-X bonds, while the remaining three bonds are compressed. This structural arrangement results in a lower energy state and can improve the overall phase stability of the material.[15]

8.2.4 Bi-Based Inorganic Perovskite

Bi^{3+} and Pb^{2+} ions have similar ionic radii and share the same electronic configuration of $6s^26p^0$. This similarity allows for the easy substitution of Pb with Bi in the perovskite lattice. Bismuth-based perovskites exhibit a wide range of structural diversity, spanning from zero-dimensional (0D) dimer units, to one-dimensional (1D) chain-like motifs, two-dimensional (2D) layered networks, and three-dimensional (3D) double perovskite structures[16]. In 0D bismuth perovskites, the structural units are discrete dimers or clusters that are stabilized within the lattice. These dimers or clusters can consist of Bi atoms coordinated with other elements or ligands. $Cs_3Bi_2I_9$ is an example of a 0D dimer species in the space group *P63/mmc* in bismuth-based perovskites. Its crystal structure is based on a hexagonal closest packing of A and X atoms, forming hexagonally stacked AX_3 layers. Within these layers, trivalent metal cations (Bi^{3+}) occupy two-thirds of the octahedral sites, and one-third of the metal sites are vacant. This arrangement results in a complex $Bi_2X_9^{3-}$ anionic cluster known as the isolated metal halide dimer unit. The anionic bi-octahedral clusters, consisting of face-sharing BiX_6 octahedra, are surrounded by monovalent cations (Cs^+) that occupy the A-site of the perovskite structure.[16] In the case of $Cs_3Bi_2I_9$, the dimeric units comprise pairs of face-sharing Bi-I octahedra surrounded by Cs^+ cations.[17–19] Compared to 3D $ABiX_3$ materials, the 0D $A_3Bi_2X_9$ systems, such as $Cs_3Bi_2I_9$, offer improved stability against phase degradation. The dimensionality of 0D $A_3Bi_2X_9$ materials can be increased by mixing nearby halides (e.g., Br-I or Cl-Br), which can modulate the crystal structure. For example, $Cs_3Bi_2Br_{9-x}I_x$ ($x > 0$) and $Cs_3Bi_2I_6Cl_3$ exhibit a 2D layered topology based on the trigonal bromide structure.[20,21]

Additionally, there are other bismuth-based perovskite compounds like $K_3Bi_2I_9$ and $Cs_3Bi_2I_9$, which have 2D structures with corrugated layers formed by corner-connected Bi-I octahedra.[22] Furthermore, black $CsBi_3I_{10}$ has a layered crystal structure that is partly broken into zero-dimensional structures, similar to $Cs_3Bi_2I_9$, between the layers.[23]

8.2.5 Sb-Based Inorganic Perovskite

Sb (antimony) and Bi (bismuth) belong to Group VA of the periodic table and share some similarities in their crystal structures and optoelectronic behavior. In antimony-based perovskites, $Sb_2X_9^{3-}$ enneahalide ions are present within the anionic sublattice, and their crystal structure can be classified into three categories.

1. Zero-Dimensional (0D) Structures: some antimony-based perovskites exhibit a 0D crystal structure, where discrete $Sb_2X_9^{3-}$ clusters are present. $Cs_3Sb_2I_9$ is an example of a zero-dimensional (0D) dimeric structure in antimony-based perovskites. It has a space group of *P63/mmc*. The crystal structure consists of isolated double octahedral structures, which are pairs of face-sharing SbX_6 octahedra. These octahedra form discrete complex anionic metal halide $Sb_2X_9^{3-}$ clusters arranged in dimer units. The clusters are surrounded by Cs atoms, and their fusion allows enough space to accommodate the Cs atoms, thereby stabilizing the dimer form.[24]

   On the other hand, $Cs_3Bi_2I_9$ has a different crystal structure. It has a *P1c1* space group with lattice parameters of $a = 14.60$ Å, $b = 8.19$ Å, $c = 25.17$ Å, and $\beta = 125.05°$. In $Cs_3Bi_2I_9$, the structure consists of fused $(Sb_2I_9)_3$-bi-octahedra clusters surrounded by Cs atoms. Unlike $Cs_3Sb_2I_9$, $Cs_3Bi_2I_9$ does not have a dimeric form but rather features these fused bismuth octahedra clusters.
2. One-Dimensional (1D) Chain Structures: in other cases, antimony-based perovskites feature 1D chain structures. In these structures, the $Sb_2X_9^{3-}$ clusters form extended chains through the linkage of octahedra. The chains can further connect or interact with each other, giving rise to unique electronic and optical properties.
3. $Cs_3Sb_2I_9$ and $Rb_3Sb_2I_9$ can also exhibit a two-dimensional (2D) layered structure. These layered forms are characterized by a $p3\bar{m}1$ space group.[16] The 2D layered forms of $Cs_3Sb_2I_9$ and $Rb_3Sb_2I_9$ provide a different arrangement of the SbX_6 octahedra compared to the 0D dimeric forms. The crystal lattice consists of corrugated double-layered polyanionic structures. These structures are based on corner-connected SbX6 octahedra, which give rise to $Sb_2X_9^{3-}$ sub-units. The polyhedral units are arranged in a 2D fashion, forming layers that extend in the plane of the crystal.

8.2.6 Double Perovskite

The halide double perovskite structure $A_2B'B''X_6$ refers to compounds where two divalent Pb(II) ions in a perovskite structure are replaced by one monovalent B(I) ion and one trivalent B(III) ion. In this structure, the A^+, B^+, B^{3+}, and X sites can be occupied by different elements, leading to a large number of possible compounds.[22] In fact, there are nearly 1,000 compounds that can be formed, and more than 600 of them follow Goldschmidt's rules on perovskite structure stability (Figure 8.2). Through first-principles calculations, 11 optimal materials have been identified as suitable bandgap absorbers for photovoltaic applications.[25] However, only a few of these compounds have been experimentally synthesized using solid-state or solution methods. Examples of such synthesized compounds include Cs_2AgBiX_6 (X = Cl, Br, I), $Cs_2AgInCl_6$, and $Cs_2AgSbBr_6$.[26–28]

For instance, $Cs_2AgBiBr_6$ has been synthesized as a red-orange truncated octahedral single crystal, crystallizing in the cubic space group Fm_3m with a lattice constant (i) of 11.25 Å at room temperature.[29] The synthesized $Cs_2AgBiCl_6$ compound is a face-centered cubic double perovskite with an indirect bandgap of 2.2 eV.[16,22,26,30–36]

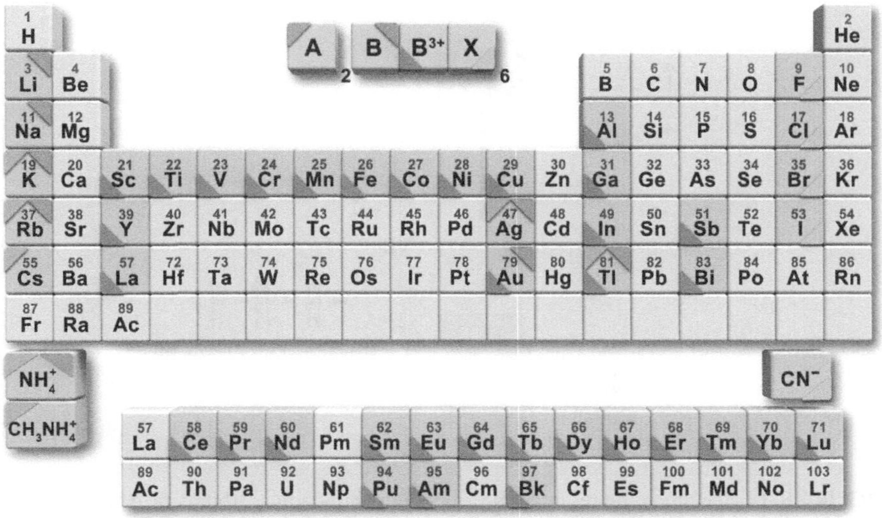

FIGURE 8.2 Elements forming halide double perovskites (elpasolites) with composition $A_2BB^{3+}X_6$. An element is colored in light blue if at least one compound with elpasolite structure containing that element has been synthesized.

8.3 OPTICAL AND ELECTRONIC PROPERTIES

8.3.1 $CsPbX_3$

The electronic structure of $CsPbX_3$ (X = Cl, Br, I) shares similar characteristics with organic-inorganic perovskites. In $CsPbX_3$, the valence band maximum (VBM) is primarily composed of antibonding hybridized Pb 6s and X np orbitals, with the X np orbitals making the major contribution. On the other hand, the conduction band minimum (CBM) is determined by the antibonding mixing of Pb 6p and X np orbitals, with the dominant contributions coming from Pb 6p orbitals, as shown in Figure 8.3. Previous studies have shown that the Cs^+ ion can negatively impact the electronic structure near the band edge, although its effect is more pronounced in the distortion of $[PbX_6]^{4-}$ octahedra when Cs^+ is substituted with other cations. For example, the different black phases of $CsPbI_3$ are attributed to variations in the Pb-I-Pb bond angles. The γ phase of $CsPbI_3$ has a bandgap of approximately 1.73 eV, while the β phase has a bandgap of 1.68 eV. However, the α-$CsPbI_3$ phase with a bond angle of 180° has not been experimentally reported due to its thermodynamic instability at room temperature.

In comparison to $CsPbI_3$, the geometric structures of $CsPbCl_3$ and $CsPbBr_3$ show minimal changes in different phases. $CsPbBr_3$ films, for instance, have a yellow color indicating a limited light absorption range. The absorption spectrum edge of $CsPbBr_3$ films is shorter than 540 nm, and an exciton-induced absorption peak is observed around 520 nm.[4] It has also been discovered that the bandgap of $CsPbBr_3$ can be tuned from 2.3 to 4.0 eV by controlling the ratio of CsBr and $PbBr_2$

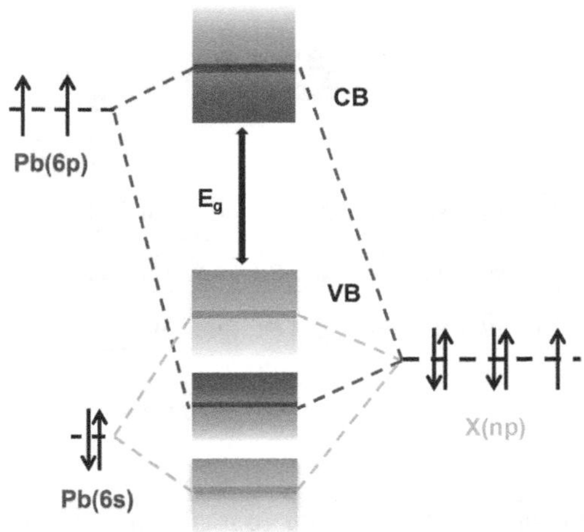

FIGURE 8.3 Schematic representation of bonding/antibonding orbitals of $APbX_3$ showing the formation of the VB and CB.

(Figure 8.4a), which allows for phase transitions from the cubic perovskite structure to derivative phases such as Cs_4PbBr_6 and $CsPb_2Br_5$.

Furthermore, $CsPbI_2Br$ exhibits a reasonable bandgap of 1.80–1.92 eV depending on the deposition technique, while $CsPbIBr_2$ achieves an acceptable bandgap of approximately 2.05 eV.[37,38] In addition, the linear trend of the absorbance plot followed Vegard's law, indicating $CsPbI_3$ and $CsPbBr_3$ have good miscibility, and thus no inhomogeneous landscape exists in the whole series of nanocomposites (Figure 8.4b).

8.3.2 Sn-Based Inorganic Perovskite

Tin-based inorganic perovskites, specifically $CsSnX_3$ (X = I, Br), have been explored as an alternative to lead-based perovskites due to their potential for achieving bandgaps closer to the ideal bandgap of the Shockley-Queisser limit. The $CsSnX_3$ family includes $CsSnI_3$, $CsSnI_2Br$, $CsSnIBr_2$, and $CsSnBr_3$. These compounds exhibit bandgaps of 1.30 eV, 1.40 eV, 1.65 eV, and 1.75 eV, respectively. The smaller gap between $CsSnI_3$ and $CsSnBr_3$ can be attributed to the larger spin-orbit coupling in iodine (I) compared to bromine (Br). Additionally, the degree of antibonding mixing of the anion p orbitals in the valence band can vary depending on the size of the X-site anion (Cl, Br, or I). As the size of the anion increases from Cl to I, the anion p levels move farther downward, leading to a decreased degree of antibonding mixing. This decrease in antibonding mixing further contributes to a smaller bandgap as the X-site anion size increases.[39] Moreover, the structural distortion, particularly the X-Sn-X bond angles, can also impact the bandgap in tin-based perovskites. The phase transition from a high-symmetry cubic phase to a lower-symmetry tetragonal phase can

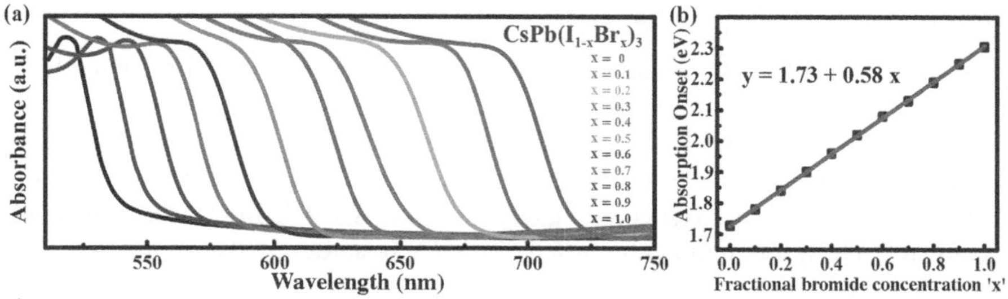

FIGURE 8.4 (a) Absorbance spectra for inorganic perovskite films of different nanocomposites, $CsPb(I_{1-x}Br_x)_3$. (b) Plots of the optical band gap of $CsPb(I_{1-x}Br_x)_3$ as a function of bromide composition (x).

induce crystal-field splitting of the CBM and further modify the bandgap in addition to the spin-orbit coupling effect.

It is interesting to note that in $CsSnI_3$, the valence band maximum (VBM) is composed of antibonding Sn 5s and I 5p orbitals, while the conduction band minimum (CBM) consists of Sn 5p and I 5p antibonding orbitals. The contribution of Cs 5d orbitals is observed at around 5 eV. Furthermore, the bandgap of these materials is direct, meaning that the minimum energy needed to excite an electron from the valence band to the conduction band is achieved at a specific crystal momentum. In the case of α-$CsSnCl_3$, it is found through density functional theory (DFT) calculations that the valence band is mainly dominated by Cl p orbitals. However, in the energy region near the band gap, the conduction band is predominantly composed of Sn p orbitals, with a minor contribution from Cl p states.[40]

In addition, Cs_2SnCl_6 and Cs_2SnBr_6 have experimentally observed band gaps of 0.6 and 0.8 eV, respectively, while theoretically, they are predicted to have band gaps of 0.2–0.3 eV.[41] Furthermore, the band gap of Cs_2SnX_6 compounds can be modulated by mixing different halide compounds. For example, $Cs_2SnI_4Cl_2$, Cs_2SnI_4ClBr, and $Cs_2SnI_4Br_2$ possess band gaps of 1.75, 1.75, and 1.56 eV, respectively.[12] In the case of Cs_2SnX_6, the valence band top is mainly composed of I-5p orbitals (or Cl-3p/Br-4p orbitals), while the conduction band bottom is dominated by Sn-5s and I-5p orbitals (or Cl-3p/Br-4p orbitals). These mixed halide compounds exhibit similar contributions from each atomic orbital to the band edges. Taking $Cs_2SnI_4Cl_2$ as an example, the valence band top is dominated by Cl-3p orbitals, while the conduction band bottom is composed of Sn-5s, I-5p, and Cl-3p orbitals. It is noteworthy that Sn-5p orbitals do not contribute to the band edges in Cs_2SnX_6 compounds, which is different from the electronic properties of $CsSnX_3$ halide perovskites.

Additionally, Cs_2SnX_6 compounds have an isolated octahedral structure, which can limit carrier mobility and result in high minority carrier-effective masses. However, if future advancements are made in increasing the carrier-diffusion lengths of Cs_2SnX_6 materials, such as producing high-quality films with larger grains and fewer defects, it could potentially lead to improved power conversion efficiency in photovoltaic applications.[42]

8.3.3 Ge-Based Inorganic Perovskite

The trend of band gap variation in $CsGeX_3$ compounds is not similar to that of $CsPbX_3$. The specific band gaps for $CsGeCl_3$, $CsGeBr_3$, and $CsGeI_3$ that you mentioned are 3.67, 2.32, and 1.6 eV respectively. [13,43] In $CsGeX_3$ compounds, the valence states are primarily determined by the combination of the Ge s- and p-states along with the p-states of the respective halogen atoms (X = Cl, Br, I). The conduction states in $CsGeX_3$ are mainly composed of the Ge p-states and the p- and d-states of the halogen atoms (X). In terms of the cesium (Cs) states, they typically appear in the higher-energy conduction and valence states.[44]

It is noted that $CsGeI_3$ has been successfully fabricated experimentally, while $CsGeF_3$, $CsGeCl_3$, and $CsGeBr_3$ have been reported theoretically. Notably, the narrower band gap (1.6 eV) of $CsGeI_3$ in the rhombohedral drives it as a promising absorber, while the cubic structure is 1.20 eV.[45] In the cubic phase, the Ge-p states located at the Brillouin zone corner experience minimal interactions with other orbitals of the conduction band minimum (CB). However, when the symmetry of the cubic phase is lowered, such as in the case of $CsGeI_3$, antibonding interactions between the Ge-p states and the I orbitals start to occur. These interactions raise the energy level of the conduction band at the Brillouin zone corner (R2) and flatten out the lowest energy level of the conduction band.[46]

8.3.4 Bi-Based Inorganic Perovskite

$Cs_3Bi_2I_9$ exhibits an excitonic absorption peak at approximately 485 nm (2.56 eV) due to the strong quantum confinement effect arising from the 0D nature of $[Bi_2I_9]^{3-}$ bio-octahedra.[47,48] The measured optical bandgap of $Cs_3Bi_2I_9$ is reported to be in the range of 2.12–2.13 eV and exhibits a relatively high optical absorption coefficient of approximately $1x10^4$ cm^{-1}.[49,50] The valence band states in $Cs_3Bi_2I_9$ are primarily determined by the I p and Bi s orbitals, while the conduction band states are composed of a combination of I sp and Bi p orbitals.[18] Mixed halide compounds based on $Cs_3Bi_2X_9$, such as $Cs_3Bi_2I_6Cl_3$ and $Cs_3Bi_2I_6Br_3$, exhibit similar band gaps to $Cs_3Bi_2I_9$, with values of 2.07 eV and 2.05 eV, respectively.[20,51] Specifically, in $Cs_3Bi_2I_6Cl_3$, it has been found that the Cl orbital in the perovskite compound has a negligible contribution to the electronic structure.[21] The absorption coefficients for $CsBi_3I_{10}$, $Cs_3Bi_2I_9$, and BiI_3 between 350 and 500 nm are reported to be approximately 1.4×10^5 cm^{-1}, 7×10^4 cm^{-1}, and 2.2×10^5 cm^{-1}, respectively.[19]

8.3.5 Sb-Based Inorganic Perovskite

The antimony halide inorganic perovskites have the basic formula $Cs_3Sb_2X_9$ and $Rb_3Sb_2X_9$. Experimental results indicate that the cesium analogue Cs3Sb2X9 has an indirect and direct band gap of approximately 2.00 eV (0D) and 2.05 eV (2D), respectively. For the rubidium analogue $Rb_3Sb_2I_9$, the Tauc plot yielded an indirect and direct band gap of approximately 2.1 eV and 2.24 eV, respectively.[24] Based on the partial density of states (PDOS) plots, the valence band maximum (VBM) is primarily determined by the Iodine (I) p states, with a smaller contribution from the Antimony (Sb) lone-pair s states. On the other hand, the conduction band is a mixture of both Iodine (I) p states and Antimony (Sb) p states. However, in $Cs_3Sb_2I_9$ perovskite, the conduction band is derived from Sb p orbitals, The p-orbital character of both the valence and conduction bands in $Cs_3Sb_2I_9$ leads to a high joint density of states, which subsequently results in a large optical absorption coefficient.[52] The availability of multiple orbitals (I p, Sb p) contributing to the electronic transitions increases the number of states available for absorption, leading to enhanced optical absorption. Both the dimer and layered structures of $Cs_3Sb_2I_9$ perovskite exhibit a similar electronic structure.[51]

$CsSbCl_4$, another perovskite compound, exhibits an optimal direct bandgap of approximately 1.4 eV with a large optical absorption coefficient of around 10^5 cm^{-1}. The bandgap of $CsSbCl_4$ can be tuned by altering its dimensionality. For example, as the layer number increases in layered $CsSbCl_4$ perovskites, the direct bandgap decreases from 1.58 eV (1-layer) to 1.57 eV (2-layer) and eventually to 1.44 eV (3-layer).[53] This change in bandgap is attributed to a weaker quantum confinement effect resulting from the increasing number of layers.

8.3.6 Double Perovskite

For $A_2M^+M^{3+}XVII_6$ perovskites, the specific electron configuration of the M+ cation can be naturally classified into three categories.

1. Type-I $A_2M^+M^{3+}X^{VII}_6$ with M^+ = Na^+, K^+, and Rb^+ (typical material: $Cs_2NaBiCl_6$): the valence band maximum (VBM) in these perovskites is derived purely from the anti-bonding state between Bi-s and Cl-p orbitals. The conduction band minimum (CBM) is a mixture of Bi-p orbitals, unoccupied s orbitals of the ionic M^+ (Na), and Cl-p states. The CBM and VBM states slightly increase or decrease as M^+ changes from Na to K to Rb, influenced by the energy of M^+ s orbitals and the size of the M^+XVII_6 and $M^{3+}XVII_6$ octahedra. The indirect band gap of $Cs_2NaBiCl_6$ is above 3.0 eV due to larger (p-s) coupling at the W and X points compared to the Γ point.
2. Type-II $A_2M^+M^{3+}X^{VII}_6$ with M^+ = Cu^+, Ag^+, and Au^+ (typical material: $Cs_2AgBiCl_6$): the VBM is mainly dominated by the anti-bonding hybridized state of Ag-d and Cl-p orbitals, with a contribution from Bi-s orbital, and The CBM is a mix of Bi-p orbital, Ag-s, and Cl-p states. The electronic symmetry reduction due to the orbital mismatch between Ag-d and Bi-s leads to a larger off-centered p-d coupling, shifting the VBM from the Γ point to the X point. The band gap variation in $Cs_2AgBiCl_6$ is mainly influenced by the change in VBM states as M^+ changes from Cu to Ag to Au, with Ag-containing compounds having the lowest VBM states and the largest band gaps due to weaker p-d repulsion.[54]
3. Type-III $A_2M^+M^{3+}X^{VII}_6$ with M^+ = In^+ and Tl^+ (typical material: $Cs_2InBiCl_6$): the band gap in these perovskites is direct, with the VBM located at the zone center Γ point. The VBM is primarily derived from the antibonding state of In-s and Cl-p orbitals due to the higher energy of In-s orbital compared to

Bi-s orbital. The CBM is composed of Bi-p, In-p, and Cl-p states. As M^+ changes from In to Tl, the lower energy of Tl-6s and the larger atomic size of Tl result in the reduced s-p coupling, leading to lower VBMs and a substantial increase in the band gaps of Tl-containing systems.

8.4 PHOTOPHYSICAL PROPERTIES

8.4.1 $CsPbX_3$

Photophysical properties encompass several key parameters, including carrier diffusion length, lifetimes, the effective mass of holes and electrons (μ), exciton binding energy (R*), and effective dielectric constants (ε_{eff}). According to the calculations, the exciton binding energies of inorganic $CsPbI_3$, $CsPbI_2Br$, and $CsPbBr_3$ are reported as 15 ± 1 mV, 22 ± 3 meV, and 33 meV, respectively. [30,55] These values suggest that as the halide composition becomes heavier (from iodine to bromine), the exciton binding energy of free carriers in the perovskite lattice increases. The effective dielectric constants (ε_{eff}) for $CsPbI_3$, $CsPbI_2Br$, and $CsPbBr_3$ perovskites are reported as 10.0, 8.6, and 7.3, respectively. The reported values for the effective carrier mobility of $CsPbI_3$, $CsPbI_2Br$, and $CsPbBr_3$ perovskites are approximately ~25 $cm^2V^{-1}s^{-1}$[55,56], ~10 $cm^2V^{-1}s^{-1}$[57], and 38 $cm^2V^{-1}s^{-1}$[58], respectively. The large lifetimes for $CsPbI_3$ perovskite were approximately 10 μs for evaporated films, which is higher compared to spin-coated films with a lifetime of around 200 ns. In the case of $CsPbBr_3$, macroscopic single crystals were reported to have charge lifetimes ranging from 2 to 7 μs. For solution-processed $CsPbI_2Br$, a shorter charge lifetime of approximately 14 ns was reported.[59] The estimated diffusion length (L_d) for solution-processed $CsPbI_3$ perovskite is reported to be approximately 1 μm.[60] For single crystal $CsPbBr_3$, separate values of diffusion length for electrons and holes were found to be around 10 μm and 12 μm, respectively.[61] However, in the case of microcrystalline films, the diffusion length for electrons was reported as 80 nm.[62] It's important to note that in general, single-crystal perovskites tend to have longer charge lifetimes and therefore longer diffusion lengths than polycrystalline films.[63] Single crystals usually exhibit lower trap densities and higher electronic quality compared to polycrystalline films. This can be connected to the presence of defects, where the low defect concentration in single crystals contributes to longer charge lifetimes and larger diffusion lengths. $CsPbX_3$ perovskites, similar to hybrid halide perovskites (HaPs), are considered defect-tolerant semiconductors. They can maintain their electronic qualities despite the presence of defects.[64] This defect tolerance is attributed to the absence of bonding-antibonding interactions between the conduction and valence bands.[65]

8.4.2 Sn-Based Inorganic Perovskite

The γ-$CsSnI_3$ perovskite exhibits a hole mobility (μ_h) of approximately 585 $cm^2\ V^{-1}\ s^{-1}$ and an electron mobility (μ_e) of around 2268 $cm^2\ V^{-1}\ s^{-1}$ at a charge density of approximately $10^{17}\ cm^{-3}$.[66] The diffusion length of $CsSnI_3$ is comparable to that of the $MAPbI_3$ perovskite. In a study on melt-synthesized $CsSnI_3$ single crystals, it was found that $CsSnI_3$ has a carrier lifetime of 6.6 ns, a carrier diffusion length of 930 nm, and a charge concentration of approximately 4.5×10^{17} cm^{-3}.[67] Furthermore, the surface recombination rate of $CsSnI_3$ was reported to be around 2×10^3 cm s^{-1}, which is comparable to that of $CsPbI_3$.[67]

The electrical properties of the Cs_2SnI_6 perovskite material vary depending on its doping and carrier concentration. In the case of undoped Cs_2SnI_6, the electrical resistivity is approximately 100 Ω cm, indicating its conductivity in the n-type semiconductor region. Hall Effect measurements have determined that the electron carrier concentration in undoped Cs_2SnI_6 is on the order of 1×10^{14} cm^{-3}. Additionally, it exhibits a high electron mobility of 310 cm^2/V.s. When Cs_2SnI_6 is doped with Sn^{2+} (using SnI_2), the resulting material retains a carrier concentration of approximately $1 \times 10^{14} cm^{-3}$ but with the opposite sign, transforming it into a p-type semiconductor. In the p-type variant, the hole mobility (the mobility of positive charge carriers) is lower than that of the n-type variant, reaching a value of 42 cm^2/V·s. The reduced hole mobility leads to a significantly increased resistivity of 780 Ω·cm. Despite the reduction in mobility, the p-type Cs_2SnI_6 still maintains a sizable hole mobility, comparable to that of $MAPbI_3$ perovskite.[68,69]

8.4.3 Ge-Based Inorganic Perovskite

$CsGeX_3$ compounds (where X represents F, Cl, Br, or I) exhibit a high absorption coefficient of around 10^7 cm^{-1} in the visible region, indicating strong light absorption capabilities. In the p-type region, the electrical conductivity (σ) of $CsGeF_3$, $CsGeCl_3$, and $CsGeBr_3$ increases to peak values. At 300 K, these peak conductivity values are approximately 18.45×10^{19} at −2.47 eV for $CsGeF_3$, 15.41^{19} at −2.26 eV for $CsGeCl_3$, and 12.78×10^{19} $(\Omega{\cdot}m{\cdot}s)^{-1}$ at −99eV for $CsGeBr_3$. These values remain constant across the entire temperature range of 0 to 1,000 K.

Conversely, in the n-type region, the electrical conductivity increases at an energy level of approximately +2.7 eV, reaching peak conductivity values. For all considered temperatures, the peak conductivity values are about 41.40×10^{19} for $CsGeF_3$, 65.56×10^{19} for $CsGeCl_3$, and 64.01×10^{19} $(\Omega{\cdot}m{\cdot}s)^{-1}$ for $CsGeBr_3$.[70] Regarding $CsGeI_3$, band structure calculations indicate an effective hole mass of 0.087. The effective mass of charge carriers is an important parameter in determining their mobility and transport properties.[14,45]

8.4.4 Bi-Based Inorganic Perovskite

In single crystal thin-film form, $Cs_3Bi_2I_9$ exhibits a carrier mobility (μ) of approximately 1.7×10^{-2} cm^2 s^{-1} V^{-1}. This value is about 3.8×10^4 times higher than the carrier mobility of a polycrystalline film, which is estimated to be 4.4×10^{-7} cm^2 s^{-1} V^{-1}.[17] Replacing Cl with I in $Cs_3Bi_2I_6Cl_3$ (known as "$Cs_3Bi_2I_9$") leads to a significant reduction in out-of-plane electron and hole effective masses. The electron effective mass decreases from 0.63 m_e to 0.39 m_e, while the hole effective mass decreases from 1.8 m_e to 0.41 m_e due to the absence of Cl.[21] In the case of 0D $Cs_3Bi_2I_9$ (presumably referring to $Cs_3Bi_2I_9$ as a zero-dimensional material), it exhibits reasonable electron mobility despite the lack of inter-octahedral connections between molecular dimers. The electron mobility is estimated to be 4.3 $cm^2V^{-1}s^{-1}$, indicating the ability of charge carriers to move through the material in response to an electric field.[71] The calculated effective masses for $Cs_3Bi_2I_9$ show the lowest in-plane effective masses for both electrons (0.34 m_0) and holes (0.89 m_0), with the highest effective mass for electrons being 9.34 m_0 in the out-of-plane direction due to the flat conduction band along the Γ-A direction. The $(\mu\tau)_h$ products, which represent the hole mobility (μ_h) multiplied by the carrier lifetime (τ), range from 1.8(2) × 10^{-5} cm2 V^{-1} for $Cs_3Bi_2I_9$ to 2.0(2) × 10^{-6} cm^2 V^{-1} for $Rb_3Bi_2I_9$. Another material, $Cs_3Bi_2I_9$ (a two-dimensional material), exhibits the best $(\mu\tau)$ product for electrons, with a value of 1.1 × 10^{-5} cm^2 V^{-1}.[71] The hole mobility of $Cs_3Bi_2I_9$ is reported to be 1.7 ± 0.1 cm^2 V^{-1} s^{-1}.

The estimated $(\mu\tau)_e$ product for electrons in these materials ranges from 1.7 × 10^{-6} cm^2 V^{-1} in $Rb_3Bi_2I_9$ to 5.4 × 10^{-5} cm^2 V^{-1} in $Cs_3Bi_2I_9$. The rise time of the α-particle generated pulse was used to estimate the electron mobility (μ_e), which ranged from 0.32 cm^2 V^{-1} s^{-1} for $Rb_3Sb_2I_9$ to 4.3 cm^2 V^{-1} s^{-1} in $Cs_3Bi_2I_9$. Similarly, the $(\mu\tau)$h values were derived from the same analysis and ranged from 1.8 × 10^{-5} cm^2 V^{-1} in $Cs_3Bi_2I_9$ to 2.0 × 10^{-6} cm^2 V^{-1} in $Rb_3Bi_2I_9$. The hole mobilities (μ_h) were reported to range from 1.7 cm^2 V^{-1} s^{-1} for $Cs_3Bi_2I_9$ to 0.14 cm^2 V^{-1} s^{-1} for $Cs_3Sb_2I_9$.[71] Additionally, it is mentioned that the electron mobility is approximately 30 times higher than the hole mobility, with the electron mobility (μ_e) being 600 cm^2 V^{-1} s^{-1} and the hole mobility (μ_h) being 20 cm^2 $V^{-1}s^{-1}$. This comparison between electron and hole mobilities indicates a significant disparity in their respective movement abilities under the influence of an electric field.[19]

8.4.5 Sb-Based Inorganic Perovskite

In the layered structure of $Cs_3Sb_2I_9$ and $Rb_3Sb_2I_9$, the valence and conduction bands exhibit dispersion, suggesting that the effective masses of electrons (m_e^*) and holes (m_h^*) should not be excessively large.[52] The electron mobility (μ_e) for $Cs_3Sb_2I_9$ is estimated to be 1.8 ± 0.1 cm^2 V^{-1} s^{-1}, while for $Rb_3Sb_2I_9$ it is estimated to be 0.32 ± 0.03 cm^2 V^{-1} s^{-1}.[71] On the other hand, the hole mobility of $Cs_3Sb_2I_9$ is reported to be 0.32 ± 0.03 cm^2 V^{-1} s^{-1}.

For few-layered $CsSbCl_4$, the carrier effective masses have been calculated to be approximately 0.1 times the free electron mass (m_0), which is smaller compared to widely used photovoltaic (PV) materials such as $MAPbI_3$ and Si but similar to that of GaAs. Moreover, the carrier mobilities in $CsSbCl_4$ can reach 103 cm^2 V^{-1} s^{-1} at room temperature, which is comparable to those observed in $MAPbI_3$, Si, and GaAs.[53]

8.4.6 Double Perovskite

The $Cs_2AgBiBr_6$ single crystal exhibits a calculated carrier lifetime of 1.4 μs, indicating long carrier lifetimes.[72] The calculated carrier diffusion length within the temperature range of room temperature (RT) to 123 K is found to be in the range of 700 nm to 2.44 μm. The corresponding carrier mobilities are reported to be 0.8 and 10 cm^2 V^{-1} s^{-1} at RT and 123 K, respectively.[73] Thin films of $Cs_2AgBiBr_6$ demonstrate carrier mobilities of up to 0.39 cm^2 V^{-1} s^{-1} (vacuum-processed) and 0.74 cm^2 V^{-1} s^{-1} (solution-processed).[74] The carrier diffusion length in $Cs_2AgBiBr_6$ films is observed to be up to 110 ± 20 nm for both electrons and holes.[75] The electron effective mass and hole effective mass of $Cs_2AgBiCl_6$ and $Cs_2AgBiBr_6$ are reported to be 0.53 m_e (0.37 m_e) and 0.15 m_e (0.14 m_e) respectively. These values indicate relatively lighter effective masses compared to lead-based analogs.[27,76] The exciton binding energy of $Cs_2AgBiBr_6$ is calculated to be 268 meV using the Elliott model. This suggests fast photon-induced carrier dissociation due to the small exciton binding energy.[77]

8.5 DEFECTS

8.5.1 $CsPbX_3$

In $CsPbX_3$ perovskite, intrinsic point defects can occur due to thermal vibrations causing atoms on the lattice nodes to deviate from their regular positions. These defects include vacancies (where lattice atoms are missing), interstitials (occupying spaces between lattice atoms), and anti-substitutions (atoms occupying incorrect positions in the lattice). Specifically, in $CsPbX_3$, vacancy defects comprise V_{Cs}, V_{Pb}, and V_I; interstitial defects include Cs_i, Pb_i, and I_i; and anti-substitution defects involve (Cs_{Pb}, Pb_{Cs}, Cs_I, Pb_I, I_{Cs}, and I_{Pb}) as per the Kroger-Vink method.[78] These defects can create shallow or deep-level trap states. Vacancy and interstitial defects are generally considered shallow-level defects due to their lower formation energies. They have minimal impact on carrier trapping or recombination. However, shallow-level defects can accelerate ion migration, leading to adverse phase separation and irreversible structural distortion, which can affect the long-term stability of the device. On the other hand, anti-substitution defects have high formation energies, making them deep-level defects. Carriers trapped by deep-level defects are challenging to release, resulting in significant non-radiative recombination, reduced carrier mobility and lifetime, and a deterioration in device performance. These deep-level defects can have a pronounced negative impact on the functionality and efficiency of the device.

In addition to point defects, perovskite films can also exhibit two other types of neutral defect pairs: Frenkel defects and Schottky defects. Frenkel defects involve the presence of ionic vacancies and interstitial ions within the crystal structure. Schottky defects, on the other hand, occur when both anionic and cationic vacancies are simultaneously present.[79] Three main types of defects can be considered in perovskite films. First, 1D defects arise from localized atomic misalignments, leading to dislocations in the crystal structure. These defects are typically observed as line defects within the material. Second, 2D defects are formed by severe atomic dislocations that occur at the crystal surface and grain boundaries (GBs) of the perovskite film. These defects are often associated with planar or interfacial regions and can have a significant influence on the properties of the material. Last, 3D defects include precipitates, large voids, and secondary phase domains. These defects can cause significant volume changes in the crystal structure and are typically observed as bulk or volumetric defects within the material. Furthermore, in perovskite films, there is a risk of irreversible decomposition due to the reduction of uncoordinated Pb^{2+} ions to Pb^0 and the oxidation of uncoordinated I- ions to volatile I_2. These processes can lead to the degradation and deterioration of the perovskite material.

8.5.2 Sn-Based Inorganic Perovskite

For the $CsSnX_3$ system, the possible intrinsic defects include Cs vacancies (V_{Cs}), Sn vacancies (V_{Sn}), iodine vacancies (V_I), Cs replacing Sn (Cs_{Sn}, also known as Cs-Sn antisite), Sn replacing Cs (Sn_{Cs}), Cs replacing I (Cs_I), iodine replacing Cs (I_{Cs}), Sn replacing I (Sn_I), iodine replacing Sn (I_{Sn}), Cs interstitial (Cs_i), Sn interstitial (Sn_i), and iodine interstitial (I_i). Under Sn-rich conditions (Figure 8.2a), V_{Cs} has the lowest formation energy among all acceptor defects, while Sn_I has the lowest formation energy among all donor defects. Over time, the formation of defect densities can be as high as approximately 10^{19} cm^{-3}. In $CsSnI_3$, under Sn-poor conditions, the facile formation of acceptor defects such as V_{Cs} and V_{Sn} contributes to high p-type electrical conductivity. In this scenario, all low-energy defects have resonant defect levels, and there are no recombination-center defects. As the Sn content increases, the concentration of acceptor defects decreases, while the concentration of donor defects increases. Consequently, the hole carrier concentration and p-type electrical conductivity decrease. Additionally, the concentration of a deep-level defect, Sn_I antisite, increases, leading to nonradiative recombination of electron-hole pairs.[80,81]

The contribution of defects to carrier concentration and conductivity in $CsSnI_3$ can be further studied by ionization, which is determined by their respective transition energy levels. Acceptor defects have transition energy levels below the valence band maximum (VBM), indicating lower ionization energy, as per Figure 8.5a. Therefore, a higher concentration of acceptor defects, such as V_{Cs} and V_{Sn}, results in a higher concentration of hole carriers and higher conductivity. Donor defects, on the other hand, mostly have shallow energy transition levels located near the conduction band minimum (CBM). However, the Sn_I defect exhibits a deep transition level within the band gap. In p-type $CsSnI_3$, the deep-level Sn_I defect acts as an electron trap and

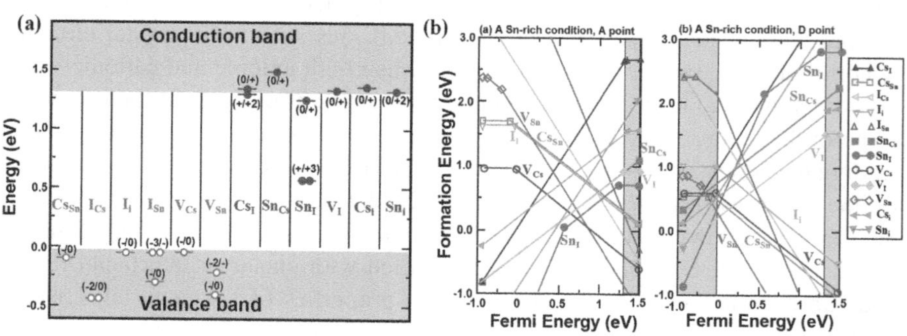

FIGURE 8.5 (a) Calculated transition energy levels for various intrinsic defects. The regions under VBM and above CBM are in shaded color. Acceptor and donor defect levels are denoted by red and blue lines, respectively. The hollow (solid) points at the levels show the number of hole (electron) carriers that can be released during the transition of the defect charge state. (b) Calculated formation energy of defects as a function of the Fermi energy EF when (μCs, μSn) is at (a) A and (b) D points.

recombination center for photogenerated electron-hole pairs, leading to decreased electrical conductivity and shorter carrier lifetime.

Controlling the Sn content is crucial for controlling the defect density in $CsSnI_3$. Under different Sn content conditions, the formation energies of acceptor defects (V_{Cs} and V_{Sn}) are much lower than those of donor defects, indicating a high concentration of acceptor defects and p-type conductivity. However, when the Fermi level is in the middle of the band gap, the electrical conductivity will be lower. Additionally, as the Sn content increases, the minimum formation energy of acceptor defects increases, indicating a decrease in hole carrier concentration during crystal growth.

8.5.3 Ge-Based Inorganic Perovskite

Germanium halide perovskites have received less experimental investigation compared to other halide perovskite materials. This lack of study is mainly attributed to the chemical instability that arises from the oxidation of the Ge^{2+} cation.[13,43] Compared to tin-based perovskites, germanium-based perovskites are more susceptible to this oxidation stability issue due to the reduced inert electron pair effect. The inert electron pair effect refers to the reluctance of elements with filled or partially filled d or f orbitals to participate in bonding. The Ge^{2+} has a completely filled $3d^{10}$ electron configuration, which makes it less prone to bond formation. This reduced chemical reactivity makes it susceptible to oxidation when exposed to air or other oxidizing agents. The oxidation of the Ge^{2+} cation leads to the generation of Ge^{4+}, which results in a change in the oxidation state of germanium from +2 to +4. This oxidation process introduces additional defects and alters the electronic structure of the perovskite material, which negatively impacts its performance as a semiconductor.

8.5.4 Double Perovskite

$A_2M^{+}M^{3+}XVII_6$ perovskites, such as $Cs_2InSbCl_6$ and $Cs_2InBiCl_6$, are mentioned as examples that exhibit similar defect-tolerant features to $CH_3NH_3PbI_3$ due to their comparable chemical bonding and band structure. This similarity arises from the presence of M^{3+} ions (Sb^{3+} or Bi^{3+}) that possess the same electron configuration (ns^2) as Sn^{2+} or Pb^{2+} ions. In the case of $Cs_2InSbCl_6$ and $Cs_2InBiCl_6$ perovskites, the p-type defects, such as $V_{Sb/Bi}$ (Sb/Bi vacancies) and V_{In} (indium vacancies), are expected to be shallow defects. The shallow nature of these defects can be attributed to the anti-bonding feature between the In^{+}/Bi^{3+}-s (valence orbitals of In^{+}/Bi^{3+} ions) and Cl-p (valence orbitals of Cl ions) orbitals present in the valence bands of these materials. The anti-bonding nature of these orbitals makes it energetically favorable for the defects to have shallow energy levels in the band gap.

Regarding the most likely n-type defects in $A_2M^{+}M^{3+}XVII_6$ perovskites, Cs_i (cesium interstitial) and V_{Cl} (chlorine vacancies), it is suggested that these defects might also be shallow. This can be attributed to the strong ionic bonding nature of the materials in this family. The strong ionic bonds result in a large energy gap between the valence and conduction bands, which leads to the formation of shallow defect levels in the band gap.[25,82]

8.6 PASSIVATION STRATEGY

Although the pristine inorganic perovskite has shown promising materials photovoltaic performance, the intrinsic layer needs to further improve its crystallinity, phase stability, free carrier transport efficiency, and photovoltaic performance.

8.6.1 ELEMENT DOPING

Indeed, ionic incorporation is an efficient strategy to optimize the tolerance factor and enhance the phase stability of inorganic perovskite materials. In the case of $CsPbI_3$, which suffers from phase instability, partially replacing I- ions with Br- ions can increase the tolerance factor and improve phase stability. This substitution can result in the formation of $CsPbI_2Br$ and $CsPbIBr_2$ compositions. Although this may lead to a loss in efficiency, it offers better stability for device operation. Additionally, other elements such as Bi^{3+}, Sn^{2+}, Ca^{2+}, Sb^{3+}, Eu^{3+}, Mn^{2+}, etc. have been successfully incorporated into the $CsPbX_3$ framework to tune crystallization kinetics, improve phase stability, and reduce the presence of point defects.

Sn-based inorganic perovskites face the challenge of Sn^{2+} oxidation to Sn^{4+}. To address this issue, tin halide additives like SnX_2 (X = F, Cl, Br) have been used. These additives not only inhibit the oxidation of Sn^{2+} but also serve as Sn compensation, increasing the formation energy of tin vacancies and reducing the background hole carrier density. Furthermore, A-site cations such as MA (methylammonium), FA (formamidinium), and Ru (rubidium) can be doped into the inorganic perovskite framework. This doping strategy offers another approach to modify the properties of inorganic perovskite materials.

8.6.2 ADDITIVE

The chapter discusses the division of additives into two categories: Lewis acids and Lewis bases, along with their respective roles in controlling perovskite crystallization and improving film quality.

Lewis acids are compounds that can accept a pair of non-bonding electrons. They are effective in passivating electron-rich defects and suppressing the oxidation reaction of halogen ions in perovskite films. Metal cations, such as Pb^{2+} and K^+, can act as Lewis acids by forming ionic bonds with uncoordinated anions at the surface and grain boundaries of the perovskite. Ammonium salts have also been reported as effective additives for passivating A-site defects and anti-site defects in inorganic perovskites.

On the other hand, Lewis bases are compounds that can donate a lone pair of non-bonding electrons. They can form coordination bonds with undercoordinated ions, such as Pb^{2+}. Lewis base additives, based on the types of electron donors, can be categorized into oxygen atom donors, sulfur atom donors, and nitrogen atom donors. By coordinating with Pb^{2+}, these Lewis bases help reduce defects in the perovskite film caused by the lack of electrons and suppress the reduction reaction from Pb^{2+} to Pb. Commonly used Lewis bases include mercaptans, urea, pyridine, thiophene, and others.

Overall, Lewis acids and Lewis bases play critical roles in manipulating the crystallization process and enhancing the quality of perovskite films through their interactions with the perovskite structure and defects.

8.6.3 Interfacial Engineering

This section focuses on three interfaces in perovskite solar cells: the ETL/perovskite interface, the perovskite/HTL interface, and the perovskite/electrode interface. The electron transport layer (ETL) is responsible for facilitating the movement of holes towards the cathode in the photoelectron transfer process. Common ETL materials include TiO_2, ZnO, SnO_2, WO_3, and PCBM. However, these ETLs often face band mismatch with the inorganic perovskite film and have intrinsic drawbacks. For example, TiO_2 has a low bulk electron mobility and requires a high-temperature sintering process. ZnO can react with perovskite at higher annealing temperatures. SnO_2, prepared through low-temperature processing, produces oxygen vacancies on the surface, leading to perovskite degradation. To improve interface contact and optimize band alignment, buffer layers like amino acids (e.g., glycine), graphene quantum dots, and trimethylammonium bromide (PMABr) can be introduced.

The mismatch in energy levels between the inorganic perovskite and the hole transport layer (HTL) can result in charge recombination. Inorganic perovskite solar cells incorporating low-dimensional perovskite materials, perovskite quantum dots, and silicon quantum dots have advantages in terms of low preparation cost and high stability. However, the performance of perovskite solar cells based on inorganic HTLs is currently lower than those based on organic HTL systems. Further investigation is needed to explore interface engineering to optimize the band alignment and improve the quality of the perovskite film.

In commercial production, noble metal electrodes, such as gold or platinum, can be expensive, making them less suitable for large-scale production of perovskite solar cells. Carbon-based electrodes have emerged as potential alternatives due to their low cost, abundance, and compatibility with solution-based processing methods. Research efforts are ongoing to optimize the performance, stability, and scalability of carbon-based, all-inorganic perovskite solar cells without HTLs. However, direct contact between the perovskite film and carbon can lead to undesirable charge recombination and poor band alignment. Addressing these issues is crucial to improving the performance of such solar cells.

8.6.4 Solvent Engineering

Perovskite precursor solvent engineering is indeed an effective approach to improving the quality of perovskite films. The choice of solvent used in the preparation of perovskite films has a significant impact on the crystallization kinetics, film morphology, and overall film quality. Here are some key points regarding perovskite precursor solvent engineering:

Solvent-Precursor Interaction: the solvent used, such as dimethyl sulfoxide (DMSO), can interact with the perovskite precursor components, such as PbI_2 and $PbBr_2$, to form Lewis base adducts. These adducts, such as PbI_2(DMSO) and

$PbBr_2$(DMSO), can modify the crystallization process of perovskite films. By slowing down the crystallization kinetics, larger-scale crystalline grains can be obtained, leading to improved film quality, long carrier lifetimes, and enhanced stability in air.

Annealing Temperature: the formation of CsI-PbX_2-DMSO intermediates can significantly reduce the annealing temperature required for the formation of inorganic perovskite films. This lowers the thermal stress on the film and facilitates the formation of high-quality perovskite films.

Solvent Evaporation: the high boiling point of DMSO allows for residual DMSO to remain in the precursor film during the annealing process. This residual solvent can improve mass transport and diffusion within the film, aiding in the formation of high-quality perovskite films. However, excessive residual DMSO may lead to significant volume shrinkage during annealing, which could affect the film morphology.

Antisolvent Method: to eliminate the solvent-edge effect and improve film morphology, the antisolvent method is often employed. Solvents such as isopropyl alcohol (IPA), methyl acetate (MeOAc), and chlorobenzene (CB) are used to control the nucleation process and enhance the formation of high-quality perovskite films.

8.6.5 Dimension Engineering

In low-dimensional inorganic perovskites, the unique structure created by the inorganic framework and long-chain molecules acts as quantum wells and barriers, respectively. This results in a quantum-well structure that enhances the stability of the material compared to traditional three-dimensional (3D) perovskites.

In the case of 0D perovskites, such as Cs_4PbX_6 (X = Cl, Br, or I), their crystal structures consist of isolated lead halide octahedra $[PbX_6]^{4-}$ surrounded by Cs^+ cations. This unique structure allows 0D perovskites to exhibit the intrinsic properties of individual octahedra. A heterostructure can be formed by tuning the stoichiometry of the precursors between CsI and PbI, resulting in the stabilization of the black-phase $CsPbI_3$. This heterostructure comprises 0D Cs_4PbI_6 and γ-$CsPbI_3$. Photovoltaic devices based on this heterostructure, known as perovskite solar cells (PSCs), have demonstrated impressive power conversion efficiency (PCE) of 16.39% and improved stability.[83] Benzyldodecyl dimethylammonium (BDA^+) has been found to act as a phase-transfer catalyst, allowing quaternary ammonium to react with the surficial layers of the $CsPbI_3$ films uniquely. This addition of cation results in the isolation of $[PbX_6]^{4-}$ octahedra and in situ formation of 0D Cs_4PbI_6 perovskite. This strategy has enabled PSCs based on $CsPbI_3$ to achieve a high PCE of 20.63% with good operational stability.[84] Furthermore, 0D $CsPbBrI_2$ and $CsPbI_3$ quantum dots (QDs) have been used as interface modifications in PSCs. These QDs introduce an electric field in the rear side of the device, promoting carrier extraction and increasing the current. The interface also allows for greater electron and hole quasi-Fermi-level splitting at the maximum power point, resulting in an increased operating voltage.[85,86]

In 1D perovskites, the charge is confined within chains of $[PbI_6]^{4-}$ octahedra. The face-sharing connections in 1D perovskite weaken the hybridization of atomic orbitals near the bandgap, creating an electronic barrier and reducing band edge dispersion. The main approach in 1D perovskite research is the formation of in-situ

1D/3D heterostructures on the surface and at the grain boundaries of $CsPbI_3$. Zhao et al. discovered that the tetrabutylammonium (TBA^+) cation can intercalate into $CsPbI_3$, effectively substituting the Cs cation. This process leads to the formation of a one-dimensional (1D) $TBAPbI_3$ layer during the post-synthesis TBAI treatment. The intercalation of TBA cations results in the in-situ formation of a protective $TBAPbI_3$ layer, which helps heal defects at the surface of the inorganic $CsPbI_3$ perovskite.[87,88] Additionally, Zhong's group reported that choline iodide (ChI) can react with $CsPbI_3$ to form a novel crystal phase called one-dimensional (1D) $ChPbI_3$. This 1D $ChPbI_3$ phase constructs a 1D/3D heterostructure with 3D $CsPbI_3$, effectively passivating defects within the $CsPbI_3$ perovskite. This process significantly improves the photoluminescence lifetime, from 20.2 ns to 49.4 ns, indicating enhanced optical properties.[89]

The formation of 2D perovskites involves the introduction of long-chain alkylammonium cations, either aliphatic or aromatic. Commonly reported types of 2D perovskites are the Ruddlesden-Popper (R-P) phases and Dion-Jacobson (D-J) phases. The interactions between the organic cations and the inorganic layers, as well as the electrostatic interactions within the layers, have a significant influence on the electronic properties of 2D perovskites. In the case of inorganic $CsPbI_3$ perovskite, the growth behavior of 2D perovskites at the 3D perovskite surface differs from that of organic perovskites due to the more challenging cation exchange between organic ligands and Cs cations.[90] For example, high-quality mixed-dimensional films of 2D ($Cs_2PbI_2Cl_2$) and 3D ($CsPbI_{2.5}Br_{0.5}$) perovskites have been successfully fabricated by carefully controlling the CsCl to PbI_2 ratios. The introduction of 2D $Cs_2PbI_2Cl_2$ significantly improves the moisture and thermal stability of perovskite solar cells (PSCs).[91] Furthermore, the use of all-inorganic 2D/3D $CsPb_2I_4Br/CsPbI_2Br$ bulk heterojunctions effectively passivates defects at the surface and grain boundaries of 3D $CsPbI_2Br$ and reduces nonradiative recombination loss in perovskite solar cells.[92] Additionally, larger cations such as phenylethylamine (PEA^+), guanidine (GA^+), n-butylammonium (PBA^+), and others are introduced into 3D inorganic perovskites or used to construct 2D/3D heterostructures by capping organic cations.[90,93–95]

8.7 ADVANTAGES OF ALL-INORGANIC PEROVSKITE

8.7.1 Crystallization

The crystallization process of perovskite films is crucial for achieving high-quality films. Nucleation and crystal growth play significant roles in determining the crystallization quality.

Nucleation can occur through two main mechanisms: homogeneous nucleation and heterogeneous nucleation. Homogeneous nucleation occurs when new phase nuclei form directly from atomic clusters and are uniformly distributed in the liquid phase without the influence of impurity particles or external surfaces. On the other hand, heterogeneous nucleation involves the formation of new phase nuclei on impurities or external surfaces present in the liquid phase. In the case of perovskite film fabrication, heterogeneous nucleation is more common due to the presence of impurities in the precursor. The nucleation process is governed by the critical nucleation radius, which is influenced by the Gibbs free energy (ΔG). ΔG is the sum of the surface free energy (γ) and the bulk free energy (ΔGV), and it represents the energy change during nucleation.[96–98]

The growth of perovskite films can be understood through the LaMer mechanism and Ostwald ripening. In the LaMer mechanism, monomer molecules accumulate in the solution, leading to a supersaturated environment. When the concentration of monomers reaches a critical limit, nuclei are formed. The monomers then diffuse and adsorb preferentially on smaller grains with high surface energy, causing them to grow faster than larger grains.[99] This stage is the growth of the nuclei by diffusion of monomers. The monomers in solution tend to preferentially adsorb onto smaller grains with high surface energy. This preferential adsorption leads to the smaller grains growing faster than the larger grains. As a result, the growth of the perovskite film is skewed towards the smaller grains, resulting in a narrow grain size distribution.

As the monomer concentration decreases, smaller grains tend to dissolve and adsorb onto the surface of larger grains, leading to the growth of larger grains by the Ostwald ripening mechanism.[100,101]

In the preparation of $CsPbX_3$ films, the LaMer mechanism is relevant. The slow volatilization rate of precursor solvents (DMF/DMSO) results in limited supersaturation of the solution, leading to a relatively fast crystal growth rate. The solute precipitates rapidly in regions with high surface energy, which can result in more pores in the as-prepared perovskite films. However, a fast nucleation process accompanied by a slow crystal growth process can produce high-quality films with large crystals and few defects, thereby improving the efficiency and stability of inorganic perovskite solar cells.[102]

Ostwald ripening involves the continuous merging of small grains, leading to the formation of larger grains with lower surface energy. The dissolved small grains can be deposited on the surface of larger grains, promoting further growth. Eventually, the merged large grains connect and form intact crystals with increased grain size.[103]

There are three types of film growth models: the 3D island growth model (Volmer-Weber), the 2D layer growth model (Frank-Vander Merwe), and the layer-island growth model (Stranski-Krastanov). In the Volmer-Weber model, the film growth process involves nucleation, island formation, network formation, and the formation of a continuous film. Small islands merge to form larger islands, and the trenches in the film fill up to create a continuous film.[104,105] This model is more favorable on substrates with poor wettability.

The Frank-Vander Merwe model describes film growth by the uniform coverage of the substrate surface with a single atomic layer, followed by layer-by-layer growth in the 3D direction. The film growth in this model can be either homostructure or heterostructure epitaxial growth, depending on the lattice type. Good substrate wettability and small lattice mismatch between the film and the substrate promote film growth according to the Frank-Vander Merwe model.

The Stranski-Krastanov model combines the characteristics of the previous two models, involving the initial formation of 2D atomic layers on the substrate surface, followed by the growth of islands and the eventual formation of continuous films.[104,106]

Substrate wettability and lattice mismatch play important roles in determining film growth according to the Stranski-Krastanov model.

8.7.2 Stability

While inorganic perovskite materials generally exhibit better stability against moisture, heat, and light compared to organic-inorganic perovskites, they do possess challenges related to phase instability.

The $CsPbI_3$ perovskite material exhibits a phase instability issue. The structural transformation of $CsPbI_3$ occurs as follows: from the black a phase (cubic, *Pm3m*) at high temperatures (around 360 °C) to the black β-phase (tetragonal, *P4/mbm*) at intermediate temperatures (around 260 °C), then to the black γ phase (orthorhombic, *Pbnm*) at lower temperatures (around 175 °C), and finally to the yellow δ phase (orthorhombic, *Pnma*) upon cooling to room temperature. The black phases are metastable and tend to convert to the yellow δ phase, which is thermodynamically stable at room temperature. Initially, it was observed that fabricating $CsPbI_3$ with smaller grain sizes could increase the phase stability due to the increased contribution of surface energy and the presence of organic ligands.[107] However, with the introduction of additives like dimethylammonium iodide (DMAI), it has been possible to fabricate distorted $CsPbI_3$ perovskite films with larger grain sizes. These distorted phases, such as the β and γ phases, have demonstrated improved stability at room temperature.

The phase instability of the black phase in $CsPbI_3$ perovskite is attributed to a lower tolerance factor, which is the ratio of the ionic radii of the constituent ions in the crystal structure. By increasing the ratio of bromine (Br) to iodine (I) in $CsPbI_3$ perovskite, the issue of phase instability can be effectively alleviated. $CsPbBr_3$ perovskite, for example, is known to be stable in the orthorhombic structure at or near room temperature.

Sn-based inorganic perovskites, such as $CsSnBr_3$ and $CsSnI_3$, have similar crystal structures to Pb-based perovskites. $CsSnBr_3$ exhibits stability at room temperature with a cubic phase (α-$CsSnBr_3$) at temperatures below 292.15 K and a tetragonal phase (β-$CsSnBr_3$) at temperatures above 292.15 K.[7] However, the large band gap of $CsSnBr_3$ limits its application in perovskite solar cells (PSCs). On the other hand, $CsSnI_3$ perovskite has a smaller band gap, making it a more suitable light absorber. However, its phase stability at room temperature is poor. It undergoes phase transitions from a pure black cubic phase (B-α) at 718 K to a black tetragonal phase (B-β) and black orthorhombic phase (B-γ) upon cooling to 431 K and 352 K, respectively. This phase instability hinders the performance of $CsSnI_3$-based perovskite solar cells.

One of the reasons for the inferior performance and poor phase stability of tin-based perovskites is the ease of Sn^{2+} oxidation. This can be attributed to the different electronic configurations of Pb^{2+} and Sn^{2+}. The lanthanide contraction effect, caused by the weak electron shielding effect of the fully occupied 4f and 5d orbitals in Pb^{2+}, makes the lone pair electrons in the 6s orbital difficult to lose after losing the 6p electrons. This phenomenon is known as the inert pair effect. In contrast, the lack of fully occupied 4f orbitals in Sn^{2+} leads to a stronger shielding effect from the 4d orbital, making the nuclear attraction of the 5s electrons weaker. As a result, the lone pair electrons in the 5s orbital of Sn^{2+} are more easily oxidized to Sn^{4+}. This difference in electron configurations between Pb^{2+} and Sn^{2+} not only causes the instability of Sn^{2+} but also increases the Lewis acidity of SnI_2 compared to PbI_2, leading to faster reaction kinetics between SnI_2 and Lewis bases.[108] Although Cs_2SnX_6 (X = I, Br) is the most stable compound among Sn-based perovskites, it does not have a fully 3D perovskite structure, which is a disadvantage.[42]

Ge-based perovskites are indeed more prone to oxidation compared to Sn-based perovskites. This is primarily due to the low binding energy of the Ge $4s^2$ electrons, making them more susceptible to oxidation and limiting the overall efficiency of Ge-based perovskite materials.[109] However, perovskite compositions that incorporate two metals, particularly Sn and Ge, have shown promise as stable narrow bandgap absorbers. The inclusion of both Sn and Ge in the perovskite structure can enhance the stability of the material compared to their monometallic counterparts.[110]

Bi- and Sb-based perovskites, such as $Cs_3Sb_2I_9$ and $Cs_3Bi_2I_9$, exhibit unique crystal structures that differ from traditional 3D perovskites. Instead, they have 2D or 0D vacant structures, which contribute to their high stability against moisture. Moreover, double perovskites, which involve the combination of different A, B^+, B^{3+}, and X elements (e.g., A = K, Rb, Cs; B^+ = Li, Na, K, Rb, Cs, Ag; B^{3+} = Al, Ga, In, Sb, Bi, Sc, Y; X = F, Cl, Br, I), have shown outstanding thermal and environmental stability. The introduction of multiple cations in the perovskite structure contributes to the increased stability by providing a greater variety of bonding interactions and structural configurations. Researchers have predicted the thermodynamic stability of over 100 double perovskites with appropriate structural tolerance and octahedra factors.[111] This suggests the vast potential for discovering new lead-free perovskite materials that meet stability requirements.

8.7.3 Quantum Dots

Perovskite quantum dots (QDs) have attracted significant attention due to their fascinating physical properties and their impact in the fields of photonics and optoelectronics (Figure 8.6a). The monodisperse $CsPbX_3$ (X = Cl, Br, I) QDs with cubic perovskite crystal structures have been successfully prepared using solution-phase synthetic methods.[107] These QDs exhibit size-tunable bandgap energies, allowing for control over their optical properties, and they demonstrate stable and spectrally narrow photoluminescence. The crystal structure of these perovskite QDs, known as inverted halide perovskite quantum dots (IHPQDs), consists of a three-dimensional framework. Divalent metal cations (such as Pb^{2+} or Sn^{2+}) and halide ions (such as Cl^-, Br^-, I^-, or mixed halides) form corner-sharing octahedra (BX_6^{4-}), while the vertexes of the octahedral cavities are occupied by monovalent inorganic cations (such as Cs^+).[112] This framework allows charge carriers to move within a limited 3D grid, resulting in relatively high carrier mobility.[113] The fabrication of these perovskite QDs is feasible at room temperature due to the use of multicomponent halide salts with ionic bond characteristics. The crystallization process is controlled by the ion coprecipitation mechanism, which is influenced by surface ligands.[114,115]

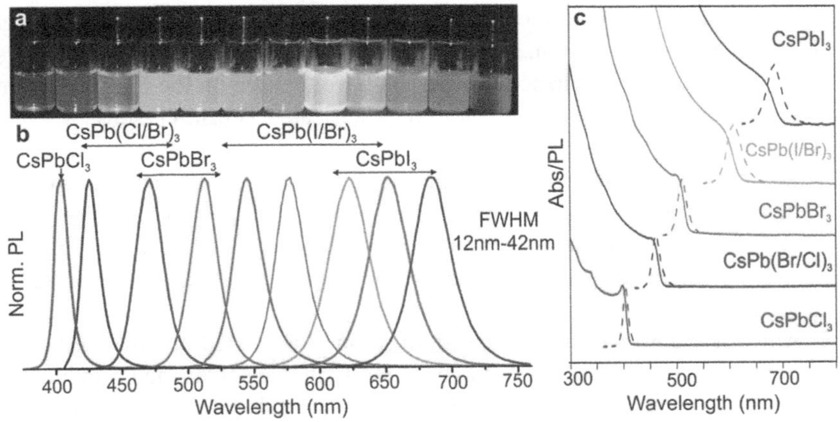

FIGURE 8.6 Colloidal perovskite $CsPbX_3$ NCs (X = Cl, Br, I) exhibit size- and composition-tunable bandgap energies covering the entire visible spectral region with narrow and bright emission: (a) colloidal solutions in toluene under UV lamp (λ = 365 nm); (b) representative PL spectra (λexc = 400 nm for all but 350 nm for $CsPbCl_3$ samples); (c) typical optical absorption and PL spectra.

Compared to traditional II-VI and III-V colloidal quantum dots like CdS, PbSe, and InP, IHPQDs (inverted halide perovskite quantum dots) exhibit several intriguing and remarkable properties[116,117]: (i) adjustable bandgap range: by adjusting the halogen ratio, the bandgap of IHPQDs can be tuned from 1.77 to 3.10 eV. This allows their luminescence spectrum to cover the entire visible light range from 405 to 700 nm (Figure 8.6b and c). (ii) Wide excitation range: IHPQDs can be excited by a single light source in the range of 350 to 400 nm, resulting in emissions with different wavelengths. (iii) Narrow-band emission: the emission peak of IHPQDs has a full width at half maximum (FWHM) of only 12 to 42 nm, which is much narrower than traditional quantum dots and organic dyes. This narrow emission spectrum enables a color gamut range that can reach 140% of the National Television System Committee (NTSC) standard. (iv) relatively high tolerance to defects, which enables the fluorescence quantum yield to reach ≈100% without passivation modification, and (v) negligible self-absorption effects and fluorescence resonance energy transfer. IHPQDs are inorganic perovskite quantum dot materials that possess high stability, ultrahigh photoluminescence quantum yield, wide absorption spectra covering the entire visible range, size-tunable bandgap, versatile chemical processability, high extinction coefficients, and long electron-hole diffusion lengths.[118–120] These unique properties make IHPQDs promising for various optoelectronic applications, including solar cells, light-emitting diodes (LEDs), optical detectors, and more.[121]

8.7.4 Single Crystal

8.7.4.1 Lead-Based Inorganic Perovskite Single Crystal

$CsPbBr_3$ single crystal is a promising material for various applications, particularly in photodetectors and high-energy radiation detectors. The crystal structure of $CsPbBr_3$

single crystal is similar to a microcrystalline film, making it an interesting material for thin-film device fabrication. However, the fabrication of $CsPbI_3$ and $CsPbCl_3$ single crystals is challenging, hence researchers have focused on the $CsPbBr_3$ counterpart. Several methods have been employed to fabricate $CsPbBr_3$ crystals:

1. Inverse Temperature Crystallization (ITC) Method: This method utilizes the abnormal solubility of $CsPbBr_3$ in specific organic solvents. The crystals' solubility decreases with increasing temperature. As the temperature rises, spontaneous nucleation occurs when supersaturation is reached, leading to crystal growth.
2. Anti-solvent Vapor-Assisted Crystallization (AVC) Method: this method is suitable when $CsPbBr_3$ has a large solubility in one solvent and a small solubility in another. By slowly diffusing the poor solvent into the solution of $CsPbBr_3$ dissolved in the good solvent, supersaturation is achieved, resulting in crystal precipitation. This method enables crystal growth at room temperature, and heating the antisolvent can accelerate the process.[122]
3. Bridgman Growth Method: this widely used method allows for the growth of large-sized bulk crystals. The raw material is melted in a crucible, and then the crucible is slowly lowered in the furnace with a controlled temperature gradient. This directional solidification crystallization minimizes the volatilization of components and improves crystal composition uniformity. However, heterogeneous nucleation and thermal stress can affect the yield and introduce dislocation defects.
4. Electronic Dynamic Gradient (EDG) Method: by setting different temperatures in the upper and lower zones of a double temperature zone tube furnace, a temperature gradient is achieved. $CsPbBr_3$ is fully melted in the region with a higher temperature, and then the temperature is slowly decreased to facilitate crystal growth.
5. Chemical Vapor Deposition (CVD) Method: this method is commonly used for preparing single crystal films. $CsPbBr_3$ can be deposited by using a mixed powder of PbX_2 and CsX as the evaporation source in the CVD reaction. By adjusting temperature, pressure, and reaction time, the perovskite deposition on the substrate can be controlled. However, at this stage, $CsPbBr_3$ prepared by CVD often has a small crystal size, limiting its use in constructing perovskite film devices. Further exploration is required to achieve larger crystal sizes.

Indeed, $CsPbBr_3$ crystal possesses several desirable characteristics that make it suitable for various applications. First, the crystal has a large average atomic number (Pb: 82, Cs: 55, Br: 35), which is beneficial for applications in radiation detection and imaging, as it enhances the crystal's ability to interact with and absorb high-energy photons. Second, $CsPbBr_3$ exhibits a suitable band gap ranging from 1.5 to 3 eV. Additionally, $CsPbBr_3$ crystal demonstrates large electrical resistivity, meaning it has a high resistance to the flow of electric current. This property is advantageous in various electronic and optoelectronic devices, as it helps minimize leakage currents and improve energy efficiency. The higher resistivity also contributes to the crystal's ability to accurately measure and detect low levels of electrical signals, resulting in higher energy resolution. Moreover, $CsPbBr_3$ exhibits high carrier mobility-lifetime product (mt), which is a crucial parameter for charge transport in semiconductor materials. The high mt value indicates that the crystal allows for efficient movement of charge carriers (electrons and holes) while preserving their lifetimes. This property is essential for applications that require fast charge transport, such as photovoltaics, light-emitting diodes (LEDs), and sensors.

8.7.4.2 Lead-Free-Based Inorganic Perovskite Single Crystal

Inorganic lead-free halide semiconductors with perovskite and perovskite-like structures have gained significant attention in recent years due to their advantageous properties and potential applications. One of the primary advantages of these materials is their nontoxicity. Unlike lead-based perovskite materials, which have raised concerns regarding environmental and human health, lead-free halide semiconductors offer a safer alternative. Furthermore, lead-free halide semiconductors exhibit excellent stability, both in ambient conditions and under operational conditions. Moreover, lead-free halide perovskites have shown promise in photodetectors and X-ray detectors. Their high carrier mobilities, low trap densities, and large absorption coefficients enable efficient charge generation and collection, resulting in enhanced sensitivity and performance in detecting photons or X-rays.

$Cs_3Cu_2I_5$: this single crystal was first synthesized in 2018 and has an orthorhombic configuration in the *Pnma* space group. It is considered a 0D material, which means it has zero-dimensional electronic properties. $Cs_3Cu_2I_5$ exhibits a large exciton binding energy of 490 meV, indicating a strong interaction between electrons and holes. The conduction band minimum (CBM) and valence band maximum (VBM) energy levels were measured to be 2.66 eV and 6.46 eV, respectively, resulting in a sizable bandgap of 3.8 eV.[123]

$CsCu_2I_3$: this single crystal has a rod-like shape, an orthorhombic structure, and is transparent. It belongs to the A_2CuX_3-type materials. $CsCu_2I_3$ has an optical bandgap of 3.78 eV, slightly smaller than $Cs_3Cu_2I_5$, and a lower exciton binding energy of 346 meV.

A_2CuX_3-Type Materials: other compounds in the A_2CuX_3 family, such as Rb_2CuX_3 and K_2CuX_3, also possess a 1D structure.[124–126]

$CsSnI_3$ Single Crystal: $CsSnI_3$ was first reported in 2012 and exhibits phase transformations during the fabrication process.[66] The α (B-α) phase at high temperatures converts to the β phase at 431 K and then to the γ (B-γ) phase at room temperature. These phases correspond to cubic, tetragonal, and orthorhombic structures respectively. $CsSnI_3$ single crystals at room temperature demonstrate p-type semiconductor behavior with a carrier concentration of around 1017 cm^{-3} and a hole mobility of approximately 585 $cm^2\ V^{-1}\ s^{-1}$. It is worth mentioning that $CsSnBr_3$ single crystals have a cubic crystal structure, a higher optical bandgap (1.68 eV) compared to CsSnI3 (1.31 eV), and different physical properties such as exciton binding energy and hole and electron effective mass.

Lead-Free Inorganic Perovskite Crystals with Bivalent Ions: apart from Sn^{2+}, other bivalent ions like Ca^{2+}, Sr^{2+}, Mn^{2+}, Zn^{2+}, Cu^{2+}, and Eu^{2+} have been utilized to synthesize lead-free inorganic perovskite crystals. Crystals such as $CsCaX_3$ and

$RbCaX_3$ can exhibit various configurations including cubic, tetragonal, and orthorhombic structures.[127] However, not all the bivalent lead-free ions form regular octahedrons with halide ions, and the configurations and dimensions of other bivalent ions based lead-free inorganic perovskite crystal single crystals are various. For instance, $CsMnCl_3(H_2O)_2$ single crystals are 1D and orthorhombic, while Cs_2ZnBr_4 single crystal with orthorhombic configuration is 0D structure and $CsCuCl_3$ single crystal with a novel DNA-like crystal configuration is 1D structure.[128]

Apart from bivalent ions, the ions of IIIA and VA elements neighboring Pb are also considered as the alternates for lead free perovskites. Generally, the M^{3+} ions can form a perovskite-like derivative – $A_3M_2X_9$, behaving as 0D or 2D crystal structures, and there are two main types: cubic close packing and hexagonal close packing of A and X ions. For instance, $Cs_3Bi_2I_9$ single crystals form a 0D configuration, while $K_3Bi_2I_9$ and $Cs_3Bi_2I_9$ tend to form layer-like 2D structures and $Cs_3Sb_2I_9$ and $Rb_3Sb_2I_9$ show 2D crystal structures.[51]

Tetravalent lead-free ions. The Sn^{4+}-based ILFP single crystals possess a cubic structure and a quasi-symmetric configuration, ensuring their crystal and chemical stability. Such Cs_2SnX_6 forms with 0D configurations.[68] In addition to Sn^{4+}, Te^{4+} and Pd^{4+} ions can also coordinate with halide ions to form Cs_2TeI_6 and Cs_2PdBr_6 single crystals, respectively.[129,130] The Cs_2PdBr_6 single crystals exhibit a n-type behavior with a direct bandgap, while Cs_2TeBr_6 single crystals have an indirect bandgap. Overall, the lead-free Cs_2MX_6 single crystals, including Cs_2SnX_6, Cs_2TeI_6, and Cs_2PdBr_6, possess cubic structures, excellent chemical stabilities, and high defect tolerance.

Inorganic lead-free double perovskite single crystals are divided into two main types: $A_2M^+M^{3+}X_6$-type and $A_4M^{2+}M^{3+}{}_2X_{12}$-type single crystals. For the form type, these single crystals involve replacing two Pb^{2+} ions in the perovskite structure with a monovalent M^+ ion (Ag^+, Na^+, K^+, Rb^+) and a trivalent M^{3+} ion (Bi^{3+}, Sb^{3+}, In^{3+}, Fe^{3+}, TI^{3+}, Tb^{3+}, Eu^{3+}) to maintain charge balance. The $Cs_2AgBiBr_6$ single crystal with a 3D cubic structure was first reported by Karunadasa and co-workers.[29] This single crystal has a 3D cubic structure and exhibits an indirect bandgap of 1.95 eV with an assisting photon energy of 0.12 eV. For $Cs_2AgBiCl_6$ single crystals, the compression of $[AgCl_6]$ and $[BiCl_6]$ octahedrons in $Cs_2AgBiCl_6$ single crystals causes crystal distortion, leading to a movement of energy levels and an increase in the bandgap. In addition, $Cs_2AgSbBr_6$ and $Cs_2AgSbCl_6$ single crystals exhibit cubic structures and indirect bandgaps.[131,132]

In comparison In^{3+}-based single crystals such as $Cs_2AgInCl_6$ and $Cs_2AgInBr_6$, these crystals have a large and direct bandgap of approximately 3.2 eV. This results in the single crystals appearing gray and transparent.[133] In recent research, there has been a focus on doping double perovskite single crystals to investigate the impact of dopants on the crystal and electronic structure. For example, when Cu^{2+} ions are incorporated into $Cs_2AgBiBr_6$ single crystals, they do not affect the cubic crystal structure but can broaden the absorption in the near-infrared range. Additionally, the substitution of alkali ions (K^+, Na^+, Rb^+) in $Cs_2AgBiBr_6$ single crystals and their impact on the crystal structures and indirect bandgaps have been confirmed through their characteristics.

The $A_4M^{2+}M^{3+}{}_2X_{12}$-type perovskites are characterized by having three octahedron centers occupied by a vacancy, a bivalent M^{2+} ion, and a trivalent M^{3+} ion. For instance, single crystals such as $Cs_4CdSb_2Cl_{12}$ and $Cs_4CdBi_2Cl_{12}$ exhibit cubic crystal structures with direct and large bandgaps. On the other hand, $Cs_4MnBi_2Cl_{12}$ single crystal has been reported to possess a 2D-layered trigonal structure. It features a special sandwich structure composed of corner-sharing octahedrons, specifically $[BiCl_6]$-$[MnCl_6]$-$[BiCl_6]$. The perovskite structure of $Cs_4MnBi_2Cl_{12}$ consists of triple-layered units, with the triple-layered structure serving as the basic unit in these perovskites.

8.7.5 Solar Cells

In recent years, there has been significant progress in the development of all-inorganic halide perovskites without any organic components, particularly in terms of their performance, which has increased by over 21%. The first all-inorganic halide perovskite solar cells (PSCs) based on $CsSnI_3$ using a Schottky junction structure (ITO/$CsSnI_3$/Au/Ti) were reported in 2012, achieving a low power conversion efficiency (PCE) of 0.88%.[134] Over time, the performance of $CsSnI_3$-based PSCs improved, and a standard device architecture (ITO/PEDOT:PSS/$CsSnI_3$-PTM/ICBC/BCP/Ag) achieved a PCE over 10%.[135] Similarly, $CsPbX_3$ (X = halide) PSCs, like $CsSnI_3$ solar cells, have also shown rapid improvements in PCE, exceeding 21%. In addition to $CsSnI_3$ and $CsPbX_3$, many other types of all-inorganic halide perovskites have been both theoretically predicted and synthesized, including $CsAgBiBr_6$ and others.

8.7.5.1 Electron/Hole Transport Layer

In all-inorganic perovskite solar cells (PSCs), electron transporting materials (ETMs) and hole transporting materials (HTMs) play crucial roles in facilitating the movement of charges and blocking undesirable charge carriers. Commonly used ETMs include titanium dioxide (TiO_2), tin dioxide (SnO_2), zinc oxide (ZnO), and phenyl-C_{61}-butyric acid methyl ester (PCBM/$PC_{61}BM$). These materials primarily function to transport electrons and block the movement of holes. On the other hand, HTMs such as Spiro-OMeTAD, poly(3,4-ethylenedioxythiophene): poly(styrenesulfonate) (PEDOT:PSS), poly[bis(4-phenyl)(2,4,6-trimethylphenyl)amine] (PTAA), and nickel oxide (NiO_x) are commonly used to transport holes and block the movement of electrons. The selection of HTL (hole transporting layer) and ETL (electron transporting layer) materials is guided by the energy band alignment of the device structure and the practicality of the precursor solution used in the fabrication process.

Spiro-OMeTAD is widely used as a molecular hole transporting material (HTM) in perovskite solar cells due to its popularity and high efficiency. However, the intrinsic Spiro-OMeTAD has low carrier mobility. To enhance its performance, two additives are frequently employed: bis(trifluoromethylsulfonyl)amine lithium salt (LiTFSI) and 4-tert-butylpyridine (tBP). LiTFSI promotes the oxidation and p-doping of Spiro-OMeTAD, thereby improving its hole transporting properties. On the other hand, tBP helps prevent the aggregation of LiTFSI, leading to better interface contact between the HTM layer and the perovskite layer. The hole mobility of doped Spiro-OMeTAD has been estimated to be 3.32×10^{-3} cm^2 V^{-1} s^{-1}.[136] Another commonly used HTM is poly(3,4-ethylenedioxythiophene): poly(styrene-sulfonate) (PEDOT:PSS). It is primarily applied in the p-i-n structure and exhibits a high hole mobility of > 10^{-2} cm^2 V^{-1} s^{-1}. Poly[bis(4-phenyl)(2,4,6-trimethylphenyl)amine] (PTAA) is a polymer HTM with conjugated chains and high stability. It is used as the main HTM in both p-i-n and n-i-p-type perovskite solar cells with high efficiencies. PTAA is often doped to improve its hole mobility, which can reach up to 10^{-2} cm^2 V^{-1} s^{-1}. NiO_x is an inorganic p-type semiconductor material with a hole mobility of 0.5 cm^2 V^{-1} s^{-1} and high chemical stability. It is also utilized as an HTM in perovskite solar cells.[137]

In perovskite solar cells (PSCs), titanium dioxide (TiO_2) is the most commonly used electron transporting layer (ETL). TiO_2 has n-type electrical conduction characteristics, high electron mobility (10^{-3} cm^2 V^{-1} s^{-1}), and good stability. The compact TiO_2 layer can be produced using various methods, including spray pyrolysis, thermal oxidation, atomic layer deposition, and electrochemical deposition.[138] Mesoporous TiO_2, on the other hand, is fabricated at higher temperatures (up to 450 °C) and is known to reduce the diffusion length of charge carriers, thereby improving the performance of solar cells. However, the use of mesoporous TiO_2 presents challenges for application on different substrates. Inorganic materials such as tin dioxide (SnO_2) and zinc oxide (ZnO) are also utilized as ETLs in PSCs. SnO_2 exhibits an electron mobility of about 100–200 cm^2 V^{-1} s^{-1}, while ZnO has an electron mobility of 26 cm^2 V^{-1} s^{-1}, both of which are excellent for facilitating electron transport in the solar cell.[139,140] In addition to inorganic metal oxides, the organic small molecule phenyl-C_{61}-butyric acid methyl ester (PCBM) is commonly used as an ETL in the p-i-n device structure.[141] PCBM has an electron mobility of 2×10^{-3} cm^2 V^{-1} s^{-1}.

8.7.5.2 Efficiency Improvement

8.7.5.2.1 $CsPbI_3$

In 2015, Snaith et al. achieved a record power conversion efficiency (PCE) of 2.9% by adding the HI solution into the $CsPbI_3$ precursor solution, resulting in the stable α-$CsPbI_3$ phase.[142] Subsequently, Luo et.al found that α-$CsPbI_3$ could transform to the δ phase in air, and by first fabricating Cs_4PbI_6 and then converting it to α-$CsPbI_3$, they obtained a PCE of 4.13%. This α-$CsPbI_3$, which underwent an intermediate phase transition, was maintained at room temperature for 72 h.[143] Ripolles et al. introduced a MoO_3 thin film between HTM and Au layers to reduce hysteresis and obtained a PCE of 4.68% in $CsPbI_3$ PSCs.[144] Yixing Zhao et.al improved the environmental stability of $CsPbI_3$ film by using ethylenediamine iodine ($EDAPbI_4$) as an additive, resulting in improved stability and a PCE of 11.8%.[145] Wang et al. explored spin coating the PEA-IPA solution on $CsPbI_3$ film, which resulted in the formation of a $CsPbI_3$ film without a two-dimensional-like structure and achieved a PCE of 13.5%.[95] Jingbi You et al. achieved a high-quality $CsPbI_3$ microcrystalline film with a uniform morphology and improved grain size (5 μm), leading to a PCE of 15.7%.[146] Our group used $HPbI_{3+x}$ and CsI precursors and PEAI as a passivator to fabricate distorted β-$CsPbI_3$ and γ-$CsPbI_3$ black films, achieving a PCE of 15.07% with improved stability. Wang et al. employed phenyl trimethylammonium bromide (PTABr) as a surface passivator, resulting in a PCE of 17.06% and improved phase stability.[147] Choline iodine (CHI) was found to enter the CsPbI3 film through pinholes or cracks, leading to well-matched energy levels and a PCE of 18.4% in β-$CsPbI_3$ PSCs.[148] The introduction of DMAI (dimethylammonium iodide) affected the phase separation of $CsPbI_3$, leading to the formation of γ-$CsPbI_3$ or β-$CsPbI_3$, each with different characteristics and performances. The use of phenyl-trimethylammonium chloride (PTACl) as the surface passivator resulted in a PCE of 19.03% in β-$CsPbI_3$ PSCs.[149] Du et al. focus on the design and fabrication of a layered structure utilizing a series of organic ligands, aimed at enhancing the stability and performance of $CsPbI_3$ solar cells. The graded alignment of the 2D quantum wells plays a crucial role in facilitating the transport of charge carriers within the solar cells. As a result, the solar cells achieved a PCE of up to 20.33% and a relatively high V_{OC} of 1.23 V.[90] Tan et al. incorporated phenyltrimethylammonium iodide (PTAI) into the precursor solution of $CsPbI_3$ perovskite solar cells (PSCs). Their findings showed the formation of low-dimensional perovskites, specifically 1D $PTAPbI_3$ and 2D PTA_2PbI_4, located at the grain boundaries and film surface of the $CsPbI_3$ film. These can not only suppress non-radiative recombination but also stabilize black-phase $CsPbI_3$ to prevent moisture intrusion. The champion PSC device, based on PTAI-$CsPbI_3$, achieved a record efficiency of 21.0% and a certified efficiency of 20.1% (reverse scanning) with a high V_{oc} of nearly 1.2 V.[150] Wang et.al used an acyloin ligand (1,2-di(thiophen-2-yl)ethane1,2-dione (DED) as a phase stabilizer and defect passivator in the perovskite precursor solution. The incorporation of DED effectively passivated Pb-related defects and suppressed ion migration within the $CsPbI_3$ perovskite. Consequently, The DED-treated $CsPbI_3$ PSCs achieved a champion PCE of 21.15%. Furthermore, the long-term stability of the DED-$CsPbI_3$ PSCs was enhanced.[151] Zhang et.al designed a passivator trifluoroacetamidine (TFA) to suppress $CsPbI_3$. The amidine group of TFA strongly chelates onto the surface of the perovskite, effectively suppressing iodide vacancies, and this interaction is further strengthened by additional hydrogen bonding. In addition, the presence of three fluorine atoms in TFA enables strong intermolecular connections through intermolecular hydrogen bonding. As a result, TFA-$CsPbI_3$ PSCs achieved PCEs of 21.35% for a 0.09 cm^2 and 17.21% for a larger 1.0 cm^2 device areas, respectively.[152] Recently, Wang et al. adopted 2,6-DAPy as a perovskite-surface passivator. The stronger nucleophilicity of the N-pyridine group in 2,6-DAPy plays a crucial role in inhibiting multiple iodine-related defects, such as I_i^+, V_I, and released I_2, through a combination of alogen-Npyridine and coordination bonds. Additionally, the two symmetric neighboring -NH_2 groups in 2,6-DAPy interact with adjacent halides within the octahedral cluster by forming hydrogen bonds. This further promotes the adsorption of 2,6-DAPy molecules onto the perovskite surface. The synergetic effects of these three chemical bonds significantly reduce the number of harmful iodine-related defects and undercoordinated Pb^{2+}, leading to prolonged carrier lifetimes and facilitating interfacial hole transfer. As a result, 2,6-DAPy based inorganic perovskite solar cells achieved a record PCE up to 21.8%. Moreover, the corresponding devices displayed higher environmental stability against moisture invasion.[153]

8.7.5.2.2 $CsPbBr_3$

In Cs-based inorganic perovskites, $CsPbBr_3$ has shown good stability to moisture, oxygen, and heat and can be synthesized using a simple solution method in air environments. Several studies have been conducted to improve the efficiency and stability of $CsPbBr_3$ perovskite solar cells (PSCs).

In 2015, Kulbak et al. successfully fabricated $CsPbBr_3$ perovskite films and incorporated them into solar cells with a specific device structure: FTO/c-TiO_2/m-TiO_2/$CsPbBr_3$/PTAA/Au. The power conversion efficiency (PCE) achieved with $CsPbBr_3$ was 5.95%.[154] In 2016, Liang et al. reported a classical structure of FTO/c-TiO_2/m-TiO_2/$CsPbBr_3$/C, which removed the expensive organic hole transport layer and used the C electrode instead of the expensive metal electrode. Finally, the device based on this structure achieved 6.7% PCE and showed good stability in air.[155] In addition to the large band gap of $CsPbBr_3$ perovskite (2.3 eV), the PCE of $CsPbBr_3$ PSCs is also limited by the low crystallinity and poor morphology when using conventional two-step solution deposition method. This is mainly due to the low solubility of the CsBr in the precursor solution, resulting in the lack of sufficient raw material and thus formation of $CsPbBr_3$ with discontinuous film morphology. To improve the crystallization process of $CsPbBr_3$ film, Duan et al. prepared high-purity $CsPbBr_3$ films by a multistep solution deposition method, and the device achieved a PCE of 7.54%. This method also lays a solid foundation for the improvement of $CsPbBr_3$PSCs performance.[156] To obtain uniform, large-area, and high-quality $CsPbBr_3$ thin films, many researchers have also tried to use other methods to prepare thin films and achieved good results. Chen et al. proposed for the first time to prepare $CsPbBr_3$ thin films by vacuum evaporation method, and the device prepared based on this method achieved a PCE of 7.78%.[157] On this basis, Li et al. proposed a two-source vacuum continuous evaporation method to fabricate $CsPbBr_3$ thin film, and the device achieved a PCE of 8.34%.[158] Due to the large energy level barrier between the $CsPbBr_3$ layer and the charge transport layer or the electrode interface, many researchers optimized the film by interface modification and successfully made the solar cells achieve PCE over 10%.[159–161] Tong et al. proposed a vacuum evaporation phase transition induction (PTI) method, based on which the n-i-p structure device achieved a PCE up to 10.91%.[16] Tang's group incorporated an inorganic 2D Cl-terminated Ti_3C_2 ($Ti_3C_2Cl_x$) MXene into the bulk and surface of $CsPbBr_3$ film to heal the defects and expanded lattices. As a result, $CsPbBr_3$ PSC achieves an efficiency as high as 11.08% with an ultrahigh Voc of 1.702 V, which are the highest PCE and voltage records for $CsPbBr_3$ solar cells.[163]

8.7.5.2.3 $CsPbI_2Br$

In 2016, Snaith et al. demonstrated that the cesium lead mixed-halide perovskites, $CsPbI_2Br$, are stable in the cubic phase at room temperature even for bulk materials. The first $CsPbI_2Br$ solar cell has shown a surprising PCE of 9.8%.[164] In 2018, Mai et al. designed an all inorganic $CsPbI_2Br$ PSCs with the structure of FTO/NiO_X/$CsPbI_2Br$/ZnO@C_{60}/Ag and delivered a PCE over 13%.[165] Due to its inorganic properties, the device showed excellent thermal stability. The unencapsulated device maintained 80% of the initial efficiency after annealing at 85°C for 15 d. A year later, Li et al. adopted the gradient annealing method to optimize the crystallization process of the $CsPbI_2Br$ film. As a result, the solar cells achieved a PCE of 16.07%.[166] In 2019, our group replaced the iodide ion with an acetate anion, and the PCE of $CsPbI_2Br$ perovskite solar cells reached 15.56%.[167] In the following year, we used $CaCl_2$ as an additive into the $CsPbI_2Br$ perovskite precursor solution, which effectively slowed down the crystallization rate of the perovskite film and enhanced the crystallization of the film, finally in a PCE of 16.79%.[168] Meanwhile, Hong et al. updated the efficiency of $CsPbI_2Br$ all-inorganic perovskite solar cells to 17.46% by Ru cation doped in A-site strategy.[169] The modification approach using polylactic acid (PLA) has shown promising results in enhancing the performance of perovskite solar cells (PSCs). The interaction between the carbonyl group (C=O) in PLA and under-coordinated Pb^{2+} ions helps to passivate surface defects and promote secondary grain growth in the perovskite film. Moreover, this modification can induce a change in the surface region of the perovskite from n-type to p-type, which facilitates better charge transport from the perovskite to the hole transport layer (HTL). As a result of these improvements, a record PCEs of 18.05% have been achieved for $CsPbI_2Br$ PSCs.[170]

8.7.5.2.4 $CsPbIBr_2$

In 2016, the first $CsPbIBr_2$ PSCs were fabricated using the dual-source co-evaporation method. The device structure was FTO/c-TiO_2/$CsPbIBr_2$/Au, and it achieved a power conversion efficiency (PCE) of 4.7%.[171] Liu et al. developed a one-step solution method to prepare $CsPbIBr_2$ films with high crystal quality and good thermal stability. They utilized MoO_X as the cathode buffer layer, which effectively reduced trap density at the interface and resulted in a PCE of 5.52%.[172] Yang's team replaced the traditional TiO_2 electron transport layer (ETL) with In_2S_3 deposited by a low-temperature chemical bath deposition (CBD) method. This modification led to a PCE of 5.59% in $CsPbIBr_2$ PSCs.[173] Subsequently, Lau et al. introduced a spray-assisted solution method for the fabrication of $CsPbIBr_2$ PSCs, achieving a PCE of 6.3%.[174] Zhu et al. demonstrated that incorporating metal cations into the $CsPbIBr_2$ framework through intermolecular exchange paths enhanced the optical response, resulting in a PCE of 9.16% and improved long-term stability.[38] In 2022, the PCE of $CsPbIBr_2$ solar cells was further enhanced to 11.31% through the use of the iodized formidine strategy.[175]

8.7.5.2.5 Carbon-Based Inorganic Perovskite Solar Cells

Carbon is considered a highly promising material for use as a counter electrode in inorganic perovskite solar cells (IPSCs) due to several advantageous properties. It is abundant, easily tunable in terms of structure, chemically stable, low-cost, resistant to water, and exhibits high carrier mobility (μ_{cm}). Additionally, carbon is inert to ion migration originating from perovskite materials. The use of carbon electrodes in IPSCs offers potential benefits such as reduced production costs, simplified fabrication processes, improved device stability, and the possibility of large-scale commercialization for carbon-based IPSCs (C-IPSCs).[176] Dong et al. predicted the S–Q limits of $CsPbI_{3-x}Br_x$ C-IPSCs.[177] The $CsPbBr_3$, $CsPbIBr_2$, $CsPbI_2Br$, and $CsPbI_3$ C-IPSCs can reach the PCEs of 13.3%, 17.1%, 20.1%, and 23.0%, respectively. Several studies have demonstrated the potential of C-IPSCs with different light absorbers. $CsPbI_3$-based C-IPSCs have achieved the highest PCE of 18.05% (certified: 17.8%),[89] while

$CsPbI_2Br$,[178] $CsPbIBr_2$,[179] and $CsPbBr_3$[180]-based C-IPSCs have achieved PCEs of 15.56%, 10.66%, and 10.13%, respectively, making them exhibit great potential for high PCE.

8.8 CONCLUSIONS AND OUTLOOK

In this chapter, we have provided an overview of the development of Cs-based inorganic perovskite materials and their applications in solar cells. The chapter discusses various Cs-based inorganic perovskite materials including $CsPbX_3$ (X = I, Br), $CsSnX_3$, $CsGeX_3$, $CsSbX_3$, and $CsBiX_3$, as well as double perovskite and their derivatives. The crystal structure, physical and chemical properties, defects, and stability of these materials are explored. The chapter clarifies the photovoltaic performance of inorganic perovskite solar cells. Specifically, it highlights the progress made in $CsPbI_3$ PSCs, which have achieved a significant increase in power conversion efficiency (PCE) from 2.9% to over 21% in recent years. $CsPbBr_3$ and $CsSnI_3$ PSCs have also demonstrated PCEs exceeding 10%.

It is noted that Sn and Pb-based perovskites can leverage technologies from each other to address stability concerns. However, Ge-based perovskites are not considered an ideal choice. Sn-Ge mixed perovskites, on the other hand, show promise in achieving efficiencies higher than 10%. There exists a trade-off between high-efficiency but poor stability (Sn^{2+}/Pb^{2+}-based) and good stability but low performance (Sn^{4+}/Sb/Bi-based, etc.). Double perovskites have a narrow light absorption window, limiting the current density in solar cells. As a result, new candidates need to be developed that combine high efficiencies with low-cost, scalable fabrication, and long operational stability.

REFERENCES

1. A. Marronnier, G. Roma, S. Boyer-Richard, L. Pedesseau, J. M. Jancu, Y. Bonnassieux, C. Katan, C. C. Stoumpos, M. G. Kanatzidis, J. Even, Anharmonicity and Disorder in the Black Phases of Cesium Lead Iodide Used for Stable Inorganic Perovskite Solar Cells, *ACS Nano*, **2018**, *12*, 3477.
2. G. Li, H. Wang, Z. Zhu, Y. Chang, T. Zhang, Z. Song, Y. Jiang, Shape and Phase Evolution from $CsPbBr_3$ Perovskite Nanocubes to Tetragonal $CsPb_2Br_5$ Nanosheets with an Indirect Bandgap, *Chem. Commun.*, **2016**, *52*, 11296.
3. K.-H. Wang, L. Wu, L. Li, H.-B. Yao, H.-S. Qian, S.-H. Yu, Large-Scale Synthesis of Highly Luminescent Perovskite-Related $CsPb_2Br_5$ Nanoplatelets and Their Fast Anion Exchange, *Angew. Chem.*, **2016**, *128*, 1.
4. G. Tong, L. K. Ono, Y. Qi, Recent Progress of All-Bromide Inorganic Perovskite Solar Cells, *Energy Technol.*, **2019**, *8*, 1900961.
5. Y. Chen, T. Shi, P. Liu, W. Xie, K. Chen, X. Xu, L. Shui, C. Shang, Z. Chen, H.-L. Yip, G. Zhou, X. Wang, The Distinctive Phase Stability and Defect Physics in $CsPbI_2Br$ Perovskite, *J. Mater. Chem. A*, **2019**, *7*, 20201.
6. J. Lin, M. Lai, L. Dou, C. S. Kley, H. Chen, F. Peng, J. Sun, D. Lu, S. A. Hawks, C. Xie, F. Cui, A. P. Alivisatos, D. T. Limmer, P. Yang, Thermochromic Halide Perovskite Solar Cells, *Nat. Mater.*, **2018**, *17*, 261.
7. D. E. Scaife, P. F. Weller, W. G. Fisher, Crystal Preparation and Properties of Cesium Tin(U) Trihalides, *J. Solid State Chem.*, **1973**, *9*, 308.

8. Y. Koji, F. Shinya, H. Hiromi, M. Takashi, O. Tsutomu, I. Sumio, Structural Phase Transitions of the Polymorphs of CsSnI3 by Means of Rietveld Analysis of the X Ray Diffraction, *Chem. Lett.*, **1991**, *20*, 801.
9. J. Jiang, C. K. Onwudinanti, R. A. Hatton, P. A. Bobbert, S. Tao, Stabilizing Lead-Free All-Inorganic Tin Halide Perovskites by Ion Exchange, *J. Phys. Chem. C*, **2018**, *122*, 17660.
10. K. P. Marshall, M. Walker, R. I. Walton, R. A. Hatton, Enhanced Stability and Efficiency in Hole-Transport-Layer-Free $CsSnI_3$ Perovskite Photovoltaics, *Nat. Energy*, **2016**, *1*, 1.
11. C. G. Ma, V. Krasnenko, M. G. Brik, First-Principles Calculations of Different (001) Surface Terminations of Three Cubic Perovskites $CsCaBr_3$, $CsGeBr_3$, and $CsSnBr_3$, *J. Phys. Chem. Solids*, **2018**, *115*, 289.
12. S. Guo, Q. Zhang, R. Sa, D. Liu, Exploring the Stability and Physical Properties of Double Perovskite Cs_2SnI_6 by Doping the Cl/Br Atom: A DFT Study, *Comput. Theor. Chem.*, **2022**, *1209*.
13. T. Krishnamoorthy, H. Ding, C. Yan, W. L. Leong, T. Baikie, Z. Zhang, M. Sherburne, S. Li, M. Asta, N. Mathews, S. G. Mhaisalkar, Lead-Free Germanium Iodide Perovskite Materials for Photovoltaic Applications, *J. Mater. Chem. A*, **2015**, *3*, 23829.
14. L.-y. Huang, W. R. L. Lambrecht, Electronic Band Structure Trends of Perovskite Halides: Beyond Pb and Sn to Ge and Si, *Phys. Rev. B*, **2016**, *93*, 195211.
15. A. C. Dias, M. P. Lima, J. L. F. Da Silva, Role of Structural Phases and Octahedra Distortions in the Optoelectronic and Excitonic Properties of $CsGeX_3$ (X = Cl, Br, I) Perovskites, *J. Phys. Chem. C*, **2021**, *125*, 19142.
16. S. F. Hoefler, G. Trimmel, T. Rath, Progress on Lead-Free Metal Halide Perovskites for Photovoltaic Applications: A Review, *Monatsh Chem.*, **2017**, *148*, 795.
17. B. W. Park, B. Philippe, X. Zhang, H. Rensmo, G. Boschloo, E. M. Johansson, Bismuth Based Hybrid Perovskites $A_3Bi_2I_9$ (A: Methylammonium or Cesium) for Solar Cell Application, *Adv. Mater.*, **2015**, *27*, 6806.
18. K.-H. Hong, J. Kim, L. Debbichi, H. Kim, S. H. Im, Band Gap Engineering of $Cs_3Bi_2I_9$ Perovskites with Trivalent Atoms Using a Dual Metal Cation, *J. Phys. Chem. C*, **2016**, *121*, 969.
19. D. H. Fabini, J. G. Labram, A. J. Lehner, J. S. Bechtel, H. A. Evans, A. Van der Ven, F. Wudl, M. L. Chabinyc, R. Seshadri, Main-Group Halide Semiconductors Derived from Perovskite: Distinguishing Chemical, Structural, and Electronic Aspects, *Inorg. Chem.*, **2017**, *56*, 11.
20. T. L. Hodgkins, C. N. Savory, K. K. Bass, B. L. Seckman, D. O. Scanlon, P. I. Djurovich, M. E. Thompson, B. C. Melot, Anionic Order and Band Gap Engineering in Vacancy Ordered Triple Perovskites, *Chem. Commun.*, **2019**, *55*, 3164.
21. K. M. McCall, C. C. Stoumpos, O. Y. Kontsevoi, G. C. B. Alexander, B. W. Wessels, M. G. Kanatzidis, From 0D $Cs_3Bi_2I_9$ to 2D $Cs_3Bi_2I_6Cl_3$: Dimensional Expansion Induces a Direct Band Gap but Enhances Electron–Phonon Coupling, *Chem. Mater.*, **2019**, *31*, 2644.
22. Z. Jin, Z. Zhang, J. Xiu, H. Song, T. Gatti, Z. He, A Critical Review on Bismuth and Antimony Halide Based Perovskites and Their Derivatives for Photovoltaic Applications: Recent Advances and Challenges, *J. Mater. Chem. A*, **2020**, *8*, 16166.
23. M. B. Johansson, H. Zhu, E. M. Johansson, Extended Photo-Conversion Spectrum in Low-Toxic Bismuth Halide Perovskite Solar Cells, *J. Phys. Chem. Lett.*, **2016**, *7*, 3467.
24. P. C. Harikesh, H. K. Mulmudi, B. Ghosh, T. W. Goh, Y. T. Teng, K. Thirumal, M. Lockrey, K. Weber, T. M. Koh, S. Li, S. Mhaisalkar, N. Mathews, Rb as an Alternative Cation for Templating Inorganic Lead-Free Perovskites for Solution Processed Photovoltaics, *Chem. Mater.*, **2016**, *28*, 7496.
25. X. Zhao, J. Yang, Y. Fu, D. Yang, Q. Xu, L. Yu, S.-H. Wei, L. Zhang, Design of Lead-Free Inorganic Halide Perovskites for Solar Cells via Cation-Transmutation, *J. Am. Chem. Soc.*, **2017**, *139*, 2630.

26. T. Xuan, R.-J. Xie, Recent Processes on Light-Emitting Lead-Free Metal Halide Perovskites, *Chem. Eng. J.*, **2020**, *393*, 124757.
27. E. T. McClure, M. R. Ball, W. Windl, P. M. Woodward, Cs_2AgBiX_6 (X = Br, Cl): New Visible Light Absorbing, Lead-Free Halide Perovskite Semiconductors, *Chem. Mater.*, **2016**, *28*, 1348.
28. M. R. Filip, S. Hillman, A. A. Haghighirad, H. J. Snaith, F. Giustino, Band Gaps of the Lead-Free Halide Double Perovskites $Cs_2BiAgCl_6$ and $Cs_2BiAgBr_6$ from Theory and Experiment, *J. Phys. Chem. Lett.*, **2016**, *7*, 2579.
29. A. H. Slavney, T. Hu, A. M. Lindenberg, H. I. Karunadasa, A Bismuth-Halide Double Perovskite with Long Carrier Recombination Lifetime for Photovoltaic Applications, *J. Am. Chem. Soc.*, **2016**, *138*, 2138.
30. M. B. Faheem, B. Khan, C. Feng, M. U. Farooq, F. Raziq, Y. Xiao, Y. Li, All-Inorganic Perovskite Solar Cells: Energetics, Key Challenges, and Strategies toward Commercialization, *ACS Energy Lett.*, **2019**, *5*, 290.
31. F. Giustino, H. J. Snaith, Toward Lead-Free Perovskite Solar Cells, *ACS Energy Lett.*, **2016**, *1*, 1233.
32. J. Tian, Q. Xue, Q. Yao, N. Li, C. J. Brabec, H. L. Yip, Inorganic Halide Perovskite Solar Cells: Progress and Challenges, *Adv. Energy Mater.*, **2020**, *10*, 2000183.
33. W. Xiang, S. Liu, W. Tress, A Review on the Stability of Inorganic Metal Halide Perovskites: Challenges and Opportunities for Stable Solar Cells, *Energy Environ. Sci.*, **2021**, *14*, 2090.
34. X. Zhou, J. Jankowska, H. Dong, O. V. Prezhdo, Recent Theoretical Progress in the Development of Perovskite Photovoltaic Materials, *J. Energy Chem.*, **2018**, *27*, 637.
35. O. O. Bello, M. E. Emetere, Progress and Limitation of Lead-Free Inorganic Perovskites for Solar Cell Application, *Solar Energy*, **2022**, *243*, 370.
36. H. Chen, S. Xiang, W. Li, H. Liu, L. Zhu, S. Yang, Inorganic Perovskite Solar Cells: A Rapidly Growing Field, *Solar RRL*, **2018**, *2*, 1700188.
37. J. K. Nam, M. S. Jung, S. U. Chai, Y. J. Choi, D. Kim, J. H. Park, Unveiling the Crystal Formation of Cesium Lead Mixed-Halide Perovskites for Efficient and Stable Solar Cells, *J. Phys. Chem. Lett.*, **2017**, *8*, 2936.
38. W. Zhu, Q. Zhang, D. Chen, Z. Zhang, Z. Lin, J. Chang, J. Zhang, C. Zhang, Y. Hao, Intermolecular Exchange Boosts Efficiency of Air-Stable, Carbon-Based All-Inorganic Planar $CsPbIBr_2$ Perovskite Solar Cells to Over 9%, *Adv. Energy Mater.*, **2018**, *8*, 1802080.
39. L.-y. Huang, W. R. L. Lambrecht, Electronic Band Structure, Phonons, and Exciton Binding Energies of Halide Perovskites $CsSnCl_3$, $CsSnBr_3$, and $CsSnI_3$, *Phys. Rev. B*, **2013**, *88*, 165203.
40. I. Borriello, G. Cantele, D. Ninno, Ab Initio Investigation of Hybrid Organic-Inorganic Perovskites Based on Tin Halides, *Phys. Rev. B*, **2008**, *77*, 235214.
41. A. Kaltzoglou, M. Antoniadou, A. G. Kontos, C. C. Stoumpos, D. Perganti, E. Siranidi, V. Raptis, K. Trohidou, V. Psycharis, M. G. Kanatzidis, P. Falaras, Optical-Vibrational Properties of the Cs2SnX6 (X = Cl, Br, I) Defect Perovskites and Hole-Transport Efficiency in Dye-Sensitized Solar Cells, *J. Phys. Chem. C*, **2016**, *120*, 11777.
42. W. Ke, M. G. Kanatzidis, Prospects for Low-Toxicity Lead-Free Perovskite Solar Cells, *Nat. Commun.*, **2019**, *10*, 965.
43. C. C. Stoumpos, L. Fraser, D. J. Clark, Y. S. Kim, S. H. Rhim, A. J. Freeman, J. B. Ketterson, J. I. Jang, M. G. Kanatzidis, Hybrid Germanium Iodide Perovskite Semiconductors: Active Lone Pairs, Structural Distortions, Direct and Indirect Energy Gaps and Strong Nonlinear Optical Properties, *J. Am. Chem. Soc.*, **2015**, *137*, 6804.
44. C. C. Stoumpos, M. G. Kanatzidis, The Renaissance of Halide Perovskites and Their Evolution as Emerging Semiconductors, *Acc. Chem. Res.*, **2015**, *48*, 2791.

45. L.-y. Huang, W. R. L. Lambrecht, Vibrational Spectra and Nonlinear Optical Coefficients of Rhombohedral $CsGeX_3$ Halide Compounds with X=I, Br, Cl, *Phys. Rev. B*, **2016**, *94*, 115202
46. S. K. Radha, W. R. L. Lambrecht, Understanding the Crystallographic Phase Relations in Alkali-Trihalogeno-Germanates in Terms of Ferroelectric or Antiferroelectric Arrangements of the Tetrahedral GeX_3 Units, *Adv. Elect. Mater.*, **2019**, *6*, 1900887.
47. H. Dammak, A. Yangui, S. Triki, Y. Abid, H. Feki, Structural Characterization, Vibrational, Optical Properties and DFT Investigation of a New Luminescent Organic–Inorganic Material: $(C_6H_{14}N)_3Bi_2I_9$, *J. Lumin.*, **2015**, *161*, 214.
48. T. Kawai, S. Shimanu, Optical Studies of $(CH_3NH_3)_3Bi_2I_9$ Single Crystals, *Phys. Stat. Sol. (b)* **1993**, *177*, K43.
49. W. G. Li, X. D. Wang, J. F. Liao, Y. Jiang, D. B. Kuang, Enhanced On–Off Ratio Photodetectors Based on Lead-Free $Cs_3Bi_2I_9$ Single Crystal Thin Films, *Adv. Funct. Mater.*, **2020**, *30*, 1909701.
50. B. Ghosh, B. Wu, H. K. Mulmudi, C. Guet, K. Weber, T. C. Sum, S. Mhaisalkar, N. Mathews, Limitations of $Cs_3Bi_2I_9$ as Lead-Free Photovoltaic Absorber Materials, *ACS Appl. Mater. Inter.*, **2018**, *10*, 35000.
51. K. M. McCall, C. C. Stoumpos, S. S. Kostina, M. G. Kanatzidis, B. W. Wessels, Strong Electron–Phonon Coupling and Self-Trapped Excitons in the Defect Halide Perovskites $A_3M_2I_9$ (A = Cs, Rb; M = Bi, Sb), *Chem. Mater.*, **2017**, *29*, 4129.
52. B. Saparov, F. Hong, J.-P. Sun, H.-S. Duan, W. Meng, S. Cameron, I. G. Hill, Y. Yan, D. B. Mitzi, Thin-Film Preparation and Characterization of $Cs_3Sb_2I_9$: A Lead-Free Layered Perovskite Semiconductor, *Chem. Mater.*, **2015**, *27*, 5622.
53. W.-h. Guo, Y.-h. Zhu, M. Zhang, J. Du, Y.-l. Cen, S.-m. Liu, Y. He, H.-x. Zhong, X. Wang, J.-j. Shi, The Dion–Jacobson Perovskite $CsSbCl_4$: A Promising Pb-Free Solar-Cell Absorber with Optimal Bandgap ~1.4 eV, Strong Optical Absorption ~105 cm−1, and Large Power-Conversion Efficiency above 20%, *J. Mater. Chem. A*, **2021**, *9*, 16436.
54. S.-H. Wei, S. B. Zhang, Chemical Trends of Defect Formation and Doping Limit in II–VI Semiconductors: The Case of CdTe, *Phy. Rev. B*, **2002**, *66*, 155211.
55. Z. Yang, A. Surrente, K. Galkowski, A. Miyata, O. Portugall, R. J. Sutton, A. Haghighirad, H. J. Snaith, D. K. Maude, P. Plochocka, R. J. Nicholas, The Impact of the Halide Cage on the Electronic Properties of Fully Inorganic Caesium Lead Halide Perovskites, *ACS Energy Lett.*, **2017**, *2*, 1621.
56. E. M. Hutter, R. J. Sutton, S. Chandrashekar, M. Abdi-Jalebi, S. D. Stranks, H. J. Snaith, T. J. Savenije, Vapour-Deposited Cesium Lead Iodide Perovskites: Microsecond Charge Carrier Lifetimes and Enhanced Photovoltaic Performance, *ACS Energy Lett.*, **2017**, *2*, 1901.
57. C. L. Kennedy, A. H. Hill, E. S. Massaro, E. M. Grumstrup, Ultrafast Excited State Transport and Decay Dynamics in Cesium Lead Mixed-Halide Perovskites, *ACS Energy Lett.*, **2017**, *2*, 1501.
58. H. Zhu, M. T. Trinh, J. Wang, Y. Fu, P. P. Joshi, K. Miyata, S. Jin, X. Y. Zhu, Organic Cations Might Not Be Essential to the Remarkable Properties of Band Edge Carriers in Lead Halide Perovskites, *Adv. Mater.*, **2017**, *29*, 1603072.
59. J. K. Nam, S. U. Chai, W. Cha, Y. J. Choi, W. Kim, M. S. Jung, J. Kwon, D. Kim, J. H. Park, Potassium Incorporation for Enhanced Performance and Stability of Fully Inorganic Cesium Lead Halide Perovskite Solar Cells, *Nano Lett.*, **2017**, *17*, 2028.
60. S. Dastidar, S. Li, S. Y. Smolin, J. B. Baxter, A. T. Fafarman, Slow Electron–Hole Recombination in Lead Iodide Perovskites Does Not Require a Molecular Dipole, *ACS Energy Lett.*, **2017**, *2*, 2239.
61. J. Song, Q. Cui, J. Li, J. Xu, Y. Wang, L. Xu, J. Xue, Y. Dong, T. Tian, H. Sun, H. Zeng, Ultralarge All-Inorganic Perovskite Bulk Single Crystal for High-Performance Visible-Infrared Dual-Modal Photodetectors, *Adv. Opt. Mater.*, **2017**, *5*, 1700157.

62. M. Kulbak, I. Levine, E. Barak-Kulbak, S. Gupta, A. Zohar, I. Balberg, G. Hodes, D. Cahen, Control over Self-Doping in High Band Gap Perovskite Films, *Adv. Energy Mater.*, **2018**, *8*, 1800398.
63. G. A. Elbaz, D. B. Straus, O. E. Semonin, T. D. Hull, D. W. Paley, P. Kim, J. S. Owen, C. R. Kagan, X. Roy, Unbalanced Hole and Electron Diffusion in Lead Bromide Perovskites, *Nano Lett.*, **2017**, *17*, 1727.
64. A. S. Thind, G. Luo, J. A. Hachtel, M. V. Morrell, S. B. Cho, A. Y. Borisevich, J. C. Idrobo, Y. Xing, R. Mishra, Atomic Structure and Electrical Activity of Grain Boundaries and Ruddlesden-Popper Faults in Cesium Lead Bromide Perovskite, *Adv. Mater.*, **2019**, *31*, 1805047.
65. J. Kang, L. W. Wang, High Defect Tolerance in Lead Halide Perovskite $CsPbBr_3$, *J. Phys. Chem. Lett.*, **2017**, *8*, 489.
66. I. Chung, J. H. Song, J. Im, J. Androulakis, C. D. Malliakas, H. Li, A. J. Freeman, J. T. Kenney, M. G. Kanatzidis, $CsSnI_3$: Semiconductor or Metal? High Electrical Conductivity and Strong Near-Infrared Photoluminescence from a Single Material. High Hole Mobility and Phase-Transitions, *J. Am. Chem. Soc.*, **2012**, *134*, 8579.
67. B. Wu, Y. Zhou, G. Xing, Q. Xu, H. F. Garces, A. Solanki, T. W. Goh, N. P. Padture, T. C. Sum, *Adv. Funct. Mater.*, Long Minority-Carrier Diffusion Length and Low Surface-Recombination Velocity in Inorganic Lead-Free $CsSnI_3$ Perovskite Crystal for Solar Cells, **2017**, *27*, 1604818.
68. C. C. Stoumpos, C. D. Malliakas, M. G. Kanatzidis, Semiconducting Tin and Lead Iodide Perovskites with Organic Cations: Phase Transitions, High Mobilities, and Near-Infrared Photoluminescent Properties, *Inorg. Chem.*, **2013**, *52*, 9019.
69. Y. Ogomi, A. Morita, S. Tsukamoto, T. Saitho, N. Fujikawa, Q. Shen, T. Toyoda, K. Yoshino, S. S. Pandey, T. Ma, S. Hayase, $CH_3NH_3SnxPb_{(1-x)}I_3$ Perovskite Solar Cells Covering up to 1060 nm, *J. Phys. Chem. Lett.*, **2014**, *5*, 1004.
70. S. Bouhmaidi, A. Marjaoui, A. Talbi, M. Zanouni, K. Nouneh, L. Setti, A DFT Study of Electronic, Optical and Thermoelectric Properties of Ge-Halide Perovskites $CsGeX_3$ (X=F, Cl and Br), *Comput. Condens. Matter*, **2022**, *31*, e00663.
71. K. M. McCall, Z. Liu, G. Trimarchi, C. C. Stoumpos, W. Lin, Y. He, I. Hadar, M. G. Kanatzidis, B. W. Wessels, α-Particle Detection and Charge Transport Characteristics in the $A_3M_2I_9$ Defect Perovskites (A = Cs, Rb; M = Bi, Sb), *ACS Photo.*, **2018**, *5*, 3748.
72. R. L. Z. Hoye, L. Eyre, F. Wei, F. Brivio, A. Sadhanala, S. Sun, W. Li, K. H. L. Zhang, J. L. MacManus-Driscoll, P. D. Bristowe, R. H. Friend, A. K. Cheetham, F. Deschler, Fundamental Carrier Lifetime Exceeding 1 μs in $Cs_2AgBiBr_6$ Double Perovskite, *Adv. Mater. Inter.*, **2018**, *5*, 1800464.
73. Y. Yin, W. Tian, J. Leng, J. Bian, S. Jin, Carrier Transport Limited by Trap State in $Cs_2AgBiBr_6$ Double Perovskites, *J. Phys. Chem. Lett.*, **2020**, *11*, 6956.
74. F. Igbari, R. Wang, Z. K. Wang, X. J. Ma, Q. Wang, K. L. Wang, Y. Zhang, L. S. Liao, Y. Yang, Composition Stoichiometry of $Cs_2AgBiBr_6$ Films for Highly Efficient Lead-Free Perovskite Solar Cells, *Nano Lett.*, **2019**, *19*, 2066.
75. W. Ning, F. Wang, B. Wu, J. Lu, Z. Yan, X. Liu, Y. Tao, J. M. Liu, W. Huang, M. Fahlman, L. Hultman, T. C. Sum, F. Gao, Long Electron-Hole Diffusion Length in High-Quality Lead-Free Double Perovskite Films, *Adv. Mater.*, **2018**, *30*, e1706246.
76. G. Volonakis, M. R. Filip, A. A. Haghighirad, N. Sakai, B. Wenger, H. J. Snaith, F. Giustino, Lead-Free Halide Double Perovskites via Heterovalent Substitution of Noble Metals, *J. Phys. Chem. Lett.*, **2016**, *7*, 1254.
77. F. Ji, G. Boschloo, F. Wang, F. Gao, Challenges and Progress in Lead-Free Halide Double Perovskite Solar Cells, *Solar RRL*, **2023**, *7*.
78. H. Jin, E. Debroye, M. Keshavarz, I. G. Scheblykin, M. B. J. Roeffaers, J. Hofkens, J. A. Steele, It's a Trap! On the Nature of Localised States and Charge Trapping in Lead Halide Perovskites, *Mater. Horiz.*, **2020**, *7*, 397.

79. J. Kim, S. H. Lee, J. H. Lee, K. H. Hong, The Role of Intrinsic Defects in Methylammonium Lead Iodide Perovskite, *J. Phys. Chem. Lett.*, **2014**, *5*, 1312.
80. P. Xu, S. Chen, H.-J. Xiang, X.-G. Gong, S.-H. Wei, Influence of Defects and Synthesis Conditions on the Photovoltaic Performance of Perovskite Semiconductor $CsSnI_3$, *Chem. Mater.*, **2014**, *26*, 6068.
81. T. C. Jellicoe, J. M. Richter, H. F. Glass, M. Tabachnyk, R. Brady, S. E. Dutton, A. Rao, R. H. Friend, D. Credgington, N. C. Greenham, M. L. Bohm, Synthesis and Optical Properties of Lead-Free Cesium Tin Halide Perovskite Nanocrystals, *J. Am. Chem. Soc.*, **2016**, *138*, 2941.
82. W.-J. Yin, J.-H. Yanga, J. Kanga, Y. Yan, S.-H. Wei, Halide Perovskite Materials for Solar Cells: A Theoretical Review, *J. Mater. Chen. A*, **2015**, *3*, 8926.
83. F. Bai, J. Zhang, Y. Yuan, H. Liu, X. Li, C. C. Chueh, H. Yan, Z. Zhu, A. K. Jen, A 0D/3D Heterostructured All-Inorganic Halide Perovskite Solar Cell with High Performance and Enhanced Phase Stability, *Adv. Mater.*, **2019**, *31*, e1904735.
84. Y. Chen, X. Wang, Y. Wang, X. Liu, Y. Miao, Y. Zhao, Functional Organic Cation Induced 3D-to-0D Phase Transformation and Surface Reconstruction of $CsPbI_3$ Inorganic Perovskite, *Sci. Bull.*, **2023**, *68*, 706.
85. J. Zhang, D. Bai, Z. Jin, H. Bian, K. Wang, J. Sun, Q. Wang, S. Liu, 3D–2D–0D Interface Profiling for Record Efficiency All-Inorganic $CsPbBrI_2$ Perovskite Solar Cells with Superior Stability, *Adv. Energy Mater.*, **2018**, *8*, 1703246.
86. K. Wang, Q. Tian, G. Zhao, J. Wen, J. Huang, C. Gao, Z. Xu, Y. Liu, L. Liang, L. Meng, L. Zhang, Z. Liu, Z. Jin, S. Olthof, S. Liu, Beach-Chair-Shaped Energy Band Alignment for High-Performance β-$CsPbI_3$ Solar Cells, *Cell Rep. Phys. Sci.*, **2020**, *1*, 100180.
87. X. Liu, X. Wang, T. Zhang, Y. Miao, Z. Qin, Y. Chen, Y. Zhao, Organic Tetrabutylammonium Cation Intercalation to Heal Inorganic $CsPbI_3$ Perovskite, *Angew. Chem. Int. Ed.*, **2021**, *60*, 12351.
88. J. Mi, Y. Chen, X. Liu, X. Wang, Y. Miao, Y. Qi, Y. Zhao, Synergistic Stabilization of $CsPbI_3$ Inorganic Perovskite via 1D Capping and Secondary Growth, *J. Energy Chem.*, **2022**, *68*, 387.
89. J. Zhang, G. Zhang, P. Y. Su, R. Huang, J. Lin, W. Wang, Z. Pan, H. Rao, X. Zhong, 1D Choline-PbI_3 -Based Heterostructure Boosts Efficiency and Stability of $CsPbI_3$ Perovskite Solar Cells, *Angew. Chem. Int. Ed.*, **2023**, *62*, e202303486.
90. Y. Du, Q. Tian, S. Wang, T. Yang, L. Yin, H. Zhang, W. Cai, Y. Wu, W. Huang, L. Zhang, K. Zhao, S. F. Liu, Manipulating the Formation of 2D/3D Heterostructure in Stable High-Performance Printable $CsPbI_3$ Perovskite Solar Cells, *Adv. Mater.*, **2023**, *35*, 2206451.
91. Z. Li, X. Liu, J. Xu, S. Yang, H. Zhao, H. Huang, S. F. Liu, J. Yao, 2D-3D CsPbICl-CsPbIBr. Mixed-Dimensional Films for All-Inorganic Perovskite Solar Cells with Enhanced Efficiency and Stability, *J. Phys. Chem. Lett.*, **2020**, *11*, 4138.
92. C. Kang, S. Xu, H. Rao, Z. Pan, X. Zhong, All-Inorganic $CsPb_2I_4Br/CsPbI_2Br$ 2D/3D Bulk Heterojunction Boosting Carbon-Based $CsPbI_2Br$ Perovskite Solar Cells with an Efficiency of Over 15%, *ACS Energy Lett.*, **2023**, *8*, 909.
93. M. Tai, Y. Zhou, X. Yin, J. Han, Q. Zhang, Y. Zhou, H. Lin, In Situ Formation of a 2D/3D Heterostructure for Efficient and Stable $CsPbI_2Br$ Solar Cells, *J. Mater. Chem. A*, **2019**, *7*, 22675.
94. Y. Jiang, J. Yuan, Y. Ni, J. Yang, Y. Wang, T. Jiu, M. Yuan, J. Chen, Reduced-Dimensional α-$CsPbX_3$ Perovskites for Efficient and Stable Photovoltaics, *Joule*, **2018**, *2*, 1356.
95. Y. Wang, T. Zhang, M. Kan, Y. Li, T. Wang, Y. Zhao, Efficient α-$CsPbI_3$ Photovoltaics with Surface Terminated Organic Cations, *Joule*, **2018**, *2*, 2065.
96. J. W. Lee, D. K. Lee, D. N. Jeong, N. G. Park, Control of Crystal Growth toward Scalable Fabrication of Perovskite Solar Cells, *Adv. Funct. Mater.*, **2018**, *29*, 1807047.
97. H. Hu, M. Singh, X. Wan, J. Tang, C.-W. Chu, G. Li, Nucleation and Crystal Growth Control for Scalable Solution-Processed Organic–Inorganic Hybrid Perovskite Solar Cells, *J. Mater. Chem. A*, **2020**, *8*, 1578.

98. W.-J. Yin, J.-H. Yang, J. Kang, Y. Yan, S.-H. Wei, Halide Perovskite Materials for Solar Cells: A Theoretical Review, *J. Mater. Chem. A*, **2015**, *3*, 8926.
99. N. T. Thanh, N. Maclean, S. Mahiddine, Mechanisms of Nucleation and Growth of Nanoparticles in Solution, *Chem. Rev.*, **2014**, *114*, 7610.
100. S. G. Kwon, T. Hyeon, Formation Mechanisms of Uniform Nanocrystals via Hot-Injection and Heat-Up Methods, *Small*, **2011**, *7*, 2685.
101. B. M. Cossairt, Shining Light on Indium Phosphide Quantum Dots: Understanding the Interplay among Precursor Conversion, Nucleation, and Growth, *Chem. Mater.*, **2016**, *28*, 7181.
102. C. Liu, Y. B. Cheng, Z. Ge, Understanding of Perovskite Crystal Growth and Film Formation in Scalable Deposition Processes, *Chem. Soc. Rev.*, **2020**, *49*, 1653.
103. W. Zhu, C. Bao, Y. Wang, F. Li, X. Zhou, J. Yang, B. Lv, X. Wang, T. Yu, Z. Zou, Coarsening of One-Step Deposited Organolead Triiodide Perovskite Films via Ostwald Ripening for High Efficiency Planar-Heterojunction Solar Cells, *Dalton. Trans.*, **2016**, *45*, 7856.
104. J. Zhao, Z. Li, M. Wang, Q. Wang, Z. Jin, Exploring the Film Growth in Perovskite Solar Cells, *J. Mater. Chem. A*, **2021**, *9*, 6029.
105. Y. C. Zheng, S. Yang, X. Chen, Y. Chen, Y. Hou, H. G. Yang, Thermal-Induced Volmer–Weber Growth Behavior for Planar Heterojunction Perovskites Solar Cells, *Chem. Mater.*, **2015**, *27*, 5116.
106. J. E. Prieto, I. Markov, Stranski–Krastanov Mechanism of Growth and the Effect of Misfit Sign on Quantum Dots Nucleation, *Surf. Sci.*, **2017**, *664*, 172.
107. L. Protesescu, S. Yakunin, M. I. Bodnarchuk, F. Krieg, R. Caputo, C. H. Hendon, R. X. Yang, A. Walsh, M. V. Kovalenko, Nanocrystals of Cesium Lead Halide Perovskites ($CsPbX_3$, X = Cl, Br, and I): Novel Optoelectronic Materials Showing Bright Emission with Wide Color Gamut, *Nano Lett.*, **2015**, *15*, 3692.
108. M. Zhang, Z. Zhang, H. Cao, T. Zhang, H. Yu, J. Du, Y. Shen, X.-L. Zhang, J. Zhu, P. Chen, M. Wang, *Mater.* Recent Progress in Inorganic Tin Perovskite Solar Cells, *Mater. Today Energy, Today Energy*, **2022**, *23*, 100891.
109. T. Miyasaka, A. Kulkarni, G. M. Kim, S. Öz, A. K. Jena, Perovskite Solar Cells: Can We Go Organic-Free, Lead-Free, and Dopant-Free?, *Adv. Energy Mater.*, **2019**, *10*, 1902500.
110. M. G. Ju, J. Dai, L. Ma, X. C. Zeng, Lead-Free Mixed Tin and Germanium Perovskites for Photovoltaic Application, *J. Am. Chem. Soc.*, **2017**, *139*, 8038.
111. H. Lei, D. Hardy, F. Gao, Lead-Free Double Perovskite $Cs_2AgBiBr_6$: Fundamentals, Applications, and Perspectives, *Adv. Funct. Mater.*, **2021**, *31*, 2105898.
112. W. Lv, L. Li, M. Xu, J. Hong, X. Tang, L. Xu, Y. Wu, R. Zhu, R. Chen, W. Huang, Improving the Stability of Metal Halide Perovskite Quantum Dots by Encapsulation, *Adv. Mater.*, **2019**, *31*, 1900682.
113. W. L. Hong, Y. C. Huang, C. Y. Chang, Z. C. Zhang, H. R. Tsai, N. Y. Chang, Y. C. Chao, Efficient Low-Temperature Solution-Processed Lead-Free Perovskite Infrared Light-Emitting Diodes, *Adv. Mater.*, **2016**, *28*, 8029.
114. Y. Yuan, Z. Liu, Z. Liu, L. Peng, Y. Li, A. Tang, Photoluminescence and Self-Assembly of Cesium Lead Halide Perovskite Nanocrystals: Effects of Chain Length Of Organic Amines and Reaction Temperature, *Appl. Surf. Sci.*, **2017**, *405*, 280.
115. B. Ai, C. Liu, Z. Deng, J. Wang, J. Han, X. Zhao, Low Temperature Photoluminescence Properties of $CsPbBr_3$ Quantum Dots Embedded in Glasses, *Phys. Chem. Chem. Phys.*, **2017**, *19*, 17349.
116. X. Li, Y. Wu, S. Zhang, B. Cai, Y. Gu, J. Song, H. Zeng, $CsPbX_3$ Quantum Dots for Lighting and Displays: Room-Temperature Synthesis, Photoluminescence Superiorities, Underlying Origins and White Light-Emitting Diodes, *Adv. Funct. Mater.*, **2016**, *26*, 2435.

117. A. Swarnkar, R. Chulliyil, V. K. Ravi, M. Irfanullah, A. Chowdhury, A. Nag, Colloidal CsPbBr3 Perovskite Nanocrystals: Luminescence beyond Traditional Quantum Dots, *Angew. Chem. Int. Ed.*, **2015**, *54*, 15424.
118. Y. Wang, X. Li, J. Song, L. Xiao, H. Zeng, H. Sun, All-Inorganic Colloidal Perovskite Quantum Dots: A New Class of Lasing Materials with Favorable Characteristics, *Adv. Mater.*, **2015**, *27*, 7101.
119. D. V. Talapin, J.-S. Lee, M. V. Kovalenko, E. V. Shevchenko, Prospects of Colloidal Nanocrystals for Electronic and Optoelectronic Applications, *Chem. Rev.*, **2009**, *110*, 389.
120. Y. F. Xu, M. Z. Yang, B. X. Chen, X. D. Wang, H. Y. Chen, D. B. Kuang, C. Y. Su, A $CsPbBr_3$ Perovskite Quantum Dot/Graphene Oxide Composite for Photocatalytic CO_2 Reduction, *J. Am. Chem. Soc.*, **2017**, *139*, 5660.
121. J. Liang, D. Chen, X. Yao, K. Zhang, F. Qu, L. Qin, Y. Huang, J. Li, Recent Progress and Development in Inorganic Halide Perovskite Quantum Dots for Photoelectrochemical Applications, *Small*, **2019**, *16*, 1903398.
122. Y. Rakita, N. Kedem, S. Gupta, A. Sadhanala, V. Kalchenko, M. L. Böhm, M. Kulbak, R. H. Friend, D. Cahen, G. Hodes, Low-Temperature Solution-Grown $CsPbBr_3$ Single Crystals and Their Characterization, *Cryst. Growth Des.*, **2016**, *16*, 5717.
123. T. Jun, K. Sim, S. Iimura, M. Sasase, H. Kamioka, J. Kim, H. Hosono, Lead-Free Highly Efficient Blue-Emitting Cs3Cu2I5 with 0D Electronic Structure, *Adv. Mater.*, **2018**, *30*, 1804547.
124 T. D. Creason, A. Yangui, R. Roccanova, A. Strom, M.-H. Du, B. Saparov, *Adv. Opt. Mater.*, Rb_2CuX_3 (X = Cl, Br): 1D All-Inorganic Copper Halides with Ultrabright Blue Emission and Up-Conversion Photoluminescence, **2020**, *8*, 1901338.
125. B. Yang, L. Yin, G. Niu, J.-H. Yuan, K.-H. Xue, Z. Tan, X.-S. Miao, M. Niu, X. Du, H. Song, E. Lifshitz, J. Tang, Lead-Free Halide Rb_2CuBr_3 as Sensitive X-Ray Scintillator, *Adv. Mater.*, **2019**, *31*, 1904711.
126. X. Zhao, G. Niu, J. Zhu, B. Yang, J.-H. Yuan, S. Li, W. Gao, Q. Hu, L. Yin, K.-H. Xue, E. Lifshitz, X. Miao, J. Tang, All-Inorganic Copper Halide as Stable and Self-Absorption Free X-Ray Scintillator, *J. Phys. Chem. Lett.*, **2020**, *11*, 1873.
127. N. V. Rebrova, A. Y. Grippa, A. S. Pushak, T. E. Gorbacheva, V. Y. Pedash, O. G. Viagin, V. L. Cherginets, V. A. Tarasov, V. V. Vistovskyy, A. P. Vas'kiv, S. V. Myagkota, Crystal Growth and Characterization of Eu2+ Doped $RbCaX_3$ (X = Cl, Br) Scintillators, *J. Cryst. Growth*, **2017**, *466*, 39.
128. S. Cui, Y. Chen, S. Tao, J. Cui, C. Yuan, N. Yu, H. Zhou, J. Yin, X. Zhang, Synthesis, Crystal Structure and Photoelectric Response of All-Inorganic Copper Halide Salts $CsCuCl_3$, *Eur. J. Inorg. Chem.*, **2020**, *2020*, 2165.
129. B. Xiao, F. Wang, M. Xu, X. Liu, Q. Sun, B.-B. Zhang, W. Jie, P. Sellin, Y. Xu, Melt-Grown Large-Sized Cs_2TeI_6 Crystals for X-ray Detection, *CrystEngComm*, **2020**, *22*, 5130.
130. N. Sakai, A. A. Haghighirad, M. R. Filip, P. K. Nayak, S. Nayak, A. Ramadan, Z. Wang, F. Giustino, H. J. Snaith, Solution-Processed Cesium Hexabromopalladate(IV), Cs_2PdBr_6, for Optoelectronic Applications, *J. Am. Chem. Soc.*, **2017**, *139*, 6030.
131. F. Wei, Z. Deng, S. Sun, N. T. P. Hartono, H. L. Seng, T. Buonassisi, P. D. Bristowe, A. K. Cheetham, Enhanced Visible Light Absorption for Lead-Free Double Perovskite $Cs_2AgSbBr_6$, *Chem. Commun.*, **2019**, *55*, 3721.
132. T. T. Tran, J. R. Panella, J. R. Chamorro, J. R. Morey, T. M. McQueen, Designing Indirect–Direct Bandgap Transitions in Double Perovskites, *Mater. Horiz.*, **2017**, *4*, 688.
133. J. Zhou, Z. Xia, M. S. Molokeev, X. Zhang, D. Peng, Q. Liu, Composition Design, Optical Gap and Stability Investigations of Lead-Free Halide Double Perovskite $Cs_2AgInCl_6$, *J. Mater. Chem. A*, **2017**, *5*, 15031.

134. K. Shum, Z. Chen, J. Qureshi, C. Yu, J. J. Wang, W. Pfenninger, N. Vockic, J. Midgley, J. T. Kenney, Synthesis and Characterization of $CsSnI_3$ Thin Films, *Appl. Phys. Lett.*, **2010**, *96*, 221903.
135. T. Ye, X. Wang, K. Wang, S. Ma, D. Yang, Y. Hou, J. Yoon, K. Wang, S. Priya, Localized Electron Density Engineering for Stabilized B-γ $CsSnI_3$-Based Perovskite Solar Cells with Efficiencies >10%, *ACS Energy Lett.*, **2021**, *6*, 1480.
136. Q. Xiao, J. Tian, Q. Xue, J. Wang, B. Xiong, M. Han, Z. Li, Z. Zhu, H. L. Yip, Z. Li, Dopant-Free Squaraine-Based Polymeric Hole-Transporting Materials with Comprehensive Passivation Effects for Efficient All-Inorganic Perovskite Solar Cells, *Angew. Chem. Int. Ed.*, **2019**, *58*, 17724.
137. S. Xiao, F. Xu, Y. Bai, J. Xiao, T. Zhang, C. Hu, X. Meng, H. Tan, H. P. Ho, S. Yang, An Ultra-low Concentration of Gold Nanoparticles Embedded in the NiO Hole Transport Layer Boosts the Performance of p-i-n Perovskite Solar Cells, *Sol. RRL*, **2018**, *3*, 1800278.
138. J. Wang, Y. Liu, X. Chen, C. Chen, P. Chen, Z. Wang, Y. Duan, Functional Metal Oxides in Perovskite Solar Cells, *Chemphyschem*, **2019**, *20*, 2580.
139. X. Zhang, J. Liang, G. Gao, S. Ding, Z. Yang, W. Yu, B. Q. Li, The Preparation of Mesoporous SnO_2 Nanotubes by Carbon Nanofibers Template and Their Lithium Storage Properties, *Electrochim. Acta*, **2013**, *98*, 263.
140. H. Katoa, M. Sanoa, K. Miyamotoa, T. Yao, Growth and Characterization of Ga-Doped ZnO Layers on A-Plane Sapphire Substrates Grown by Molecular Beam Epitaxy, *J. Cryst. Growth*, **2002**, *237*, 538.
141. M. Morana, P. Koers, C. Waldauf, M. Koppe, D. Muehlbacher, P. Denk, M. Scharber, D. Waller, C. Brabec, Organic Field-Effect Devices as Tool to Characterize the Bipolar Transport in Polymer-Fullerene Blends: The Case of P3HT-PCBM, *Adv. Funct. Mater.*, **2007**, *17*, 3274.
142. G. E. Eperon, G. M. Paternò, R. J. Sutton, A. Zampetti, A. A. Haghighirad, F. Cacialli, H. J. Snaith, Inorganic Caesium Lead Iodide Perovskite Solar Cells, *J. Mater. Chem. A*, **2015**, *3*, 19688.
143. P. Luo, W. Xia, S. Zhou, L. Sun, J. Cheng, C. Xu, Y. Lu, Solvent Engineering for Ambient-Air-Processed, Phase-Stable $CsPbI_3$ in Perovskite Solar Cells, *J. Phys. Chem. Lett.*, **2016**, *7*, 3603.
144. T. S. Ripolles, K. Nishinaka, Y. Ogomi, Y. Miyata, S. Hayase, Efficiency Enhancement by Changing Perovskite Crystal Phase and Adding a Charge Extraction Interlayer in Organic Amine Free-Perovskite Solar Cells Based on Cesium, *Sol. Energy Mat. Sol. C.*, **2016**, *144*, 532.
145. T. Zhang, M. I. Dar, G. Li, F. Xu, N. Guo, M. Grätzel, Y. Zhao, Bication Lead Iodide 2D Perovskite Component to Stabilize Inorganic α-$CsPbI_3$ Perovskite Phase for High-Efficiency Solar Cells, *Sci. Adv.*, **2017**, *3*, 1700841.
146. P. Wang, X. Zhang, Y. Zhou, Q. Jiang, Q. Ye, Z. Chu, X. Li, X. Yang, Z. Yin, J. You, Solvent-Controlled Growth of Inorganic Perovskite Films in Dry Environment for Efficient and Stable Solar Cells, *Nat. Commun.*, **2018**, *9*, 2225.
147. K. Wang, Z. Jin, L. Liang, H. Bian, D. Bai, H. Wang, J. Zhang, Q. Wang, S. Liu, All-Inorganic Cesium Lead Iodide Perovskite Solar Cells with Stabilized Efficiency Beyond 15%, *Nat. Commun.*, **2018**, *9*, 4544.
148. Y. Wang, M. I. Dar, L. K. Ono, T. Zhang, M. Kan, Y. Li, L. Zhang, X. Wang, Y. Yang, X. Gao, Y. Qi, M. Grätzel, Y. Zhao, Thermodynamically Stabilized β-$CsPbI_3$-Based Perovskite Solar Cells with Efficiencies >18%, *Science*, **2019**, *365*, 591.
149. Y. Wang, X. Liu, T. Zhang, X. Wang, M. Kan, J. Shi, Y. Zhao, The Role of Dimethylammonium Iodide in $CsPbI_3$ Perovskite Fabrication: Additive or Dopant?, *Angew. Chem. Int. Ed.*, **2019**, *58*, 16691.

150. S. Tan, B. Yu, Y. Cui, F. Meng, C. Huang, Y. Li, Z. Chen, H. Wu, J. Shi, Y. Luo, D. Li, Q. Meng, Temperature-Reliable Low-Dimensional Perovskites Passivated Black-Phase $CsPbI_3$ toward Stable and Efficient Photovoltaics, *Angew. Chem. Int. Ed.*, **2022**, *61*, 202201300.
151. J. Wang, Y. Che, Y. Duan, Z. Liu, S. Yang, D. Xu, Z. Fang, X. Lei, Y. Li, S. F. Liu, 21.15%-Efficiency and Stable gamma -$CsPbI_3$ Perovskite Solar Cells Enabled by an Acyloin Ligand, *Adv. Mater.*, **2023**, *35*, 2210223.
152. H. Zhang, W. Xiang, X. Zuo, X. Gu, S. Zhang, Y. Du, Z. Wang, Y. Liu, H. Wu, P. Wang, Q. Cui, H. Su, Q. Tian, S. F. Liu, Fluorine-Containing Passivation Layer via Surface Chelation for Inorganic Perovskite Solar Cells, *Angew. Chem. Int. Ed.*, **2023**, *62*, 202216634.
153. Z. Wang, Q. Tian, H. Zhang, H. Xie, Y. Du, L. Liu, X. Feng, A. Najar, X. Ren, S. F. Liu, Managing Multiple Halide-Related Defects for Efficient and Stable Inorganic Perovskite Solar Cells, *Angew. Chem. Int. Ed.*, **2023**202305815.
154. M. Kulbak, D. Cahen, G. Hodes, How Important Is the Organic Part of Lead Halide Perovskite Photovoltaic Cells? Efficient $CsPbBr_3$ Cells, *J. Phys. Chem. Lett.*, **2015**, *6*, 2452.
155. J. Liang, C. Wang, Y. Wang, Z. Xu, Z. Lu, Y. Ma, H. Zhu, Y. Hu, C. Xiao, X. Yi, G. Zhu, H. Lv, L. Ma, T. Chen, Z. Tie, Z. Jin, J. Liu, All-Inorganic Perovskite Solar Cells, *J. Am. Chem. Soc.*, **2016**, *138*, 15829.
156. J. Duan, Y. Zhao, B. He, Q. Tang, High-Purity Inorganic Perovskite Films for Solar Cells with 9.72 % Efficiency, *Angew. Chem. Int. Ed.*, **2018**, *57*, 3787.
157. W. Chen, J. Zhang, G. Xu, R. Xue, Y. Li, Y. Zhou, J. Hou, Y. Li, A Semitransparent Inorganic Perovskite Film for Overcoming Ultraviolet Light Instability of Organic Solar Cells and Achieving 14.03% Efficiency, *Adv. Mater.*, **2018**, *30*, 1800855.
158. H. Li, G. Tong, T. Chen, H. Zhu, G. Li, Y. Chang, L. Wang, Y. Jiang, Interface Engineering Using a Perovskite Derivative Phase for Efficient and Stable $CsPbBr_3$ Solar Cells, *J. Mater. Chem. A*, **2018**, *6*, 14255.
159. J. Duan, Y. Zhao, X. Yang, Y. Wang, B. He, Q. Tang, Lanthanide Ions Doped $CsPbBr_3$ Halides for HTM-Free 10.14%-Efficiency Inorganic Perovskite Solar Cell with an Ultrahigh Open-Circuit Voltage of 1.594 V, *Adv. Energy Mater.*, **2018**, *8*, 1802346.
160. J. Duan, Y. Zhao, Y. Wang, X. Yang, Q. Tang, Hole-Boosted Cu(Cr,M)O_2 Nanocrystals for All-Inorganic $CsPbBr_3$ Perovskite Solar Cells, *Angew. Chem. Int. Ed.*, **2019**, *58*, 16147.
161. J. Duan, Y. Wang, X. Yang, Q. Tang, Alkyl-Chain-Regulated Charge Transfer in Fluorescent Inorganic $CsPbBr_3$ Perovskite Solar Cells, *Angew. Chem. Int. Ed.*, **2020**, *59*, 4391.
162. G. Tong, T. Chen, H. Li, L. Qiu, Z. Liu, Y. Dang, W. Song, L. K. Ono, Y. Jiang, Y. Qi, Phase Transition Induced Recrystallization and Low Surface Potential Barrier Leading to 10.91%-Efficient $CsPbBr_3$ Perovskite Solar Cells, *Nano Energy*, **2019**, *65*, 104015.
163. Q. Zhou, J. Duan, J. Du, Q. Guo, Q. Zhang, X. Yang, Y. Duan, Q. Tang, Tailored Lattice "Tape" to Confine Tensile Interface for 11.08%-Efficiency All-Inorganic $CsPbBr_3$ Perovskite Solar Cell with an Ultrahigh Voltage of 1.702 V, *Adv. Sci.*, **2021**, *8*, 2101418.
164. R. J. Sutton, G. E. Eperon, L. Miranda, E. S. Parrott, B. A. Kamino, J. B. Patel, M. T. Hörantner, M. B. Johnston, A. A. Haghighirad, D. T. Moore, H. J. Snaith, Bandgap-Tunable Cesium Lead Halide Perovskites with High Thermal Stability for Efficient Solar Cells, *Adv. Energy Mater.*, **2016**, *6*, 1502458.
165. C. Liu, W. Li, C. Zhang, Y. Ma, J. Fan, Y. Mai, All-Inorganic $CsPbI_2Br$ Perovskite Solar Cells with High Efficiency Exceeding 13, *J. Am. Chem. Soc.*, **2018**, *140*, 3825.
166. W. Chen, H. Chen, G. Xu, R. Xue, S. Wang, Y. Li, Y. Li, Precise Control of Crystal Growth for Highly Efficient $CsPbI_2Br$ Perovskite Solar Cells, *Joule*, **2019**, *3*, 191.
167. H. Zhao, Y. Han, Z. Xu, C. Duan, S. Yang, S. Yuan, Z. Yang, Z. Liu, S. Liu, A Novel Anion Doping for Stable $CsPbI_2Br$ Perovskite Solar Cells with an Efficiency of 15.56% and an Open Circuit Voltage of 1.30 V, *Adv. Energy Mater.*, **2019**, *9*, 1902279.

168. Y. Han, H. Zhao, C. Duan, S. Yang, Z. Yang, Z. Liu, S. Liu, Controlled n-Doping in Air-Stable $CsPbI_2Br$ Perovskite Solar Cells with a Record Efficiency of 16.79%, *Adv. Funct. Mater.*, **2020**, *30*, 1909972.
169. S. S. Mali, J. V. Patil, P. S. Shinde, G. de Miguel, C. K. Hong, Fully Air-Processed Dynamic Hot-Air-Assisted M:$CsPbI_2Br$ (M: Eu2+, In3+) for Stable Inorganic Perovskite Solar Cells, *Matter*, **2021**, *4*, 635.
170. H. Xiao, C. Zuo, K. Yan, Z. Jin, Y. Cheng, H. Tian, Z. Xiao, F. Liu, Y. Ding, L. Ding, Highly Efficient and Air-Stable Inorganic Perovskite Solar Cells Enabled by Polylactic Acid Modification, *Adv. Energy Mater.*, **2023**, *13*, 2300738.
171. Q. Ma, S. Huang, X. Wen, M. A. Green, A. W. Y. Ho-Baillie, Hole Transport Layer Free Inorganic $CsPbIBr_2$ Perovskite Solar Cell by Dual Source Thermal Evaporation, *Adv. Energy Mater.*, **2016**, *6*, 1502202.
172. C. Liu, W. Li, J. Chen, J. Fan, Y. Mai, R. E. I. Schropp, Ultra-Thin MoO_x as Cathode Buffer Layer for the Improvement of All-Inorganic $CsPbIBr_2$ Perovskite Solar Cells, *Nano Energy*, **2017**, *41*, 75.
173. B. Yang, M. Wang, X. Hu, T. Zhou, Z. Zang, Highly Efficient Semitransparent $CsPbIBr_2$ Perovskite Solar Cells via Low-Temperature Processed In_2S_3 as Electron-Transport-Layer, *Nano Energy*, **2019**, *57*, 718.
174. C. F. J. Lau, X. Deng, Q. Ma, J. Zheng, J. S. Yun, M. A. Green, S. Huang, A. W. Y. Ho-Baillie, $CsPbIBr_2$ Perovskite Solar Cell by Spray-Assisted Deposition, *ACS Energy Lett.*, **2016**, *1*, 573.
175. J. He, J. Su, J. Di, Z. Lin, S. Zhang, J. Ma, J. Zhang, S. Liu, J. Chang, Y. Hao, Surface Reconstruction Strategy Improves the All-Inorganic $CsPbIBr_2$ Based Perovskite Solar Cells and Photodetectors Performance, *Nano Energy*, **2022**, *94*.
176. X. Zhang, Z. Yu, D. Zhang, Q. Tai, X. Z. Zhao, Recent Progress of Carbon-Based Inorganic Perovskite Solar Cells: From Efficiency to Stability, *Adv. Energy Mater.*, **2022**, *13*, 2201320.
177. C. Dong, B. Xu, D. Liu, E. G. Moloney, F. Tan, G. Yue, R. Liu, D. Zhang, W. Zhang, M. I. Saidaminov, Carbon-Based All-Inorganic Perovskite Solar Cells: Progress, Challenges and Strategies Toward 20% Efficiency, *Mater. Today*, **2021**, *50*, 239.
178. G. Zhang, J. Zhang, Z. Pan, H. Rao, X. Zhong, Enhancing Hole Extraction via Carbon Nanotubes/Poly(3-Hexylthiophene) Composite for Carbon-Based $CsPbI_2Br$ Solar Cells with a New Record Efficiency, *Sci. China Mater.*, **2023**, *66*, 1727.
179. Z. Du, Y. Xu, Y. Jing, R. Li, J. Lin, J. Wu, Z. Lan, Thick Film High-Performance Carbon-Based $CsPbIBr_2$ Perovskite Solar Cells vis Nano-Perovskite Modified Hot Casting Method, *J. Alloy. Compd.*, **2023**, *966*, 171583.
180. M. Wei, B. He, J. Mao, W. Liu, Z. Wang, H. Chen, Q. Tang, Back Interface Management by Multifunctional Composite Carbon Nanospheres Electrode for Efficient and Stable HTL-Free Carbon-Based Perovskite Solar Cells, *Carbon*, **2023**, *215*, 118482.

9 Fully Printable Mesoscopic Perovskite Solar Cells

Wenjun Wu

9.1 INTRODUCTION

The potential of perovskite solar cells in converting sunlight into electricity has been widely recognized. However, as with any new technology, it faces certain challenges that need to be addressed to make it practical for widespread use. One such challenge is the high equipment requirement in the evaporation process, which can lead to increased costs and slow down production. Additionally, the hole transport layer, Spiro-OMeTAD, and counter electrodes made of noble metals such as gold and silver can add to the expense and lead to materials and solvent waste. These challenges have created a need for perovskite solar cells that are low-cost, easy to assemble, and highly stable. To address these needs, printable mesoscopic perovskite solar cells (p-MPSCs) have been developed.

In 2014, Prof. Han H. and his co-workers introduced a groundbreaking structure for perovskite solar cells.[1] This innovative design features a scaffold consisting of three layers of mesoporous TiO_2, ZrO_2, and carbon, which is then infused with perovskite precursor (Figure 9.1). What sets this approach apart is that it eliminates the need of a hole-conducting layer, simplifying the device architecture and reducing production costs. To create the perovskite layer, a solution of PbI_2, methylammonium iodide (MAI), and 5-ammoniumvaleric acid iodide (5-AVAI) is drop-cast onto a carbon film. The addition of 5-AVAI templating generates mixed-cation perovskite $(5\text{-AVA})_x(MA)_{1-x}PbI_3$ crystals, which exhibit lower defect concentration and superior pore filling. Consequently, the scaffold achieves more complete contact with the perovskite layer, resulting in a longer exciton lifetime and higher quantum yield for photoinduced charge separation.

In comparison to the frequently utilized $MAPbI_3$, the 5-AVA templated perovskite yielded a superior power conversion efficiency (PCE) of 12.8%. Moreover, even when exposed to full sunlight in regular air for over 1,000 h, the device exhibited remarkable stability with minimal PCE degradation. These results demonstrate the vast potential of perovskite solar cells and pave the way for further research in this captivating field.

In the initial stage of developing p-MPSCs, the focus was on optimizing the technology by exploring alternative materials to replace 5-AVAI. One promising avenue of research involves the use of bifunctional organic molecules, such as 4-(aminomethyl)

DOI: 10.1201/9781003400493-9

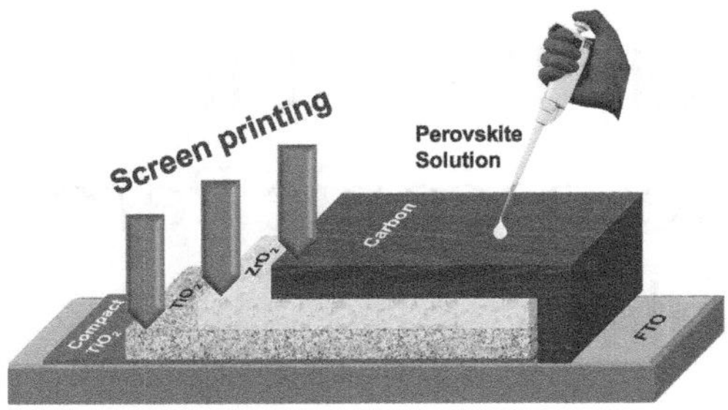

FIGURE 9.1 Schematic diagram of structure of printable mesoscopic perovskite solar cell.

benzoic acid hydroiodide (AB), as organic cations in organic-inorganic halide perovskite materials.[2] In a study conducted by Hu Y. et al., AB was compared to the monofunctional cation benzylamine hydroiodide (BA) and the non-conjugated bifunctional organic molecule 5-AVAI. The results showed that devices based on AB-$MAPbI_3$ exhibited superior PCE of 15.6% and high stability. This is attributed to the bifunctional conjugated cation promoting the growth of perovskite crystals in the mesoporous network and facilitating charge transport. The study provides valuable insights into the rational design of novel organic cations for p-MPSCs.

In summary, it is evident that the assembly process of p-MPSCs boasts various benefits that have caught the attention of researchers in the perovskite solar cell field. These advantages include their low fabrication cost, ability to assemble in air, and use of screen-printing equipment for mass production. Since 2014, several studies have been conducted to investigate different aspects of p-MPSCs such as their stability, electrode materials, insulating and electron transport layers, compact layer, passivation techniques (including co-crystallization and post-treatment), and perovskite materials. These comprehensive studies have demonstrated the potential of p-MPSCs and opened up new avenues for further research in this field.

9.2 WORKING PRINCIPLE OF FULLY PRINTABLE MESOSCOPIC PEROVSKITE SOLAR CELLS

As illustrated in Figure 9.1, a typical device structure of p-MPSCs comprises fluorine-doped tin oxide (FTO) glass acting as a transparent front contact, coated with a thin and compact layer of titanium dioxide (c-TiO_2) as a blocking layer. Electronic extraction is achieved through mesoporous titanium dioxide (m-TiO_2), while holes are transported from the perovskite absorption layer to the counter electrode made of nanocrystalline graphite and carbon black. The titanium dioxide and carbon electrode are electrically insulated from each other by an interlayer of mesoporous zirconium dioxide (m-ZrO_2).

Figure 9.2a illustrates the photovoltaic power conversion mechanism in p-MPSCs. When photons are absorbed by the perovskite material (1), they produce electron-hole pairs. These pairs can undergo either radiative (2) or non-radiative recombination (6a and 6b), depending on whether the circuit is open or closed, respectively.

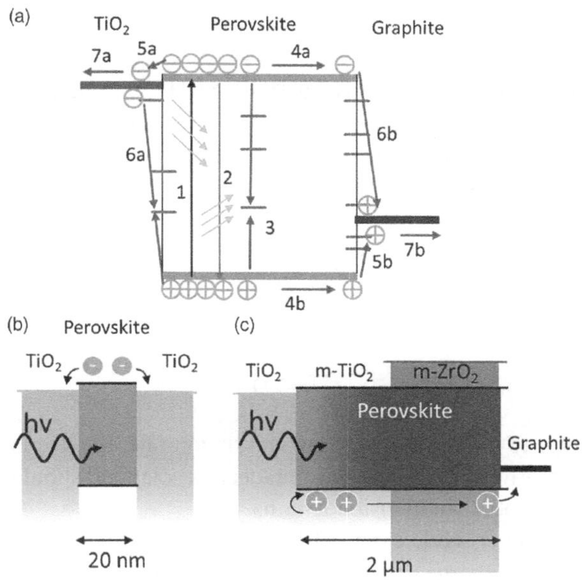

FIGURE 9.2 a) Schematic illustration of the basic generation, recombination and transport processes in a mesoscopic graphite-based PSC. Charge extraction at the respective contacts is depicted for b) electrons and c) holes.

Non-radiative recombination can occur within the bulk absorber material (3) or through surface recombination if the charge carriers are injected into the charge transport layers (CTL; 5). Due to the close proximity of the electron-hole pairs to the front TiO_2 electrode, a diffusion-assisted movement of charge carriers towards the carbon (Graphite) back electrode (4) takes place. Finally, in the presence of an external circuit, the charges can also be extracted.

The open-circuit voltage (V_{oc}) can be described by a simple equation in the limit where the external quantum efficiency approaches the internal quantum efficiency of luminescence.

$$V_{oc} = V_{oc}^{rad} + \frac{kT}{q}\ln\left(\frac{R_{rad}}{R_{rad}+R_{nrad}}\right) \tag{1}$$

Where V_{oc}^{rad} is the radiative limit of the open-circuit voltage (1.34 V for $MAPbI_3$), k is the Boltzmann constant, T is the temperature, and q is the elementary charge. R_{rad} and R_{nrad} represent the radiative and non-radiative recombination rates, respectively.

As shown in Equation 2, the effective lifetime τ_{eff}^{nrad-1} of non-radiative recombination is inversely proportional to R_{nrad}. For perovskites, it includes both bulk (τ_b) and surface (τ_s) recombination lifetimes. The latter is the limiting factor for most PSC devices, as perovskites typically exhibit very high bulk recombination lifetimes.

$$\tau_{\text{eff}}^{\text{nrad}-1} = \tau_b^{-1} + \tau_s^{-1} \tag{2}$$

τ_b is a material-specific Shockley-Read-Hall (SRH) recombination lifetime that essentially depends on the trap density of perovskites. The bulk lifetime of negative (n) and positive (p) charge carriers is related to their respective diffusion lengths L and diffusion coefficients D (Equations 3a and b).

$$\tau_{\text{b,n}} = \frac{L_{\text{n}}^2}{D_{\text{n}}} \tag{3a}$$

$$\tau_{\text{b,p}} = \frac{L_{\text{p}}^2}{D_{\text{p}}} \tag{3b}$$

Therefore, in order to reduce the bulk recombination rate, the average transport distance of each carrier to their respective contacts in a solar cell should be significantly shorter than their respective diffusion lengths.

On the other hand, τ_s is determined by recombination at the interface between the perovskite and the charge extraction layer (CEL), which strongly depends on the "selectivity" of the CEL or its ability to accept one type carrier while repelling the another one. Therefore, the generally drawn conclusion is that in order to maximize the V_{oc} of perovskite devices, high crystallinity of the bulk perovskite material, selectivity of the CELs in solar cell, and quality of the interfaces are all indispensable.

As depicted in Figure 9.2b, the perovskite absorber in these devices is intricately embedded in nanopores formed by mesoporous nano-particles (20 nm) of either m-TiO_2 or m-ZrO_2. Within the m-TiO_2 region, photogenerated electrons are efficiently extracted to TiO_2, with an average transport distance within the absorber of less than 10 nm. This results in significantly lower bulk and surface recombination rates for electrons, as the transport distance is an order of magnitude smaller than the electron diffusion length, and TiO_2 serves as a highly selective electron extraction material (EEM). In contrast, if electrons are generated within the m-ZrO_2 region, the average free path is much longer but not necessarily greater than L_n. Importantly, the electron density produced within the perovskite-filled pores of m-ZrO_2 primarily depends on the thickness of the m-TiO_2 layer and the distance that the absorption spectrum penetrates into the cell.

Referring to Figure 9.2c, the transport of photogenerated holes in p-MPSCs differs from that of electrons. The holes must travel through the porous structure and reach the carbon electrode, with their average transport path determined by the electrode spacing d, which represents the combined thickness of m-TiO_2 and m-ZrO_2.

Han's group conducted a study[3] to investigate the effect of different thicknesses of m-ZrO_2 and m-TiO_2 on the V_{oc}. The results showed that as the thickness of the m-ZrO_2 layer increased, the V_{oc} also increased (Figure 9.3a). For a 0.5-μm-thick m-ZrO_2 layer, the light-induced voltage was below 0.81 V but rapidly increased with the thickness of m-ZrO_2, plateauing at 0.91 V for a 2.2-μm layer. The highest V_{oc} value of 0.93 V was recorded with a 3.7-μm-thick m-ZrO_2 layer. However, the performance of devices with 3.0-μm and 4.4-μm m-ZrO_2 layers was lower due to the less controllable pore filling with increasing layer thickness. It is worth noting that with a m-ZrO_2 layer thickness of less than 1 μm, there was a higher probability of direct electrical contact points between TiO_2 and graphite during screen printing, which further reduced the V_{oc}.

The steady-state V_{oc} of samples with varying thicknesses of m-TiO_2 is presented in Figure 9.3c (with a fixed m-ZrO_2 thickness of 2.2 μm) and Figure 9.3d (with a fixed m-ZrO_2 thickness of 0.79 or 1.7 μm). The hollow frames depicted in Figure 9.3a and c represent the overlapping samples with an m-TiO_2 thickness of 350 nm and m-ZrO_2 thickness of 2.2 μm. It was observed that as the thickness of m-TiO_2 increased from 120 nm to 560 nm, the V_{oc} exhibited a corresponding increase from 0.88 V to 0.93 V. However, when the thickness of m-TiO_2 reached 6 μm, the photocurrent voltage dropped below 0.57 V. Intriguingly, it should be noted that for samples devoid of m-TiO_2 ("0 nm," consisting only of compact TiO_2), the highest V_{oc} recorded was a meager 0.41 V.

The differences in V_{oc} for various thicknesses of m-ZrO_2 and m-TiO_2 can primarily be attributed to variations in quasi-Fermi level splitting (QFLS). In the bulk, there are two recombination processes, namely radiative and non-radiative bulk recombination. At the interface, during transient states, two processes occur:

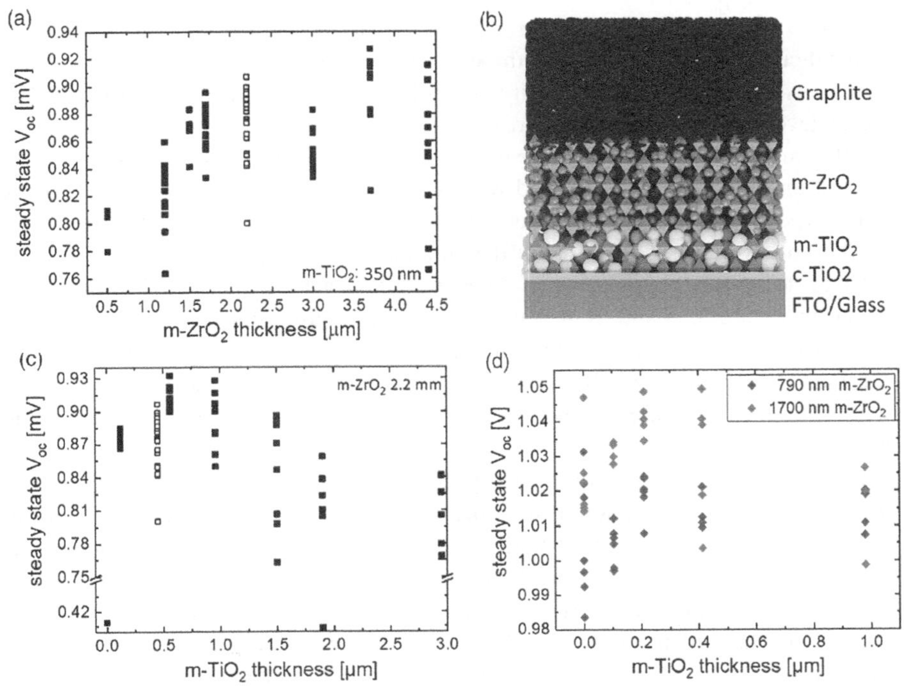

FIGURE 9.3 Steady-state V_{oc} values for a variation of the (a) m-ZrO_2 and (c, d) m-TiO_2. The layer structure is sketched in part (b). The devices of (a, c) have been fabricated at HUST with 10–20 devices for each variation. The empty-box data-points refer to identical samples in each plot. Devices of (d) were fabricated at ISE with seven samples per variation for devices with 0.79 (blue) and 1.7-μm (red)-thick m-ZrO_2 layers. Measurements were carried out under simulated AM1.5G light.

charge extraction to CELs (leading to charge separation and suppressing photoluminescence (PL)), and interface recombination. However, during steady states, the latter determines the interface recombination rate. It is noteworthy that this process is nonlinear, and changes in carrier concentration affect the interdependence of these processes. Charge extraction can be considered the fastest process, followed by interface recombination.

In the realm of p-MPSCs, the role of the electron transport layer TiO_2 cannot be overstated. It is responsible for light absorption and generation, with a depth of approximately 1 μm. As the electrons are extracted into m-TiO_2, they can spread to the "dark" region. Therefore, as the thickness of m-TiO_2 increases, the likelihood of interface recombination with the holes in the perovskite also increases.

Moreover, the compact TiO_2 (c-TiO_2) blocking layer plays a pivotal role in preventing charge recombination at the FTO front electrode. Comparison of devices with and without m-TiO_2 ("0 nm" and 120 nm) has revealed an increase in V_{oc}. This can be attributed to the fact that the charge extraction rate from the perovskite to TiO_2 remains the same for both compact and mesoporous cases, yet the addition of m-TiO_2 layers significantly reduces surface recombination at these interfaces.

Figure 9.4a presents a simulation of the porosity of m-TiO_2 and the perovskite using two distinct layered structures placed adjacent to each other in a direction perpendicular to the thickness of the cell. In this regard, mesoporous ZrO_2 solely serves as a geometric spacer, and it is regarded as an inert electrical insulator. The energy levels of the effective semiconducting layer, where the perovskite infuses into m-ZrO_2, are established by the pure perovskite layer's band structure. The minimum conduction band of TiO_2 is aligned with that of the perovskite at -4.0 eV, ensuring that charges can be effectively extracted towards both m-TiO_2 and c-TiO_2 layers.

The photovoltages derived from the simulations are depicted in Figure 9.4b. In order to accurately reproduce the experimental data from the HUST samples (as shown in Figure 9.3a and c), it was necessary to assume trap-assisted recombination at the m-TiO_2 interface, while initially, no traps were introduced at the c-TiO_2 interface (represented by filled symbols). Consequently, both the rise in V_{oc} with increasing m-ZrO_2 thickness and the fall in V_{oc} with higher m-TiO_2 thickness could be precisely replicated.

Figure 9.4c and d depicts the energy diagrams of 0.2 and 2.2 μm-ZrO_2 devices under open-circuit illumination, with the green lines representing QFLS values of 0.85 and 0.91 eV, respectively. The diffusion of both carriers to the graphite contact is delayed by the increased electrode spacing d. However, the dominant effect is that of diffusion limitations, as carrier generation primarily occurs near the FTO contact, from where they diffuse outwards. Moreover, the layer thickness not only impacts the photovoltage but also significantly affects the short-circuit current density, owing to an increase in surface recombination. The increased electrode spacing results in greater charge transport losses, leading to a decrease in the fill factor.

9.3 COMPACT LAYER

For conventional processes, a mixture of isopropanol and ethanol containing a certain concentration of titanium isopropoxide, commonly known as titanium

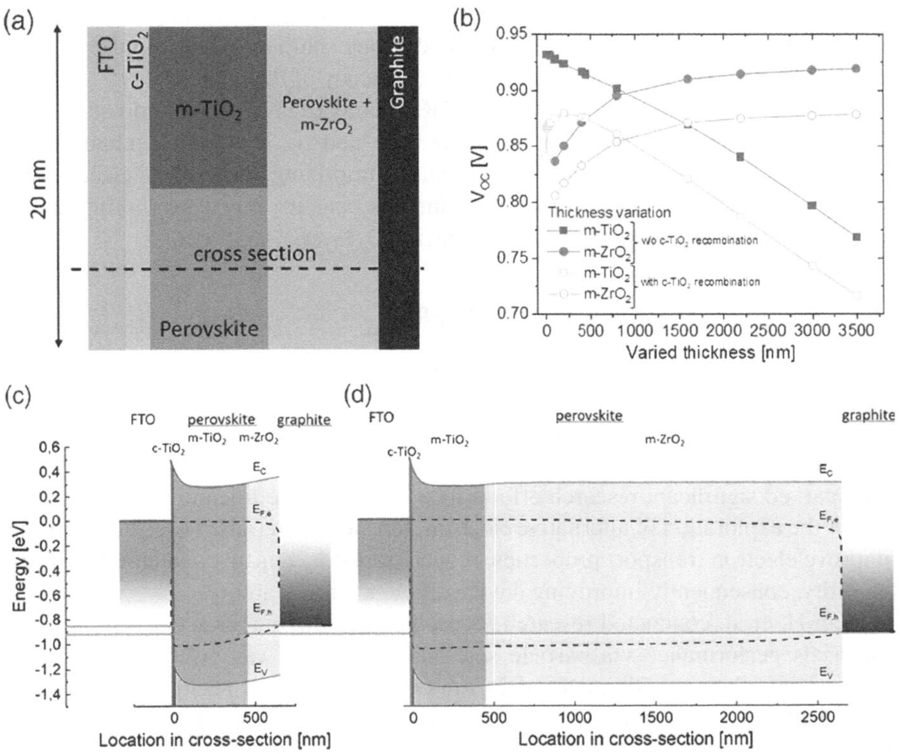

FIGURE 9.4 Results of the 2D simulation. a) Illustration of a cross-section through the modeled unit cell of the 2D simulation, representing the porous structures inside the device. The dashed line marks the cross section for the following figures, whereas the origin is set to the FTO/c-TiO_2 interface. b) Calculated V_{oc} for different layer thicknesses. Band energies along the cross section for a device with both 0.45 μm m-TiO_2 and an m-ZrO_2 thickness of c) 0.2 and d) 2.2 μm in V_{oc} under illumination. The QFLS is marked by horizontal green lines.

acetylacetonate, is typically used. The substrate is heated to around 500°C, and a certain thickness of compact TiO_2 layer is prepared on the transparent conductive substrate using a spray coating technique.

In the pursuit of enhancing the efficiency of p-MPSCs, the use of the TiO_2 compact layer (c-TiO_2) has become a popular choice. However, the relatively low electron mobility of c-TiO_2 may hinder device performance. In this regard, novel compact layers have been investigated. One such study by Liu Q. et al. involved the development of a SnO_2 compact layer (c-SnO_2) through spray pyrolysis using a solution of dibutyltin bisacetylacetonate in ethanol. The champion device incorporating c-SnO_2 exhibited a less pronounced hysteresis effect and achieved an impressive efficiency of 13.77%. This remarkable performance was attributed to the higher electron mobility of SnO_2 and less charge accumulation at interfaces.[4]

In 2016, Han's group published a report on compact-layer-free p-MPSC devices that exhibited an efficiency exceeding 13%, which is comparable to that of devices equipped with a TiO_2 compact layer. Compact-layer-free p-MPSCs offer a technologically simpler and more cost-effective photovoltaic solution, which effectively combines the benefits of low cost with the high efficiency of the device.[5]

In current designs of p-MPSCs, the TiO_2 compact layer plays a pivotal role in improving the contact between the perovskite and the electrode, enhancing the extraction of photo-generated electrons and minimizing the risk of electron-hole recombination. While this layer offers numerous benefits, it is worth noting that the majority of p-MPSCs still feature a compact TiO_2 layer.

9.4 ELECTRON TRANSPORT LAYERS

To date, mesoporous TiO_2 has demonstrated the most efficient electron transport layer (ETL) for p-MPSCs. Nevertheless, the relatively low electron mobility of TiO_2 may result in charge accumulation and recombination at the TiO_2/perovskite interface, which sometimes limits the device performance. Overcoming this challenge has sparked significant research efforts focused on the modification of TiO_2 properties or the exploration of alternative electron transport materials. These efforts aim to improve electron transport properties, reduce trap-state density, and enhance charge mobility, consequently improving device efficiency and stability.

Han H. et al. conducted research focused on optimizing TiO_2 electron transport materials performance via particle size adjustments. It was discovered that TiO_2 nanoparticles having a diameter of 25 nm exhibited the best performance.[6] In their continued efforts to optimize the performance of TiO_2 electron transport materials, Han H. et al. made significant progress in 2018 by introducing a pristine C60 layer onto the TiO_2 ETL layer via a spray method.[7] This modification effectively improved charge transfer at the interface between the ETL and perovskite, resulting in a remarkable increase in PCE from 13.7% to 15.4% and the TiO_2 (C60)-based device showed reduced hysteresis effect.

By incorporating 100-nm-sized TiO_2 nanoparticles into the conventional TiO_2 scaffold layer, researchers have developed a dual-sized TiO_2 scaffold layer that builds upon the foundation of the 25 nm TiO_2 ETL layer. This innovative approach not only enhances the light scattering ability of the scaffold layer – resulting in higher light-harvesting efficiency – but also facilitates excellent infiltration of perovskite due to the larger voids created within the dual-sized scaffold layer. These advantages translate to a significantly higher short-circuit photocurrent density and, ultimately, a higher PCE when compared to conventional ETLs.[8]

Through the implementation of zinc tin oxide as an electron transport layer (ETL) between TiO_2 and FTO, a gradient energy level was constructed, resulting in a remarkable increase in the V_{oc} of p-MPSCs to 1.02 V.[9] This groundbreaking discovery represents a simple yet effective strategy for fine-tuning the band alignment, which successfully suppresses interfacial trap-assisted recombination, ultimately leading to enhanced voltage and charge extraction simultaneously. This innovative approach has the potential to revolutionize the design of ETLs and significantly improve the performance of p-MPSCs in the future.

Recently, Han H. et al. conducted optimization of the preparation process for barium stannate $BaSnO_3$ (BSO) to enhance its performance as a mesoporous electron transport layer (m-ETL) in p-MPSCs. The novel process involves initial heating of BSO in nitrogen to achieve better crystallization and dispersion, followed by cooling in oxygen to reduce surface defects. This optimization strategy resulted in a significant improvement in the PCE of p-MPSCs featuring the BSO m-ETL, achieving an impressive 14.77% PCE, clearly demonstrating the effectiveness of this approach.[10]

Liu Q. et al. demonstrated the synthesis of porous nano-TiO_2 with a large specific surface area using titanium metal organic framework (MOF) MIL-125, as illustrated in Figure 9.5. The cake-like morphology composed of nanocrystals exhibited enhanced perovskite crystallization and reduced recombination of photo-generated electron-hole pairs. The optimized nano-TiO_2-based device achieved a PCE of 13.42%, with a fill factor (*FF*) and open circuit voltage (V_{oc}) of 68.14% and 0.907 V, respectively. This cost-effective and easily prepared Nano-TiO_2 can be printed by screen-printing, making it a promising strategy for the development of high-performance, low-cost porous TiO_2 for fully printable hole-conductor-free carbon-based perovskite solar cells.[11] Furthermore, the research group successfully prepared an anatase TiO_2 electron transport layer with high specific surface area and superior wettability using NH_2-MIL-125 metal framework material, which was applied in p-MPSCs with improved performance.[12] The long-term stability of the device was also impressive, maintaining 94% of its original efficiency for 600 h under a relative humidity of 30–35%.

Recently, Han's group reported a novel oxygen vacancy (OV) detailing management approach for the electron transport layer, achieving a high PCE of 18.96% for this type of device. Initially, a reducing agent-mediated reduction process was employed to introduce an abundance of OVs into both the surface and sub-surface of SnO_2 nanocrystals. Next, a hydrogen peroxide treatment was implemented to remove surface OVs while retaining sub-surface OVs, forming a layered OV structure. This layered OV structure was shown to improve the conductivity of the SnO_2 ETL by increasing electron donation and reducing surface defects. A significant synergistic effect was also observed, with the layered OVs contributing to improving the energy level matching, suppressing interface recombination, and enhancing charge

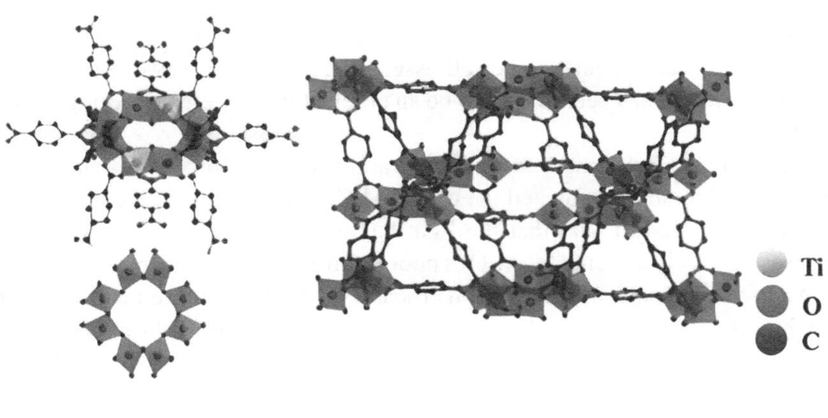

FIGURE 9.5 Schematic structure of MIL-125.

extraction and transfer. These factors collectively contributed to the mp-SnO_2's efficacy as a high-performance ETL, ultimately resulting in a breaking power conversion efficiency.[13] To mitigate non-radiative recombination at the interface of titanium dioxide and perovskite, the electron transport layer (TiO_2) was further refined with the addition of ammonium phosphate. This strategic modification led to a significant boost in the power conversion efficiency of p-MPSCs, setting a new record at 22.2%. Impressively, these cells demonstrated robust performance, maintaining 97% of their initial efficiency even after 750 hours of maximum power point tracking under conditions of 55 ± 5 °C.[14]

Moreover, researchers have conducted a series of studies suppressing non-radiative recombination within the device, further enhancing the photoelectric conversion performance. To this end, dual-sided passivation between the electronic transport layer and perovskite,[15] TiO_2 nanosheets,[16] and other methods[17,18] have also been explored.

9.5 INSULATING LAYERS

In p-MPSCs, ZrO_2 is utilized as both a spacer and supported layer due to its high-band gap width of 5 eV.[19] This material serves as a site for the separation of electrons and holes, effectively inhibiting charge recombination and resulting in an increase in the V_{oc}.[20] According to recent research, the particle size of the ZrO_2 spacer layer has a significant impact on the time-resolved photoluminescence and band-edge emission spectrum of the perovskite. Furthermore, there is a clear correlation between the insulating performance of the spacer layer and the V_{oc}.

In order to address concerns of pollution and waste associated with the use of large amounts of perovskite, Han H. et al.[21,22] devised a solution by introducing a thin layer of Al_2O_3 between TiO_2 and ZrO_2 through the use of a spraying pyrolysis method. This technique effectively reduced the effective thickness of mesoporous ZrO_2. Consequently, charge recombination between TiO_2 and carbon electrode was suppressed, resulting in an increase in open circuit voltage. Additionally, the conduction band of perovskite shifted downwards, thereby improving the overall charge transfer properties. The effectiveness of this approach was demonstrated by a reduction in effective thickness of mesoporous ZrO_2 from 3 μm to 1.2 μm.

Furthermore, optimizing the interface band alignments and charge lifetime is essential for enhancing the performance of p-MPSCs. In this regard, researchers have introduced an organic silane self-assembled monolayer between the TiO_2 and ZrO_2 layers, which has been shown to be an efficient interface-engineering approach for p-MPSCs.[23]

In an effort to develop more cost-effective and novel-structure perovskite solar cells (PSCs), Fan K. et al. investigated the use of Al_2O_3 in place of ZrO_2 as a spacer layer. However, their study found that the light harvesting efficiency of $(5\text{-AVA})_x(MA)_{1-x}PbI_3$ was adversely affected by Al_2O_3's poor infiltration and filling properties, resulting in a lower photoelectric conversion efficiency than that achieved with ZrO_2. This study highlights the critical role played by spacer layers in p-MPSCs' performance, providing valuable insights and guidance for the rational design of future PSCs with higher efficiency and lower costs.[24]

In 2022, Wu W. et al.[25] utilized spray pyrolysis to deposit an ultrathin ZrO_2 layer onto the mesoporous TiO_2 surface. Their study found that the surface lattice perturbation of Zr^{4+} led to a reduction in surface oxygen vacancies of TiO_2, decreasing the density of defective states at the TiO_2-perovskite interface and inhibiting Shockley-Read-Hall recombination (also known as nonradiative recombination) as shown in Figure 9.6. The team utilized perovskite $Cs_{0.05}(FA_{0.92}MA_{0.08})_{0.95}Pb(I_{0.92}Br_{0.08})_3$ in their study and achieved a high PCE of 17.81% with high open-circuit voltage and short-circuit current density. These findings have important implications for the future development of PSCs by identifying promising approaches for enhancing device performance through effective interface engineering.

9.6 CARBON ELECTRODE COMPONENTS

The work function of carbon counter electrodes (CEs) is approximately 5.0 eV, which is only slightly higher than the valence band edge of the perovskite. Consequently, the energy level matching between the carbon electrode and perovskite is the primary factor restricting the photogenerated hole extraction of p-MPSCs at the perovskite–carbon interface.[26,27] Thus, to enhance the performance of p-MPSCs, structural modification or screening of alternative carbon materials for the counter electrode have become essential focal points. These efforts are targeted at further optimizing the square resistance and work function of the carbon electrode, which could ultimately lead to more efficient and cost-effective PSCs with improved stability and durability.

In a study conducted by Han H. et al.,[28] mesoporous graphite/carbon black counter electrodes with different sizes of flake graphite were fabricated for p-MPSCs, resulting in varying FF and PCE due to differences in square resistance. By manipulating the graphite size, the team achieved PCEs of more than 11%. The use of mesoporous graphite and carbon black counter electrodes contributed to their low block resistance and large aperture sizes, which are key factors in enhancing the performance of

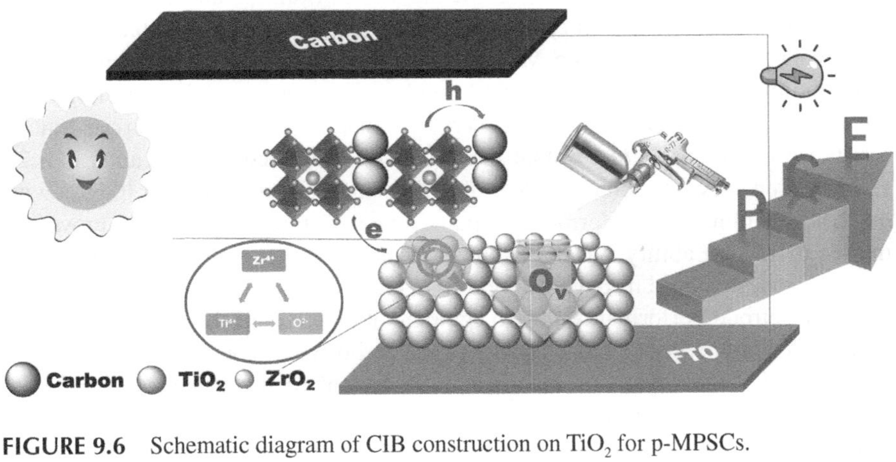

FIGURE 9.6 Schematic diagram of CIB construction on TiO_2 for p-MPSCs.

p-MPSCs. These findings provide valuable insights into the development of efficient and cost-effective PSCs with the potential for large-scale production.

To further enhance the performance of carbon electrodes, Mathews N. et al.[29] investigated the influence of three commonly used carbon blacks, namely vulcan carbon black (VC), meso carbon black (MC), and super P carbon black (SP) on the performance of p-MPSCs. Their results showed that VC-based carbon electrodes possessed good wetting properties for perovskite precursors and strong adhesion within the device stack due to their high porosity and pore volume. Additionally, the highly conductive properties of VC-based counter electrodes resulted in a thin layer resistance as low as 11.0 Ohm/sq, which was lower than that of MC (15.3 Ohm/sq) and SP (22.0 Ohm/sq). Based on VC, the team achieved a record champion power conversion efficiency of 12.55% with almost no reduction in efficiency over 30 d under ambient conditions (approximately 75% relative humidity) without encapsulation. These findings have important implications for the future development of high-performance PSCs, particularly in terms of improving the stability and durability of carbon-based electrodes.

Furthermore, needle coke (NC) – a new carbon material obtained from coal, petroleum, or pitch coke (PC), synthetic graphite (SG), as well as natural graphite (NG) – is widely employed in the production of various carbon electrodes for batteries and electrolysis-based devices.[30] This is due to its excellent characteristics such as ease of graphitization, high electrical conductivity, and low cost. Wu W. et al.[31] took an innovative approach, introducing NC into carbon electrodes to replace conventional carbon black in p-MPSCs. This modification resulted in a substantial increase in the champion PCE, from 7.49% for carbon black to 11.66% for NC due to its high charge extraction capability; morphology-assisted perovskite injection and crystallization, which is beneficial for charge extraction, suppression of current leakage, and charge recombination. These findings highlight the potential of NC as a promising new material for the production of efficient and cost-effective p-MPSCs, with important implications for the future manufacture of advanced energy conversion devices.

The use of liquid metal (LM), particularly Galinstan, as an interface modifier material in combination with carbon electrodes has shown considerable promise in improving the performance of p-MPSCs. LM's high conductivity and room-temperature fluidity make it an ideal material for enhancing hole extraction and improving contact properties between ZrO_2 and carbon. This, in turn, reduces carrier transfer resistance at the carbon/perovskite interface. An optimized concentration of 1.2% LM has been shown to achieve a PCE of 13.51% in PSCs, compared to 10.69% without LM addition. The performance enhancement is attributed to the improvement in hole extraction ability and the engineered interface between perovskite and carbon electrode achieved through LM addition. However, excessive LM can negatively impact electrode performance due to agglomeration. Overall, LM holds significant promise for improving the conductivity of the carbon electrode and optimizing contact properties, making it a promising material for enhancing mesoscopic PSCs.[32]

In 2019, Prajongtat P. et al. utilized electrostatic spinning technique to introduce a polymer film onto the surface of carbon electrodes, aiming to enhance the stability of the device under humid conditions.[33]

The development of high-efficiency low-temperature carbon paste is becoming increasingly important in terms of energy-saving and production efficiency. Han H. et al.[34] utilized isopropanolic titanium (IV) and acetic acid as functional additives and successfully obtained a novel, highly conductive, low-temperature, calcite-phase compatible carbon paste. The functional additives in the carbon paste generated new complex polymeric Ti-O-Ti species in situ, acting as a binder and plasticizer. This approach resulted in an increase in the conductivity of the carbon film to 1.13×10^4 S m^{-1}, surpassing that of transparent FTO or ITO electrodes. The low-temperature carbon film was then employed as a carbon electrode in p-MPSCs, achieving a champion efficiency of 14.04%. By surface modifying PEDOT:PSS, the PCE can be further increased to 14.55%.[35] These findings reveal great potential for the development of highly efficient, low-temperature carbon paste for future applications in the field of solar energy conversion.

A recent study has reported the development of a phenol-formaldehyde resin-based carbon (PFc) material as counter electrodes for p-MPSCs, utilizing an improved Stöber strategy. By modulating the solvent polarity (EtOH/H_2O ratio) during synthesis, PFc materials with various morphologies were effectively obtained. At the optimal ratio (EtOH/H_2O = 4/3), the perovskite precursor infiltration in carbon electrodes was significantly improved due to the interaction between the surface functional groups (C–O/C–N and C=O) of the prepared carbon material and Pb^{2+} ions, as well as the homogeneous porous structure. Moreover, open-circuit voltage (V_{oc}) of PFc-based p-MPSCs increased from 0.97 V to 1.03 V (without any passivation treatment) and a maximum power conversion efficiency (PCE) of 17.64% was achieved based on the champion device.[36] These results indicate that the phenol-formaldehyde resin-based carbon material possesses excellent potential for use as a counter electrode in p-MPSCs, and its morphology can be effectively tailored by adjusting solvent polarity during synthesis.

Carbon electrodes serve a dual purpose as a hole transport layer and a current collector in p-MPSCs. In addition to the aforementioned strategies, various other approaches have been employed to optimize the performance of carbon electrodes. These techniques include the use of ultrathin graphite,[37] graphite chloride,[38] introducing boron atoms into the graphite lattice,[39] oxygen management strategies,[40] naturally derived cross-linked carbon nanoparticles,[41] oxygen-rich activated carbon,[42] multi-walled carbon nanotubes,[43] and waste toner carbon.[44] These investigations have provided insights into the utilization and development of carbon electrodes for p-MPSCs, emphasizing the importance of such research for enhancing the overall efficiency and stability of these devices.

9.7 STRATEGIES FOR PERFORMANCE IMPROVEMENT

9.7.1 Passivation

Passivation is a highly effective technique used to suppress nonradiative recombination in perovskite solar cells, and it is also widely employed in the study of p-MPSCs. Currently, there are two main approaches to passivation: co-crystalization and post-treatment. In the co-crystalization method, the passivator is blended with the perovskite precursor and the resulting mixture is simultaneously crystallized. The second method involves post-treating the prepared perovskite film with a passivator solution, using an antisolvent of the perovskite as the solvent. Both approaches have demonstrated promising results in enhancing the performance of p-MPSCs.

9.7.1.1 Co-Crystalization

Highly crystalline perovskite films with large grains and few grain boundaries are essential for achieving highly efficient and stable PSCs. However, current methods for preparing perovskite films rely on a fast crystallization process, which leads to rapid nucleation and insufficient growth. Recently, researchers have introduced N-methylformamide in $MAPbI_3$ solution to inhibit nucleation and promote growth in the TiO_2/ZrO_2/carbon triple mesoscopic scaffold. Additionally, methylammonium chloride was added to decrease the formation of large colloids, which can serve as nucleation sites. During the crystallization process, the solvent was slowly and carefully removed to prevent rapid nucleation and allow sufficient time for crystal growth. Consequently, highly oriented $MAPbI_3$ crystals with suppressed non-radiative recombination and promoted charge transport were obtained in the triple mesoscopic layer with disordered pores. The resulting p-MPSCs displayed a highest PCE of 18.82% and demonstrated promising long-term operational stability of 1,000 h under continuous illumination at the maximum power point at 55 ± 5 °C, as well as damp-heat stability of 1,340 h aging at 85 °C and 85% relative humidity.[45]

In 2017, Chen W. et al. [46] introduced an environmentally friendly approach to enhance the performance and stability of p-MPSCs. The researchers incorporated strontium chloride ($SrCl_2$) into the precursor solution of $CH_3NH_3PbI_3$ to form $CH_3NH_3PbI_3(SrCl_2)_x$ perovskite with lower defect concentration and better pore filling. The optimized $CH_3NH_3PbI_3(SrCl_2)_{0.1}$ perovskite composition substantially improved the PCE of the device from 13.0% to 15.9%, compared to the pristine $CH_3NH_3PbI_3$ device. Additionally, the stability of the device under ambient illumination was also improved. This study provides a promising pathway for the development of efficient and stable p-MPSCs with reduced environmental impact.

In 2017, Hu Y. et al. [47] presented a novel approach to improve the performance of hybrid perovskite solar cells by incorporating $MABF_4$ into the precursor solution of $(5\text{-AVA})_{0.034}MA_{0.966}PbI_3$ perovskite. The resulting $(5\text{-AVA})_{0.034}MA_{0.966}PbI_{2.95}(BF_4)_{0.05}$ perovskite exhibited reduced charge recombination, leading to a champion PCE of 15.5% with a remarkable V_{oc} of 0.97 V. The addition of $MABF_4$ effectively passivated the defect states in the perovskite film and enhanced the hole transport, resulting in improved device performance. These findings provide a promising strategy for achieving high-performance and stable hybrid PSCs.

In recent years, there has been increasing interest in passivating defects in PSCs to improve their performance and stability. In this regard, Wu W. et al.[48] have introduced imidazolium-based ionic liquids, 1-allyl-3-methylimidazole tetrafluoroborate ($[AMIm]BF_4$) and 1-benzyl-3-methylimidazole tetrafluoroborate ($[BzMIm]BF_4$), as passivants for p-MPSCs. The authors found that the positive ion at the near end N atom of the conjugated side chain plays a key role in dual-site simultaneous passivation. As the conjugation degree of the side chain increases from allyl to benzyl, it not only effectively inhibits non-radiative charge recombination but also reduces the passivation adsorption energy and increases the charge transfer rate. Consequently, a novel concept of "tautomeric dual-site passivation" is proposed (shown in Figure 9.7). The authors achieved a high efficiency of 16.87% with a high *FF* (> 0.7), a J_{sc} of 23.88 mA cm^{-2}, and a V_{oc} up to 1.0 V. This work presents an innovative approach to passivating defects in p-MPSCs, which could lead to the development of more efficient and stable solar cells.

A novel passivator, N,1-fluoroformamidinium iodide (F-FAI), was also synthesized and applied to p-MPSCs. The passivator was designed by replacing an amino group of guanidinium with a fluorine atom, resulting in enhanced molecular polarity and interactions with $MAPbI_3$. F-FAI improved the crystallinity of $MAPbI_3$, passivated defects, and downshifted the Fermi level. The PCE of p-MPSCs was improved from 15.24% with guanidinium iodide to 17.01% with F-FAI.[49] These results demonstrate the promising potential of F-FAI as an effective passivator for p-MPSCs.

The presence of shallow defects in PSCs causes charge recombination, which limits their photocurrent. However, suppressing these defects, particularly methylamine vacancies, is challenging in p-MPSCs owing to their large perovskite thickness and uncontrollable crystallization processes. Wu W. et al. addressed this issue by introducing N719@TiO_2 nanoparticles, which incorporate the commercial N719 dye onto the surface of TiO_2, to improve the passivation effect of MA vacancies and enhance the charge extraction. The unbound carboxyl group of N719 serves as a shell on the surface of TiO_2, while the core TiO_2 binds with the dyes, reinforcing the passivation effect. As a result, the charge extraction capacity for the devices based on N719@TiO_2 was elevated to 1.8007×10^9 C, exceeding that of the control group (1.5507×10^9 C), and the J_{sc} was significantly enhanced to 23.58 mA cm^{-2}, surpassing that of the control device (21.95 mA cm^{-2}).[50] These results demonstrate the potential of N719@

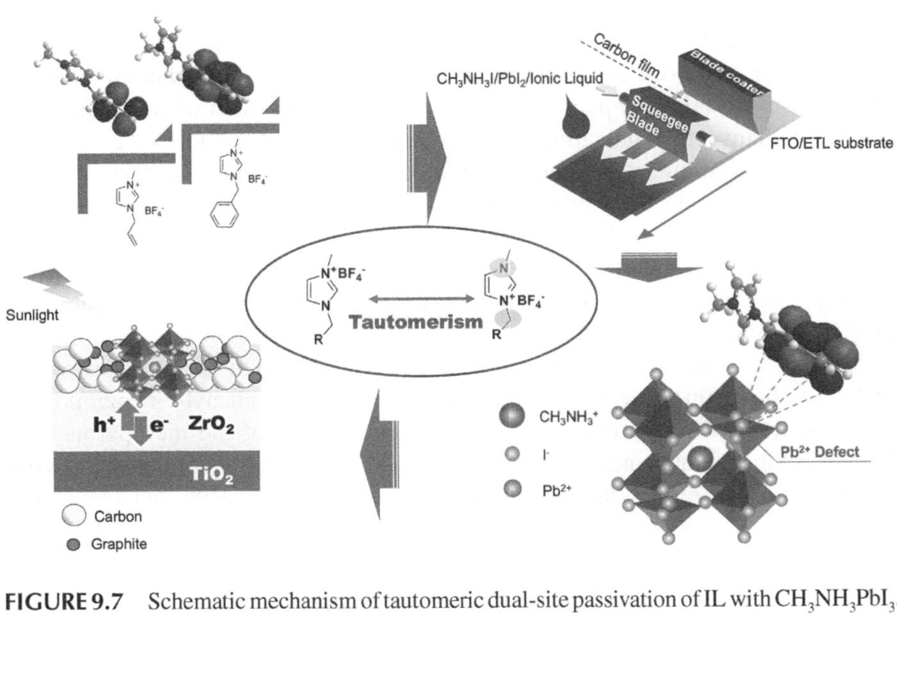

FIGURE 9.7 Schematic mechanism of tautomeric dual-site passivation of IL with $CH_3NH_3PbI_3$.

TiO_2 nanoparticles in mitigating the deleterious effects of shallow defects in p-MPSCs, thereby advancing the development of more efficient and stable perovskite-based solar cells.

To address the issue of poor mesopore filling by the perovskite precursor solution throughout the triple-mesoporous scaffold for p-MPSCs, a bifunctional additive, biuret, was designed to facilitate controllable crystallization of the precursor solution. Biuret contains both carbonyl and amino groups that strongly coordinate with perovskite components, resulting in improved regulation of crystallization kinetics and under-coordinated ion passivation. A champion PCE of 13.42% (J_{sc} = 19.49 mA cm^{-2} and V_{oc} = 0.96 V) was achieved with only 3 mol % biuret, representing a 26% improvement compared to the control device (10.66%).[51]

Additionally, other passivators, such as thiourea,[52] biguanide hydrochloride (BH),[53] bis-adduct 2,5-(dimethyl ester) C60 fullerpyrrolidine (bis-DMEC60),[54] 1-(2-chloroethyl) piperidine hydrochloride (ClEP),[55] 1-ethylpyridine chloride (1-EC),[56] LiCl,[57] ethylene carbonate (EC),[58] melamine hydroiodide (MLAI),[59] D-sorbitan hexaacetate (DSHA),[60] octyltrimethylammonium chloride (OTAC),[61] polyoxometalate (POM),[62] phenethylamine (PEA),[63] ammonium chloride (NH_4Cl),[64] trivalent europium cations,[65,66] ammonium methyl acetate (MAAc),[67] thiopheneformamidine hydrochloride,[68] CuSCN,[69] and amide additives (formamide, acetamide, and urea)[70] have also been used to achieve co-crystallization in p-MPSCs, leading to significant advancements in scientific research efforts.

9.7.1.2 Post-Treatment

The interfacial recombination of PSCs, particularly the loss associated with recombination on the surface of the perovskite/carbon layer, presents a significant bottleneck to further improve the PCE of p-MPSCs. To address this issue, Han's group introduced 2-bromo-6-fluoronaphthalene as an interfacial regulator for p-MPSCs through post-processing. The terminal bromine of the molecule interacts with the iodide ion in the perovskite through a non-covalent halogen bond, acting as the electrophilic side. Additionally, the fused ring of naphthalene accommodates the electron cloud density absorbed by perovskite, improving the hole extraction whilst suppressing non-radiative recombination. The molecule's electron-withdrawing and electron-storage capabilities alter the energy levels of the perovskite surface, resulting in improved hole selectivity and accelerated interfacial charge transfer. The pristine device achieved a PCE of 15.5%, with an increase to 16.77% after the introduction of 2-bromo-6-fluoronaphthalene.[71] These findings underscore the potential of interfacial regulators in mitigating interfacial recombination losses, significantly improving PCE in p-MPSCs.

The low V_{oc} of p-MPCs represents a significant limitation to further improve their PCEs. To address this issue, Han's group introduced sodium 2-phenyl-5-benzimidazole sulfonate through post-processing. The conduction band of the mesoporous TiO_2 layer was significantly increased from −4.22 eV to −4.11 eV, making the electron transfer from the perovskite to the electron transport layer easier and suppressing surface charge recombination. The champion device achieved an V_{oc} of 1.02V, a J_{sc} of 23.06 mA cm^{-2}, an *FF* of 0.70, and a total efficiency of 16.51%. Furthermore, the device maintained 91.7% of its initial efficiency after continuous operation for

1,000 h at the maximum power point under a 1.5G solar simulator.[72] These results underscore the potential of sodium 2-phenyl-5-benzimidazole sulfonate as an effective approach to address the low V_{oc} limitation in p-MPCs, paving the way towards more efficient and stable perovskite-based solar cells.

p-MPSC fabrication processes typically involve high temperatures and can limit synchronization of the passivation step with device fabrication. To address this issue, a post-decoration process utilizing PCBM in chlorobenzene as anti-solvent has been developed. This process induces the regrowth of perovskite crystals – reducing grain boundaries – and enables the relaying delivery of excited state electrons from perovskite to TiO_2 via the ultra-thin PCBM/perovskite heterojunction. The champion device achieved a power conversion efficiency of 14.12% with an V_{oc} of 0.91 V, a J_{sc} of 24.03 mA/cm^2 and an *FF* of 0.64. This creative approach offers a simple solution to achieving high efficiency without compromising the long-term stability of p-MPSCs.[73] These findings highlight the significance of the post-decoration process in achieving high efficiency and stable perovskite-based solar cells and pave the way for further advancements in the field.

One of the main challenges faced by p-MPSCs is the potential loss of voltage caused by the lack of a hole transport layer (HTM). To address this issue, low-dimensional perovskites have been introduced to provide wider band gaps and promote type II band alignment, which enhances directional charge transfer, leading to improved device efficiency. Introducing PEA (phenethylamine) series compounds through post-processing has been shown to effectively construct low-dimensional perovskites between perovskite and carbon electrodes. Through this approach, it was found that the V_{oc} of the device using $MAPbI_3$ as a light absorber was increased from 0.92 to 0.98 V, and the efficiency was increased to 16.24%, with the $FAPbI_3$-based device reaching 17.47% and an V_{oc} of 1.02 V.[74] These findings highlight the importance of low-dimensional perovskite structures in improving voltage loss in p-MPSCs and pave the way for further research on effective post-processing approaches towards achieving higher efficiencies.

The absence of a hole transport layer often results in a low work function of the carbon layer used in p-MPSCs, typically within a range of 4.8–5.0 eV, leading to reduced photovoltage (V_{oc}) and overall efficiency. To address this issue, the introduction of vanadium oxide to the surface of carbon/perovskite via post-processing has been shown to effectively improve the photovoltage and PCE of these devices. Vanadium oxide has a high work function, approximately 5.39 eV, which promotes the charge transfer from perovskite to carbon while maintaining the conductivity of the carbon layer. As a result, the V_{oc} of the device increased from 892.73 to 922.86 mV, the average PCE reached 14.47%, with the champion efficiency reaching 15.77%. Furthermore, the introduction of vanadium oxide was found to suppress the hysteresis effect, with the hysteresis index reduced from 0.025 to -0.001.[75] These results demonstrate the potential of vanadium oxide as a promising post-processing approach for enhancing the performance and stability of p-MPSCs and pave the way towards realizing more efficient and stable perovskite-based solar cells.

In recent research, the use of a p-type molecule, 2, 3, 5, 6-tetrafluoro-7, 7, 8, 8-tetracyanoquinodimethane, was found to effectively modify the perovskite/carbon surface through post-processing, resulting in optimized charge transfer rates and suppressed charge recombination. This approach was found to enhance the PCE of the treated device up to 18.05%, with the V_{oc} notably enhanced to 1,044 mV. The band alignment of the perovskite/carbon interface was successfully tuned, leading to a larger upward band bending that facilitated the transfer of photogenerated holes.[76] These findings suggest that post-processing modifications using p-type molecules could be a promising strategy to enhance the performance of perovskite-based solar cells, and further investigations could lead to the development of more efficient and stable devices.

9.7.2 Other Works

To enhance the efficiency of photoelectric conversion in p-MPSCs, researchers have extensively studied and refined their assembly process and device structure, building upon the foundational structure shown in Figure 9.1. These efforts have aimed to optimize the functioning of the cells and improve their overall performance, bringing them closer to being suitable for practical applications.

Effective control of crystallization is crucial for achieving high-performing printable mesoscopic perovskite solar cells, but the complicated mesoporous structure of these cells makes it challenging to study their crystallization mechanisms. To address this challenge, a solvent evaporation-controlled crystallization method has been proposed. This approach involves adjusting the evaporation rate of solvent during the annealing process to control the rate of perovskite crystallization. By implementing this method, researchers were able to obtain a homogeneous pore filling mesoscopic structure, leading to a stabilized power conversion efficiency of 16.26%.[77] These results demonstrate the potential of solvent evaporation-controlled crystallization as a promising strategy for optimizing the performance and stability of printable mesoscopic perovskite solar cells.

Similar to planar and mesoporous devices, two-step methods have also been investigated for p-MPSCs. In this approach, a DMF solution of PbI_2 is dropwise added onto the carbon electrode at 70 °C and subsequently soaked in an isopropanol solution of MAI or FAI for 15 min. The unreacted MAI or FAI is rinsed with isopropanol, and the mixture is subsequently heated at 70 °C for 30 min. Based on $FAPbI_3$, a PCE of 11.9% was achieved, which is higher than that of $MAPbI_3$ (11.4%). When using an optimized mixing ratio of 3:2 for formamidinium and methylammonium cations, a PCE of 12.9% was achieved.[78,79] Using this optimized two-step method, the built-in electric field is strengthened, voltage loss is reduced, and the PCE is further increased to 17.68%.[80]

In an effort to improve hole transport properties in perovskite-based solar cells, Wang M. et al. introduced a NiO_x layer between the carbon and ZrO_2 layer of p-MPSCs for the first time, forming a p-i-n architecture. This approach led to an appreciated PCE of 14.9% under AM 1.5G illumination. It is believed that the NiO_x layer effectively accelerated charge extraction to the external circuit by inducing a downward shift of the Fermi level.[81] Subsequently, other researchers investigated the use of Ni metal [82] and Ni oxides [83–85] in p-MPSCs. These investigations collectively highlight the potential of incorporating Ni-based materials to improve the performance and stability of p-MPSCs.

To enhance the PCE of solar cells, reduction of optical losses such as reflection, absorption, and transmission losses is a commonly employed approach. For p-MPSCs, researchers have successfully employed a mesoporous SiO_2 antireflection coating at the air/glass interface to enhance the transmittance of the fluorine-doped tin oxide (FTO) glass substrate. The mesoporous SiO_2 layer allows for a graded refractive index interface, resulting in a 2%–4% increase in transmittance of FTO glass in the 350–800 nm region at normal incident angle, while also exhibiting wide-angle and broadband antireflection properties. The resulting SiO_2 coatings have successfully increased the average J_{sc} and PCE by approximately 3% for p-MPSCs.[86] These findings highlight the potential of using mesoporous SiO_2 layers as an effective means to reduce optical losses in perovskite-based solar cells, leading to enhanced efficiency and performance.

To further highlight the importance of interface engineering, Wang's group has made significant progress in studying the crystallization of $CH_3NH_3PbI_3$ and $CH_3NH_3PbI_3$/C surfaces using in situ spatial confinement-induced fluorescence for the first time. By analyzing the derived reaction kinetics, they were able to quantitatively describe the perovskite formation process. Additionally, the group revealed that the interfacial contact between carbon and perovskite plays a vital role in hole extraction efficiency and short-circuit current density. As a result, they introduced a solvent vapor-assisted PbI_2 diffusion method to carefully control the $CH_3NH_3PbI_3$/C surface and reduce the unreacted PbI_2 barrier. Improving surface conditions led to enhanced hole extraction efficiency by altering charge extraction resistance and fluorescence lifetime, ultimately increasing the short-circuit current density.[87]

The use of amino acids as multisite passivators for perovskite solar cells (PSCs) has been proposed due to their unique molecular makeup containing amino, carboxyl, and other functional groups. However, little research has been conducted on the effects of chirality on the photovoltaic properties of PSCs. Wu W. et al. introduced L- and D-cysteines as additives into p-MPSCs, achieving power conversion efficiencies of 17.41% and 15.12%, respectively. Through density-functional theory analysis and photophysical characterization, the researchers deeply explored the underlying causes for the differences in photoelectric conversion performances. DFT analysis showed that the chiral molecular environment of L-cysteine enhanced the negative surface electrostatic potential around the carboxyl group, thereby promoting intramolecular charge transfer and strengthening the coordination to under-coordinated Pb^{2+} (halide vacancy) defects. The chiral environment of L-cysteine also inhibited nonradiative recombination, perovskite crystallization, stability, and light capture, highlighting a novel research pathway extending passivation mechanisms from functional groups to the molecular environment.[88] These findings underscore the potential of utilizing chiral compounds in modulating the performance and stability of perovskite-based solar cells.

Yang S. et al. have developed and characterized a perovskite/carbon bulk heterojunction (BHJ) for carbon-based perovskite solar cells (C-PSCs). The first step involved developing a two-step orthogonal solvent method to deposit high-quality perovskite films directly onto a pre-formed perovskite film, providing a cost-effective and facile means of manipulating the compositions, band alignment, and charge transfer of the perovskite junction. In the second step, this method was applied to a porous carbon electrode to fabricate a perovskite/carbon BHJ with perovskite crystals seamlessly filling in the porous carbon film, resulting in a high contact area of the perovskite/carbon heterojunction. This BHJ accelerated hole collection of the carbon electrode from the perovskite layer, significantly boosting the performance and stability of p-MPSCs with methylammonium lead triiodide ($MAPbI_3$) as the active layer.[89]

Furthermore, the researchers conducted studies such as in situ crystal transfer[90] and solvent engineering[91,92] to improve the fabrication of high-performance p-MPSCs.

9.8 STABILITY

Perovskite solar cells have great potential due to their high-power conversion efficiency, low-cost, and ease of production. Despite these advantages, stability remains a major obstacle to large-scale commercialization. While numerous methods have been developed to improve perovskite stability, they have not yet reached the standard set by mature PV technologies according to IEC61215:2016. However, researchers have successfully improved the stability of p-MPSCs by using bifunctional organic molecules of 5-ammoniumvaleric acid to suppress $MAPbI_3$ perovskite decomposition, reconstruction, and irreversible ionic migration at grain boundaries. Through this method, the p-MPSCs were able to pass key IEC qualification tests, including damp heat, thermal cycling, and ultraviolet preconditioning and demonstrated over 9,000 h of operational stability.[93] These findings represent a significant step towards the large-scale application of stable perovskite solar cells.

The International Electrotechnical Commission (IEC) is a globally recognized organization that develops international standards for electric technologies based on extensive field experience and regular updates. The latest version of the IEC 61215 standard, which has been refined over almost 40 years, is primarily designed to test industrial modules. However, research labs often focus on small-area, single solar cells. To bridge this gap, we have concentrated on all parameters of the IEC test that are also applicable to single cells. However, for perovskites to become a mainstream technology, they will have to transition to modules. In this regard, module stability will become a critical area of research focus. This will require a thorough evaluation of upscaling methods for PSCs, including the encapsulation of large-scale perovskite modules, which is a significant challenge due to the high temperatures involved that can damage the perovskite absorber.

Official testing for the IEC 61215 standard begins with ten modules that are handled according to the flowchart in Figure 9.8. All modules are initially tested to confirm the producer's specifications, such as efficiency. After the first five tests, the modules are separated and subjected to one of the sequences called A, B, C, D, or E and then tested again for efficiency and safety. Each sequence contains several stress tests targeting specific degradation causes that commonly occur. These are briefly:

- Sequence A: mild conditions for basic characterization.
- Sequence B: tests hot-spots and outdoor behavior.
- Sequence C: combines several stress tests where the modules are first preconditioned with UV light and then subjected to 50 thermal and 10 humidity freeze cycles.
- Sequence D: contains 200 thermal cycles.
- Sequence E: damp-heat test together with mechanical stability tests.

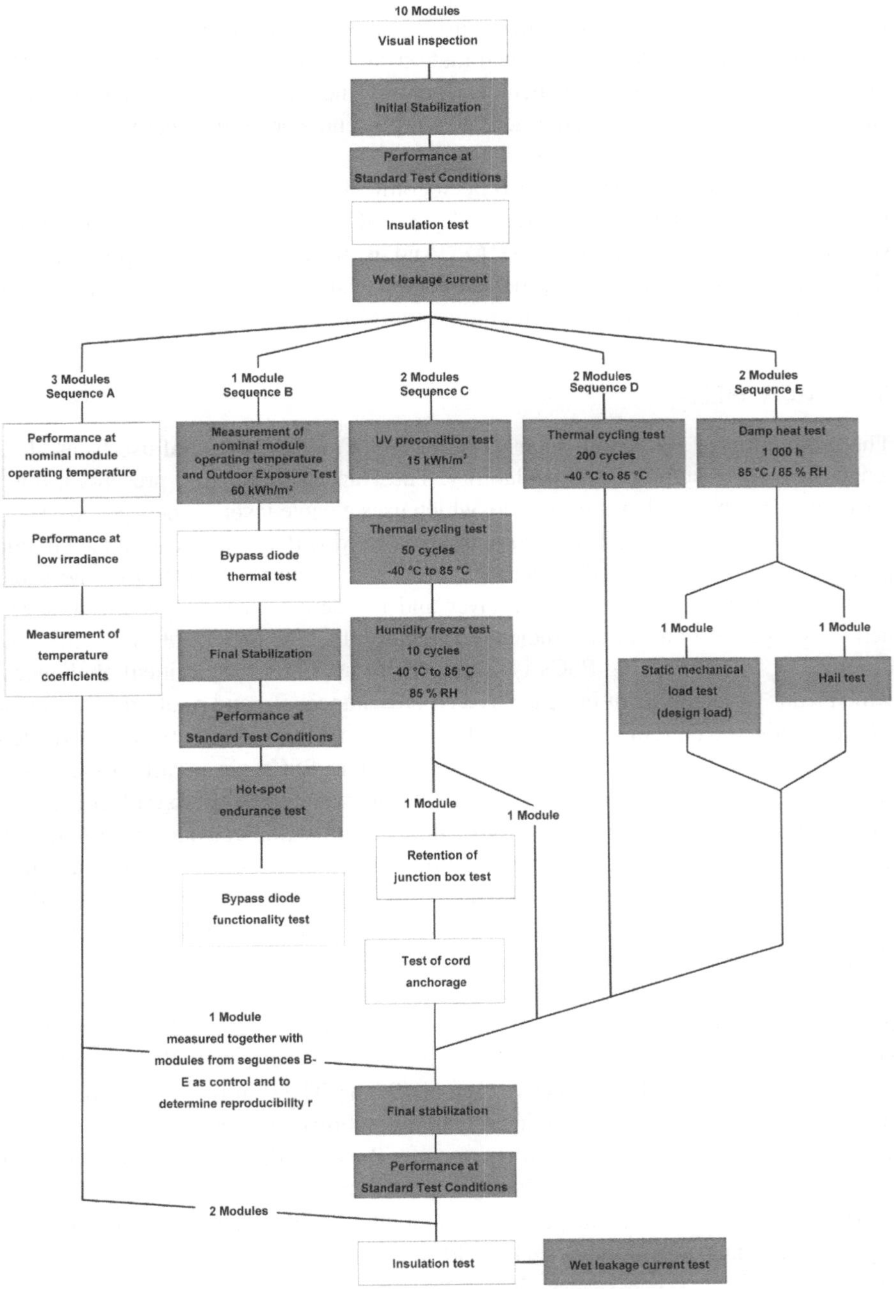

FIGURE 9.8 Full test flow for design qualification and type approval of PV modules. There are five testing sequences called A, B, C, D, and E.

The standard is approved for the module series if all modules pass the requirements of each sequence. If one module fails, two additional new modules are tested again for the respective sequence. In general, after each test sequence (or for test Sequence B after the bypass diode test), the degradation should not exceed approximately 5%. Other general passing criteria include repeating the wet leakage current test after each test and ensuring no visual defects are present. For more specific testing and descriptions, please refer to the literature.[94]

The stability of p-MPSC has also been confirmed by other works. Encapsulated with UV-curable adhesive, the prepared p-MPSC devices could maintain thermal stability for more than 1500 h at 100 °C.[95] And using hot melt adhesive polyurethane (PU) to encapsulate the devices, the submodules can maintain 97.52% of the initial efficiency after 2,136 h under outdoor conditions.[96]

9.9 MODULES

The production of perovskite solar modules (PSM) for commercial use faces challenges related to stability and scalability. The Han H. group first presents a hole-conductor-free printable solar module, which uses a triple layer of mesoporous TiO_2/ZrO_2/carbon as a p-MPSC framework and is infused with a mixed cation lead halide perovskite $(5\text{-AVA})_x(MA)_{1-x}PbI_3$ as a light absorber. By optimizing the thickness and alignment of the triple mesoscopic layer and the active area design, the unit cell achieves a power conversion efficiency (PCE) of 14.02% under 100 mW cm^2 conditions. For industrializing PSCs is scaling them up, one of the critical challenges, which leads to a significant increase in series resistance (R_S) and a subsequent loss of *FF* that limits the PCE of larger-area PSCs. To address this issue, a study investigated the modulation of R_S in fully printable mesoscopic PSCs via a carbon electrode, whereby device geometry was optimized, busbars were prepared to collect currents from electrodes, and the conductivity of the carbon back electrode was improved. The use of a rectangular device shape and tin busbars on conductive substrate effectively reduced the R_S, while an additional highly conductive low-temperature carbon layer hot-pressed onto the back carbon electrode further lowered the R_S. As a result, R_S modulation led to an enhanced PCE of 13.99% for 1 cm^2 PSCs, whereas the control device exhibited a PCE of 7.74%. This increase in PCE can be attributed to the improvement of *FF* from 0.372 to 0.638, with reduced R_S from 27.13 to 9.66 $\Omega\ cm^2$.[97] For a larger area of a module consisting of ten serially connected cells (10 × 10 cm^2), it shows a PCE of 10.4% on an active area of 49 cm^2. The stability tests demonstrate light-soaking stability of 1,000 h, local outdoor stability of 1 month, and a shelf-life stability of over 1 year. These findings open up the potential for the development of efficient and stable large-area PSMs for industrial use.[98]

To realize the commercial application of photovoltaic devices, fabrication of high-performance mini-modules (10–200 cm^2), submodules (200–800 cm^2), and modules (≥ 800 cm^2) is necessary. But for p-MPSCs, limited by the uniformity defect of the mesoporous membrane of screen-printing technique for scaling-up devices, Han H. et al. employed slot-die coating technology to prepare p-MPSCs with an active area of 60.08 cm^2 and obtained the highest PCE of 12.87% for a mini-module.[99]

FIGURE 9.9 p-MPSCs power station of Wuhan Wondersolar Photovoltaic Energy Co., Ltd. in China.

For larger scale research and production, such as Hubei Wonder Solar Technology Co., Ltd., established in 2016, focuses on developing printable mesoscopic solar cells and optoelectronic devices using a triple-layer mesoporous membrane structure. They are exploring breakthrough solutions for single-junction solar cells to surpass the Shockley-Queisser limit and are integrating the entire industry chain, from fundamental research to industrial promotion, to achieve low-cost photovoltaic power generation and related high-performance optoelectronic devices. They have constructed a small-scale power station based on p-MPSCs modules, as illustrated in Figure 9.9. In 2021, Wuhan Wonder Solar Photovoltaic Research Institute was established, and Wonder Solar Photovoltaic Technology (Gedian) Industrial Park was built to promote the "industry-university-research-application" chain. Construction began on a 200 MW large-scale production line for printable mesoscopic perovskite solar cell components in April 2022, and the second 200 MW production line for printable mesoscopic perovskite solar cell components will commence construction in June 2023.

9.10 CONCLUSION AND OUTLOOK

To fabricate p-MPSCs, a triple mesoscopic scaffold comprising mesoporous TiO_2, mesoporous ZrO_2, and a carbon layer is a commonly used approach. Perovskite is integrated within this three-layer mesoporous structure through diffusion crystallization, where TiO_2 serves for electron collection and ZrO_2 separates irradiated excitons and facilitates the migration of electrons and holes towards the TiO_2 and carbon layers, respectively, to accomplish photoelectric conversion. However, the controllability of perovskite crystal through diffusion crystallization poses significant challenges, leading to more surface and internal defects compared to conventional devices, ultimately resulting in lower photoelectric conversion efficiency.

Despite limitations in practical applications involving long-term stability and large-scale manufacturing processes, printable mesoscopic perovskite solar cells exhibit great potential in the photovoltaic industry. Significant research has been carried out in photovoltaic conversion mechanism development, performance improvement, stability enhancement, and large-scale production. Continued efforts aim to refine performance, enhance stability, and reduce manufacturing costs through advances in perovskite drop-casting processes, electrodes, and dielectric materials. As a result, printable mesoscopic perovskite solar cells are poised to become a significant contender in the future photovoltaic market.

ACKNOWLEDGMENTS

This work was supported financially by NSFC/China (grant nos. 22075083), the Programme of Introducing Talents of Discipline to Universities (grant no. B16017). Thanks to Professor He Tian from East China University of Science and Technology for his valuable support and assistance.

REFERENCES

1. A. Y. Mei, X. Li, L. F. Liu, Z. L. Ku, T. F. Liu, Y. G. Rong, M. Xu, M. Hu, J. Z. Chen, Y. Yang, M. Gratzel, H. W. Han, A hole-conductor–free, fully printable mesoscopic perovskite solar cell with high stability, *Science* **2014**, *345*, 295.
2. Y. Hu, Z. H. Zhang, A. Y. Mei, Y. Y. Jiang, X. M. Hou, Q. F. Wang, K. Du, Y. G. Rong, Y. H. Zhou, G. Z. Xu, H. W. Han, Improved performance of printable perovskite solar cells with bifunctional conjugated organic molecule, *Adv. Mater.* **2018**, *30*, 1705786.
3. L. Wagner, C. Qiu, M. Unmüssig, D. Bogachuk, S. Mastroianni, U. Würfel, Y. Hu, H. W. Han, A. Hinsch, A 2D model for interfacial recombination in mesoscopic perovskite solar cells with printed back contact, *Sol. RRL* **2021**, *5*, 2000595.
4. X. X. Jiang, Y. L. Xiong, Z. H. Zhang, Y. G. Rong, A. Y. Mei, C. B. Tian, J. Zhang, Y. M. Zhang, Y. X. Jin, H. W. Han, Q. J. Liu, Efficient hole-conductor-free printable mesoscopic perovskite solar cells based on SnO_2 compact layer, *Electrochim. Acta* **2018**, *263*, 134.
5. X. X. Jiang, Y. L. Xiong, A. Y. Mei, Y. G. Rong, Y. Hu, L. Hong, Y. X. Jin, Q. J. Liu, H. W. Han, Efficient compact-layer-free, hole-conductor-free, fully printable mesoscopic perovskite solar cell, *J. Phys. Chem. Lett.* **2016**, *7*, 4142.
6. Y. Yang, K. Ri, A. Y. Mei, L. F. Liu, M. Hu, T. F. Liu, X. Li, H. W. Han, The size effect of TiO_2 nanoparticles on a printable mesoscopic perovskite solar cell, *J. Mater. Chem. A* **2015**, *3*, 9103.
7. C. B. Tian, S. J. Zhang, S. Li, A. Y. Mei, D. Li, S. Liu, D. Y. Zhang, Y. Hu, Y. G. Rong, H. W. Han, A C_{60} modification layer using a scalable deposition technology for efficient printable mesoscopic perovskite solar cells, *Sol. RRL* **2018**, *2*, 1800174.
8. Z. H. Liu, S. Bi, G. L. Hou, C. Z. Ying, X. J. Su, Dual-sized TiO_2 nanoparticles as scaffold layers in carbon-based mesoscopic perovskite solar cells with enhanced performance, *J. Power Sources* **2019**, *430*, 12.
9. J. H. Zhao, Y. M. Zhang, X. B. Zhao, J. Zhang, H. Wang, Z. Q. Zhu, Q. J. Liu, Band alignment strategy for printable triple mesoscopic perovskite solar cells with enhanced photovoltage, *ACS Appl. Energy Mater.* **2019**, *2*, 2034.
10. J. L. Liu, Y. J. Guan, S. Liu, S. Li, C. X. Gao, J. K. Du, C. Qiu, D. Y. Li, D. Y. Zhang, X. D. Wang, Y. F. Wang, Y. Hu, Y. G. Rong, A. Y. Mei, H. W. Han, Modulating oxygen vacancies in $BaSnO_3$ for printable carbon-based mesoscopic perovskite solar cells, *ACS Appl. Energy Mater.* **2021**, *4*, 11032.

11. X. B. Zhao, J. H. Zhao, J. C. He, B. Li, Y. M. Zhang, J. C. Hu, H. P. Wang, D. M. Zhang, Q. J. Liu, Porous anatase TiO_2 nanocrystal derived from the metal–organic framework as electron transport material for carbon-based perovskite solar cells, *ACS Appl. Energy Mater.* **2020**, *3*, 6180.
12. B. Li, J. Zhao, Q. Lu, S. Zhou, H. Wei, T. Lv, Y. Zhang, J. Zhang, Q. Liu, Carbon-based printable perovskite solar cells with a mesoporous TiO_2 electron transporting layer derived from metal–organic framework NH_2-MIL-125, *Energy Technol.* **2021**, *9*, 2000957.
13. J. L. Liu, S. Li, S. Liu, Y. M. Chu, T. Ye, C. Qiu, Z. X. Qiu, X. D. Wang, Y. F. Wang, Y. Q. Su, Y. Hu, Y. G. Rong, A. Y. Mei, H. W. Han, Oxygen vacancy management for high-temperature mesoporous SnO_2 electron transport layers in printable perovskite solar cells, *Angew. Chem. Int. Ed.* **2022**, *61*, e202202012.
14. J. L. Liu, X. Y. Chen, K. Z. Chen, W. M. Tian, Y. S. Sheng, B. She, Y. Y. Jiang, D. Y. Zhang, Y. Liu, J. H. Qi, K. Chen, Y. M. Ma, Z. X. Qiu, C. Y. Wang, Y. F. Yin, S. L. Zhao, J. Leng, S. Y. Jin, W. S. Zhao, Y. Y. Qin, Y. Q. Su, X. Y. Li, X. J. Li, Y. Zhou, Y. H. Zhou, F. R. Ling, A. Y. Mei, H. W. Han, Electron injection and defect passivation for high-efficiency mesoporous perovskite solar cells, *Science* **2024**, *383*, 1198–120.
15. J. H. Zhao, Y. M. Zhang, Q. Zhang, X. B. Zhao, B. Li, J. Zhang, Z. Q. Zhu, J. Liu, Q. J. Liu, Efficient bifacial passivation enables printable mesoscopic perovskite solar cells with improved photovoltage and fill factor, *Sol. RRL* **2020**, 2000288.
16. Y. G. Rong, Z. L. Ku, A. Y. Mei, T. F. Liu, M. Xu, S. Ko, X. Li, H. W. Han, Hole-conductor-free mesoscopic $TiO_2/CH_3NH_3PbI_3$ heterojunction solar cells based on anatase nanosheets and carbon counter electrodes, *J. Phys. Chem. Lett.* **2014**, *5*, 2160.
17. J. L. Liu, S. Li, S. Liu, Y. M. Chu, T. Ye, C. Qiu, Z. X. Qiu, X. D. Wang, Y. F. Wang, Y. Q. Su, Y. Hu, Y. G. Rong, A. Y. Mei, H. W. Han, Oxygen vacancy management for high-temperature mesoporous SnO_2 electron transport layers in printable perovskite solar cells, *Angew. Chem. Int. Ed.* **2022**, *61*, e202202012.
18. Z. N. Wan, M. Xu, Z. Y. Fu, D. Li, A. Y. Mei, Y. Hu, Y. G. Rong, H. W. Han, Screen printing process control for coating high throughput titanium dioxide films toward printable mesoscopic perovskite solar cells, *Front. Optoelectron.* **2019**, *12*, 344.
19. D. Bi, S. J. Moon, L. Häggman, G. Boschloo, L. Yang, E. M. Johansson, A. Hagfeldt, Using a two-step deposition technique to prepare perovskite ($CH_3NH_3PbI_3$) for thin film solar cells based on ZrO_2 and TiO_2 mesostructures, *RSC Adv.* **2013**, *3*, 18762.
20. Z. A. Xie, S. F. Liu, L. X. Qin, S. P. Pang, W. Wang, Y. Yan, L. Yao, Z. J. Chen, S. F. Wang, H. L. Du, M. H. Yu, G. G. Qin, Refractive index and extinction coefficient of $CH_3NH_3PbI_3$ studied by spectroscopic ellipsometry, *Opt. Mater. Express* **2015**, *5*, 29.
21. Q. F. Wang, S. Liu, Y. Ming, Y. J. Guan, D. Y. Li, C. C. Zhang, Z. K. Wang, Y. G. Rong, Y. Hu and H. W. Han, Sustain. Improvements in printable mesoscopic perovskite solar cells via thinner spacer layers, *Energy Fuels* **2018**, *2*, 2412.
22. T. Liu, L. Liu, M. Hu, Y. Yang, L. Zhang, A. Mei, H. Han, Critical parameters in TiO_2/ZrO_2/Carbon-based mesoscopic perovskite solar cell, *J. Power Sources* **2015**, *293*, 533.
23. L. F. Liu, A. Y. Mei, T. F. Liu, P. Jiang, Y. S. Sheng, L. J. Zhang, H. W. Han, Fully printable mesoscopic perovskite solar cells with organic silane self-assembled monolayer, *J. Am. Chem. Soc.* **2015**, *137*, 1790.
24. Z. Meng, D. P. Guo, J. G. Yu, K. Fan, Investigation of Al_2O_3 and ZrO_2 spacer layers for fully printable and hole-conductor-free mesoscopic perovskite solar cells, *Appl. Surf. Sci.* **2018**, *430*, 632.
25. J. S. He, T. Ding, W. J. Wu, Surface lattice perturbation of electron transport layer reducing oxygen vacancies for positive photovoltaic effect, *Sol. RRL* **2022**, 2200226.
26. Z. H. Wei, K. Y. Yan, H. N. Chen, Y. Yi, T. Zhang, X. Long, J. K. Li, L. X. Zhang, J. N. Wang and S. H. Yang, Cost-efficient clamping solar cells using candle soot for hole extraction from ambipolar perovskites, *Energy Environ. Sci.* **2014**, *7*, 3326.

27. Z. Li, S. A. Kulkarni, P. P. Boix, E. Z. Shi, A. Y. Cao, K. W. Fu, S. K. Batabyal, J. Zhang, Q. H. Xiong, L. H. Wong, N. Mathews, S. G. Mhaisalkar, Laminated carbon nanotube networks for metal electrode-free efficient perovskite solar cells, *ACS Nano.* **2014**, *8*, 6797.
28. Y. G. Rong, Z. L. Ku, A. Y. Mei, T. F. Liu, M. Xu, S. Ko, X. Li, H. W. Han, Hole-conductor-free mesoscopic $TiO_2/CH_3NH_3PbI_3$ heterojunction solar cells based on anatase nanosheets and carbon counter electrodes, *J. Phys. Chem. Lett.* **2014**, *5*, 2160.
29. P. Kajal, J. H. Lew, A. Kanwat, P. J. S. Rana, G. V. Nutan, T. M. Koh, S. G. Mhaisalkar, S. Powar, N. Mathews, Unveiling the role of carbon black in printable mesoscopic perovskite solar cells, *J. Power Sources* **2021**, *501*, 230019.
30. W. F. Ren, Z. L. Zhang, Y. H. Wang, G. W. Kan, Q. Q. Tan, Z. Y. Zhong and F. B. Su. Preparation of porous carbon microspheres anode materials from fine needle coke powders for lithium-ion batteries, *RSC Advances* **2015**, *5*, 11115.
31. Y. Zhong, L. Xu, C. Li, B. Zhang, W. J. Wu, Needle coke: A predominant carbon black alternative for printable triple mesoscopic perovskite solar cells, *Carbon* **2019**, *153*, 602.
32. Y. M. Zhang, J. H. Zhao, J. Zhang, X. X. Jiang, Z. Q. Zhu, Q. J. Liu, Interface engineering based on liquid metal for compact-layer-free, fully printable mesoscopic perovskite solar cells, *ACS Appl. Mater. Interfaces* **2018**, *10*, 15616.
33. P. Prajongtat, C. Sriprachuabwong, R. Wongkanya, D. Dechtrirat, J. Sudchanham, N. Srisamran, W. Sangthong, P. Chuysinuan, A. Tuantranont, S. Hannongbua, N. Chattham, Moisture-resistant electrospun polymer membranes for efficient and stable fully printable perovskite solar cells prepared in humid air, *ACS Appl. Mater. Interfaces* **2019**, *11*, 27677.
34. P. Jiang, T. W. Jones, N. W. Duffy, K. F. Anderson, R. Bennett, M. Grigore, P. Marvig, Y. Xiong, T. Liu, Y. Sheng, L. Hong, X. Hou, M. Duan, Y. Hu, Y. Rong, G. J. Wilson, H. Han, Fully printable perovskite solar cells with highly-conductive, low-temperature, perovskite-compatible carbon electrode, *Carbon* **2018**, *129*, 830.
35. S. Y. Wang, P. Jiang, W. J. Shen, A. Y. Mei, S. X. Xiong, X. S. Jiang, Y. G. Rong, Y. W. Tang, Y. Hu and H. W. Han, A low-temperature carbon electrode with good perovskite compatibility and high flexibility in carbon based perovskite solar cells, *Chem. Commun.* **2019**, *55*, 2765.
36. J. Sheng, X. T. Zhu, X. L. Xu, J. S. He, D. Ma, J. L. Liu and W. J. Wu, EtOH/H_2O ratio modulation on carbon for high-V_{oc} (1.03 V) printable mesoscopic perovskite solar cells without any passivation, *Mater. Adv.* **2023**, *4*, 1534.
37. M. Duan, Y. G. Rong, A. Y. Mei, Y. Hu, Y. S. Sheng, Y. J. Guan, H. W. Han, Efficient hole-conductor-free, fully printable mesoscopic perovskite solar cells with carbon electrode based on ultrathin graphite, *Carbon* **2017**, *120*, 71.
38. X. M. Chen, L. Lu, D. W. Gu, X. Y. Zhang, H. Yu, F. Chen, Y. J. Rui, J. W. Hou and Y. Yang, Chlorine management of a carbon counter electrode for high performance printable perovskite solar cells, *J. Mater. Chem. C* **2021**, *9*, 8615.
39. M. Duan, C. B. Tian, Y. Hu, A. Y. Mei, Y. G. Rong, Y. L. Xiong, M. Xu, Y. S. Sheng, P. Jiang, X. M. Hou, X. T. Zhu, F. Qin, H. W. Han, Boron-doped graphite for high work function carbon electrode in printable hole-conductor-free mesoscopic perovskite solar cells, *ACS Appl. Mater. Interfaces* **2017**, *9*, 31721.
40. C. B. Tiana, A. Y. Mei, S. J. Zhang, H. R. Tian, S. Liu, F. Qin, Y. L. Xiong, Y. G. Rong, Y. Hu, Y. H. Zhou, S. Y. Xie, H. W. Han, Oxygen management in carbon electrode for high-performance printable perovskite solar cells, *Nano Energy* **2018**, *53*, 160.
41. S. S. Mali, H. Kim, J. V. Patil, C. K. Hong, Bio-inspired carbon hole transporting layer derived from aloe vera plant for cost-effective fully printable mesoscopic carbon perovskite solar cells, *ACS Appl. Mater. Interfaces* **2018**, *10*, 31280.
42. C. Liu, C. X. Gao, W. Wang, X. D. Wang, Y. F. Wang, W. J. Hu, Y. G. Rong, Y. Hu, L. B. Guo, A. Y. Mei, H. W. Han, Cellulose-based oxygen-rich activated carbon for printable mesoscopic perovskite solar cells, *Sol. RRL* **2021**, *5*, 2100333.

43. F. R. Li, Y. Xu, W. Chen, S. H. Xie, J. Y. Li, Nanotube enhanced carbon grids as top electrodes for fully printable mesoscopic semitransparent perovskite solar cells, *J. Mater. Chem. A* **2017**, *5*, 10374.
44. D. Ma, J. Sheng, Y. Z. Hao, J. S. He, Y. Zhong, W. J. Wu, Photovoltaic green application of waste toner carbon on fully printable mesoscopic perovskite solar cells, *Sol. Energy* **2021**, *228*, 439.
45. S. Liu, D. Y. Zhang, Y. S. Sheng, W. H. Zhang, Z. T. Qin, M. C. Qin, S. Li, Y. F. Wang, C. X. Gao, Q. F. Wang, Y. Ming, C. Liu, K. Yang, Q. Y. Huang, J. H. Qi, Q. J. Gao, K. Chen, Y. Hu, Y. G. Rong, X. H. Lu, A. Y. Mei, H. W. Han, Highly oriented $MAPbI_3$ crystals for efficient hole-conductor-free printable mesoscopic perovskite solar cells, *Fundam. Res.* **2022**, *2*, 276.
46. H. Zhang, H. Wang, S. T. Williams, D. H. Xiong, W. J. Zhang, C. C. Chueh, W. Chen, A. K. Y. Jen, $SrCl_2$ derived perovskite facilitating a high efficiency of 16% in hole-conductor-free fully printable mesoscopic perovskite solar cells, *Adv. Mater.* **2017**, *29*, 1606608.
47. Y. S. Sheng, A. Y. Mei, S. Liu, M. Duan, P. Jiang, C. B. Tian, Y. L. Xiong, Y. G. Rong, H. W. Han, Y. Hu, Mixed $(5\text{-AVA})_xMA_{1-x}PbI_{3-y}(BF_4)_y$ perovskites enhance the photovoltaic performance of hole-conductor-free printable mesoscopic solar cells, *J. Mater. Chem. A* **2018**, *6*, 2360.
48. Y. Chen, D. Ma, Z. Q. Wang, J. S. He, X. Q. Gong, W. J. Wu, Tautomeric dual-site passivation for carbon-based printable mesoscopic perovskite solar cells, *Adv. Mater. Interfaces* **2022**, *9*, 2200326.
49. C. Liu, S. Liu, Y. F. Wang, Y. M. Chu, K. Yang, X. D. Wang, C. X. Gao, Q. F. Wang, J. K. Du, S. Li, Y. Hu, Y. G. Rong, L. B. Guo, A. Y. Mei, H. W. Han, Improving the performance of perovskite solar cells via a novel additive of N, 1-Fluoroformamidinium iodide with electron-withdrawing fluorine group, *Adv. Funct. Mater.* **2021**, *31*, 2010603.
50. L. Xu, Y. M. Li, J. J. Shi, N. Robertson, W. J. Wu, Q. B. Meng, H. Tian, Suppressing shallow defect of printable mesoscopic perovskite solar cells with a N719@TiO_2 inorganic–organic core–shell structured additive, *Sol. RRL* **2020**, 2000042.
51. D. J. Wang, Z. L. Zhang, J. Liu, Y. Zhang, K. Chen, B. She, B. C. Liu, Y. Huang, J. Xiong, J. Zhang, Synergistic effect of defect passivation and crystallization control enabled by bifunctional additives for carbon-based mesoscopic perovskite solar cells, *ACS Appl. Mater. Interfaces* **2021**, *13*, 45435.
52. S. G. Ko, G. I. Ryu, B. Kim, G. J. Cha, J. H. Ri, G. S. Sonu, U. C. Kim, Effects of thiourea on the perovskite crystallization for fully printable solar cells, *Sol. Energy Mater. Sol. Cells* **2019**, *196*, 105.
53. J. W. Wu, W. H. Zhang, Q. F. Wang, S. Liu, J. K. Du, A. Y. Mei, Y. G. Rong, Y. Hu, H. W. Han, A favored crystal orientation for efficient printable mesoscopic perovskite solar cells, *J. Mater. Chem. A* **2020**, *8*, 11148.
54. C. Tian, S. Zhang, A. Mei, Y. Rong, Y. Hu, K. Du, M. Duan, Y. Sheng, P. Jiang, G. Xu, H. Han, A multifunctional bis-adduct fullerene for efficient printable mesoscopic perovskite solar cells, *ACS Appl. Mater. Interfaces* **2018**, *10*, 10835.
55. J. Yang, S. Li, X. Xiao, J. Du, M. Xia, X. Xiao, W. Wang, W. Hu, A. Mei, Y. Hu, H. Han, A multifunctional piperidine-based modulator for printable mesoscopic perovskite solar cells, *Chem. Eng. J.* **2022**, *446*, 136967.
56. W. Zhang, J. Du, C. Qiu, K. Yang, Q. Huang, Q. Wang, W. Zhang, H. Han, X. Gao, Y. Hu, Enhanced efficiency of printable mesoscopic perovskite solar cells using ionic liquid additives, *Chem. Commun.* **2021**, *57*(33), 4027.
57. Y. Sheng, Y. Hu, A. Y. Mei, P. Jiang, X. Hou, M. Duan, L. Hong, Y. Guan, Y. G. Rong, Y. Xiong, W. H. Han, Enhanced electronic properties in $CH_3NH_3PbI_3$ *via* LiCl mixing for hole-conductor-free printable perovskite solar cells, *J. Mater. Chem. A* **2016**, *4*, 16731.
58. K. Chen, Z. Zhang, J. Liu, Y. Huang, D. Wang, B. She, B. Liu, X. Zhang, J. Zhang, Enhanced performance of carbon-based, fully printed mesoscopic perovskite solar cells through defects passivation, *Adv. Mater. Interfaces* **2022**, *9*, 2100395.

59. J. Liu, D. Wang, K. Chen, J. Kang, J. Yang, J. Zhang, H. Zhang, Enhanced photovoltaic performance and reduced hysteresis in hole-conductor-free, printable mesoscopic perovskite solar cells based on melamine hydroiodide modified $MAPbI_3$, *Sol. Energy* **2020**, *206*, 548.
60. J. Liu, D. Wang, K. Chen, B. She, B. Liu, Y. Huang, W. Xie, J. Zhang, H. Zhang, Improved pore-filling and passivation of defects in hole-conductor-free, fully printable mesoscopic perovskite solar cells based on D-sorbitol hexaacetate-modified $MAPbI_3$, *ACS Appl. Mater. Interfaces* **2020**, *12*, 47677.
61. K. Yang, S. Liu, J. Du, W. Zhang, Q. Huang, W. Zhang, W. Hu, Y. Hu, Y. G. Rong, A. Y. Mei, H. W. Han, Improving hole-conductor-free fully printable mesoscopic perovskite solar cells' performance with enhanced open-circuit voltage via the octyltrimethylammonium chloride additive, *Sol. RRL* **2021**, *5*, 2000825.
62. Y. Zhang, Y. Wang, Z. Sun, F. Li, R. Tao, Z. Jin, L. Xu, Large grain growth for hole-conductor-free fully printable perovskite solar cells *via* polyoxometalate molecular doping, *Chem. Commun.* **2017**, *53*, 2290.
63. C. Xu, Z. Zhang, Y. Hu, Y. Sheng, P. Jiang, H. W. Han, J. Zhang, Printed hole-conductor-free mesoscopic perovskite solar cells with excellent long-term stability using PEAI as an additive, *J. Energy Chem.* **2018**, *27*, 764.
64. Y. G. Rong, X. Hou, Y. Hu, A. Y. Mei, L. Liu, P. Wang, H. W. Han, Synergy of ammonium chloride and moisture on perovskite crystallization for efficient printable mesoscopic solar cells, *Nat. Commun.* **2017**, *8*, 14555.
65. J. Liu, S. Wang, M. Chen, P. Jiang, L. Cao, W. Li, H. Ran, Y. Hu, H. W. Han, Y. Tang, Trivalent europium ions doped $CsPbBr_3$ for highly efficient and stable printable mesoscopic perovskite solar cells and driving water electrolysis, *Sol. RRL* **2022**, 2200361.
66. S. Valastro, E. Smecca, C. Bongiorno, C. Spampinato, G. Mannino, S. Biagi, I. Deretzis, F. Giannazzo, A. K. Jena, T. Miyasaka, A. L. Magna, A. Alberti, Out-of-glovebox integration of recyclable europium-doped $CsPbI_3$ in triple-mesoscopic carbon-based solar cells exceeding 9% efficiency, *Sol. RRL* **2022**, *6*, 2200267
67. D. Wang, Z. Zhang, T. Huang, B. She, B. Liu, Y. Chen, L. Wang, C. Wu, J. Xiong, Y. Huang, J. Zhang, Crystallization kinetics control enabled by a green ionic liquid additive toward efficient and stable carbon-based mesoscopic perovskite solar cells, *ACS Appl. Mater. Interfaces* **2022**, *14*, 9161.
68. J. Wu, Y. Li, Y. Zhang, Y. Li, Y. Huang, Z. Jiang, Q. Ai, Y. Liu, L. Zhang, Y. Peng, X. Wang, B. Xu, C. Cheng, Highly orientational order perovskite induced by in situ-generated 1D perovskitoid for efficient and stable printable photovoltaics, *Small* **2022**, *18*, 2200130.
69. I. Zimmermann, P. Gratia, D. Martineau, G. Grancini, J. N. Audinot, T. Wirtz, M. K. Nazeeruddin, Improved efficiency and reduced hysteresis in ultra-stable fully printable mesoscopic perovskite solar cells through incorporation of CuSCN into the perovskite layer, *J. Mater. Chem. A* **2019**, *7*, 8073.
70. S. Liu, S. Li, J. Wu, Q. Wang, Y. Ming, D. Zhang, Y. Sheng, Y. Hu, Y. Rong, A. Mei, H. Han, Amide additives induced a fermi level shift to improve the performance of hole-conductor-free, printable mesoscopic perovskite solar cells, *J. Phys. Chem. Lett.* **2019**, *10*, 6865.
71. D. Y. Zhang, P. Jiang, D. Y. Li, S. Li, J. H. Qi, X. D. Wang, Y. Hu, Y. G. Rong, A. Y. Mei, H. W. Han, Halogen bond involved post-treatment for improved performance of printable hole-conductor-free mesoscopic perovskite solar cells, *Sol. RRL* **2022**, *6*, 2100851.
72. Y. S. Sheng, W. X. Ji, Y. M. Chu, Y. Ming, A. Y. Mei, Y. Hu, Y. G. Rong, H. W. Han, Post-treatment of mesoporous scaffolds for enhanced photovoltage of triple-mesoscopic perovskite solar cells, *Sol. RRL* **2020**, *4*, 2000185.
73. L. Xu, C. Li, Y. Zhong, Z. H. Pang, W. J. Wu, Relaying delivery of excited state electrons for fully printable perovskite solar cells via ultra-thin gradient PCBM/perovskite heterojunction, *Sol. Energy* **2019**, *187*, 352.

74. X. Y. Chen, Y. K. Xia, Q. Y. Huang, Z. Li, A. Y. Mei, Y. Hu, T. Wang, R. Cheacharoen, Y. G. Rong, H. W. Han, Tailoring the dimensionality of hybrid perovskites in mesoporous carbon electrodes for type-II band alignment and enhanced performance of printable hole-conductor-free perovskite solar cells, *Adv. Energy Mater.* **2021**, *11*, 2100292.
75. D. Li, C. H. Tong, W. X. Ji, Z. Y. Fu, Z. N. Wan, Q. Y. Huang, Y. Ming, A. Y. Mei, Y. Hu, Y. G. Rong, H. W. Han, Vanadium oxide post-treatment for enhanced photovoltage of printable perovskite solar cells, *ACS Sustain. Chem. Eng.* **2019**, *7*, 2619.
76. Y. K. Xia, X. Y. Chen, Z. W. Zheng, X. Xiao, C. X. Ling, M. H. Xia, J. W. Gong, L. Y. Gao, J. W. Xiang, Y. Hu, A. Y. Mei, Y. G. Rong, H. W. Han, Interfacial energy band alignment enables the reduction of potential loss for hole-conductor-free printable mesoscopic perovskite solar cells, *J. Phys. Chem. Lett.* **2022**, *13*, 2144.
77. Q. Wang, W. Zhang, Z. Zhang, S. Liu, J. Wu, Y. Guan, A. Mei, Y. Rong, Y. Hu, H. Han, Crystallization control of ternary-cation perovskite absorber in triple-mesoscopic layer for efficient solar Cells, *Adv. Energy Mater.* **2020**, *10*, 1903092
78. M. Hu, L. Liu, A. Mei, Y. Yang, T. Liu, H. Han, Efficient hole-conductor-free, fully printable mesoscopic perovskite solar cells with a broad light harvester $NH_2CH{=}NH_2PbI_3$, *J. Mater. Chem. A* **2014**, *2*, 17115
79. K. Cao, H. Li, S. Liu, J. Cui, Y. Shen, M. Wang, $MAPbI_{3-x}Br_x$ mixed halide perovskites for fully printable mesoscopic solar cells with enhanced efficiency and less hysteresis, *Nanoscale* **2016**, *8*, 8839
80. J. Du, C. Qiu, S. Li, W. Zhang, W. Zhang, Y. Wang, Z. Qiu, Q. Wang, K. Yang, A. Mei, Y. Rong, Y. Hu, H. Han, Minimizing the voltage loss in hole-conductor-free printable mesoscopic perovskite solar cells, *Adv. Energy Mater.* **2022**, *12*, 2102229.
81. X. B. Xu, Z. H. Liu, Z. X. Zuo, M. Zhang, Z. X. Zhao, Y. Shen, H. P. Zhou, Q. Chen, Y. Yang, M. K. Wang, Hole selective NiO contact for efficient perovskite solar cells with carbon electrode, *Nano Lett.* **2015**, *15*, 2402.
82. Z. Ku, X. Xia, H. Shen, N. H. Tiep, H. J. Fan, A mesoporous nickel counter electrode for printable and reusable perovskite solar cells, *Nanoscale* **2015**, *7*, 13363.
83. D. Guo, J. Yu, K. Fan, H. Zou, B. He, Nanosheet-based printable perovskite solar cells, *Sol. Energy Mater. Sol. Cells* **2017**, *159*, 518.
84. A. Bashir, J. H. Lew, S. Shukla, D. Gupta, T. Baikie, S. Chakraborty, R. Patidar, A. Bruno, S. Mhaisalkar, Z. Akhter, Cu-doped nickel oxide interface layer with nanoscale thickness for efficient and highly stable printable carbon-based perovskite solar cell, *Sol. Energy* **2019**, *182*, 225.
85. F. Behrouznejad, C.-M. Tsai, S. Narra, E. W. G. Diau, N. Taghavinia, Interfacial investigation on printable carbon-based mesoscopic perovskite solar cells with NiO_x/C back electrode, *ACS Appl. Mater. Interfaces* **2017**, *9*, 25204
86. X. Wang, W. Wang, J. Liu, J. Qi, Y. He, Y. Wang, W. Hu, Y. Cheng, K. Chen, Y. Hu, A. Mei, H. Han, Reducing optical reflection loss for perovskite solar cells via printable mesoporous SiO_2 antireflection coatings, *Adv. Funct. Mater.* **2022**, *32*, 2203872.
87. J. W. Li, G. D. Niu, W. Z. Li, K. Cao, M. K. Wang, L. D. Wang, Insight into the $CH_3NH_3PbI_3$/C interface in hole-conductor-free mesoscopic perovskite solar cells, *Nanoscale* **2016**, *8*, 14163.
88. J. L. Liu, J. W. He, D. Ma, J. S. He, W. J. Wu, Chiral molecular environment determining selective coordination of cysteine to perovskite halogen vacancies, *Sol. RRL* **2023**, 2200935.
89. T. F. Liu, Y. Li, R. G. Xu, L. Y. Lou, J. Q. Gao, K. Zhang, M. Z. Su, W. Qian, G. P. Wang, S. Xiao, S. H. Yang, Building bulk heterojunction to enhance hole extraction for high-performance printable carbon-based perovskite solar cells, *Sol. RRL* **2022**, *6*, 2200206.
90. Y. Guan, M. Xu, W. Zhang, D. Li, X. Hou, L. Hong, Q. Wang, Z. Zhang, A. Mei, M. Chen, Y. Zhou, N. P. Padture, Y. Hu, Y. Rong, H. Han, In situ transfer of $CH_3NH_3PbI_3$ single crystals in mesoporous scaffolds for efficient perovskite solar cells, *Chem. Sci.* **2020**, *11*, 474.

91. J. Liu, D. Wang, Y. Zhang, K. Chen, B. She, B. Liu, Z. Zhang, Y. Huang, J. Xiong, H. Zhang, J. Zhang, Room-temperature-processed, carbon-based fully printed mesoscopic perovskite solar cells with 15% efficiency, *Sol. RRL* **2021**, *5*, 2100274.
92. J. Chen, Y. Xiong, Y. Rong, A. Mei, Y. Sheng, P. Jiang, Y. Hu, X. Li, H. Han, Solvent effect on the hole-conductor-free fully printable perovskite solar cells, *Nano Energy* **2016**, *27*, 130.
93. A. Mei, Y. Sheng, Y. Ming, Y. Hu, Y. Rong, W. Zhang, S. Luo, G. Na, C. Tian, X. Hou, Y. Xiong, Z. Zhang, S. Liu, S. Uchida, T.-W. Kim, Y. Yuan, L. Zhang, Y. Zhou, H. Han, Stabilizing perovskite solar cells to IEC61215:2016 standards with over 9,000-h operational tracking, *Joule* **2020**, *4*, 2646.
94. P. Holzhey, M. Saliba, A full overview of international standards assessing the long-term stability of perovskite solar cells, *J. Mater. Chem. A* **2018**, *6*, 21794.
95. A. K. Baranwal, S. Kanaya, T. A. N. Peiris, G. Mizuta, T. Nishina, H. Kanda, T. Miyasaka, H. Segawa, S. Ito, 100 °C thermal stability of printable perovskite solar cells using porous carbon counter electrodes, *ChemSusChem* **2016**, *9*, 2604.
96. Z. Fu, M. Xu, Y. Sheng, Z. Yan, J. Meng, C. Tong, D. Li, Z. Wan, Y. Ming, A. Mei, Y. Hu, Y. Rong, H. Han, Encapsulation of printable mesoscopic perovskite solar cells enables high temperature and long-term outdoor stability, *Adv. Funct. Mater.* **2019**, *29*, 1809129.
97. D. Y. Li, P. Jiang, W. H. Zhang, J. K. Du, C. Qiu, J. L. Liu, Y. Hu, Y. G. Rong, A. Y. Mei, H. W. Han, Series resistance modulation for large-area fully printable mesoscopic perovskite solar cells, *Sol. RRL* **2022**, *6*, 2100554.
98. Y. Hu, S. Si, A. Y. Mei, Y. G. Rong, H. W. Liu, X. Li, H. W. Han, Stable large-area (10 × 10 cm2) printable mesoscopic perovskite module exceeding 10% efficiency, *Sol. RRL* **2017**, *1*, 1600019.
99. M. Xu, W. Ji, Y. Sheng, Y. Wu, H. Cheng, J. Meng, Z. Yan, J. Xu, A. Mei, Y. Hu, Y. Rong, H. Han, Efficient triple-mesoscopic perovskite solar mini-modules fabricated with slot-die coating, *Nano Energy* **2020**, *74*, 104842.

10 Electron/Hole Transport Layer Free Perovskite Solar Cells

Pengpeng Teng, Nian Cheng, Jun Zhang, Shaowen Tang, Mengran Chen, and Peng-an Zong

10.1 INTRODUCTION

Perovskite solar cells (PSCs) have garnered significant attention in recent years due to their impressive power conversion efficiency (PCE) and the potential for cost-effective production. Nevertheless, the commercial viability of PSCs heavily relies on addressing issues related to device stability and fabrication costs, which are closely linked to various components such as metal oxides (TiO_2 or ZnO) in the electron transport layer (ETL),[1–3] organic materials (Spiro-OMeTAD or PTAA) in the hole transport layer (HTL),[4–6] and metal electrodes (Au or Ag).[7–8] One pragmatic approach to tackle these challenges is to eliminate the charge transport layers (CTL) in PSCs. This not only reduces the manufacturing cost of PSCs but also enhances their overall stability. In pursuit of improved performance and stability, researchers have explored various strategies, including innovative designs that obviate the need for ETL or HTL. These design innovations aim to simplify the device architecture and streamline manufacturing processes. Currently, ETL/HTL-free PSCs offer the advantage of a simplified device structure, potentially lowering production costs and enhancing device stability. However, they also present challenges related to optimizing charge transport, achieving high efficiency, and ensuring long-term stability.

10.1.1 Device Structure

Following more than a decade of continuous advancement, ETL/HTL-free PSCs have achieved an impressive PCE exceeding 20%.[9–11] These devices are typically assembled using a transparent conducting oxide (TCO) substrate, an ETL or HTL, a perovskite active layer, and a metal counter electrode. The structural configuration of ETL/HTL-free PSCs falls into two categories based on the inclusion of a mesoscopic scaffold layer: the mesoporous structure and the planar structure.

1. Mesoporous ETL/HTL-Free PSCs: within this design, a mesoporous framework is integrated into the perovskite layer to improve the carrier transmission efficiency which can reduce the recombination of electrons

DOI: 10.1201/9781003400493-10

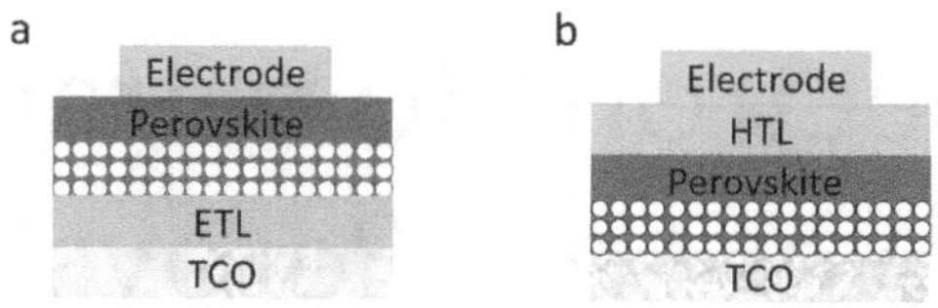

FIGURE 10.1 The schematic diagrams of the mesoporous HTL-free (a) and ETL-free (b) structure.

and holes, and cause light scattering to increase the absorption of photons. The mesoporous structure serves as an efficient conduit for charge transport, eliminating the necessity for distinct ETL and HTL layers. There are two types of mesoporous structures in ETL/HTL-free PSCs. The first type is a mesoporous HTL-free device using semiconductor nanoparticles (TiO_2, ZnO, NiO, etc.) as ETL and the device structure is TCO/ETL/Mesoporous ETL (m-ETL)/Perovskite/Electrode (Figure 10.1a). Another type is ETL-free device with the structure of TCO/Mesoporous insulating nanoparticles (Al_2O_3, ZrO_2, SiO_2, etc.)/Perovskite/HTL/Electrode (Figure 10.1b). However, the preparation of mesoporous layers usually requires high-temperature annealing treatment at 400–500 °C, which increases the cost of PSCs.

2. Planar ETL/HTL-Free PSCs: in this configuration, the structural design of PSCs has been greatly expanded without mesoporous framework. The PSCs with planar structure do not use mesoporous materials and deposit perovskite film on substrate directly. Currently, the ETL/HTL-free PSCs have developed normal and inverted structures based on traditional n-i-p and p-i-n devices. The difference between the inverted structure and normal structure is that the positions of ETL and HTL are opposite. As shown in Figure 10.2a and b, the normal planner HTL-free and ETL-free structures are TCO/ETL/Perovskite/Electrode and TCO/Perovskite/HTL/Electrode, respectively. Figure 10.2c and d also illustrate the schematic inverted structures for CTL-free PSCs. The inverted HTL-free structure is shown in Figure 10.2c, which consists of a TCO, a perovskite active layer, an ETL layer, and an electrode. Accordingly, Figure 10.2d shows the inverted ETL-free PSCs, which consist of a TCO layer, an HTL layer, a perovskite active layer, and an electrode.

10.1.2 Working Principles

The perovskite active layer exhibits characteristics of bipolar transportation, which indicates the perovskite film can transport electron and hole simultaneously.[12] When the counter electrode with a suitable work function is directly contacted with the perovskite active layer, the photo-generated carriers can efficiently transport from the perovskite layer to the counter electrode, as shown in Figure 10.3. Therefore, ETL/HTL-free PSCs are proposed, which are much easier to fabricate and exhibit

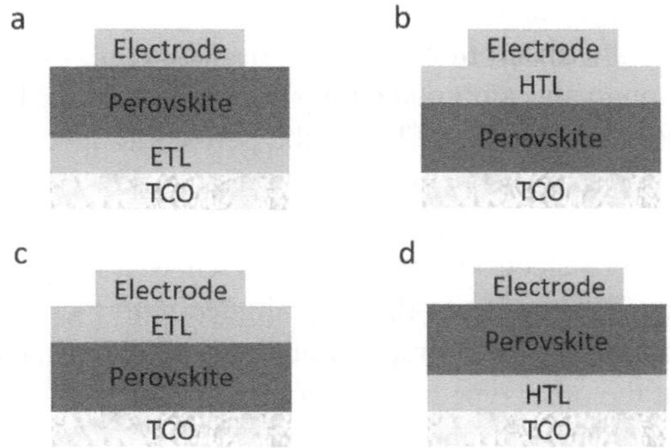

FIGURE 10.2 The schematic diagrams of four planner CTL-free PSCs: (a) regular HTL-free structure, (b) regular ETL-free structure, (c) inverted HTL-free structure, and (d) inverted ETL-free structure.

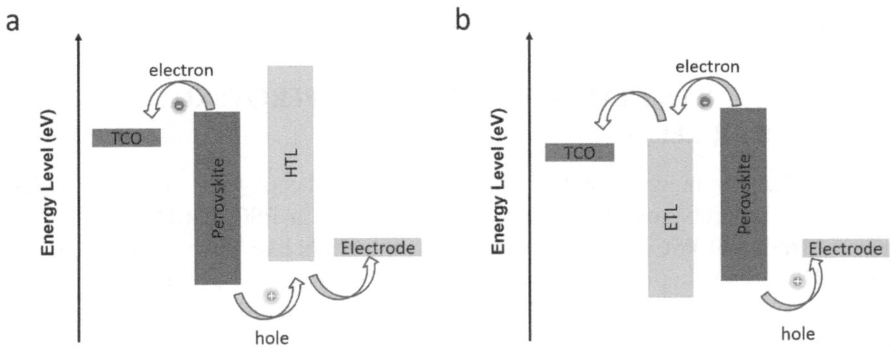

FIGURE 10.3 Energy level alignment of the ETL-free (a) and HTL-free (b) device.

superior device stability. The ETL/HTL-free PSCs are generally composed of a TCO substrate, a carrier transport layer, perovskite, and electrodes. The CTLs provide a selective contact in PSCs. The ETL should selectively permit the passage of electrons and block holes, whereas the HTL serves the opposite function. Like traditional solar cells, the working process of devices can be roughly divided into: (i) generation and separation charge carrier (electrons and holes); (ii) transport of charge carriers; (iii) carrier collection and current generation.

1. Generation and separation charge carriers
   Perovskites are direct bandgap semiconductors that have high light absorption coefficients. Typically, perovskite with a thickness of ~300 nm can almost fully absorb incident photons. The perovskite layer absorbs photons to produce excitons, which are then rapidly separated into free electrons and holes due to very weak Coulomb force binding.

2. Transport of charge carriers
   Specially, the light absorption layer can simultaneously transport electrons and holes through the ambipolar transfer characteristic. Furthermore, the heterojunction between the CTL and the perovskite layer will serve as the driving force for charge separation and transport. For ETL/HTL-free PSCs the free carriers can be transported to the interfaces of perovskite/CTL and perovskite/TCO or electrode respectively. Hence, it is crucial for the work function of the TCO or electrode to be aligned with the energy level of the perovskite, facilitating the extraction of one type of carrier while impeding the transfer of the other type of free carrier.
3. Carrier collection and current generation
   The electrons and holes transferred through the CTL are collected by the electrodes to form current and voltage. The selection of electrodes is also very important and the appropriate electrode work function can reduce the energy level barrier injected by carriers and form good ohmic contact. When a counter electrode with a suitable work function is directly contacted with the perovskite active layer, the photo-generated holes can efficiently transport from the perovskite layer to the counter electrode.

10.2 ELECTRON TRANSPORT LAYER FREE PEROVSKITE SOLAR CELLS

Despite the advantages that ETLs bring to PSCs, the oxygen vacancies in ETL introduce the problems of instability and hysteresis in PSCs, hindering the commercialization of PSCs.[1,13] Electron transport materials like TiO_2 and ZnO can promote the degradation of perovskite, which contributes to the instability of PSCs.[2,3] Additionally, the development of ETL-free structures simplifies the fabrication process and decreases the manufacturing costs. In this respect, substantial efforts have been devoted in recent years to develop ETL-free structures for PSCs.

10.2.1 Regular ETL-Free Structure

The traditional ETL-free devices are composed of TCO substrate, perovskite, HTL (usually Spiro-OMeTAD), and metal electrode. In 2014, Kelly et al. first fabricated ETL-free PSCs, as shown in Figure 10.4a, with a structure of ITO/perovskite/Spiro-OMeTAD/Ag. The direct contact between ITO and perovskite resulted in contact resistance and energy level mismatch, limiting charge carrier transport. This drawback could be compensated for by reducing interface recombination. Furthermore, the effective extraction of holes by HTL alleviated electron-hole radiative recombination on perovskite, ultimately achieving a higher PCE of 13.5%.[14]

The disparity in energy levels and suboptimal interface contact between the perovskite layer and the TCO lead to a sluggish electron-injection rate at the ETL-free interface, resulting in diminished device performance. To tackle this issue, various strategies have been implemented to enhance the perovskite/anode interface. Hu and their team achieved a substantial improvement in carrier lifetimes within the perovskite films, yielding ETL-free devices with an outstanding PCE of 19.5%.

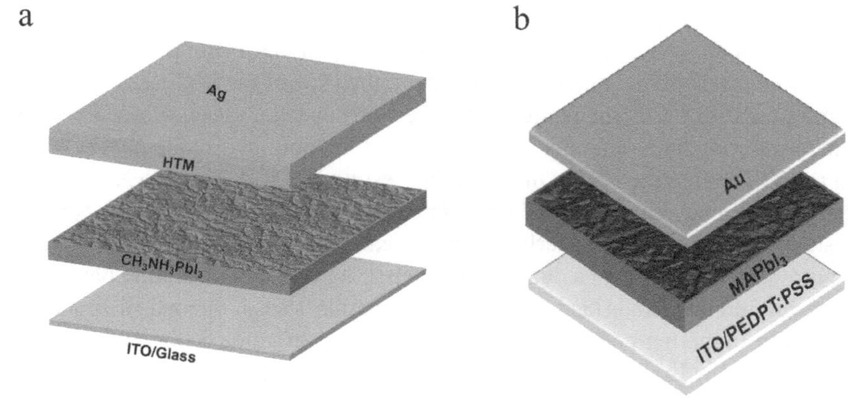

FIGURE 10.4 The device structure of (a) the regular ETL-free PSCs and (b) the inverted ETL-free structure.

This advancement nearly eradicated hysteresis and ensured robust stability.[15] Akin and colleagues introduced sodium fluoride (NaF) as an interfacial layer to align the energy levels between the perovskite and the fluorine-doped tin oxide (FTO). This interface modification induced band bending at the FTO/perovskite interface, promoting effective charge collection. Furthermore, they observed that certain Na^+ ions could migrate into the absorber layer, contributing to the enlargement of perovskite grain size and self-passivation. As a result, the efficiency of NaF-treated conventional ETL-free PSCs reached 20%, a performance on par with ETL-based PSCs.[16] In a recent development, Hui and their team employed 1,4-butanediammonium diiodide (BDADI) to form a low-dimensional perovskite layer between the perovskite film and the FTO. This innovation yielded high-quality perovskite films and substantially elevated the PCE of regular ETL-free PSCs to over 22%.[10]

10.2.2 Inverted ETL-Free Structure

The main hurdle in achieving high-efficiency inverted ETL-free PSCs is the comparatively low charge extraction efficiency at the interface between the metal electrode and the perovskite layer. In the absence of an ETL, electrons have a propensity to swiftly migrate towards the electrode, rendering them vulnerable to capture by defects, which ultimately leads to efficiency losses. Therefore, it is essential to find effective ways to promote efficient electron move towards the electrode. Huang et al. reported an inverted ETL-free perovskite device with the structure of ITO/PEDOT:PSS/$MAPbI_3$/Au in 2014,[17] as shown in Figure 10.4b.

On this basis, Wang et al. inserted a Ti layer between the perovskite and Au, which obtained better contact and improved interface morphology of perovskite/Au. The device presented a PCE to 13% with hysteresis-free photocurrent curves.[18] In 2018, Xiao and colleagues presented a noteworthy development by introducing inverted ETL-free PSCs with a composite metal electrode composed of titanium (Ti) and gold (Au). In this configuration, polytriarylamine (PTAA) served as the HTL to enhance the perovskite grain size. By incorporating an ultra-thin titanium interlayer, measuring a mere 10 nanometers in thickness, the PCE of the leading device reached an impressive 18.1%.[19] Huang and colleagues introduced a noteworthy enhancement by incorporating an atomically thin amorphous niobium hydroxide (a-NbOH) layer into the TCO substrate.[11] This modification resulted in ETL-free PSCs with an outstanding efficiency exceeding 21%, accompanied by notable improvements in working stability. The a-NbOH interlayer, strategically positioned between the TCO and the perovskite layer, serves a dual purpose. First, it acts as an insulating barrier layer, effectively reducing interface charge recombination losses. Second, it functions as an electron tunneling contact, facilitating electron extraction while enhancing hole blocking. The hydroxyl-rich surface of the a-NbOH layer promotes crystal growth, leading to high-quality perovskite thin films characterized by improved crystallinity, larger grain size, and reduced trap density. This breakthrough offers promising prospects for the fabrication of large-area ETL-free PSCs, with an impressive efficiency of approximately 20.0% achieved for ETL-free PSCs possessing an effective area of 1.1 cm^2.

10.3 HOLE TRANSPORT LAYER FREE PEROVSKITE SOLAR CELLS

HTL-free PSCs are typically composed of an electron-transporting layer, a perovskite active layer, and a metal counter electrode. Depending on the PSC's design, a mesoscopic scaffold layer may also be incorporated. It is noteworthy that traditional organic hole-transporting layers suffer from instability and are unable to effectively impede the diffusion of metal atoms and halide ions.[4,5] Furthermore, these organic hole-transport materials tend to be relatively expensive, which escalates the overall fabrication cost of PSCs.[20–22] However, HTL-free PSCs have made significant strides and hold great promise for various applications in photovoltaic devices.

10.3.1 Regular HTL-Free Structure

The traditional HTL-free PSCs are composed of a TCO substrate, an ETL layer (usually mesoporous TiO_2 layer), a perovskite active layer, and a suitable electrode layer. In 2012, Grätzel et al. reported the traditional HTL-free PSCs for the first time with a device structure of FTO/TiO_2 nanosheets/$MAPbI_3$/Au.[23] The schematic device structure and corresponding energy level alignment are presented in Figure 10.5. The $MAPbI_3$ layer absorbs the incident light and transports the photo-generated holes at the same time. The work-function of Au (-5.1 eV) is slightly larger than the valence band level of $MAPbI_3$ (-5.43 eV), which can facilitate the transportation of photo-generated holes from $MAPbI_3$ to Au electrode. Therefore, this prototype HTL-free PSC can demonstrate a PCE of 5.5% under simulated AM 1.5 G irradiation. After optimizing the composition of perovskite film using Br- alloying, Etgar et al. reported a PCE of 8.54% for the traditional HTL-free PSCs with $CH_3NH_3PbI_xBr_{3-x}$ perovskite layer.[24] In 2015, the influence of film thickness of the mesoporous TiO_2 skeleton layer on the PCEs of the HTL-free PSCs were studied by Xiao et al. The optimized film thickness for TiO_2 skeleton layer is found to be 800 nm for HTL-free

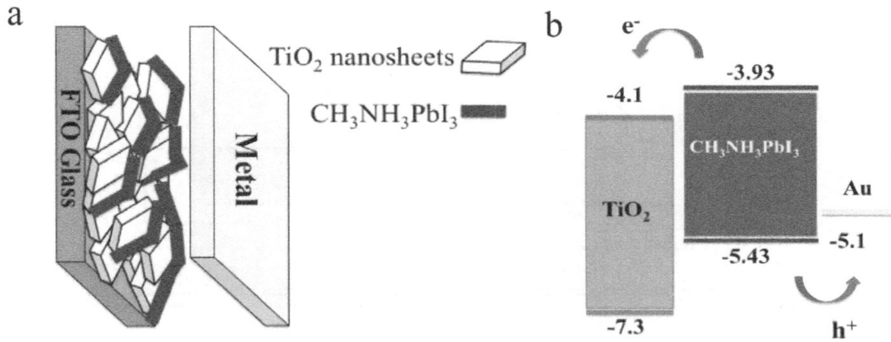

FIGURE 10.5 (a) Schematic device structure. (b) Energy level diagram of the traditional HTL-free PSCs.[23] Reprinted with permission from the American Chemical Society.

PSCs, which can provide a PCE of 12.22%.[25] Although the Au electrode is widely employed in the early studies of the traditional HTL-free PSCs, the direct contact of Au electrode with the perovskite layer will induce easy diffusion of the Au atom into the perovskite layer, which deteriorates the performance of the PSCs.

To improve the device stability of the traditional HTL-free PSCs, carbon electrode (CE), which is low cost, highly stable, hydrophobic, and highly conductive, is widely employed in subsequent studies.[26] Carbon-based materials, typically, carbon nanotubes (CNTs) and graphene, own the capability to function as both ETL and HTL in PSCs. For instance, CNTs can be uniformly distributed as a thin film on a substrate, with the perovskite layer subsequently deposited atop them. This CNT layer is proficient at effectively transporting both electrons and holes.[27,28] Similarly, graphene, owing to its exceptional electrical conductivity and transparency, can serve as a versatile charge transport layer within ETL/HTL-free configurations.[29,30] At the same time, CE is insensitive to ion migration, which plays a vital role in enhancing the device stability of HTL-free PSCs. In 2014, Han et al. reported the HTL-free PSCs with mesoporous TiO_2/ZrO_2 bilayer structure and mesoporous CE.[31] The schematic device structure is illustrated in Figure 10.6, which can deliver an excellent PCE of 12.8%. Since then, CEs have almost replaced metal electrodes in HTL-free PSCs. In a recent development, Zhong and collaborators introduced a moisture-induced secondary crystal growth approach that has demonstrated notable enhancements in the crystallinity, grain size, and overall morphology of perovskite films. This technique has led to a reduction in the density of defect states and has significantly improved the interface contact between perovskite films and CEs. As a result of these advancements, the PCE of regular HTL-free PSCs based on CEs has now reached an impressive 19.52%.[32]

10.3.2 Inverted HTL-Free Structure

In comparison with traditional HTL-Free devices, the inverted structure of PSCs is composed of four parts: TCO, perovskite absorbing layer, ETL and electrode. For

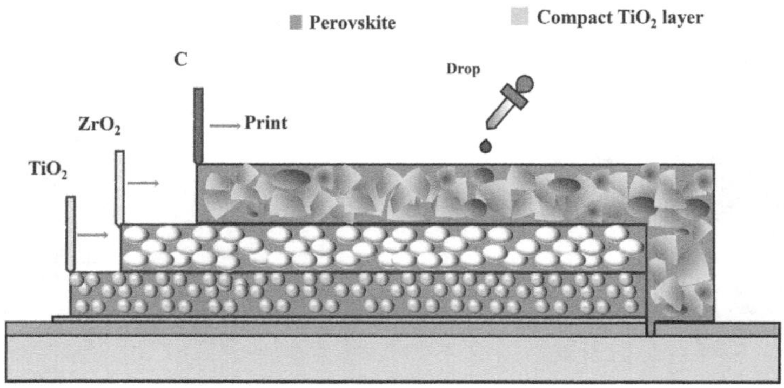

FIGURE 10.6 Cross-section of TiO_2 and ZrO_2 mesoporous bilayer PSCs.

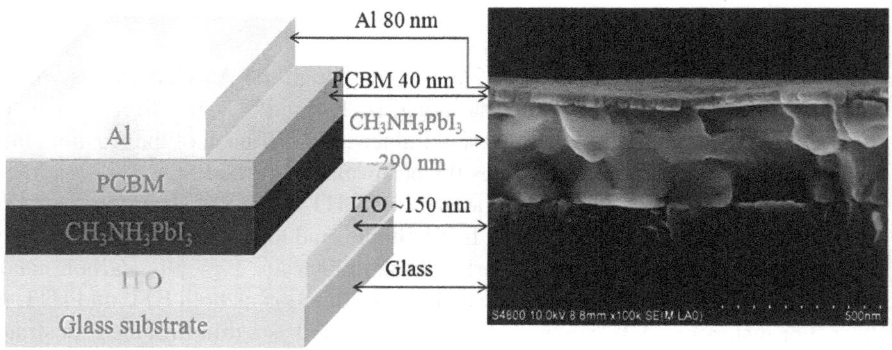

FIGURE 10.7 The structure and SEM cross-section of inverted HTL-free device.[35] Reprinted with permission from the Royal Society of Chemistry.

the regular HTL-free structure, the mesoporous TiO_2 is usually applied as ETL. However, the formation of mesoporous TiO_2 layers requires treatment at high temperatures ~500 °C to improve charge mobility, which increases the production cost and complexity. In the inverted hole-free structure, the use of mesoporous TiO_2 layer can be avoided. The perovskite layer contacts with the ITO or FTO transparent electrode directly, and these metal oxide transparent electrodes are very stable and do not react with perovskite and destroy the device. Hence, the adoption of the inverted HTL-free structure appears to be a more rational choice for enhancing the stability of PSCs.

In 2014, Cui et al. reported the first inverted HTL-free PSC with a structure of ITO/$MAPbI_3$/C60/Ag and achieved a PCE of 5.4%.[33] Next year, Tsai et al. reported a perovskite/fullerene (PCBM) heterojunction thin-film PSC without HTL and a PCE of 11.02% were achieved.[34] Yang et al. obtained PCE of 12.78% by effectively controlling the morphology of perovskite films. The device structure is ITO/$CH_3NH_3PbI_3$/PCBM/Al and SEM cross-section are shown in Figure 10.7.[35] In 2018, the highest PCE based on this structure has surpassed 20%.[9] Since the first report of inverted HTL-free PSCs, this structure has attracted widespread attention and achieved significant progress.

10.4 PERFORMANCE OPTIMIZATION OF ETL/HTL FREE PSCs

Despite the numerous advantages of ETL/HTL-free devices in terms of cost, process, and device stability, their performance is still inferior to PSCs with both ETL and HTL. So far, a lot of effort has been devoted to optimizing device performance. In the following, we demonstrate the performance optimization of ETL/HTL-free PSCs, including crystallization engineering, composition engineering, additive engineering, surface passivation, and perovskite heterojunction.

10.4.1 Crystallization Engineering

The optimization of perovskite crystal growth processes for perovskite thin films is a crucial aspect in the development of high-efficiency solar cells. Herein, we focus on HTL-free and ETL-free PSCs. Several methods are employed to optimize these processes. The one-step method is a straightforward approach where all the precursor materials necessary for perovskite formation are mixed in a single solution. Typically, this solution is spin-coated or deposited onto a substrate, and then it is heated to facilitate the crystallization of the perovskite. The key to optimization in this method lies in finding the right solvent, precursor ratios, and temperature conditions to achieve the desired perovskite morphology and crystal quality.[36,37] For instance, N-methyl-2-pyrrolidone (NMP) was adopted as a precursor solvent to slow down the crystallization of perovskite, which obtained PSCs with high performance and excellent stability.[38] Pan and colleagues employed formamide as a dual-role solvent to expedite the nucleation process of formamidinium and cesium-based perovskite. As depicted in Figure 10.8a, this approach resulted in a notable enhancement in the crystal quality of the produced perovskite film. Consequently, the power conversion efficiency of HTL-free PSCs with a carbon-based electrode surpassed 19%.[39]

In two-step method, the perovskite precursor materials are first deposited as a thin film of intermediate phases and then subjected to a second step where a specific solvent or additive is applied to induce perovskite formation. The two-step method allows for better control over the perovskite formation and morphology, leading to enhanced device performance.[40,41] Yu et al. improved the traditional dipping method and fabricated high-quality $CsPbBr_3$ perovskite films in HTL-free PSCs.[42] After that, they introduced 2-methoxyethanol as dipping solution instead of commonly used methanol to enhance the crystallinity and purity of perovskite.[43] Safari et al. reported that an improved two-step deposition route and a dense perovskite film with large grains is formed.[44] The presence of large crystals forming a dense perovskite layer inhibits the penetration of oxygen or moisture into the film, thereby improving the stability of the device.

The multi-step solution method involves multiple deposition and treatment steps to achieve precise control over perovskite layer thickness, composition, and crystallinity. By carefully adjusting each step's conditions, researchers can tailor the perovskite film properties to optimize solar cell efficiency. By tuning the number of deposition cycles of CsBr solution, Duan et al. achieved vertical- and monolayer-aligned perovskite grains on mesoporous TiO_2 substrate. The carbon-based $CsPbBr_3$ PSCs achieved a PCE of 9.72%.[45] In gas-phase transport method, gaseous precursors are used to grow perovskite crystals. It offers the advantage of better control over crystal growth and uniformity. Vapor deposition methods, such as chemical vapor deposition (CVD) and vapor-assisted solution deposition (VASP), are commonly employed.[46–48] Optimization here involves precise control of precursor flow rates, temperatures, and deposition conditions. Yang et al. introduced vapor-assisted solution technique to prepare high-quality perovskite films and fabricated HTL-free PSCs with improved PCE.[49]

10.4.2 Composition Optimization

The optimization of perovskite composition for perovskite thin films plays a crucial role in advancing the development of high-efficiency solar cells, particularly in the context of HTL-free (hole-transporting layer-free) and ETL-free (electron-transporting layer-free) solar cells. This optimization encompasses various approaches.

Single-Component $MAPbI_3$: single-component perovskite materials like methylammonium lead iodide ($MAPbI_3$) are commonly used. To optimize their composition, researchers focus on controlling precursor ratios, crystallization conditions, and deposition techniques. This helps achieve the desired crystal structure, bandgap, and optical properties for efficient photovoltaic performance. Li et al. achieved component management of $MAPbI_3$ perovskite by increasing the annealing temperature from 60 °C to 150 °C. The results showed that n-p heterojunctions were spontaneously formed at the interface of $MAPbI_3$/Spiro-OMeTAD, effectively promoting charge separation and extraction, inhibiting charge recombination, and resulting in a 15.69% PCE of ETL-free PSC.[50]

Two-Component FA-MA Systems: two-component systems, combining formamidinium (FA) and methylammonium (MA) with lead halides, offer flexibility in tuning perovskite properties. Optimization involves finding the optimal ratio of FA to MA and controlling the composition of lead halides. This fine-tuning can enhance stability, bandgap engineering, and overall solar cell efficiency. Zhu et al. fabricated the $FAPbI_3$/$MAPbI_3$ CTL-free PSCs by depositing $MAPbI_3$ onto the as-fabricated $FAPbI_3$ (Figure 10.8b) and successfully obtained a high-quality $FAPbI_3$/$MAPbI_3$ hetero-bilayer and a record PCE of 16.57%.[51]

Multicomponent Systems: (i) introducing additional elements or ions (e.g., cesium, rubidium) into the perovskite lattice is a strategy to optimize composition. This approach enhances the structural stability, tunability of bandgaps, and overall performance of perovskite materials, leading to improved solar cell efficiency.[52] In this respect, researchers have demonstrated that appropriate concentrations of Cs can enhance the stability of $MAPbI_3$. Wang et al. prepared a mixed organic-inorganic perovskite active layer $Cs_XMA_{1-X}PbI_3$ in air atmosphere, in which MAI was partially replaced by CsI.[53] They showed that moderate doping of Cs in perovskite membranes can significantly improve the morphology and crystallinity of the perovskite films. Then Guo et al. reported that HL-free $CsPbI_2Br$ solar cells have enhanced stability and performance by incorporating rubidium (Rb) cations. Compared with pure

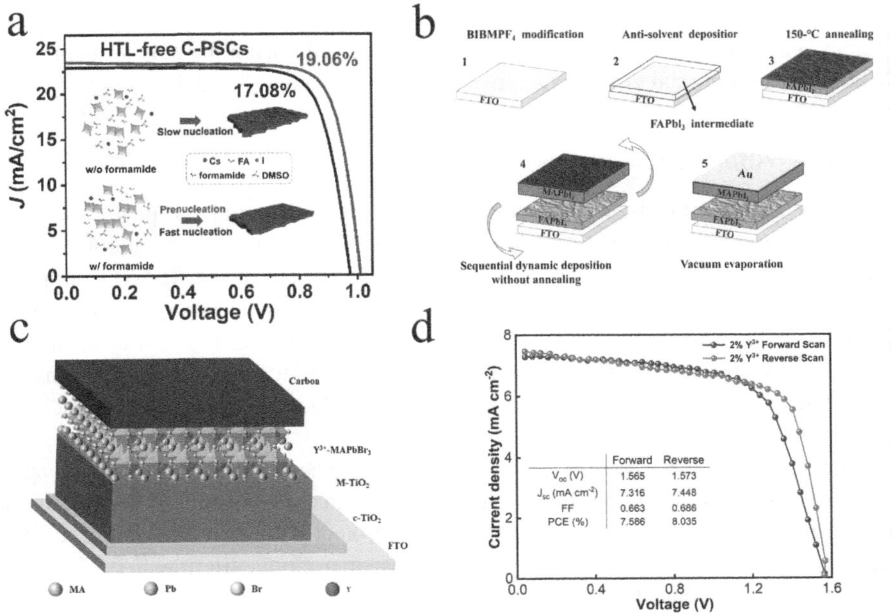

FIGURE 10.8 (a) The crystallization engineering and J-V curve of the HTL-free PSCs.[39] Reprinted with permission from John Wiley and Sons, Inc. (b)The fabrication procedure of the CTL-free PSC.[51] (c) Schematic diagram of Y^{3+}-$MAPbBr_3$ PSC and (d) J-V curve of the best device of $MAPb_{0.98}Y_{0.02}Br_3$.[57]

$CsPbI_2Br$ film, Rb-doped $CsPbI_2Br$ film has excellent crystallinity, pinhole-free surface morphology and enhanced light absorption.[54]

(ii) Pb-Site Doping: doping the perovskite lattice by substituting lead (Pb) ions with other elements is a technique used to optimize performance. Pb-site doping can enhance stability, charge carrier mobility, and optical properties.[55,56] Careful selection of dopants and doping concentrations is critical. Zhao et al. successfully introduced yttrium (Y^{3+}) ions into the $MAPbBr_3$ lattice, and $Pb^{2+}$was replaced by Y^{3+} resulting in lattice shrinkage and stronger reciprocity between $Pb^{2+}$and Br-, which improved phase stability of perovskite. By optimizing the doping process, the maximum PCE of $MAPb_{0.98}Y_{0.02}Br_3$ PSC reached ~8.3%.[57] The device structure and J-V curve are shown in Figure 10.8c and d. In addition, the unpackaged device exhibits excellent long-term stability within 80 days under dry air of 20–25 °C and relative humidity of 20–30%.

(iii) X-Position Component Variations: modifying the halide composition at the "X" position within perovskite materials is another optimization strategy. This variation allows fine-tuning of bandgap, stability, and optical properties.[58,59] Researchers explore mixed-halide perovskite to achieve desired performance characteristics. Yin et al. investigated the doping of $CsPbI_2Br$ perovskite precursors with $PbCl_2$. The role of chlorine (Cl) atoms in modulating the crystallization of organic-inorganic perovskite and passivating bulk defects is crucial. They demonstrated the incorporation of Cl ions into the $CsPbI_2Br$ perovskite lattice, forming $CsPbI_2Br$-Cl perovskite with a larger bandgap. Rapid and uniform nucleation of $CsPbI_2Br$-Cl occurred during the annealing process, leading to slow crystal growth, increased crystal size, and reduced defect density.[60]

10.4.3 Additive Engineering

Additive engineering is an important strategy for optimizing the performance of HTL-free and ETL-free PSCs. Additives are specific materials or compounds introduced into the perovskite layer or other functional layers of the device to improve various aspects of device performance.[61,62] Additives can be incorporated into the perovskite layer to improve the mobility of charge carriers. For instance, organic additives like fullerenes or small molecules may be used to facilitate charge transport within the perovskite material.[63–65] This results in reduced charge recombination and higher device efficiency.

Some additives are chosen for their ability to enhance the stability of the perovskite layer. PSCs are known to be sensitive to moisture and oxygen, leading to degradation over time. Additives can act as stabilizers to protect the perovskite material from environmental factors and extend the device's operational lifetime.[66,67] Moreover, additives can influence the morphology and crystallinity of perovskite film. They can help in achieving uniform and dense perovskite layers, reducing defects, and improving light absorption.[68,69] This, in turn, leads to enhanced device performance. Liu and colleagues reported the preparation of high-quality perovskite films by adding 2,6-di-tert-butyl-4-methylphenol (BHT) into the precursor solution. Compared to the original device with a PCE of 12.16%, the champion device based on BHT-$MAPbI_3$ exhibited a significantly improved PCE of 16.88% and enhanced stability.[70]

Furthermore, additive engineering can mitigate hysteresis, resulting in more reliable and consistent device performance.[71] Huang and their team incorporated 2,3,5,6-tetrafluoro-7,7,8,8-tetracyanoquinodimethane (F4TCNQ) into the perovskite layer of HTL-free PSCs. This additive effectively enhanced the conductivity of the perovskite films and improved their electronic contact with the TCO, thereby reducing the series resistance. The device, featuring the structure of ITO/Perovskite/ETL/Cu, achieved an exceptional PCE exceeding 20.0% and exhibited negligible hysteresis.[9] In 2018, Zhou et al. achieved high-quality perovskite films by co-doping the perovskite precursor solution with Cu(thiourea)Cl and thiourea sulfur compound.[72] In comparison to the control device, which displayed an efficiency of 13.10%, the HTL-free PSC, prepared via additive engineering, attained an impressive PCE of 22%, and it exhibited virtually no hysteresis, as illustrated in Figure 10.9a. Moreover, the optimized devices created through this co-doping strategy demonstrated significantly improved environmental stability, thermal stability, and long-term operational stability when compared to the control devices.

In addition, additives can be introduced to the hole or ETLs to improve light harvesting and charge generation. This can involve the use of luminescent materials or nanostructured additives that scatter or trap light, increasing its interaction with the perovskite layer.[73–75] It is crucial that the chosen additives do not interfere with the simplified HTL-free and ETL-free device architecture. Additives should be compatible with the charge transport and extraction mechanisms of these designs.

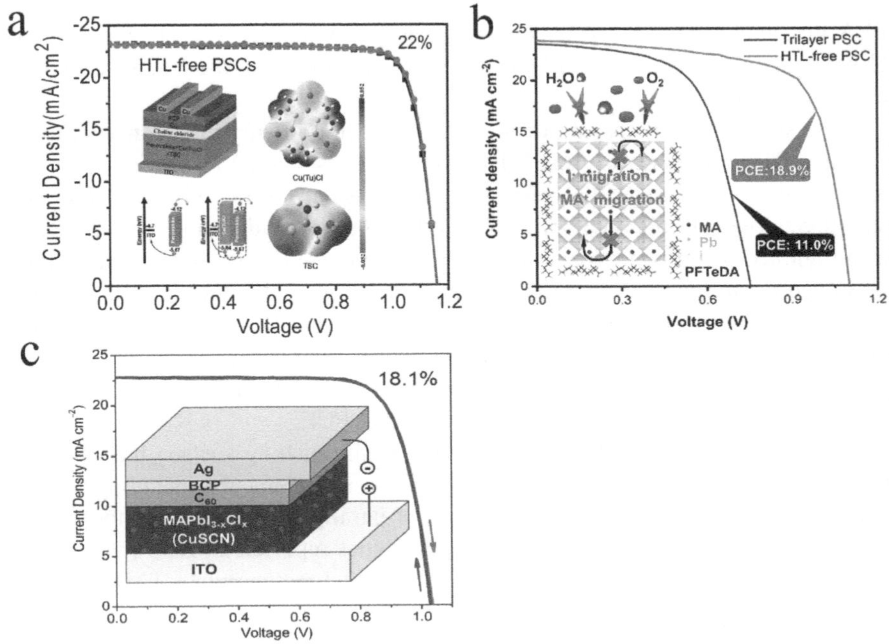

FIGURE 10.9 (a) The additive engineering and J-V curve of the HTL-free PSCs.[72] Reprinted with permission from Elsevier. (b) The surface passivation and J-V curve of the HTL-free PSCs.[85] Reprinted with permission from the American Chemical Society. (c) The schematic diagram and J-V curve of the HTL-free PSCs optimized by Heterojunction engineering.[91] Reprinted with permission from John Wiley and Sons.

10.4.4 Surface Passivation

Surface passivation is a crucial technique in improving the performance and stability of PSCs, especially in the context of HTL-free and ETL-free configurations. Surface passivation aims to reduce defects, trap states, and incite ionic migration at the perovskite surface, which can significantly enhance device efficiency and longevity.[76] Passivation is achieved by depositing a thin layer of passivation material onto the surface of the perovskite layer. Common passivation materials include organic molecules (such as long-chain organic ligands), inorganic materials (like metal oxides), and polymers.[77–80]

Passivation materials help reduce surface defects and trap states, which can lead to recombination of charge carriers (electrons and holes) and lower device efficiency.[81] By capping these defects, passivation materials enable more efficient charge transport within the perovskite layer. Zhang et al. presented an effective surface crystal reconstruction method, involving the coating of $CsPbI_2Br$ film with Phenethylamine iodide (PEAI) followed by annealing at 280 °C, to remove defects within $CsPbI_2Br$ and enhance the efficiency of solar cells. The PEAI treatment significantly improved the crystal quality of $CsPbI_2Br$, narrowing the bandgap and reducing defect density.

Simultaneously, PEAI reacted with $CsPbI_2Br$ to sequentially form two-dimensional PEA_2PbI_4 and quasi-two-dimensional $PEA_2PbI_4 \cdot (CsPbI_{2+x}Br_{1-x})_{n-1}$. The HTL-free CE $CsPbI_2Br$ PSC achieved a high PCE of 13.38%, compared to the control group with a PCE of 11.26%.[82]

Perovskite materials are susceptible to ionic migration, which can lead to performance degradation over time.[83] Passivation layers can act as barriers to prevent ionic movement between the perovskite and adjacent layers, enhancing long-term stability.[84] Ye and collaborators introduced perfluorotetradecanoic acid (PFTeDA) as an additive to mitigate ion migration and passivate crystal defects, resulting in improved environmental, thermal, and light stability of the devices. Ultimately, as depicted in Figure 10.9b, the HTL-free PSCs with carbon-based electrodes attained a notable PCE of 18.9%.[85] Moreover, the passivation layer can also smooth the surface of the perovskite film, reducing surface roughness and enhancing light absorption. This results in improved light trapping and higher PCE.

Passivation materials play a crucial role in diminishing surface recombination by forming a protective shield against environmental elements, including moisture and oxygen.[86,87] This protective function is of paramount significance for ensuring the long-term stability of PSCs. Wu and colleagues reported an ETL-free PSC that was enhanced with the incorporation of poly(methyl methacrylate; PMMA). When compared to the reference PSC without PMMA treatment, it exhibited improved photovoltaic performance and the moisture resistance of the film is significantly improved.[88] This finding underscores the benefits of surface treatment for perovskite films in ETL-free devices. In efforts to enhance the performance of the perovskite film and alleviate damage resulting from moisture exposure, Grätzel and colleagues introduced a thin layer of a fullerene derivative, specifically α-Bis-PCBM, onto the surface of the perovskite film. The α-Bis-PCBM surface passivation layer effectively served as a barrier against moisture infiltration, further optimizing the perovskite interface.[89]

Additionally, passivation layers must be chosen or engineered to be compatible with HTL-free and ETL-free designs. They should not interfere with charge carrier extraction and transport, ensuring the simplified device architecture remains effective. Ideally, passivation layers should be optically transparent in the relevant wavelength range to allow incident light to penetrate and be absorbed by the perovskite layer.

10.4.5 Perovskite Heterojunction

Perovskite heterojunction engineering is a critical strategy for optimizing the performance of HTL-free and ETL-free PSCs. Heterojunctions are interfaces between two different materials within the device structure, and their design and engineering play a key role in improving device efficiency, stability, and other important characteristics.[90] PSCs typically contain multiple heterojunctions, including the perovskite/HTL interface, perovskite/ETL interface, perovskite/perovskite interface, and the interfaces between various charge extraction layers (e.g., transparent electrodes).

Heterojunction engineering aims to achieve proper alignment of energy levels (band alignment) between the perovskite layer and adjacent layers. For HTL-free designs, achieving good alignment with the hole extraction materials is crucial, while for ETL-free designs, alignment with electron extraction materials is essential. Proper band alignment reduces energy barriers for charge carrier transport and extraction. Engineering the heterojunctions can help reduce charge carrier recombination at these interfaces. By carefully selecting and designing the adjacent materials, researchers can minimize the loss of electrons and holes, leading to improved overall device efficiency. Huang and colleagues achieved the inverted PSCs without an ETL by incorporating CuSCN into the perovskite precursor. Thanks to the favorable energy level alignment between CuSCN and the $MAPbI_{3-x}Cl_x$ perovskite, the incorporation of CuSCN served a dual purpose. It not only effectively passivated defects and grain boundaries within the perovskite but also accelerated the transfer of holes from the perovskite to the ITO substrate. Time-resolved photoluminescence (TRPL) further confirmed the accelerated hole transfer and collection brought about by CuSCN. These results demonstrate that carriers are more evenly transported to their respective electrode contacts, reducing charge accumulation while inhibiting charge recombination. Figure 10.9c shows an impressive PCE of 18.1% was achieved in the HTL-free PSCs.[91]

Heterojunction engineering can involve the introduction of passivating materials or thin layers at the interfaces to reduce defects and trap states. These passivation layers improve charge carrier lifetime and reduce non-radiative recombination, enhancing device performance and stability. Huo and colleagues have introduced a synergistic modification strategy for all-inorganic $CsPbI_2Br$ perovskite. This approach utilizes solutions of ethylammonium iodide (EAI) and polyethyleneamine iodide (PAI), both at a concentration of 10 mg/mL, with a 1:1 concentration ratio of EAI to PAI. This strategy operates on two key principles: first, the ethylammonium ion (EA^+) from EAI infiltrates the internal lattice of the perovskite, passivating the bulk defects. Second, PAI contributes to passivation by forming local two-dimensional perovskite structures, coordination bonds, and molecular interactions on the surface or within grain boundaries of the perovskite film. The synergy of these effects significantly enhances the performance of the passivated PSCs. Following this modification, the PSCs exhibit remarkable improvements in stability and PCE. Additionally, the refined energy level alignment enhances charge extraction efficiency at the perovskite-to-CE interface while suppressing charge recombination. These modified PSCs showcase outstanding performance, including a PCE of up to 13.76%, a high V_{OC} of 1.17 V, and a substantial FF of up to 80.32%. Encouragingly, even without encapsulation, these modified PSCs maintain significant long-term stability when stored in a glovebox for over 1,000 h, with only an 8% loss in PCE. This method provides a novel approach to enhancing the performance and stability of all-inorganic $CsPbI_2Br$ PSCs.[92]

The heterojunctions play a critical role in charge extraction from the perovskite layer to the respective electrodes (HTL or ETL). Engineering these interfaces can enhance the efficiency of charge collection and minimize charge losses during transport. Careful heterojunction design can also mitigate hysteresis effects commonly observed in PSCs. This leads to more stable and predictable device behavior. Zhang and his research team employed an in-situ reaction method to synthesize a hybrid perovskite material, $MAPbI_3$-$MAPbI_xBr_{3-x}$. This was achieved by initially depositing a layer of MABr onto the surface of $MAPbI_3$, triggering the in-situ growth of a mixed-halide $MAPbI_xBr_{3-x}$ thin layer. This innovative approach enabled the fabrication of HTL-free PSCs with a CE directly applied to the perovskite layer. Upon the introduction of the $MAPbI_xBr_{3-x}$ perovskite layer, a rapid photoluminescence intensity decay was observed, signifying efficient extraction of photogenerated holes from the $MAPbI_3$ layer to the CE. In the absence of modification, both photo-induced electrons and holes could freely migrate to the CE, leading to recombination due to the mismatched energy levels between $MAPbI_3$ and carbon. In contrast, the surface modification with the $MAPbI_xBr_{3-x}$ layer not only improved hole extraction but also hindered electron flow from the perovskite layer to the CE. This modification enhanced interfacial charge transport and suppressed interfacial charge recombination, resulting in a notable increase in PCE from 11.3% to 16.2%, all while maintaining a consistent and high level of reproducibility.[93]

In advanced designs, tandem and ternary heterojunctions may be employed to further optimize the device's light absorption, charge separation, and extraction processes. Yin and colleagues introduced the concept of a multi-gradient heterojunction (GHJ), comprising multiple layers of inorganic $CsPbBr_xI_{3-x}$ nanocrystals with a gradient bandgap superimposed on $MAPbI_3$. By adjusting the halide composition rationally, the best energy level alignment was achieved between $MAPbI_3$ and $CsPbBr_xI_{3-x}$, promoting hole extraction and transport from $MAPbI_3$ to the electrode. The valence band maximum shift between $MAPbI_3$ and several layers of $CsPbBr_xI_{3-x}$ introduced an intrinsic electric field, which facilitated charge separation and extraction at the $MAPbI_3$/CE interface. Notably, the 3-level GHJ exhibited the most efficient charge extraction, evident from the rapid TRPL decay. Furthermore, the multiple layers of $CsPbBr_xI_{3-x}$ served as robust electron barriers, effectively impeding the electron flow from the perovskite to the CE, thereby delaying interfacial charge recombination. Therefore, this well-designed three-level $MAPbI_3$-$CsPbBr_xI_{3-x}$ GHJ can construct highly efficient HTL-free PSCs.[94] Moreover, the introduction of an organic $CsPbBr_xI_{3-x}$ multi-gradient layer effectively prevents direct contact between $MAPbI_3$ and infiltrating water molecules, while enhancing the thermal stability of the perovskite film. These more complex structures can enhance overall device efficiency. Heterojunction engineering should be compatible with the simplified HTL-free and ETL-free device architectures.

10.5 CONCLUSION AND OUTLOOK

In response to challenges concerning PSC device stability and manufacturing costs, researchers have explored various strategies, including the removal of CTL in PSCs. The elimination of either the ETL or HTL streamlines device architecture, offering the prospect of cost reduction and enhanced stability. The structures of ETL/HTL-free PSCs can be categorized into two types, depending on the presence of a mesoscopic scaffold layer: the mesoporous structure and the planar structure. Given that the ETL/HTL-free design poses specific challenges related to optimizing charge transport, achieving high efficiency, and ensuring long-term stability, the establishment of efficient charge transport pathways becomes essential in the absence of ETL and HTL. This might involve modifications to the perovskite layer itself and the interfaces to facilitate charge transport. The maintenance of a high PCE is critical for the commercial success of PSCs, necessitating a strong focus on refining ETL/HTL-free designs to ensure they do not compromise device efficiency. Long-term stability remains a paramount concern within the domain of PSCs, and ETL/HTL-free designs must comprehensively address this issue to compete with traditional PSC configurations. Current strategies encompass techniques such as crystallization engineering, composition optimization, additive engineering, surface passivation, and heterojunction formation, among others. Sustained research and development efforts are imperative to confront these challenges and position ETL/HTL-free PSCs as a competitive and viable technology in the solar energy market.

REFERENCES

1. T. Leijtens, G. E. Eperon, S. Pathak, A. Abate, M. M. Lee, H. J. Snaith, Overcoming ultraviolet light instability of sensitized TiO_2 with meso-superstructured organometal tri-halide perovskite solar cells. *Nature Communications* **2013**, 4, 2885.
2. Q. Jiang, X. Zhang, J. You, SnO_2: a wonderful electron transport layer for perovskite solar cells. *Small* **2018**, 14, 1801154.
3. J. Cao, B. Wu, R. Chen, Y. Wu, Y. Hui, B. W. Mao, N. Zheng, Efficient, hysteresis-free, and stable perovskite solar cells with ZnO as electron-transport layer: effect of surface passivation. *Advanced Materials* **2018**, 30, 1705596.
4. J. Liu, Y. Wu, C. Qin, X. Yang, T. Yasuda, A. Islam, K. Zhang, W. Peng, W. Chen, L. Han, A dopant-free hole-transporting material for efficient and stable perovskite solar cells. *Energy & Environmental Science* **2014**, 7, 2963.
5. Z. H. Bakr, Q. Wali, A. Fakharuddin, L. Schmidt-Mende, T. M. Brown, R. Jose, Advances in hole transport materials engineering for stable and efficient perovskite solar cells. *Nano Energy* **2017**, 34, 271.
6. T. Zhang, F. Wang, H.-B. Kim, I.-W. Choi, C. Wang, E. Cho, R. Konefal, Y. Puttisong, K. Terado, L. Kobera, M. Chen, M. Yang, S. Bai, B. Yang, J. Suo, S.-C. Yang, X. Liu, F. Fu, H. Yoshida, W. M. Chen, J. Brus, V. Coropceanu, A. Hagfeldt, J.-L. Brédas, M. Fahlman, D. S. Kim, Z. Hu, F. Gao, Ion-modulated radical doping of spiro-OMeTAD for more efficient and stable perovskite solar cells. *Science* **2022**, 377, 495.
7. K. Domanski, J.-P. Correa-Baena, N. Mine, M. K. Nazeeruddin, A. Abate, M. Saliba, W. Tress, A. Hagfeldt, M. Grätzel, Not all that glitters is gold: metal-migration-induced degradation in perovskite solar cells. *ACS Nano* **2016**, 10, 6306.
8. J. Li, Q. Dong, N. Li, L. Wang, Direct evidence of ion diffusion for the silver-electrode-induced thermal degradation of inverted perovskite solar cells. *Advanced Energy Materials* **2017**, 7, 1602922.
9. W.-Q. Wu, Q. Wang, Y. Fang, Y. Shao, S. Tang, Y. Deng, H. Lu, Y. Liu, T. Li, Z. Yang, Molecular doping enabled scalable blading of efficient hole-transport-layer-free perovskite solar cells. *Nature Communications* **2018**, 9, 1.
10. W. Hui, X. Kang, B. Wang, D. Li, Z. Su, Y. Bao, L. Gu, B. Zhang, X. Gao, L. Song, W. Huang, Stable electron-transport-layer-free perovskite solar cells with over 22% power conversion efficiency. *Nano Letters* **2023**, 23, 2195.
11. W. Q. Wu, J. F. Liao, J. X. Zhong, Y. F. Xu, L. Wang, J. Huang, Suppressing interfacial charge recombination in electron-transport-layer-free perovskite solar cells to give an efficiency exceeding 21%. *Angewandte Chemie International Edition* **2020**, 132, 21166.
12. S. Guarnera, A. Abate, W. Zhang, J. M. Foster, G. Richardson, A. Petrozza, H. J. Snaith, Improving the long-term stability of perovskite solar cells with a porous Al_2O_3 buffer layer. *The Journal of Physical Chemistry Letters* **2015**, 6, 432.
13. L. Huang, Z. Ge, Simple, robust, and going more efficient: recent advance on electron transport layer-free perovskite solar cells. *Advanced Energy Materials* **2019**, 9, 1900248.

14. D. Liu, J. Yang, T. L. Kelly, Compact layer free perovskite solar cells with 13.5% efficiency. *Journal of the American Chemical Society* **2014**, 136, 17116.
15. Q. Han, J. Ding, Y. Bai, T. Li, J.-Y. Ma, Y.-X. Chen, Y. Zhou, J. Liu, Q.-Q. Ge, J. Chen, J. T. Glass, M. J. Therien, J. Liu, D. B. Mitzi, J.-S Hu, Carrier dynamics engineering for high-performance electron-transport-layer-free perovskite photovoltaics. *Chem* **2018**, 4, 2405.
16. F. Sadegh, E. Akman, D. Prochowicz, M. M. Tavakoli, P. Yadav, S. Akin, Facile NaF treatment achieves 20% efficient ETL-free perovskite solar cells. *ACS Applied Materials & Interfaces* **2022**, 14, 38631.
17. Z. Xiao, Y. Yuan, Y. Shao, Q. Wang, Q. Dong, C. Bi, P. Sharma, A. Gruverman, J. Huang, Giant switchable photovoltaic effect in organometal trihalide perovskite devices. *Nature Materials* **2015**, 14, 193.
18. T. Shi, J. Chen, J. Zheng, X. Li, B. Zhou, H. Cao, Y. Wang, Ti/Au cathode for electronic transport material-free organic-inorganic hybrid perovskite solar cells. *Scientific Reports* **2016**, 6, 39132.
19. T. Chen, T. Shi, X. Li, J. Zheng, W. Fan, B. Ni, Y. Wang, J. Dai, Z. Xiao, Efficient perovskite solar cells with titanium cathode interlayer. *Solar RRL* **2018**, 2, 1800167.
20. J. Liang, C. Wang, Y. Wang, Z. Xu, Z. Lu, Y. Ma, H. Zhu, Y. Hu, C. Xiao, X. Yi, G. Zhu, H. Lv, L. Ma, T. Chen, Z. Tie, Z. Jin, J. Liu, All-inorganic perovskite solar cells. *Journal of the American Chemical Society* **2016**, 138, 15829.
21. H. Chen, X. Zheng, Q. Li, Y. Yang, S. Xiao, C. Hu, Y. Bai, T. Zhang, K. S. Wong, S. Yang, An amorphous precursor route to the conformable oriented crystallization of $CH_3NH_3PbBr_3$ in mesoporous scaffolds: toward efficient and thermally stable carbon-based perovskite solar cells. *Journal of Materials Chemistry A* **2016**, 4, 12897.
22. L. Fagiolari, F. Bella, Carbon-based materials for stable, cheaper and large-scale processable perovskite solar cells. *Energy & Environmental Science* **2019**, 12, 3437.
23. L. Etgar, P. Gao, Z. Xue, Q. Peng, A. K. Chandiran, B. Liu, M. K. Nazeeruddin, M. Grätzel, Mesoscopic $CH_3NH_3PbI_3/TiO_2$ heterojunction solar cells. *Journal of the American Chemical Society* **2012**, 134, 17396.
24. S. Aharon, B. E. Cohen, L. Etgar, Hybrid lead halide iodide and lead halide bromide in efficient hole conductor free perovskite solar cell. *The Journal of Physical Chemistry C* **2014**, 118, 17160.
25. Y. Xiao, G. Han, Y. Chang, Y. Zhang, Y. Li, M. Li, Investigation of perovskite-sensitized nanoporous titanium dioxide photoanodes with different thicknesses in perovskite solar cells. *Journal of Power Sources* **2015**, 286, 118.
26. H. Chen, S. Yang, Carbon-based perovskite solar cells without hole transport materials: The front runner to the market? *Advanced Materials* **2017**, 29, 1603994.
27. X. Zheng, H. Chen, Q. Li, Y. Yang, Z. Wei, Y. Bai, Y. Qiu, D. Zhou, K. S. Wong, S. Yang, Boron doping of multiwalled carbon nanotubes significantly enhances hole extraction in carbon-based perovskite solar cells. *Nano Letters* **2017**, 17, 2496.
28. S. N. Habisreutinger, R. J. Nicholas, H. J. Snaith, Carbon nanotubes in perovskite solar cells. *Advanced Energy Materials* **2017**, 7, 1601839.
29. A. Agresti, S. Pescetelli, B. Taheri, A. E. Del Rio Castillo, L. Cinà, F. Bonaccorso, A. Di Carlo, Graphene–perovskite solar cells exceed 18% efficiency: a stability study. *ChemSusChem* **2016**, 9, 2609.
30. P. You, Z. Liu, Q. Tai, S. Liu, F. Yan, Efficient semitransparent perovskite solar cells with graphene electrodes. *Advanced Materials* **2015**, 27, 3632.
31. A. Mei, X. Li, L. Liu, Z. Ku, T. Liu, Y. Rong, M. Xu, M. Hu, J. Chen, Y. Yang, A hole-conductor-free, fully printable mesoscopic perovskite solar cell with high stability. *Science* **2014**, 345, 295.
32. L. Li, H. Rao, Z. Wu, J. Hong, J. Zhang, Z. Pan, X. Zhong, Moisture induced secondary crystal growth boosting the efficiency of hole transport layer-free carbon-based perovskite solar cells beyond 19.5%. *Advanced Functional Materials* **2023**, 2308428.

33. H. Hu, D. Wang, Y. Zhou, J. Zhang, S. Lv, S. Pang, X. Chen, Z. Liu, N. P. Padture, G. Cui, Vapour-based processing of hole-conductor-free $CH_3NH_3PbI_3$ perovskite/C_{60} fullerene planar solar cells. *RSC Advances* **2014**, 4, 28964.
34. K.-W. Tsai, C.-C. Chueh, S. T. Williams, T.-C. Wen, A. K. Jen, High-performance hole-transporting layer-free conventional perovskite/fullerene heterojunction thin-film solar cells. *Journal of Materials Chemistry A* **2015**, 3, 9128.
35. X. Bao, Q. Zhu, M. Qiu, A. Yang, Y. Wang, D. Zhu, J. Wang, R. Yang, High-performance inverted planar perovskite solar cells without a hole transport layer via a solution process under ambient conditions. *Journal of Materials Chemistry A* **2015**, 3, 19294.
36. H.-H. Huang, Q.-H. Liu, H. Tsai, S. Shrestha, L.-Y. Su, P.-T. Chen, Y.-T. Chen, T.-A. Yang, H. Lu, C.-H. Chuang, K.-F. Lin, S.-P. Rwei, W. Nie, L. Wang, A simple one-step method with wide processing window for high-quality perovskite mini-module fabrication. *Joule* **2021**, 5, 958.
37. Y. Zhao, K. Zhu, CH_3NH_3Cl-assisted one-step solution growth of $CH_3NH_3PbI_3$: structure, charge-carrier dynamics, and photovoltaic properties of perovskite solar cells. *The Journal of Physical Chemistry C* **2014**, 118, 9412.
38. C.-M. Tsai, G.-W. Wu, S. Narra, H.-M. Chang, N. Mohanta, H.-P. Wu, C.-L. Wang, E. W.-G. Diau, Control of preferred orientation with slow crystallization for carbon-based mesoscopic perovskite solar cells attaining efficiency 15%. *Journal of Materials Chemistry A* **2017**, 5, 739.
39. Y. Wang, L. Li, Z. Wu, R. Zhang, J. Hong, J. Zhang, H. Rao, Z. Pan, X. Zhong, Self-driven prenucleation-induced perovskite crystallization enables efficient perovskite solar cells. *Angewandte Chemie International Edition* **2023**, 62, e202302342.
40. J.-H. Im, H.-S. Kim, N.-G. Park, Morphology-photovoltaic property correlation in perovskite solar cells: one-step versus two-step deposition of $CH_3NH_3PbI_3$. *APL Materials* **2014**, 2, 081510.
41. D. Bi, S.-J. Moon, L. Häggman, G. Boschloo, L. Yang, E. M. Johansson, M. K. Nazeeruddin, M. Grätzel, A. Hagfeldt, Using a two-step deposition technique to prepare perovskite ($CH_3NH_3PbI_3$) for thin film solar cells based on ZrO_2 and TiO_2 mesostructures. *RSC Advances* **2013**, 3, 18762.
42. P. Teng, X. Han, J. Li, Y. Xu, L. Kang, Y. Wang, Y. Yang, T. Yu, Elegant face-down liquid-space-restricted deposition of $CsPbBr_3$ films for efficient carbon-based all-inorganic planar perovskite solar cells. *ACS Applied Materials & Interfaces* **2018**, 10, 9541.
43. J. Feng, X. Han, H. Huang, Q. Meng, Z. Zhu, T. Yu, Z. Li, Z. Zou, Curing the fundamental issue of impurity phases in two-step solution-processed $CsPbBr_3$ perovskite films. *Science Bulletin* **2020**, 65, 726.
44. Z. Safari, M. B. Zarandi, M. R. Nateghi, Improved environmental stability of HTM free perovskite solar cells by a modified deposition route. *Chemical Papers* **2019**, 73, 2667.
45. J. Duan, Y. Zhao, B. He, Q. Tang, High-purity inorganic perovskite films for solar cells with 9.72% efficiency. *Angewandte Chemie International Edition* **2018**, 57, 3787.
46. M. M. Tavakoli, L. Gu, Y. Gao, C. Reckmeier, J. He, A. L. Rogach, Y. Yao, Z. Fan, Fabrication of efficient planar perovskite solar cells using a one-step chemical vapor deposition method. *Scientific Reports* **2015**, 5, 14083.
47. M. R. Leyden, M. V. Lee, S. R. Raga, Y. Qi, Large formamidinium lead trihalide perovskite solar cells using chemical vapor deposition with high reproducibility and tunable chlorine concentrations. *Journal of Materials Chemistry A* **2015**, 3, 16097.
48. Q. Chen, H. Zhou, Z. Hong, S. Luo, H.-S. Duan, H.-H. Wang, Y. Liu, G. Li, Y. Yang, Planar heterojunction perovskite solar cells via vapor-assisted solution process. *Journal of the American Chemical Society* **2014**, 136, 622.
49. X. Li, Y. Tan, H. Lai, S. Li, Y. Chen, S. Li, P. Xu, J. Yang, All-inorganic $CsPbBr_3$ perovskite solar cells with 10.45% efficiency by evaporation-assisted deposition and setting intermediate energy levels. *ACS Applied Materials & Interfaces* **2019**, 11, 29746.

50. P. Cui, D. Wei, J. Ji, D. Song, Y. Li, X. Liu, J. Huang, T. Wang, J. You, M. Li, Highly efficient electron-selective layer free perovskite solar cells by constructing effective p-n heterojunction. *Solar RRL* **2017**, 1, 1600027.
51. X. Yang, Q. Li, Y. Zheng, D. Luo, Y. Zhang, Y. Tu, L. Zhao, Y. Wang, F. Xu, Q. Gong, R. Zhu, Perovskite hetero-bilayer for efficient charge-transport-layer-free solar cells. *Joule* **2022**, 6, 1277.
52. D. P. McMeekin, G. Sadoughi, W. Rehman, G. E. Eperon, M. Saliba, M. T. H.rantner, A. Haghighirad, N. Sakai, L. Korte, B. Rech, M. B. Johnston, L. M. Herz, H. J. Snaith, A mixed-cation lead mixed-halide perovskite absorber for tandem solar cells. *Science* **2016**, 351, 151.
53. S. Wang, H. Liu, H. Bala, B. Zong, L. Huang, Z.-a. Guo, W. Fu, B. Zhang, G. Sun, J. Cao, Z. Zhang, A highly stable hole-conductor-free $Cs_XMA_{1-X}PbI_3$ perovskite solar cell based on carbon counter electrode. *Electrochimica Acta* **2020**, 335, 135686.
54. Y. Guo, F. Zhao, J. Tao, J. Jiang, J. Zhang, J. Yang, Z. Hu, J. Chu, Efficient and hole-transporting-layer-free $CsPbI_2Br$ planar heterojunction perovskite solar cells through rubidium passivation. *ChemSusChem* **2019**, 12, 983.
55. A. Swarnkar, W. J. Mir, A. Nag, Can B-site doping or alloying improve thermal-and phase-stability of all-inorganic $CsPbX_3$ (X = Cl, Br, I) perovskites? *ACS Energy Letters* **2018**, 3, 286.
56. B. Luo, F. Li, K. Xu, Y. Guo, Y. Liu, Z. Xia, J. Z. Zhang, B-Site doped lead halide perovskites: synthesis, band engineering, photophysics, and light emission applications. *Journal of Materials Chemistry C* **2019**, 7, 2781.
57. Y. Du, X. Zhang, Y. Shi, X. Hou, F. Li, Q. Zhang, Q. Tai, P. Liu, X.-Z. Zhao, Optimized crystallization and defect passivation with Yttrium (III) doped $MAPbBr_3$ film for highly efficient and stable hole-transport-layer-free carbon-based perovskite solar cells. *Journal of Alloys and Compounds* **2022**, 890, 161909.
58. S. Colella, E. Mosconi, P. Fedeli, A. Listorti, F. Gazza, F. Orlandi, P. Ferro, T. Besagni, A. Rizzo, G. Calestani, G. Gigli, F. D. Angelis, R. Mosca, $MAPbI_{3-X}Clx$ mixed halide perovskite for hybrid solar cells: the role of chloride as dopant on the transport and structural properties. *Chemistry of Materials* **2013**, 25, 4613.
59. R. K. Misra, L. Ciammaruchi, S. Aharon, D. Mogilyansky, L. Etgar, I. Visoly-Fisher, E. A. Katz, Effect of halide composition on the photochemical stability of perovskite photovoltaic materials. *ChemSusChem* **2016**, 9, 2572.
60. K. Wang, W. Sun, W. Liu, X. Huo, R. Yin, J. Liu, Y. Gao, T. You, P. Yin, Mitigating interfacial and bulk defects via chlorine modulation for HTL-free all-inorganic $CsPbI_2Br$ carbon-based perovskite solar cells with efficiency over 14%. *Chemical Engineering Journal* **2022**, 445, 136781.
61. F. Zhang, K. Zhu, Additive engineering for efficient and stable perovskite solar cells. *Advanced Energy Materials* **2020**, 10, 1902579.
62. S. Bai, P. Da, C. Li, Z. Wang, Z. Yuan, F. Fu, M. Kawecki, X. Liu, N. Sakai, J. T.-W. Wang, S. Huettner, S. Buecheler, M. Fahlman, F. Gao, H. J. Snaith, Planar perovskite solar cells with long-term stability using ionic liquid additives. *Nature* **2019**, 571, 245.
63. Y. Fang, C. Bi, D. Wang, J. Huang, The functions of fullerenes in hybrid perovskite solar cells. *ACS Energy Letters* **2017**, 2, 782.
64. C. Park, H. Ko, D. H. Sin, K. C. Song, K. Cho, Organometal halide perovskite solar cells with improved thermal stability via grain boundary passivation using a molecular additive. *Advanced Functional Materials* **2017**, 27, 1703546.
65. C.-C. Chen, Z. Hong, G. Li, Q. Chen, H. Zhou, Y. Yang, One-step, low-temperature deposited perovskite solar cell utilizing small molecule additive. *Journal of Photonics for Energy* **2015**, 5, 057405.
66. R. Ranjan, S. Ranjan, M. Monalisa, K. S. Nalwa, A. Singh, A. Garg, R. K. Gupta, Enhanced thermal and moisture stability via dual additives approach in methylammonium lead iodide based planar perovskite solar cells. *Solar Energy* **2021**, 225, 200.

67. Y. Chen, X. Yang, P. Liu, W. Wang, R. Ran, W. Zhou, Z. Shao, Improving moisture/thermal stability and efficiency of $CH_3NH_3PbI_3$-based perovskite solar cells via gentle butyl acrylate additive strategy. *Solar RRL* **2021**, 5, 2000621.
68. P. W. Liang, C. Y. Liao, C. C. Chueh, F. Zuo, S. T. Williams, X. K. Xin, J. Lin, A. K. Y. Jen, Additive enhanced crystallization of solution-processed perovskite for highly efficient planar-heterojunction solar cells. *Advanced Materials* **2014**, 26, 3748.
69. X. Gong, M. Li, X. B. Shi, H. Ma, Z. K. Wang, L. S. Liao, Controllable perovskite crystallization by water additive for high-performance solar cells. *Advanced Functional Materials* **2015**, 25, 6671.
70. Y. Cheng, Q. Wei, N. Wang, Z. Ye, Y. Zhao, Q. Wang, D. Chu, L. Zan, F. Fu, Y. Liu, Antioxidant induced bulk passivation for efficient and stable hole transport layer-free carbon electrode perovskite solar cells. *Chinese Chemical Letters* **2023**, 34, 107933.
71. S. Liu, Y. Guan, Y. Sheng, Y. Hu, Y. Rong, A. Mei, H. Han, A review on additives for halide perovskite solar cells. *Advanced Energy Materials* **2020**, 10, 1902492.
72. X. Zhou, L. Zhang, H. Hu, Z. Jiang, D. Wang, J. Chen, Y. Li, J. Wu, Y. Zhang, M. Zhang. C. Liu, Y. Peng, X. Wang, B. Xu, Highly efficient and stable hole-transport-layer-free inverted perovskite solar cells achieved 22% efficiency through p-type molecular synergistic doping. *Nano Energy* **2022**, 104, 107988.
73. Q. He, H. Zhang, S. Han, R. Wang, Y. Li, X. Zhang, Y. Xing, Improvement of thiourea (Lewis base)-modified SnO_2 electron-transport layer for carbon-based $CsPbIBr_2$ perovskite solar cells. *ACS Applied Energy Materials* **2021**, 4, 10958.
74. Y. Qiang, Y. Xie, Y. Qi, P. Wei, H. Shi, C. Geng, H. Liu, Enhanced performance of carbon-based perovskite solar cells with a Li+-doped SnO_2 electron transport layer and Al_2O_3 scaffold layer. *Solar Energy* **2020**, 201, 523.
75. X. Liu, Z. Liu, B. Sun, X. Tan, H. Ye, Y. Tu, T. Shi, Z. Tang, G. Liao, 17.46% efficient and highly stable carbon-based planar perovskite solar cells employing Ni-doped rutile TiO_2 as electron transport layer. *Nano Energy* **2018**, 50, 201.
76. Q. Jiang, Y. Zhao, X. Zhang, X. Yang, Y. Chen, Z. Chu, Q. Ye, X. Li, Z. Yin, J. You, Surface passivation of perovskite film for efficient solar cells. *Nature Photonics* **2019**, 13, 460.
77. C.-H. Chiang, Z.-L. Tseng, C.-G. Wu, Planar heterojunction perovskite/$PC_{71}BM$ solar cells with enhanced open-circuit voltage via a (2/1)-step spin-coating process. *Journal of Materials Chemistry A* **2014**, 2, 15897.
78. G. Yang, P. Qin, G. Fang, G. Li, A Lewis base-assisted passivation strategy towards highly efficient and stable perovskite solar cells. *Solar RRL* **2018**, 2, 1800055.
79. M. Kim, S. G. Motti, R. Sorrentino, A. Petrozza, Enhanced solar cell stability by hygroscopic polymer passivation of metal halide perovskite thin film. *Energy & Environmental Science* **2018**, 11, 2609.
80. A. Mingorance, H. Xie, H. S. Kim, Z. Wang, M. Balsells, A. Morales-Melgares, N. Domingo, N. Kazuteru, W. Tress, J. Fraxedas, N. Vlachopoulos, A. Hagfeldt, M. Lira-Cantu, Interfacial engineering of metal oxides for highly stable halide perovskite solar cells. *Advanced Materials Interfaces* **2018**, 5, 1800367.
81. E. Aydin, M. De Bastiani, S. De Wolf, Defect and contact passivation for perovskite solar cells. *Advanced Materials* **2019**, 31, 1900428.
82. Y. Wu, Q. Zhang, L. Fan, C. Liu, M. Wu, D. Wang, T. Zhang, Surface reconstruction-induced efficient $CsPbI_2Br$ perovskite solar cell using phenylethylammonium iodide. *ACS Applied Energy Materials* **2021**, 4, 5583.
83. Y. Yuan, J. Huang, Ion migration in organometal trihalide perovskite and its impact on photovoltaic efficiency and stability. *Accounts of Chemical Research* **2016**, 49, 286.
84. B. Chen, P. N. Rudd, S. Yang, Y. Yuan, J. Huang, Imperfections and their passivation in halide perovskite solar cells. *Chemical Society Reviews* **2019**, 48, 3842.
85. T. Ye, Y. Hou, A. Nozariasbmarz, D. Yang, J. Yoon, L. Zheng, K. Wang, K. Wang, S. Ramakrishna, S. Priya, Cost-effective high-performance charge-carrier-transport-layer-free perovskite solar cells achieved by suppressing ion migration. *ACS Energy Letters* **2021**, 6, 3044.

86. G. Wu, R. Liang, M. Ge, G. Sun, Y. Zhang, G. Xing, Surface passivation using 2D perovskites toward efficient and stable perovskite solar cells. *Advanced Materials* **2022**, 34, 2105635.
87. H. Zhang, M. K. Nazeeruddin, W. C. Choy, Perovskite photovoltaics: the significant role of ligands in film formation, passivation, and stability. *Advanced Materials* **2019**, 31, 1805702.
88. X. Zhang, Y. Zhou, Y. Li, J. Sun, X. Lu, X. Gao, J. Gao, L. Shui, S. Wu, J.-M. Liu, Efficient and carbon-based hole transport layer-free $CsPbI_2Br$ planar perovskite solar cells using PMMA modification. *Journal of Materials Chemistry C* **2019**, 7, 3852.
89. F. Zhang, W. Shi, J. Luo, N. Pellet, C. Yi, X. Li, X. Zhao, T. J. S. Dennis, X. Li, S. Wang, Y. Xiao, S. M. Zakeeruddin, D. Bi, M. Grätzel, Isomer-pure bis-PCBM-assisted crystal engineering of perovskite solar cells showing excellent efficiency and stability. *Advanced Materials* **2017**, 29, 1606806.
90. X. Wu, B. Li, Z. Zhu, C.-C. Chueh, A. K.-Y. Jen, Designs from single junctions, heterojunctions to multijunctions for high-performance perovskite solar cells. *Chemical Society Reviews* **2021**, 50, 13090.
91. S. Ye, H. Rao, W. Yan, Y. Li, W. Sun, H. Peng, Z. Liu, Z. Bian, Y. Li, C. Huang, A strategy to simplify the preparation process of perovskite solar cells by co-deposition of a hole-conductor and a perovskite layer. *Advanced Materials* **2016**, 28, 9648.
92. X.-N. Huo, K.-X. Wang, R. Yin, W.-W. Sun, Y.-S. Sun, Y.-K. Gao, T.-T. You, P.-G. Yin, High-performance carbon-based all-inorganic $CsPbI_2Br$ perovskite solar cells via ethylammonium iodide and phenethylammonium iodide synergistic passivation. *Solar Energy Materials and Solar Cells* **2022**, 247, 111963.
93. J. Liu, Q. Zhou, N. K. Thein, L. Tian, D. Jia, E. M. Johansson, X. Zhang, In situ growth of perovskite stacking layers for high-efficiency carbon-based hole conductor free perovskite solar cells. *Journal of Materials Chemistry A* **2019**, 7, 13777.
94. B. Li, Y. Zhang, L. Zhang, L. Yin, Graded heterojunction engineering for hole-conductor-free perovskite solar cells with high hole extraction efficiency and conductivity. *Advanced Materials* **2017**, 29, 1701221.

11 Semitransparent Perovskite Solar Cells for Building-Integrated Photovoltaics (BIPV) and Tandem Solar Cells

Yiqiang Zhang, Ge Chen, and Pengwei Li

11.1 INTRODUCTION

As a result of the depletion of conventional energy sources and the undeniable shifts in global climate patterns, nations across the globe find themselves facing mounting pressure to embark on a journey of energy transformation and adopt alternative energy strategies.[1] One such measure entails harnessing solar energy, which offers a multitude of advantages, including sustainability, minimal pollution, and an abundance of resources. The significance of solar power in reshaping the landscape of energy infrastructure competition is consequently on the rise.[2] At present, silicon (Si) solar cells occupy most of the global photovoltaic market. In this regard, perovskite solar cells (PSCs) offer a viable and low-cost solution compared to Si solar cells.[3] Over the past decade, the power conversion efficiencies (PCEs) of PSCs have been remarkably increased from 3.8% to 26.1%, making them an attractive alternative. Additionally, perovskite materials have other merits such as low exciton binding energy, long carrier diffusion length, and high carrier mobility, facilitating the dissociation of the photogenerated exciton and the transport of the charge carriers, and thus enhancing the performance of PSCs.[4–6]

Semi-transparent perovskite solar cells (ST-PSCs) hold great promise in building integrated photovoltaics (BIPV) due to their tunable band gaps and excellent light transmittance.[7] Applying photovoltaics to the windows and roofs of buildings can meet the advantages of building shading and energy supply at the same time.[8] BIPV is an ideal new building material in terms of energy saving. The application of silicon-based solar cells in photovoltaic windows has been widely studied before.[9] However, owing to the low absorption coefficient and rigid unbendable feature of silicon materials, alternative photovoltaic materials were fully explored for lowering the manufacturing cost and improving the device performance, such as CdTe, dye-sensitized solar cells and PSCs. The application of photovoltaic devices in buildings needs to be analyzed from three aspects: power generation, thermal performance,

DOI: 10.1201/9781003400493-11

and optical performance. Therefore, in addition to considering the power conversion efficiency (PCE), heat-insulating ability, average visible light transmittance (AVT), and color properties, integrability are key in the application of ST-PSCs in BIPV.[10] In addition, ST-PSCs can serve as a top cell for making tandem cells, binding with the bottom cell to allow full utilization of the solar spectrum and achieve higher PCE, as compared to the single-junction solar cells.[11] Theoretical calculations indicate that double-junction tandem solar cells (TSCs) can overcome the Shockley-Queisser limit of single-junction solar cells, with a theoretical efficiency of about 46%. TSCs have thus been proven to be a successful way to overcome the Shockley-Queisser limit of single-junction solar cells.[12] Therefore, it is imperative to streamline the prospective market applications of ST-PSCs for BIPV and TSCs.[13,14]

The performance of ST-PSCs is evaluated based on two main parameters: power conversion efficiency (PCE) and average visible light transmittance (AVT). These two parameters interact with each other, posing a challenge when attempting to optimize both simultaneously. Achieving high AVT invariably results in a trade-off in PCE and vice versa. For instance, in order to obtain high light transmittance of the device, Eperon et al.[15] formed an island-like perovskite microstructure layer through the dehumidification of the film, making the device obtain an AVT of 30%. Another effective way to achieve high light transmittance of the PSCs is reducing the thickness of the perovskite active layer.[16] However, these methods inevitably cause insufficient light absorption of the device, thereby sacrificing the PCE of the solar cells. To solve this problem, light utilization efficiency (LUE) is proposed, which is the product of PCE and AVT.[17] Balancing PCE and AVT is crucial for achieving high-performance in ST-PSCs. To improve the insufficient light absorption in ST-PSCs, strategies, such as device structure optimization,[18–20] transparent cell electrode design,[21–24] and anti-reflection coating,[25–27] are constantly being studied to maximize their performance. These endeavors will serve as crucial stepping stones towards further practical utility of ST-PSCs in various domains, including BIPV, TSCs, and associated areas like displays, automotive technologies, and sensor applications.

Photon propagation management related to AVT and carrier dynamics control associated with PCE are crucial in regulating LUE of ST-PSCs. The optimization of these two aspects can promote ST-PSCs to the best trade-off in terms of optical transmission and electricity generation, thus maximizing the performance of ST-PSCs. In addition,[28] maintaining the aesthetics of ST-PSCs is important for their commercialization, particularly when applied to windows, roofs, and other areas that require vibrant colors,[29] At present, photonic crystals, band gap control, and other strategies are commonly used ways to achieve color control of ST-PSCs.

ST-PSCs are also used as top cells of TSCs series with Si cells,[30] CIGS,[31] organic solar cells (OSCs),[32] and PSCs.[33] The certified efficiency of perovskite/silicon tandem devices has reached to 33.2%, indicating the remarkable potential for application in TSCs. However, due to the phase separation of perovskites, textured surface of silicon cells, and poor conductivity of interconnect layers, the measures of optimizing performance of ST-PSCs are critical in further advancing TSCs, including stability enhancement of perovskite materials, conformal perovskite coating on textured surfaces, and design of high-quality interconnect layers.

This chapter provides an in-depth analysis of recent advances in the application of ST-PSCs in BIPV and TSCs. First, the theoretical knowledge of photon transmission path, carrier dynamics, and color rendering of ST-PSCs is systematically introduced. Second, control strategies for photon propagation, carrier dynamics, and color of ST-PSCs in BIPV are summarized. Then, the selection of different bottom cells in perovskite-based tandem devices and the design of interconnect layers are investigated. Finally, aiming for the commercial application of perovskite-based BIPV and TSCs, the development and prospects of environmental management, stability, and large-area regulation of Pb are profoundly prospected.

11.2 THEORETICAL ANALYSIS ON ST-PSCs (FIGURE 11.1)

11.2.1 Photon Propagation

Light exhibits wave-particle duality. When the light demonstrates particle properties, it is referred to as photons. The energy of a photon ($E = h\nu$) and bandgap energy (E_g) of semiconductors determine the relationship between photons and semiconductor media. If $E = h\nu < E_g$, photons pass through the semiconductor and the semiconductor remains transparent. However, if $E = h\nu > E_g$, photons will elevate valence electrons onto the conduction band and produce excitons (electron-hole pairs). The light intensity passing through an absorption semiconductor is denoted as I(x) with the unit of energy $cm^{-2}\ s^{-1}$. The absorption coefficient represents the relative decrease in light intensity as it passes through the absorption semiconductor, denoted by α with the unit of cm^{-1}. The relationship between I(x) and α can be expressed by the formula (1):

$$I(x) = I_0 e^{-\alpha x} \tag{1}$$

where I_0 is the intensity of the incident light, and x is the thickness of the semiconductor medium. Apparently, the intensity of light passing through a semiconductor medium is affected by the absorption coefficient and thickness of the material. In perovskite materials, the high absorption coefficient (10^4–10^5 cm^{-1}) and suitable bandgap (1.17–2.3 eV) in the visible light range reflects their excellent absorption capacity (Figure 11.2).

Transmittance (T) is defined as the ratio of the intensity of the incident light (I_0) to that of the transmitted light intensity (I_d). For an object with uniform thickness d and uniform absorption coefficient α, considering the primary transmission, the transmittance of incident light at the rear interface can be expressed as formula (2):

$$T = I_d / I_0 = (1 - R)^2 e^{-\alpha d} \tag{2}$$

Besides, the overall light transmittance is also affected by refractive index coupling among each functional layer. Considering the spectral response in the visible wavelength range, the AVT can be further calculated using formula (3)[42]:

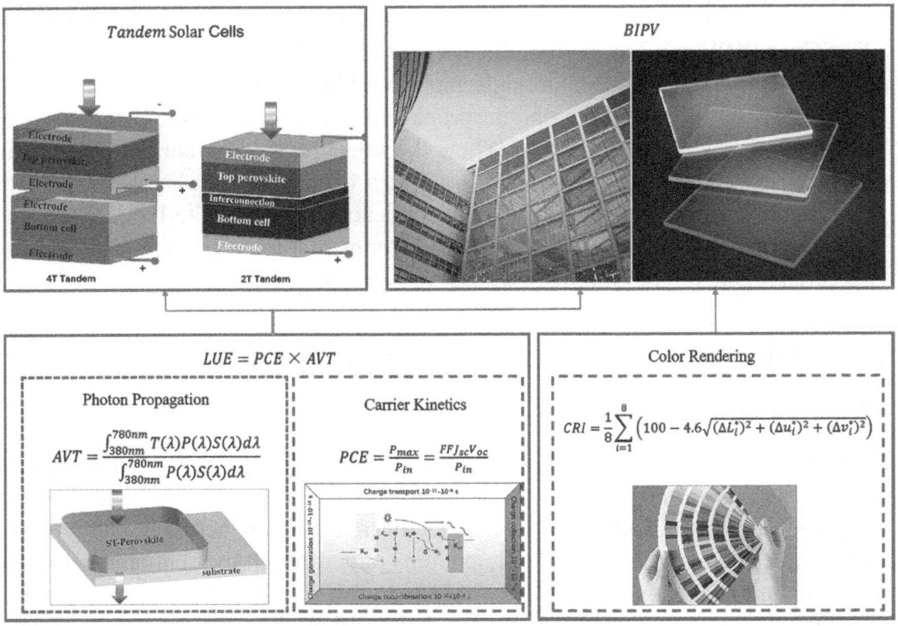

FIGURE 11.1 A brief summary of theoretical analysis on ST-PSCs.

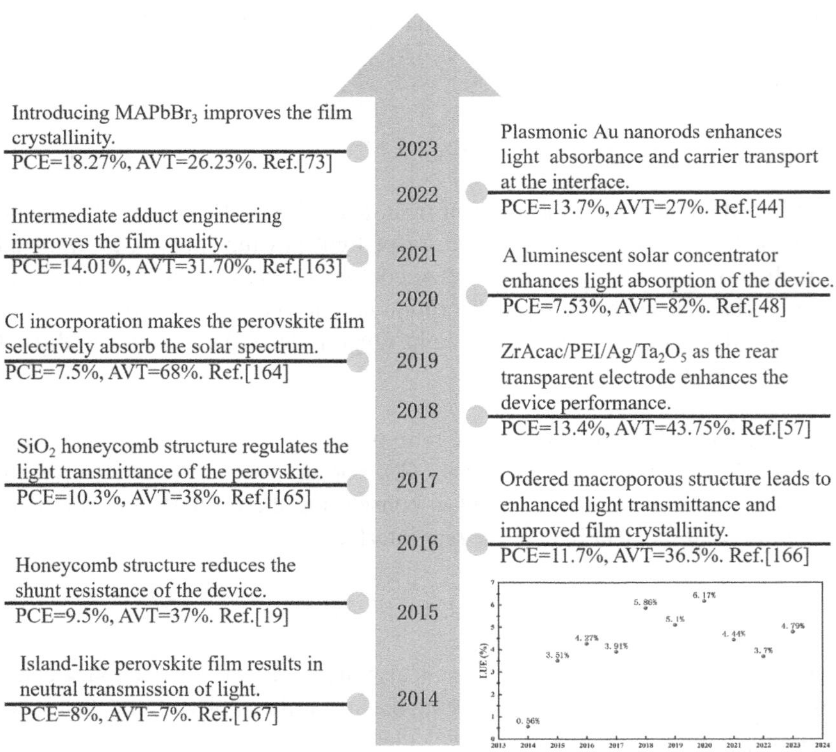

FIGURE 11.2 A brief figure of the efficiency progress of semi-transparent perovskite solar cells.[19,34–41]

$$AVT = \frac{\int_{380nm}^{780nm} T(\lambda)P(\lambda)S(\lambda)d\lambda}{\int_{380nm}^{780nm} P(\lambda)S(\lambda)d\lambda} \tag{3}$$

where λ represents wavelength, T represents transmittance, P reflects the human eye's response to visible light, and S is the solar photon flux under AM1.5G light illumination. Based on the principles of energy conservation, the correlations of absorption rate (A), reflectivity (R), and transmittance (T) can be derived as formula (4):

$$A + R + T = 1 \tag{4}$$

Adjusting the propagation path of photons and thus tuning A, R, and T can improve the device performance, as the absorptivity, reflectivity, and transmittance of the stacked films significantly impact the final performance of ST-PSCs.

11.2.2 Carrier Kinetics

In an ideal photovoltaic device, photons are converted into electron-hole pairs. However, defects in the light absorption layer cause non-radiative recombination loss and open circuit voltage loss (V_{loss}). The carrier recombination dynamics process is expressed by formula (5):

$$\frac{d_n}{d_t} = -k_1 n - k_2 n^2 - k_3 n^3 \tag{5}$$

Where n is the photogenerated carrier density and k_1, k_2, and k_3 are the first-, second-, and third-order rate constants of Shockley–Read–Hall defect-assisted (monomolecular), radiative (bimolecular), and Auger (three-body) recombination processes, respectively. Nonradiative recombination losses mainly result from Shockley–Read–Hall defect-assisted recombination and Auger recombination.

PCE is a key index to evaluate the photoelectric conversion capability of solar cells, which can be expressed as (6)[43]:

$$PCE = \frac{P_{max}}{P_{in}} = \frac{FF \times J_{sc} \times V_{oc}}{P_{in}} \tag{6}$$

where P_{max} and P_{in} are the maximum output power and incident light power, respectively. FF is the fill factor, J_{SC} is the short-circuit current, and V_{OC} is the open-circuit voltage. Jsc is an important parameter to characterize the current output, which can be expressed as (7):

$$J_{sc} = \int qQE(E)b_s dE \tag{7}$$

where q is the elementary charge (q = 1.6 × 10^{-9} C). b_s is the incident light intensity, which describes the number of solar radiation photons within the energy range between E and E + dE per unit time and unit area. Quantum efficiency (QE) depicts the probability that the carriers excited by a photon with energy (E) entering an external circuit. FF is mainly affected by the series resistance (R_s) and shunt resistance (R_{sh}) of the entire device, which is an important index related to the quality of photovoltaic devices. Reducing R*s* or increasing R_{sh} enhances FF.

Open circuit voltage (V_{oc}) is an important index to evaluate the energy loss of devices, expressed as (8):

$$V_{oc} = \frac{KT}{q} In\left(\frac{J_{sc} N_D \tau_{eff}}{q n_i^2 d}\right) \tag{8}$$

where k is the Boltzmann constant, T is the temperature, q is the elementary charge, J_{SC} is the short-circuit current density under AM1.5 light, J_0 is the initial current density, N_D is the doping concentration, τ_{eff} is the radiative recombination lifetime of photogenerated carriers, n_i is the intrinsic carrier concentration, and d is the thickness of the absorption layer. It is worth noting that τ_{eff} as a key parameter exhibits a proportional correlation with V_{OC}, having a predominant impact on the device performance. The total energy loss (E_{loss}) of PSCs is expressed by the following formula (9):

$$E_{loss} = qV_{loss} = E_g - qV_{oc} = \Delta E_1 + \Delta E_2 + \Delta E_3 \tag{9}$$

where ΔE_1 is radiation recombination loss, ΔE_2 is thermodynamic loss, and ΔE_3 is nonradiative recombination loss. ΔE_1 and ΔE_2 mainly depend on the intrinsic material properties, while ΔE_3 is the dominant source for V_{loss}. ST-PSCs typically have higher V_{loss} than opaque PSCs, improving the performance of ST-PSCs requires reducing V_{loss} while considering light transmittance. There is a seesaw pattern between *PCE* and *AVT*. To unify the evaluation indicators of ST-PSCs, researchers proposed *LUE*, which can be expressed in formula (10):

$$LUE = PCE \times AVT \tag{10}$$

11.2.3 Color Rendering

The transparent photovoltaics used in architecture, automobiles and other fields have specific requirements for visible light transmittance and color. The CRI and CIELAB color coordinates (a^*, b^*) are used to determine the device aesthetic standards. The color rendering index (CRI) is used to determine the range of visible light and evaluate the color hue that the human eye can perceive. CRIs are calculated based on ideal transmission profiles (step-functions) in combination with CIE 1976 three-dimensional (3D) uniform color space (CIELUV), CIE 1974 test-color samples, and correction for chromatic adaptation (non-planckian locus), when necessary. CRI LUE can be expressed by the following formula (11)[44]:

$$CRI = \frac{1}{8}\sum_{i=1}^{8}\left(100 - 4.6\sqrt{\left(\Delta L_i^*\right)^2 + \left(\Delta u_i^*\right)^2 + \left(\Delta v_i^*\right)^2}\right) \tag{11}$$

where ΔL_i^*, Δu_i^*, and Δv_i^* are the differences in lightness (L^*) and chromaticity coordinates (u^*, v^*) between each color sample, i (8 in total) "illuminated" with a fixed reference solar spectrum (AM1.5G) and the transmission sources (T(λ), AM1.5(λ)). The CRI index is used to find the visible range of ST-PSCs to maximize the efficiency of solar devices. A high CRI value indicates that the device has better color rendering ability. CRI is related to the device's structure, optical interference, and other factors. Researchers have used various methods to regulate the CRI of the device to improve the color rendering ability.

The CIE color coordinates are calculated from the corresponding transmitted light and expressed by the product of the incident spectrum and the transmission spectrum of the device.[45] The standard sunlight source D65 and AM 1.5G solar spectrum are usually selected as the reference light source (incident light) to determine the color parameters of ST-PSCs. The XY chromaticity diagram is designed specifically to represent the human eye's perception of color and the relative color relative to the reference lighting source. CIE color coordinates are used to evaluate the transparent color characteristics of ST-PSCs.

11.3 ST-PSCs FOR BIPV

Perovskite solar cells (PSCs) have gained significant attention due to their low manufacturing cost and high photovoltaic performance. The use of solar photovoltaics on buildings has become an imperative trend.[46] PSCs collect solar energy and convert it into electrical energy to fulfill the requirements of lighting and aesthetics. ST-PSCs feature lightweight construction, specific light transmittance and adjustable color,[13] making them suitable for application not only on traditional reinforced roofs but also on windows and facades.[47] However, the efficiency of ST-PSCs is lower than that of opaque PSCs. Improving the performance of ST-PSCs requires a certain trade-off between PCE and AVT. In addition to PCE and AVT, color also plays a pivotal role in influencing the suitability of ST-PSCs for applications within the construction industry.[48] Therefore, this section will discuss the regulation of translucent perovskite solar cells from three aspects: photon propagation management, dynamic carrier regulation strategy and color regulation to promote the better use of ST-PSCs in the field of BIPV.

11.3.1 Photon Propagation Management

11.3.1.1 Light Capture Engineering

Surface plasmon is an essential strategy for enhancing the light capture of ST-PSCs. When fast electrons pass through the metal, they will experience characteristic energy loss. These losses are attributed to the collective oscillations of free electrons defined as "plasma" by David Pines.[49] Surface plasmon, on the other hand, is an electromagnetic excitation that occurs at the interface between metal and dielectric materials. There are two mechanisms for improving the performance of perovskite through plasma nanostructures[50]: (i) far-field scattering effect. Surface plasmons capture and propagate light as a scattering element, which increases the effective optical length and the probability of light being absorbed by solar cells. (ii) Near-field antenna effect. This allows plasma particles to be coupled with semiconductors, amplifying the electromagnetic field and expanding their effective absorption cross-sectional area. Surface plasmon nanostructures have been applied in PSCs in three configurations: on the cell surface, in the semiconductor active layer, and at the bottom interface between the semiconductor and the metal (Figure 11.3a).[51] Each configuration improves the performance of solar cells through different physical principles.

The commonly used metals in plasma metals are gold (Au) and silver (Ag). Lie et al.[34] added gold nanoplasma between the perovskite layer and the functional layer. The resonance peak of gold nanoplasma overlapped with the perovskite bandgap, and tetraoctylammonium bromide (TOAB) ligand was added to prevent the agglomeration of Au nanoplasma (Figure 11.3b). The addition of Au nanoplasma reduces the formation of PbI_2 phase, thus passivating the surface of perovskite layer and enhancing the stability of device. ST-PSCs with PCE of 13.7% and AVT of 27% were prepared. Kim et al.[52] used electrode-coupled plasmons (ECPs) to improve the performance of ST-PSCs. ECPs are the plasmonic coupling between Ag nanoparticles (NPs) and a metal film coated as a top-electrode on the NPs with a thin intervening semiconductor layer between them (Figure 11.3c). By changing the particle size and the thickness of the semiconductor layer, the wavelength of the ECPs is adjusted in the deep red region, thereby improving the PCE and photocurrent of the ST-PSCs. By reducing the thickness of the perovskite layer and introducing ECPs, the PCE

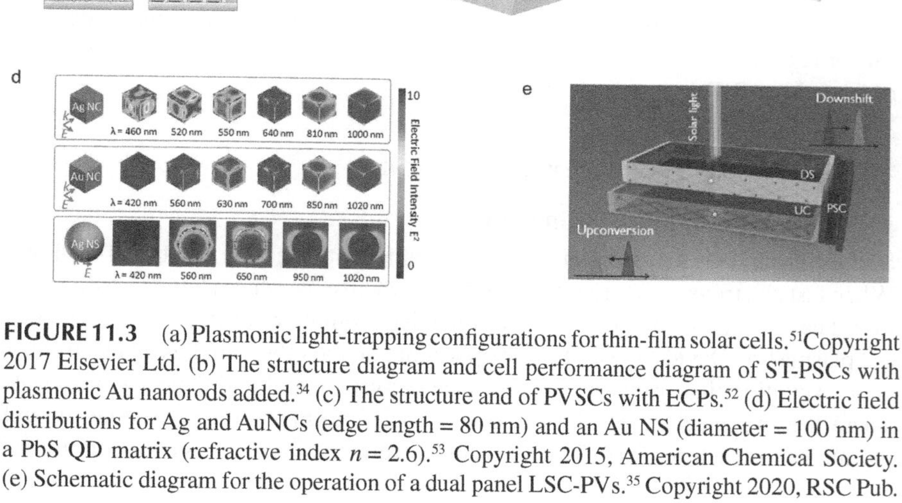

FIGURE 11.3 (a) Plasmonic light-trapping configurations for thin-film solar cells.[51] Copyright 2017 Elsevier Ltd. (b) The structure diagram and cell performance diagram of ST-PSCs with plasmonic Au nanorods added.[34] (c) The structure and of PVSCs with ECPs.[52] (d) Electric field distributions for Ag and AuNCs (edge length = 80 nm) and an Au NS (diameter = 100 nm) in a PbS QD matrix (refractive index $n = 2.6$).[53] Copyright 2015, American Chemical Society. (e) Schematic diagram for the operation of a dual panel LSC-PVs.[35] Copyright 2020, RSC Pub.

and visual transparency index (VTI) values of ST-PSCs were 6% and 28% higher than the initial values, respectively. The enhancement of thin-film solar cells by Ag nanoplasma depends on the oscillating electric field intensity near the metal nanoplasma. The electric field intensity depends on the metal type, size, shape, proximity of adjacent particles, and dielectric environment. Compared with the Ag nanoplasma structure of the sphere, the Ag cubic nanoplasma exhibits a strong electric field in a wider wavelength range, and its far-field light scattering effect is more effective in improving the performance of thin-film solar cells (Figure 11.3d).[53] Compared with precious metals such as Au and Ag, copper (Cu) has the advantage of low price. Rahman et al.[54] found that the Cu cubic nanoplasma has similar local field enhancement with the Ag nanoplasma, resulting in similar absorption enhancement and transparency. The use of Cu nanoplasma can effectively compensate for the decrease in light absorption performance caused by the decrease of active layer thickness. The 120-nm active layer with 70-nm Cu nanoplasma has the same light absorption and transparency characteristics as the 270-nm active layer without nanoplasma. Cu nanoplasma can be a substitute for Ag and Au plasma to enhance the performance of perovskite solar cells.

In addition to surface plasmon nanostructures, luminescent solar concentrators have shown excellent use in light capture. Solar concentrators made of upconversion (UC) and downconversion (DC) materials can convert infrared and ultraviolet light into photon energy effectively utilized by perovskites. Kim et al.[35] impregnated fluorescent dye molecules into polyurethane (PU) films to prepare solid films of DC and UC luminescent solar concentrator (LSC) plates (Figure 11.3e). The sunlight first enters the DC, and the dye molecules on the DC absorb 375–450-nm purple light and convert it into 450–550-nm blue light. The 635-nm red light entering UC is converted to blue light by the bimolecular organic dye on UC. The configuration of the DC/UC double-sided panel has the advantage of collecting dual wavelengths, and the direct and delayed fluorescence emission of the double LSC panel is incident on the PSC. Specifically, PL emission is guided to the LSC edge by total internal reflection, and then strong photons are collected by the attached PSC. Although up-down conversion materials can enhance the sunlight capture of ST-PSCs, research is still in its infancy and requires further study.

11.3.1.2 Photon Reflection Engineering

The incident light entering the perovskite absorption layer is inevitably limited by the Fresnel reflection at the air-substrate interface. This reflection of some photons is one of the reasons for the poor performance of ST-PSCs. To reduce light reflection, an anti-reflection (AR) layer can be added to the surface of the device. The AR film with different refractive indexes from the device increases the light transmission path and reduces light reflection through the optical interference principle, which promotes the efficiency and transmittance of ST-PSCs. In 2014, Bolink group[55] used a thin perovskite absorption layer, a thin metal electrode, and a 100-nm LiF AR layer to achieve PCE with an efficiency of more than 6% and ST-PSCs with a full device transparency of 30% (Figure 11.4a). The LiF AR layer shields the thin metal electrode layer while reducing the energy loss through the device's crystal surface reflection, maintaining the device's high transparency and high J_{sc}. In addition to

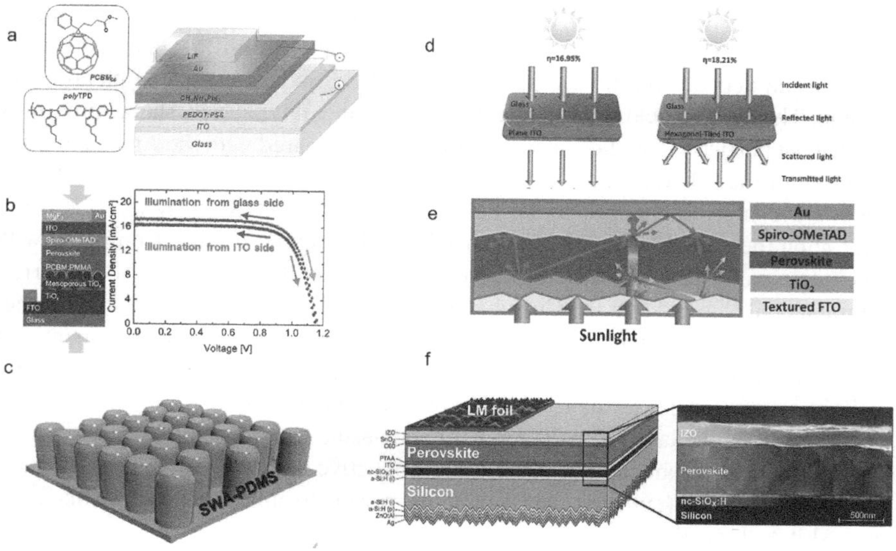

FIGURE 11.4 (a) Schematic layout of the semitransparent solar cell and chemical structures of the organic hole and electron blocking materials.[55] Copyright 2014, RSC Publishing. (b) Schematic structures of the perovskite solar cells with a MgF_2 antireflection coating and no contact grid in the active cell area.[56] (c) Schematic of the SWA-PDMS.[57] Copyright 2018, Royal Society of Chemistry. (d) Plane ITO and hexagonal-tiled ITO light absorption diagrams.[58] Copyright 2018, American Chemical Society. (e) Schematics of possible incident light paths within perovskite solar cells with textured substrate. The concept of light trapping in solar cell is illustrated by the arrows representing incoming and scattered sunlight.[59] Copyright 2017 Elsevier B.V. (f) Tandem solar cell device schematics of the experimentally realized architecture and SEM cross section image of the top cell with the layers as indicated.[60] Copyright 2018, RSC Publishing.

the LiF AR layer, the MgF_2 AR layer has also received extensive attention due to its excellent transmittance in a wide spectral range (120 nm–750 nm). In 2019, Bett et al.[56] added a MgF_2 AR layer to the ITO layer, effectively reducing the current loss of the ITO layer. The device's efficiency using MgF_2 AR layer is increased to 13.4% (Figure 11.4b). Some organic materials can also be used as AR layers to reduce light reflection. Dudem et al.[57] proposed to use a subwavelength architecture of polydimethylsiloxane (SWA-PDMS; Figure 11.4c) as the AR layer of organic solar cells. The light transmission rate and efficiency of the device are greatly improved. The AR layer has outstanding chemical properties and corrosion resistance, which can protect the device from damage in the external environment.

The texture structure is another important strategy to improve the light absorption performance of ST-PSCs by adjusting the reflection/propagation ratio of the incident light on the texture surface. The rough texture surface makes the incident light reflect multiple times in the active layer, prolonging the optical path length and promoting the reflected light to re-enter the active layer. Huang et al.[58] found that ST-PSCs have a higher efficiency using textured substrate structure. They have prepared hexagonal indium tin oxide (ITO) substrates using polystyrene templates and sputtering techniques. The hexagonal ITO substrate changes the transmission path of light and produces scattered light, reducing optical loss (Figure 11.4d). Compared with the traditional ITO substrate, the J_{SC} and PCE of PSCs prepared with hexagonal ITO substrate increased by 7.2% and 7.4%, respectively. Shi et al.[59] used textured fluorine-doped SnO_2 (FTO) substrates to prepare PSCs. The rough FTO surface has strong light scattering, promoting stronger absorption of light by the absorption layer (Figure 11.4e). The rough FTO surface also promotes the growth of high-quality crystallization of the perovskite layer. Compared with the smooth substrate, the J_{SC} and PCE of the thin PSC based on textured substrate are increased by 14.5% and 22%, respectively. Jost et al.[60] added a layer of textured light management foil to the perovskite/silicon tandem device (Figure 11.4f). The light management foil enhances the light capture of the bottom layer through the scattering effect and reduces the light reflection. The efficiency of the stacked device with textured light management foil is significantly increased from 23.4% to 25.5%. AR layers and texture structures are effective strategies to improve light capture, not only in ST-PSCs but also in tandem devices. However, the type of AR layer and the precipitation in the upper layer of the texture structure hinder their wide application. Further research is needed.

11.3.1.3 Transparent Electrode Engineering

When applied to building photovoltaics, ST-PSCs require not only a certain level of PCE but also a minimum level for AVT, which is typically between 20% and 30% for photovoltaic windows. Therefore, it is necessary to optically control the substrate of the electrode, active layer, and interfacial layer to meet the device's requirements for AVT. The transparent electrode (TE) is an important component of ST-PSCs that influences both PCE and AVT.[61] A good TE should have high transparency and conductivity, excellent stability, low cost, and be well-matched with adjacent layers in the manufacturing process and energy band. At present, thin-layer metal electrodes, conductive metal oxides, nanomaterials, graphene, and conductive polymers are mainly studied.

Ultrathin metal films have garnered significant interest thanks to their remarkable electrical conductivity, high light transmittance, straightforward deposition procedures, and robust bonding properties. However, they do face certain challenges. One primary concern is their vulnerability to damage from ion bombardment during the sputtering deposition process. Additionally, the smooth surfaces of these metal films can lead to optical losses, and there is a propensity for the formation of discontinuous films when applied to high-performance substrates. Furthermore, it's crucial for ultrathin metal electrodes to be carefully matched to the underlying energy levels to prevent the creation of barriers that could impede the transmission of charge carriers. Therefore, optimizing the ultrathin metal electrode is key to improve ST-PSC performance. Researchers often add interface layers on both sides of the metal electrode, such as MoO_x and PEIE (polyethyleneimine ethoxylate),[62] to optimize its energy level and act as a reflective layer to capture more light and improve device performance. For example, Ying et al.[63] prepared a $Zr(acac)_4$/PEI (polyetherimide)/Ag/Ta_2O_5 ultrathin Ag metal electrode (Figure 11.5a). $Zr(acac)_4$/PEI was the interface between the functional layer and the electrode, optimizing the energy level of the electrode (with

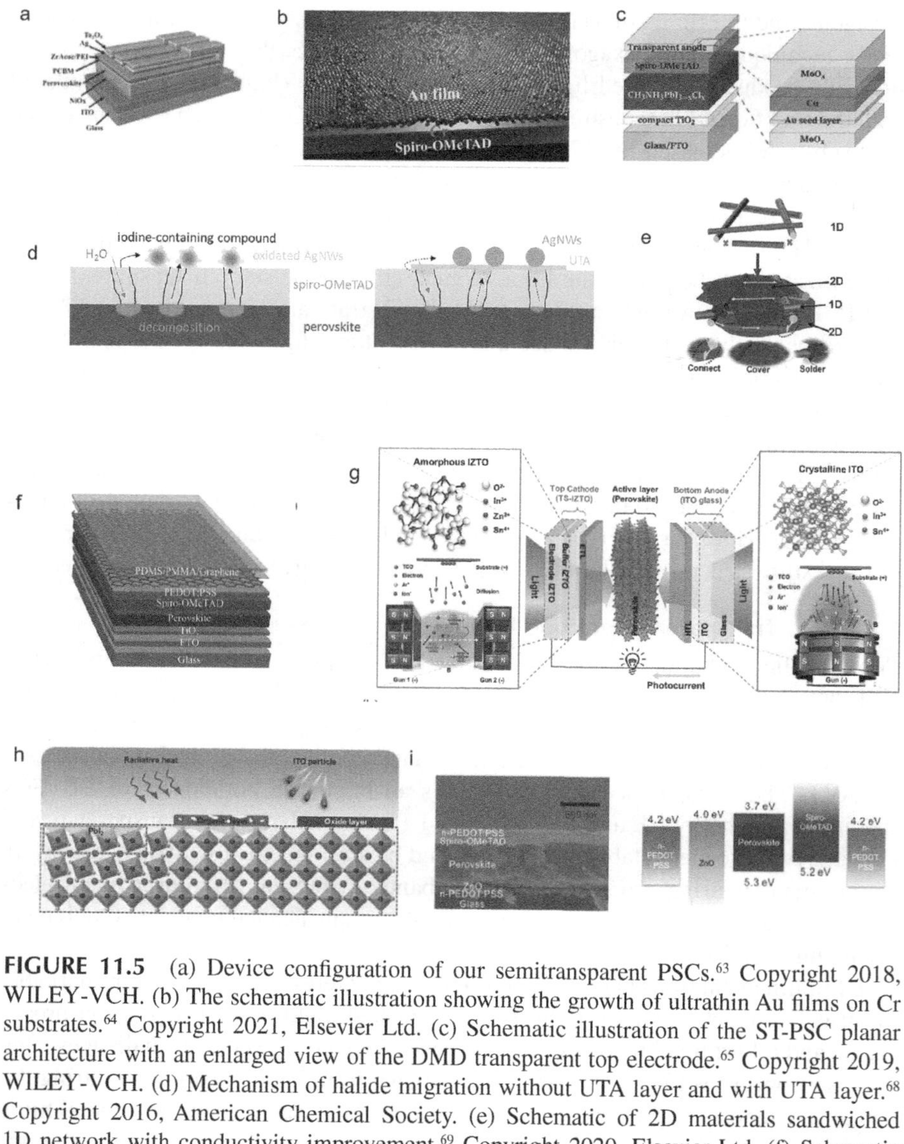

FIGURE 11.5 (a) Device configuration of our semitransparent PSCs.[63] Copyright 2018, WILEY-VCH. (b) The schematic illustration showing the growth of ultrathin Au films on Cr substrates.[64] Copyright 2021, Elsevier Ltd. (c) Schematic illustration of the ST-PSC planar architecture with an enlarged view of the DMD transparent top electrode.[65] Copyright 2019, WILEY-VCH. (d) Mechanism of halide migration without UTA layer and with UTA layer.[68] Copyright 2016, American Chemical Society. (e) Schematic of 2D materials sandwiched 1D network with conductivity improvement.[69] Copyright 2020, Elsevier Ltd. (f) Schematic diagram of a semitransparent perovskite solar cell.[70] Copyright 2015, WILEY-VCH. (g) Schematic structure of semi-transparent PSC with crystalline ITO anode grown by DC magnetron sputtering and amorphous IZTO cathode grown by cosputtering of ITO and IZO targets using linear FTS.[71] Copyright 2020, Elsevier Ltd. (h) A schematic illustration of the mechanism for damage-free semi-transparent perovskite solar cells.[72] Copyright 2022, Wiley-VCH. (i) SEM cross-sectional image (left) and energy-level diagram (right) of the TCO-free ST-PSC.[73] Copyright 2017, WILEY-VCH.

a work function of −4.04 eV). The Ag layer deposited on PEI exhibits low roughness and continuity, Additionally, the Ta_2O_5 coating with high refractive index (n = 2.1) significantly improved the transmittance of the whole device in the red infrared region, increasing the device's transmittance from 12.47% to 43.75%.

The metal island has local surface plasmon resonance (SPR), leading to a decrease in the transmittance of the device compared with the ideal thin gold metal film. To obtain continuous and uniform thin metal electrodes, metal seeds are often introduced to promote the continuous growth of metal films. For example, Yang et al.[64] proposed the use of cromium (Cr) as the seed layer with large surface energy (2.62 J/m^2) to induce the growth of continuous ultrathin gold layers. By inducing the growth of the gold layer (with a surface energy of 1.51 J/m^2) according to the Frank-van der Merwe principle, Au forms a continuous ultra-thin film with good transmittance and conductivity (Figure 11.5b), improving the PCE of the ST-PSCs up to 19.8%. Cu metal can also be used as a TE in ST-PSCs, but it tends to self-aggregate and diffuse into dielectric thin-film oxides, which leads to instability. Giuliano et al.[65] found that using gold as a seed layer can address this problem by acting as an effective wetting inducer and diffusion barrier to prevent the diffusion of Cu to oxides and the formation of continuous Cu metal thin layers. They prepared ST-PSCs with MoO_x/Au-seed/Cu/MoO_x as electrodes (Figure 11.5c), and the efficiency reached 12.5%.

The thin layer metal electrode has attractive advantages, while its shortcomings also promote researchers to continuously explore new TE to meet the requirements of device development. Ag nanowires (Ag NWs) electrodes have excellent photoelectric and mechanical properties and can be used as transparent electrodes for flexible devices. However, it's challenging to balance the relationship between haze, transmittance, and conductivity of the electrode to achieve high device efficiency. Some new preparation techniques, including rapid radial electrochemical etching[66] and inkjet printing,[67] have been utilized to prepare Ag nanowire electrodes and achieve a balance between PCE and AVT of the device. Direct deposition on the hole transport layer will lead to the decomposition of the Ag NWs electrode, and the halide in the perovskite could migrate to the electrode interface and irreversibly react with the electrode, leading to the formation of silver halide. To solve this problem, Dai et al.[68] inserted an Au layer between the Ag NWs electrode and the hole transport layer. As a physical separation layer, the intermediate layer prevents unnecessary electrode degradation and significantly enhances Ohmic contact. The Au layer also prevents external water molecules from entering the perovskite layer, inhibiting the decomposition of the perovskite layer (Figure 11.5d). The overall PCE measured under front illumination is more than 11%, while the PCE measured under back illumination is 7.53%. Dou et al.[69] used two-dimensional (2D) graphene and one-dimensional (1D) Ag NW materials to fabricate 2D/1D/2D electrodes (Figure 11.5e). The graphene layer blocked the diffusion of I ions; improved the blocking of iodide ions; prevented the corrosion of electrodes in water, oxygen and iodine; and enhanced the efficiency and stability of the device. A PCE of 14.69 % was obtained (increased from 12.28%) with an AVT of 20.11%.

Graphene materials, due to their high carrier mobility, excellent photoelectric properties, mechanical flexibility, and convenient preparation, show great potential as transparent electrodes for ST-PSCs. In 2015, You et al.[70] introduced multilayer graphene stacks (PDMS (poly(dimethylsiloxane))/PMMA (poly(methyl methacrylate))/graphene) prepared by chemical vapor deposition as transparent electrodes for ST-PSCs, attaining a PCE of 11.65% (Figure 11.5f).

Transparent conductive oxides, such as indium-doped tin oxide (ITO), aluminum-doped zinc oxide (AZO), and zinc oxide indium (IZO) are frequently used TEs in ST-PSCs due to their environmental stability, uniform structure, and good interface contact. However, the typical magnetron sputtering process used to prepare these TEs can cause damage to the active layer at the bottom, thereby limiting their use. Sang-Hwi Lim et al.[71] prepared an InZnSnO (IZTO) electrode (Figure 11.5g) by a two-step linear facing target sputtering (FTS) process that resulted in a high-quality transparent electrode without any plasma damage. The fabrication process was composed of the slow sputtering buffer IZTO layer and the fast sputtering dense IZTO layer. Thus, a thick and dense amorphous IZTO electrode layer was obtained without grain boundaries, preventing water and air from diffusing from the surface to the active layer. The device with this electrode prepared by FTS process achieved a PCE of 15.72%, which was significantly improved compared to the device with an ITO electrode deposited by magnetron sputtering (PCE = 3.34%). NiO_x nanoparticle coating layer can be used a protective layer for the deposition of TEs ST-PSCs. Jeong et al.[72] introduced a p-type nickel oxide (NiO_x) nanoparticle coating layer on the perovskite layer, which serves as both a hole transport layer and a buffer layer of the electrode (Figure 11.5h). The addition of this NiO_x buffer layer prevented the sputtering damage during the deposition of the ITO transparent electrode, increasing the PCE of ST-PSCs to 19.5%.

The transparent electrode of ST-PSCs is one of the key factors in adjusting the PCE and AVT. However, the preparation of a suitable TE often requires specific vacuum techniques and time-consuming etching technology. Therefore, TE prepared using simple and inexpensive techniques is desirable. Zhang et al.[73] prepared highly conductive PEDOT:PSS as the electrode material for ST-PSCs using a low-temperature and complete solution treatment (Figure 11.5i). This electrode material has reasonable light transmittance and can be prepared via printing technology, achieving a PCE of 13.9% and making it a viable option for large-scale production of ST-PSCs.

TE is a critical component of ST-PSCs that impacts the efficiency and transmittance of the device, constituting a significant portion of the manufacturing cost of ST-PSCs. While transparent conductive oxides are commonly used TEs, recent studies have explored other materials such as graphene, Ag NWs, NiO_x, and PEDOT:PSS. Future research should focus on finding more TE materials with simple preparation processes and excellent photoelectric properties to promote the development of ST-PSCs.

11.3.2 Carrier Kinetics Regulation

LUE is an important indicator for measuring the performance of ST-PSCs and represents the balance between PCE and AVT. To achieve large-scale application of photovoltaic materials in the construction field, ST-PSCs need to maximize their LUE. However, the recombination process of photogenerated carrier in the device limits the improvement of the LUE. The carrier dynamics control strategy is an effective approach for improving LUE. By passivating the defects in the perovskite layer and optimizing the energy level alignment between the functional layers of ST-PSCs, it is possible to reduce the non-radiative recombination of photogenerated carriers, to improve light utilization rate of the device, and to enhance the efficiency of ST-PSCs. Therefore, the progress of carrier dynamics regulation strategy is summarized in the following section.

11.3.2.1 Perovskite Layer Regulation

Additive engineering is one of the common strategies for defect passivation of perovskite layers. The additive interacts with Pb^{2+} in the perovskite layer and Sn^{4+} in the electron transport layer to passivate the defects of the perovskite functional layer. This promotes charge transfer between the perovskite layer and the functional layer, reducing the non-radiative recombination at the interface, and obtaining ST-PSC with excellent performance. Gunes et al.[74] used thieno[3, 2-b]thiophen-2-ylmethanaminium iodide (TTMAI) to treat the surface of FAMA-based perovskite (Figure 11.6a). The low concentration of TTMAI tends to form a 2D structure on the 3D surface, while the high concentration of TTMAI exists on the 3D perovskite surface in the form of a salt. There is a strong π-π and intermolecular interaction between sulfur atoms in TTMAI, which promotes hole transport between the perovskite layer and the hole transport layer, thereby improving the FF of the device. The as-prepared ST-PSCs exhibited a PCE of 17.9%. Passivating the defects of the buried interface of the perovskite layer using derivatives with functional groups is another effective approach to improve the performance of the PSC. Wang et al. [75] used three derivatives with functional Lewis base groups to passivate the defects at the buried interface. The catechol group in 3, 4-Dihydroxyphenylalanine (DOPA) can react with the uncoordinated Sn atom in SnO_2, while the carboxyl and amino groups on the other side of the molecule react with the uncoordinated Pb^{2+} in the perovskite layer, realizing the bidirectional defect passivation between the buried interface of the perovskite layer and the SnO_2 layer (Figure 11.6b). Dong et al.[76] introduced chlorobenzenesulfonic potassium salts as a passivation agent for the buried interface. The as-fabricated ST- PSCs demonstrated an efficiency of 12.83% and an AVT of over 27%. Zhang et al.[77] used derivatives with pyridine groups 1-propyl-[4, 4'-bipyridin]-1-ium iodide (BiPy-I) as a passivator (Figure 11.6c). The pyridyl group passivates the Pb-I antisite defect through electrostatic interaction, and BiPy-I helps to enhance the transparency of the device. The PCE of the prepared inverted ST-PSC is increased by 11.74% at 23% AVT and maintains approximately 91% of the original efficiency after 240 h storage in an ambient atmosphere.

High-quality crystalline thin films promote carrier transport and improve device performance. The ion interaction between the additive and the perovskite can regulate the process of crystallization growth in the perovskite film. Dokkhan et al.[78] used submicron sponge polymer colloidal particles of poly(N-vinyl formamide-co-2-(N-vinylformamido)ethyl ether) (PNVF-NVEE) as additives, which delayed the crystallization process, improved the film coverage, and prepared high-quality films with minimal pinholes. Yang et al.[36] added $MAPbBr_3$ to the perovskite precursor solution, changing the colloid type of the nucleation site during the film crystallization process, increasing the nucleation size and grains size of the

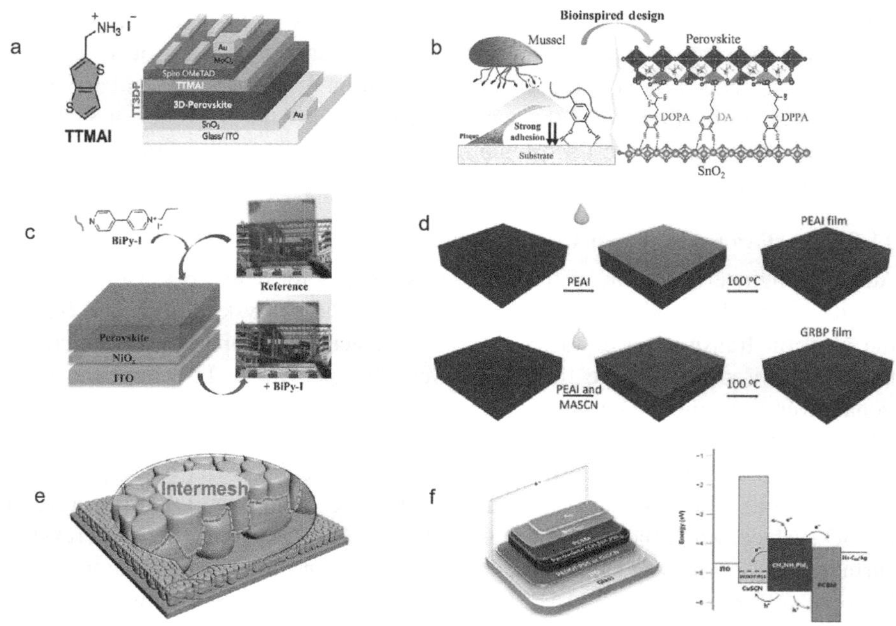

FIGURE 11.6 (a) Schematics of the thieno[3,2-b]thiophen-2-ylmethanaminium iodide (TTMAI) structure and illustration of the solar cell architecture used in this study.[74] Copyright 2021, Wiley-VCH. (b) Schematic illustration of mussel-inspired design of DOPA, DA and DPPA as crosslinkers between the SnO_2 and perovskite buried interface.[75] Copyright 2021, part of Springer Nature. (c) Graphs of the semi-transparent N-methyl pyrrolidone processed perovskite films with/without BiPy-I modification.[77] Copyright 2020, Elsevier B.V. (d) Schematics of the two posttreatment processes preparing the wide-bandgap perovskite films.[79] Copyright 2022, Wiley-VCH. (e) Schematic illustration of the combination process of NP-SnO_2 and Col-SnO_2 via the intermeshing SnO_2 ETL strategy.[81] Copyright 2020, Royal Society of Chemistry. (f) Device architecture of PHJ PVSCs fabricated in this study (left) and corresponding energy band diagram relative to vacuum level (right).[81] Copyright 2015 Wiley-VCH.

film, and reducing the presence of grain boundaries and defects. As a consequence, ST-PSCs with high-quality ultra-thin films were obtained, featuring an increased FF from 70% to 75% and achieving a PCE of 18.27% and an AVT of 26.23%. Liu et al.[79] used methyl ammonium thiocyanate (MASCN) and phenethyl ammonium iodide (PEAI) as post-treatment additives in the film to reduce the non-radiative recombination loss of the bulk and interface of the perovskite layer (Figure 11.6d). The MASCN additive promotes the growth of the film grains and the crystal preferred orientation. PEAI exists in the bulk phase of the perovskite film and the interface with HTL, passivating the film defects. The ST-PSC exhibited a PCE of 19.9% (V_{OC} = 1.189 V, J_{SC} = 20.3 mA cm^{-2}, FF = 82.0%).

Overall, these additives and methods have helped to improve the efficiency and stability of ST-PSCs, making them a promising candidate for future renewable energy applications.

11.3.2.2 Regulation of Functional Layer

The non-radiative recombination in ST-PSCs is influenced by the energy level mismatch at the contact interface between the hole transport layer (HTL) or electron transport layer (ETL) and the perovskite film. Therefore, the selection and optimization of ETL and HTL are crucial to enhance the performance of ST-PSCs. A functional layer with an ideal surface will enhance the uniform density and low-defect growth of the perovskite layer. At the same time, the favorable energy level alignment between the functional layer and the perovskite layer will promote charge transport and suppress charge recombination. Researchers have explored various ways to optimize the functional layer of ST-PSCs. For instance, Xue et al.[80] used modified NiO_x as an efficient hole transport layer and fullerene diadduct C_{60} (CH_2) (Ind) to replace the commonly used PCBM as ETL. The addition of dihydromethane and indene groups improved the energy level of C_{60}, while NiO_x promoted high smoothness of perovskite films. Consequently, high-performance ST-PSCs with AVT of 21.5% and PCE of 12.6% were obtained. Liu et al.[81] combined nano-SnO_2 with colloidal SnO_2 to construct an intermeshing SnO_2 (ImSnO_2) ETL with high transmittance in the visible region (Figure 11.6e). The intermeshing structure of ImSnO_2 realized the smooth surface of SnO_2, which provides conditions for the growth of a compact perovskite layer and a large grain film, resulting in a good energy level alignment with the perovskite layer to enhance electron transport. The improvement of charge extraction efficiency and the inhibition of interface recombination substantially reduced the E_{loss} of $CsPbI_2Br$ PSCs to less than 0.6 eV, while increasing the V_{oc} to as high as 1.31 V. CuSCN exhibits high hole mobility (0.1 $cm^2\ V^{-1}\ s^{-1}$), suitable energy levels ($E_{VB} = -5.4$ eV and $E_{CB} = -1.8$ eV) and excellent transparency in the UV-vis-NIR spectrum, making it an attractive material for HTL of ST-PSCs (Figure 11.6f). However, the surface of CuSCN prepared by the solution method is rough, thus affecting device performance. Jung et al.[82] used diethylsulfide (DES) as a solvent to obtain a thin and compact film through the spin-coating process, leading to ST-PSC with an AVT of 25% and a PCE of 10%. Sun et al.[83] optimized the p-conductivity in CuSCN HTL by using butylammonium iodide as an additive, promoting the formation of Cu^{2+} ions, thereby improving the conductivity of the HTL. The additive improves the solubility of the CuSCN precursor solution, resulting in a uniform coverage of the perovskite layer and a PCE of 19.24% for ST-PSCs.

Apparently, regulating the functional layer is essential for achieving high-performance ST-PSCs. By controlling the thickness, composition, and morphology of the functional layers, it is possible to optimize the optical and electrical properties of the cells, in turn enhancing their performance. Moreover, regulating the functional layer can also impact the transparency and color of the cells, which is vital for applications. Therefore, careful control over the functional layer is essential for the development of efficient and stable semi-transparent perovskite solar cells that are suitable for practical applications in various fields.

11.3.3 Color Control

Perovskite films that are several hundred nanometers thick can fully absorb light energy, convert it into electrical energy, and attain high energy conversion efficiency.

However, their dark color may not meet the aesthetic requirements of the BIPV field. Researchers have attempted to reduce the thickness of the perovskite layer to obtain lighter brown and red devices, but it remains difficult to meet the color specifications needed for real-life building materials. To address this issue, there are two main approaches to controlling the color of ST-PSCs. First, the color of perovskite devices can be changed by regulating the perovskite materials themselves. This involves modifying the chemical composition or doping of perovskite materials to alter their optical properties and achieve the desired color. Second, physical structures such as photonic crystals and optical microcavities can be utilized to adjust the color of perovskite devices. These structures can control the reflection, transmission, and absorption of light in the perovskite layer, resulting in a specific color. Overall, the development of ST-PSCs with vibrant colors, excellent transparency, and high efficiency is a key objective in the BIPV field.

11.3.3.1 Regulation of Perovskite Materials

Unique characteristics of perovskite materials, such as adjustable bandgaps, distinctive compound arrays, and phase transitions are closely related to the observed color changes in perovskite films. Various studies have investigated these color changes and their underlying mechanisms. Wheeler et al.[84] found that the dissociation of CH_3NH_2 induced by low formation/dissociation energy of the methylammonium lead iodide-methylamine complex ($CH_3NH_3PbI_3{\cdot}xCH_3NH_2$) during photo-heating caused perovskite devices to change from a transparent state (with an AVT of 68%) to a colored state (Figure 11.7a). The PCE of the device in the colored state is as high as 11.3%. Halder et al.[85] found that $MAPbI_3$ undergoes a phase transition from tetragonal to cubic phase accompanied by a change in the perovskite bandgap and film color when hydrated or heated (Figure 11.7b). Noh et al.[86] demonstrated that the bandgap (from 1.58 eV to 2.28 eV) and film color (from dark brown to brown/red to yellow) of $MAPbX_3$ can be regulated by changing the content of Br and I elements (Figure 11.7c). Jiang et al.[87] prepared color devices by adjusting the reflection spectrum of perovskite films using various thicknesses of transparent conductive PEDOT:PSS electrode material. A series of devices with bright colors were fabricated by coherently superposing of the reflected and transmitted electromagnetic fields among the electrode material, the hole transport layer, and the perovskite layer (Figure 11.7d). Although the color of the device can be adjusted by regulating the perovskite material itself, altering the perovskite material to achieve specific colors may affect device efficiency. While there are few studies in this field, more research on color control is being conducted on photonic crystals.

11.3.3.2 Physical Structures

Photonic crystals and dielectric microscopes stacked with materials with different refractive indices show wavelength-dependent reflectivity as a function of their structure and material refractive index. The light interference effect between materials with alternating refractive index between photonic crystals makes the film show different colors. Photonic crystals can be mechanically stacked on thin-film devices without sacrificing device performance, compared with bandgap control strategies to change color. Their optical response is controlled by varying the thickness, number

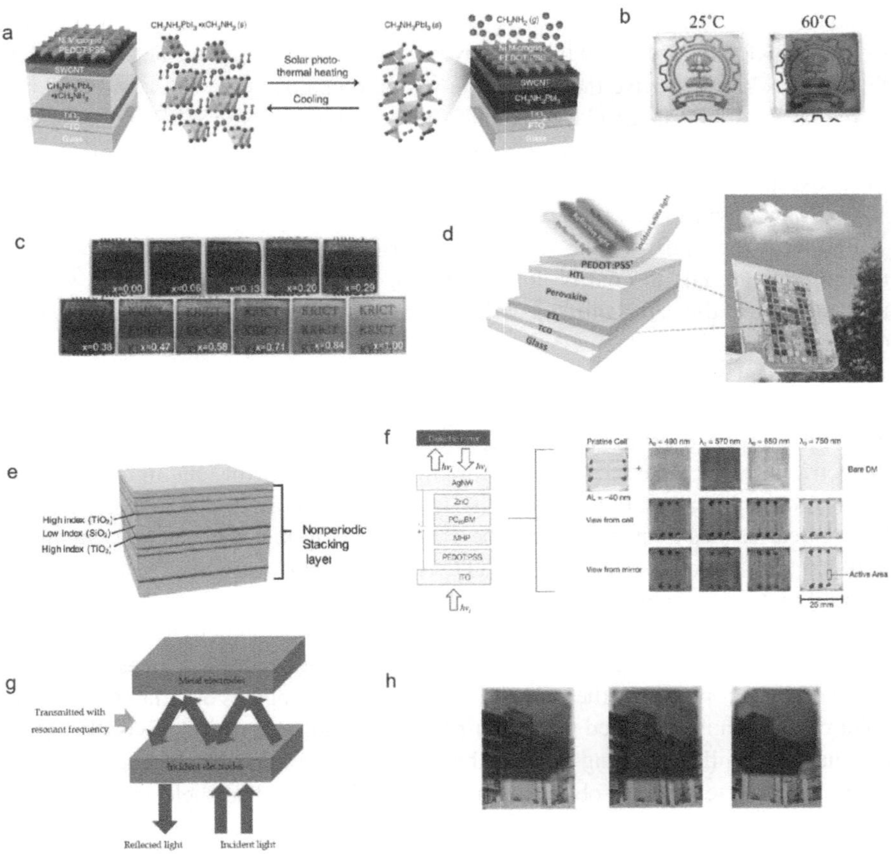

FIGURE 11.7 (a) Schematic of PV window device architecture and switching process.[84] (b) The corresponding image of the thin film of dihydrated perovskite at 25°C and at 60 °C is presented.[85] Copyright 2015, American Chemical Society. (c) Photographs of 3D TiO_2/$MAPb(I_{1-x}Br_x)_3$ bilayer nanocomposites on FTO glass substrates.[86] Copyright 2013, American Chemical Society. (d) Device architecture of the cells with conducting polymer PEDOT:PSST as the top electrode where the PEDOT:PSST means the PEDOT:PSS electrode was prepared by transfer printing and photographic image of a colored schematic "H" assembled by colorful perovskite solar cells.[87] Copyright 2016, American Chemical Society. (e) Schematic diagram of the structure of NBRF formed by nonperiodic stacking of alternatively high-refractive index (TiO_2) and low-refractive-index (SiO_2) layers.[88] Copyright 2019, American Chemical Society. (f) Digital images of a pristine cell with a 40-nm-thick mixed halide perovskite (MHP) as active layer (AL), four bare dielectric mirrors, and a full-device stack comprising perovskite solar cells and dielectric mirrors, including a view from the solar cell and from the mirror and total transmittance spectra of pristine cells and full stack including dielectric mirrors.[89] Copyright 2016, American Chemical Society. (g) Optical microcavity configuration in solar cells.[89] (h) Images of fabricated colored PV devices. The background image can be clearly seen through our fabricated samples with various colors.[92] Copyright 2016, American Chemical Society.

of layers, and refractive index. Yoo et al.[88] prepared narrow-bandwidth reflective filters (NBRFs; Figure 11.7e) by multilayering TiO_2 with a high refractive index and SiO_2 with a low refractive index in a non-periodic manner. The order and thickness of the layers affect the displayed colors. Compared to periodic NBRFs, ripple-free, aperiodic NBRFs provide a wider color gamut and higher color purity, with a small PCE loss and enhanced device stability. The PCE of red, green, and blue devices is 18.0%, 18.6%, and 18.9%, respectively, compared to that of the black control group of 20.1%. Ramírez Quiroz et al.[89] created dielectric mirrors using high and low refractive index materials to control color of PSCs. Provskite active layer with variable thicknesses and dielectric mirrors with different wavelengths provides specific colors (Figure 11.7f).

The optical microcavity structure (Figure 11.7g) has two planar electrodes, which reflect the incident light and induce the microcavity resonance in the intermediate medium to produce strong interference.[90] Changing the thickness of the intermediate dielectric layer regulates the color of the film without affecting the light absorption and photocurrent generation of the active layer.[91] The optical microcavity is an excellent optical strategy for regulating the color of perovskite devices and enhancing light capture. Lu et al.[92] used an Ag/ITO/Ag microcavity structure to obtain films with vibrant colors ranging from red-orange to blue (Figure 11.7h). The PCE of the yellow ST-PSCs is 7.2%. Lee et al.[93] introduced a phase compensation microcavity structure consisting of Ag/SiO_2/ZnS/Ag/ZnS, which reduced the devices, sensitivity of the optical properties to the angle. When the incident angle is 60°, the transmission color of the film is generated by collecting a very small proportion of visible light in the sunlight, while capturing most of the light in the visible light for the absorption of the perovskite layer to obtain high device efficiency. The dielectric functional layer is placed on the traditional optical microcavity to achieve impedance matching, thereby enhancing the efficiency of translucent color. The device efficiency for red, green and blue color reaches 10.47%, 10.66%, and 11.18%, respectively.

In conclusion, various structures such as photonic crystals, dielectric microscopes, and optical microcavities can regulate and enhance the color of ST-PSCs. These structures offer a wide range of colors, high color purity, and improved efficiency compared to traditional methods.

11.4 ST-PSCs FOR TSCs

The tandem device is a type of solar cell that consists of two subcells connected in series, each with different band gaps. The top cell has a larger band gap of approximately 1.5–1.9 eV, while the bottom cell has a smaller band gap of about 0.9–1.3 eV. There are two types of tandem devices, namely the 2-terminal (2-T) and 4-terminal (4-T) structure (Figure 11.8a, b).[94] The 4-T structure is relatively simple to manufacture and operate, but it has higher costs due to the additional two electrodes and the resulting optical loss. On the other hand, the 2T structure has fewer electrodes, but it requires an intermediate interconnect layer that is crucial for the efficiency of the device. Current matching should also be established between the subcells in series. Perovskite material is an ideal material for subcell in tandem devices because of its adjustable band gap and fast PCE growth. Tandem solar devices composed of

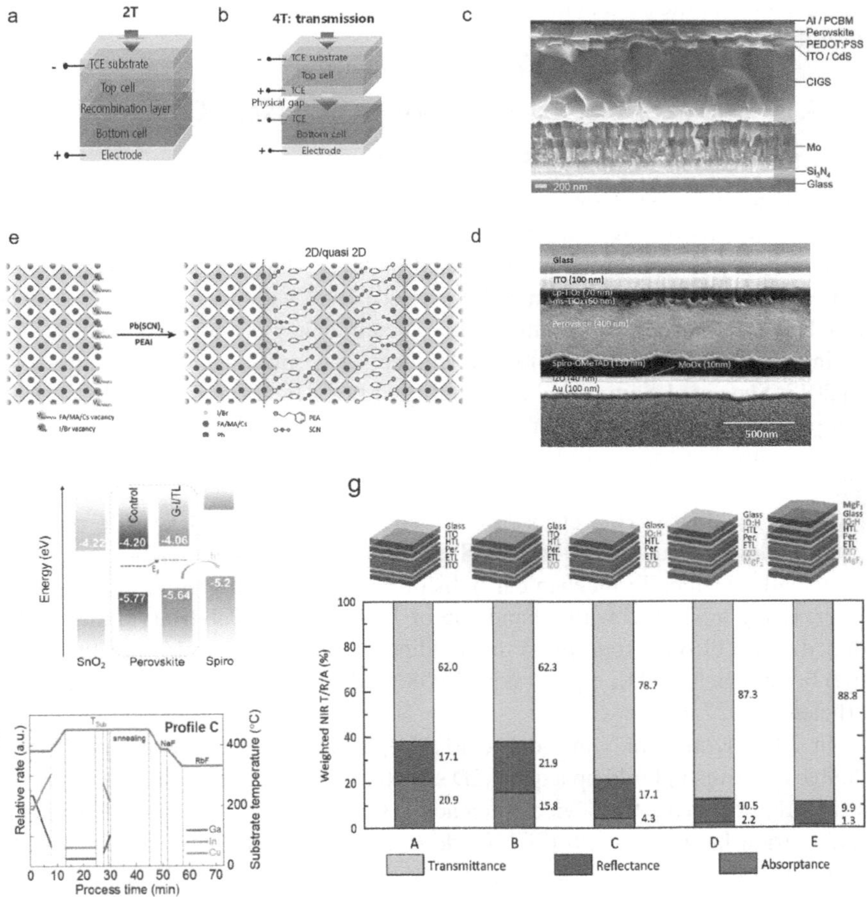

FIGURE 11.8 Series tandem solar cells: (a) 2T, (b) 4T transmission.[94] Copyright 2020, Wiley-VCH. (c) SEM image of the perovskite-CIGS tandem solar cell structure, Al/PCBM/perovskite/PEDOT:PSS/ITO/CdS/CIGS/Mo/Si_3N_4/glass.[31] Copyright 2015 Wiley-VCH. (d) SEM cross-sectional image of the semi-transparent cell based on $Cs_{0.05}Rb_{0.05}FA_{0.765}MA_{0.135}PbI_{2.55}Br_{0.45}$ with a bandgap of −1.62 eV.[95] (e) Schematic illustration of defect passivation at perovskite grain boundaries (or surfaces) by PEA^+ from PEAI and SCN^- from $Pb(SCN)_2$.[97] Copyright 2019, Elsevier Inc. (f) Energy level diagrams of control and G-I/TL treated films.[99] (g) Schematic illustration of semitransparent perovskite solar cells (PSCs) with various layer configurations.[100] Copyright 2022, Wiley-VCH. (h) Metal flux rates and sequences for a growth profile with Ga and Cu co-evaporated in the second step.[101]

perovskite materials and silicon, CIGS (Cu(In, Ga)Se_2), organic and perovskite solar materials have made plenty of progress. The development and challenges of each stacked device are summarized later.

However, there are still challenges in the development of each stacked device. For instance, the stability and durability of perovskite materials need to be improved to enhance the long-term performance of the tandem device. Moreover, further optimization of the current matching between the subcells is required to improve the overall PCE of the device.

11.4.1 Choice of Cell Configuration of the Tandem Cells

11.4.1.1 CIGS/Perovskite TSCs

CIGS material has a narrow band gap (1.1 eV), making it an ideal absorption material for bottom cells in TSCs. In 2015, Todorov et al.[31] changed the band gap range of perovskite materials through a gas-based halide exchange reaction reactor to prepare a perovskite/CIGS tandem cell device structure without the intrinsic ZnO layer of CIGS cell, as shown in Figure 11.8c. The efficiency of the TSCs reaches 10.9%, laying the foundation for future development of perovskite/CIGS tandem devices. In 2018, Shen et al.[95] improved the stability of PSCs by using a Cs/Rb/FA/MA four-ion perovskite absorption layer instead of the $MAPbI_3$ perovskite absorption layer, attaining an efficiency of 18.1% for ST-PSCs at the top of tandem cells (Figure 11.8d). The ST-PSCs and CIGS cells are mechanically stacked to form a TSC, achieving an efficiency of 23.9%. Han et al.[96] preserved the B-doped ZnO layer to protect the original device structure of the CIGS cell and deposited a sufficiently smooth ITO layer with chemically mechanical polishing to ensure a high-quality interconnect layer. Using a ST-PSC with a band gap of 1.59 eV and a Polytriarylamine (PTAA) doped with TPFB (4-isopropyl-4'-methyldiphenyliodonium tetrakis (pentafluorophenyl) borate) as an HTL, a perovskite/CIGS tandem cell was prepared, achieving an efficiency of 22.43%.

In 2019, Kim et al.[97] introduced PEAI (phenethylamine iodide) and $Pb(SCN)_2$ double additives to build up a quasi-2D structure (Figure 11.8e) with a 3D perovskite structure, improving the device efficiency. As a 2D cationic additive, PEAI can passivate grain boundaries. Pb $(SCN)_2$ additive can increase the grain size and form excessive PbI_2. However, when the two additives are used simultaneously, the formation of excess PbI_2 is reduced, while high-quality film morphology with fewer defects is obtained. This, in turn, improves carrier mobility and lifetime. The as-fabricated perovskite/CIGS TSCs exhibited a PCE of 25.9 %. In 2021, Chen et al.[98] used 2-thienylethylammonium chloride (TEACl) as an additive to create a 2D/3D heterojunction structure as an interface. The sulfur atom in thiophene can interact with Pb^{2+}, and the amino group interacts with the anion. The 2D layer passivates the defects on the surface and grain boundaries of the perovskite layer, prolongs the carrier lifetime, and improves the stability of the device. A perovskite/CIGS TSC with an efficiency of 24.0% was prepared.

In 2022, Tang et al.[99] utilized guanidine bromide (GABr) for interface treatment between the perovskite layer and the hole transport layer to prevent the dissolution of the perovskite layer by the IPA solvent. The film improved by a mixed solvent shows better absorbance and energy level alignment (Figure 11.8f). Additionally, a ST-PSC with an efficiency of 18.3 % was prepared. Combined with a CIGS bottom cell with an efficiency of 17.5 %, a 4T perovskite/CIGS TSC with an efficiency of 25.5% was realized. Feeney et al.[100] used in-house sputtered indium-doped zinc oxide (IZO) and internally sputtered indium oxide (IO:H) as ITO front-end transparent conductive oxide. In addition, anti-reflection layers of MgF_2 were coated on both sides of the cell to optimize the structure of perovskite/CIGS tandem cell devices for light management (Figure 11.8g). The device efficiency of the optimal light management structure reaches 27.3%. At present, most of the research focuses on rigid perovskite/CIGS TSC, and there are few studies on flexible TSC. In 2023, Luo et al.[101] obtained three different growth profiles of Ga by varying the evaporation rate of Ga, In, Cu, and Se in the three-step co-evaporation process. From the corresponding scanning electron microscopy, it was found that the best CIGS absorption layer was obtained with a large grain size and the best GGI ([Ga]/([Ga]+[In]) double-graded structure (Figure 11.8h). In addition, the deposition of Ag on the Mo layer promotes the crystallization of the CIGS film. A fully flexible 4T perovskite/CIGS TSC with an efficiency of more than 21.5% was fabricated on a PI substrate. To sum up, CIGS material is a highly promising bottom cell material for perovskite-based TSCs, owing to its low toxicity, appropriate band gap, and impressive efficiency and stability.

11.4.1.2 Si/Perovskite TSCs

The photovoltaic market is dominated by efficient and stable silicon cells, but the efficiency of silicon cells has grown slowly in recent years. To meet people's growing energy demand, it is necessary to continuously improve the efficiency of solar cells. The utilization of wide bandgap and narrow bandgap semiconductors connected in series or parallel to form TSCs is considered an optimal strategy to break through the silicon Shockley-Queisser (S-Q) limit and obtain enhanced efficiency. Although the efficiency of perovskite/Si TSCs has increased to 32.5% in less than a decade, showing great potential of application,[102] some shortcomings hinder its further development. Specifically, wide bandgap perovskite materials exhibit significant open circuit voltage losses due to poor crystal quality, defects in grain boundaries, and bulk. In addition, the conformal growth of perovskite layer on textured silicon is a huge challenge. To overcome these limitations, researchers have proposed several strategies to improve the efficiency of perovskite/Si TSCs. In this section, we will summarize these strategies from the aspects of conformal growth, fabrication technology advancement, and additive engineering/interface engineering of wide bandgap perovskite layer.

11.4.1.2.1 Conformal Growth of Perovskite Layer

In the perovskite/Si TSC, the photoluminescence response of perovskite changes due to the different morphology of the bottom silicon cell. The photoluminescence distribution of perovskite deposited a textured substrate is narrower than that on a plane[103] (Figure 11.9a). Therefore, modifying the pyramid structure of the bottom silicon cell during different perovskite film manufacturing processes can lead to improved film formation and enhanced device performance. In 2016, Werner et al.[104] prepared top PSCs using the same manufacturing process and single-chip TSCs using silicon cells with a single-sided bottom texture as shown in Figure 11.9b. The efficiency increased to 20.5% compared with the previous double-sided polished bottom silicon cells (19.2%). The polished silicon cell surface causes light reflection and loss of light capture. The use of textured silicon bottom cells improves the spectral response of the edge band of the silicon cell and increases the current density of the bottom cell. However, preparing high-quality perovskite top cells on textured surfaces is

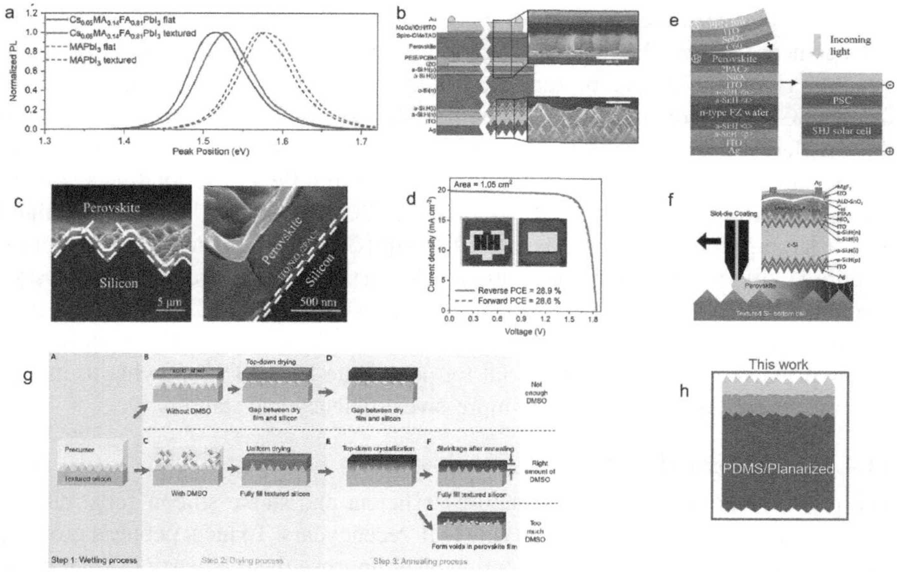

FIGURE 11.9 (a) PL emission of iodide-only perovskite films (triple cation $Cs_{0.05}MA_{0.14}FA_{0.81}PbI_3$ solid line, and MAPbI3 dashed line) on textured (red) and flat (black) silicon substrates.[103] Copyright 2022, Wiley-VCH GmbH. (b) Schematic drawings of monolithic tandem devices without and with rear-side texture and the corresponding SEM cross-section images for the front and rear side.[104] (c) SEM cross-section images of a perovskite/silicon tandem device, showing complete perovskite and functional layers.[106] Copyright 2022, Wiley-VCH GmbH. (d) J–V curves of the tandem device (1.05 cm^2 aperture area), the digital photo of a device is shown in the inset.[107] Copyright 2023, Wiley-VCH GmbH. (e) Illustration of the lamination process and device architecture of the laminated monolithic perovskite/silicon tandem solar cells.[108] (f) Device architecture schematic of perovskite/silicon tandem devices wherein the wide-band-gap perovskite absorber layer ($MAPb(I_{0.75}Br_{0.25})_{3_}$1.68 eV) is deposited by slot-die-coating.[110] Copyright 2020, American Chemical Society. (g) Mechanism of dense perovskite film formation.[112] Copyright 2020, Elsevier Inc. (h) Double-side textured silicon cell with a planarizing perovskite top cell and a PDMS light-scattering layer.[112] Copyright 2020, Elsevier Inc.

very challenging. In 2018, Sahli et al.[105] used a hybrid two-step deposition method combined with sequential co-evaporation and spin-coating to fabricate the top perovskite layer. The perovskite layer completely encapsulates the pyramid structure of the silicon cell and is well deposited on the micron-sized textured monocrystalline silicon pyramid. The silicon cell with a pyramid structure reduces reflection losses and enhances light capture ability. A full-texture perovskite/silicon tandem cell with a current density of 19.5 mA cm^{-2} and a certified efficiency of 25.2% was obtained. Mao et al.[106] designed NiO_x/2PACz ([2-(9H-carbazol-9-yl)ethyl] phosphonic acid) as an ultrathin hybrid HTL on an ITO recombination junction. A strong P-O···Ni or P = O···Ni bond is formed between the hydroxyl group and the phosphonic acid group in NiO_x. The NiO_x treated by 2PACz has uniform surface potential and an increased water contact angle. These factors promote the growth of large grains in the perovskite layer and the conformal coverage of the HTL layer and the perovskite layer in the textured silicon bottom cell (Figure 11.9c). The HTL of NiO_x treated by 2PACz reduces the shunt loss caused by direct contact between the ITO and the perovskite layer. A full-textured perovskite/silicon tandem cell with a masked area of 1.2 cm^2 and a PCE of 28.84% was obtained. The mixed two-step method can deposit a conformal perovskite layer on the textured silicon bottom cell, but the perovskite film exhibits low quality and poor photovoltaic performance. To improve the film quality of the perovskite layer, Luo et al.[107] introduced an anionic additive composed of ammonium chloride (MACl) and methyl ammonium thiocyanate (MASCN) to control the crystallization process of the wide band gap perovskite film. The MACl additive promotes the reaction and crystallization of perovskite, without altering the growth orientation of perovskite crystal. The MASCN additive promotes the rapid nuclei growth and has a strong preferred orientation. The simultaneous addition of MACl and MASCN forms a perovskite layer with larger grains, lower defect density, and suppressed phase segregation. The perovskite layer is deposited conformally on the textured silicon bottom cell, resulting in a stable perovskite/silicon TSC with a 1 cm^2 certification efficiency of 27.9%. The J-V curve is shown in Figure 11.9d.

11.4.1.2.2 Fabrication Technology Advancement of Perovskite/Si TSC

The deposition of the perovskite layer using sequential deposition technology is influenced by several factors, such as material, temperature, solvent, and process. Roger et al.[108] proposed a lamination process that combined the two parts of A and B in Figure 11.9e through heat and pressure to connect the perovskite layer and the C_{60} layer. This process promotes the growth of larger grains in the perovskite layer, reduces roughness, and decreases the surface area of the perovskite/C_{60} interface, thereby increasing the V_{oc} of the PSC. The lamination technology was successfully used to prepare a device with a PCE of 20%, providing a novel method for the preparation of laminated cells. In 2022, Chen et al.[109] prepared TSC by a two-step solvent method. In the first step, FAI was added to a mixed solution of $PbBr_2$ and PbI_2 to suppress excessive PbI_2, and rubidium acetate (RbAc) was added to reduce the hysteresis and roughness of the top perovskite cell. This resulted in a perovskite/Si TSC with a certified PCE of 27.20%. Although PSCs prepared by spin coating have made remarkable progress, they are not suitable for large-scale production of large-area PSCs due to centrifugal force. Therefore, slot-die coating, blade coating, inkjet printing, and other methods are being explored extensively to promote the commercialization of PSCs. In 2022, Subbiah et al.[110] added the surfactant LP (L-α-phosphatidylcholine) to PTAA to obtain better wettability and promote better interaction between ink and substrate. A solvent system consisting of a mixture of low boiling point acetonitrile (ACN) and methylamine (MA (MeOH)) dissolved in methanol was introduced to ensure the rapid evaporation of the solvent. Using cysteine hydrochloride as a passivation additive, the first perovskite/textured silicon TSC (Figure 11.9f) was prepared by slot-die coating, with a PCE of 23.8%. Xu et al.[111] achieved a stable output power of 19.4% using a halide perovskite single-junction solar cell with an annealing temperature of 150°C and a slot-die coating process. A 2-terminal perovskite/Si TSC with a PCE of 25.2% on a 1 cm^2 active area was achieved. Chen et al.[112] deposited a hole transport layer and a perovskite layer on a substrate at 70 °C and a rate of 1.5 m/min using a (N_2)-knife-assisted blading process. The perovskite precursor solution was added with 25 mol% DMSO and 2-ME as solvents. The addition of DMSO changed the film drying and crystallization process, resulting in a dense perovskite layer that completely covered the full-textured silicon cell (Figure 11.9g). The planarization of the perovskite layer was realized on the silicon cell with a submicron pyramid structure (Figure 11.9h), and a monolithic perovskite/Si TSC with an efficiency of 26% was prepared.

11.4.1.2.3 Additive Engineering and Interface Engineering of Perovskite/Si TSCs

Improving the efficiency of wide bandgap PSCs is crucial for achieving high-performance perovskite/Si TSC. However, these types of PSCs often have poor film quality due to defects in the crystallization process and encounter issues such as halide phase separation and interface recombination at the interface under light and atmospheric environments, resulting in reduced efficiency.

To address these challenges, additive engineering and interface engineering are used to improve the performance of wide bandgap PSCs. For instance, Chen et al.[113] added MACl and MAH_2PO_2 to the perovskite precursor solution, which changed the crystal quality of the wide band gap perovskite film, leading to a large grain size, reduced defect density, and smoother perovskite film. The open-circuit voltage deficit of the device is reduced to 0.49–0.51 V, and the PCE is 25.4%. Similarly, TaO et al.[114] used pseudohalogen ions of PF_6^- and BF_4^- to replace I^-/Br^- in perovskite (Figure 11.10a) and found that this addition resulted in high-quality crystal films with less carrier recombination, lower charge transfer loss, and inhibited phase separation. The PCE of the 4T perovskite/Si TSC reached 27.35% (PF_6^-) and 27.11% (BF_4^-), respectively. Phenformin hydrochloride (PhenHCl) with electron-rich and electron-poor regions was introduced as a multifunctional passivator,[115] which effectively passivated the perovskite grain boundaries and surface dangling bonds (Figure 11.10b), inhibited the phase separation of halides, and reduced ion migration. The PCE of the perovskite/Si TSC with this passivator increased from 25.4% to 27.4%. Duong et al.[116] found that a 2D perovskite phase could be formed on the surface of 3D perovskite. This 2D perovskite phase can passivate surface defects and change the electronic structure of the 3D perovskite surface, thereby obtaining better carrier lifetime and higher efficiency. The n-BABr (n-butyl ammonium bromide) cationic surface coating strategy was used to prepare a ST-PSC with a high efficiency of 17.1%. When combined with the silicon cell, the 4T perovskite/Si TSC achieves an efficiency of 27.7%. Interface engineering is also essential for improving the efficiency of perovskite/Si TSCs. The textured surface of the silicon cell enhances the light absorption capacity, but it also poses a challenge to the deposition of the top perovskite layer. Perovskite layers deposited on flat surface can resulting in optical loss. The top PSCs were prepared by vacuum evaporation of inorganic metal salts and spin-coated organic salts. The lack of chemical bonds and van der Waals gaps at the interface will lead to uneven perovskite at the top and bottom. Li et al.[117] introduced a CsBr thin layer (Figure 11.10c) between the perovskite layer and the HTL, which interacted with the PbI_2 at the bottom to form a gradient perovskite, optimizing the energy level matching between the layers of the tandem cell. The

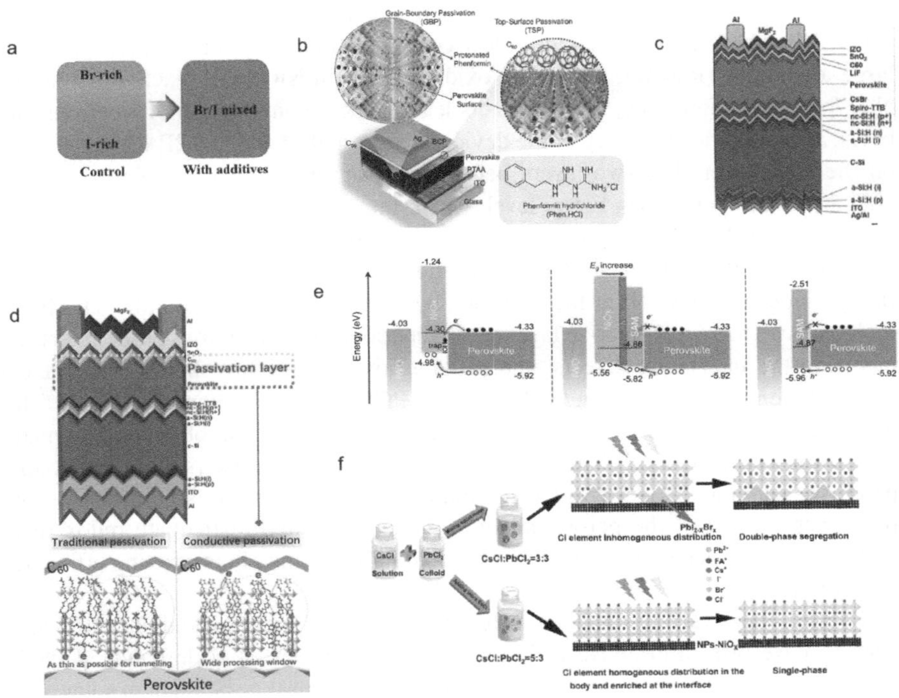

FIGURE 11.10 (a) Schematic diagram of the phase separation of perovskite without and with additives.[114] Copyright 2022, American Chemical Society. (b) Illustration of the p-I-n device structure, GBP and TSP strategies, and the molecular structure of the PhenHCl passivation molecule used in this study.[115] Copyright 2021, Elsevier Inc. (c) Schematic illustration of a fully textured perovskite/c-Si TSC.[117] Copyright 2021, Wiley-VCH GmbH. (d) Schematic of conformal perovskite/textured silicon TSC architecture. The following diagram shows the comparison between conductive passivation and traditional passivation.[118] Copyright 2022, Wiley-VCH GmbH. (e) Energy-level alignment of the IWO/HTL/perovskite interfaces with NiO_x, SAM, and NiO_x/SAM as the HTL, respectively.[119] Copyright 2022, Royal Society of Chemistry. (f) Schematic of the influence of the addition of excess CsCl to the perovskite precursor, preventing the formation of $PbI_{2-x}Br_x$ at the interface.[121] Copyright 2022, Wiley-VCH GmbH.

efficiency of the perovskite/Si TSC is 27.8%. The interface recombination between the perovskite layer and C_{60} functional layer leads to the loss of open circuit voltage and fill factor. Traditional interface passivators usually only reduce the interface recombination by about 30 eV. Xu et al.[118] used a conductive passivator organic amine salt 2-thiophenethylamine bromide (TEABr) as an interfacial passivation molecule (Figure 11.10d). The π electron conjugation on the aromatic ring structure of the passivator enhances its electrical conductivity. In addition, the delocalized lone pair electrons of the heteroatoms of the heteroaromatics promote the charge transfer at the interface. Using TEABr as an interfacial passivation molecule, the V_{OC} of the device was significantly increased by approximately 100 mV, with a higher FF. The efficiency of the perovskite/Si TSC was 28.51%.

Nickel oxide (NiO_x) has become an attractive hole transport material for perovskite solar cells due to its good stability, low cost, and suitability for the preparation of large-area cells. However, the highly oxidative Ni^{3+} in NiO_x will react with I^- at the NiO_x/perovskite interface, where defects lead to the recombination of electrons and holes, affecting the efficiency of the device. Zheng et al.[119] added a 2PACz monolayer on the NiO_x layer. The two layers between the self-assembled monolayer and the NiO_x layer have a uniform Fermi level, which improves the valence band maximum (VBM) of the NiO_x layer, passivates the defects of the NiO_x layer (Figure 11.10e), and improves the efficiency of PSCs. Zhumagali et al.[120] introduced organometallic dye molecules (N719) at the interface between NiO_x and perovskite. The N719 molecule can form a protective layer to prevent poor redox reactions at the interface, passivate the interfacial defects, and promote the migration of holes. An efficient textured monolithic perovskite/Si TSC with a PCE of 26.2% (compared to 23.5% for dye-free device) was achieved. Li et al.[121] changed the band gap of perovskite materials from 1.64 eV to 1.67 eV by adding inorganic $CsPbCl_3$ clusters to the perovskite precursor solution. This led to Cl distribution at the perovskite/NiO_x interface and the upper surface of the perovskite film, which prevented inadequate redox reactions at the interface and inhibited phase separation, ultimately improving efficiency (Figure 11.10f).

Overall, these various methods including additive engineering and interface engineering have proven effective in improving the efficiency of wide bandgap PSCs, enabling them to be more competitive in the field of perovskite/Si tandem solar cells.

11.4.1.3 OPV/Perovskite TSCs

With the discovery of A-DAD-A structured small molecule acceptor materials, such as Y6, Y18, and BTPV-4F, the efficiency of organic solar cells has significantly improved. However, organic solar cells are prone to decompose under ultraviolet (UV) light. PSCs have emerged as a promising alternative due to their semitransparent nature, which blocks UV damage to bottom organic solar cells. Perovskite/organic tandem cells can fully utilize the solar spectrum and demonstrate excellent PCE.

In 2018, Chen et al.[122] used a dual-source vacuum co-evaporation method to prepare a $CsPbBr_3$ perovskite photoactive layer as the top cell of TSC. A tandem device with an efficiency of 14.03% was fabricated by using an organic solar cell with wide band gap polymer, PBDB-T-SF, as a donor and narrow band gap n-type organic semiconductor, IT-4F, as an acceptor (Figure 11.11a). The device maintained high stability of over 99.4% of its initial PCE after 120 h of UV irradiation. In 2019, Zeng et al.[123] prepared a 2T inorganic perovskite/organic TSC with an efficiency of 15.04% using a 1.92 eV band gap $CsPbI_2Br$ perovskite absorber layer and a PTB_7-Th:COi_8DFIC:$PC_{71}BM$ ternary cell as an organic bottom cell. However, there are still many shortcomings in the preparation of perovskite/organic TSCs, such as poor film quality and phase segregation in the perovskite layer, resulting in low device efficiency.

Additive engineering is an effective strategy for passivating film defects and improving efficiency. In 2021, Wu et al.[124] replaced the commonly used spiro-OMeTAD in the perovskite layer with the polymer PBDB-T, and used benzyl

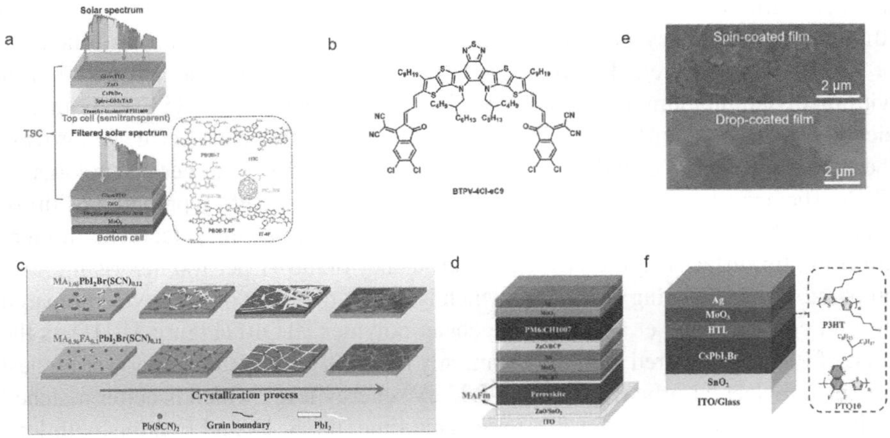

FIGURE 11.11 (a) Device structure of four-terminal TSCs: semitransparent top cell with a structure of glass/ITO/ZnO/$CsPbBr_3$/spiro-OMeTAD/transfer laminated PH1000; bottom OSCs with a structure of glass/ITO/ZnO/organic photoactive layer/MoO3/Al.[122] Copyright 2018, Wiley-VCH. (b) Chemical structures of BTPV-4Cl-eC9.[125] Copyright 2022, Wiley-VCH GmbH. (c) Schematic illustration of the film formation processes for the $MA_{1.06}PbI_2Br(SCN)_{0.12}$ film and the $MA_{0.96}FA_{0.1}PbI_2Br(SCN)_{0.12}$ film.[126] Copyright 2022, Wiley-VCH GmbH. (d) Tandem device structure adopted in this work.[127] Copyright 2023, Wiley-VCH GmbH. (e) SEM images for inorganic perovskite films made by spin-coating and drop-coating.[129] (f) Device structure of $CsPbI_2Br$ PSCs and molecule structures of PTQ10 and P3HT.[32] Copyright 2022, Elsevier Ltd.

ammonium chloride (PMACl) to passivate the perovskite layer, inhibiting light-induced phase separation. The perovskite/organic TSC with an efficiency of 18.06% was prepared, which showed excellent stability even after 150 h of UV irradiation, 250 h of primary sunlight irradiation, and 100 h of heating at 80 °C. In 2022, Qin et al.[125] used chlorformamidine ($ClFA^+$) to passivate the defects in the perovskite layer. The chloride ion added to the formamidine is partially combined with the Pb^{2+} cation between the iodine vacancy defects, resulting in significant passivation these defects. In addition, the organic small molecule acceptor, BTPV-4Cl-eC9 (Figure 11.11b), broadens the absorption spectrum range of organic cells, achieving a PCE of perovskite/organic TSC up to 22.0%. Xie et al.[126] introduced a small amount of formamidine (FA^+) cations into the basic composition of $MA_{1.06}PbI_2Br(SCN)_{0.12}$ to regulate the crystallization process of perovskite films. The addition of FA^+ increased the solubility of $Pb(SCN)_2$ in the precursor solution, inhibiting the formation of PbI_2 aggregates. The composition of the film was optimized to form a high-quality film with uniformly passivated PbI_2 at the grain boundaries (Figure 11.11c). The optimized perovskite film was combined with organic (PM6:CH1007) solar cells, leading to a high efficiency of 21.2%.

The ionic liquid is an effective passivator for passivating perovskite defects. Sun et al.[127] used methylammonium formate (MAFm) ionic liquid to modify the bottom and top of the wide bandgap $CsPbI_2Br$ inorganic perovskite. The formic acid group interacts with Pb^{2+}, resulting in significant passivation of the defects in the modified film. A high-efficiency perovskite/organic (PM_6:CH1007) TSC with an efficiency of 23.21% was prepared (Figure 11.11d). Yao et al.[128] modified the perovskite layer with methylammonium acetate (MAAc) ionic liquid, obtaining a TSC with an efficiency of 22.43% (certified 21.42%). In addition to additive engineering, improving the preparation process and functional layers also promotes the perovskite/organic TSCs efficiency. Liu et al.[129] found that the PCE of inorganic PSCs prepared by drop-coating (self-spreading) is much higher than that of spin-coating, featuring a more uniform film surface (Figure 11.11e). The efficiency of the 4T perovskite/organic TSC prepared by drop coating is 22.34%, which is higher than that of the device prepared by spin coating. Ding et al.[32] used the cheap polymer PTQ10 (Figure 11.11f) as the HTL of PSCs. Compared with the commonly used P3HT, PTQ10 has a lower highest occupied molecular orbital (HOMO; −5.23 eV) energy level, which is better matched with the perovskite energy level. At the same time, the S, N, and F atoms with lone pair electrons on PTQ10 interact with uncoordinated Pb^{2+} ions, which passivates the surface of perovskite and further improves the performance of PSCs. A perovskite/organic TSC with an open circuit voltage of 2.22 V and a PCE of 21.4% was prepared.

11.4.1.4 Perovskite/Perovskite TSCs

All-perovskite TSC is a promising technology due to its advantages of low cost, flexibility, and solution processing. The band gap of perovskite materials can be adjusted to 1.2 eV by adding the Sn element, which can be combined with wide-bandgap perovskite to make full use of sunlight and thus produce high-efficiency all-perovskite TSCs. In this section, we will focus on various strategies that have been used to enhance the efficiency and stability of Sn-Pb mixed perovskite materials.

In 2016, Eperon et al.[130] added 50% Sn to adjust the band gap of the perovskite material and replaced a certain proportion of FA ions with Cs ions to increase the device's stability. A $FA_{0.75}Cs_{0.25}Sn_{0.5}Pb_{0.5}I_3$ PSC with a V_{oc} of 1.2 V and PCE of 14.8% was obtained. By combining it with wide-bandgap $FA_{0.83}Cs_{0.17}Pb(I_{0.5}Br_{0.5})_3$ material, they achieved a monolithic 2T tandem device with an efficiency of 17.0% and a mechanically stacked 4T tandem device with a PCE of 20.3%, respectively. Sn-Pb mixed perovskite materials have a low bandgap suitable for tandem devices, but the efficiency of Sn-Pb mixed perovskite materials is generally lower than that of pure Pb-based perovskite materials. Sn^{2+} in Sn-based perovskite materials is easily oxidized, resulting in Sn vacancy defects, which reduces the efficiency and stability of the device. Additives and interface engineering are effective strategies to solve these problems. Tong et al.[131] used guanidinium thiocyanate (GuaSCN) as an additive to make grain boundaries more dense and minimize the appearance of pinholes in the film. The addition of GuaSCN forms a 2D structure at the grain boundary, which passivates the defects formed by Sn oxidation and enhances the carrier lifetime (> 1 microsecond). The 4T and 2T all-perovskite TSCs with efficiencies of 25% and 23.1% were fabricated, respectively. Lin et al.[132] added trifluoromethylphenylammonium (CF3-PA) as a passivator to the Sn-Pb perovskite precursor solution and interacted with the receptor-like defects on the surface of the Pb-Sn perovskite through ionic bonds and hydrogen bonds. At high temperature, the formation of surface iodine vacancies is inhibited, and iodine defects are reduced (Figure 11.12a). The

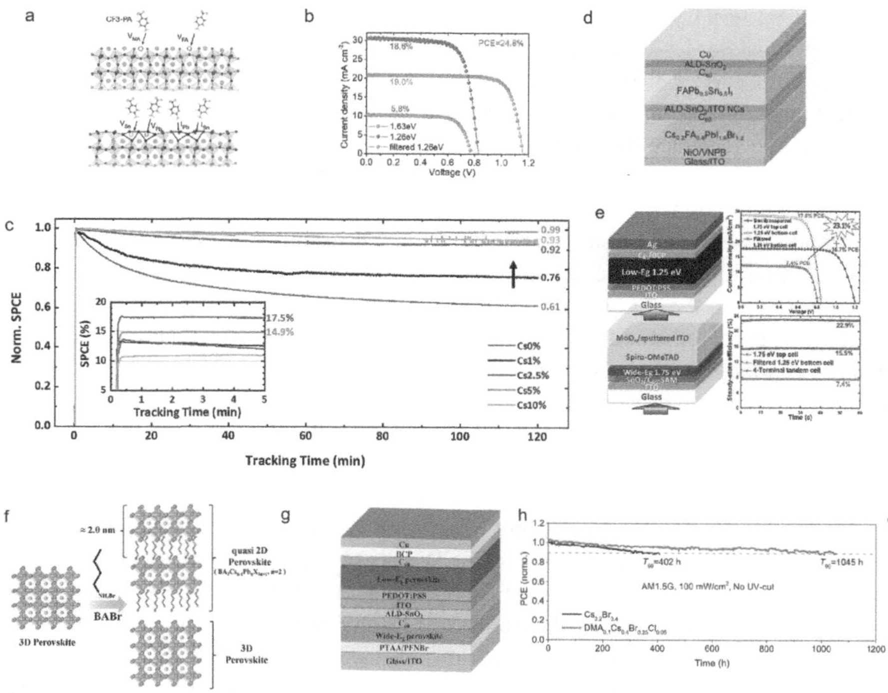

FIGURE 11.12 (a) Schematic diagram of the interaction between ammonium cations with the acceptor-like defects, that is formamidinium (FA) vacancy (V_{FA}), methylammonium (MA) vacancy (V_{MA}), Sn vacancy (V_{Sn}), Pb vacancy (V_{Pb}), I substituted at Sn site (I_{Sn}), and I substituted at Pb site (I_{Pb}). Both the A-I and B-I terminated (001) perovskite surfaces are considered.[132] Copyright 2022, Springer Nature. (b) J–V characteristics of the best-performing single-junction NBG bottom solar cell (Eg = 1.26 eV) with IPH-interlayer, WBG top solar cell (E_g = 1.63 eV), and filtered bottom solar cell measured from backward scan under AM 1.5 G (1,000 W m^{-2}) irradiation.[134] (c) Stabilized power conversion efficiency (SPCE) of the best performing $Cs_x(FA_{0.8}MA_{0.2})_{(1-x)}Sn_{0.5}Pb_{0.5}I_3$ LBG PSCs with different Cs concentrations varying from x = 0%, to 1%, 2.5%, 5%, and 10% (denoted as Cs 0%, Cs 1%, Cs 2.5%, Cs 5%, and Cs 10%, respectively) under AM 1.5G illumination at MPP tracking conditions at 25°C for 120 min.[136] (d) Schematic structure of monolithic all-perovskite tandem solar cells.[137] Copyright 2022, Wiley-VCH GmbH. (e) Performance and structure of tandem solar cells.[138] (f) Schematic representations of the quasi-2D perovskite and relevant spacing.[140] Copyright 2022, Royal Society of Chemistry. (g) Schematic structure of a 2T TSC device.[141] Copyright 2018, American Chemical Society. Copyright 2023, RSC Publishing. (h) MPP tracking measured with the encapsulated $Cs_{0.2}Br_{0.4}$ and $DMA_{0.1}Cs_{0.4}Br_{0.25}Cl_{0.05}$ devices under full solar illumination (AM 1.5G, 100 mW cm^{-2}) in ambient conditions.[142] Copyright 2022, Wiley-VCH GmbH.

carrier diffusion length is doubled to more than 5 μm, and the efficiency of Pb-Sn perovskite solar cells is increased to more than 22%. The certification efficiency of full perovskite TSC is 26.4%. The packaged series device can maintain more than 90% of its initial performance after 600 h of operation at the maximum power point under environmental conditions and one solar illumination. Yu et al.[133] introduced the textured ternary Sn (II) alloy SnOCl as a hole transport material (HTM) for Sn-Pb perovskite devices, which enhanced the optical absorption of the device and increased the carrier diffusion length to 3.63 μm. Datta et al.[134] used a solution-treated quaternary ammonium salt passivation layer to suppress the non-radiative recombination at the interface and improve the V_{oc} of the cell, obtaining an efficiency of over 23% for TSCs. Hu et al.[135] proposed solution-treated fullerene derivative indene-C60-propionic acid hexyl ester (IPH) as an intermediate layer between the perovskite layer and C60. The addition of the intermediate layer has a higher conduction band offset, which decreases the defect density at the interface, inhibiting non-radiative recombination of the charge. A 4T all-perovskite tandem device with an efficiency of 24.8% was prepared (Figure 11.12b). The stability of Sn-Pb mixed perovskite is also a major challenge in the commercialization of PSCs. It has been found that the stability of perovskite devices can be adjusted by changing the composition of perovskite. Moghadamzadeh et al.[136] introduced trace Cs into $FA_{0.8}MA_{0.2}Sn_{0.5}Pb_{0.5}I_3$ perovskite semiconductor. Cs reacts with the residual $Sn_yPb_{(1-y)}I_2$ ($0 < y < 1$) on the surface, which improves the operational stability of the device. The narrow bandgap PSC can maintain an initial power output of up to 92% after two hours under simulated one-time solar irradiation (Figure 11.12c). Wu et al.[137] obtained a stable all-perovskite solar tandem device with a PCE of 26.3% by using all-FA-based narrow-bandgap perovskite cells and replacing the traditional PEDOT:PSS with solution-treated inorganic ITO nanocrystals as the charge transport layer (Figure 11.12d).

The wide bandgap perovskite is used as the front cell in the all-perovskite TSC, which needs to further improved while the improvement strategy for the wide band gap perovskite is also abundant. In 2018, Zhao et al.[138] prepared a highly efficient translucent $FA_{0.8}Cs_{0.2}Pb(I_{0.7}Br_{0.3})_3$ perovskite top cell with a bandgap of 1.75 eV, and the electrode was fabricated by more transparent MoO_x/ITO. For the first time, the efficiency of the 4T all-perovskite TSC (Figure 11.12e) is higher than that of single-junction PSCs, exceeding 23%. However, phase segregation and non-radiative recombination in wide bandgap perovskite cause large V_{oc} loss and poor device stability. Researchers have found that the 2D structure can improve device stability and inhibit interface charge recombination.[139] Yan et al.[140] obtained a 2D/3D heterojunction structure (Figure 11.12f) by coating n-butylammonium bromide (BABr) onto a 3D $CsPbI_2Br$ perovskite film. The quasi-2D structure modified the interface defects and optimized the band matching. In addition, Zhao et al.[141] used lead chloride ($PbCl_2$) and phenethylammonium chloride (PMACl) to introduce Cl into the bulk membrane and form a 2D phase on the surface. Cl in $PbCl_2$ enters the perovskite film lattice, reducing halide vacancies and inhibiting ion migration. PMACl promotes the formation of a 2D perovskite phase on the grain surface and reduces non-radiative recombination. The stability of the device is enhanced, with a reduction of the V_{oc} loss from 0.558 V to 0.418 V, and the efficiency of the prepared 2T all-perovskite tandem device (Figure 11.12g) has reached 26.68%.

Composition regulation engineering is another effective strategy to suppress phase segregation. Wen et al.[142] reduced the Br content in the precursor solution to 25 mol%, added dimethylammonium (DMAI) and $MAPbCl_3$ to the precursor solution, and added 10 mol% DMA and 5 mol% Cl to $Cs_{0.4}FA_{0.6}Pb(I_{0.75}Br_{0.25})_3$ to obtain a perovskite film with a bandgap of 1.8 eV. The defect density of the film decreases, and strain relaxation inhibits light-induced halide segregation. The wide bandgap PSCs exhibit excellent performance and stability, maintaining an initial efficiency of 90% after 1,000 h of operation at the maximum power point (Figure 11.12h), and the efficiency of the all-perovskite TSC is 26.0%. Jiang et al.[143] combined the characteristics of rapid crystallization of Br with the gas quenching method to prepare a film with a gradient distribution of Br content on the surface and in the body. Compared with the film prepared by the anti-solvent method, the film prepared by the gas-quench method has fewer defects, inhibits phase segregation, and improves stability. An efficient all-perovskite 2T TSC with an efficiency of 27.1% was fabricated by mixing WBG perovskite film with 1.75 V Br-I and a 1.25 V narrow band gap PSC. Under continuous 0.8 solar illumination, the maximum PCE of 81% was maintained within 1,825 h at 35 °C.

In conclusion, various strategies have been used to enhance the efficiency and stability of both narrow and wide bandgap perovskite materials. These strategies include bandgap engineering, additives, interface engineering, and composition engineering. Further research is needed to commercialize these all-perovskite TSCs.

11.4.2 Design of the Interconnect Layer

The interconnect layer (ICL) in a 2T TSC is where the charges of two subcells are recombined; it should have low resistivity, high transmittance, and good compatibility with the subcells to effectively extract one carrier while preventing another carrier.[144] There are three types of interconnect layers: tunnel diode, recombination layer, and mechanical stack.[145] The ICL affects the optical and electrical properties between subcells. The following describes the progress of the interconnect layer from three aspects: preparation technology, performance improvement, and defect passivation.

11.4.2.1 Choice of the Process Techniques

It is important to choose the appropriate process techniques for preparing the interconnect layer. The interconnect layer is an intermediate layer connecting two subcells, and it must be a protective barrier to prevent damage from the upper cell's preparation process on the lower cell. Common processes for preparing interconnect layers include thermal evaporation, solution preparation, atomic layer deposition (ALD), and sputtering technology. Each technology has its own advantages and disadvantages.

Magnetron sputtering is recognized as the most advanced technology for preparing transparent conductive oxide (TCO). However, high-energy sputtering ions during the sputtering process may cause certain damage to other functional layers. ALD is a self-limiting surface reaction that enables layer-by-layer conformal growth in the nanometer range and reduces damage to other layers. Thiesbrummel et al.[146] used ALD to deposit a quasi-nondestructive thin layer of indium oxide and tin oxide on the Al-doped ZnO nanoparticle layer. Brinkmann et al.[147] controlled the thickness of the grown InO_x layer by ALD technology. ALD-grown InO_x layers with a thickness of 1.5 nm inserted into SnO_x and MoO_x significantly improved the J-V characteristics of TSC (Figure 11.13a). In addition, Zhou et al.[148] found that the thin SnO_2

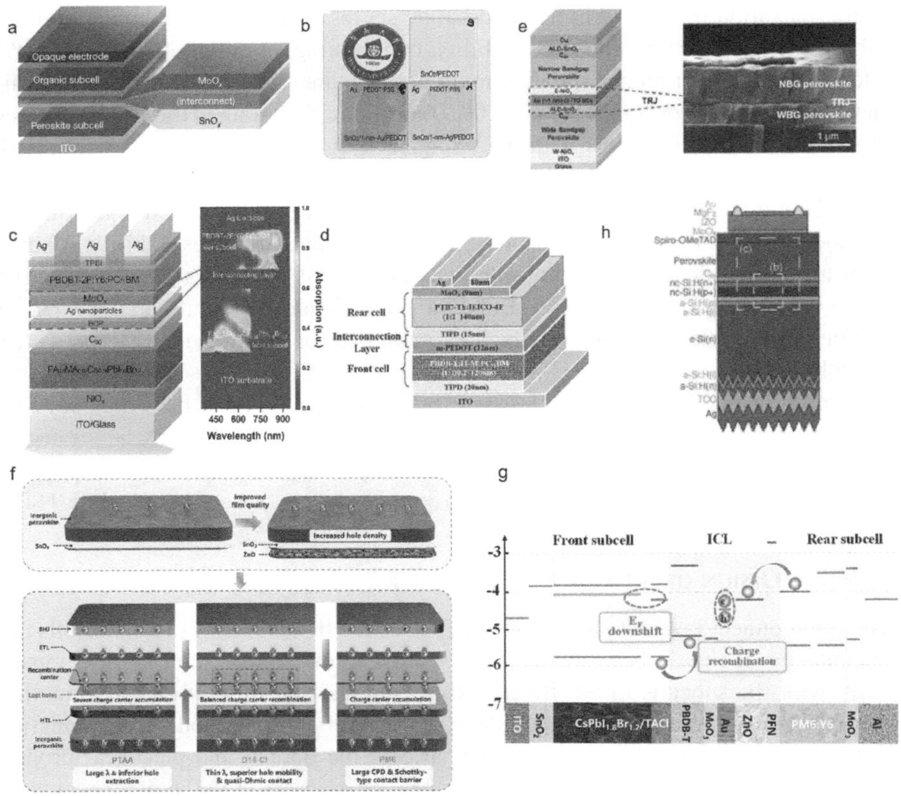

FIGURE 11.13 (a) Schematic of a perovskite–organic tandem cell with InO_x or Ag as interconnect.[147] Copyright 2022, Springer Nature. (b) Digital photos of the three prepared interconnect layers.[148] Copyright 2022, American Chemical Society. (c) Tandem cell structure with a highlight on ICL design and the distributions of photon abortions in tandem, simulated by transfer matrix method.[149] Copyright 2020, Elsevier Inc. (d) ITO/TIPD/PBDB-T:ITM:PC71BM/mPEDOT/TIPD/PTB7-Th:IEICO-4F/MoO_3/Ag.[150] Copyright 2021, Wiley-VCH GmbH. (e) The device structure and cross-sectional SEM image of an all-perovskite tandem solar cell.[153] Copyright 2021, Wiley-VCH GmbH. (f) Schematic of the hole transport and charge carrier recombination in the ICLs with different HTLs.[154] Copyright 2022, Wiley-VCH GmbH. (g) The energy-level diagrams of the 2T TSC.[155] Copyright 2021, Wiley-VCH GmbH. (h) Schematic view of the perovskite/SHJ monolithic tandem cell with a nc-Si:H recombination junction.[157] Copyright 2017, Wiley-VCH.

layer prepared by ALD is a useful ETL with excellent solvent resistance and good transparency (Figure 11.13b). Direct contact with PEDOT:PSS promotes the effective extraction of carriers between the two subcells. Chen et al.[149] prepared an interconnect layer composed of MoO_x, Ag nanoparticles and BCP by total thermal evaporation (Figure 11.13c). Ag nanoparticles promote the carriers recombination and reduce electrical loss. MoO_x prevents damage to the bottom cell caused by chloroform.

All-solution fabrication of tandem devices is challenging, and the ICL is the key to fabricating TSCs. Zeng et al.[150] used a solution composed of poly (3,4-ethylene dioxythiophene):poly(styrenesulfonate) (m-PEDOT) and titanium (diisopropoxide) bis(2,4-pentanedionate) (TIPD) to treat ICL (Figure 11.13d). The whole solution preparation process avoids the damage from complex vacuum evaporation and sputtering technology to other layers, ensuring low cost and high reproducibility. The prepared ICL also has sufficient physical robustness to prevent the damage of solvents such as chlorobenzene to the underlying films. It is found that when the interconnect layer is prepared by solution method, the formation of a three-dimensional polymer network under heating or illumination can increase the solvent resistance of the material and prevent the damage to the bottom layer when depositing the upper layer. Chang et al.[151] used (1,2-bis[4- (azido-methyl)phenyl]-1,2-diphenylethene) (TPE-MN_3) as crosslinking agent and molybdenum tris-[1-(trifluoroethanoyl)-2-(trifluoromethyl)ethane-1,2-dithiolene] as p-type dopant to improve the PTAA hole transport layer. They found that the improved interconnect layer had good solvent resistance, appropriate energy level alignment, and good conductivity, preparing a monolithic all-perovskite TSC with an efficiency of 18.69%. The encapsulated TSC maintained an initial efficiency of about 91% after exposure to air for 9,300 h.

In summary, the choice of process techniques is critical for preparing the interconnect layer, which affects the optical and electrical properties between subcells. Different technologies have their advantages and disadvantages, and researchers are exploring new methods to improve the performance of the interconnect layer, such as ALD, total thermal evaporation, and full-solution deposition.

11.4.2.2 Improving of Photovoltaic Performance

The improvement of the interconnect layer is crucial for enhancing the thermal stability and overall performance of TSCs. The commonly used PEDOT:PSS in HTL and BCP in ETL have been found to be unstable under certain conditions, leading to perovskite decomposition and device defects. Therefore, selecting an appropriate interconnect layer with high transparency, stability, good electrical contact, and carrier mobility is essential for maintaining device stability. Chou et al.[152] used MoO_3/Al/ZnO layer to replace the PEDOT:PSS layer as ICL. MoO_3 and ZnO layers are used for effective hole and electron collection, while the Al layer is inserted in the middle to ensure optimal electrical contact and processing performance. An ICL with high transparency and stability was thus obtained. However, the metal layer between ETL and HTL will diffuse into perovskite under light and heat conditions, forming insulating metal halide defects and inducing the decomposition of the perovskite. Gao et al.[153] used nickel oxide in ethanol, atomic layer deposited SnO_2, and ITO nanoparticles as tunnel junctions (ALD-SnO_2/ITO NCs/E-NiO*x*) (Figure 11.13e). The thermal stability of all-perovskite TSC has been significantly improved. The encapsulated device can maintain 85% initial performance after aging at 85 °C for 2500 h and can maintain 80% initial efficiency after continuous irradiation at 65 °C for 900 h.

An excellent interconnect layer requires good Ohmic contact with the subcell, high light transmittance, high carrier mobility, and low lateral conductivity, ensuring that the carriers within the subcell reach the interconnect layer, complete the recombination, and prevent the shunt of the interconnect layer. The balance between extracting holes and electron density from the subcell will promote effective recombination at the interconnect layer. However, the energy level mismatch between the perovskite layer and the functional layer (ETL and HTL) and the defect state of the perovskite layer affect the mobility of electrons and holes to the interconnect layer. Yang et al.[154] used nanoporous ZnO films to decrease the roughness of ZnO/SnO_2 double-layer ETL, reduce the volume defects of perovskite films, and improve the electron mobility to the interconnect layer. Using organic semiconductor D18-Cl as the HTL, a quasi-Ohmic contact can be established, reducing hole accumulation between the perovskite layer and HTL, achieving a balance between hole and electron density extracted by the interconnect layer (Figure 11.13f). A perovskite/organic TSC with a significant PCE of 23.17% was prepared. Chen et al.[155] used trimethylammonium chloride (TACl) to reconstruct the surface of all-inorganic perovskite, resulting in high-quality perovskite films with low surface ion defects and well-aligned Fermi level, promoting high-quality growth of the interconnect layer, efficient charge extraction, and recombination at the interconnect layer (Figure 11.13g).

In summary, an excellent interconnect layer plays a vital role in improving photovoltaic performance, and careful design is required to achieve optimal stability and efficiency.

11.4.2.3 Passivating Defect of the Interconnect Layer

Improving the interconnect layer is crucial for enhancing the performance of TSC. Nevertheless, the existing interconnect layer has numerous defects that need to be addressed. Previously, the interconnect layer was composed of PEDOT:PSS and n-type metal oxides. The use of PEDOT:PSS biased towards neutrality can negatively affect the V_{oc} of the device by preventing the decomposition of the ETL. Zhou et al.[156] demonstrated that utilizing pH-neutral conjugated polyelectrolytes (CPEs) as p-type HTLs and metal oxides as new interconnect layers for n-type ETLs played a vital role in improving device efficiency. The TCO material used in the interconnect layer has parasitic absorption that produces optical loss, which adversely affects the performance of the tandem device. To address this, Sahli et al.[157] proposed using a nanocrystalline silicon composite layer to replace the TCO. This alternative solution effectively reduces parasitic absorption and shunt loss due to its low transverse conductivity. Furthermore, a perovskite/silicon TSC with an impressive efficiency of 22.7% was fabricated using nanocrystalline silicon as a recombination layer (Figure 11.13h).

The interconnect layer is critical for tandem devices, but there are still many defects in the current technology, making it challenging to create an ideal interconnect layer. Significant effort is required to overcome these challenges.

11.5 PROSPECTS OF ST-PSCs FOR THE APPLICATIONS OF BIPV AND TSCs

ST-PSCs have the potential to promote the practical application and large-scale production of BIPV and TSCs. However, to achieve real commercialization, certain conditions must be met, including maximizing device lifetime, reducing harm to the environment and creatures, and reproducible large-scale production. Therefore, this section aims to explore the application prospects of ST-PSCs in BIPV and TSCs.

11.5.1 Maximizing Device Lifetime of ST-PSCs

The long service life of ST-PSCs is a prerequisite for their large-scale production. Maximizing their service life requires high stability. Unfortunately, when ST-PSCs are exposed to the atmospheric environment, light, heat, and humidity can cause chemical decomposition, which affects the practical application of PSCs. Currently, few studies focus on improving the stability of ST-PSCs; however, their stability control is crucial for commercialization. Doping cations in perovskite materials improves the stability of PSCs. For instance, Niu et al.[158] found that doping a small amount of Cs ions in $MAPbI_3$ could significantly improve the thermal stability of the device (Figure 11.14a). Zhang et al.[159] doped Rb into the perovskite precursor solution, extending the shelf life of the unpackaged $FA_{0.95}Rb_{0.05}PbI_3$ device under

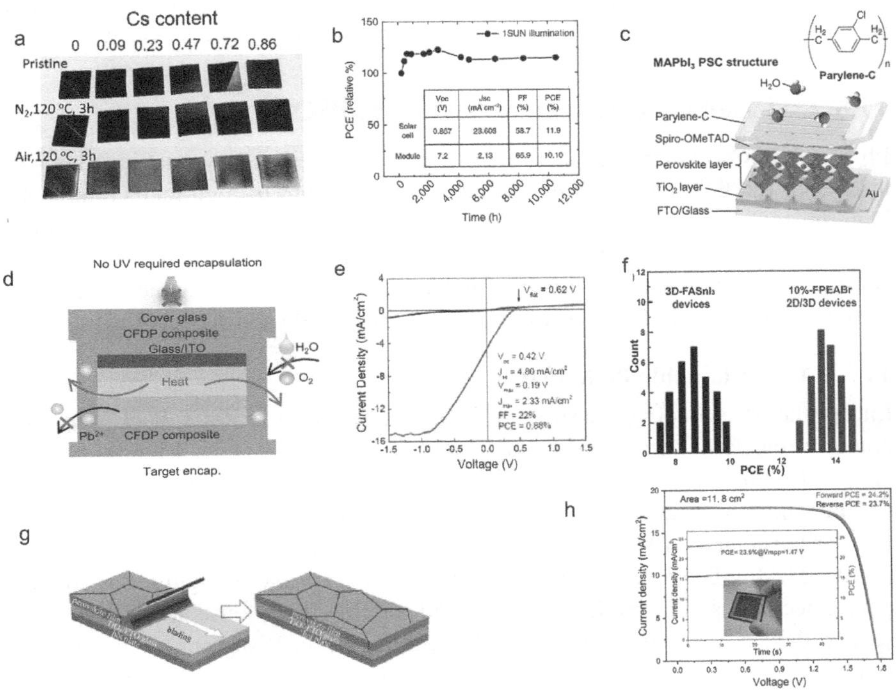

FIGURE 11.14 (a) Photographs of perovskite films exposing to different atmosphere.[158] (b) Typical module stability test under 1 sun AM 1.5 G conditions at stabilized temperature of 55° and at short circuit conditions.[160] (c) Schematic of the $MAPbI_3$ PSC structure with Parylene-C encapsulation and the chemical structure of Parylene-C.[161] (d) Scheme of the PSCs with target encapsulation.[162] (e) Photocurrent vs. voltage in dark (black) and under light illumination (blue) with a power density of 50 mW/cm^2, equivalent to the half of the intensity of AM 1.5 standard radiations.[163] Copyright 2012, AIP Publishing. (f) The reproducibility of the control cells and 2D/3D-based PSCs from 30 devices.[164] Copyright 2021, Wiley-VCH GmbH. (g) Descriptive illustration of the blade-coating process.[166] Copyright 2020, Higher Education Press. (h) Performance of champion tandem devices of 11.8 cm^2 sizes.[167] Copyright 2023, RSC Publishing.

environmental conditions to one month, while also improving its wet stability. In addition to doping ions in the perovskite precursor solution, adding molecules with amino functional groups to the solution has formed a 2D perovskite structure that enhances device stability. Grancini et al.[160] formed a 2D/3D ($(HOOC(CH_2)_4NH_3)_2PbI_4$/$CH_3NH_3PbI_3$ perovskite junction, combining the high stability of 2D with high efficiency of 3D. The hydrophobic carbon electrodes and hole-transporting material-free solar devices maintained significant stability for over 400 d without any efficiency loss (Figure 11.14b). In addition, the metal electrode of ST-PSCs may react with the decomposed iodine in perovskite to produce an irreversible iodide, which affects the stability of devices. However, the decomposition of perovskite can be minimized by incorporating a buffer layer with high transparency and hydrophobicity between the electrode and the functional layer. Improving the stability of ST-PSCs is critical for the application of BIPV and TSCs. The expected service life of windows and buildings in the construction industry should be around 5 to 25 years, which is far from the current service life of the device. Therefore, further research is necessary.

11.5.2 Tackling Environmental Issue of ST-PSCs

Pb-based PSCs exhibit excellent energy conversion efficiency, making them the leaders in the field of photovoltaics. However, exposure to high temperatures and humidity can lead to the decomposition of perovskite films into toxic heavy metal compounds such as PbI_2, Pb, or PbO. These compounds pose a risk to human health, causing diseases such as hypertension and nephrotoxicity. Ensuring the commercialization of Pb-based PSCs requires measures to reduce Pb leakage and the use of Pb.

11.5.2.1 Sustainable Pb Management

Encapsulation is an effective strategy for Pb management in PSCs. Packaging technology protects devices from damage caused by light, heat, and humidity while preventing Pb from harming the environment and human health. At present, both organic and inorganic packaging materials have been used for this purpose. Kim et al.[161] encapsulated $MAPbI_3$ solar cells using poly(p-chloro-xylylene) (Parylene-C) (Figure 11.14c). The results show that Parylene-C encapsulation significantly reduces device decomposition, resulting in improved stability. As a consequence, the encapsulated device maintains the initial PCE within 196 h. Wang et al.[162] used a self-synthesized crosslinked fluoropropyl methylsiloxane-dimethylsiloxane multiblock polymer (CFDP) as the encapsulation layer. CFDP was found to be effective in maintaining device efficiency, with good light transmittance, UV stability, water and oxygen resistance, and improved heat dissipation capacity (Figure 11.14d). The obtained device maintained an initial efficiency of 98% and 95% under hygrothermal and thermal cycling tests, respectively, which meets the International Electrotechnical Commission (IEC) IEC61215 standard. In addition, chemical absorption materials with functional chemical groups such as carboxyl, sulfhydryl, or sulfinyl groups are used for Pb management. These materials capture Pb ions through strong chelation with the functional groups, reducing Pb leakage.

Effective management of Pb leakage and toxicity is vital for the commercialization of Pb-based PSCs, which hold great potential in photovoltaics. Encapsulation technologies that utilize organic and inorganic materials, as well as chemical absorption materials with functional groups, are effective strategies to reduce environmental damage and human health risks associated with Pb-based PSCs.

11.5.2.2 Development of Lead-Free or Lead-Less ST-PSCs

Packaging technology plays an important role in extending the use time of devices and preventing Pb leakage. However, current packaging technologies are expensive and have cumbersome processes that make them unsuitable for large-scale production. Consequently, it is becoming popular to substitute Pb ions with other materials in perovskite materials, with Sn being the most commonly used material as its electronic structure and ionic radius are similar to Pb. Furthermore, the bandgap of Sn-based perovskite materials can be adjusted between 1.2–1.4 eV, which is very close to the optimal band gap (1.34 eV) of the Shockley-Queisser limit. Chen et al.[163] synthesized $CsSnI_3$ (1.3 eV) thin films on glass substrates by alternate deposition of $SnCl_2$ and CsI layers on glass substrates followed by thermal annealing. The efficiency of Schottky solar cells is 0.9% (Figure 11.14e). However, there are two major problems in the development of Sn-based perovskites. First, Sn^{2+} is easily oxidized to Sn^{4+}, resulting in poor photoelectric performance and stability. Second, the rapid crystallization of Sn-based perovskites leads to poor film quality.

From 2012 to present, strategies such as crystallization regulation engineering and defect passivation engineering have been adopted to optimize the efficiency of pure Sn-based PSCs. He's group[164] added 4-fluoro-phenethylammonium bromide (FPEABr) to the precursor solution of $FASnI_3$ to form a 2D/3D mixed tin perovskite film. The 2D phase distributed at the grain boundary and surface of the 3D phase reduces the defect density and inhibits the oxidation of Sn. The 2D/3D Sn-based PSCs with an efficiency of 14.81% were obtained (Figure 11.14f). Although pure Sn-based perovskite has made some progress, its efficiency is still lower than that of Pb-based perovskite. To enhance the efficiency of Sn-based perovskite, researchers have initiated the development of Sn-Pb mixed perovskite materials. In 2014, Ogomi et al.[165] synthesized a $CH_3NH_3Sn_{0.5}Pb_{0.5}I_3$ perovskite film by adding PbI_2 to the SnI_2 precursor solution, which exhibited photovoltaic properties that tin halide perovskite films do not possess. Moreover, the film has a broader range of light absorption than the pure Pb-based perovskite film, expanding to an absorption edge of 1060 nm. The device achieved an efficiency of 4.18%. This study illustrates the reliability of perovskite films with partial Sn substitution for Pb. To date, the efficiency of Sn-Pb mixed perovskite has reached 23.7%, which is beneficial to the subsequent development of Pb-free and Pb-less PSCs.

11.5.3 Evaluating Large-Scale Manufacturing of Perovskite-Based BIPV and TSCs

The laboratory has made some progress in ST-PSCs. However, for them to be commercialized in the field of BIPV and TSCs, large-scale production is necessary. The spin-coating method, commonly used in the laboratories, is only suitable for preparing cells with an area of up to 1 cm^2. Therefore, it is not appropriate for producing large-area devices. Ying et al.[166] used a doctor-blade coating method with

CTAB surfactant as an additive to prepare $FAPbBr_3$ thin films with uniform pore-free feature (Figure 11.14g). An AVT and PCE of 25.42% and 5.11%, respectively, were achieved for the 5*5 cm ST-PSCs with silver nanowires as the top electrode. Zheng et al.[167] added an ultra-thin layer of ITO between perovskite and silicon cells. They discovered that adding a 1.7 nm ultrathin ITO layer reduces the valence band offset between the SnO_2 layer and promotes the electron capture of the SnO_2 layer. Additionally, the ultra-thin ITO layer has limited lateral conductivity, which is advantageous to reducing the local shunt of large-area tandem devices. They fabricated a tandem device with a PCE of 24.2% and an area of 11.8 cm^2 (Figure 11.14h). These works demonstrate the feasibility of large-scale fabrication of ST-PSCs. However, large-scale manufacturing of ST-PSCs still requires further research. For example, when printing, blade coating and other technologies are used, appropriate inks must be selected to prepare high-quality films. The expansion of the device area is accompanied by an increase in defect density. Therefore, appropriate additives should be selected to passivate film defects and improve the efficiency of the device. In the preparation of large-area tandem devices, the appropriate preparation process and solvent must be chosen to reduce the damage to the bottom cell caused by the top cell. In addition, the proper interconnect layer can promote charge recombination and minimize interface shunt.

In summary, the prospect of ST-PSCs in BIPV and TSCs applications is promising but requires further research and development to address technical, environmental, and economic challenges. With continued efforts, ST-PSCs have great potential for revolutionizing the solar industry and contributing to a sustainable future (Figure 11.15).

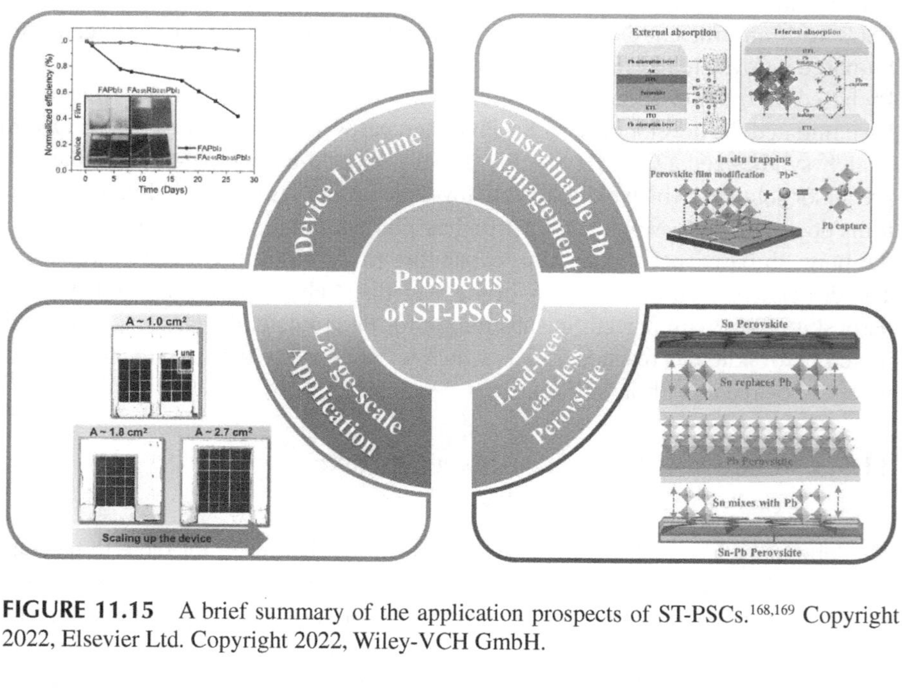

FIGURE 11.15 A brief summary of the application prospects of ST-PSCs.[168,169] Copyright 2022, Elsevier Ltd. Copyright 2022, Wiley-VCH GmbH.

REFERENCES

1. Green M A. Commercial progress and challenges for photovoltaics. *Nature Energy* **2016**, *1*(1), 15015.
2. Kurtz S R, Leilaeioun A, King R R, et al. Revisiting the Terawatt challenge. *MRS Bulletin* **2020**, *45*(3), 159–164.
3. Lou J, Feng J, Liu S, et al. Semitransparent perovskite solar cells for photovoltaic application. *Solar RRL* **2022**, *7*(1), 2200708.
4. Galkowski K, Mitioglu A, Miyata A, et al. Determination of the exciton binding energy and effective masses for methylammonium and formamidinium lead tri-halide perovskite semiconductors. *Energy & Environmental Science* **2016**, *9*(3), 962–970.
5. Dong Q, Fang Y, Shao Y, et al. Electron-hole diffusion lengths> 175 μm in solution-grown CH3NH3PbI3 single crystals. *Science* **2015**, *347*(6225), 967–970.
6. Zhumekenov A A, Saidaminov M I, Haque M A, et al. Formamidinium lead halide perovskite crystals with unprecedented long carrier dynamics and diffusion length. *ACS Energy Letters* **2016**, *1*(1), 32–37.
7. Shi B, Duan L, Zhao Y, et al. Semitransparent perovskite solar cells: From materials and devices to applications. *Advanced Materials* **2020**, *32*(3): e1806474.
8. Shukla A K, Sudhakar K, Baredar P. Recent advancement in BIPV product technologies: A review. *Energy and Buildings* **2017**, *140*, 188–195.
9. Yu G, Yang H, Luo D, et al. A review on developments and researches of building integrated photovoltaic (BIPV) windows and shading blinds. *Renewable and Sustainable Energy Reviews* **2021**, *149*, 111355.
10. Zhou Z, Yuan Z, Yin Z, et al. Progress of semitransparent emerging photovoltaics for building integrated applications. *Green Energy & Environment* **2023**, 22.
11. Hossain M I, Qarony W, Ma S, et al. Perovskite/silicon tandem solar cells: From detailed balance limit calculations to photon management. *Nano-Micro Letters* **2019**, *11*(1), 58.
12. De Vos A. Detailed balance limit of the efficiency of tandem solar cells. *Journal of Physics D: Applied Physics* **1980**, *13*(5), 839.
13. Ritzer D B, Abdollahi Nejand B, Ruiz-Preciado M A, et al. Translucent perovskite photovoltaics for building integration. *Energy & Environmental Science* **2023**, *16*(5), 2212–2225.
14. Koh T M, Wang H, Ng Y F, et al. Halide Perovskite solar cells for building integrated photovoltaics: Transforming building facades into power generators. *Advanced Materials* **2022**, *34*(25): e2104661.
15. Eperon G E, Burlakov V M, Goriely A, Snaith H J. Neutral color semitransparent microstructured perovskite solar cells. *ACS Nano* **2014**, *8*(1), 591–598.
16. Della Gaspera E, Peng Y, Hou Q, et al. Ultra-thin high efficiency semitransparent perovskite solar cells. *Nano Energy* **2015**, *13*, 249–257.
17. Yang C, Liu D, Bates M, et al. How to accurately report transparent solar cells. *Joule* **2019**, *3*(8), 1803–1809.
18. Yang W, Yang Z, Shou C, et al. Optical design and optimization for back-contact perovskite solar cells. *Solar Energy* **2020**, *201*, 84–91.
19. Hörantner M T, Zhang W, Saliba M, et al. Templated microstructural growth of perovskite thin films via colloidal monolayer lithography. *Energy & Environmental Science* **2015**, *8*(7), 2041–2047.
20. Hörantner M T, Nayak P K, Mukhopadhyay S, et al. Shunt-blocking layers for semitransparent perovskite solar cells. *Advanced Materials Interfaces* **2016**, *3*(10), 1500837.
21. Li F R, Xu Y, Chen W, et al. Nanotube enhanced carbon grids as top electrodes for fully printable mesoscopic semitransparent perovskite solar cells. *Journal of Materials Chemistry A* **2017**, *5*(21), 10374–10379.

22. Tavakoli M M, Nasilowski M, Zhao J, et al. Efficient semitransparent $CsPbI_3$ quantum dots photovoltaics using a graphene electrode. *Small Methods* **2019**, *3*(12), 1900449.
23. Tran V-D, Pammi S V N, Park B-J, et al. Transfer-free graphene electrodes for super-flexible and semi-transparent perovskite solar cells fabricated under ambient air. *Nano Energy* **2019**, *65*, 104018.
24. Shen J-J. Recently-explored top electrode materials for transparent organic solar cells. *Synthetic Metals* **2021**, *271*, 116582.
25. Li Y, Ji C, Qu Y, et al. Enhanced light utilization in semitransparent organic photovoltaics using an optical outcoupling architecture. *Advanced Materials* **2019**, *31*(40): e1903173.
26. Hassan S, Dhimish M. Review of current state-of-the-art research on photovoltaic soiling, anti-reflective coating, and solar roads deployment supported by a pilot experiment on a PV road. *Energies* **2022**, *15*(24), 764132.
27. Shanmugam N, Pugazhendhi R, Madurai Elavarasan R, et al. Anti-reflective coating materials: A holistic review from PV perspective. *Energies* **2020**, *13*(10), 2631.
28. Sun J, Li Y, Sun J, et al. Reversible self-powered fluorescent electrochromic windows driven by perovskite solar cells. *Chemical Communications (Cambridge)* **2019**, *55*(80), 12060–12063.
29. Boopathi K M, Hanmandlu C, Singh A, et al. UV- and NIR-protective semitransparent smart windows based on metal halide solar cells. *ACS Applied Energy Materials* **2018**, *1*(2), 632–637.
30. Loper P, Moon S J, de Nicolas S M, et al. Organic-inorganic halide perovskite/crystalline silicon four-terminal tandem solar cells. *Physical Chemistry Chemical Physics* **2015**, *17*(3), 1619–1629.
31. Todorov T, Gershon T, Gunawan O, et al. Monolithic perovskite-CIGS tandem solar cells via in situ band gap engineering. *Advanced Energy Materials* **2015**, *5*(23), 1500799.
32. Ding Y, Guo Q, Geng Y, et al. A low-cost hole transport layer enables CsPbI2Br single-junction and tandem perovskite solar cells with record efficiencies of 17.8% and 21.4%. *Nano Today* **2022**, *46*, 101586.
33. Lai H, Luo J, Zwirner Y, et al. High-performance flexible all-perovskite tandem solar cells with reduced VOC -deficit in wide-bandgap subcell. *Advanced Energy Materials* **2022**, *12*(45), 2202438.
34. Lie S, Bruno A, Wong L H, et al. Semitransparent perovskite solar cells with > 13% efficiency and 27% transparency using plasmonic Au nanorods. *ACS Applied Materials & Interfaces* **2022**, *14*(9), 11339–11349.
35. Kim G M, Tatsuma T. Semi-transparent perovskite solar cells developed by considering human luminosity function. *Scientific Reports* **2017**, *7*(1), 10699.
36. Yang Z, Niu Y, Zhang X, et al. Efficiency improvement of semi-transparent perovskite solar cells via crystallinity enhancement. *Journal of Materials Chemistry A* **2023**, *11*(6), 3070–3079.
37. Wang M, Cao F, Wang M, et al. Intermediate-adduct-assisted growth of stable CsPbI2Br inorganic perovskite films for high-efficiency semitransparent solar cells. *Advanced Materials* **2021**, *33*(10), 2006745.
38. Zuo L, Shi X, Fu W, et al. Highly efficient semitransparent solar cells with selective absorption and tandem architecture. *Advanced Materials* **2019**, *31*(36), 1901683.
39. Zhang L, Hörantner M T, Zhang W, et al. Near-neutral-colored semitransparent perovskite films using a combination of colloidal self-assembly and plasma etching. *Solar Energy Materials and Solar Cells* **2017**, *160*, 193–202.
40. Chen B-X, Rao H-S, Chen H-Y, et al. Ordered macroporous CH3NH3PbI3 perovskite semitransparent film for high-performance solar cells. *Journal of Materials Chemistry A* **2016**, *4*(40), 15662–15669.
41. Eperon G E, Burlakov V M, Goriely A, et al. Neutral color semitransparent microstructured perovskite solar cells. *ACS Nano* **2014**, *8*(1), 591–598.

42. Mujahid M, Chen C, Zhang J, et al. Recent advances in semitransparent perovskite solar cells. *InfoMat* **2020**, *3*(1), 101–124.
43. Hu Z, Wang J, Ma X, et al. A critical review on semitransparent organic solar cells. *Nano Energy* **2020**, *78*, 105376.
44. Lunt R R. Theoretical limits for visibly transparent photovoltaics. *Applied Physics Letters* **2012**, *101*(4).
45. Tai Q, Yan F. Emerging semitransparent solar cells: Materials and device design. *Advanced Materials* **2017**, *29*(34), 1700192.
46. Chen K-S, Salinas J-F, Yip H-L, et al. Semi-transparent polymer solar cells with 6% PCE, 25% average visible transmittance and a color rendering index close to 100 for power generating window applications. *Energy & Environmental Science* **2012**, *5*(11), 9551–9557.
47. Xia Y, Liang X, Jiang Y, et al. High-efficiency and reliable smart photovoltaic windows enabled by multiresponsive liquid crystal composite films and semi-transparent perovskite solar cells. *Advanced Energy Materials* **2019**, *9*(33), 1900720.
48. Cannavale A, Eperon G E, Cossari P, et al. Perovskite photovoltachromic cells for building integration. *Energy & Environmental Science* **2015**, *8*(5), 1578–1584.
49. Pines D. Collective energy losses in solids. *Reviews of Modern Physics* **1956**, *28*(3), 184–198.
50. Kim G M, Tatsuma T. Photocurrent enhancement of perovskite solar cells at the absorption edge by electrode-coupled plasmons of silver nanocubes. *The Journal of Physical Chemistry C* **2017**, *121*(21), 11693–11699.
51. Enrichi F, Quandt A, Righini G C. Plasmonic enhanced solar cells: Summary of possible strategies and recent results. *Renewable and Sustainable Energy Reviews* **2018**, *82*, 2433–2439.
52. Kawawaki T, Wang H, Kubo T, et al. Efficiency enhancement of PbS quantum dot/ZnO nanowire bulk-heterojunction solar cells by plasmonic silver nanocubes. *ACS Nano* **2015**, *9*(4), 4165–4172.
53. Rahman M S, Ibrahim M, Munira S, et al. Plasmon enhanced semitransparent planar perovskite solar cells with copper nanocubes: FDTD study[C]//2019 IEEE PES Asia-Pacific Power and Energy Engineering Conference (APPEEC). *IEEE*, 2019, 1–6.
54. Kim K, Nam S K, Cho J, et al. Photon upconversion-assisted dual-band luminescence solar concentrators coupled with perovskite solar cells for highly efficient semi-transparent photovoltaic systems. *Nanoscale* **2020**, *12*(23), 12426–12431.
55. Roldán-Carmona C, Malinkiewicz O, Betancur R, et al. High efficiency single-junction semitransparent perovskite solar cells. *Energy & Environmental Science* **2014**, *7*(9), 2968–2973.
56. Bett A J, Winkler K M, Bivour M, et al. Semi-transparent perovskite solar cells with ITO directly sputtered on spiro-OMeTAD for tandem applications. *ACS Applied Materials Interfaces* **2019**, *11*(49), 45796–45804.
57. Dudem B, Jung J W, Yu J S. Improved light harvesting efficiency of semitransparent organic solar cells enabled by broadband/omnidirectional subwavelength antireflective architectures. *Journal of Materials Chemistry A* **2018**, *6*(30), 14769–14779.
58. Huang Y, Jin R, Xiong Z, et al. Hexagonal-tiled indium tin oxide electrodes to enhance light trapping in perovskite solar cells. *ACS Applied Nano Materials* **2018**, *1*(11), 6159–6167.
59. Shi B, Liu B, Luo J, et al. Enhanced light absorption of thin perovskite solar cells using textured substrates. *Solar Energy Materials and Solar Cells* **2017**, *168*, 214–220.
60. Jošt M, Köhnen E, Morales-Vilches A B, et al. Textured interfaces in monolithic perovskite/silicon tandem solar cells: Advanced light management for improved efficiency and energy yield. *Energy & Environmental Science* **2018**, *11*(12), 3511–3523.
61. Hecht D S, Hu L, Irvin G. Emerging transparent electrodes based on thin films of carbon nanotubes, graphene, and metallic nanostructures. *Advanced Materials* **2011**, *23*(13), 1482–1513.

62. Pang S, Chen D, Zhang C, et al. Efficient bifacial semitransparent perovskite solar cells with silver thin film electrode. *Solar Energy Materials and Solar Cells* **2017**, *170*, 278–286.
63. Ying Z, Chen W, Lin Y, et al. Supersmooth Ta2O5/Ag/polyetherimide film as the rear transparent electrode for high performance semitransparent perovskite solar cells. *Advanced Optical Materials* **2018**, *7*(4), 1801409.
64. Yang D, Zhang X, Hou Y, et al. 28.3%-efficiency perovskite/silicon tandem solar cell by optimal transparent electrode for high efficient semitransparent top cell. *Nano Energy* **2021**, *84*, 105934.
65. Giuliano G, Cataldo S, Scopelliti M, et al. Nonprecious copper-based transparent top electrode via seed layer–assisted thermal evaporation for high-performance semi-transparent n-i-p perovskite solar cells. *Advanced Materials Technologies* **2019**, *4*(5), 1800688.
66. Fang Y, Wu Z, Li J, et al. High-performance hazy silver nanowire transparent electrodes through diameter tailoring for semitransparent photovoltaics. *Advanced Functional Materials* **2018**, *28*(9), 1705409.
67. Xie M, Lu H, Zhang L, et al. Fully solution-processed semi-transparent perovskite solar cells with ink-jet printed silver nanowires top electrode. *Solar RRL* **2018**, *2*(2), 1700184.
68. Dai X, Zhang Y, Shen H, et al. Working from both sides: Composite metallic semitransparent top electrode for high performance perovskite solar cells. *ACS Applied Materials Interfaces* **2016**, *8*(7), 4523–45231.
69. Dou Y, Liu Z, Wu Z, et al. Self-augmented ion blocking of sandwiched 2D/1D/2D electrode for solution processed high efficiency semitransparent perovskite solar cell. *Nano Energy* **2020**, *71*, 104567.
70. You P, Liu Z, Tai Q, et al. Efficient semitransparent perovskite solar cells with graphene electrodes. *Advanced Materials* **2015**, *27*(24), 3632–3638.
71. Lim S-H, Seok H-J, Kwak M-J, et al. Semi-transparent perovskite solar cells with bidirectional transparent electrodes. *Nano Energy* **2021**, *82*, 105703.
72. Jeong M J, Lee J H, You C H, et al. Oxide/halide/oxide architecture for high performance semi-transparent perovskite solar cells. *Advanced Energy Materials* **2022**, *12*(31), 2200661.
73. Zhang Y, Wu Z, Li P, et al. Fully solution-processed TCO-free semitransparent perovskite solar cells for tandem and flexible applications. *Advanced Energy Materials* **2018**, *8*(1), 1701569.
74. Gunes U, Bag Celik E, Akgul C C, et al. A thienothiophene-based cation treatment allows semitransparent perovskite solar cells with improved efficiency and stability. *Advanced Functional Materials* **2021**, *31*(42), 2103130.
75. Wang B, Ma J, Li Z, et al. Bioinspired molecules design for bilateral synergistic passivation in buried interfaces of planar perovskite solar cells. *Nano Research* **2021**, *15*(2), 1069–1078.
76. Dong Y, Shen W, Dong W, et al. Chlorobenzenesulfonic potassium salts as the efficient multifunctional passivator for the buried interface in regular perovskite solar cells. *Advanced Energy Materials* **2022**, *12*(20), 2200417.
77. Zhang Y-W, Cheng P-P, Tan W-Y, et al. Balance the thickness, transparency and stability of semi-transparent perovskite solar cells by solvent engineering and using a bifunctional additive. *Applied Surface Science* **2021**, *537*, 147908.
78. Dokkhan C, Mokhtar M Z, Ke C-R, et al. Modulating crystallization in semitransparent perovskite films using submicrometer spongelike polymer colloid particles to improve solar cell performance. *ACS Applied Energy Materials* **2019**, *2*(9), 6624–6633.
79. Liu Z, Zhu C, Luo H, et al. Grain regrowth and bifacial passivation for high-efficiency wide-bandgap perovskite solar cells. *Advanced Energy Materials* **2022**, *13*(2), 2203230.

80. Xue Q, Bai Y, Liu M, et al. Dual interfacial modifications enable high performance semi-transparent perovskite solar cells with large open circuit voltage and fill factor. *Advanced Energy Materials* **2017**, *7*(9), 1602333.
81. Liu S, Chen W, Shen Y, et al. An intermeshing electron transporting layer for efficient and stable CsPbI2Br perovskite solar cells with open circuit voltage over 1.3 V. *Journal of Materials Chemistry A* **2020**, *8*(29), 14555–614555.
82. Jung J W, Chueh C-C, Jen A K Y. High-performance semitransparent perovskite solar cells with 10% power conversion efficiency and 25% average visible transmittance based on transparent CuSCN as the hole-transporting material. *Advanced Energy Materials* **2015**, *5*(17), 1500486.
83. Sun J, Zhang N, Wu J, et al. Additive engineering of the CuSCN hole transport layer for high-performance perovskite semitransparent solar cells. *ACS Applied Materials Interfaces* **2022**, *14*(46), 52223–52232.
84. Wheeler L M, Moore D T, Ihly R, et al. Switchable photovoltaic windows enabled by reversible photothermal complex dissociation from methylammonium lead iodide. *Nature Communications* **2017**, *8*(1), 1722.
85. Halder A, Choudhury D, Ghosh S, et al. Exploring thermochromic behavior of hydrated hybrid perovskites in solar cells. *The Journal of Physical Chemistry Letters* **2015**, *6*(16), 3180–3184.
86. Noh J H, Im S H, Heo J H, et al. Chemical management for colorful, efficient, and stable inorganic-organic hybrid nanostructured solar cells. *Nano Letters* **2013**, *13*(4), 1764–1749.
87. Jiang Y, Luo B, Jiang F, et al. Efficient colorful perovskite solar cells using a top polymer electrode simultaneously as spectrally selective antireflection coating. *Nano Letters* **2016**, *16*(12), 7829–7835.
88. Yoo G Y, Azmi R, Kim C, et al. Stable and colorful perovskite solar cells using a nonperiodic SiO(2)/TiO(2) multi-nanolayer filter. *ACS Nano* **2019**, *13*(9), 10129–10139.
89. Ramirez Quiroz C O, Bronnbauer C, Levchuk I, et al. Coloring semitransparent perovskite solar cells via dielectric mirrors. *ACS Nano* **2016**, *10*(5), 5104–5112.
90. He F, You H, Li X, et al. Transparent ultrathin metal electrode with microcavity configuration for highly efficient TCO-free perovskite solar cells. *Materials* **2020**, *13*(10), 2328.
91. Lee K-T, Fukuda M, Joglekar S, et al. Colored, see-through perovskite solar cells employing an optical cavity. *Journal of Materials Chemistry C* **2015**, *3*(21), 5377–5382.
92. Lu J-H, Yu Y-L, Chuang S-R, et al. High-performance, semitransparent, easily tunable vivid colorful perovskite photovoltaics featuring Ag/ITO/Ag microcavity structures. *The Journal of Physical Chemistry C* **2016**, *120*(8), 4233–4239.
93. Lee K-T, Jang J-Y, Ha N Y, et al. High-performance colorful semitransparent perovskite solar cells with phase-compensated microcavities. *Nano Research* **2018**, *11*(5), 2553–2561.
94. Yeom K M, Kim S U, Woo M Y, et al. Recent progress in metal halide perovskite-based tandem solar cells. *Advanced Materials* **2020**, *32*(51): e2002228.
95. Shen H, Duong T, Peng J, et al. Mechanically-stacked perovskite/CIGS tandem solar cells with efficiency of 23.9% and reduced oxygen sensitivity. *Energy & Environmental Science* **2018**, *11*(2), 394–406.
96. Han Q, Hsieh Y T, Meng L, et al. High-performance perovskite/Cu (In, Ga) Se2 monolithic tandem solar cells. *Scienc* 2018, *361*(6405), 904–908.
97. Kim D H, Muzzillo C P, Tong J, et al. Bimolecular additives improve wide-band-gap perovskites for efficient tandem solar cells with CIGS. *Joul* 2019, *3*(7), 1734–1745.
98. Chen C, Liang J, Zhang J, et al. Interfacial engineering of a thiophene-based 2D/3D perovskite heterojunction for efficient and stable inverted wide-bandgap perovskite solar cells. *Nano Energy* **2021**, *90*, 106608.

99. Tang L, Wang X, Liu X, et al. Mixed solvents assisted post-treatment enables high-efficiency single-junction perovskite and 4T perovskite/CIGS tandem solar cells. *Advanced Science (Weinheim)* **2022**, *9*(23): e2201768.
100. Feeney T, Hossain I M, Gharibzadeh S, et al. Four-Terminal perovskite/copper indium gallium selenide tandem solar cells: Unveiling the path to >27% in power conversion efficiency. *Solar RRL* **2022**, *6*(12), 2200662.
101. Luo J, Tang L, Wang S, et al. Manipulating Ga growth profile enables all-flexible high-performance single-junction CIGS and 4T perovskite/CIGS tandem solar cells. *Chemical Engineering Journal* **2023**, 455.
102. Li Z, Zhao Y, Wang X, et al. Cost analysis of perovskite tandem photovoltaics. *Joule* **2018**, *2*(8), 1559–1572.
103. De Bastiani M, Jalmood R, Liu J, et al. Monolithic perovskite/silicon tandems with >28% efficiency: Role of silicon-surface texture on perovskite properties. *Advanced Functional Materials* **2022**, *33*(4), 2205557.
104. Werner J, Barraud L, Walter A, et al. Efficient near-infrared-transparent perovskite solar cells enabling direct comparison of 4-terminal and monolithic perovskite/silicon tandem cells. *ACS Energy Letters* **2016**, *1*(2), 474–480.
105. Sahli F, Werner J, Kamino B A, et al. Fully textured monolithic perovskite/silicon tandem solar cells with 25.2% power conversion efficiency. *Nature Materials* **2018**, *17*(9), 820–826.
106. Mao L, Yang T, Zhang H, et al. Fully textured, production-line compatible monolithic perovskite/silicon tandem solar cells approaching 29% efficiency. *Advanced Materials* **2022**, *34*(40): e2206193.
107. Luo X, Luo H, Li H, et al. Efficient perovskite/silicon tandem solar cells on industrially compatible textured silicon. *Advanced Material* **2023**, *35*(9): e2207883.
108. Roger J, Schorn L K, Heydarian M, et al. Laminated monolithic perovskite/silicon tandem photovoltaics. *Advanced Energy Materials* **2022**, *12*(27), 2200961.
109. Chen B, Wang P, Li R, et al. A two-step solution-processed wide-bandgap perovskite for monolithic silicon-based tandem solar cells with >27% efficiency. *ACS Energy Letters* **2022**, *7*(8), 2771–2780.
110. Subbiah A S, Isikgor F H, Howells C T, et al. High-performance perovskite single-junction and textured perovskite/silicon tandem solar cells via slot-die-coating. *ACS Energy Letters* **2020**, *5*(9), 3034–3040.
111. Xu K, Al-Ashouri A, Peng Z W, et al. Slot-die coated triple-halide perovskites for efficient and scalable perovskite/silicon tandem solar cells. *ACS Energy Letters* **2022**, *7*(10), 3600–3611.
112. Chen B, Yu Z J, Manzoor S, et al. Blade-coated perovskites on textured silicon for 26%-efficient monolithic perovskite/silicon tandem solar cells. *Joule* **2020**, *4*(4), 850–864.
113. Chen B, Yu Z, Liu K, et al. Grain engineering for perovskite/silicon monolithic tandem solar cells with efficiency of 25.4%. *Joule* **2019**, *3*(1), 177–190.
114. Tao J, Liu X, Shen J, et al. F-type pseudo-halide anions for high-efficiency and stable wide-band-gap inverted perovskite solar cells with fill factor exceeding 84. *ACS Nano* **2022**, *16*(7), 10798–10810.
115. Isikgor F H, Furlan F, Liu J, et al. Concurrent cationic and anionic perovskite defect passivation enables 27.4% perovskite/silicon tandems with suppression of halide segregation. *Joule* **2021**, *5*(6), 1566–1586.
116. Duong T, Pham H, Kho T C, et al. High efficiency perovskite-silicon tandem solar cells: Effect of surface coating versus bulk incorporation of 2D perovskite. *Advanced Energy Materials* **2020**, *10*(9), 1903553.
117. Li Y, Shi B, Xu Q, et al. Wide bandgap interface layer induced stabilized perovskite/silicon tandem solar cells with stability over ten thousand hours. *Advanced Energy Materials* **2021**, *11*(48), 2102046.

118. Xu Q, Shi B, Li Y, et al. Conductive passivator for efficient monolithic perovskite/silicon tandem solar cell on commercially textured silicon. *Advanced Energy Materials* **2022**, *12*(46), 2202404.
119. Zheng L, Xuan Y, Wang J, et al. Inverted perovskite/silicon V-shaped tandem solar cells with 27.6% efficiency via self-assembled monolayer-modified nickel oxide layer. *Journal of Materials Chemistry A* **2022**, *10*(13), 7251–7262.
120. Zhumagali S, Isikgor F H, Maity P, et al. Linked nickel oxide/perovskite interface passivation for high-performance textured monolithic tandem solar cells. *Advanced Energy Materials* **2021**, *11*(40), 2101662.
121. Li R, Chen B, Ren N, et al. CsPbCl(3) -cluster-widened bandgap and inhibited phase segregation in a wide-bandgap perovskite and its application to NiO(x) -based perovskite/silicon tandem solar cells. *Advanced Materials* **2022**, *34*(27): e2201451.
122. Chen W, Zhang J, Xu G, et al. A semitransparent inorganic perovskite film for overcoming ultraviolet light instability of organic solar cells and achieving 14.03% efficiency. *Advanced Materials* **2018**, *30*(21): e1800855.
123. Zeng Q, Liu L, Xiao Z, et al. A two-terminal all-inorganic perovskite/organic tandem solar cell. *Science Bulletin (Beijing)* **2019**, *64*(13), 885–887.
124. Wu X, Liu Y, Qi F, et al. Improved stability and efficiency of perovskite/organic tandem solar cells with an all-inorganic perovskite layer. *Journal of Materials Chemistry A* **2021**, *9*(35), 19778–19787.
125. Qin S, Lu C, Jia Z, et al. Constructing monolithic perovskite/organic tandem solar cell with efficiency of 22.0% via reduced open-circuit voltage loss and broadened absorption spectra. *Advanced Materials* **2022**, *34*(11): e2108829.
126. Xie Y M, Yao Q, Zeng Z, et al. Homogeneous grain boundary passivation in wide-bandgap perovskite films enables fabrication of monolithic perovskite/organic tandem solar cells with over 21% efficiency. *Advanced Functional Materials* **2022**, *32*(19), 2112126.
127. Sun S Q, Xu X, Sun Q, et al. All-inorganic perovskite-based monolithic perovskite/organic tandem solar cells with 23.21% efficiency by dual-interface engineering. *Advanced Energy Materials* **2023**, *13*(16), 2204347.
128. Yao Q, Xie Y M, Zhou Y, et al. Dual sub-cells modification enables high-efficiency n–i–p type monolithic perovskite/organic tandem solar cells. *Advanced Functional Materials* **2023**, *33*(8), 2212599.
129. Liu L, Xiao H, Jin K, et al. 4-Terminal inorganic perovskite/organic tandem solar cells offer 22% efficiency. *Nanomicro Letters* **2022**, *15*(1), 23.
130. Eperon G E, Leijtens T, Bush K A, et al. Perovskite-perovskite tandem photovoltaics with optimized band gaps. *Science* **2016**, *354*(6314), 861–865.
131. Tong J, Song Z, Kim D H, et al. Carrier lifetimes of> 1 μs in Sn-Pb perovskites enable efficient all-perovskite tandem solar cells. *Science* **2019**, *364*(6439), 475–479.
132. Lin R, Xu J, Wei M, et al. All-perovskite tandem solar cells with improved grain surface passivation. *Nature* **2022**, *603*(7899), 73–78.
133. Yu Z, Wang J, Chen B, et al. Solution-processed ternary tin (II) alloy as hole-transport layer of Sn-Pb perovskite solar cells for enhanced efficiency and stability. *Advanced Materials* **2022**, *34*(49): e2205769.
134. Datta K, Wang J, Zhang D, et al. Monolithic all-perovskite tandem solar cells with minimized optical and energetic losses. *Advanced Materials* **2022**, *34*(11): e2110053.
135. Hu H, Moghadamzadeh S, Azmi R, et al. Sn-Pb mixed perovskites with fullerene-derivative interlayers for efficient four-terminal all-perovskite tandem solar cells. *Advanced Functional Materials* **2021**, *32*(12), 2107650.
136. Moghadamzadeh S, Hossain I M, Duong T, et al. Triple-cation low-bandgap perovskite thin-films for high-efficiency four-terminal all-perovskite tandem solar cells. *Journal of Materials Chemistry A* **2020**, *8*(46), 24608–24619.

137. Wu P, Wen J, Wang Y, et al. Efficient and thermally stable all-perovskite tandem solar cells using all-FA narrow-bandgap perovskite and metal-oxide-based tunnel junction. *Advanced Energy Materials* **2022**, *12*(48), 2202948.
138. Zhao D, Wang C, Song Z, et al. Four-terminal all-perovskite tandem solar cells achieving power conversion efficiencies exceeding 23%. *ACS Energy Letters* **2018**, *3*(2), 305–306.
139. He R, Yi Z, Luo Y, et al. Pure 2D perovskite formation by interfacial engineering yields a high open-circuit voltage beyond 1.28 V for 1.77-eV wide-bandgap perovskite solar cells. *Advanced Science (Weinheim)* **2022**, *9*(36): e2203210.
140. Yan L, Li Y, Li S, et al. 2D/3D heterostructured CsPbI2Br solar cells: A choice for a monolithic all-perovskite tandem device. *Journal of Materials Chemistry A* **2022**, *10*(28), 14799–14809.
141. Zhao Y, Wang C, Ma T, et al. Reduced 0.418 V VOC-deficit of 1.73 eV wide-bandgap perovskite solar cells assisted by dual chlorides for efficient all-perovskite tandems. *Energy & Environmental Science* **2023**, *5*(16), 2080–2089.
142. Wen J, Zhao Y, Liu Z, et al. Steric engineering enables efficient and photostable wide-bandgap perovskites for all-perovskite tandem solar cells. *Advanced Materials* **2022**, *34*(26): e2110356.
143. Jiang Q, Tong J, Scheidt R A, et al. Compositional texture engineering for highly stable wide-bandgap perovskite solar cells. *Science* **2022**, *378*(6626), 1295–1300.
144. Zhang M, Lin Z. Efficient interconnecting layers in monolithic all-perovskite tandem solar cells. *Energy & Environmental Science* **2022**, *15*(8), 3152–3170.
145. Ko Y, Park H, Lee C, et al. Recent progress in interconnection layer for hybrid photovoltaic tandems. *Advanced Materials* **2020**, *32*(51): e2002196.
146. Thiesbrummel J, Peña-Camargo F, Brinkmann K O, et al. Understanding and minimizing VOC losses in all-perovskite tandem photovoltaics. *Advanced Energy Materials* **2022**, *13*(3), 2202674.
147. Brinkmann K O, Becker T, Zimmermann F, et al. Perovskite-organic tandem solar cells with indium oxide interconnect. *Nature* **2022**, *604*(7905), 280–286.
148. Zhou X, Lai H, Huang T, et al. Suppressing nonradiative losses in wide-band-gap perovskites affords efficient and printable all-perovskite tandem solar cells with a metal-free charge recombination layer. *ACS Energy Letters* **2022**, *8*(1), 502–512.
149. Chen X, Jia Z, Chen Z, et al. Efficient and reproducible monolithic perovskite/organic tandem solar cells with low-loss interconnecting layers. *Joule* **2020**, *4*(7), 1594–1606.
150. Zeng R, Yu R, Jin S, et al. Efficient organic tandem solar cells enabled by solution-processed interconnection layer and fine-tuned active layer. *Advanced Optical Materials* **2021**, *9*(21), 2101246.
151. Chang C-Y, Tsai B-C, Hsiao Y-C, et al. Solution-processed conductive interconnecting layer for highly-efficient and long-term stable monolithic perovskite tandem solar cells. *Nano Energy* **2019**, *55*, 354–367.
152. Chou C H, Kwan W L, Hong Z, et al. A metal-oxide interconnection layer for polymer tandem solar cells with an inverted architecture. *Advanced Materials* **2011**, *23*(10), 1282–1286.
153. Gao H, Lu Q, Xiao K, et al. Thermally stable all-perovskite tandem solar cells fully using metal oxide charge transport layers and tunnel junction. *Solar RRL* **2021**, *5*(12), 2100814.
154. Yang H, Chen W, Yu Y, et al. Regulating charge carrier recombination in the interconnecting layer to boost the efficiency and stability of monolithic perovskite/organic tandem solar cells. *Advanced Materials* **2023**, *35*(6): e2208604.
155. Chen W, Li D, Chen X, et al. Surface reconstruction for stable monolithic all-inorganic perovskite/organic tandem solar cells with over 21% efficiency. *Advanced Functional Materials* **2021**, *32*(5), 2109321.

156. Zhou H, Zhang Y, Mai C K, et al. Polymer homo-tandem solar cells with best efficiency of 11.3%. *Advanced Materials* **2015**, *27*(10), 1767–1773.
157. Sahli F, Kamino B A, Werner J, et al. Improved optics in monolithic perovskite/silicon tandem solar cells with a nanocrystalline silicon recombination junction. *Advanced Energy Materials* **2018**, *8*(6), 1701609.
158. Niu G, Li W, Li J, et al. Enhancement of thermal stability for perovskite solar cells through cesium doping. *RSC Advances* **2017**, *7*(28), 17473–17479.
159. Zhang M, Yun J S, Ma Q, et al. High-efficiency rubidium-incorporated perovskite solar cells by gas quenching. *ACS Energy Letters* **2017**, *2*(2), 438–444.
160. Grancini G, Roldan-Carmona C, Zimmermann I, et al. One-Year stable perovskite solar cells by 2D/3D interface engineering. *Nature Communications* **2017**, *8*, 15684.
161. Kim H, Lee J, Kim B, et al. Enhanced stability of MAPbI(3) perovskite solar cells using poly(p-chloro-xylylene) encapsulation. *Scientific Reports* **2019**, *9*(1), 15461.
162. Wang T, Yang J, Cao Q, et al. Room temperature nondestructive encapsulation via self-crosslinked fluorosilicone polymer enables damp heat-stable sustainable perovskite solar cells. *Nature Communications* **2023**, *14*(1), 1342.
163. Chen Z, Wang J J, Ren Y, et al. Schottky solar cells based on CsSnI3 thin-films. *Applied Physics Letters* **2012**, *101*(9).
164. Yu B B, Chen Z, Zhu Y, et al. Heterogeneous 2D/3D tin-halides perovskite solar cells with certified conversion efficiency breaking 14%. *Advanced Materials* **2021**, *33*(36): e2102055.
165. Ogomi Y, Morita A, Tsukamoto S, et al. CH3NH3SnxPb(1-x)I3 perovskite solar cells covering up to 1060 nm. *The Journal of Physical Chemistry Letters* **2014**, *5*(6), 1004–1011.
166. Ying H, Liu Y, Dou Y, et al. Surfactant-assisted doctor-blading-printed FAPbBr3 films for efficient semitransparent perovskite solar cells. *Frontiers of Optoelectronics* **2020**, *13*(3), 272–281.
167. Zheng J, Duan W, Guo Y, et al. Efficient monolithic perovskite–Si tandem solar cells enabled by an ultra-thin indium tin oxide interlayer. *Energy & Environmental Science* **2023**, *16*(3), 1223–1233.
168. Tyagi B, Lee H B, Kumar N, et al. High-performance, large-area semitransparent and tandem perovskite solar cells featuring highly scalable a-ITO/Ag mesh 3D top electrodes. *Nano Energy* **2022**, *95*, 106978.
169. Luo H, Li P, Ma J, et al. Sustainable Pb management in perovskite solar cells toward eco-friendly development. *Advanced Energy Materials* **2022**, *12*(30), 2201242.

12 Perovskite Indoor Photovoltaics

Chun-Hao Chen, Xin Chen, and Zhao-Kui Wang

12.1 INTRODUCTION

The Internet of Things (IoTs) is a rapidly evolving network that connects physical objects, enabling them to communicate and interact through the Internet and other information channels.[1,2] With the increasing number of self-powered IoTs devices, such as sensors and health monitoring equipment, the demand for energy has become more pronounced. To improve the stability and autonomy of the IoTs system, there is a necessity for off-grid power systems that seamlessly integrate with wireless networks. Currently, portable batteries serve as the primary power source for IoTs devices. However, their low power-to-mass ratio and the need for frequent battery replacements can lead to instability and high energy consumption within the IoTs system. Consequently, there is a higher demand for off-grid power supply equipment that is lightweight, energy-efficient, and cost-effective.[3] Indoor photovoltaics (IPVs) have emerged as a promising solution to meet these requirements, becoming a critical energy support technology for the IoTs.[4–6]

For indoor applications, photovoltaic devices (PVs) must be designed differently from those optimized for outdoor use. Indoor light sources have significantly different intensity and wavelengths compared to outdoor conditions, making PVs optimized for standard outdoor conditions (AM 1.5G) less suitable for indoor use. IPVs require an absorption layer with a band gap of around 1.8~2.0 eV to cover the indoor light spectrum and produce high optical voltage, making perovskite solar cells (PSCs) an ideal choice for IPVs.[7,8] PSCs offer several advantages over other mainstream IPVs. They can achieve high optical voltage output (above 1.2 V) under typical indoor lighting conditions (~1000 lux),[9,10] surpassing the voltage outputs of amorphous silicon solar cells (a-Si),[11] dye-sensitized solar cells (DSSCs),[12] and organic photovoltaics (OPVs).[13] The tunable band gap of perovskite materials plays a key role in achieving high power conversion efficiency (PCE) in both indoor and outdoor conditions.[14,15] Moreover, perovskite indoor photovoltaics (PIPVs) benefit from low production costs and flexible fabrication methods, similar to their outdoor counterparts.[16,17] To establish a competitive edge in terms of energy cost when compared to conventional fossil fuel sources, PIPVs must amalgamate the benefits of affordability, exceptional performance, and prolonged stability. Based on initial life cycle assessment findings, PSC modules are anticipated to manifest reduced environmental impact and a shorter energy payback time (EPBT) metric.[18] Anticipated reductions in PSC production costs are expected due to extended operational durations, which can significantly curtail maintenance

 DOI: 10.1201/9781003400493-12

and replacement expenditures linked to photovoltaic modules. Furthermore, various research teams have meticulously evaluated the production expenses associated with PSC modules, thereby deriving precise projections for the levelized cost of power (LCOE). According to calculations by Egan et al., the manufacturing expenses for PSCs span a range from $87 to $140 per square meter, yielding LCOE values ranging from 9 to 19 cents per kilowatt-hour.[19] This data underscores the potential for PSCs to offer a cost-effective energy generation solution when compared to traditional fossil fuels. However, the use of lead in perovskite materials raises concerns about toxicity in indoor photovoltaic applications.[20,21] Addressing this issue is crucial before mass-market adoption. Importantly, perovskite photovoltaics exhibit improved indoor stability compared to outdoor PSCs, as indoor conditions avoid the strong temperature fluctuations and adverse environmental conditions experienced outdoors. In conclusion, PIPVs hold significant promise as an essential energy support technology for the IoTs. Their ability to generate high optical voltage, combined with low cost and improved indoor stability, makes them a competitive choice for powering self-sustained IoTs devices.[22] Addressing toxicity concerns will be pivotal for their successful commercialization and widespread use in various IoTs applications.

12.2 ACCURATE MEASUREMENT OF IPVs

The fundamental difference between indoor and outdoor photovoltaic power generation lies in the significant disparity in the intensity and wavelength of the received light source. IPVs operate under low-intensity diffuse light primarily in the visible spectrum (350–750 nm) provided by artificial light sources such as compact fluorescent lamps (CFL) and light-emitting diodes (LEDs), as shown in Figure 12.1.[23]

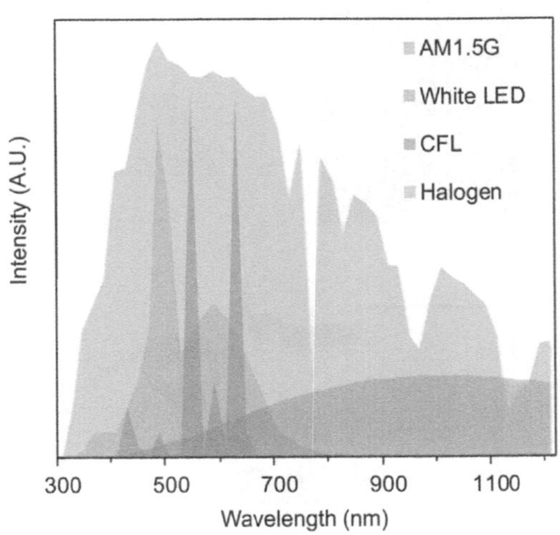

FIGURE. 12.1 The different light spectra of AM 1.5G, white LED, CFL, and halogen. Reprinted with permission from Mathews et al., Joule **3**, 1415 (2019). Copyright 2019, Elsevier Inc.[23]

In contrast, outdoor PSCs function under AM 1.5G conditions, with a spectrum ranging from ultraviolet (150 nm) to mid-infrared (4,000 nm). Accurately evaluating the photovoltaic performance of IPVs becomes crucial due to these differences.[24] The following formula can be used to calculate the illuminance of the indoor light source using the corresponding incident power.[25]

$$E_v = K_m \times \int_0^{\infty} E_\lambda \times V(\lambda) d_\lambda$$

where $V(\lambda)$ indicates CIE spectral brightness factor of human light vision, K_m equals 683 lm/W. To calibrate the indoor light source, researchers have proposed using a universal LED light meter with National Institute of Standards and Technology (NIST)-traceable calibration.[25] Accurate testing of IPVs, as suggested by Hou et al. in Figure 12.2,[26] involves several key elements, such as ensuring the instability of indoor photovoltaic measuring light source intensity is less than 2%, using an optical mask for testing, utilizing a spectrometer for light source calibration to improve accuracy, and limiting the difference between EQE integrated current density and actual current density to within 5%. The incident power density of indoor light sources (CFL and LED) at 1,000 lux illuminance is about 300 μW/cm², resulting in an estimated current density (J_{sc}) limit of approximately 140 μA/cm² for IPVs.[27] In

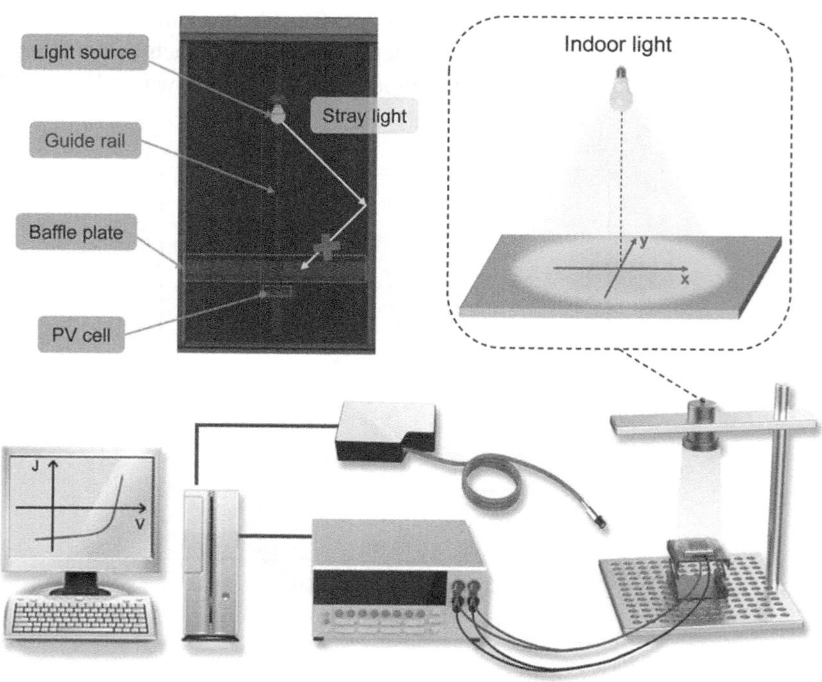

FIGURE 12.2 The typical setup for indoor PCE measurement and schematic illustrations. Reprinted with permission from Cui et al., Joule 5, 1016 (2021). Copyright 2021, Elsevier Inc.

some cases, PIPVs may exhibit current density exceeding this limit, which could be attributed to inaccurate testing methods.

12.3 TUNABLE BAND GAP OF PIPVs

The difference in spectra between indoor and outdoor lighting poses a challenge for mature perovskite photovoltaic devices optimized for outdoor conditions. The band gap of high-performance outdoor PSCs typically ranges from 1.1 to 1.6 eV, while the optimal band gap for PIPVs is around 1.8 to 2.0 eV. This specific band gap allows PIPVs to fully absorb the indoor full-spectrum light source and generate high optical voltage. Unlike other photovoltaic technologies with restrictedly adjustable band gaps (e.g., a-Si, DSSCs, and OPVs), perovskite materials offer a wide and precisely tunable range of band gaps, making them particularly suitable for IPVs.

The crystal structure of perovskite is described as ABX_3, where A is a monovalent cation (such as $CH_3NH_3^+$ (MA), $HC(NH_2)_2^+$ (FA), or Cs), B is a divalent metal cation (Pb or Sn), and X is a halide anion (I, Br, Cl). Figure 12.3 illustrates how the band gap of perovskite can be continuously adjusted from infrared (1.15 eV) to ultraviolet (up to 3 eV) through the engineering of mixed cations or anions.[28] While the mainstream system for outdoor PSCs is typically based on $FAPbI_3$ with a narrow band gap of about 1.5 eV, researchers have applied photovoltaic devices of this system into IPVs and achieved high PCE.[29] However, the narrow band gap results in low optical voltage for indoor photovoltaic applications. Therefore, perovskite materials with appropriate band gap widths need to be developed and utilized for indoor photovoltaic scenarios.

By considering the match between the band gap and the indoor light source spectrum (350–750 nm), $CsPbBrI_2$ (with a band gap of 1.89 eV) and $CsPbI_3$ (with a band gap of 1.7 eV) can be used to prepare high-performance PIPVs.[30,31] Compared with other IPV devices like a-Si (0.73 V, 36% PCE),[11] DSSCs (0.98 V, 35% PCE),[12] and OPVs (1.10 V, 26.1% PCE),[13] PIPVs achieve an open-circuit voltage of more than 1.20 V due to the matched band gap width under 1,000 lux indoor illumination. For instance, Guo et al. employed the perovskite system of $CsPbI_{2.25}Br_{0.75}$ for indoor

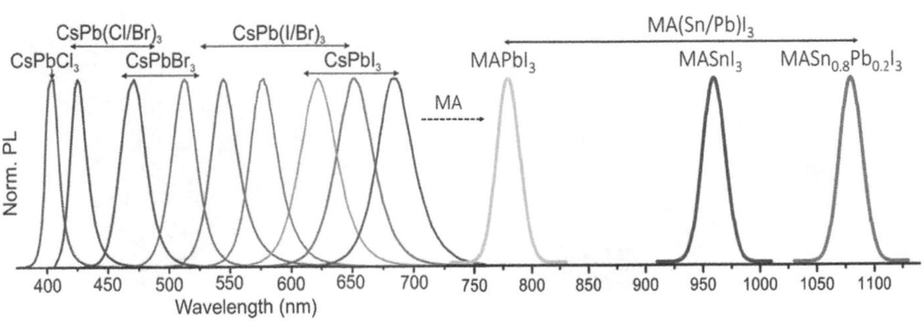

FIGURE 12.3 The tunable band gap of perovskite from infrared to ultraviolet through the composition engineering of mixed cations or anions. Reprinted with permission from Saliba et al., Angew. Chem. Int. Ed. **57**, 2554 (2018). Copyright 2018, Wiley-VCH.

applications and achieved a recorded photovoltage of 1.25 V under 1,000 lux illuminance.[10] In 2023, Wang et al. utilized the polarity of amino naphthalene sulfonates to functionalize TiO_2, anchoring $CsPbI_3$ perovskite crystals onto the substrate through strong ionic dipole interactions between the molecular-level polar layer and the ionic perovskite film.[32] The resulting high-quality $CsPbI_3$ thin-film exhibits advantages of trap tolerance and high shunt resistance under low-light conditions. As a result, a recorded indoor PCE of 41.2% was obtained under 1,062 lux (2,956 K, P_{in}: 334.11 μW cm^{-2}), which is sufficient for practical production and application.

12.4 FLEXIBLE AND LARGE-AREA PIPVs

PIPVs offer a significant advantage in that perovskite thin films can be deposited using a solution method, allowing for low-cost and high-efficiency production on flexible polymer substrates. This flexibility opens up new application possibilities within the IoTs system. Flexible perovskite solar cells (FPSCs) have several unique advantages, including their light weight, high power-to-mass ratio, and high PCE, making them highly suitable for powering self-powered IoTs devices.[33,34] Unlike rigid devices, FPSCs provide a more adaptable and customizable surface for various IoTs applications, catering to specific customer needs. Chen et al. conducted pioneering research on flexible PIPVs and introduced a three-dimensional crosslinking agent called borax to achieve comprehensive stress release for perovskite films (as shown in Figure 12.4).[35] This improved the mechanical stability of the PIPVs under large temperature changes and external forces, leading to a recorded PCE of 31.85% for flexible PIPVs under 1,062 lux illuminance from an LED source with a color temperature of 2,956 K.

To further reduce the cost of flexible perovskite photovoltaic devices, the development of large-area flexible manufacturing technologies is crucial. The continuous roll-to-roll (R2R) process presents a promising approach, characterized by fast fabrication speed and low cost, making it highly suitable for IoTs system applications.[36–38]

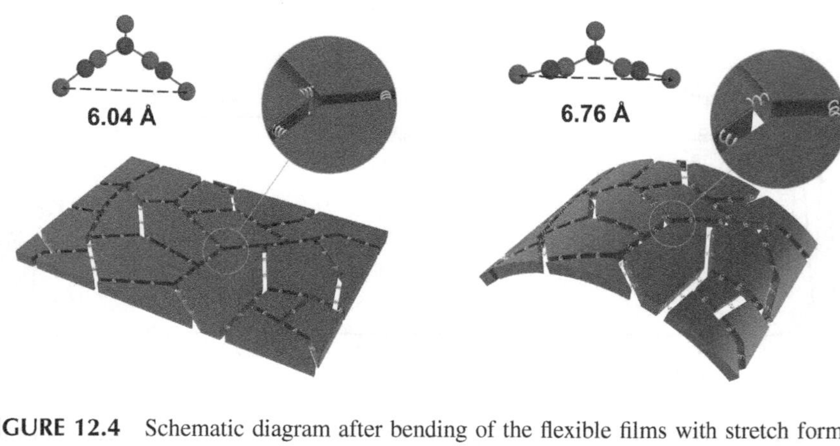

FIGURE 12.4 Schematic diagram after bending of the flexible films with stretch forms of borax. Reprinted with permission from Chen et al., Adv. Mater. **34**, 2200320 (2022). Copyright 2022, Wiley-VCH.

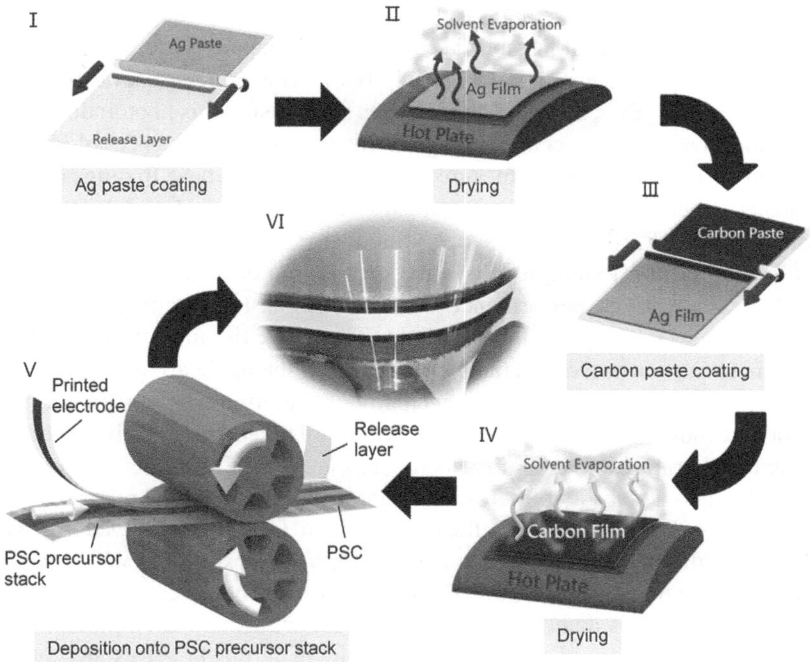

FIGURE 12.5 Flow-chart illustrating the fabrication steps of the flexible printed electrodes. Reprinted with permission from Sutherland et al., Adv. Energy Mater. **12**, 2202142 (2022). Copyright 2022, Wiley-VCH.

In 2022, Sutherland et al. introduced a new vacuum-free R2R compatible method to deposit electrodes based on conductive pastes (Figure 12.5).[39] This approach allowed all functional layers, except for transparent conductive electrodes, to be produced through vacuum-free deposition. The resulting devices achieved a record PCE of up to 16.7% and exhibited outstanding operational and mechanical stability. This large-area flexible PIPV preparation process offers competitive cost advantages for IoTs system applications. In summary, the combination of perovskite's solution-based thin-film deposition, flexibility in manufacturing, and outstanding performance makes flexible PIPVs a promising technology for meeting the energy needs of IoTs devices. The ability to customize and adapt these devices to specific IoTs applications, coupled with their low cost and high efficiency, positions flexible PIPVs as a key player in advancing the integration of IoTs systems with renewable energy sources.

12.5 STABILITY ISSUES OF PIPVs

Indeed, the stability of PSCs is a critical area of study and optimization for their widespread commercialization. While advanced encapsulation technologies can effectively mitigate the impact of water and oxygen on PSCs,[40] photo-induced damage caused by sunlight remains a challenging issue. Brennan et al. identified that the light-induced structural instabilities of $APb(I_{1-x}Br_x)_3$ perovskites are attributed to reversible halide phase segregation (Figure 12.6(a)).[41] However, the use of all-inorganic perovskite $CsPb(I_{1-x}Br_x)_3$ has shown promise in suppressing this photoinduced segregation. Additionally, Liang et al. investigated the origins and influences of Pb^0, which is induced by light and X-ray exposure (Figure 12.6(b)).[42] These findings contribute to a deeper understanding of the mechanisms behind photo-induced damage in perovskite materials. Furthermore, the operating temperature of perovskite devices can increase sharply under sunlight exposure, leading to a decline in dynamic stability and a decrease in device performance. However, indoor light sources typically have significantly lower intensities than sunlight. Therefore, the impact of light on perovskite materials varies considerably depending on the application scenario. Under indoor conditions, PIPVs experience much milder operating environments compared to outdoor conditions. There is no exposure to solar radiation, such as ultraviolet light, and temperature and humidity fluctuations are typically less severe indoors. As a result, the indoor stability of perovskite photovoltaics is greatly improved, and the working time of PIPVs can be significantly extended. Guo et al. observed that intense redshift in both photo-luminescence (PL) and UV-vis absorption, which appeared under 1 sun irradiation for 20 min (Figure 12.7(a)), did not occur under LED light (Figure 12.7(b)).[22] This demonstrates that the mild application environment indoors contributes to the enhanced stability of perovskite materials in PIPVs, making them more suitable for meeting market requirements.

In conclusion, while stability remains an important challenge for PSCs, particularly under intense sunlight exposure, the application environment plays a crucial role in influencing the performance and stability of perovskite materials. The mild conditions experienced indoors offer significant advantages for PIPVs, making them more viable for commercial deployment and mass-market integration in IoTs and other indoor applications. Nonetheless, ongoing research and optimization efforts

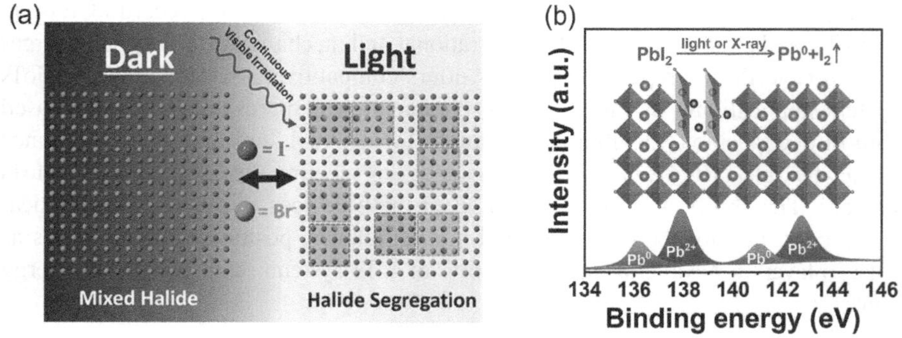

FIGURE 12.6 a) Schematic diagram of the phase segregation of perovskite under illumination. Reprinted with permission from Brennan et al., ACS Energy Lett. **3**, 204 (2018). Copyright 2017, American Chemical Society. b) Schematic diagram on the origins and influences of Pb^0 for PSCs. Reprinted with permission from Liang et al., Joule **6**, 816 (2022). Copyright 2022, Elsevier Inc.

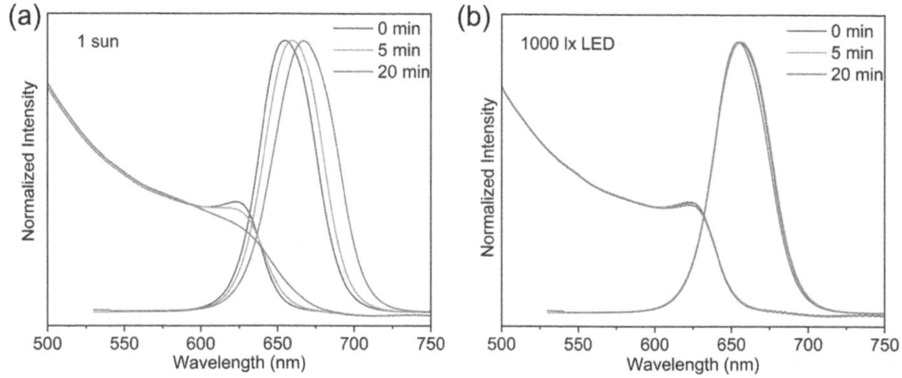

FIGURE 12.7 PL and UV-vis spectra variation of perovskite films under the illumination of a) sunlight and b) LED light. Reprinted with permission from Guo et al., ACS Energy Lett. **8**, 90 (2023). Copyright 2022, American Chemical Society.

are essential to further enhance the stability and reliability of perovskite-based photovoltaic technologies.

12.6 TOXICITY OF LEAD IN PIPVs

Indeed, the toxicity of lead in PSCs is a critical consideration, especially when used in indoor applications like PIPVs, where they may be in closer contact with humans. Lead exposure can have harmful effects on human health, particularly for young children and fetuses. International efforts to prevent and control lead poisoning often involve monitoring lead concentration in blood. Elevated blood lead concentrations in children above safety thresholds, such as 100 μg/L for hyperlipemia and 200 μg/L for lead poisoning, can be of concern. It has been calculated that without proper encapsulation, the lead leakage from a 1.4 cm^2 PSC devices could cause the blood lead concentration of young children to exceed the safety threshold.[43,44] Considering that perovskite photovoltaic modules have much larger surface areas, effective isolation strategies and packaging methods should be developed to minimize human exposure to lead from PIPVs. Researchers have been investigating various encapsulation methods to address lead leakage from PSC modules. Chen et al. used low-cost cation-exchange resins (CERs) to trap lead in PSCs (Figure 12.8(a)).[45] This treatment significantly reduced the Pb leakage rate, offering a promising solution to mitigate lead exposure risks. Additionally, Jiang et al. explored three encapsulation methods (Figure 12.8(b)) and found that epoxy resin-based polymers added between the perovskite layer and the top glass cover were effective in significantly reducing lead leakage rates.[46]

Furthermore, researchers have been exploring lead-free alternatives for PSCs. Yang et al. introduced catechin into perovskite layers to inhibit the oxidation of Sn^{2+}, resulting in $FA_{0.75}MA_{0.25}SnI_2Br$ films with a lower Sn^{4+} ratio.[47] By employing lead-free tin-based perovskites, they achieved a maximum PCE of 12.81% under

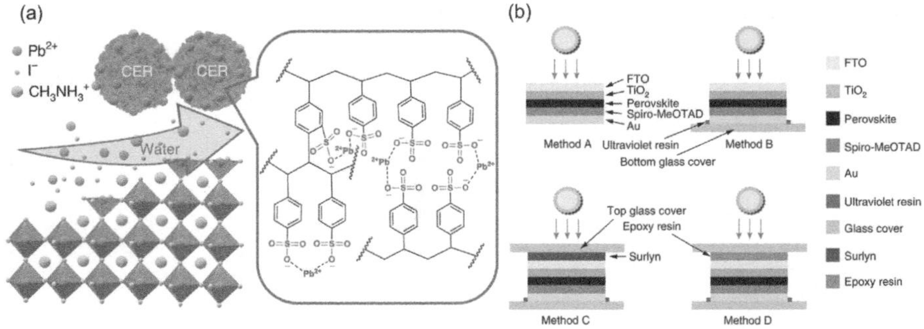

FIGURE 12.8 a) Schematic of how CERs prevent lead leakage. Reprinted with permission from Chen et al., Nat. Energy **5**, 1003 (2020). Copyright 2020, Springer Nature Limited. b) Schematic showing the four different encapsulation methods. Reprinted with permission from Jiang et al., Nat. Energy **4**, 585 (2019). Copyright 2019, Springer Nature Limited.

1,000 lux. This work represents a significant step towards advancing the potential of lead-free PSCs for indoor photovoltaic applications like PIPVs. The development of effective encapsulation methods, lead trapping techniques, and the exploration of lead-free perovskite materials are essential to ensure the safety of PIPVs and their suitability for mass-market adoption in indoor settings. It is crucial to prioritize human health and minimize potential risks associated with lead exposure in PSC-based technologies. Continued research and innovation in these areas will play a vital role in shaping the future of PIPVs and their safe integration into the IoTs and other indoor applications.

12.7 CONCLUSIONS AND OUTLOOK

The rapid development of the IoTs has created a significant energy demand for self-powered devices within the system. To address this energy gap and support the IoTs ecosystem, IPVs have emerged as a crucial technology. Among the various options, PIPVs have gained attention due to their tunable band gap, high optical voltage, and flexible fabrication methods. To ensure the future development and widespread adoption of this promising technology, the following points should be considered and optimized:

Optimize Band Gap: leveraging the tunable band gap properties of perovskite materials is essential to control and tailor it within the range of 1.8–2.0 eV. This adjustment will enable PIPVs to generate higher indoor optical voltage, enhancing the energy-harvesting efficiency under various indoor lighting conditions.

Flexible Large-Area R2R Printing: continuously advancing flexible large-area R2R printing technology will be crucial to reduce the cost of PIPV production. Scalable and cost-effective manufacturing methods will make PIPVs more competitive and accessible for widespread implementation.

Customized Service System: developing a customized service system for flexible PIPV devices is essential to cater to specific customer needs and applications.

Offering tailored solutions will enable the seamless integration of PIPVs into various IoTs devices and systems, further expanding their utility and versatility.

Lead Toxicity Mitigation: addressing the issue of lead toxicity in perovskite materials is of paramount importance. Investing in research and development to enhance encapsulation technologies that effectively reduce lead leakage is vital. Additionally, exploring the development of lead-free perovskite materials can offer similar performance without the potential risks associated with lead exposure.

By focusing on these key areas, the future of PIPVs appears promising, offering a sustainable and efficient energy supply for IoTs devices. Advancements in these aspects will accelerate the integration of PIPVs into the IoTs ecosystem, contributing to a more connected and energy-efficient future.

REFERENCES

1. L. Atzori, A. Iera, G. Morabito, The internet of things: A survey, *Comput. Networks* **2010**, *54*, 2787.
2. F. Xia, L. T. Yang, L. Wang, A. Vinel, Internet of things, *Int. J. Commun. Syst.* **2012**, *25*, 1101.
3. S. Li, L. D. Xu, S. Zhao, The internet of things: a survey, *Inf. Syst Front.* **2015**, *17*, 243.
4. I. Mathews, S. N. Kantareddy, T. Buonassisi, I. M. Peters, Technology and market perspective for indoor photovoltaic cells, *Joule* **2019**, *3*, 1415.
5. J. K. W. Ho, H. Yin, S. K. So, From 33% to 57%—an elevated potential of efficiency limit for indoor photovoltaics, *J. Mater. Chem. A* **2020**, *8*, 1717.
6. M. Freitag, J. Teuscher, Y. Saygili, X. Zhang, F. Giordano, P. Liska, A. Hagfeldt, Dye-sensitized solar cells for efficient power generation under ambient lighting, *Nat. Photonics* **2017**, *11*, 372.
7. A. Abbas, A. Abdollahinia, A. G. Aberle, S. Adachi, J. Adams, J. G. J. Adams, D. Aiken, 2013 index, *IEEE J. Photovoltaics* **2013**, *3*, 1465.
8. M. Li, F. Igbari, Z. K. Wang, L. S. Liao, Indoor thin-film photovoltaics: progress and challenges, *Adv. Energy Mater.* **2020**, *10*, 2000641.
9. Z. Guo, A. K. Jena, I. Takei, M. Ikegami, A. Ishii, Y. Numata, N. Shibayama, T. Miyasaka, Dopant-free polymer HTM-based $CsPbI_2Br$ solar cells with efficiency over 17% in sunlight and 34% in indoor light, *Adv. Funct. Mater.* **2021**, *31*, 2103614.
10. Z. Guo, S. Zhao, N. Shibayama, A. Kumar Jena, I. Takei, T. Miyasaka, A universal method of perovskite surface passivation for $CsPbX_3$ solar cells with VOC over 90% of the S-Q limit, *Adv. Funct. Mater.* **2022**, *32*, 2207554.
11. G. Kim, J. W. Lim, J. Kim, S. J. Yun, M. A. Park, Transparent thin-film silicon solar cells for indoor light harvesting with conversion efficiencies of 36% without photodegradation, *ACS Appl. Mater. Interfaces* **2020**, *12*, 27122.
12. D. Zhang, M. Stojanovic, Y. Ren, Y. Cao, F. T. Eickemeyer, E. Socie, N. Vlachopoulos, J. E. Moser, S. M. Zakeeruddin, A. Hagfeldt, M. Grätzel, A molecular photosensitizer achieves a V_{OC} of 1.24 V enabling highly efficient and stable dye-sensitized solar cells with copper(II/I)-based electrolyte, *Nat. Commun.* **2021**, *12*, 1777.
13. Y. Cui, Y. Wang, J. Bergqvist, H. Yao, Y. Xu, B. Gao, C. Yang, S. Zhang, O. Inganäs, F. Gao, J. Hou, Wide-gap non-fullerene acceptor enabling high-performance organic photovoltaic cells for indoor applications, *Nat. Energy* **2019**, *4*, 768.
14. C. Zhao, W. Tian, Q. Sun, Z. Yin, J. Leng, S. Wang, S. Jin, Trap-enabled long-distance carrier transport in perovskite quantum wells, *J. Am. Chem. Soc.* **2020**, 142, 15091.
15. A. Karmakar, M. S. Dodd, S. Agnihotri, E. Ravera, V. K. Michaelis, Cu(II)-doped Cs2SbAgCl6 double perovskite: a lead-free, low-bandgap material, *Chem. Mater.* **2018**, *30*, 8280.

16. T. Y. Yang, Y. Y. Kim, J. Seo, Roll-to-roll manufacturing toward lab-to-fab-translation of perovskite solar cells, *APL Mater.* **2021**, *9*, 110901.
17. K. Wojciechowski, D. Forgács, Commercial applications of indoor photovoltaics based on flexible perovskite solar cells, *ACS Energy Lett.* **2022**, *7*, 3729.
18. J. Gong, S. B. Darling, F. You, Perovskite photovoltaics: life-cycle assessment of energy and environmental impacts, *Energy Environ. Sci.* **2015**, 8, 1953.
19. N. L. Chang, A. W. Yi Ho-Baillie, P. A. Basore, T. L. Young, R. Evans, R. J. Egan, A manufacturing cost estimation method with uncertainty analysis and its application to perovskite on glass photovoltaic modules, *Prog. Photovolt.* **2017**, *25*, 390.
20. W. Ke and M. G. Kanatzidis, Prospects for low-toxicity lead-free perovskite solar cells, *Nat. Commun.* **2019**, 10, 965.
21. A. Abate, Perovskite solar cells go lead free, *Joule* **2017**, *1*, 659.
22. Z. Guo, A. K. Jena, T. Miyasaka, Halide perovskites for indoor photovoltaics: the next possibility, *ACS Energy Lett.* **2023**, *8*, 90.
23. I. Mathews, S. N. Kantareddy, T. Buonassisi, I. M. Peters, Technology and market perspective for indoor photovoltaic cells, *Joule* **2019**, *3*, 1415.
24. K. L. Wang, Y. H. Zhou, Y. H. Lou, Z. K. Wang, Perovskite indoor photovoltaics: opportunity and challenges, *Chem. Sci.* **2021**, *12*, 11936.
25. A. Venkateswararao, J. K. W. Ho, S. K. So, S. W. Liu, K. T. Wong, Small molecules for vacuum-processed organic photovoltaics: Past, current status, prospect, *Mater. Sci. Eng. R Rep.* **2020**, *139*, 100517.
26. Y. Cui, L. Hong, T. Zhang, H. Meng, H. Yan, F. Gao, J. Hou, Accurate photovoltaic measurement of organic cells for indoor applications, *Joule* **2021**, *5*, 1016.
27. H. Y. Lin, C. Y. Chen, B. W. Hsu, Y. L. Cheng, W. L. Tsai, Y. C. Huang, C. S. Tsao, H. W. Lin, Efficient cesium lead halide perovskite solar cells through alternative thousand-layer rapid deposition, *Adv. Funct. Mater.* **2019**, 29, 1905163.
28. M. Saliba, J. P. Correa-Baena, M. Grätzel, A. Hagfeldt, A. Abate, Perovskite solar cells: from the atomic level to film quality and device performance, *Angew. Chem. Int. Ed.* **2018**, *57*, 2554.
29. C. H. Chen, F. Hu, Z. H. Su, Y. J. Yu, K. L. Wang, Y. R. Shi, J. Chen, Y. Xia, X. Y. Gao, Z. K. Wang, L. S. Liao, Spring-like ammonium salt assisting stress release for low-temperature deposited $FAPbI_3$ films toward flexible photovoltaic application, *Adv. Funct. Mater.* **2023**, *33*, 2213661.
30. M. A. Green, Commercial progress and challenges for photovoltaics, *Nat. Energy* **2016**, *1*, 15015.
31. C. C. Chueh, C. Z. Li, A. K. Y. Jen, Recent progress and perspective in solution-processed Interfacial materials for efficient and stable polymer and organometal perovskite solar cells, *Energy Environ. Sci.* **2015**, *8*, 1160.
32. K. L. Wang, H. Lu, M. Li, C. H. Chen, D. B. Zhang, J. Chen, J. J. Wu, Y. H. Zhou, X. Q. Wang, Z. H. Su, Y. R. Shi, Q. S. Tian, Y. X. Gao, S. M. Zakeeruddin, M. Grätzel, Z. K. Wang, L. S. Liao, Ion–dipole interaction enabling highly efficient CsPbI3 perovskite indoor photovoltaics, *Adv.* **2023**, *Mater.* 2210106.
33. Z. Chen, Q. Cheng, H. Chen, Y. Chen, Y. Wu, J. Ding, X. Wu, H. Yang, H. Liu, W. Chen, X. Tang, X. Lu, Y. Li, Y. Li, Perovskite grain-boundary manipulation using room-temperature dynamic self-healing "Ligaments" for developing highly stable flexible perovskite solar cells with 23.8% efficiency, *Adv. Mater.* **2023**, *35*, 2300513.
34. L. Li, Y. Wang, X. Wang, R. Lin, X. Luo, Z. Liu, K. Zhou, S. Xiong, Q. Bao, G. Chen, Y. Tian, Y. Deng, K. Xiao, J. Wu, M. I. Saidaminov, H. Lin, C. Q. Ma, Z. Zhao, Y. Wu, L. Zhang, H. Tan, Flexible all-perovskite tandem solar cells approaching 25% efficiency with molecule-bridged hole-selective contact, *Nat. Energy* **2022**, *7*, 708.
35. C. H. Chen, Z. H. Su, Y. H. Lou, Y. J. Yu, K. L. Wang, G. L. Liu, Y. R. Shi, J. Chen, J. J. Cao, L. Zhang, X. Y. Gao, Z. K. Wang, Full-dimensional grain boundary stress release for flexible perovskite indoor photovoltaics, *Adv. Mater.* **2022**, *34*, 2200320.

36. C. Zuo, D. Vak, D. Angmo, L. Ding, M. Gao, One-step roll-to-roll air processed high efficiency perovskite solar cells, *Nano Energy* **2018**, *46*, 185.
37. J. E. Kim, S. S. Kim, C. Zuo, M. Gao, D. Vak, D. Y. Kim, Humidity-tolerant roll-to-roll fabrication of perovskite solar cells via polymer-additive-assisted hot slot die deposition, *Adv. Funct. Mater.* **2019**, *29*, 1809194.
38. Y. Y. Kim, T. Y. Yang, R. Suhonen, M. Välimäki, T. Maaninen, A. Kemppainen, N. J. Jeon, J. Seo, Gravure-printed flexible perovskite solar cells: toward roll-to-roll manufacturing, *Adv. Sci.* **2019**, *6*, 1802094.
39. L. J. Sutherland, D. Vak, M. Gao, T. A. N. Peiris, J. Jasieniak, G. P. Simon, H. Weerasinghe, Vacuum-free and solvent-free deposition of electrodes for roll-to-roll fabricated perovskite solar cells, *Adv. Energy Mater.* **2022**, *12*, 2202142.
40. J. Li, R. Xia, W. Qi, X. Zhou, J. Cheng, Y. Chen, G. Hou, Y. Ding, Y. Li, Y. Zhao, X. Zhang, Encapsulation of perovskite solar cells for enhanced stability: Structures, materials and characterization, *J. Power Sources* **2021**, *485*, 229313.
41. M. C. Brennan, S. Draguta, P. V Kamat, M. Kuno, Light-induced anion phase segregation in mixed halide perovskites, *ACS Energy Lett.* **2018**, *3*, 204.
42. J. Liang, X. Hu, C. Wang, C. Liang, C. Chen, M. Xiao, J. Li, C. Tao, G. Xing, R. Yu, W. Ke, G. Fang, Origins and influences of metallic lead in perovskite solar cells, *Joule* **2022**, *6*, 816.
43. RoHS, https://ec.europa.eu/growth/single-market/european-standards/.
44. C. H. Chen, S. N. Cheng, L. Cheng, Z. K. Wang, L. S. Liao, Toxicity, leakage, and recycling of lead in perovskite photovoltaics, *Adv. Energy Mater.* **2023**, *13*, 2204144.
45. S. Chen, Y. Deng, H. Gu, S. Xu, S. Wang, Z. Yu, V. Blum, J. Huang, Trapping lead in perovskite solar modules with abundant and low-cost cation-exchange resins, *Nat. Energy* **2020**, *5*, 1003.
46. Y. Jiang, L. Qiu, E. J. Juarez-Perez, L. K. Ono, Z. Hu, Z. Liu, Z. Wu, L. Meng, Q. Wang, Y. Qi, Reduction of lead leakage from damaged lead halide perovskite solar modules using self-healing polymer-based encapsulation, *Nat. Energy* **2019**, *4*, 585.
47. W. F. Yang, J. J. Cao, C. Dong, M. Li, Q. S. Tian, Z. K. Wang, L. S. Liao, Suppressed oxidation of tin perovskite by Catechin for eco-friendly indoor photovoltaics, *Appl. Phys. Lett.* **2021**, *118*, 023501.

13 Inverted Perovskite Solar Cells

Xiaodong Li and Junfeng Fang

13.1 DEVICE CONFIGURATION AND ADVANTAGES OF INVERTED PEROVSKITE SOLAR CELLS

Perovskite solar cells (PSCs) usually adopt a sandwich-like device configuration, where the perovskite active layer is sandwiched between hole and electron transport layers. According to the relative position of hole and electron transport layers, PSCs can be divided into regular n-i-p devices and inverted p-i-n devices (Figure 13.1).[1] Among regular PSCs, two types of device structures exist: mesoporous and planar structures, depending on whether a mesoporous scaffold is applied (usually mesoporous TiO_2 or Al_2O_3). In regular PSCs, TiO_2 and Spiro-OMeTAD are the most commonly used electron and hole transport layers. Despite the high efficiency, TiO_2 can induce device degradation especially under illumination due to the intrinsically photocatalytic property of TiO_2. Besides, the doping process of Spiro-OMeTAD is also an inevitable issue for device degradation. In comparison, inverted PSCs avoid the use of TiO_2 and Spiro-OMeTAD, exhibiting huge potential in device stability. Inverted PSCs can be fabricated in low-temperature condition and easy to realize flexible devices. In addition, due to the absence of parasitic absorption from Spiro-OMeTAD, inverted PSCs are much more suitable to fabricate tandem solar cells (perovskite/perovskite or c-Si/perovskite tandem devices), and most perovskite-related tandem solar cells have adopted inverted PSCs now.

13.2 TRANSPORTING MATERIALS IN INVERTED PEROVSKITE SOLAR CELLS

Throughout the development of inverted PSCs, transporting materials have played an important role in the performance of inverted PSCs, which not only affect carrier extraction and transport but also affect device stability.[2] According to their functions in solar cells, transporting materials include hole transport material (HTM) and electron transport material (ETM). In inverted PSCs, perovskite is directly deposited on HTM. Therefore, HTM greatly affects perovskite crystalline and growth process, apart from the ability of hole transporting. Similarly, ETM depositing on top of perovskite can protect perovskite films and help reduce the harmful impacts of external atmosphere (O_2, H_2O, etc.), apart from its electron transporting function. Now, various transporting materials are reported for application in inverted PSCs, which will be described in detail in the following sections.

 DOI: 10.1201/9781003400493-13

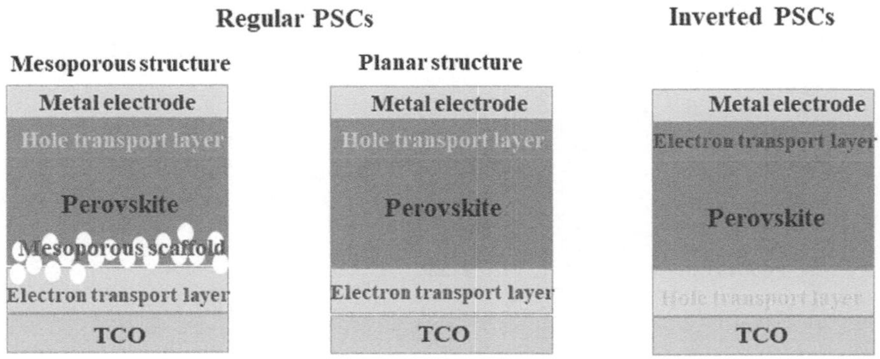

FIGURE 13.1 Device configuration of PSCs.

13.2.1 Hole Transport Materials (HTMs; (Figure 13.2))

13.2.1.1 PEDOT:PSS

PEDOT:PSS is a typical HTM widely used in organic photoelectric devices due to its good transmittance and high conductivity. The first inverted PSCs was fabricated with PEDOT:PSS as the HTM despite its low efficiency of 3.9%.[3] The low efficiency is mainly ascribed to the relatively low work function of PEDOT:PSS (~4.9 eV) in comparison with the valance band of perovskite layer (~5.4 eV). Therefore, many works have been focused on the modification of PEDOT:PSS. For example, Ding et al. increased the work function to 5.2 eV by adding a polymer electrolyte PSS-Na into PEDOT:PSS, thus enhancing device V_{oc} from 0.96 V to 1.11 V and realizing over 15% device efficiency.[4] Lee et al. obtained a self-organized PEDOT:PSS with work function of 5.4 eV through introducing perfluorinated ionomer (PFI) additive, also greatly improving device efficiency.[5] Besides work function, Gong et al. reported a highly conductive PEDOT:PSS through PEO modification, thus increasing device J_{sc} and efficiency.[6] Although great efforts have been made in PEDOT:PSS for application in PSCs, the devices performance still lags behind, which promotes researchers to develop other new HTMs.

13.2.1.2 PTAA

PTAA, poly(bis(4-phenyl)(2,4-bimethylphenyl)amine, is initially adopted in regular PSCs to replace Spiro-OMeTAD as doping-free HTM due to its well-matched energy level. Then Huang et al. introduced PTAA into inverted PSCs as HTM and obtained a high efficiency of 18% in 2015.[7] The hydrophobic nature of PTAA will form a non-wetting surface with perovskite solution, inducing high-aspect-ratio perovskite grain growth, thus benefiting device performance. From then on, PTAA starts to be widely applied in inverted PSCs and the device efficiency based on PTAA HTM has reached 23.6% in 2021.[8] The non-wetting property of PTAA also makes it challenging to uniformly deposit perovskite solution on it, which is the main obstacle limiting

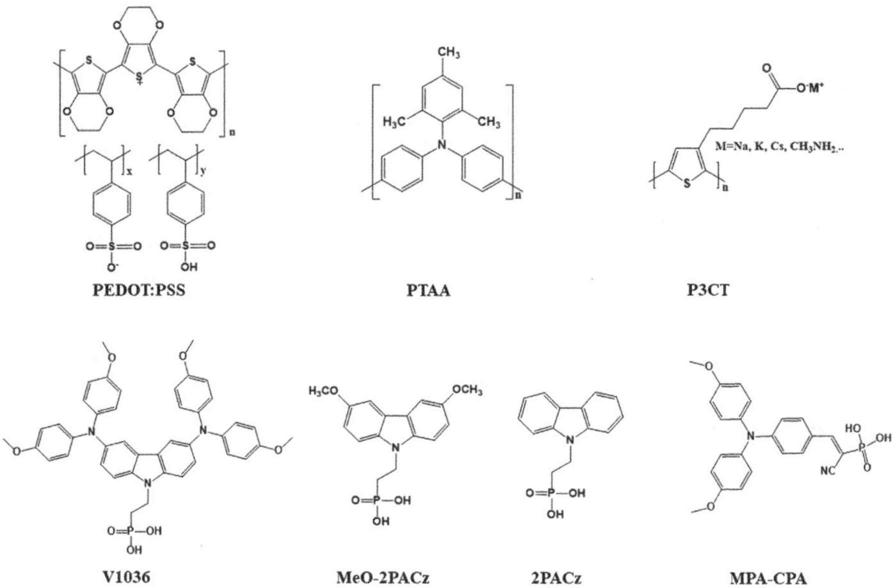

FIGURE 13.2 Typical organic HTMs in inverted PSCs.

PTAA application in inverted PSCs. To address this issue, several approaches like prewashing PTAA layer with polar solvents, oxygen plasma, and UV treatment have been reported[7] to improve its wettability. Another method is to introduce an extra interlayer between PTAA and perovskite as surface linker. For example, PFN and PMMA have been reported to help perovskite deposition on PTAA and result in much better devices performance.[9, 10] In addition, anionic conjugated polyelectrolytes are also a good choice, and large area (1 cm^2) inverted PSCs based on PTAA are obtained with an efficiency of over 18%.[11]

13.2.1.3 P3CT

P3CT (Poly [3-(4-carboxybutyl)thiophene-2,5-diyl]) can react with alkali and form a series of water or alcohol soluble polyelectrolytes HTM for application in inverted PSCs. The sodium salt of P3CT (P3CT-Na) was first reported by Li et al. as an efficient HTM in 2015 and a high efficiency of over 16% is realized owing to its deep work function, high conductivity, and good wetting property with perovskite solution.[12] However, the strong aggregation of P3CT-Na in water limits its film uniformness and devices reproducibility. Li et al. further resolved this issue through solvent optimization from water to methanol and counterions tailoring from Na^+ to $CH_3NH_2^+$ and obtained a more efficient P3CT-based HTM (P3CT-CH_3NH_2 or P3CT-N).[13] From then on, P3CT-based HTM starts to attract much attention and lots of other P3CT-based HTM has been developed (P3CT-K, P3CT-Cs et al). In 2022, Li et al. first realized an over 24% efficiency in inverted PSCs by using P3CT-N HTM.[14]

13.2.1.4 SAM

SAMs (self-assembled monolayers) are ordered organic molecules with a few molecules' thickness that spontaneously assembles on substrates by adsorption of the anchoring group.[15] Generally, SAM molecules consist of three parts including anchoring group, linker, and terminal groups. The anchoring group helps SAMs to bind on the surface of substrate and strongly affects the interfacial dipole, work function, and coverage degree of substrate; the linker connects the anchoring group and terminal group, which will affect the packing structure of molecules during assembling process thus determining carriers' extracting efficiency at perovskite/SAMs interface; the terminal group is in direct contact with the upper perovskite and will work as a template for perovskite growth. In 2018, Getautis et al. first introduced a SAM HTM called V1036 with a phosphonic acid anchoring group, ethyl linker, and carbazole terminal group in inverted PSCs despite its moderate efficiency of ~17%.[16] Then they developed two typical SAMs of 2PACz and MeO-2PACz HTM and further increased efficiency to over 20%, which was even higher than that of PTAA devices.[17] Subsequently, various SAMs are reported with different anchoring groups of -SO_3H, -COOH, and -PO_3H_2 or a linking and terminal group. In 2022, Wolf et al. reported high efficiency of over 24% in inverted PSCs by using 2PACz SAMs as HTM.[18] Wu et al. developed a new SAM of MPA-CPA and directly spin-coated onto ITO substrate as HTM in inverted PSCs. They found that MPA-CPA would form a bilayer stack with chemically anchored SAM and an extra un-adsorbed, disordered overlayer. Just to this un-adsorbed overlayer, the obtained HTM layer exhibited superwetting properties with perovskite solution and promoting device efficiency to over 25%.[19]

13.2.1.5 Inorganic HTM

Inorganic HTM also shows great potential in inverted PSCs owing to their low cost and intrinsic stability. Among various inorganic HTMs, NiOx is the most widely used in inverted PSCs. In 2015, Han et al. reported the first certified large area (1 cm^2) PSCs with efficiency of over 15% by using doped NiOx HTM (NiMgLiO).[20] You and Yang et al. also reported efficient and stable inverted PSCs with NiOx HTM.[21] The main obstacle for NiOx HTM lies in its low conductivity and high temperature process (> 200 °C). To overcome this issue, Jen et al. reported a low-temperature processed and Cu-doped NiOx with high conductivity by using a combustion method, which can even be used to fabricate ITO-free PSCs with an efficiency of 13%.[22] In addition, Li and Fang et al. reported a photoconductive NiOx HTM (NiOx-Pc) through combining NiOx with phthalocyanine molecule.[23] In NiOx-Pc, the Pc molecule can absorb light and transfer the photogenerated charge NiOx, thus increasing its hole density and conductivity. As a result, high efficiency of over 21% is obtained in inverted PSCs with NiOx. Besides NiOx, Anthopoulos et al. reported solution-process ultrathin CuSCN in inverted PSCs in 2017.[24] Electrodeposited or successive ionic layer adsorption and reaction processed CuSCN are also reported for application in inverted PSCs.[25]

13.2.2 Electron Transport Material (ETM)

The most widely used ETMs in inverted PSCs are fullerene derivatives, like PCBM and C60 owing to their good electron extracting ability, matched energy level, and potential traps passivation.[26] Apart from PCBM or C60 themselves, some works on fullerene modifications are also reported. For example, Huang et al. reported cross-linkable silane-functionalized fullerene to increase its water-resistance in PSCs, thus enhancing device stability.[27] Jiu, Fang and Li et al. reported a graphdiyne doped PCBM to improve its conductivity and thus increased device efficiency from 13.6% to 14.8% in 2015.[28] In addition, benefiting from the non-fullerene development in organic solar cells, many other non-fullerene ETMs are synthesized and applied into inverted PSCs, like HATNA-derived ETM, PFN-2TNDI and so on.[29,30]

13.3 EFFICIENCY IMPROVEMENT OF INVERTED PEROVSKITE SOLAR CELLS

Apart from the optimizations of transporting materials and perovskite films, interfacial modification also plays an important role in efficiency improvement of inverted PSCs. Huang et al. reported that in comparison with bulk perovskite, perovskite surface exhibited 1~2 orders of magnitude more traps density in inverted PSCs.[31] Neher et al. also found severe quasi-Fermi-level splitting losses (~80mV) at the perovskite/transporting layer interface induced by interfacial recombination, which was considered one major reason for Voc loss in inverted PSCs (Figure 13.3).[32]

To suppress the interfacial recombination, Zhu et al. proposed a solution-processed secondary growth (SSG) method to re-grow perovskite surface and thus improved devices efficiency to over 21% with a high V_{oc} of > 1.2V.[33] Huang et al. reported the peeling-off method to remove the defective layer on perovskite surface with adhesive tapes, thus improving device efficiency to 22%.[34] Surface passivation has been demonstrated to be a very effective method to inhibit interfacial traps in regular PSCs. Therefore, various surface passivating approaches have also been introduced into inverted PSCs, among which using organic ammonium salt to form 2D perovskite on 3D bulk perovskite is most widely used. For example, Vaynzof et al. passivated both top and buried perovskite surface with PEAI and realized > 23% efficiency in inverted PSCs.[35] Sargent et al. passivated perovskite surface with

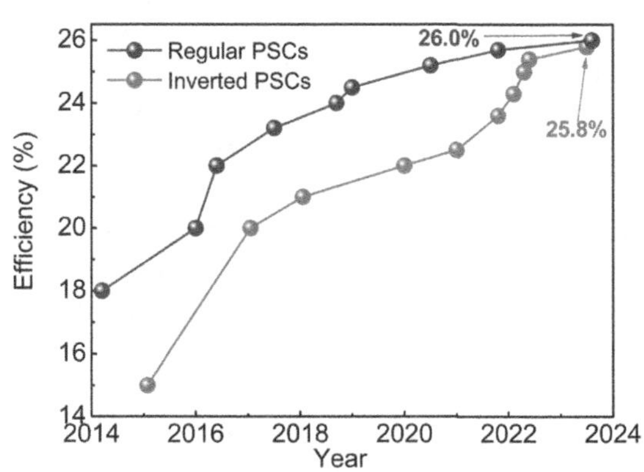

FIGURE 13.3 Efficiency evolution of Inverted PSCs with respect to regular PSCs.

MAI and 2D ligand mixture to fine tune the quantum-size and energy level at perovskite interface, enhancing device efficiency to 23.9%.[36] Wolf et al. treated the 3D perovskite surface with oleylammonium iodide and formed a higher-dimensionality 2D perovskite capping layer on top, reducing possible energy loss and improving efficiency to 24.3%.[18]

Recently, from the point of interfacial heterojunction, Fang and Li et al. proposed a surface sulfidation method (SST) and constructed perovskite/PbS interfacial heterojunction in inverted PSCs.[14] With this method, perovskite surface tends to be more n-type with up-shifted Fermi level, thus generating an extra back-surface field at perovskite/PbS interface, thus inhibiting recombination and enhancing the whole V_{bi} of inverted PSCs. As a result, device V_{oc} is greatly enhanced to 1.19 V with a low voltage loss of 0.36 V. And device efficiency is increased to over 24% for the first time. Zhu et al. reported a reactive surface engineering approach through 3-(aminomethyl)pyridine (3-APy) treatment on perovskite surface. 3-APy would react with FA^+ ions on perovskite surface and induced a n-doping of perovskite top region, thus increasing device efficiency to over 25%.[37] Zhu et al. reported a new passivator of ferrocenyl-bis-thiophene-2-carboxylate (FcTc2) for application in inverted PSCs and realized high efficiency of 25%.[38]

Besides top interface, the buried interface is also important for efficiency improvement in inverted PSCs. Zang et al. introduced 2-aminoindan hydrochloride additive into perovskite solution and formed a bottom-up 2D/3D heterojunction at buried interface, thus inhibiting interfacial recombination and resulting in over 25% device efficiency.[39] Xu et al. reported a porous insulator contact (PIC) model in inverted PSCs by depositing thick Al_2O_3 insulating layer with random nanoscale openings on PTAA HTM, reducing contact area between perovskite and ITO electrode, and thus inhibiting interfacial recombination. As a result, ~25% efficiency in small-area devices and over 23% in 1 cm^2 devices are obtained in inverted PSCs.[40] He et al. introduced dimethylacridine molecule into perovskite solution to induce p-doping perovskite at the buried interface and constructed a well-matched p-perovskite/ITO contact, thus improving device efficiency to 25.8%, which is the highest efficiency of inverted PSCs up to now.[41]

13.4 STABILITY ISSUES OF INVERTED PEROVSKITE SOLAR CELLS

The stability of inverted PSCs relates to both carrier transporting layer and perovskite layer. In 2015, You and Yang et al. tried to use inorganic NiO_x and ZnO as HTM and ETM respectively to improve device air stability.[21] Han et al. used doped HTM (NiMgLiO) and ETL ($Ti(Nb)O_x$) and also greatly improved device stability.[20] Besides HTM or ETL-related degradation, ion migration in perovskite layer is another important reason for device degradation, especially under illumination. Li and Fang et al. reported an in-situ crosslinking strategy to block ion migration along grain boundaries by introducing cross-linkable additives of TMTA into perovskite films.[42] TMTA bonds to grain boundaries through coordination interaction and then crosslinks after thermal treatment, thus blocking the channel of ions migration. To further block potential ion diffusion towards ETM, Han et al. reported diffusion engineering by introducing graphene into PCBM ETM.[43] Jen et al. developed a cross-linkable ETM of c-HATNA to block ions diffusion and greatly improved device stability.[44]

Lee et al. found that the corrosion of metal electrodes is a major origin for intrinsic device degradation through exploring the self-degradation process of PSCs. The decomposition products of perovskite (e.g., HI and I_2) can react with commonly used metal electrodes in inverted PSCs, thus accelerating the decomposition of perovskite.[45] On the other hand, ion migration in PSCs is a reversible process and may not cause permanent device degradation under illumination as the efficiency can be fully recovered after being left in the dark. However, such a reversible process will be broken once the migrated ions react with metal electrodes. As a result, irreversible degradation of PSCs occurs far ahead of the degradation of perovskite layers. To inhibit electrode corrosion, Huang et al. tried to use Cu to replace commonly used Ag electrode and realized improved device stability.[46] Li and Fang et al. further proposed a chemical anti-corrosion strategy to inhibit Cu corrosion in PSCs by introducing typical organic inhibitor of BTA before the Cu electrode.[47] BTA will interact with Cu atom and form a compact [BTA-Cu] polymer at the Cu surface, thus inhibiting chemical or electrochemical corrosion under perovskite atmosphere. As a result, air, thermal, and operational stability of PSCs are improved simultaneously, retaining ~90% of initial efficiency after MPP tracking for 1,000 h. Chen et al. also reported chemically inert bismuth interlayer before electrode to inhibit possible Ag corrosion in PSCs and retained over 95% of initial efficiency after light illumination for 500 h.[48]

Recently, interfacial degradation in inverted PSCs has attracted much attention and many efforts have been made on this point. For example, Huang et al. stabilized the perovskite surfaces with wide-bandgap lead oxysalts, retaining over 95% of initial efficiency after MPP tracking for 1,200 h.[49] Wolf et al. reported a tailored-dimensionality 2D/3D heterojunction in PSCs to inhibit interfacial degradation and thus improve the damp heat stability of PSCs.[18] Yan et al. proposed phosphorus stabilization of perovskites owing to its strongest bonding with lead. The resulting PSCs maintained ~23% efficiency for more than 1,500 h under open-circuit conditions at 85 °C.[50] Li and Abate et al. tried to buffer the strain and stabilize perovskite phase through introducing polymer dipoles on perovskite surface, thus releasing the lattice strain created by temperature changes during PSCs aging. Resulting PSCs retained over 90% of initial efficiency after aging for 3,000 h and after repeated cycling between -60° and 80 °C.[51] Sargent et al. constructed a robust 2D/3D heterostructure at perovskite surface with limited penetration into bulk perovskite by using anilinium and its fluorinated derivatives, thus enhancing devices stability and showing T_{85} of ~1,560 h under 1 sun illumination at 50% RH and 85 °C.[52]

13.5 OUTLOOK OF INVERTED PEROVSKITE SOLAR CELLS

In the past two years, device performance of inverted PSCs has been increasing rapidly. Considering the advantages inverted PSCs in simple processing, high stability, and compatibility with tandem devices, more efforts should be made to promote the further development of inverted PSCs. Further improvements should lie in the interfacial optimization, especially the energy level alignment at perovskite/transporting layer interface, which is critical to the charge extraction and benefits device Voc. In addition, complete passivation of film defects at buried, top surface as well as bulk perovskite should also be noted, which will help to increase device J_{sc} and FF. Based on these achievements, efficiency of 26% or even higher can be anticipated in inverted PSCs.

REFERENCES

1. L. Meng, J. You, T. F. Guo, Y. Yang, Recent advances in the inverted planar structure of perovskite solar cells. *Acc. Chem. Res.* **2016**, *49*, 155.
2. W. Yan, S. Ye, Y. Li, W. Sun, H. Rao, Z. Liu, Z. Bian, C. Huang, Hole-transporting materials in inverted planar perovskite solar cells. *Adv. Energy Mater.* **2016**, *6*, 1600474.
3. J.-Y. Jeng, Y.-F. Chiang, M.-H. Lee, S.-R. Peng, T.-F. Guo, P. Chen, T.-C. Wen, CH3NH3PbI3 perovskite/fullerene planar-heterojunction hybrid solar cells. *Adv. Mater.* **2013**, *25*, 3727.
4. C. Zuo, L. Ding, Modified PEDOT layer makes a 1.52 VVocfor perovskite/PCBM solar cells. *Adv. Energy Mater.* **2017**, *7*, 1601193.
5. K. G. Lim, H. B. Kim, J. Jeong, H. Kim, J. Y. Kim, T. W. Lee, Boosting the power conversion efficiency of perovskite solar cells using self-organized polymeric hole extraction layers with high work function. *Adv. Mater.* **2014**, *26*, 6461.
6. X. Huang, K. Wang, C. Yi, T. Meng, X. Gong, Efficient perovskite hybrid solar cells by highly electrical conductive PEDOT:PSS hole transport layer. *Adv. Energy Mater.* **2016**, *6*, 1501773.
7. C. Bi, Q. Wang, Y. Shao, Y. Yuan, Z. Xiao, J. Huang, Non-wetting surface-driven high-aspect-ratio crystalline grain growth for efficient hybrid perovskite solar cells. *Nat. Commun.* **2015**, *6*, 7747.
8. S. Chen, X. Dai, S. Xu, H. Jiao, L. Zhao, J. Huang, Stabilizing perovskite-substrate interfaces for high-performance perovskite modules. *Science* **2021**, *373*, 902.
9. J. Lee, H. Kang, G. Kim, H. Back, J. Kim, S. Hong, B. Park, E. Lee, K. Lee, Achieving large-area planar perovskite solar cells by introducing an interfacial compatibilizer. *Adv. Mater.* **2017**, *29*, 1606363.
10. J. Wang, J. Xu, Z. Li, X. Lin, C. Yu, H. Wu, H.-l. Wang, Front-contact passivation of PIN MAPbI3 solar cells with superior device performances. *ACS Appl. Energy Mater.* **2020**, *3*, 6344.
11. E. D. Jung, A. K. Harit, D. H. Kim, C. H. Jang, J. H. Park, S. Cho, M. H. Song, H. Y. Woo, Multiply charged conjugated polyelectrolytes as a multifunctional interlayer for efficient and scalable perovskite solar cells. *Adv. Mater.* **2020**, *32*, 2002333.
12. X. Li, X. Liu, X. Wang, L. Zhao, T. Jiu, J. Fang, Polyelectrolyte based hole-transporting materials for high performance solution processed planar perovskite solar cells. *J. Mater. Chem. A* **2015**, *3*, 15024.
13. X. Li, Y. C. Wang, L. Zhu, W. Zhang, H. Q. Wang, J. Fang, Improving efficiency and reproducibility of perovskite solar cells through aggregation control in polyelectrolytes hole transport layer. *ACS Appl. Mater. Interfaces* **2017**, *9*, 31357.
14. X. Li, W. Zhang, X. Guo, C. Lu, J. Wei, J. Fang, Constructing heterojunctions by surface sulfidation for efficient inverted perovskite solar cells. *Science* **2022**, *375*, 434.
15. Y. Yao, C. Cheng, C. Zhang, H. Hu, K. Wang, S. De Wolf, Organic hole-transport layers for efficient, stable, and scalable inverted perovskite solar cells. *Adv. Mater.* **2022**, *34*, 2203794
16. A. Magomedov, A. Al-Ashouri, E. Kasparavičius, S. Strazdaite, G. Niaura, M. Jošt, T. Malinauskas, S. Albrecht, V. Getautis, Self-assembled hole transporting monolayer for highly efficient perovskite solar cells. *Adv. Energy Mater.* **2018**, *8*, 1870139.

17. A. Al-Ashouri, A. Magomedov, M. Roß, M. Jošt, M. Talaikis, G. Chistiakova, T. Bertram, J. A. Márquez, E. Köhnen, E. Kasparavičius, S. Levcenco, L. Gil-Escrig, C. J. Hages, R. Schlatmann, B. Rech, T. Malinauskas, T. Unold, C. A. Kaufmann, L. Korte, G. Niaura, V. Getautis, S. Albrecht, Conformal monolayer contacts with lossless interfaces for perovskite single junction and monolithic tandem solar cells. *Energy Environ. Sci.* **2019**, *12*, 3356.
18. R. Azmi, E. Ugur, A. Seitkhan, F. Aljamaan, A. S. Subbiah, J. Liu, G. T. Harrison, M. I. Nugraha, M. K. Eswaran, M. Babics, Y. Chen, F. Xu, T. G. Allen, A. u. Rehman, C.-L. Wang, T. D. Anthopoulos, U. Schwingenschlögl, M. De Bastiani, E. Aydin, S. De Wolf, Damp heat-stable perovskite solar cells with tailored-dimensionality 2D/3D heterojunctions. *Science* **2022**, *376*, 73.
19. S. Zhang, F. Ye, X. Wang, R. Chen, H. Zhang, L. Zhan, X. Jiang, Y. Li, X. Ji, S. Liu, M. Yu, F. Yu, Y. Zhang, R. Wu, Z. Liu, Z. Ning, D. Neher, L. Han, Y. Lin, H. Tian, W. Chen, M. Stolterfoht, L. Zhang, W.-H. Zhu, Y. Wu, Minimizing buried interfacial defects for efficient inverted perovskite solar cells. *Science* **2023**, *380*, 404.
20. W. Chen, Y. Wu, Y. Yue, J. Liu, W. Zhang, X. Yang, H. Chen, E. Bi, I. Ashraful, M. Grätzel, L. Han, Efficient and stable large-area perovskite solar cells with inorganic charge extraction layers. *Science* **2015**, *350*, 944.
21. J. You, L. Meng, T. B. Song, T. F. Guo, Y. M. Yang, W. H. Chang, Z. Hong, H. Chen, H. Zhou, Q. Chen, Y. Liu, N. De Marco, Y. Yang, Improved air stability of perovskite solar cells via solution-processed metal oxide transport layers. *Nat. Nanotechnol.* **2016**, *11*, 75.
22. J. W. Jung, C. C. Chueh, A. K. Jen, Cu-doped nickel oxide hole-transporting layer via the combustion method for high-performance thin-film perovskite solar cells. *Adv. Mater.* **2015**, *27*, 7874.
23. X. Zhao, W. Zhang, X. Feng, X. Guo, C. Lu, X. Li, J. Fang, Photoconductive NiOx hole transport layer for efficient perovskite solar cells. *Chem. Eng. J.* **2022**, *435*, 135140.
24. N. Wijeyasinghe, A. Regoutz, F. Eisner, T. Du, L. Tsetseris, Y.-H. Lin, H. Faber, P. Pattanasattayavong, J. Li, F. Yan, M. A. McLachlan, D. J. Payne, M. Heeney, T. D. Anthopoulos, Copper(I) thiocyanate (CuSCN) hole-transport layers processed from aqueous precursor solutions and their application in thin-film transistors and highly efficient organic and organometal halide perovskite solar cells. *Adv. Funct. Mater.* **2017**, *27*, 1701818.
25. S. Ye, W. Sun, Y. Li, W. Yan, H. Peng, Z. Bian, Z. Liu, C. Huang, CuSCN-based inverted planar perovskite solar cell with an average PCE of 15.6%. *Nano. Lett.* **2015**, *15*, 3723.
26. Y. Shao, Z. Xiao, C. Bi, Y. Yuan, J. Huang, Origin and elimination of photocurrent hysteresis by fullerene passivation in CH3NH3PbI3 planar heterojunction solar cells. *Nat. Commun.* **2014**, *5*, 5784.
27. Y. Bai, Q. Dong, Y. Shao, Y. Deng, Q. Wang, L. Shen, D. Wang, W. Wei, J. Huang, Enhancing stability and efficiency of perovskite solar cells with crosslinkable silane-functionalized and doped fullerene. *Nat. Commun.* **2016**, *7*, 12806.
28. C. Kuang, G. Tang, T. Jiu, H. Yang, H. Liu, B. Li, W. Luo, X. Li, W. Zhang, F. Lu, J. Fang, Y. Li, Highly efficient electron transport obtained by doping pcbm with graphdiyne in planar-heterojunction perovskite solar cells. *Nano. Lett.* **2015**, *15*, 2756.
29. C. Sun, Z. Wu, H.-L. Yip, H. Zhang, X.-F. Jiang, Q. Xue, Z. Hu, Z. Hu, Y. Shen, M. Wang, F. Huang, Y. Cao, Amino-functionalized conjugated polymer as an efficient electron transport layer for high-performance planar-heterojunction perovskite solar cells. *Adv. Energy Mater.* **2016**, *6*, 1501534.
30. D. Zhao, Z. Zhu, M. Y. Kuo, C. C. Chueh, A. K. Jen, Hexaazatrinaphthylene derivatives: efficient electron-transporting materials with tunable energy levels for inverted perovskite solar cells. *Angew. Chem. Int. Ed. Engl.* **2016**, *55*, 8999.
31. Z. Ni, C. Bao, Y. Liu, Q. Jiang, W.-Q. Wu, S. Chen, X. Dai, B. Chen, B. Hartweg, Z. Yu, Z. Holman, J. Huang, Resolving spatial and energetic distributions of trap states in metal halide perovskite solar cells. *Science* **2020**, *367*, 1352.

32. M. Stolterfoht, C. M. Wolff, J. A. Márquez, S. Zhang, C. J. Hages, D. Rothhardt, S. Albrecht, P. L. Burn, P. Meredith, T. Unold, D. Neher, Visualization and suppression of interfacial recombination for high-efficiency large-area pin perovskite solar cells. *Nat. Energy* **2018**, *3*, 847.
33. D. Luo, W. Yang, Z. Wang, A. Sadhanala, Q. Hu, R. Su, R. Shivanna, G. F. Trindade, J. F. Watts, Z. Xu, T. Liu, K. Chen, F. Ye, P. Wu, L. Zhao, J. Wu, Y. Tu, Y. Zhang, X. Yang, W. Zhang, R. H. Friend, Q. Gong, H. J. Snaith, R. Zhu, Enhanced photovoltage for inverted planar heterojunction perovskite solar cells. *Science* **2018**, *360*, 1442.
34. S. Chen, Y. Liu, X. Xiao, Z. Yu, Y. Deng, X. Dai, Z. Ni, J. Huang, Identifying the soft nature of defective perovskite surface layer and its removal using a facile mechanical approach. *Joule* **2020**, *4*, 2661.
35. M. Degani, Q. An, M. Albaladejo-Siguan, Y. J. Hofstetter, C. Cho, F. Paulus, G. Grancini, Y. Vaynzof, 23.7% Efficient inverted perovskite solar cells by dual interfacial modification. *Sci. Adv.* **2021**, *7*, eabj7930.
36. H. Chen, S. Teale, B. Chen, Y. Hou, L. Grater, T. Zhu, K. Bertens, S. M. Park, H. R. Atapattu, Y. Gao, M. Wei, A. K. Johnston, Q. Zhou, K. Xu, D. Yu, C. Han, T. Cui, E. H. Jung, C. Zhou, W. Zhou, A. H. Proppe, S. Hoogland, F. Laquai, T. Filleter, K. R. Graham, Z. Ning, E. H. Sargent, Quantum-size-tuned heterostructures enable efficient and stable inverted perovskite solar cells. *Nat. Photonics* **2022**, *16*, 352.
37. Q. Jiang, J. Tong, Y. Xian, R. A. Kerner, S. P. Dunfield, C. Xiao, R. A. Scheidt, D. Kuciauskas, X. Wang, M. P. Hautzinger, R. Tirawat, M. C. Beard, D. P. Fenning, J. J. Berry, B. W. Larson, Y. Yan, K. Zhu, Surface reaction for efficient and stable inverted perovskite solar cells. *Nature* **2022**, *611*, 278.
38. Z. Li, B. Li, X. Wu, S. A. Sheppard, S. Zhang, D. Gao, N. J. Long, Z. Zhu, Organometallic-functionalized interfaces for highly efficient inverted perovskite solar cells. *Science* **2022**, *376*, 416.
39. H. Li, C. Zhang, C. Gong, D. Zhang, H. Zhang, Q. Zhuang, X. Yu, S. Gong, X. Chen, J. Yang, X. Li, R. Li, J. Li, J. Zhou, H. Yang, Q. Lin, J. Chu, M. Grätzel, J. Chen, Z. Zang, 2D/3D Heterojunction engineering at the buried interface towards high-performance inverted methylammonium-free perovskite solar cells. *Nat. Energy* **2023**, *8*, 946.
40. W. Peng, K. Mao, F. Cai, H. Meng, Z. Zhu, T. Li, S. Yuan, Z. Xu, X. Feng, J. Xu, M. D. McGehee, J. Xu, Reducing nonradiative recombination in perovskite solar cells with a porous insulator contact. *Science* **2023**, *379*, 683.
41. Q. Tan, Z. Li, G. Luo, X. Zhang, B. Che, G. Chen, H. Gao, D. He, G. Ma, J. Wang, J. Xiu, H. Yi, T. Chen, Z. He, Inverted perovskite solar cells using dimethylacridine-based dopants. *Nature* **2023**, *620*, 545.
42. X. Li, W. Zhang, Y.-C. Wang, W. Zhang, H.-Q. Wang, J. Fang, In-situ cross-linking strategy for efficient and operationally stable methylammoniun lead iodide solar cells. *Nat. Commun.* **2018**, *9*, 3806.
43. E. Bi, H. Chen, F. Xie, Y. Wu, W. Chen, Y. Su, A. Islam, M. Grätzel, X. Yang, L. Han, Diffusion engineering of ions and charge carriers for stable efficient perovskite solar cells. *Nat. Commun.* **2017**, *8*, 15330.
44. Z. Zhu, D. Zhao, C.-C. Chueh, X. Shi, Z. Li, A. K. Y. Jen, Highly efficient and stable perovskite solar cells enabled by all-crosslinked charge-transporting layers. *Joule* **2018**, *2*, 168.
45. H. Back, G. Kim, J. Kim, J. Kong, T. K. Kim, H. Kang, H. Kim, J. Lee, S. Lee, K. Lee, Achieving long-term stable perovskite solar cells via ion neutralization. *Energy Environ. Sci.* **2016**, *9*, 1258.
46. J. Zhao, X. Zheng, Y. Deng, T. Li, Y. Shao, A. Gruverman, J. Shield, J. Huang, Is Cu a stable electrode material in hybrid perovskite solar cells for a 30-year lifetime? *Energy Environ. Sci.* **2016**, *9*, 3650.
47. X. Li, S. Fu, W. Zhang, S. Ke, W. Song, J. Fang, Chemical anti-corrosion strategy for stable inverted perovskite solar cells. *Sci. Adv.* **2020**, *6*, eabd1580.

48. S. Wu, R. Chen, S. Zhang, B. H. Babu, Y. Yue, H. Zhu, Z. Yang, C. Chen, W. Chen, Y. Huang, S. Fang, T. Liu, L. Han, W. Chen, A chemically inert bismuth interlayer enhances long-term stability of inverted perovskite solar cells. *Nat. Commun.* **2019**, *10*, 1161.
49. S. Yang, S. Chen, E. Mosconi, Y. Fang, X. Xiao, C. Wang, Y. Zhou, Z. Yu, J. Zhao, Y. Gao, F. De Angelis, J. Huang, Stabilizing halide perovskite surfaces for solar cell operation with wide-bandgap lead oxysalts. *Science* **2019**, *365*, 473.
50. C. Li, X. Wang, E. Bi, F. Jiang, S. M. Park, Y. Li, L. Chen, Z. Wang, L. Zeng, H. Chen, Y. Liu, C. R. Grice, A. Abudulimu, J. Chung, Y. Xian, T. Zhu, H. Lai, B. Chen, R. J. Ellingson, F. Fu, D. S. Ginger, Z. Song, E. H. Sargent, Y. Yan, Rational design of lewis base molecules for stable and efficient inverted perovskite solar cells. *Science* **2023**, *379*, 690.
51. G. Li, Z. Su, L. Canil, D. Hughes, M. H. Aldamasy, J. Dagar, S. Trofimov, L. Wang, W. Zuo, J. J. Jerónimo-Rendon, M. M. Byranvand, C. Wang, R. Zhu, Z. Zhang, F. Yang, G. Nasti, B. Naydenov, W. C. Tsoi, Z. Li, X. Gao, Z. Wang, Y. Jia, E. Unger, M. Saliba, M. Li, A. Abate, Highly efficient p-i-n perovskite solar cells that endure temperature variations. *Science* **2023**, *379*, 399.
52. S. M. Park, M. Wei, J. Xu, H. R. Atapattu, F. T. Eickemeyer, K. Darabi, L. Grater, Y. Yang, C. Liu, S. Teale, B. Chen, H. Chen, T. Wang, L. Zeng, A. Maxwell, Z. Wang, K. R. Rao, Z. Cai, S. M. Zakeeruddin, J. T. Pham, C. M. Risko, A. Amassian, M. G. Kanatzidis, K. R. Graham, M. Grätzel, E. H. Sargent, Engineering ligand reactivity enables high-temperature operation of stable perovskite solar cells. *Science* **2023**, *381*, 209.

Index

Note: First use only recorded in the index.

For Product Safety Concerns and Information please contact our
EU representative GPSR@taylorandfrancis.com Taylor & Francis
Verlag GmbH, Kaufingerstraße 24, 80331 München, Germany